W0091471

Hypersingular Integral Equations and Their Applications

Differential and Integral Equations and Their Applications

A series edited by:
A.D. Polyanin
Institute for Problems in Mechanics, Moscow, Russia

Volume 1
Handbook of First Order Partial Differential Equations
A.D. Polyanin, V.F. Zaitsev and A. Moussiaux

Volume 2
Group-Theoretic Methods in Mechanics and Applied Mathematics
D.M. Klimov and V.Ph. Zhuravlev

Volume 3
Quantization Methods in the Theory of Differential Equations
V.E. Nazaikinskii, B.-W. Schulze and B.Yu. Sternin

Volume 4
Hypersingular Integral Equations and Their Applications
I.K. Lifanov, L.N. Poltavskii and G.M. Vainikko

This book is part of a series. The publisher will accept continuation orders which may be cancelled at any time and which provide for automatic billing and shipping of each title in the series upon publication. Please ask for details.

Hypersingular Integral Equations and Their Applications

I.K. Lifanov, L.N. Poltavskii
and G.N. Vainikko

CRC Press
Taylor & Francis Group
Boca Raton London New York

CRC Press is an imprint of the
Taylor & Francis Group, an **informa** business

A TAYLOR & FRANCIS BOOK

First published 2004 by Taylor & Francis

Published 2023 by CRC Press
Taylor & Francis Group
6000 Broken Sound Parkway NW, Suite 300
Boca Raton, FL 33487-2742

© 2004 by Taylor & Francis Group, LLC
CRC Press is an imprint of Taylor & Francis Group, an Informa business

No claim to original U.S. Government works

ISBN 13: 978-0-415-30998-1 (hbk)

This book contains information obtained from authentic and highly regarded sources. Reasonable efforts have been made to publish reliable data and information, but the author and publisher cannot assume responsibility for the validity of all materials or the consequences of their use. The authors and publishers have attempted to trace the copyright holders of all material reproduced in this publication and apologize to copyright holders if permission to publish in this form has not been obtained. If any copyright material has not been acknowledged please write and let us know so we may rectify in any future reprint.

Except as permitted under U.S. Copyright Law, no part of this book may be reprinted, reproduced, transmitted, or utilized in any form by any electronic, mechanical, or other means, now known or hereafter invented, including photocopying, microfilming, and recording, or in any information storage or retrieval system, without written permission from the publishers.

For permission to photocopy or use material electronically from this work, please access www.copyright.com (http://www.copyright.com/) or contact the Copyright Clearance Center, Inc. (CCC), 222 Rosewood Drive, Danvers, MA 01923, 978-750-8400. CCC is a not-for-profit organization that provides licenses and registration for a variety of users. For organizations that have been granted a photocopy license by the CCC, a separate system of payment has been arranged.

Trademark Notice: Product or corporate names may be trademarks or registered trademarks, and are used only for identification and explanation without intent to infringe.

**Visit the Taylor & Francis Web site at
http://www.taylorandfrancis.com**

**and the CRC Press Web site at
http//www.crcpress.com**

Publisher's Note
The publisher has gone to great lengths to ensure the quality of this reprint
but points out that some imperfections in the original copies may be
apparent.

Every effort has been made to ensure that the advice and information
in this book is true and accurate at the time of going to press.
However, neither the publisher nor the authors can accept any legal
responsibility or liability for any errors or omissions that may be made.
In the case of drug administration, any medical procedure or
the use of technical equipment mentioned within this book,
you are strongly advised to consult the manufacturer's guidelines.

British Library Cataloguing in Publication Data
A catalogue record for this book is available
from the British Library

Library of Congress Cataloging in Publication Data
A catalog record for this book has been requested

Contents

Preface

Hypersingular integral equations, i.e., integral equations whose kernel has a singularity of an order greater than one, are a convenient tool for studying spatial problems in air and fluid dynamics, elasticity, the theory of diffraction of electromagnetic and acoustic waves, ecology, etc. Usually, hypersingular integral equations are obtained as a result of reducing Neumann boundary value problems for the Laplace or the Helmholtz equation to integral equations by means of the double-layer potential.

In this book, exact analytical solutions of some two-dimensional hypersingular integral equations are constructed for the first time. An analytical solution in quadratures is obtained for the hypersingular equation on the sphere to which the Neumann problem for the Laplace equation on the sphere is reduced.

The book also contains an original exposition of some topics in the theory of the double-layer and single-layer potentials.

For the numerical solution of two-dimensional hypersingular integral equations the method of closed discrete vortex frameworks is developed (a version of this method was previously used for the numerical analysis of spatial problems in aerodynamics). In order to justify this method, the authors study the convergence of quadrature sums replacing hypersingular integrals to the exact values of those integrals. For the analysis of the convergence of numerical solutions, spaces of fractional quotients and pseudodifferential operators are used.

Theorems of the existence and the uniqueness of solutions are established for hypersingular integral equations. The theory developed in the monograph can be applied to the investigation of the convergence of approximate solutions obtained by finite-difference methods for partial differential equations of elliptic type.

Applications of hypersingular integral equations are demonstrated by numerous examples. Thus, solutions are obtained for the spatial nonstationary problem of flow past the deck of a ship, with the deck superstructure and vortex traces taken into account; the problem of diffraction of acoustic waves on a cube, the problem of radiation be an antenna with current sources on its surface, and also the problem of flow past an airfoil with suction on its surface.

The authors hope that the book will be useful for scientists and engineers in various areas of mathematics, physics, and mechanics.

Ivan K. Lifanov
Lev N. Poltavskii
Gennadi M. Vainikko

Authors

Ivan K. Lifanov, Professor, D.Sc., Ph.D., is a prominent scientist in the fields of numerical analysis, integral equations, and aerodynamics.

Ivan Lifanov graduated from the Department of Mechanics and Mathematics of the Moscow State University in 1965. He received his Ph.D. degree in 1968 at the Moscow State University and his D.Sc. degree in 1981 at the Computing Center of the Russian Academy of Sciences. Since 1968 Ivan Lifanov has been staff member at the Zhukovsky Air Force Engineering Academy, where he is at present Head of the Department of Mathematics. He is also Chief Researcher at the Institute of Computing Mathematics of the Russian Academy of Sciences and is Professor of the Orel State University. Since 1996 he has the title of Honoured Researcher of the Russian Federation and since 2000 is Vice-chairman of the Scientific Methodical Council on Mathematics of the Ministry of Education.

Professor Lifanov is a member of the Editorial Board of the journals *Differential Equations* and *Electromagnetic Waves and Electronic Systems*. He is also an author of over 230 scientific publications and 8 monographs.

Address: Zhukovsky Air Force Engineering Academy, Department of Mathematics, Planetnaya 3, 125190 Moscow, Russia.

Lev N. Poltavskii, Professor, D.Sc., Ph.D., is Professor at the Department of Mathematics of the Zhukovsky Air Force Engineering Academy.

Lev Poltavskii graduated from the Department of Mechanics and Mathematics of the Moscow State University in 1965. He received his Ph.D. degree in 1970 at the Moscow Institute of Physics and Technology and his D.Sc. degree in 1993 at the Moscow State University. Professor Poltavskii is an author of about 50 scientific publications.

Address: Zhukovsky Air Force Engineering Academy, Department of Mathematics, Planetnaya 3, 125190 Moscow, Russia.

Gennadi M. Vainikko, Professor, D.Sc., Ph.D., is a noted scientist in the fields of applied mathematics, differential and integral equations, ill-posed and inverse problems.

Gennadi M. Vainikko graduated from the University of Tartu in 1961 and received his Ph.D. degree there in 1964. He received his D.Sc. degree in 1969 at the Voronezh State University. At present he is Professor at the Helsinki University of Technology. Since 1986 he is Full Member of the Estonian Academy of Sciences.

Professor Vainikko is an author of about 200 scientific publications and 12 monographs.

Address: Helsinki University of Technology, Otakaari 1M, 02150 Espoo, Finland.

Chapter 1

Singular Integrals and Integral Equations

1.1. Some Facts from the Theory of One-Dimensional Integrals

All lines considered below belong to a plane with a Cartesian right-handed coordinate frame OXY. Sometimes, the points of the plane will be regarded as complex numbers and denoted by $t = x + iy$, where i is the imaginary unit.

We say that L *is a smooth open-ended curve (arc)*, if L can be defined by the parametric relations (see *Muskhelishvili* (1968)):

$$x = x(s), \qquad y = y(s), \qquad s_a \leq s \leq s_b, \tag{1.1.1}$$

where s_a and s_b are finite constants; $x(s)$, $y(s)$ are continuously differentiable functions on $[s_a, s_b]$; the derivatives $x'(s)$, $y'(s)$ cannot be both equal to zero at the same point; and it is assumed that different values of the parameter s correspond to different points of the curve L.

The relation $t = x(s) + iy(s)$ for the points of the curve L establishes a one-to-one correspondence between $t \in L$ and $s \in [s_a, s_b]$, and we have $t'_s = x'(s) + iy'(s)$.

In some cases, the curve L will be denoted by ab, with $a = t(s_a)$ and $b = t(s_b)$.

We say that L *is a smooth closed curve*, if L is a smooth curve such that

$$x(s_b) = x(s_a), \quad y(s_b) = y(s_a),$$
$$x'(s_b - 0) = x'(s_a + 0), \quad y'(s_b - 0) = y'(s_a + 0).$$

In this case, the functions $x(s)$, $y(s)$ and $x'(s)$, $y'(s)$ can be regarded as periodic with period $T = s_b - s_a$.

A *smooth (simple) line* is defined as a union of finitely many mutually disjoint closed or open-ended curves (in particular, these cannot have common end-points).

A curve is said to be *piecewise smooth* if it consists of finitely many smooth open-ended curves having no points in common except, possibly, end-points. Such a curve is said to have only *angular nodes* if the angle between any two curves entering each node is different from zero, i.e., the node cannot be a cuspidal point.

In Appendix 1 to *Muskhelishvili* (1968), the following result is proved. Let L be a simple piecewise smooth curve (i.e., it consists of finitely many smooth open-ended curves a_1a_2, $a_2a_3, \ldots, a_{n-1}a_n$ such that the last point of the preceding curve coincides with the first point of the next curve) whose nodes are all of angular type. Then, for any two points t_1, t_2 of the curve L, the following inequality holds:

$$K_0\sigma(t_1, t_2) \leq r(t_1, t_2) \leq \sigma(t_1, t_2), \tag{1.1.2}$$

where $\sigma(t_1, t_2)$ is the length of the part of L between the points t_1 and t_2; if L is closed, $\sigma(t_1, t_2)$ is the length of the smaller part; $r(t_1, t_2) = |t_1 - t_2|$ is the distance between the points t_1 and t_2 on the plane OXY; $K_0 \in (0, 1)$ is a constant that does not depend on the points t_1, t_2 of the curve L.

Note that the inequality (1.1.2) is also valid for a piecewise smooth curve L which has only angular nodes.

Definition 1.1.1. A function $\varphi(t)$ defined on a set D (on the complex plane, in general) *is of class $H(\mu)$ on D*, in other words, *satisfies the Hölder condition with exponent μ*, if for any $t_1, t_2 \in D$, the inequality

$$|\varphi(t_1) - \varphi(t_2)| \le A|t_1 - t_2|^\mu \tag{1.1.3}$$

holds with constants $A \ge 0$ and $0 < \mu \le 1$. These constants are respectively called the *coefficient* and the *exponent* in the Hölder condition. If the exponent μ is of no interest, we simply say that *the function $\varphi(t)$ satisfies the H-condition or belongs to the class H on the set D*. We write $\varphi(t) \in H(\mu)$ or $\varphi(t) \in H$.

Note that the inclusion $\varphi(t) \in H(\mu)$ implies that $|\varphi(t)| \in H(\mu)$.

The *H*-condition can be extended to functions of several variables as follows. A function $\varphi(t_1, \ldots, t_n)$ defined for $(t_1, \ldots, t_n) \in D$ is said *to be of class $H(\mu_1, \ldots, \mu_n)$* (or *satisfy the H-condition*) on the set D, if for any points $(t'_1, \ldots, t'_n)$, $(t''_1, \ldots, t''_n) \in D$, the inequality

$$|\varphi(t''_1, \ldots, t''_n) - \varphi(t'_1, \ldots, t'_n)| \le A_1|t''_1 - t'_1|^{\mu_1} + \cdots + A_n|t''_1 - t'_n|^{\mu_n} \tag{1.1.4}$$

holds with constants $A_j \ge 0$, $0 < \mu_j \le 1$, $j = 1, \ldots, n$.

From (1.1.4), it follows that if $\varphi(t_1, \ldots, t_n) \in H$, then for any t_k, $k = 1, \ldots, n$, the function φ belongs to the class $H(\mu_k)$, uniformly with respect to the rest of the variables, i.e.,

$$|\varphi(t_1, \ldots, t''_k, \ldots, t_n) - \varphi(t_1, \ldots, t'_k, \ldots, t_n)| \le A|t''_k - t'_k|^{\mu_k},$$

where A is a constant independent of $t_1, \ldots, t_n$. The converse statement is also true.

In what follows, when saying that a function $\varphi(t_1, \ldots, t_n)$ satisfies the *H*-condition with respect to each variable separately, we mean that this condition holds uniformly with respect to the rest of the variables.

A smooth open-ended curve L is called a *Lyapunov curve*, if the derivative $t'(s)$ satisfies the $H(\alpha)$-condition on $[s_a, s_b]$. As shown by *Muskhelishvili* (1968), in this case, the function

$$f(s, s_0) = \frac{t - t_0}{s - s_0} = \frac{t(s) - t(s_0)}{s - s_0}$$

satisfies the $H(\alpha)$-condition with respect to both variables s and s_0 and does not vanish on the segment $[s_a, s_b]$. Moreover, if $t^{(n)}(s) \in H(\alpha)$, then all partial derivatives of the function $f(s, s_0)$ of the orders $\le n - 1$ belong to $H(\alpha)$ with respect to both variables.

A piecewise smooth curve L is called *a piecewise Lyapunov curve*, if its smooth pieces are Lyapunov curves.

Definition 1.1.2. A function $\varphi(t)$ belongs to the *class H^* on a piecewise smooth curve L*, if

$$\varphi(t) = \frac{\varphi^*(t)}{P_L^\nu(t)}, \qquad P_L^\nu(t) = \prod_{k=1}^{p} |t - c_k|^{\nu_k}, \tag{1.1.5}$$

where $\varphi^*(t) \in H_0$ on L, i.e., it belongs to the class H on every smooth piece of the curve L; $0 \le \nu_k < 1$; and c_k, $k = 1, \ldots, p$, are the nodes of the curve L.

Without loss of generality, we can assume that $\varphi^*(t) \in H$ on L.

Now, we recall the definition of a singular integral of Cauchy type over a piecewise smooth curve.

Definition 1.1.3. Let t_0 be a point of the curve L outside its nodes (i.e., its interior point). Consider a circular curve with center at t_0 whose radius $\varepsilon > 0$ is so small that its intersection with L consists of two points t' and t''. Denote by l the arc $t't'' \subset L$. If the integral

$$\int_{L \setminus l} \frac{\varphi(t)\, dt}{t - t_0}$$

has a finite limit $\Phi(t_0)$ as $\varepsilon \to 0$, this limit is called *the Cauchy principal value of the integral,*

$$\Phi(t_0) = \lim_{\varepsilon \to 0} \int_{L \setminus l} \frac{\varphi(t)\, dt}{t - t_0} = \int_L \frac{\varphi(t)\, dt}{t - t_0}. \tag{1.1.6}$$

As shown by *Muskhelishvili* (1968), the class H^* of functions on a piecewise smooth curve L is invariant with respect to the integral in the sense of the Cauchy principal value (singular integral). In other words, if $\varphi(t) \in H^*$ on L, then $\Phi(t_0) \in H^*$ on L.

In some problems it is convenient to replace an integration curve L by another curve Λ (to change variables in a singular integral). In such cases, it is assumed that between the points t of smooth pieces of the curve L and the points τ of smooth pieces of the curve Λ there is a one-to-one correspondence $t = t(\tau)$ such that the derivative $t'(\tau) = dt/d\tau$ exists, is different from zero, and $t'(\tau) \in H(\alpha)$ on Λ. Now, suppose that $\varphi(t) \in H$ in a neighborhood of a point t_0 (different from the nodes) and $\varphi(t)$ is integrable on L. Then the following formula holds:

$$\int_L \frac{\varphi(t)\, dt}{t - t_0} = \int_\Lambda \frac{\psi(\tau, \tau_0)\, d\tau}{\tau - \tau_0}, \tag{1.1.7}$$

$$\psi(\tau, \tau_0) = \frac{(\tau - \tau_0) t'(\tau)}{t(\tau) - t(\tau_0)} \varphi(t(\tau)), \tag{1.1.8}$$

where $t_0 = t(\tau_0)$, $\tau_0 \in \Lambda$.

For singular integrals over piecewise smooth curves, the order of integration can be changed according to the following Poincaré–Bertrand formula (see *Lifanov* (1996) and *Muskhelishvili* (1968)):

$$\int_L \frac{dt}{t - t_0} \int_L \frac{\varphi(t, \tau)\, d\tau}{\tau - t} = \int_L d\tau \int_L \frac{\varphi(t, \tau)\, dt}{(t - t_0)(\tau - t)} - \pi^2 \varphi(t_0, t_0), \qquad t_0 \in L. \tag{1.1.9}$$

Here, L is a piecewise smooth curve, t_0 is outside the nodes of L, and the function $\varphi(t, \tau)$ has the form

$$\varphi(t, \tau) = \frac{\varphi^*(t, \tau)}{\Pi(t, \tau)}, \tag{1.1.10}$$

where $\varphi^*(t, \tau)$ is a function of class H_0 on L, i.e., it satisfies the H-condition on every smooth arc of L, and

$$\Pi(t, \tau) = \prod_{k=1}^{n} |t - C_k|^{\alpha_k} |\tau - C_k|^{\beta_k},$$

with C_k, $k = 1, \ldots, n$, being the nodes of L, α_k, $\beta_k \geq 0$, $\alpha_k + \beta_k < 1$.

Consider the special case of a singular integral on the segment $L = [a, b]$ of the real axis OX. Then, formula (1.1.6) reads

$$\Phi(x_0) = \lim_{\varepsilon \to 0} \int_{L \setminus l} \frac{\varphi(x)\, dx}{x - x_0} = \tag{1.1.11}$$

$$= \lim_{\varepsilon \to 0} \left[\int_a^{x_0 - \varepsilon} \frac{\varphi(x)\, dx}{x - x_0} + \int_{x_0 - \varepsilon}^b \frac{\varphi(x)\, dx}{x - x_0} \right] = \int_a^b \frac{\varphi(x)\, dx}{x - x_0}, \qquad x_0 \in (a, b).$$

It is convenient to reduce the problem of circulation-free flow of an ideal incompressible fluid past the segment $[a, b]$ to a strongly singular integral of the form

$$F(x_0) = \int_a^b \frac{\varphi(x)\,dx}{(x - x_0)^2}, \qquad x_0 \in (a, b). \tag{1.1.12}$$

This integral is understood as *a finite part in the sense of Hadamard* (see *Hadamard* (1932)),

$$F(x_0) = \lim_{\varepsilon \to 0} \left[\int_a^{x_0 - \varepsilon} \frac{\varphi(x)\,dx}{(x - x_0)^2} + \int_{x_0 + \varepsilon}^b \frac{\varphi(x)\,dx}{(x - x_0)^2} - \frac{2\varphi(x_0)}{\varepsilon} \right].$$

Theorem 1.1.1. *Let $\varphi(x)$ be a function defined on $[a, b]$ such that $\varphi'(x) \in H(\alpha)$ on $[a, b]$, i.e., $\varphi(x) \in H_1(\alpha)$ on $[a, b]$. Then the integral $F(x_0)$ exists for any $x_0 \in (a, b)$.*

Proof. First, we take $\varphi(x) \equiv 1$ on $[a, b]$. Then

$$\lim_{\varepsilon \to 0} \left[\int_a^{x_0 - \varepsilon} \frac{dx}{(x - x_0)^2} + \int_{x + \varepsilon}^b \frac{dx}{(x - x_0)^2} - \frac{2}{\varepsilon} \right] =$$

$$= \lim_{\varepsilon \to 0} \left[\frac{1}{a - x_0} - \frac{1}{b - x_0} \right] = \frac{1}{a - x_0} - \frac{1}{b - x_0} = \int_a^b \frac{dx}{(x - x_0)^2}. \tag{1.1.13}$$

Now, we have

$$\int_a^b \frac{\varphi(x)\,dx}{(x - x_0)^2} = \int_a^b \frac{1}{x - x_0} \frac{\varphi(x) - \varphi(x_0)}{x - x_0}\,dx + \varphi(x_0) \int_a^b \frac{dx}{(x - x_0)^2}.$$

The second integral on the right-hand side of this formula exists as a strongly singular integral, by virtue of (1.1.13). As shown by *Muskhelishvili* (1968), for $\varphi(x) \in H_1(\alpha)$, we have $(\varphi(x) - \varphi(x_0))/(x - x_0) \in H(\alpha)$ as a function of two variables. Therefore, the first integral on the right-hand side exists, too.

Remark 1.1.1. From the definition of a strongly singular integral $F(x_0)$ on the segment, we see that if $\varphi'(x) \in H(\alpha)$, then

$$F(x_0) = \frac{\varphi(a)}{a - x_0} - \frac{\varphi(b)}{b - x_0} + \int_a^b \frac{\varphi'(x)\,dx}{x - x_0}, \tag{1.1.14}$$

which means that the integral $F(x_0)$ can be understood as a result of formal integration by parts.

Remark 1.1.2. Similarly, the expressions

$$\Phi(\theta_0) = \int_0^{2\pi} \cot \frac{\theta - \theta_0}{2} \varphi(\theta)\,d\theta, \tag{1.1.15}$$

$$F(\theta_0) = \int_0^{2\pi} \frac{\varphi(\theta)\,d\theta}{\sin^2 \frac{\theta - \theta_0}{2}}, \qquad \theta_0 \in [0, 2\pi], \tag{1.1.16}$$

can be regarded as singular and strongly singular integrals, respectively.

Integrals of the type (1.1.15) are called *Hilbert integrals*, while those of the type (1.1.16) we call *hypersingular Hilbert integrals*.

For the Hilbert integrals, the following formulas should be mentioned:

$$\int_0^{2\pi} \cot \frac{\theta_0 - \theta}{2}\,d\theta \int_0^{2\pi} \cot \frac{\theta - \theta_1}{2} \varphi(\theta_1)\,d\theta_1 = -4\pi^2 \varphi(\theta_0) + 2\pi \int_0^{2\pi} \varphi(\theta)\,d\theta, \tag{1.1.17}$$

$$\int_0^{2\pi} \cot \frac{\theta - \theta_0}{2} \cot \frac{\theta_1 - \theta}{2}\,d\theta = 2\pi, \qquad \theta_0, \theta_1 \in [0, 2\pi]. \tag{1.1.18}$$

Note that for $\theta_0 = \theta_1$ in (1.1.18), we obtain a hypersingular integral.

1.2. One-Dimensional Equations

In this section, we give some facts pertaining to the solution of one-dimensional singular and hypersingular integral equations to be used in the sequel for solving applied problems. Most frequently, one has to deal with singular integral equations of the first kind on a segment. The solutions of such an equation in complete form,

$$\frac{1}{\pi} \int_{-1}^{1} \frac{\varphi(x)\,dx}{x - x_0} + \int_{-1}^{1} K(x_0, x)\varphi(x)\,dx = f(x_0), \qquad x_0 \in (-1, 1), \tag{1.2.1}$$

have the same specific features as the solutions of the corresponding characteristic equation

$$\frac{1}{\pi} \int_{-1}^{1} \frac{\varphi(x)\,dx}{x - x_0} = f(x_0), \qquad x_0 \in (-1, 1). \tag{1.2.2}$$

Equation (1.2.2) may have solutions of index $\varkappa = 1, 0, -1$.

For $\varkappa = 1$, all solutions of equation (1.2.2) (in this case, we say that the solutions are of class (∞, ∞), i.e., may be unbounded at both end-points of the segment) are given by the formula

$$\varphi(x_0) = -\frac{1}{\pi\sqrt{1 - x_0^2}} \int_{-1}^{1} \frac{\sqrt{1 - x^2}\,f(x)\,dx}{x - x_0} + \frac{C}{\pi\sqrt{1 - x_0^2}}, \tag{1.2.3}$$

where C is an arbitrary constant and

$$\int_{-1}^{1} \varphi(x)\,dx = C. \tag{1.2.4}$$

For $\varkappa = 0$, the equation admits only one solution (we say, that the solution is class $(\infty, 0)$, i.e., is bounded at the point $x = 1$),

$$\varphi(x_0) = -\frac{1}{\pi}\sqrt{\frac{1 - x_0}{1 + x_0}} \int_{-1}^{1} \sqrt{\frac{1 + x}{1 - x}}\,\frac{f(x)\,dx}{x - x_0}, \tag{1.2.5}$$

and only one solution (of class $(0, \infty)$)

$$\varphi(x_0) = -\frac{1}{\pi}\sqrt{\frac{1 + x_0}{1 - x_0}} \int_{-1}^{1} \sqrt{\frac{1 - x}{1 + x}}\,\frac{f(x)\,dx}{x - x_0}. \tag{1.2.6}$$

For $\varkappa = -1$, there is only one solution (of class $(0, 0)$),

$$\varphi(x_0) = -\frac{\sqrt{1 - x_0^2}}{\pi} \int_{-1}^{1} \frac{f(x)\,dx}{\sqrt{1 - x^2}\,(x - x_0)}, \tag{1.2.7}$$

provided that

$$\int_{-1}^{1} \frac{f(x)\,dx}{\sqrt{1 - x^2}} = 0.$$

In particular, if we take $f(x) \equiv 1$, $C = 0$ in (1.2.3), we get

$$\varphi(x_0) = \frac{x_0}{\sqrt{1 - x_0^2}}. \tag{1.2.8}$$

Taking $f(x) \equiv -1$ in (1.2.5), we get

$$\varphi(x_0) = \sqrt{\frac{1 - x_0}{1 + x_0}}, \tag{1.2.9}$$

and, for $f(x) \equiv -x$ in (1.2.7), we have

$$\varphi(x_0) = \sqrt{1 - x_0^2}.$$

As shown by *Lifanov* (1996), for equations (1.2.1) and (1.2.2) one can construct solutions of the form

$$\varphi(x_0) = \frac{\psi(x)}{x - q}, \qquad q \in (-1, 1), \tag{1.2.10}$$

where $\psi(x) \in H^*$ on $[-1, 1]$. These we call *singular solutions*. Let us find all solutions of equation (1.2.2) of the form (1.2.10).

The solutions of class (∞, ∞) of the form (1.2.10) are given by

$$\varphi(x_0) = -\frac{1}{\pi \sqrt{1 - x_0^2}\,(x_0 - q)} \int_{-1}^{1} \frac{\sqrt{1 - x^2}\,(x - q)f(x)\,dx}{x - x_0} + \frac{A + Bx_0}{\sqrt{1 - x_0^2}\,(x_0 - q)} =$$

$$= -\frac{1}{\pi \sqrt{1 - x_0^2}} \int_{-1}^{1} \frac{\sqrt{1 - x^2}\,f(x)\,dx}{x - x_0} + \frac{C}{\pi \sqrt{1 - x_0^2}} + \frac{B_1}{\sqrt{1 - x_0^2}\,(x_0 - q)}, \tag{1.2.11}$$

where A, B, C, B_1 are arbitrary constants.

The solutions of class $(\infty, 0)$ of the form (1.2.10) are given by

$$\varphi(x_0) = -\frac{1}{\pi} \sqrt{\frac{1 - x_0}{1 + x_0}} \frac{1}{x - q} \int_{-1}^{1} \sqrt{\frac{1 + x}{1 - x}} \frac{f(x)(x - q)\,dx}{x - x_0} - \sqrt{\frac{1 - x_0}{1 + x_0}} \frac{A}{x_0 - q} =$$

$$= -\frac{1}{\pi} \sqrt{\frac{1 - x_0}{1 + x_0}} \int_{-1}^{1} \sqrt{\frac{1 + x}{1 - x}} \frac{f(x)\,dx}{x - x_0} + \frac{B}{x_0 - q} \sqrt{\frac{1 - x_0}{1 + x_0}}. \tag{1.2.12}$$

The solutions of class $(0, 0)$ of the form (1.2.10) are given by

$$\varphi(x_0) = -\frac{\sqrt{1 - x_0^2}}{\pi(x_0 - q)} \int_{-1}^{1} \frac{(x - q)f(x)\,dx}{\sqrt{1 - x^2}\,(x - x_0)} =$$

$$= -\frac{\sqrt{1 - x_0^2}}{\pi} \int_{-1}^{1} \frac{f(x)\,dx}{\sqrt{1 - x^2}\,(x - x_0)} - \frac{\sqrt{1 - x_0^2}}{\pi(x_0 - q)} \int_{-1}^{1} \frac{f(x)\,dx}{\sqrt{1 - x^2}}. \tag{1.2.13}$$

Next, consider the characteristic singular integral equation of the second kind with constant coefficients, namely,

$$a\varphi(x_0) + \frac{b}{\pi} \int_{-1}^{1} \frac{\varphi(x)\,dx}{x - x_0} = f(x_0), \qquad x_0(-1, 1), \tag{1.2.14}$$

where a and b are real numbers such that $b \neq 0$, $a \pm b \neq 0$, $a^2 + b^2 = 1$, and $f(x) \in H(\alpha)$ on $[-1, 1]$. Let us briefly recall some results for equation (1.2.14) (see *Muskhelishvili* (1968)).

The index of this equation can take the values $1, 0, -1$.

The respective solutions for $\varkappa = 1, 0$ have the form

$$\varphi(x_0) = a f(x_0) - \frac{b}{\pi} \omega_\varkappa(x_0) \int_{-1}^1 \frac{f(x)\,dx}{\omega_\varkappa(x)(x - x_0)} + A\omega_\varkappa(x_0), \qquad (1.2.15)$$

where A is an arbitrary constant for $\varkappa = 1$; $A = 0$ for $\varkappa = 0$;

$$\omega_1(x) = (1 + x)^\alpha (1 - x)^{-1-\alpha}, \qquad \omega_0(x) = (1 + x)^\alpha (1 - x)^{-\alpha}, \qquad (1.2.16)$$

and the number α is a root of the equation

$$a + b \cot \pi\alpha = 0 \qquad (1.2.17)$$

such that

$$-1 < \alpha < 0.$$

For $\varkappa = -1$, the solution is unique and it exists if and only if

$$\int_{-1}^1 \frac{f(x)\,dx}{\omega_{-1}(x)} = 0, \qquad (1.2.18)$$

where $\omega_{-1}(x) = (1 + x)^\alpha (1 - x)^{1-\alpha}$, $0 < \alpha < 1$, α is a solution of equation (1.2.17) and is given by (1.2.15) with $A = 0$.

As shown by *Lifanov* (1996), it is convenient to reduce the problem of circulation-free flow past the segment $[-1, 1]$ to the hypersingular integral equation (see *Lifanov* (1996))

$$\int_{-1}^1 \frac{g(x)\,dx}{(x - x_0)^2} = f(x_0), \qquad x_0 \in (-1, 1), \qquad (1.2.19)$$

which admits a unique solution in the class of functions vanishing at the end-points of the segment,

$$\varphi(-1) = \varphi(1) = 0. \qquad (1.2.20)$$

Then, in view of (1.1.14), equation (1.2.19) is equivalent to the equation

$$\int_{-1}^1 \frac{g'(x)\,dx}{x - x_0} = f(x_0), \qquad x_0 \in (-1, 1), \qquad (1.2.21)$$

provided that

$$\int_{-1}^1 g'(x)\,dx = 0. \qquad (1.2.22)$$

Now, using (1.2.3) for the solution of equation (1.2.21) with the condition (1.2.22), we get

$$g'(x_0) = -\frac{1}{\pi^2 \sqrt{1 - x_0^2}} \int_{-1}^1 \frac{\sqrt{1 - x^2}\, f(x)\,dx}{x - x_0} \qquad (1.2.23)$$

or

$$g(x_0) = -\frac{1}{\pi^2} \int_{-1}^{x_0} \frac{d\tau}{\sqrt{1 - \tau^2}} \int_{-1}^1 \frac{\sqrt{1 - x^2}\, f(x)\,dx}{\tau - x} =$$

$$= \frac{1}{\pi^2} \int_{-1}^1 f(x) \ln \left| \frac{\sqrt{(1-x)(1+x_0)} - \sqrt{(1+x)(1-x_0)}}{\sqrt{(1-x)(1+x_0)} + \sqrt{(1+x)(1-x_0)}} \right| \, dx. \qquad (1.2.24)$$

Condition (1.2.20) can be verified for the solution (1.2.24) of equation (1.2.19).

Next, we turn to the characteristic equations of the second kind with Hilbert kernel and constant coefficients,

$$a\varphi(\theta_0) + \frac{b}{2\pi} \int_0^{2\pi} \varphi(\theta) \cot \frac{\theta - \theta_0}{2} \, d\theta = f(\theta_0), \tag{1.2.25}$$

where $\theta_0 \in [0, 2\pi]$, $a \neq 0$, a and b are real numbers.

The index of equation (1.2.25) with $a \neq 0$ is equal to zero and this equation admits only one solution (see *Gakhov* (1966), *Lifanov* (1996)), which is given by

$$\varphi(\theta_0) = a f(\theta_0) - \frac{b}{2\pi} \int_0^{2\pi} f(\theta) \cot \frac{\theta - \theta_0}{2} \, d\theta + \frac{b^2}{2\pi a} \int_0^{2\pi} f(\theta) \, d\theta, \qquad \theta_0 \in [0, 2\pi]. \tag{1.2.26}$$

The fact that $\varphi(\theta_0)$ defined by (1.2.26) is a solution of equation (1.2.25) can be verified directly by means of (1.1.17).

For $a = 0$, $b = 1$, we obtain an equation of the first kind with Hilbert kernel, namely,

$$\frac{1}{2\pi} \int_0^{2\pi} \varphi(\theta) \cot \frac{\theta - \theta_0}{2} \, d\theta = f(\theta_0), \qquad \theta_0 \in [0, 2\pi]. \tag{1.2.27}$$

Its solution is non-unique and has the form

$$\varphi(\theta_0) = -\frac{1}{2\pi} \int_0^{2\pi} f(\theta) \cot \frac{\theta - \theta_0}{2} \, d\theta + C, \qquad \theta_0 \in [0, 2\pi], \tag{1.2.28}$$

where C is an arbitrary constant, provided that the condition

$$\int_0^{2\pi} f(\theta) \, d\theta = 0 \tag{1.2.29}$$

is satisfied. Note that if (1.2.29) does not hold, i.e., $\int_0^{2\pi} f(\theta) \, d\theta \neq 0$, it is still possible to find a solution of equation (1.2.27), but this solution is outside the class of absolutely integrable functions. To be more precise, the function

$$\varphi(\theta_0) = -\frac{1}{2\pi} \int_0^{2\pi} f(\theta) \cot \frac{\theta - \theta_0}{2} \, d\theta + C + \frac{1}{2\pi} \cot \frac{\theta - \theta_q}{2} \int_0^{2\pi} f(\theta) \, \theta, \tag{1.2.30}$$

where θ_q is an arbitrarily fixed point of $[0, 2\pi]$, is a solution of equation (1.2.28) (see *Lifanov* (1996)).

Just as in the case of the equation on a segment, we can consider a hypersingular integral equation with a kernel of Hilbert type (see *Lifanov* (1996)),

$$\frac{1}{4\pi} \int_0^{2\pi} g(\theta) \left[\sin \frac{\theta - \theta_0}{2} \right]^{-2} d\theta = f(\theta_0), \qquad \theta_0 \in [0, 2\pi]. \tag{1.2.31}$$

For a periodic $g(\theta) \in H_1(\alpha)$ on $[0, 2\pi]$, equation (1.2.31) is equivalent to

$$\frac{1}{2\pi} \int_0^{2\pi} g'(\theta) \cot \frac{\theta - \theta_0}{2} \, d\theta = f(\theta_0), \qquad \theta_0 \in [0, 2\pi]. \tag{1.2.32}$$

Therefore, the solution of equation (1.2.31) is given by the formula

$$g(\theta_0) = -\frac{1}{\pi} \int_0^{2\pi} f(\theta) \ln \left| \sin \frac{\theta - \theta_0}{2} \right| d\theta + C, \tag{1.2.33}$$

where C is an arbitrary constant.

The following relations should also be mentioned (see *Lifanov* (1996)):

$$\frac{1}{4\pi} \int_0^{2\pi} \left[\sin \frac{\theta_0 - \theta}{2} \right]^{-2} (a_n \cos n\theta + b_n \sin n\theta) \, d\theta = -n(a_n \cos n\theta_0 + b_n \sin n\theta_0), \tag{1.2.34}$$

where $\theta_0 \in [0, 2\pi]$.

1.3. Some Facts from the Theory of Multi-Dimensional Integrals

While singular integrals usually considered in the one-dimensional case are mostly of the same type, in the case of several dimensions (≥ 2), there exist many more types of singular integrals. We start with a most natural generalization of the concept of one-dimensional singular integral.

Let $L_1, \ldots, L_n$ be piecewise smooth plane curves. Following *Gakhov* (1966), we introduce a *framework* as the topological product of these curves and denote it by $L = L_1 \times \cdots \times L_n$. Let $\varphi(t) = \varphi(t^1, \ldots, t^n)$ be a function defined on the framework L. A point $t = (t^1, \ldots, t^n)$ is called *interior* if t^k does not coincide with any of the nodes of L_k, $k = 1, \ldots, n$. Let t_0 be an interior point of the framework L. In the plane of the curve L_k, consider the circular curve of radius ε_k centered at t_0^k and denote by l^k the piece of L_k inside the circle (ε_k is assumed so small that l_k is a smooth open-ended arc).

Definition 1.3.1. *The Cauchy integral (multiple singular integral) of the function* $\varphi(t^1, \ldots, t^n)$ *over the framework* L *at the point* $t_0 = (t_0^1, \ldots, t_0^n)$ *is the limit*

$$\Phi(t_0) = \lim_{\varepsilon_1, \ldots, \varepsilon_n \to 0} \int_{L_*} \frac{\varphi(t^1, \ldots, t^n)\, dt^1 \cdots dt^n}{(t^1 - t_0^1) \cdots (t^n - t_0^n)}, \qquad L_* = (L_1 \setminus l_1) \times \cdots \times (L_n \setminus l_n), \quad (1.3.1)$$

where $\varepsilon_1, \ldots, \varepsilon_n$ *tend to zero independently. This limit is denoted by*

$$\Phi(t_0) = \int_{L_1 \times \cdots \times L_n} \frac{\varphi(t^1, \ldots, t^n)\, dt^1 \cdots dt^n}{(t^1 - t_0^1) \cdots (t^n - t_0^n)} = \int_L \frac{\varphi(t)\, dt}{((t - t_0))}, \qquad (1.3.2)$$

where $((t - t_0)) = (t - t_0^1) \cdots (t^n - t_0^n)$, $dt = dt^1 \cdots dt^n$.

Just as for the one-dimensional singular integral, it can be shown that $\Phi(t_0)$ exists at the point t_0 if the function $\varphi(t) = \varphi(t^1, \ldots, t^n)$ satisfies the H-condition in a neighborhood of $t_0 = (t_0^1, \ldots, t_0^n) \in L$. On the other hand, formula (1.3.1) defining $\Phi(t_0)$ allows us to consider the multiple Cauchy integral as an iterated integral,

$$\int_{L_1 \times \cdots \times L_n} \frac{\varphi(t^1, \ldots, t^n)\, dt^1 \cdots dt^n}{(t^1 - t_0^1) \cdots (t^n - t_0^n)} = \int_{L_1} \frac{dt^1}{t^1 - t_0^1} \left(\cdots \left(\int_{L_n} \frac{\varphi(t^1, \ldots, t^n)\, dt^n}{t^n - t_0^n} \right) \cdots \right). \quad (1.3.3)$$

Recall that (see *Gakhov* (1966), *Muskhelishvili* (1968)) the one-dimensional singular integral over a closed smooth curve l (as well as the integral with Hilbert kernel) preserves the regularity properties of the density function: if $\varphi(t) \in H_m(\alpha)$ ($\varphi^{(m)} \in H(\alpha)$) on L, then $\Phi(t) \in H_m(\alpha)$ on L. For multiple integrals this is not true. The following result holds (see *Lifanov* (1996)).

Theorem 1.3.1. *Let* $\varphi(t^1, \ldots, t^n)$ *be a function of class* $H(\mu_1, \ldots, \mu_n)$ *on the* n*-dimensional torus* $L = L_1 \times \cdots \times L_n$, *where* L_k, $k = 1, \ldots, n$, *are closed smooth curves. Then the singular integral* $\Phi(t_0^1, \ldots, t_0^n)$ *belongs to the class* $H(\mu_1 - \varepsilon, \ldots, \mu_n - \varepsilon)$ *on* L, *where* $\varepsilon > 0$ *is an arbitrarily small constant.*

The function $\varphi(t) = \varphi(t^1, \ldots, t^n)$ is said to be of class H^* *on the framework* $L = L_1 \times \cdots \times L_n$ (L_k, $k = 1, \ldots, n$, are plane piecewise smooth curves), if

$$\varphi(t) = \frac{\varphi^*(t)}{P_{L_1}^{\nu^1}(t^1) \cdots P_{L_n}^{\nu^n}(t^n)}, \qquad (1.3.4)$$

where

$$\varphi^* \in H \quad \text{on} \quad L, \qquad P_{L_k}^{\nu^k}(t^k) = \prod_{i_k = 1}^{m_k} \left| t^k - C_{i_k}^k \right|^{\nu_{i_k}^k},$$

$C_{i_k}^k$, $i_k = 1, \ldots, m_k$, are the nodes of the curve L_k, and $\nu^k = \left(\nu_1^k, \ldots, \nu_{m_k}^k \right)$, $0 \leq \nu_{i_k}^k < 1$, $i_k = 1, \ldots, m_k$; $k = 1, \ldots, n$.

Theorem 1.3.2. *The class H^* on the framework $L = L_1 \times \cdots \times L_n$ of piecewise smooth curves L_k, $k = 1, \ldots, n$, is invariant with respect to the operator associated with the multiple Cauchy integral, i.e., if $\varphi(t) \in H^*$ on L, then $\Phi(t_0) \in H^*$ on L.*

It should be mentioned that the order of integration in multiple integrals over a framework can be changed according to a generalization of the Poincaré–Bertrand formula (see *Lifanov* (1996)). Thus, for any smooth closed curve L we have the identity

$$\int_L \frac{dt_0}{(t - t_0)(t_0 - \tau)} \equiv 0, \tag{1.3.5}$$

and therefore, the following inversion formula holds for multiple integrals.

Theorem 1.3.3. *Let $\varphi(t^1, \ldots, t^n)$ be of class H on the framework L of closed smooth curves $L_1, \ldots, L_n$. Then*

$$\int_L \frac{dt_0}{((t_0 - \tau))} \int_L \frac{\varphi(t)\, dt}{((t - t_0))} \equiv (-\pi^2)^n \varphi(\tau), \qquad \tau \in L_0. \tag{1.3.6}$$

Another natural generalization of the one-dimensional singular integral is (see *Mikhlin* (1962))

$$V(x_0) = \int_D \frac{f(x_0, \theta)}{r^2} u(x)\, dx. \tag{1.3.7}$$

Here we consider in more detail the case of a plane closed domain $D \subset \mathbb{R}_2$, which may be bounded or unbounded. The two-dimensional Jordan measure of its boundary is equal to zero. The points of D are denoted by $x = (x^1, x^2)$ and $x_0 = (x_0^1, x_0^2)$, $r = |x - x_0|$, and $\theta = (x - x_0)/r$ is a point of the unit circle S in $\mathbb{R}_2$. We say that $u(x)$ is the *density function* of the singular integral and assume that $u(x)$ is absolutely integrable in D and on any closed set F inside D it satisfies the $H(\alpha)$-condition. The function $f(x_0, \theta)$ is called the *characteristic function*. It is assumed that $f(x_0, \theta)$ is bounded and, for any fixed x_0, is continuous in θ.

Let x_0 be an interior point of the domain D. The integral (1.3.7) is understood in the following sense:

$$V(x_0) = \lim_{\varepsilon \to 0} \int_{D \setminus O(x_0, \varepsilon)} \frac{f(x_0, \theta)}{r^2} u(x)\, dx, \tag{1.3.8}$$

where $O(x_0, \varepsilon)$ is an ε-neighborhood of the point x_0.

As shown by *Mikhlin* (1962) and *Stein* (1970), the necessary and sufficient condition for the existence of the integral (1.3.7) in the sense of *the principal value* (1.3.8) is

$$\int_S f(x_0, \theta)\, dS = 0. \tag{1.3.9}$$

The integral (1.3.7) and many of its properties have been studied in the cited monographs, especially in the case of $D = \mathbb{R}_2$. However, in order to have a unified approach to the solution of exterior problems for the Laplace equation with Neumann boundary conditions on closed surfaces (i.e., surfaces without border) or non-closed surfaces (i.e., surfaces with border), it is convenient to use the double layer potential. In this connection, to facilitate the construction of the numerical method of closed discrete vortex frameworks for problems in aerodynamics and a similar method for problems in electrodynamics (see *Anfinogenov and Lifanov* (1992), *Lifanov* (1996)), it is preferable to pass to hypersingular integral equations. In this case, the crucial role belongs to integrals of the form

$$I(\nu) = \frac{1}{4\pi} \int_\sigma \frac{\nu(M_0)\, d\sigma_{M_0}}{|M_0 M|^2}. \tag{1.3.10}$$

Here, σ is a surface (a differentiable manifold with boundary); $M_0, M \in \sigma$. The function $\nu(M_0)$ is of class $C^2(\sigma^*)$, where σ^* is the interior of the surface σ, i.e., $\sigma^* = \sigma \setminus \partial\sigma$, $\partial\sigma$ is the boundary of σ.

Further exposition in this section follows *Lifanov and Poltavskii* (1998).

Assume that there is a Cartesian coordinate system $OX_0Y_0Z_0$ such that the plane OX_0Y_0 is non-orthogonal to the tangential planes to σ at the points $M \in \sigma$, and the equation of σ in these coordinates is $z = f(x, y)$. A coordinate system of this type will be called a *normal coordinate system*. The coordinates of the point M are denoted by $(x, y, f(x, y))$ and those of the point M_0 by $(x_0, y_0, f(x_0, y_0))$. Thus,

$$|M_0M| = \sqrt{(x - x_0)^2 + (y - y^0)^2 + |f(x, y) - f(x_0, y_0)|^2} \ .$$

Let D be the projection of the surface σ to the plane OX_0Y_0, and

$$\nu(M_0) = \nu(x_0, y_0, f(x_0, y_0)) = g(x_0, y_0), \qquad d\sigma_{M_0} = \sqrt{1 + f_{x_0}'^2(x_0, y_0) + f_{y_0}'^2(x_0, y_0)} \, dx_0 \, dy_0.$$

Then the integral (1.3.10) can be written in the form

$$I(\nu) = I^*(g) = \frac{1}{4\pi} \iint_D K(x, y, x_0, y_0)g(x_0, y_0) \, dx_0 \, dy_0, \qquad (1.3.11)$$

$$K(x, y, x_0, y_0) = \frac{F_1(x_0, y_0)}{|F_2(x, y, x_0, y_0)|^3},$$

$$F_1(x_0, y_0) = \sqrt{1 + f_{x_0}'^2(x_0, y_0) + f_{y_0}'^2(x_0, y_0)},$$

$$F_2(x, y, x_0, y_0) = \sqrt{(x - x_0)^2 + (y - y_0)^2 + |f(x, y) - f(x_0, y_0)|^2} \ .$$

The integral (1.3.11) will be understood in the following sense. Fix a point $M^*(x, y) \in D$ outside the boundary of the domain D, $M^* \notin \partial D$, so that $M(x, y, f(x, y)) \in \sigma$ and $M(x, y, f(x, y)) \notin \partial\sigma$. Consider an ellipse on the plane OX_0Y_0 with center at $M^*(x, y)$ and the following structure. The focal axis of that ellipse coincides with the straight line l passing through the point $M^*(x, y)$ and parallel to the tangential plane P to the surface σ at the point $M(x, y, f(x, y))$. Let $a = \varepsilon$ be the length of the focal semi-axis on that line. The other axis of symmetry of the ellipse passes through the point $M^*(x, y)$ in the direction orthogonal to l, and the length of the focal semi-axis on that line is equal to $b = \varepsilon|\cos\beta|$, where β is the angle between the plane OXY and the tangential plane P. Consider the cylindrical surface whose generatrix is parallel to OZ and the directrix coincides with the ellipse. This cylindrical surface cuts from the plane P a circle of radius ε with center at $M(x, y, f(x, y))$. The normal vector to the plane P at the point $M^*(x, y)$ is $N = f_x'(x, y)i + f_y'(x, y)j - k$, and

$$|\cos\beta| = \frac{|(N, k)|}{|N||k|} = \frac{1}{\sqrt{1 + f_x'^2(x, y) + f_y'^2(x, y)}} \ .$$

Denote by U_ε the interior of the said ellipse. Then the integral (1.3.11) is defined as follows *Poltavskii* (1993):

$$I(\nu) = \lim_{\varepsilon \to 0} \left[\frac{1}{4\pi} \iint_{D \setminus U_\varepsilon} K(x, y, x_0, y_0)g(x_0, y_0) \, dx_0 \, dy_0 - \frac{g(x, y)}{2\varepsilon} \right]. \qquad (1.3.12)$$

Consider polar coordinates with the pole at $M^*(x, y)$ and the polar axis parallel to OX_0 and having the same direction. Then

$$x_0 - x = r\cos\varphi, \qquad y_0 - y = r\sin\varphi,$$

where $r = \left| \overline{M^* M_0^*} \right|$ and φ is the polar angle of the point $M_0^*(x_0, y_0)$ in these coordinates. Assuming that $f(x_0, y_0)$ is sufficiently smooth and applying the Taylor formula to $f(x_0, y_0)$ and $F_1(x_0, y_0)$ (see (1.3.11)) at the point $M^*(x, y)$, we obtain

$$F_1(x_0, y_0) = F_1(x, y) + rA(x, y, \varphi) + r^2 O_1(x, y, t, \varphi), \tag{1.3.13}$$

where

$$A(x, y, \varphi) = A_1(x, y) \cos \varphi + A_2(x, y) \sin \varphi, \qquad |O_1(x, y, r, \varphi)| < C,$$
$$F_2^{-3}(x, y, x_0, y_0) = \frac{1}{r^3 L^{3/2}(x, y, \varphi)} + \frac{B(x, y, \varphi)}{r^2 L^{5/2}(x, y, \varphi)} + \frac{O_2(x, y, r, \varphi)}{t}, \tag{1.3.14}$$

where

$$B(x, y, \varphi) = B_1(x, y) \cos^3 \varphi + B_2(x, y) \cos^2 \varphi \sin \varphi + B_3(x, y) \cos \varphi \sin^2 \varphi + B_4(x, y) \sin^3 \varphi,$$
$$|O_2(x, y, r, \varphi)| < C, \qquad L(x, y, \varphi) = 1 + \left[f'_{x_0}(x, y) \cos \varphi + f'_{y_0}(x, y) \sin \varphi \right]^2.$$

Remark 1.3.1. We say that a function $\lambda(x, y, \varphi)$ defined on the unit circle *is even with respect to* φ, if $\lambda(x, y, \varphi + \pi) = \lambda(x, y, \varphi)$ for any φ; and $\lambda(x, y, \varphi)$ *is odd with respect to* φ, if $\lambda(x, y, \varphi + \pi) = -\lambda(x, y, \varphi)$. For an odd function, its integral in φ over any segment of length 2π is equal to zero.

Now, we note that the function $L(x, y, \varphi)$ is even with respect to φ and the functions $A(x, y, \varphi)$ and $B(x, y, \varphi)$ are odd in the above sense. Therefore,

$$\int_0^{2\pi} A(x, y, \varphi) \, d\varphi = \int_0^{2\pi} B(x, y, \varphi) \, d\varphi = 0. \tag{1.3.15}$$

Let us represent the kernel $K(x, y, x_0, y_0)$ of the integral (1.3.11) in the form

$$K(x, y, x_0, y_0) = \frac{K_1(x, y, \varphi)}{r^3} + \frac{K_2(x, y, \varphi)}{r^2} + \frac{K_3(x, y, \varphi)}{r}, \tag{1.3.16}$$

where

$$K_1(x, y, \varphi) = \frac{F_1(x, y)}{L^{3/2}(x, y, \varphi)},$$
$$K_2(x, y, \varphi) = \frac{A(x, y, \varphi)}{L^{3/2}(x, y, \varphi)} + \frac{F_1(x, y) B(x, y, \varphi)}{L^{5/3}(x, y, \varphi)},$$
$$|K_3(x, y, r, \varphi)| < C.$$

Thus, the function $K_2(x, y, \varphi)$ is odd with respect to φ in the above sense, and therefore,

$$\int_0^{2\pi} K_2(x, y, \varphi) = 0. \tag{1.3.17}$$

Let us examine the integral (1.3.11), using the expansion of its kernel in terms of (1.3.16). We assume that $g(x_0, y_0) \in C^2[D]$, which means that $g(x_0, y_0)$ is twice continuously differentiable in D and vanishes on ∂D. Then, it follows from (1.3.16) that the integral

$$I_3(\nu) = \iint_D \frac{1}{r} K_3(x, y, r, \varphi) g(x_0, y_0) \, dx_0 \, dy_0 = \iint_{D^*} K_3(x, y, r, \varphi) b(x, y, r, \varphi) \, dr \, d\varphi, \tag{1.3.18}$$

with $b(x, y, r, \varphi) = g(x_0, y_0) = g(x + r \cos \varphi, \, y + r \sin \varphi)$, is absolutely convergent.

Lemma 1.3.1. *Let $g(x_0, y_0) \in C^2[D]$. Then*

$$\lim_{\varepsilon \to 0} \iint_{D \setminus U_\varepsilon} \frac{g(x_0, y_0)}{r^2} K_2(x, y, \varphi) \, dx_0 \, dy_0 = \lim_{\varepsilon \to 0} \iint_{D \setminus U_\varepsilon^*} \frac{g(x_0, y_0)}{r^2} K_2(x, y, \varphi) \, dx_0 \, dy_0, \quad (1.3.19)$$

where U_ε^* is the interior of the circle of radius ε with center at $M^*(x, y)$ on the plane OXY.

Proof. By virtue of (1.3.17), the limit on right-hand side of (1.3.19) exists. Therefore, in order to prove our statement, it suffices to show that

$$\lim_{\varepsilon \to 0} \iint_{U_\varepsilon^* \setminus U_\varepsilon} \frac{K_2(x, y, \varphi)}{r} b(x, y, r, \varphi) \, dr \, d\varphi = 0. \quad (1.3.20)$$

Let us change the variables: $\varphi = \psi + \alpha$, where the angle α is defined by

$$f'_{y_0}(x, y) \cos \alpha - f'_{x_0}(x, y) \sin \alpha = 0. \quad (1.3.21)$$

If $f'_{x_0}(x, y) = f'_{y_0}(x, y) = 0$, then α can be taken arbitrary, say, $\alpha = 0$.

Consider Cartesian coordinates with the origin at $M^*(x, y)$ and the axes parallel to the original ones, $M^*\xi \parallel OX_0$ and $M^*\eta \parallel OY_0$. Turning this coordinate frame by the angle α about the point $M^*(x, y)$, we obtain another coordinate frame $M^*\xi_1\eta_1$. Thus, we have

$$\begin{aligned} x_0 &= x + \xi = x + \xi_1 \cos \alpha - \eta_1 \sin \alpha, \\ y_0 &= y + \eta = y + \xi_1 \sin \alpha + \eta_1 \cos \alpha. \end{aligned} \quad (1.3.22)$$

The first partial derivatives of the function $f(x, y)$ with respect to the new variables have the form

$$\begin{aligned} \frac{\partial f}{\partial \xi_1} &= \frac{\partial f}{\partial x_0} \frac{\partial x_0}{\partial \xi_1} + \frac{\partial f}{\partial y_0} \frac{\partial y_0}{\partial \xi_1} = f'_{x_0}(x_0, y_0) \cos \alpha + f'_{y_0}(x_0, y_0) \sin \alpha, \\ \frac{\partial f}{\partial \eta_1} &= \frac{\partial f}{\partial x_0} \frac{\partial x_0}{\partial \eta_1} + \frac{\partial f}{\partial y_0} \frac{\partial y_0}{\partial \eta_1} == -f'_{x_0}(x_0, y_0) \sin \alpha + f'_{y_0}(x_0, y_0) \cos \alpha. \end{aligned} \quad (1.3.23)$$

In view of (1.3.21), we have $f'_{\eta_1} = 0$ at the point $M^*(x, y)$, and therefore, the axes η_1 and ξ_1 are the symmetry axes of the ellipse U_ε, and η_1 is its focal axis. By the definition of the ellipse U_ε, its equation has the form

$$\frac{\eta_1^2}{\varepsilon^2} + \frac{\xi_1^2}{\varepsilon^2 \cos^2 \beta} = 0. \quad (1.3.24)$$

In terms of the polar radius r, we can write $\xi_1 = r \cos \psi$, $\eta_1 = r \sin \psi$, and therefore, the equation of the ellipse becomes

$$\frac{r^2 \sin^2 \psi}{\varepsilon^2} + \frac{r^2 \cos^2 \psi}{\varepsilon^2 \cos^2 \beta} = 1. \quad (1.3.25)$$

From the definition of the angle β, we get

$$\frac{1}{\cos^2 \beta} = 1 + \left[f'^2_{x_0}(x, y) + f'^2_{y_0}(x, y) \right] = 1 + A_\varepsilon^2(x, y). \quad (1.3.26)$$

Using (1.3.25) and (1.3.26), we find that

$$r = \frac{\varepsilon}{\sqrt{1 + \cos^2 \psi A_\varepsilon^2(x, y)}} = r_0(\psi, \varepsilon). \quad (1.3.27)$$

Remark 1.3.2. If we take $\varphi = \psi + \alpha$ in the function $L(x, y, \varphi)$, which is even with respect to φ in the above sense, then, using (1.3.21), we get

$$L(x, y, \varphi) = L_\varepsilon(x, y, \psi) = 1 + \cos^2 \psi A_\varepsilon^2(x, y),$$

where $A_\varepsilon^2(x, y)$ is defined by (1.3.26). Therefore, we can write

$$r_0(\psi, \varepsilon) = r_0^*(\varphi, \varepsilon) = \frac{\varepsilon}{\sqrt{L(x, y, \varphi)}} .$$

Let $b(x, y, 0, \varphi) = b(x, y) = g(x, y)$ (see (1.3.18)) and consider the following transformations:

$$\iint_{U_\varepsilon^* \setminus U_\varepsilon} \frac{K_2(x, y, \varphi)}{r} b(x, y, r, \varphi)\, dr\, d\varphi = \iint_{U_\varepsilon^* \setminus U_\varepsilon} \frac{b(x, y, r, \varphi) - b(x, y)}{r} K_2(x, y, \varphi)\, dr\, d\varphi +$$

$$+\, b(x, y) \int_0^{2\pi} K_2(x, y, \varphi)\, d\varphi \int_{r_0^*(\varphi, \varepsilon)}^\varepsilon \frac{dr}{r} = I_1(\varepsilon) + I_2(\varepsilon). \tag{1.3.28}$$

Since $b(x, y) = g(x, y) \in C^2[D]$, the integral $I_1(\varepsilon)$ is absolutely convergent, and therefore,

$$\lim_{\varepsilon \to 0} I_1(\varepsilon) = 0. \tag{1.3.29}$$

Further, we have

$$I_2(\varepsilon) = b(x, y) \int_0^{2\pi} K_2(x, y, \varphi) \ln \sqrt{L(x, y, \varphi)}\, d\varphi, \tag{1.3.30}$$

since the integrand is odd with respect to φ. Relations (1.3.29) and (1.3.30) imply (1.3.20). This completes the proof of Lemma 1.3.1.

Now, consider the first term in (1.3.16),

$$A_1(\varepsilon) = \frac{1}{4\pi} \iint_{D \setminus U_\varepsilon} \frac{b(x, y, r, \varphi)}{r^2} K_1(x, y, \varphi)\, dr\, d\varphi - \frac{g(x, y)}{2\varepsilon} . \tag{1.3.31}$$

We assume that the boundary Γ of the domain D is three times continuously differentiable and that the function $g(x_0, y_0)$ has been extended to the plane $\mathbb{R}_2$ as $\tilde{g}(x_0, y_0) \in C^2[\mathbb{R}_2]$, so that $\tilde{g}(x_0, y_0) = 0$ outside an ellipse $U_R \supset D$. Assuming that $g(x_0, y_0)$ is of class C^2 inside D, we transform the integral (1.3.31) as follows:

$$A_1(\varepsilon) = \frac{1}{4\pi} \iint_{U_R \setminus U_\varepsilon^*} \frac{b(x, y, r, \varphi) - g(x, y)}{r^2} K_1(x, y, \varphi)\, dr\, d\varphi +$$

$$+ \left[g(x, y) \frac{1}{4\pi} \iint_{U_R \setminus U_\varepsilon^*} \frac{K_1(x, y, r, \varphi)}{r^2}\, dr\, d\varphi - \frac{g(x, y)}{2\varepsilon} \right] =$$

$$= B_1 + B_2 . \tag{1.3.32}$$

Taking the Taylor expansion of the function $b(x, y, r, \varphi) = g(x_0, y_0)$ at the point $M^*(x, y)$, we get

$$B_1 = \frac{1}{4\pi} \int_0^{2\pi} K_1(x, y, \varphi)\, d\varphi \int_{r_0^*}^\varepsilon \frac{g_{x_0}'(x, y) r \cos \varphi + g_{y_0}'(x, y) r \sin \varphi}{r^2}\, dr +$$

$$+ \frac{1}{4\pi} \iint_{U_\varepsilon^* \setminus U_\varepsilon} K_1(x, y, \varphi) \frac{r^2 A_\varepsilon(x, y, r, \varphi)}{r^2}\, dr\, d\varphi = M_1(\varepsilon) + M_2(\varepsilon). \tag{1.3.33}$$

Here, $|A_\varepsilon(x, y, r, \varphi)| < C$, meas $(U_\varepsilon^* \setminus U_\varepsilon) \to 0$ as $\varepsilon \to 0$, and therefore,

$$\lim_{\varepsilon \to 0} M_2(\varepsilon) = 0. \tag{1.3.34}$$

Further, we have

$$M_1(\varepsilon) = \frac{1}{4\pi} \int_0^{2\pi} K_1(x, y, \varphi) \left[g'_{x_0}(x, y) \cos \varphi + g'_{y_0}(x, y) \sin \varphi \right] \ln L^{1/2}(x, y, \varphi) \, d\varphi = 0, \tag{1.3.35}$$

since the integrand is odd with respect to φ.

It follows from (1.3.33)–(1.3.35) that

$$\lim_{\varepsilon \to 0} B_1(\varepsilon) = 0. \tag{1.3.36}$$

Using the properties of integrals of periodic functions, we find that

$$\iint_{U_R \setminus U_\varepsilon} \frac{K_1(x, y, \varphi)}{r^2} \, dr \, d\varphi = \int_0^{2\pi} \frac{\left[1 + A_*^2(x, y) \right]^{1/2}}{L_*^{3/2}(x, y, \psi)} \, d\psi \int_{r_0(\psi, \varepsilon)}^{r_0(\psi, R)} \frac{dr}{r^2} =$$

$$= \left(\frac{1}{\varepsilon} - \frac{1}{R} \right) \int_0^{2\pi} \frac{\left[1 + A_*^2(x, y) \right]^{1/2}}{1 + A_*^2(x, y) \cos^2 \psi} \, d\psi = 2\pi \left(\frac{1}{\varepsilon} - \frac{1}{R} \right). \tag{1.3.37}$$

From (1.3.37), it follows that

$$B_2 = -\frac{g(x, y)}{2R}. \tag{1.3.38}$$

Further, we get

$$\iint_{U_R \setminus U_\varepsilon^*} \frac{K_1(x, y, \varphi)}{r^2} \, dr \, d\varphi = \int_0^{2\pi} \frac{\left[1 + A_*^2(x, y) \right]^{1/2}}{L_*^{3/2}(x, y, \psi)} \, d\psi \int_\varepsilon^{r_0(\psi, R)} \frac{dr}{r^2} =$$

$$= \frac{\left[1 + A_*^2(x, y) \right]^{1/2}}{\varepsilon} \int_0^{2\pi} \frac{d\psi}{L^{3/2}(x, y, \psi)} - \frac{1}{R} \int_0^{2\pi} \frac{\left[1 + A_*^2(x, y) \right]^{1/2} \, d\psi}{1 + A_*^2(x, y) \cos^2 \psi} =$$

$$= \frac{1}{\varepsilon} G(x, y) - \frac{2\pi}{R}, \tag{1.3.39}$$

$$G(x, y) = \left[1 + A_*^2(x, y) \right]^{1/2} \int_0^{2\pi} \frac{d\psi}{L^{3/2}(x, y, \psi)}.$$

From (1.3.32), using (1.3.36)–(1.3.39), we obtain

$$A_1(\varepsilon) = \frac{1}{4\pi} \iint_{D \setminus U_\varepsilon^*} \frac{K_1(x, y, \varphi)}{r^2} b(x, y, r, \varphi) \, dr \, d\varphi - \frac{g(x, y)}{4\pi\varepsilon} G(x, y). \tag{1.3.40}$$

On the basis of the above arguments, we can make the following conclusion.

Theorem 1.3.4. *If $g(M) \in C^2(\sigma^*)$, then for any point $M \in \sigma^*$, the integral (1.3.10) exists and can be represented in the form*

$$I(\nu) = \frac{1}{4\pi} \int_\sigma \frac{\nu(M_0) \, d\sigma_{M_0}}{|M_0 M|^3} = \frac{1}{4\pi} \iint_D \frac{g(x_0, y_0) \left[1 + f_{x_0}'^2(x_0, y_0) + f_{y_0}'^2(x_0, y_0) \right]^{1/2} dx_0 \, dy_0}{\left[(x - x_0)^2 + (y - y_0)^2 + (f(x, y) - f(x_0, y_0))^2 \right]^{3/2}} =$$

$$= \lim_{\varepsilon \to 0} \left[\frac{1}{4\pi} \iint_{D \setminus U_\varepsilon^*} \frac{K_1(x, y, \varphi) b(x, y, r, \varphi)}{r^2} \, dr \, d\varphi - \frac{g(x, y)}{4\pi\varepsilon} G(x, y) \right] +$$

$$+ \frac{1}{4\pi} \iint_D \frac{K_2(x, y, \varphi)}{r^2} g(x_0, y_0) \, dx_0 \, dy_0 + \frac{1}{4\pi} \iint_D \frac{K_3(x, y, \varphi)}{r} g(x_0, y_0) \, dx_0 \, dy_0 =$$

$$= I_1 + I_2 + I_3, \tag{1.3.41}$$

where $G(x, y)$ is defined in (1.3.39).

Remark 1.3.3. Suppose that there exist partial derivatives $g'_{x_0}(x_0, y_0)$ and $g'_{y_0}(x_0, y_0)$ for which the Hölder condition holds with exponent α, $0 < \alpha \leq 1$. Then, we can write

$$g(x_0, y_0) - g(x, y) = g(x_0, y_0) - g(x_0, y) + g(x_0, y) - g(x, y) =$$
$$= g'_{y_0}(x, y_0 + \theta_1(y_0 - y))(y_0 - y) + g'_{x_0}(x_0 + \theta_2(x_0 - x), y)(x_0, x) =$$
$$= g'_{x_0}(x, y)(x_0 - x) + g'_{y_0}(x, y)(y_0 - y) + A(x_0, y_0, x, y). \tag{1.3.42}$$

In view of the Hölder continuity of the partial derivatives of $g(x_0, y_0)$, we get

$$|A(x_0, y_0, x, y)| \leq |x_0 - x| \, \left| g'_{x_0}(x_0 + \theta_2(x_0 - x), y) - g'_{x_0}(x, y) \right| +$$
$$+ |y_0 - y| \, \left| g'_{y_0}(x_0 + \theta_1(y_0 - y)) - g'_{y_0}(x, y) \right| \leq$$
$$\leq C_1 |x_0 - x| \, |1 + \theta_2|^\alpha |x_0 - x|^\alpha + C_2 |y_0 - y| \, |1 + \theta_1|^\alpha \, |y_0 - y|^\alpha \leq C_3 r^{1+\alpha}. \tag{1.3.43}$$

where $r = \left[(x_0 - x)^2 + (y_0 - y)^2 \right]^{1/2}$.

We write $g(M) \in C^1_\alpha(\sigma^*)$, if the first partial derivatives of this function locally satisfy the Hölder condition with exponent α (the constant in the Hölder inequality may become arbitrarily large near the boundary).

Now, the assumptions of Theorem 1.3.4 can be weakened.

Theorem 1.3.5. *If $g(M) \in C^1_\alpha(\sigma^*)$, then for any point $M \in \sigma^*$, the Hadamard integral* (1.3.10) *exists and can be represented in the form* (1.3.41).

1.4. Two-Dimensional Equations

The development of a general theory of the existence of solutions for two-dimensional singular and hypersingular integral equations on piecewise smooth surfaces encounters serious difficulties. In subsequent chapters, the existence of solutions is established for certain hypersingular integral equations on smooth surfaces or a union of disjoint smooth surfaces. In order to work out various numerical methods, it would be reasonable to start with the examination of some two-dimensional singular and hypersingular integral equations whose solutions can be represented by analytical formulas.

As an example, consider the problem of flow past a rectangular airfoil. *Bisplinghoff, Ashley, and Halfman* (1958) have shown that for an inviscid incompressible fluid, this problem can be reduced to a bisingular integral equation of the first kind. This equation has the form

$$\frac{1}{\pi^2} \int_{-1}^{1} \int_{-1}^{1} \frac{\left[(x - x_0)^2 + (y - y_0)^2 \right]^{1/2}}{(x - x_0)(y - y_0)} \gamma(x, y) \, dx \, dy = f(x_0, y_0), \tag{1.4.1}$$

where $(x_0, y_0) \in \tilde{I}_0^2 = (-1, 1) \times (-1, 1)$. No analytical solution is known for this problem, and it has to be solved by approximate methods. As shown by *Bisplinghoff, Ashley, and Halfman* (1958), for a rectangular airfoil of a large span, equation (1.4.1) can be replaced by

$$\frac{1}{\pi^2} \int_{-1}^{1} \int_{-1}^{1} \frac{\gamma(x, y) \, dx \, dy}{(x - x_0)(y - y_0)} = f(x_0, y_0), \tag{1.4.2}$$

where $(x_0, y_0) \in \tilde{I}_0^2$. Equation (1.4.2) can be solved analytically. Indeed, for a Hölder continuous $f(x_0, y_0)$ on I^2, all solutions of this equation in the class of absolutely integrable functions on $I_0^2 = [-1, 1] \times [-1, 1]$ have been constructed in *Belotserkovskii and Lifanov* (1993) and *Lifanov* (1996). Now, we are going to consider analytical solutions of equation (1.4.2) in a wider functional class (see *Elliot, Lifanov, and Litvinchuk* (1997)). This corresponds to the *problem of flow with suction*, in which case the solution has a singularity of the type $1/x$ with respect to one of the variables. Let us describe the problem in more precise terms. For a given point $q \in (-1, 1) \subset OX$, by $H_q^*(x, y)$ we denote the class of functions which can be written in the form

$$\varphi(x) + \frac{\psi(x, y)}{q - x},$$

where φ and ψ are Hölder continuous on any compact subdomain of $\widetilde{I_0^2}$ and are absolutely integrable on I_0^2. Our immediate aim is to construct an analytical solution of equation (1.4.2) under various conditions. In order to avoid the necessity of choosing a solution, we make additional assumptions ensuring its uniqueness.

Theorem 1.4.1. *Let f be a Hölder continuous function on I^2. A solution of equation* (1.4.2), *which is bounded on the lines $x = \pm 1$ and $y = \pm 1$ and belongs to the class $H_q^*(x, y)$ on I_0^2, is unique and is given by the formula*

$$\gamma(x_0, y_0) = \frac{(1 - x_0^2)^{1/2}(1 - y_0^2)^{1/2}}{\pi^2} \int_{-1}^{1} \int_{-1}^{1} \frac{f(x, y)\, dx\, dy}{(1 - x^2)^{1/2}(1 - y^2)^{1/2}(x - x_0)(y - y_0)} -$$
$$- \frac{(1 - x_0^2)^{1/2}(1 - y_0^2)^{1/2}}{\pi^2(q - x_0)} \int_{-1}^{1} \frac{dx}{(1 - x^2)^{1/2}} \left(\frac{1}{\pi^2} \int_{-1}^{1} \frac{f(x, y)\, dx}{(1 - y^2)^{1/2}(y - y_0)} \right), \quad (1.4.3)$$

provided that

$$\int_{-1}^{1} \frac{f(x_0, y)\, dy}{(1 - y^2)^{1/2}} = 0, \qquad -1 \le x_0 \le 1. \quad (1.4.4)$$

Proof. We can rewrite (1.4.2) as

$$\frac{1}{\pi} \int_{-1}^{1} \frac{dy}{y - y_0} \left(\frac{1}{\pi} \int_{-1}^{1} \frac{\gamma(x, y)\, dx}{x - x_0} \right) = f(x_0, y_0), \quad (1.4.5)$$

which in turn can be written as

$$\frac{1}{\pi} \int_{-1}^{1} \frac{\psi(x_0, y)}{y - y_0}\, dy = f(x_0, y_0), \quad (1.4.6)$$

where

$$\psi(x_0, y) = \frac{1}{\pi} \int_{-1}^{1} \frac{\gamma(x, y)\, dx}{x - x_0}. \quad (1.4.7)$$

Equation (1.4.6) can be regarded as a one-dimensional singular integral equation for ψ with the parameter x_0. From (1.2.7), we have

$$\psi(x_0, y_0) = -\frac{(1 - y_0^2)^{1/2}}{\pi} \int_{-1}^{1} \frac{f(x_0, y)\, dy}{(1 - y^2)^{1/2}(y - y_0)}, \quad (1.4.8)$$

provided that

$$\int_{-1}^{1} \frac{f(x_0, y)\, dy}{(1 - y^2)^{1/2}} = 0. \quad (1.4.9)$$

This condition coincides with (1.4.4). That ψ in (1.4.8) is Hölder continuous on I^2 can be established by arguments similar to those used in *Muskhelishvili* (1968). Returning to (1.4.7), let us consider ψ

as a known function and y as a parameter. Then, using (1.2.13), we solve this equation with respect to γ in the class $H_q^*(x, y)$. From (1.2.13), it follows that

$$\gamma(x_0, y) = -\frac{(1 - x_0^2)^{1/2}}{\pi} \int_{-1}^{1} \frac{\psi(x, y) \, dx}{(1 - x^2)^{1/2}(x - x_0)} + \frac{(1 - x_0^2)^{1/2}}{q - x_0} \frac{1}{\pi} \int_{-1}^{1} \frac{\psi(x, y) \, dx}{(1 - x^2)^{1/2}}. \qquad (1.4.10)$$

Hence, taking ψ of the form (1.4.8), we obtain (1.4.3).

Now, let us show that the result of solving equation (1.4.2) will be the same, if we initially write this equation in the form

$$\frac{1}{\pi} \int_{-1}^{1} \frac{\varphi(x, y_0) \, dx}{x - x_0} = f(x_0, y_0), \qquad (1.4.11)$$

where

$$\varphi(x, y_0) = \frac{1}{\pi} \int_{-1}^{1} \frac{\gamma(x, y) \, dy}{y - y_0}. \qquad (1.4.12)$$

Solving equation (1.4.11) in the class $H_q^*(x, y)$ with the help of (1.2.13), we get

$$\varphi(x_0, y_0) = -\frac{(1 - x_0^2)^{1/2}}{\pi} \int_{-1}^{1} \frac{f(x, y_0) \, dx}{(1 - x^2)^{1/2}(x - x_0)} + \frac{(1 - x_0^2)^{1/2}}{q - x_0} \frac{1}{\pi} \int_{-1}^{1} \frac{f(x, y_0) \, dx}{(1 - x^2)^{1/2}}. \qquad (1.4.13)$$

Now, taking φ given by (1.4.13), let us solve equation (1.4.2) with respect to γ, assuming that γ is bounded on the edges. From (1.2.17), we again obtain (1.4.3). But the condition in (1.2.7) shows that this function will be a solution if

$$\int_{-1}^{1} \frac{dy}{(1 - y^2)^{1/2}} \left\{ -\frac{(1 - x_0^2)^{1/2}}{\pi} \int_{-1}^{1} \frac{f(x, y) \, dx}{(1 - x^2)^{1/2}(x - x_0)} + \right.$$
$$\left. + \frac{(1 - x_0^2)^{1/2}}{q - x_0} \frac{1}{\pi} \int_{-1}^{1} \frac{f(x, y) \, dx}{(1 - x^2)^{1/2}} \right\} = 0, \qquad (1.4.14)$$

for all $x_0 \in [-1, 1]$. Now (1.4.14) can be written in the form

$$-F(x_0) + \frac{A}{q - x_0} = 0 \qquad (1.4.15)$$

for all $x_0 \in (-1, 1)$, $x_0 \neq q$, where

$$F(x_0) = \frac{1}{\pi} \int_{-1}^{1} \frac{dx}{(1 - x^2)^{1/2}(x - x_0)} \left(\int_{-1}^{1} \frac{f(x, y) \, dy}{(1 - y^2)^{1/2}} \right), \qquad (1.4.16)$$

$$A = \int_{-1}^{1} \int_{-1}^{1} \frac{f(x, y) \, dx \, dy}{(1 - x^2)^{1/2}(1 - y^2)^{1/2}}. \qquad (1.4.17)$$

To ensure (1.4.15), we should have $A = 0$ and $F(x_0) = 0$ on $(-1, 1)$. If $F(x_0) = 0$ on $(-1, 1)$, then

$$\int_{-1}^{1} \frac{f(x, y) \, dy}{(1 - y^2)^{1/2}} = B \qquad (1.4.18)$$

with some constant B. In this case, (1.4.17) yields

$$0 = B \int_{-1}^{1} \frac{dx}{(1 - x^2)^{1/2}}, \qquad (1.4.19)$$

and thus $B = 0$. Thereby, we have obtained condition (1.4.4) and the proof is complete.

Theorem 1.4.2. *Let $\gamma(x, y)$ be a solution of equation* (1.4.2) *of class $H_q^*(x, y)$ on I_0^2. Suppose that $\gamma(x, y)$ satisfies the additional condition*

$$\int_{-1}^{1} dx \frac{1}{\pi} \int_{-1}^{1} \frac{\gamma(x, y)\, dy}{y - y_0} = A_1(y_0), \tag{1.4.20}$$

is bounded on the lines $x = 1$, $y = 1$, $y = -1$, and is absolutely integrable on the line $x = -1$. Then, the solution $\gamma(x, y)$ is given by the formula

$$\gamma(x_0, y_0) = \frac{\left(1 - y_0^2\right)^{1/2}}{\pi^2} \left(\frac{1 - x_0}{1 + x_0}\right)^{1/2} \int_{-1}^{1} \int_{-1}^{1} \left(\frac{1 + x}{1 - x}\right)^{1/2} \frac{f(x, y)\, dx\, dy}{(1 - y^2)^{1/2}(x - x_0)(y - y_0)} -$$

$$- \frac{1}{\pi} \left(\frac{1 - x_0}{1 + x_0}\right)^{1/2} \frac{\left(1 - y_0^2\right)^{1/2}}{q - x_0} \int_{-1}^{1} \left[A_1(y) + \int_{-1}^{1} \left(\frac{1 + x}{1 - x}\right)^{1/2} f(x, y)\, dx\right] \frac{dy}{(1 - y^2)^{1/2}(y - y_0)}, \tag{1.4.21}$$

provided that

$$\int_{-1}^{1} \frac{f(x_0, y)\, dy}{(1 - y^2)^{1/2}} = 0, \tag{1.4.22}$$

for all $x \in (-1, 1)$, and

$$\int_{-1}^{1} \frac{A_1(y)\, dy}{(1 - y^2)^{1/2}} = 0. \tag{1.4.23}$$

Proof. Let us rewrite (1.4.2) and (1.4.20) as

$$\frac{1}{\pi} \int_{-1}^{1} \frac{dy}{y - y_0} \left(\frac{1}{\pi} \int_{-1}^{1} \frac{\gamma(x, y)\, dx}{x - x_0}\right) = f(x_0, y_0), \tag{1.2.24}$$

$$\frac{1}{\pi} \int_{-1}^{1} \frac{dy}{y - y_0} \left(\int_{-1}^{1} \gamma(x, y)\, dx\right) = A_1(y_0), \tag{1.4.25}$$

respectively. From these equations, with x_0 regarded as a parameter, taking into account that the solutions should be bounded on the lines $y = 1$, $y = -1$, we find, by virtue of (1.2.7), that

$$\frac{1}{\pi} \int_{-1}^{1} \frac{\gamma(x, y_0)\, dx}{x - x_0} = -\frac{\left(1 - y_0^2\right)^{1/2}}{\pi} \int_{-1}^{1} \frac{f(x_0, y)\, dy}{(1 - y^2)(y - y_0)}, \tag{1.4.26}$$

provided that condition (1.4.22) holds. Again, because of (1.4.23), we have from (1.4.25)

$$\int_{-1}^{1} \gamma(x, y_0)\, dx = -\frac{\left(1 - y_0^2\right)^{1/2}}{\pi} \int_{-1}^{1} \frac{A_1(y)\, dy}{(1 - y^2)(y - y_0)}. \tag{1.4.27}$$

Viewing (1.4.26) as equation (1.2.2) with the parameter y_0 and solving this equation in the class $H_q^*(x, y)$, we find from (1.2.12) that

$$\gamma(x_0, y_0) = \frac{\left(1 - y_0^2\right)^{1/2}}{\pi^2} \left(\frac{1 - x_0}{1 + x_0}\right)^{1/2} \int_{-1}^{1} \int_{-1}^{1} \left(\frac{1 + x}{1 - x}\right)^{1/2} \frac{f(x, y)\, dx\, dy}{(1 - y^2)(x - x_0)(y - y_0)} +$$

$$+ \frac{A(y_0)}{\pi} \left(\frac{1 - x_0}{1 + x_0}\right)^{1/2} \frac{1}{q - x_0}. \tag{1.4.28}$$

Hence, using the relation

$$\frac{1}{\pi} \int_{-1}^{1} \left(\frac{1-x}{1+x} \right)^{1/2} \frac{dx}{x-x_0} = -1 \tag{1.4.29}$$

for all $x_0 \in (-1, 1)$, we find that

$$\int_{-1}^{1} \gamma(x_0, y_0) \, dx_0 = A(y_0) + \frac{(1-y_0^2)^{1/2}}{\pi} \int_{-1}^{1} \left(\frac{1-x}{1+x} \right)^{1/2} \left(\int_{-1}^{1} \frac{f(x,y) \, dy}{(1-y^2)^{1/2}(y-y_0)} \right) dx. \tag{1.4.30}$$

Relation (1.4.27) shows that the unknown function $A(y_0)$ can be expressed through $A_1(y_0)$, and by (1.4.28) we obtain (1.4.21).

Again, as in the proof of the preceding theorem, we obtain the same solution if we change the order of iterated integration in (1.4.24) and (1.4.25). We omit the details. Theorem 1.4.2 is proved.

In a similar way, one can obtain other analytic formulas, depending on whether the solution is sought in the class of bounded or unbounded functions on the lines $x = \pm 1$ and $y = \pm 1$. Let us just formulate one more result.

Theorem 1.4.3. *The solution of equation* (1.4.2) *satisfying the additional condition*

$$\frac{1}{\pi} \int_{-1}^{1} \left(\int_{-1}^{1} \frac{\gamma(x,y) \, dx}{x-x_0} \right) dy = C(x_0), \tag{1.4.31}$$

belonging to the class $H_q^*(x, y)$ on I_0^2, bounded on the lines $x \pm 1$, and absolutely integrable on the lines $y = \pm 1$ (on which it may be unbounded) is given by the formula

$$
\gamma(x_0, y_0) = \frac{(1-x_0^2)^{1/2}}{(1-y_0^2)^{1/2} \pi^2} \int_{-1}^{1} \int_{-1}^{1} \frac{(1-y^2)^{1/2} f(x,y) \, dx \, dy}{(1-x^2)^{1/2}(x-x_0)(y-y_0)} -
$$
$$
- \frac{(1-x_0^2)^{1/2}}{(q-x_0)(1-y_0^2)^{1/2} \pi^2} \int_{-1}^{1} \int_{-1}^{1} \frac{(1-y^2)^{1/2} f(x,y) \, dx \, dy}{(1-x^2)^{1/2}(x-x_0)(y-y_0)} -
$$
$$
- \frac{(1-x_0^2)^{1/2}}{(1-y_0^2)^{1/2} \pi^2} \int_{-1}^{1} \frac{C(x) \, dx}{(1-x^2)^{1/2}(x-x_0)} +
$$
$$
+ \frac{(1-x_0^2)^{1/2}}{(q-x_0)(1-y_0^2)^{1/2} \pi^2} \int_{-1}^{1} \frac{C(x) \, dx}{(1-x^2)^{1/2}} . \tag{1.4.32}
$$

The results described so far pertain to two-dimensional operators which can be represented as a composition of two one-dimensional singular integral operators. Next, we consider another type of two-dimensional singular integral equations with analytical solutions.

Let σ be a closed smooth surface. With the help of the simple layer potential, the interior and the exterior Dirichlet problems for the Laplace equation can be reduced to the following weakly singular integral equation of the first kind:

$$\frac{1}{\pi} \int_{\sigma} \frac{\psi(M) d\sigma_M}{r_{MM_0}} = f(M_0), \qquad M \in \sigma, \tag{1.4.33}$$

where $d\sigma_M$ is the area element on the surface σ with respect to the coordinates of the point $M \in \sigma$.

Suppose now that σ is a sphere of radius R centered at the origin (we denote it by σ_R). If $f(M_0)$ is a spherical function, then the Dirichlet boundary value problems admit solutions which can also be expressed in terms of spherical functions (see *Lifanov* (1996) and *Schwartz* (1967)). Thus, using the relation between the limit values of the normal derivative of the simple layer potential and its density, we obtain the following relations (see *Lifanov* (1996)):

$$\frac{1}{4\pi} \int_{\sigma_R} \frac{Y_n(\theta, \varphi)\, d\sigma_{RM}}{r_{MM_0}} = \frac{R}{2n+1} Y_n(\theta_0, \varphi_0), \quad 0 \le \theta_0 \le \pi, \quad 0 \le \varphi_0 \le 2\pi, \quad n = 0, 1, \ldots, \quad (1.4.34)$$

where $Y_n(\theta, \varphi)$ are spherical functions of the order n,

$$Y_n(\theta, \varphi) = \sum_{m=0}^{n} (A_{nm} \cos m\varphi + B_{mn} \sin m\varphi) P_n^m(\cos \theta), \qquad (1.4.35)$$

where A_{nm} and B_{nm} are arbitrary constants;

$$P_n^m(\mu) = \left(1 - \mu^2\right)^{m/2} P_n^{(m)}(\mu), \quad n = 0, 1, \ldots; \quad m = 0, 1, \ldots, n, \qquad (1.4.36)$$

are the associated Legendre functions,

$$P_n(\mu) = \frac{1}{2^n n!} \frac{d^n}{d\mu^n} (\mu^2 - 1)^n, \qquad n = 0, 1, \ldots, \qquad (1.4.37)$$

are the Legendre polynomials of degree n.

In particular, if we take $n = 1$ in (1.4.34) and $A_{10} = 1$, $A_{11} = B_{11} = B_{10} = 0$ in (1.4.35), we obtain the following result: the function $\psi(M) = 3 \cos \theta$ is an explicit solution of equation (1.4.33) for the unit sphere σ_1 with $f(M_0) = \cos \theta_0$.

In a similar way, we find that for a closed smooth surface σ, the Dirichlet problem for the Helmholtz equation is reduced, by means of the simple layer potential for that equation, to the following weakly singular integral equation (see *Lifanov* (1996)):

$$\frac{1}{4\pi} \int_{\sigma} \frac{\exp\left(-ikr_{MM_0}\right)}{r_{MM_0}} \psi(M) d\sigma_M = f(M_0), \qquad M_0 \in \sigma. \qquad (1.4.38)$$

Then, for the sphere σ_R we obtain (see *Lifanov* (1996)):

$$\frac{1}{4\pi} \frac{\exp(-ikr_{MM_0})}{r_{MM_0}} Y(\theta, \varphi)\, d\sigma_{R_M} = B_n^{-1} Y_n(\theta_0, \varphi_0), \qquad n = 0, 1, \ldots. \qquad (1.4.39)$$

Here,

$$B_n = k \left[\frac{\eta_n'(kR)}{\eta_n(kR)} - \frac{\xi_n^{(2)'}(kR)}{\xi_n^{(2)}(kR)} \right], \qquad (1.4.40)$$

$$\eta_n(kR) = \frac{1}{\sqrt{kR}} J_{n+1/2}(KR), \qquad (1.4.41)$$

$$\xi_n^{(2)}(kR) = \frac{1}{\sqrt{kR}} H_{n+1/2}^{(2)}(kR), \qquad (1.4.42)$$

where $J_{n+1/2}(kR)$ is the Bessel function and $H_{n+1/2}^{(2)}(kR)$ is the Hankel function of the second kind.

Remark 1.4.1. It can be shown that $B_n \rightarrow (2n+1)/R$ as $k \rightarrow 0$ (see (1.4.34)).

For a closed smooth surface, the exterior and the interior Neumann problems for the Laplace equation can be reduced (by means of the double layer potential; see *Lifanov* (1996)) to the following hypersingular integral equation of the first kind:

$$\frac{1}{4\pi} \int_{\sigma} \frac{\partial}{\partial n_{M_0}} \frac{\partial}{\partial n_{M_0}} \left(\frac{1}{r_{MM_0}} \right) g(M) d\sigma_M = f(M_0), \qquad M_0 \in \sigma. \tag{1.4.43}$$

For the sphere σ_R, equation (1.4.43) admits exact solutions given by the formula

$$\frac{1}{4\pi} \int_{\sigma_R} \frac{\partial}{\partial n_{M_0}} \frac{\partial}{\partial n_{M_0}} \left(\frac{1}{r_{MM_0}} \right) Y_n(\theta, \varphi) d\sigma_{R_M} = -\frac{n(n+1)}{R(2n+1)} Y_n(\theta_0, \varphi_0), \quad n = 1, 2, \ldots . \tag{1.4.44}$$

For a closed smooth surface, the exterior and the interior Neumann problems for the Helmholtz equation are reduced (by means of the double layer potential; see *Lifanov* (1996)) to the following hypersingular integral equation of the first kind:

$$\frac{1}{4\pi} \int_{\sigma} \frac{\partial}{\partial n_{M_0}} \frac{\partial}{\partial n_{M_0}} \left(\frac{\exp(-ikr_{MM_0})}{r_{MM_0}} \right) g(M) \, d\sigma_M = f(M_0), \qquad M_0 \in \sigma. \tag{1.4.45}$$

If $\sigma = \sigma_R$ is a sphere, then the following relations hold:

$$\frac{1}{4\pi} \int_{\sigma_R} \frac{\partial}{\partial n_{M_0}} \frac{\partial}{\partial n_{M_0}} \left(\frac{\exp(-ikr_{MM_0})}{r_{MM_0}} \right) Y_n(\theta, \varphi) \, d\sigma_{R_M} = C_n^{-1} Y_n(\theta_0, \varphi_0), \tag{1.4.46}$$

where

$$C_n = \frac{1}{k} \left[\frac{\xi_n^{(2)}(kR)}{\xi^{(2)'}(kR)} - \frac{\eta_n(kR)}{\eta_n'(kR)} \right]. \tag{1.4.47}$$

Remark 1.4.2. It can be shown that $C_n \rightarrow -R(2n+1)/(n(n+1))$ as $k \rightarrow 0$ in (1.4.47).

Finally, let us show that for the sphere $\sigma = \sigma_R$, equation (1.4.43) can be solved in quadratures, i.e., we can find the inverse operator which has the form of an integral of the function on the left-hand side of (1.4.43)). Our exposition here follows *Bitsadze* (1986), *Lifanov* (1988).

Let D_0 be a simply connected domain in the n-dimensional Euclidean space $\mathbb{R}_n$, $n \geq 2$. The points $x \in \mathbb{R}_n$ have coordinates $x_1, \ldots, x_n$ in an orthogonal Cartesian frame, and x is regarded as the radius-vector of the point. Suppose that the domain D_0 is star-shaped with respect to the origin, and its boundary σ is an $(n-1)$-dimensional Lyapunov surface. Let $D_1 = \mathbb{R}_n \setminus (D_0 \cup \sigma)$. Suppose that $U_i(x) \in C^{1,0}(D_i \cup \sigma)$ is a harmonic function in the domain D_i, $i = 0, 1$. Then the function

$$x \cdot \operatorname{grad} U_i = V_i, \qquad i = 0, 1 \tag{1.4.48}$$

is harmonic in D_i.

Consider the first boundary value problem (the Dirichlet problem) in D_i for $U_i(x)$ with the boundary condition

$$\lim_{\substack{x \rightarrow x_0 \\ x \in D_i}} V_i(x) = f(x_0), \qquad x_0 \in \sigma, \quad i = 0, 1. \tag{1.4.49}$$

We introduce the notation

$$\lim_{\substack{x \rightarrow x_0 \\ x \in D_0}} V_0(x) = V_0^+(x_0), \qquad \lim_{\substack{x \rightarrow x_0 \\ x \in D_1}} V_1(x) = V_1^-(x_0).$$

It is well known that for any $f(x) \in C^{0,0}$, this problem has one and only one solution. Moreover, if there is a Green function $G(x, y)$, this solution can be expressed by quadrature formulas

$$V_0(x) = -\frac{1}{\omega_n} \int_\sigma \frac{\partial G(x, y)}{\partial n_y^+} f(y) \, d\sigma_y,$$

$$V_1(x) = -\frac{1}{\omega_n} \int_\sigma \frac{\partial G(x, y)}{\partial n_y^-} f(y) \, d\sigma_y,$$

(1.4.50)

where ω_n is the area of the unit sphere in $\mathbb{R}_n$; n_y^+ and n_y^- are, respectively, the unit vectors of the exterior and the interior normals to σ at the point y.

The Green function $G(x, y)$ for the Dirichlet problem (1.4.49) can be explicitly found only for a very narrow class of domains. If σ is the sphere of unit radius with center at the origin, the Green function has the form

$$G_i(x, y) = (-1)^i \left(E(x, y) - E\left(|x|y, \frac{x}{|x|} \right) \right),$$

where $E(x, y)$ is the fundamental solution of the Laplace equation. In this case, (1.4.50) turns into the Poisson formula

$$U_i(x) = \frac{(-1)^i}{\omega_n} \int_\sigma \frac{1 - |x|^2}{|y - x|^2} f(y) \, d\sigma_y,$$

(1.4.51)

where $x^2 = |x|^2$ and $(y-x)^2 = |y-x|^2$ are the scalar products xx and $(y-x)(y-x)$, respectively. The function $(-1)^i(|x|^2 - 1)/|y - x|^2$ is called the *Poisson kernel* for the interior ($i = 0$) and the exterior ($i = 1$) Dirichlet problem.

In what follows, we assume that D_0 is the unit ball $|X| < 1$. Since in this case, x_0 coincides with the exterior normal n_{x_0} to the unit sphere $\sigma = \{|x_0| = 1\}$, the boundary value problem (1.4.49) coincides with the following Neumann problem: *find a harmonic function* $U_i(x) \in C^{1,0}(D_i \cup \sigma)$ (*inside* σ ($i = 0$) *or outside* σ ($i = 1$)) *satisfying the boundary condition*

$$(-1)^i n_{x_0} \cdot \lim_{\substack{x \to x_0 \\ x \in D_i}} \operatorname{grad} U_i(x) \equiv \left(\frac{\partial U_i}{\partial (-1)^i n_{x_0}} \right)^{+(i=0)}_{-(i=1)} = f(x_0), \quad x_0 \in \sigma.$$

(1.4.52)

The notion of Green function can also be introduced for the Neumann problem, but its construction requires fairly difficult calculations (see *Prandtl* (1939), *Bouligand et al.* (1935), *Kellog* (1970)). Since $V_0(0) = 0$ (see (1.4.48)), the condition

$$\int_\sigma f(y) \, d\sigma_y = 0$$

(1.4.53)

is necessary and sufficient for the solvability of the Neumann problem (1.4.52), and the solution of this problem can be written in quadratures,

$$U_i(x) = \frac{1}{\omega_n} \int_\sigma N_i(x, y) f(t) \, d\sigma_y + C,$$

(1.4.54)

where $N_i(x, y)$ is the *Neumann kernel* of the corresponding problem. Because of (1.4.53), the Neumann kernel is defined to within an arbitrary additive constant. Below, we describe a simple method for the construction of $N_i(x, y)$ for any $n \geq 2$.

It is easy to see (*Bitsadze* (1986)) that the solution of the interior problem (1.4.52) has the form

$$U_0(x) = \int_0^1 V_0(tx) \frac{dt}{t} + C,$$

(1.4.55)

where $V_0(x)$ is given by (1.4.51), and we have $U_0(0) = 0$ in view of (1.4.53).

Indeed, form (1.4.55), changing the variables $tx = \xi$, we find that

$$U'_{0x_k} = \int_0^1 V'_{0\xi_k}\, dt, \qquad U''_{0x_k x_k} = \int_0^1 V''_{0\xi_k \xi_k} t\, dt,$$

which means that

$$\Delta U_0 = \int_0^1 \Delta V_0 t\, dt, \qquad x \cdot \mathrm{grad}\, U_0(x) = V_0(x). \tag{1.4.56}$$

Since $V(x)$ is a harmonic function satisfying the Dirichlet boundary condition (1.4.49), the desired statement follows from (1.4.56).

In a similar way (see *Chang* (1970)) it can be shown that the solution of the exterior problem (1.4.52) has the form

$$U_1(x) = -\int_1^{+\infty} V_1(tx)\, \frac{dt}{t} + C, \tag{1.4.57}$$

where $V_1(x)$ is the function given by (1.4.48) and, for $n > 2$, we have

$$V_1(+\infty) = 0, \tag{1.4.58}$$

whereas for $n = 2$ relation (1.4.58) is ensured by (1.4.53).

Since relations (1.4.54) and (1.4.55) for $i = 0$, as well as (1.4.54) and (1.4.57), are equivalent, we have

$$N_0(x, y) = \int_0^1 \left(\frac{1 - |tx|^2}{|y - tx|^n} - 1 \right) \frac{dt}{t}, \tag{1.4.59}$$

$$N_1(x, y) = -\int_1^{+\infty} \frac{|tx|^2 - 1}{|y - tx|^n} \frac{dt}{t}, \qquad n > 2, \tag{1.4.60}$$

$$N_1(x, y) = \int_1^{+\infty} \left(\frac{1 - |tx|^2}{|y - tx|^2} + 1 \right) \frac{dt}{t}, \qquad n = 2,$$

because in this case,

$$V_1(tx) \to 0, \qquad \frac{1 - |tx|^2}{|y - tx|^2} + 1 \to 0 \quad \text{as} \quad t \to +\infty$$

and the second expression tends to zero as $1/t$. Now, utilizing calculations from *Bitsadze* (1986), we can express $N_0(x, y)$ and $N_1(x, y)$ through elementary functions. In the cases of greatest interest, $n = 2$ and $n = 3$, we get

$$n = 2: \qquad N_0(x, y) = -2 \ln |y - x|, \tag{1.4.61}$$

$$n = 3: \qquad N_0(x, y) = \frac{2}{|y - x|} + \ln \frac{2}{1 - x \cdot y + |y - x|}, \tag{1.4.62}$$

$$n = 2: \qquad N_1(x, y) = 2 \ln |y - x| - 2 \ln |x|, \tag{1.4.63}$$

$$n = 3: \qquad N_1(x, y) = -\frac{2}{|y - x|} + \ln \frac{1 - x \cdot y + |y - x|}{|x| - x \cdot y}. \tag{1.4.64}$$

Let us seek a solution of the Neumann problem in the form of the double layer potential on the surface σ with density $g(x)$,

$$n = 2: \qquad U_i(x) = \frac{1}{\omega_n} \int_\sigma g(y) \frac{\partial}{\partial n_y} \ln \frac{1}{|y - x|}\, d\sigma_y, \qquad x \in D_i, \tag{1.4.65}$$

$$n = 3: \qquad U_i(x) = \frac{1}{\omega_n} \int_\sigma g(y) \frac{\partial}{\partial n_y} \frac{1}{|y - x|}\, d\sigma_y, \qquad x \in D_i. \tag{1.4.66}$$

It is assumed here that relation (1.4.53) holds for the function

$$f(x) = \left.\frac{\partial u_i}{\partial n_x}\right|_{x \in \sigma}^{+(i=0)\;-(i=1)}$$

in the boundary condition.

Since the normal derivative of the double layer potential is continuous on the surface σ, the boundary conditions for the Neumann problems take the form ($x \in \sigma$)

$$n = 2: \qquad \frac{1}{2\pi} \int_\sigma g(y) \frac{\partial}{\partial n_x} \frac{\partial}{\partial n_y} \ln \frac{1}{|y-x|} \, d\sigma_y = f(x), \qquad (1.4.67)$$

$$n = 3: \qquad \frac{1}{4\pi} \int_\sigma g(y) \frac{\partial}{\partial n_x} \frac{\partial}{\partial n_y} \frac{1}{|y-x|} \, d\sigma_y = f(x). \qquad (1.4.68)$$

For the double layer potential, we have

$$U_1(x) - U_0(x) = g(x), \qquad x \in \sigma. \qquad (1.4.69)$$

Therefore, each integral equation (1.4.67) and (1.4.68) admits a solution defined to within a constant, provided that σ is a closed surface for which there exist solutions of the interior and the exterior Neumann problems. Thus, using (1.4.54), (1.4.69) and (1.4.61)–(1.4.64), we find that the solutions of equations (1.4.67) and (1.4.68) are given by the formulas ($x \in \sigma$)

$$n = 2: \qquad g(x) = \frac{2}{\pi} \int_\sigma \ln|y-x| f(y) \, d\sigma_y + C, \qquad (1.4.70)$$

$$n = 3: \qquad g(x) = -\frac{1}{4\pi} \int_\sigma \left[\frac{4}{|y-x|} + \ln \frac{2(|x|-y \cdot x)}{(1-y \cdot x + |y-x|)^2} \right] f(y) \, d\sigma_y + C. \qquad (1.4.71)$$

Chapter 2

Sobolev–Slobodetskii Spaces

2.1. Generalized Functions

Definition 2.1.1. The *space $S = S(\mathbb{R}_n)$* consists of all infinitely differentiable complex-valued functions $\varphi(x)$ defined in the n-dimensional space $\mathbb{R}_n$ and decaying at infinity, together with all their derivatives, faster than any negative power of $|x| = \left(\sum_{i=1}^{n} x_i^2\right)^{1/2} \to \infty$.

For $\varphi(x) \in S$, we use the notation

$$|[\varphi]|_m = \sup_x (1 + |x|)^m \sum_{|p| \leq m} \left| \frac{\partial^p \varphi(x)}{\partial x^p} \right|, \qquad 0 \leq m < \infty, \tag{2.1.1}$$

where $p = (p_1, \ldots, p_n)$ is a multi-index with integer components, $|p| = p_1 + \cdots + p_n$, $\forall\, p_i \geq 0$,

$$\frac{\partial^p \varphi(x)}{\partial x^p} = \frac{\partial^{p_1 + \cdots + p_n} \varphi(x)}{\partial x_1^{p_1} \ldots \partial x_n^{p_n}}.$$

The *topology in S* is defined by the norms (2.1.1). Accordingly, a sequence $\varphi^k(x) \in S$ converges to $\varphi(x) \in S$ as $k \to \infty$, if $\left|[\varphi - \varphi^k]\right|_m \to 0$ for every $m = 1, 2, \ldots$.

Let $C_0^\infty(\mathbb{R}_n)$ be the set of all compactly supported infinitely differentiable functions in $\mathbb{R}_n$. Obviously, $C_0^\infty(\mathbb{R}_n) \subset S$, and the set $C_0^\infty(\mathbb{R}_n)$ is dense in S in the topology of S.

Definition 2.1.2. The *convolution* of the functions $\varphi(x) \in S$ and $\psi(x) \in S$ is the function

$$\varphi \times \psi = \int_{\mathbb{R}_n} \varphi(x - y)\psi(y)\, dy, \qquad dy = dy_1 \cdots dy_n. \tag{2.1.2}$$

Clearly, the following relations hold:

$$\varphi \times \psi = \psi \times \varphi, \qquad \frac{\partial^p}{\partial x^p}(\varphi \times \psi) = \frac{\partial^p \varphi}{\partial x^p} \times \psi = \varphi \times \frac{\partial^p \psi}{\partial x^p}. \tag{2.1.3}$$

For any real t, we have the inequality

$$(1 + |x - y|)^{-|t|} \leq \frac{(1 + |x|)^t}{(1 + |y|)^t}. \tag{2.1.4}$$

Indeed, for the sake of definiteness, let $t > 0$. Then

$$1 + |y| \leq 1 + |y - x| + |x| \leq (1 + |x - y|)(1 + |x|). \tag{2.1.5}$$

Raising (2.1.4) to the power t, we obtain (2.1.5). In order to obtain (2.1.4) for $t < 0$, we should raise the inequality $1 + |x| \leq (1 + |x - y|)(1 + |y|)$ to the power $|t|$.

Consider a function $f(x) \in C^\infty(\mathbb{R}_n)$ such that $|f(x)| \leq C(1 + |x|)^t$, where t is a real number. It follows from (2.1.4) that for $\varphi(x) \in S$, the convolution $f \times \varphi$ makes sense. Moreover, relations (2.1.3) hold in this case.

Proposition 2.1.1. *Let $f(x)$ be a locally integrable (in the sense of Lebesgue) function in $\mathbb{R}_n$ such that $|f(x)| \le C(1 + |x|)^t$ for some real t. Then, for $\varphi(x) \in S$, we have $f \times \varphi \in C^\infty(\mathbb{R}_n)$ and the following estimate holds:*

$$\left| \frac{\partial^p}{\partial x^p}(f \times \varphi) \right| \le C|[\varphi]|_m (1 + |x|)^t, \qquad 0 \le |p| < \infty, \tag{2.1.6}$$

where the constant C does not depend on $\varphi(x)$; $m = \max(|p|, |t| + n + 1)$.

Proof. From the inequality (2.1.4) and the theorem about differentiation of integrals with respect to a parameter (see *Sobolev* (1947)), it follows that

$$\frac{\partial^p}{\partial x^p}(f \times \varphi) = \int_{\mathbb{R}_n} \frac{\partial^p \varphi(x-y)}{\partial x^p} f(y)\, dy. \tag{2.1.7}$$

Since $\varphi(x) \in S$, we have

$$\left| \frac{\partial^p \varphi(x)}{\partial x^p} \right| \le |[\varphi]|_m (1 + |x|)^{-|t|-n-1}, \qquad m = \max\{|p|, t| + n + 1\}.$$

In view of (2.1.4) and (2.1.7), we get

$$\left| \frac{\partial^p}{\partial x^p}(f \times \varphi) \right| \le C \int_{\mathbb{R}_n} \frac{(1 + |y|)^t |[\varphi]|_m\, dy}{(1 + |x - y|)^{|t|+n+1}} \le C_1 |[\varphi]|_m (1 + |x|)^t, \qquad 0 \le p \le \infty.$$

Hence follows the desired result.

Corollary 2.1.1. *If $\varphi \in S$ and $\psi \in S$, then $\varphi \times \psi \in S$.*

Indeed, in this case we can take $t = -1, -2, \ldots$ in (2.1.6), and therefore, $\varphi \times \psi \in S$. Moreover,

$$|[\varphi \times \psi]|_m \le C_m |[\varphi]|_{m+n+1}, \qquad m = 0, 1, \ldots.$$

Definition 2.1.3. Let f be a functional on S and let (f, φ) denote its value on the element $\varphi \in S$. We say that f *is a linear continuous functional on S*, if the following conditions hold:

   (i) $(f, \alpha_1 \varphi_1 + \alpha_2 \varphi_2) = \alpha_1(f, \varphi_1) + \alpha_2(f, \varphi_2)$ for any $\varphi_1, \varphi_2 \in S$ and any complex α_1, α_2;

  (ii) $(f, \varphi_n) \to (f, \varphi)$ for any sequence $\varphi_n \in S$, $\varphi_n \to \varphi$ in S.

Definition 2.1.4. A linear continuous functional on $S(\mathbb{R}_n)$ is called a *generalized function.*

Example 2.1.1. Let $f(x)$ be locally integrable in the sense of Lebesgue and for some $N > 0$,

$$\int_{\mathbb{R}_n} |f(x)|(1 + |x|)^{-N}\, dx < \infty. \tag{2.1.8}$$

Then $f(x)$ can be associated with a linear functional, namely,

$$(f, \varphi) = \int_{\mathbb{R}_n} \overline{f(x)} \varphi(x)\, dx. \tag{2.1.9}$$

From (2.1.1), it follows that $|(f, \varphi)| \le C|[\varphi]|_N$, and therefore, the functional f is continuous. The functional (2.1.9) is called a *regular functional*. Quite often, no distinction is made between a regular functional and the function $f(x)$ associated with it.

Example 2.1.2. The *delta-function* (δ-function) is the generalized function defined by the relation $(\delta, \varphi) = \varphi(0)$ for all $\varphi \in S$.

The space of generalized functions is denoted by $S' = S'(\mathbb{R}_n)$.

Proposition 2.1.2. *Let* $f \in S'$. *Then, there exists* $m \geq 0$, $m = m(f)$, *such that for all* $\varphi \in S$ *the following inequality holds:*

$$|f(\varphi)| \leq C_m |[\varphi]|_m. \tag{2.1.10}$$

Proof. Assume the contrary. Then there is a sequence $\varphi_m \in S$ such that $|(f, \varphi_m)| \geq m |[\varphi_m]|_m$. Let $\psi_m = \varphi_m(x)\left(m |[\varphi_m]|_m\right)^{-1}$. Then, we have

$$|[\psi_m]|_i \leq \frac{|[\varphi_m]|_i}{m |[\varphi_m]|_m} \leq \frac{1}{m}, \qquad \forall\, i \leq m,$$

and therefore, $|[\psi_m]|_i \to 0$ as $m \to \infty$, $i = 0, 1, \dots$. Thus, $\psi_m \to 0$ in S. On the other hand, we have $|(f, \psi_m)| \geq 1$, which is in contradiction with the definition of generalized functions. Proposition 2.1.2 is proved.

Definition 2.1.5. For a generalized function $f \in S'$, *its derivative* $\partial^k f / \partial x^k$ is the generalized function defined by the relation

$$\left(\frac{\partial^k f}{\partial x^k}, \varphi\right) = (-1)^{|k|}\left(f, \frac{\partial^k \varphi}{\partial x^k}\right), \qquad \forall\, \varphi \in S. \tag{2.1.11}$$

The generalized function $g = \partial^k f / \partial x^k$ is a linear functional for which the estimate (2.1.10) holds with m replaced by $m + |k|$, and therefore, Definition 2.1.5 is correct.

Example 2.1.3. Let $\theta(t) = 1$ for $t \geq 0$, and $\theta(t) = 0$ for $t \leq 0$. The function $\theta(t)$ defines a regular functional on $S(\mathbb{R}_n)$. Let us find its derivative $d\theta/dt$. By (2.1.11) we have

$$\left(\frac{d\theta}{dt}, \varphi\right) = -(\theta(t), \varphi'(t)) = -\int_0^\infty \frac{d\varphi}{dt} = \varphi(0) = (\delta, \varphi),$$

and therefore, $d\theta/dt = \delta$ is the delta-function.

Let $a(x) \in C^\infty(\mathbb{R}_n)$ such that

$$\left|\frac{\partial^p a(x)}{\partial x^p}\right| \leq C_p (1 + |x|)^{t_p}, \qquad 0 \leq |p| < \infty. \tag{2.1.12}$$

Definition 2.1.6. For $f \in S'$, *the product* $a(x) \cdot f$ is the generalized function defined by the relation

$$(af, \varphi) = (f, \bar{a}(x) \cdot \varphi), \tag{2.1.13}$$

where $\bar{a}(x)$ is the complex conjugate of $a(x)$.

If $\varphi \in S$ and the estimates (2.1.12) hold for $a(x)$, then $\bar{a}\varphi \in S$ and $|[\bar{a}\varphi]|_m \leq C_m |[\varphi]|_{N_m}$, where $N_m = \max_{0 \leq p \leq m} (m + |t_p|)$. Therefore, relation (2.1.13) defines a continuous linear functional on S.

Definition 2.1.7. *The shift of a generalized function* f *by a vector* $\vec{a} \in \mathbb{R}_n$ *is the generalized function* $f^{\vec{a}} = f(x - \vec{a})$ *defined by the relation*

$$\left(f^{\vec{a}}, \varphi\right) = (f, \varphi(x + \vec{a})).$$

If f is a regular functional, then $f^{\vec{a}}$ is also a regular functional associated with the function $f(x - \vec{a})$.

Definition 2.1.8. A sequence $f_n \in S'$ is said to *converge in* S' *to* $f' \in S'$ as $n \to \infty$, if

$$(f_n, \varphi) \to (f, \varphi), \qquad \forall \, \varphi \in S. \tag{2.1.14}$$

Note that the space S' is complete in the following sense: if $f_n \in S'$ is a sequence such that the numerical sequence (f_n, φ) is convergent for any $\varphi \in S$, then there is $f \in S'$ such that $f_n \to f$ as $n \to \infty$ (see *Vladimirov* (1976)).

Example 2.1.4. Suppose that for all $n = 1, 2, \ldots$, we have $|f_n(x)| \le f_0(x)$ and for $f_0(x)$ the estimate (2.1.8) holds. Let $f_n(x) \to f(x)$ almost everywhere as $n \to \infty$. Then

$$(f_n, \varphi) = \int_{\mathbb{R}_n} \overline{f_n(x)} \varphi(x)\, dx \to \int_{\mathbb{R}_n} \overline{f(x)} \varphi(x)\, dx = (f, \varphi)$$

by the Lebesgue theorem on dominated convergence.

The *support of a continuous function* $\varphi(x)$ is defined as the closure of the set of all x at which $\varphi(x) \ne 0$ and is denoted by $\operatorname{supp} \varphi$. Let U be an open domain in $\mathbb{R}_n$, which may be unbounded; and let $C_0^\infty(U)$ be the set of all infinitely differentiable functions with a compact support in U. By $S(U)$ we denote the closure of $C_0^\infty(U)$ in the topology of the space $S(\mathbb{R}_n)$.

Definition 2.1.9. Any continuous linear functional on $S(U)$ is called *a generalized function in the domain* U. The space of generalized functions in U is denoted by $S'(U)$.

It follows from the Hahn–Banach theorem that any continuous linear functional f on $S(U)$ can be extended as a continuous linear functional, denoted by lf, on $S(\mathbb{R}_n)$.

Let $F \in S'(\mathbb{R}_n)$. A functional $f \in S'(U)$ is called the *restriction* of F to the domain U, if $(F, \varphi) = (f, \varphi)$, $\forall \varphi \in S(U)$. The restriction operator is denoted by $p_U \colon p_U F = f$.

Definition 2.1.10. We say that a generalized function $f \in S'(\mathbb{R}_n)$ *vanishes on* U, if the restriction of f to U coincides with the null functional on $S(U)$.

Let $U_{\max}$ be the maximal open set on which $f = 0$. Then the complement of $U_{\max}$ is called the *support of the functional* f and is denoted by $\operatorname{supp} f$. For a regular functional f associated with a locally integrable function $f(x)$, the set $\operatorname{supp} f$ is the complement of the largest open set on which $f(x) = 0$ almost everywhere. The following inclusion holds: $\operatorname{supp} \partial^p f / \partial x^p \subseteq \operatorname{supp} f$.

Example 2.1.5. The generalized functions $\delta^k = \partial^k \delta / \partial x^k$ vanish everywhere outside the origin, so that $\operatorname{supp} \delta^k = \{0\}$, $k = 0, 1, 2, \ldots$.

Theorem 2.1.1. *For any continuous linear functional f on S there exist p and N such that f can be represented as*

$$f = \sum_{|k| \le p} \frac{\partial^k f_k}{\partial x^k}, \tag{2.1.15}$$

where f_k are regular functionals associated with continuous functions of power growth,

$$|f_k(x)| \le C(1 + |x|)^N, \qquad 0 \le |k| \le p. \tag{2.1.16}$$

Proof. By Proposition 2.1.2, there exists $m = m(f)$ such that $|(f, \varphi)| \le C |[\varphi]|_m$. We have

$$(1 + |x|)^m \frac{\partial^k \varphi}{\partial x^k} = \int_{-\infty}^{x_1} \cdots \int_{-\infty}^{x_n} \frac{\partial}{\partial y_1} \cdots \frac{\partial}{\partial y_n} \left[(1 + |y|)^m \frac{\partial^k \varphi(y)}{\partial y^k} \right] dy.$$

Calculating the derivatives in the integrand and estimating the resulting terms, we obtain

$$\|[\varphi]\|_m \le C_1 \left(\sum_{|k| \le m+n+n} \int_{\mathbb{R}_n} (1+|y|) \left| \frac{\partial^k \varphi(y)}{\partial y^k} \right| dy \right)^2 \le$$

$$\le C_2 \int_{\mathbb{R}_n} (1+|y|)^{-n-1} \left(\sum_{|k| \le m+n} \int_{\mathbb{R}_n} (1+|y|)^{2m+n+1} \left| \frac{\partial^k \varphi(y)}{\partial y^k} \right|^2 dy \right) \le$$

$$\le C_3 \left(\sum_{|k| \le m+n} \int_{\mathbb{R}_n} (1+|y|)^{2m+n+1} \left| \frac{\partial^k \varphi(y)}{\partial y^k} \right|^2 dy \right). \tag{2.1.17}$$

Let $L_2(\mathbb{R}_n)$ be the Lebesgue space of square summable functions in $\mathbb{R}_n$, and let

$$\tilde{L}_2(\mathbb{R}_n) = \bigoplus_{|k|=0}^{m+n} L_2(\mathbb{R}_n)$$

be the direct sum of copies of $L_2(\mathbb{R}_n)$ with the scalar product

$$(\vec{u}, \vec{v}) = \sum_{|k| \le m+n} \int_{\mathbb{R}_n} \bar{u}_k(x) \cdot v_k(x) \, dx,$$

where $\vec{u} = \{u_k(x)\} \in \tilde{L}_2(\mathbb{R}_n)$, $\vec{v} = \{v_k(x)\} \in \tilde{L}_2(\mathbb{R}_n)$. Consider the linear subspace $L \subset \tilde{L}_2(\mathbb{R}_n)$ that consists of all vectors $\vec{v}$ of the form

$$\vec{v} = \left\{ (1+|y|)^{m+\frac{n+1}{2}} \frac{\partial^k \varphi(y)}{\partial y^k} \right\}.$$

We define a linear functional g on L by letting $g(\vec{v}) = (f, \varphi)$, $\vec{v} \in L$. It follows from the inequality (2.1.17) that the linear functional g is bounded on $L \subset \tilde{L}_2(\mathbb{R}_n)$. Therefore, by the Hahn–Banach theorem, it can be extended as a continuous linear functional to the entire $\tilde{L}_2(\mathbb{R}_n)$. By the Riesz theorem about the general form of a linear functional in Hilbert space, there is a vector $\{g_k(y)\}$, $0 \le |k| \le m+n$, in $\tilde{L}_2(\mathbb{R}_n)$ such that

$$l_g(\vec{v}) = \sum_{|k|=0}^{m+n} \int_{\mathbb{R}_n} \overline{g_k(y)} v_k(y) \, dy,$$

where $l_g(\vec{v})$ is the extension of the linear functional g to $\tilde{L}_2(\mathbb{R}_n)$. For $\vec{v} \in L$, we have

$$l_g(\vec{v}) = g(\vec{v}) = (f, \varphi) = \sum_{|k|=0}^{m+n} \int_{\mathbb{R}_n} \overline{g_k(y)} (1+|y|)^{m+\frac{m+1}{2}} \frac{\partial^k \varphi}{\partial y^k} \, dy. \tag{2.1.18}$$

The function

$$f_k(x) = \int_0^{x_1} \cdots \int_0^{x_n} g_k(y)(1+|y|)^{m+\frac{n+1}{2}} \, dy$$

is absolutely integrable, and therefore,

$$(1+|x|)^{m+\frac{n+1}{2}} g_k(x) = \frac{\partial}{\partial x_1} \cdots \frac{\partial}{\partial x_n} f_k(x). \tag{2.1.19}$$

For the functions $f_k(x)$ the following inequality holds:

$$|f_k(x)|^2 \le \left| \int_0^{x_1} \cdots \int_0^{x_n} (1+|y|)^{2m+n+1} dy \right| \left| \int_0^{x_1} \cdots \int_0^{x_n} |g_k(y)|^2 dy \right| \le C(1+|x|)^{2m+2n+1}.$$

Substituting (2.1.19) into (2.1.18) and integrating by parts, we get

$$(f, \varphi) = \sum_{|k|=0}^{m+n} (-1)^n \int_{\mathbb{R}_n} \overline{f_k(x)} \frac{\partial}{\partial y_1} \cdots \frac{\partial}{\partial y_n} \frac{\partial^k \varphi}{\partial y^k} \, dy.$$

Theorem 2.1.1 is proved.

Definition 2.1.11. A functional $f \in S'(\mathbb{R}_n)$ is called *compactly supported*, if its support lies in a finite ball $|x| < R$.

Proposition 2.1.3. *Any compactly supported functional $f \in S'(\mathbb{R}_n)$ can be represented in the form*

$$f = \sum_{|k|=0}^{p} \frac{\partial^k f_k}{\partial x^k}, \qquad (2.1.20)$$

where f_k are regular functionals associated with continuous compactly supported functions $f_k(x)$.

Proof. Let $\operatorname{supp} f \subset U_R$, $U_R = \{x : |x| < R\}$. Consider a function $\chi(x) \in C_0^\infty(\mathbb{R}_n)$ such that $\chi(x) = 1$ for $x \in U_R$. Then $\big(f, (1 - \chi(x))\varphi(x)\big) = 0$, and therefore, $(f, \varphi) = (f, \chi\varphi)$. From Theorem 2.1.1, it follows that $f = \sum\limits_{|k|=0}^{p} \dfrac{\partial^k f_k}{\partial x^k}$, where f_k are functionals associated with continuous functions of polynomial growth. Further, we have

$$(f, \varphi) = (f, \chi\varphi) = \left(\sum_{|k|=0}^{p} \frac{\partial^k f}{\partial x^k}, \chi\varphi \right) = (-1)^{|k|} \sum_{|k|=0}^{p} \left(f_k, \frac{\partial^k \chi\varphi}{\partial x^k} \right) = \sum_{|k|=0}^{p} \left(h_k, \frac{\partial^k \varphi}{\partial x^k} \right)(-1)^{|k|},$$

where h_k are regular functionals associated with continuous compactly supported functions $h_k(x)$. Proposition 2.1.3 is proved.

Definition 2.1.12. *The convolution of a generalized function $f \in S'$ and a test function $\varphi \in S$ is defined by the relation*

$$(f \times \varphi)(x) = \overline{(f(y), \overline{\varphi(x - y)})}, \qquad (2.1.21)$$

where $\overline{\varphi(x - y)}$ is regarded as a function of y with x fixed, and we write $f(y)$ to indicate that the functional f is acting on the test function with respect to the variable y.

It is easy to see that the function $f \times \varphi$ is defined for any $x \in \mathbb{R}_n$.

Let $f \in S'$ be a regular functional. Then, using (2.1.9), we get

$$f \times \varphi = \overline{(f(y), \overline{\varphi(x - y)})} = \int_{\mathbb{R}_n} \overline{\overline{f(y)}\varphi(x - y)}\, dy = \int_{\mathbb{R}_n} f(y)\varphi(x - y)\, dy.$$

This relation shows that Definition 2.1.12 applied to regular functions coincides with the common definition of convolution (see (2.1.2)).

Proposition 2.1.4. *The function $f \times \varphi$ is infinitely differentiable and satisfies the inequalities*

$$\left| \frac{\partial^p}{\partial x^p}(f \times \varphi) \right| \le C_p (1 + |x|)^N, \qquad 0 \le |p| < \infty. \qquad (2.1.22)$$

Proof. From (2.1.21) and Theorem 2.1.1, it follows that

$$f \times \varphi = \sum_{|k| \le m} \left(\frac{\partial^k f_k}{\partial x^k} \times \varphi \right) = \sum_{|k| \le m} \overline{\left(\frac{\partial^k f_k(y)}{\partial y^k}, \overline{\varphi(x - y)} \right)} = \qquad (2.1.23)$$

$$= \sum_{|k| \le m} \overline{\left(f_k(y), (-1)^{|k|} \frac{\partial^k}{\partial y^k} \overline{\varphi(x - y)} \right)} = \sum_{|k| \le m} \int_{\mathbb{R}_n} f_k(y) \frac{\partial^k}{\partial y^k} \varphi(x - y)\, dy,$$

where $f_k(y)$ satisfies the inequality $|f_k(y)| \le C(1 + |y|)^N$. Now, Proposition 2.1.4 follows from (2.1.23) and Proposition 2.1.1.

Example 2.1.6. Let $\varphi \in S$. Then $\delta \times \varphi = \overline{(\delta(y), \overline{\varphi(x - y)})} = \overline{\overline{\varphi(x)}} = \varphi(x)$. If $P(D) = \sum\limits_{|k| \le m} a_k \dfrac{\partial^k}{\partial x^k}$ is a differential operator with constant coefficients, then

$$(P(D)\delta) \times \varphi = \sum_{|k| \le m} a_k \overline{\left(\frac{\partial^k}{\partial x^k} \delta(y), \overline{\varphi(x - y)} \right)} = \sum_{|k| \le m} a_k \frac{\partial^k \varphi}{\partial x^k} = P(D)\varphi.$$

Proposition 2.1.5. *For* $f \in S'$, $\varphi, \psi \in S$, *we have*

$$(f \times \varphi, \psi) = (f, \varphi_1 \times \psi), \qquad \varphi_1(x) = \overline{\varphi(-x)}.$$

Proof. By virtue of (2.1.23), we have

$$f \times \varphi = \sum_{|k| \le m} \int_{\mathbb{R}_n} \int_{\mathbb{R}_n} f_k(y) \frac{\partial^k}{\partial x^k} \varphi(x - y) \, dy,$$

where $|f_k(y)| \le C(1 + |y|)^N$. Therefore,

$$(f \times \varphi, \psi) = \sum_{|k| \le m} \int_{\mathbb{R}_n} \overline{f_k(y) \frac{\partial^k \varphi(x - y)}{\partial x^k}} \psi(x) \, dx \, dy. \tag{2.1.24}$$

Theorem 2.1.1 yields

$$(f, \varphi_1 \times \psi) = \sum_{|k| \le m} \left(\frac{\partial^k}{\partial y^k} f_k(y), \varphi_1 \times \psi \right) = \sum_{|k| \le m} \int_{\mathbb{R}_n} \overline{f_k(y)} (-1)^{|k|} \frac{\partial^k}{\partial y^k} (\varphi_1 \times \psi) \, dy. \tag{2.1.25}$$

By the definition of convolution of test functions, we have

$$\frac{\partial^k}{\partial y^k} (\varphi_1 \times \psi) = \int_{\mathbb{R}_n} \frac{\partial^k}{\partial y^k} \varphi_1(y - x) \psi(x) \, dx = (-1)^{|k|} \int_{\mathbb{R}_n} \frac{\partial^k}{\partial y^k} \overline{\varphi(x - y)} \psi(x) \, dy. \tag{2.1.26}$$

Proposition 2.1.5 follows from (2.1.24)–(2.1.26).

Proposition 2.1.6. *Let* $f \in S'$ *be a compactly supported functional and* $\varphi \in S$. *Then* $f \times \varphi \in S$ *and the operator of convolution with* f *is continuous on* S.

Proof. Proposition 2.1.3 implies that

$$f \times \varphi = \sum_{|k| \le m} \int_{\mathbb{R}_n} f_k(y) \frac{\partial^k}{\partial x^k} \varphi(x - y) \, dy \tag{2.1.27}$$

with compactly supported continuous functions $f_k(y)$. It follows from Proposition 2.1.1 that $f \times \varphi \in S$; moreover,

$$\|[f \times \varphi]\|_p \le C_p |[\varphi]|_{p+m}, \qquad p = 0, 1, 2, \dots.$$

This inequality implies that the convolution operator is continuous in S. Proposition 2.1.6 is proved.

Consider a compactly supported functional $g \in S'$ and define the functional g_1 by

$$(g_1, \varphi) = \overline{(g, \overline{\varphi(-x)})}.$$

Obviously, $g_1 \in S'$ and has a compact support. The following relations hold:

$$g_1 \times \varphi = \left(g_1(y), \overline{\varphi(x - y)} \right) = (g(y), \varphi(x + y)). \tag{2.1.28}$$

According to Proposition 2.1.6, we have $a(x) = g_1 \times \varphi \in S$, and therefore, $(g(y), \varphi(x + y)) \in S$.

Definition 2.1.13. For $f, g \in S'$ and g compactly supported, *the convolution $f \times g$ is defined* as a functional h such that

$$(h, \varphi) = (f \times g, \varphi) = \big(f(x), (g(y), \varphi(x+y))\big). \tag{2.1.29}$$

From (2.1.28) and Proposition 2.1.6, it follows that $h \in S'$.
For absolutely integrable functions $f(x)$ and $g(x)$, we have

$$(f \times g, \varphi) = \int_{\mathbb{R}_n} \left(\int_{\mathbb{R}_n} \overline{g(x-y)f(y)}\, dy \right) \varphi(x)\, dx = \int_{\mathbb{R}_n} \overline{f(y)} \left(\int_{\mathbb{R}_n} \overline{g(x-y)} \varphi(x)\, dx \right) dy =$$

$$= \int_{\mathbb{R}_n} \overline{f(y)} \left(\int_{\mathbb{R}_n} \overline{g(t)} \varphi(t+y)\, dt \right) dx = \big(f(x), (g(y), \varphi(x+y))\big).$$

Example 2.1.7. If $f \in S'$, then $(f \times \delta, \varphi) = (f(x), (\delta(y), \varphi(x+y))) = (f(x), \varphi(x))$, i.e., $f \times \delta = f$, where δ is the delta-function.

2.2. Fourier Transformation

Definition 2.2.1. *The Fourier transform of a test function $\varphi(x) \in S$ is defined by the formula*

$$\widehat{\varphi}(\xi) = \int_{\mathbb{R}_n} \varphi(x)\, e^{i(x,\xi)}\, dx, \qquad (x, \xi) = \sum_{i=1}^{n} x_i \xi_i. \tag{2.2.1}$$

The inverse Fourier transform is (see *Shilov* (1961))

$$\varphi(x) = \frac{1}{(2\pi)^n} \int_{\mathbb{R}_n} \widehat{\varphi}(\xi)\, e^{-i(x,\xi)}\, d\xi. \tag{2.2.2}$$

The operator of the Fourier transformation will be denoted by F and the operator of the inverse Fourier transformation by F^{-1}. The Fourier transformation of the functions in S has the following properties (see *Shilov* (1965)):

1) $F\left(\dfrac{\partial^k}{\partial x^k} \varphi(x) \right) = \xi^k \widehat{\varphi}(\xi)(-i)^{|k|}$, where $\xi^k = \xi_1^{k_1} \cdots \xi_n^{k_n}$;

2) $F(\varphi \times \psi) = \widehat{\varphi}(\xi)\widehat{\psi}(\xi)$;

3) *the Parseval identity,*

$$\int_{\mathbb{R}_n} \varphi(x)\overline{\psi}(x)\, dx = \frac{1}{(2\pi)^n} \int_{\mathbb{R}_n} \widehat{\varphi}(\xi)\overline{\widehat{\psi}(\xi)}\, d\xi;$$

4) $F(\varphi(x-a)) = e^{i(a,\xi)}\widehat{\varphi}(\xi)$, $a = (a_1, \ldots, a_n) \in \mathbb{R}_n$.

The following result is proved in *Shilov* (1961).

Proposition 2.2.1. *The Fourier transformation is a one-to-one mapping of the space $S(\mathbb{R}_n)$ onto itself.*

Denote by S^+ the subspace of S that consists of all functions $\varphi(x) = \varphi(x', x_n) \in S$ vanishing for all $x_n < 0$, where $x' = (x_1, \ldots, x_{n-1}) \in \mathbb{R}_{n-1}$.

Proposition 2.2.2. *The Fourier image $\widetilde{S}^+ = FS^+$ of the space S^+ consists of the functions $\widehat{\varphi}_+(\xi', \alpha_n)$ which can be extended as analytic functions of $\alpha_n = \xi_n + i\tau$ to the halfplane $\tau > 0$, so that the extensions are continuously differentiable in (ξ', ξ_n), $\tau \geq 0$, and satisfy the inequalities*

$$\left(1 + |\xi'| + |\xi_n| + \tau\right)^m \left| \frac{\partial^{k'}}{\partial \xi'^{k'}} \frac{\partial^{k_n}}{\partial \alpha_n^{k_n}} \widehat{\varphi}_+(\xi', \alpha_n) \right| < C_{m,|k|}, \qquad 0 \leq m, \quad |k| < \infty, \quad k = (k', k_n). \tag{2.2.3}$$

Proof. Let $\varphi_+(x', x_n) \in S^+$. Then

$$\widehat{\varphi}_+(\xi', \xi_n) = \int_0^\infty \left\{ \int_{\mathbb{R}_{n-1}} \varphi_+(x', x_n) \, e^{ix_n \xi_n + i(x', \xi')} \, dx \right\} dx_n. \tag{2.2.4}$$

Since $\varphi_+(x', x_n) \in S^+$, direct differentiation shows that the right-hand side of (2.2.4) can be analytically extended in $\alpha_n = \xi_n + i\tau$ to the halfplane $\tau > 0$,

$$\widehat{\varphi}_+(\xi', \alpha_n) = \int_0^\infty \left\{ \int_{\mathbb{R}_{n-1}} \varphi_+(x', x_n) \, e^{ix_n \alpha_n + i(x', \xi')} dx' \right\} dx_n. \tag{2.2.5}$$

Since

$$\sum_{|p|=0}^m |\xi'|^{k'} |\alpha_n|^{k_n} \geq C\left(1 + |\xi'| + |\xi_n| + \tau\right)^m, \qquad |p| = |k'| + k_n,$$

$$|\xi'|^{k'} = |\xi_1|^{k_1} \cdots |\xi_{n-1}|^{k_{n-1}}, \qquad k' = (k_1, \ldots, k_{n-1}),$$

we have

$$\left(1 + |\xi'| + |\xi_n| + \tau\right)^m \left| \frac{\partial^{k'}}{\partial \xi'^{k'}} \frac{\partial^{k_n}}{\partial \alpha_n^{k_n}} \widehat{\varphi}_+(\xi', \alpha_n) \right| \leq$$

$$\leq C_1 \int_0^\infty dx_n \int_{\mathbb{R}_{n-1}} \sum_{|p|=0}^m \left| \frac{\partial^p}{\partial x^p} \left(x^k \varphi_+(x', x_n)\right) \right| dx' \leq C_{m,|k|}.$$

Proposition 2.2.2. is proved.

Definition 2.2.2. *The Fourier transform of a functional $f \in S'$ is the generalized function $\widehat{f} \in S'(\mathbb{R}_n)$ such that*

$$(\widehat{f}, \widehat{\varphi}) = (2\pi)^n (f, \varphi), \quad \forall \varphi \in S(\mathbb{R}_n), \tag{2.2.6}$$

where $\widehat{\varphi} = F\varphi \in S(\mathbb{R}_n)$.

By Proposition 2.2.2, the functional $\widehat{f}$ is defined on the whole space S and is continuous.

If f is a regular functional associated with an absolutely integrable function $f(x)$, then $\widehat{f}$ is also a regular functional corresponding to $\widehat{f}(\xi) = Ff(x)$, and thus, (2.2.6) follows from the Parseval identity.

For the Fourier transforms of generalized functions the following relations hold:

$$F\left(\frac{\partial^k f}{\partial x^k}\right) = (-i\xi)^k \widehat{f}, \qquad F(f(x - \vec{a})) = e^{i(\vec{a}, \xi)} \widehat{f}, \tag{2.2.7}$$

where $f \in S'$ is an arbitrary generalized function, $\vec{a} \in \mathbb{R}_n$ is an arbitrary vector.

Indeed, using (2.2.6) and (2.1.13), we get

$$\left(F\left(\frac{\partial^k f}{\partial x^k}\right), \widehat{\varphi}\right) = (2\pi)^n \left(\frac{\partial^k f}{\partial x^k}, \varphi\right) =$$
$$= (2\pi)^n (-1)^{|k|} \left(f, \frac{\partial^k \varphi}{\partial x^k}\right) = (-1)^{|k|} \left(\widehat{f}, (-i\xi)^k \widehat{\varphi}\right) = \left((-i\xi)^k \widehat{f}, \widehat{\varphi}\right).$$

Hence we obtain the first relation in (2.2.7).

Using Definition 2.1.7 and relation (2.2.6), we find that

$$(Ff(x-\bar{a}), \widehat{\varphi}) = (2\pi)^n (f(x-\bar{a}), \varphi) =$$
$$= (2\pi)^n (f, \varphi(x+\bar{a})) = \left(\widehat{f}, e^{(\bar{a},\xi)} \widehat{\varphi}\right) = \left(e^{(\bar{a},\xi)} \widehat{f}, \widehat{\varphi}\right),$$

which proves the second relation in (2.2.7).

The operator F is continuous on S', which means that if $f_n \to f$ in S', then $\widehat{f} \to \widehat{f}$ on S' as $n \to \infty$. Indeed, by (2.2.6), we have

$$(\widehat{f_n}, \widehat{\varphi}) = (2\pi)^n (f_n, \varphi) \to (2\pi)^n (f, \varphi) = (\widehat{f}, \widehat{\varphi}).$$

Proposition 2.2.3. *The Fourier transform of a compactly supported functional $f \in S'$ is an entire analytic function $\widehat{f}(\xi + i\eta)$ satisfying the inequalities*

$$\left|\frac{\partial^k}{\partial \alpha^k} \widehat{f}(\xi + i\eta)\right| \le C_k (1 + |\xi| + |\eta|)^m e^{b|\eta|}, \quad \alpha = \xi + i\eta, \quad b > 0, \quad m \in \mathbb{N}, \qquad (2.2.8)$$

where $\mathbb{N}$ is the set of positive integers.

Proof. From Proposition 2.1.3, it follows that the functional f can be represented as

$$f = \sum_{|p|=0}^{m} \frac{\partial^p}{\partial x^p} f_p(x), \qquad (2.2.9)$$

where $f_p(x)$ are continuous compactly supported functions. Suppose that the support of every $f_p(x)$ belongs to one and the same ball $S_b = \{x : |x| < b, \ b > 0\}$. Then the Fourier transform

$$\widehat{f_p}(\xi + i\eta) = \int_{S_b} f_p(x) e^{i(x,\xi+i\eta)} dx$$

is an entire analytic function of $\alpha = \xi + i\eta$ and

$$\left|\frac{\partial^k}{\partial \alpha^k} \widehat{f_p}(\xi + i\eta)\right| \le \int_{S_b} |x|^{|k|} |f_p(x)| e^{-(x,\eta)} dx \le C_k e^{b|\eta|}. \qquad (2.2.10)$$

It follows from (2.2.9) that

$$\widehat{f}(\xi + i\eta) = \sum_{|p|=0}^{m} (-i)^{|p|} (\xi + i\eta)^p \widehat{f_p}(\xi + i\eta).$$

This relation, together with (2.2.10), proves Proposition 2.2.3.

Proposition 2.2.4. *If* $f \in S'$, $\varphi \in S$, *then*

$$F(f \times \varphi) = \widehat{\varphi}(\xi)\widehat{f}. \tag{2.2.11}$$

Proof. It follows from Proposition 2.1.5 that

$$\left(F(f \times \varphi), \widehat{\psi}\right) = (2\pi)^n (f \times \varphi, \psi) = (2\pi)^n (f, \varphi_1 \times \psi), \tag{2.2.12}$$

where $\varphi_1(x) = \overline{\varphi(-x)}$. We have

$$F\varphi_1 = \int_{\mathbb{R}_n} \varphi_1(x) \, e^{i(x,\xi)} \, dx = \int_{\mathbb{R}_n} \overline{\varphi(-x)} \, e^{i(x,\xi)} \, dx = \overline{\widehat{\varphi}(\xi)}. \tag{2.2.13}$$

From (2.2.12) and (2.2.13) we get

$$\left(F(f \times \varphi), \widehat{\psi}\right) = \left(\widehat{f}, \overline{\widehat{\varphi}(\xi)} \cdot \widehat{\psi}(\xi)\right) = \left(\widehat{\varphi}(\xi)\widehat{f}, \widehat{\psi}(\xi)\right).$$

Proposition 2.2.4 is proved.

Proposition 2.2.5. *If* $f \in S'$ *and* $g \in S'$ *is compactly supported, then*

$$F(f \times g) = \widehat{g}(\xi)\widehat{f}. \tag{2.2.14}$$

Proof. From (2.1.28) and (2.1.29), it follows that

$$(F(f \times g), \widehat{\varphi}) = (2\pi)^n (f \times g, \varphi) = (2\pi)^n (f, g_1 \times \varphi) = \left(Ff, F(g_1 \times \varphi)\right), \tag{2.2.15}$$

where g_1 is given by $(g_1, \varphi) = \overline{\left(g, \overline{\varphi(-x)}\right)}$. From the definition of the functional g_1, we get

$$(\widehat{g}_1, \widehat{\varphi}) = (2\pi)^n (g_1, \varphi) = (2\pi)^n \overline{\left(g, \overline{\varphi(-x)}\right)} = \overline{\left(\widehat{g}, \overline{\widehat{\varphi}(\xi)}\right)} = \int_{\mathbb{R}_n} \widehat{g}(\xi) \cdot \widehat{\varphi}(\xi) \, d\xi. \tag{2.2.16}$$

Here, we have used the regularity of the functional $\widehat{g}$ which follows from Proposition 2.2.3. Relation (2.2.16) implies that $\widehat{g}_1 = \overline{\widehat{g}(\xi)}$. From (2.2.15), using Proposition 2.2.4, we get

$$(F(f \times g), \widehat{\varphi}) = \left(Ff, \overline{\widehat{g}(\xi)} \cdot \widehat{\varphi}(\xi)\right) = \left(\widehat{g}(\xi) \cdot \widehat{f}, \widehat{\varphi}(\xi)\right).$$

This completes the proof of Proposition 2.2.5.

Example 2.2.1. Let $f = \delta$. Then $Ff = 1$. Indeed,

$$(F\delta, F\varphi) = (2\pi)^n (\delta, \varphi) = (2\pi))^n \varphi(0) = \int_{\mathbb{R}_n} 1 \cdot F(\varphi) \, d\xi = (1, F(\varphi)).$$

If $f = \delta^k = \partial^k \delta / \partial x^k$, then by (2.2.7) we have $F(\delta^k) = (-\xi)^{|k|} F(\delta) = (-i\xi)^{|k|}$.

Example 2.2.2. Suppose that $f \in S'$ and its Fourier transform Ff are known. Let us find $F(P(x)f)$, where $P(x)$ is a polynomial. For a test function φ, we have $\partial^k \widehat{\varphi}(\xi) / \partial \xi^k = F\left((ix)^k \varphi(x)\right)$. Hence,

$$F\left(x^k \varphi(x)\right) = (-i)^{|k|} \frac{\partial^k}{\partial \xi^k} \widehat{\varphi}(\xi) = \left(-i\frac{\partial}{\partial \xi}\right)^k \widehat{\varphi}(\xi).$$

Further, we have

$$\left(F(P(x)f), F\varphi\right) = (P(x)f, \varphi)(2\pi)^n = (2\pi)^n \left(f, \overline{P(x)}\varphi\right) = \left(Ff, F\left(\overline{P(x)}\varphi\right)\right) =$$

$$= \left(Ff, \overline{P\left(i\frac{\partial}{\partial\xi}\right)\widehat{\varphi}(\xi)}\right) = \left(P\left(-i\frac{\partial}{\partial\xi}\right)\widehat{f}, \widehat{\varphi}(\xi)\right).$$

It follows that

$$F(P(x)f) = P\left(-i\frac{\partial}{\partial\xi}\right)\widehat{f}(\xi).$$

Example 2.2.3. Consider the linear functional $f = 1/x$ defined by

$$(f, \varphi) = \left(\frac{1}{x}, \varphi\right) = \lim_{\varepsilon\to 0}\left(\int_{-\infty}^{-\varepsilon}\frac{\varphi(x)}{x}\,dx + \int_{\varepsilon}^{\infty}\frac{\varphi(x)}{x}\,dx\right), \qquad \varphi \in S. \tag{2.2.17}$$

Thus, the right-hand side is a singular integral. As shown in *Vladimirov* (1976), the functional f defined by (2.2.17) belongs to S'. Let us find $F(1/x)$.

A generalized function f is called *odd* if $(f, \varphi(-x)) = -(f, \varphi)$. Let us show that the Fourier transform of an odd function is an odd function. Indeed,

$$\left(\widehat{f}, \widehat{\varphi}(-\xi)\right) = (2\pi)^n(f, \varphi(-x)) = -(2\pi)^n(f, \varphi) = -(\widehat{f}, \widehat{\varphi}).$$

Since $f = 1/x$ is odd, $F(1/x)$ is also odd. Let us show that $xf = 1$. Indeed,

$$(xf, \varphi) = (f, x\varphi) = \lim_{\varepsilon\to 0}\int_{\mathbb{R}\setminus U_\varepsilon}\frac{x\varphi(x)}{x}\,dx = \int_{\mathbb{R}} 1\cdot\varphi(x)\,dx = (1, \varphi),$$

where $U_\varepsilon = \{x : |x| < \varepsilon,\ \varepsilon > 0\}$. Using the result established in Example 2.2.2, we get

$$F(xf) = -i\frac{d}{d\xi}Ff = F(1). \tag{2.2.18}$$

Let us calculate $F(1)$. We have

$$(F(1), F(\varphi)) = 2\pi(1, \varphi) = 2\pi\int_{-\infty}^{\infty}\varphi(x)\,dx = 2\pi\widehat{\varphi}(0) = 2\pi(\delta(\xi), \varphi(\xi)),$$

which means that $F(1) = 2\pi\delta(\xi)$. Letting $g(\xi) = Ff$, from (2.2.18) we get $dg(\xi)/d\xi = 2\pi i\delta(\xi)$. On the basis of Example 2.1.3, we see that

$$g(\xi) = 2\pi i[\theta(\xi) + C].$$

Since $g(\xi)$ is an odd function, we have $C = 1/2$, and therefore,

$$F\left(\frac{1}{x}\right) = g(\xi) = \pi i\,\text{sign}\,\xi.$$

Let f_λ be a family of generalized functions depending on the complex parameter λ varying in an open domain Λ. For the sake of brevity, this family will be referred to as the generalized function f_λ.

Definition 2.2.3. A generalized function f_λ is called *analytic* in the domain Λ, if for each $\varphi \in S$, the scalar function (f_λ, φ) is analytic with respect to $\lambda \in \Lambda$ (see *Zygmund* (1965)).

Two analytic functions f_λ and g_λ defined in a domain Λ and coinciding on a subset $\Lambda_0 \subset \Lambda$ with a limit point inside Λ coincide for all $\lambda \in \Lambda$. Indeed, for any test function φ, the scalar functions (f_λ, φ) and (g_λ, φ) coincide in the domain Λ, as claimed by the uniqueness theorem for scalar analytic functions.

This property is fundamental for the method of analytic extension of the functional f_λ with respect to the parameter λ. Suppose that the functional f_λ is analytic in Λ and all scalar functions of the form (f_λ, φ) can be analytically extended to a larger domain Λ_1. Then, the values (f_λ, φ), for each $\lambda \in \Lambda_1$, define a continuous linear functional f_λ on the space S.

Example 2.2.4. Consider the function x_+^λ which is equal to x^λ for $x > 0$ and identically vanishes for $x < 0$. For $\operatorname{Re} \lambda > -1$, this function defines the functional

$$\left(x_+^\lambda, \varphi\right) = \int_0^\infty x^\lambda \varphi(x)\, dx. \tag{2.2.19}$$

The scalar function (2.2.19) is analytic with respect to λ, because it admits the derivative in λ equal to

$$\int_0^\infty x^\lambda (\ln x) \varphi(x)\, dx.$$

This means that the functional x_+^λ is analytic in λ for $\operatorname{Re} \lambda > -1$. Let us construct its analytic extension to the entire plane of the variable λ. To this end, we write (2.2.19) in the form

$$\left(x^\lambda, \varphi\right) = \int_0^\infty x^\lambda \varphi(x)\, dx = \int_0^1 x^\lambda [\varphi(x) - \varphi(0)]\, dx + \int_1^\infty x^\lambda \varphi(x)\, dx + \frac{\varphi(0)}{\lambda + 1}. \tag{2.2.20}$$

The first term on the right-hand side is defined for $\operatorname{Re} \lambda > -2$, the second term is defined for all λ, and the third for $\lambda \neq -1$. Therefore, the functional x_+^λ can be analytically extended to the region $\operatorname{Re} \lambda > 2$, $\lambda \neq -1$.

In a similar way, we construct the analytic extension of the functional x_+^λ to the region $\operatorname{Re} \lambda > -n - 1$, $\lambda \neq -1, \ldots, -n$,

$$\left(x_+^\lambda, \varphi\right) = \int_0^1 x^\lambda \left[\varphi(x) - \varphi(0) - x\varphi'(0) - \cdots - \frac{x^{n-1}}{(n-1)!} \varphi^{(n-1)}(0) \right] dx +$$
$$+ \int_1^\infty x^\lambda \varphi(x)\, dx + \sum_{k=1}^n \frac{\varphi^{(k-1)}(0)}{(k-1)!(\lambda + k)}. \tag{2.2.21}$$

Thereby, the generalized function x_+^λ is defined for all $\lambda \neq -1, -2, \ldots$. The functional x_-^λ for $\operatorname{Re} \lambda > -1$ is defined by the relation

$$\left(x_-^\lambda, \varphi(x)\right) = \int_{-\infty}^0 |x|^\lambda \varphi(x)\, dx. \tag{2.2.22}$$

Relation (2.2.22) can be written in the form

$$\left(x_-^\lambda, \varphi(x)\right) = \int_0^\infty x^\lambda \varphi(-x)\, dx = \left(x_+^\lambda, \varphi(-x)\right). \tag{2.2.23}$$

Formula (2.2.23) automatically allows us to transfer the results about the analytic extension of the functional x_+^λ to the case of x_-^λ.

Example 2.2.5. Consider the functional r^λ whose action, for $\text{Re}\,\lambda > -n$, is given by

$$\left(r^\lambda, \varphi\right) = \int_{\mathbb{R}_n} r^\lambda \varphi(x)\,dx, \tag{2.2.24}$$

where $r = \left(\sum_{i=1}^n x_i^2\right)^{1/2}$, $\varphi(x) \in S$. Since

$$\frac{d}{d\lambda}(r^\lambda, \varphi) = \int_{\mathbb{R}_n} r^\lambda \ln r \varphi(x)\,dx,$$

we see that the functional r^λ is an analytic function of λ in the region $\text{Re}\,\lambda > -n$. For $\text{Re}\,\lambda \le -n$, the function r^λ is locally non-integrable. Let us define the functional r^λ by the method of analytic extension. Passing to spherical coordinates in the integral (2.2.24), we transform it to

$$(r^\lambda, \varphi) = \Omega_n \int_0^\infty r^{\lambda+n-1} S_\varphi(r)\,dr, \tag{2.2.25}$$

where Ω_n is the area of the unit sphere in $\mathbb{R}_n$,

$$S_\varphi(r) = \frac{1}{\Omega_n} \int_\Omega \varphi(r\omega)\,d\omega, \tag{2.2.26}$$

where $d\omega$ is the area element on the unit sphere Ω, and ω is a unit vector. Using basic properties of test functions φ and differentiating, one can see that $S_\varphi(r) \in S^+(\mathbb{R}_1)$ and the operator $S_\varphi(r)$ is continuous from $S(\mathbb{R}_n)$ to $S^+(\mathbb{R}_1)$.

From the Taylor expansion of the function $\varphi(x)$ we find that

$$S_\varphi(r) = \frac{1}{\Omega_n} \int_\Omega \left[\varphi(0) + \sum \frac{\partial \varphi(0)}{\partial x_j} x_j + \frac{1}{2} \sum \frac{\partial^2 \varphi(0)}{\partial x_i \partial x_j} x_i x_j + \cdots \right] d\omega.$$

Each term of the integrand with an odd number of x_j annihilates after the integration and we get

$$S_\varphi(r) = \varphi(0 + a_1 r^2 + a_2 r^4 + \cdots + a_k r^{2k} + o(r^{2k}), \qquad \lim_{r\to 0} \frac{o(r^{2k})}{r^{2k}} = 0. \tag{2.2.27}$$

It follows that

$$\frac{\partial^{2m+1} S_\varphi(0)}{\partial r^{2m+1}} = 0, \quad m = 0, 1, 2, \ldots.$$

The integral in (2.2.25) can be understood as the result of applying the functional $\Omega_n x_+^\mu$ ($\mu = \lambda + n - 1$). Example 2.2.4 shows that the function x_+^μ is analytic for $\text{Re}\,\mu > -1$ (i.e., for $\text{Re}\,\lambda > -n$) and can be extended as an analytic function to the entire plane with the points $\mu = -1, \ldots, -m$ ($\lambda = -n, -n-1, \ldots$) excluded. Since the odd order derivatives of the function $S_\varphi(x)$ vanish at $x = 0$, there are no poles corresponding to the even values of m (see Example 2.2.4). Thus, there remains the sequence of poles $\lambda = -n, -n-2, \ldots$.

Example 2.2.6. Let us find the Fourier transform of the functional r^λ. The generalized function r^λ is defined for $\lambda \ne -n, n-2, \ldots$, and is spherically symmetric, i.e., $\left(r^\lambda, \varphi(U(x)) = (r^\lambda, \varphi(x)\right)$, where U is any orthogonal transformation of $\mathbb{R}_n$. Therefore, its Fourier transform $g_\lambda(\xi)$ is a spherically symmetric generalized function. The Fourier integral

$$g_\lambda(\xi) = \int_{\mathbb{R}_n} r^\lambda\, e^{i(\xi,x)}\,dx$$

is convergent for $-n < \operatorname{Re} \lambda < 0$ and is a function of $\rho = \left(\sum \xi_i^2 \right)^{1/2}$. For any $t > 0$, we have

$$g_\lambda(t\xi) = \int_{\mathbb{R}_n} r^\lambda \, e^{i(t\xi,x)} \, dx.$$

Changing the variables: $y = tx$, $x = t^{-1}y$, $dx = t^{-n}dy$, $r = |x| = t^{-1}|y|$, we get

$$g_\lambda(t\xi) = t^{-\lambda-n} \int_{\mathbb{R}_n} |y|^\lambda \, e^{i(\xi,y)} dy = t^{-\lambda-n} g_\lambda(\xi).$$

This means that $g_\lambda(\xi)$ is a homogeneous function of degree $(-\lambda - n)$, and therefore, has the form

$$g_\lambda(\xi) = C_\lambda \rho^{-\lambda-n}. \tag{2.2.28}$$

Let us calculate the constant C_λ. We have

$$g_{1\lambda} = \int_{\mathbb{R}_n} \overline{r^\lambda} \, e^{i(\xi,x)} \, dx = \overline{\int_{\mathbb{R}_n} r^\lambda \, e^{-i(\xi,x)} \, dx} = \overline{g_\lambda(-\xi)} = \overline{g_\lambda(\xi)}.$$

Hence, we get

$$(2\pi)^n \int_{\mathbb{R}_n} r^\lambda \varphi(x) \, dx = \int_{\mathbb{R}_n} \overline{g_{1\lambda}(\xi)} \widehat{\varphi}(\xi) \, d\xi = \int_{\mathbb{R}_n} g_\lambda(\xi) \widehat{\varphi}(\xi) \, d\xi = C_\lambda \int_{\mathbb{R}_n} \rho^{-\lambda-n} \widehat{\varphi}(\xi) \, d\xi. \tag{2.2.29}$$

Take $\varphi(x) = e^{-r^2/2}$. Then, using the formula $F\left(e^{-x^2/2} \right) = \sqrt{2} e^{-\xi^2/2}$ (see *Ryzhik and Gradshteyn* (1951)), we get $\widehat{\varphi}(\xi) = (2\pi)^{n/2} e^{-\rho^2/2}$. Substituting $\varphi(x)$ and $\widehat{\varphi}(\xi)$ into (2.2.29), we find that

$$C_\lambda \int_{\mathbb{R}_n} e^{-\frac{\rho^2}{2}} \rho^{-n-\lambda} \, d\xi = (2\pi)^{\frac{n}{2}} \int_{\mathbb{R}_n} r^\lambda \, e^{-\frac{r^2}{2}} \, dx.$$

Passing to spherical coordinates, we obtain

$$C_\lambda \int_0^\infty e^{-\frac{\rho^2}{2}} \rho^{-\lambda-1} \, d\rho = \int_0^\infty e^{-\frac{r^2}{2}} r^{\lambda+n-1} \, dr. \tag{2.2.30}$$

The integrals obtained can be expressed through the gamma-function

$$\Gamma(\lambda) = \int_0^\infty x^{\lambda-1} \, e^{-x} \, dx.$$

Hence,

$$\int_0^\infty e^{-\frac{\rho^2}{2}} \rho^{-\lambda-1} \, d\rho = 2^{-\frac{\lambda}{2}-1} \Gamma\left(-\frac{\lambda}{2} \right), \qquad \int_0^\infty e^{-\frac{r^2}{2}} r^{\lambda+n-1} \, dr = 2^{\frac{\lambda+n-2}{2}} \Gamma\left(\frac{\lambda+n}{2} \right).$$

Using these relations, we find that

$$C_\lambda = 2^{\lambda+n} \pi^{\frac{n}{2}} \frac{\Gamma\left(\dfrac{\lambda+n}{2} \right)}{\Gamma\left(-\dfrac{\lambda}{2} \right)},$$

and therefore,

$$F(r^\lambda) = g_\lambda = 2^{\lambda+n} \pi^{\frac{n}{2}} \frac{\Gamma\left(\dfrac{\lambda+n}{2} \right)}{\Gamma\left(-\dfrac{\lambda}{2} \right)} \rho^{-\lambda-n}.$$

Hence we see that for $-n < \operatorname{Re} \lambda < 0$, the functional $F(r^\lambda)$ has the form

$$\left(F(r^\lambda), \psi(\xi) \right) = 2^{\lambda+n} \pi^{\frac{n}{2}} \frac{\Gamma\left(\dfrac{\lambda+n}{2} \right)}{\Gamma\left(-\dfrac{\lambda}{2} \right)} \int_{\mathbb{R}_n} \rho^{-\lambda-n} \widehat{\psi}(\xi) \, d\xi. \tag{2.2.31}$$

This relation remains valid in the whole region in which the analytic function r^λ exists, i.e., for all $\lambda \neq -n, -n-2, \dots$. The regularization of the integral is done in the same way as in Example 2.2.5.

2.3. Sobolev–Slobodetskii Spaces

Definition 2.3.1. *The Sobolev–Slobodetskii space* $H_s(\mathbb{R}_n)$, *for any real* s, *consists of all generalized functions* u *whose Fourier transform* $\hat{u}(\xi)$ *is locally integrable in the sense of Lebesgue and*

$$\|u\|_s^2 = \int_{\mathbb{R}_n} |\hat{u}(\xi)|^2 (1 + |\xi|)^{2s}\, dx < \infty, \tag{2.3.1}$$

where $|\xi| = \left(\sum_{i=1}^n \xi_i^2\right)^{1/2}$.

The Fourier image of the space $H_s = H_s(\mathbb{R}_n)$ is denoted by $\tilde{H}_s = \tilde{H}_s(\mathbb{R}_n)$. Formula (2.3.1) defines a norm in the spaces H_s and $\tilde{H}_s$. The space $\tilde{H}_s$ is a Hilbert space with respect to the scalar product

$$\langle u, v \rangle_s = \int_{\mathbb{R}_n} \hat{u}(\xi)\overline{\hat{v}(\xi)}(1 + |\xi|)^{2s}\, d\xi. \tag{2.3.2}$$

It follows that H_s is a Hilbert space, too.

Let $u \in H_s$ and $\varphi \in S$. Then, by (2.2.6), we have

$$(u, \varphi) = \frac{1}{(2\pi)^n} \int_{\mathbb{R}_n} \overline{\hat{u}(\xi)}\hat{\varphi}(\xi)\, d\xi. \tag{2.3.3}$$

Applying the Cauchy inequality to (2.3.3), we get

$$|(u, \varphi)| \le \frac{1}{(2\pi)^n} \int_{\mathbb{R}_n} |\hat{u}(\xi)| \cdot |\hat{\varphi}(\xi)| \frac{(1 + |\xi|)^s}{(1 + |\xi|)^s}\, d\xi \le \frac{1}{(2\pi)^n} \|u\|_s \cdot \|\varphi\|_{-s}. \tag{2.3.4}$$

From (2.1.1) and (2.3.1), it follows that if $\varphi_n \to \varphi$ in S, then $\|\varphi_n - \varphi\|_s \to 0$. On the other hand, by virtue of (2.3), the convergence $\|u_n - u\|_s \to 0$ implies that $(u_n, \varphi) \to (u, \varphi)$, for any $\varphi \in S$, i.e., $u_n \to u$ in S'.

For $s = 0$, the space $\tilde{H}_0$ coincides with $L_2(\mathbb{R}_n)$ — the Lebesgue space of square summable functions. By the Plancherel theorem (see *Shilov* (1961)), we have $H_0 = F^{-1}(\tilde{H}_0) = L_2(\mathbb{R}_n)$.

Proposition 2.3.1. *For integer* $s = m > 0$, *the norm* (2.3.1) *is equivalent to*

$$\|u\|_m'^2 = \sum_{|k| \le m} \int_{\mathbb{R}_n} \left| \frac{\partial^k u(x)}{\partial x^k} \right|^2 dx. \tag{2.3.5}$$

Proof. For $\hat{u}(\xi) \in \tilde{H}_m$, we have $(-i\xi)^k \hat{u}(\xi) \in L_2(\mathbb{R}_n)$ for $|k| \le m$, and it follows from the Plancherel theorem that

$$\frac{\partial^k u(x)}{\partial x^k} = F^{-1}\left((-i\xi)^k \hat{u}(\xi)\right) \in L_2(\mathbb{R}_n).$$

Thus, $H_m(\mathbb{R}_n)$ consists of square summable functions $u(x)$ whose generalized derivatives of the orders $1 \le |k| \le m$ are square summable. The following inequality holds

$$C_1(1 + |\xi|)^m \le \sum_{|k| \le m} |\xi|^{|k|} \le C_2(1 + |\xi|)^m. \tag{2.3.6}$$

Proposition 2.3.1 is proved.

Proposition 2.3.2. *For integer* $m > 0$, *any generalized function in* H_{-m} *can be represented as a sum of derivatives of functions in* $L_2(\mathbb{R}_n)$, *the order of these derivatives being* $\leq m$.

Proof. For $\widehat{u}(\xi) \in \widetilde{H}_{-m}$, let $\widehat{V}(\xi) = (1 + |\xi|)^{-m}\widehat{u}(\xi)$. Clearly, $\widehat{V}(\xi) \in \widetilde{H}_0 = L_2(\mathbb{R}_n)$. Since $|\xi| = \sum_{k=1}^{n} \xi_k(\xi_k/|\xi|)$, we can represent $\widehat{u}(\xi)$ in the form

$$\widehat{u}(\xi) = (1 + |\xi|)^m \widehat{V}(\xi) = \sum_{|k| \leq m} (-i\xi)^k \widehat{V}_k(\xi), \tag{2.3.7}$$

where $\widehat{V}(\xi) \in L_2(\mathbb{R}_n)$. Applying the inverse Fourier transformation to both sides of (2.3.7), we get

$$u(x) = \sum_{|k| \leq m} \frac{\partial^k V_k(x)}{\partial x^k}, \qquad V_k(x) \in L_2(\mathbb{R}_n).$$

Proposition 2.3.2 is proved.

Example 2.3.1. The function $\delta^k(x) = \partial^k \delta/\partial x^k$ belongs to $H_s(\mathbb{R}_1)$ for $s < -|k| - 1/2$, but does not belong to $H_{-|k|-1/2}(\mathbb{R}_1)$, since $F(\delta^k(x)) = (-i\xi)^k$, as shown in Example 2.2.1.

Theorem 2.3.1. *The functions of class* $C_0^\infty(\mathbb{R}_n)$ *are dense in* $H_s(\mathbb{R}_n)$ *with respect to the norm* (2.3.1).

Proof. Let $\alpha(x) \in C_0^\infty(\mathbb{R}_n)$, $\alpha(x) \geq 0$, $\alpha(x) = 0$ for $|x| > 1$, and $\int_{\mathbb{R}_n} \alpha(x)\,dx = 1$. Take $\alpha_\varepsilon(x) = \varepsilon^{-n}\alpha(\varepsilon^{-1}x)$. For an arbitrary function $u(x) \in H_s$, let $u_\varepsilon = u \times \alpha_\varepsilon$. By Propositions 2.1.4 and 2.2.4, we have $u_\varepsilon \in C^\infty$ and $\widehat{u}_\varepsilon(\xi) = \widehat{\alpha}_\varepsilon(\xi)\widehat{u}$. Changing the variables, $x = \varepsilon y$, we get

$$\widehat{\alpha}_\varepsilon(\xi) = \int_{\mathbb{R}_n} \alpha(\varepsilon^{-1}x)\,e^{i(x,\xi)}\varepsilon^{-n}\,dx = \int_{\mathbb{R}_n} \alpha(y)\,e^{i(y,\varepsilon\xi)}\,dy = \widehat{\alpha}(\varepsilon\xi), \tag{2.3.8}$$

and

$$|\widehat{\alpha}(\xi)| \leq \int_{\mathbb{R}_n} \alpha(y)\,dy = \widehat{\alpha}(0) = 1. \tag{2.3.9}$$

Therefore, $\widehat{u}_\varepsilon \in \widetilde{H}_s$. Moreover,

$$\|u - u_\varepsilon\|_s^2 \leq \int_{\mathbb{R}_n} |\widehat{u}(\xi)|^2 \,|1 - \widehat{\alpha}(\varepsilon\xi)|\,(1 + |\xi|)^{2s}\,d\xi. \tag{2.3.10}$$

For any fixed ξ, we have the convergence $1 - \widehat{\alpha}(\varepsilon\xi) \to 0$ as $\varepsilon \to 0$ and also the inequality $|1 - \widehat{\alpha}(\varepsilon\xi)| < 2$ (see (2.3.9)). Now, from (2.3.10) and the Lebesgue theorem, it follows that $\|u - u_\varepsilon\|_s \to 0$ as $\varepsilon \to 0$. Thus, for any $\delta > 0$ there is $\varepsilon_1 > 0$ such that for all $0 < \varepsilon < \varepsilon_1$ we have

$$\|u - u_\varepsilon\|_s < \frac{\delta}{2}. \tag{2.3.11}$$

The inclusion $\widehat{\alpha}(\varepsilon\xi) \in S(\mathbb{R}_n)$ means that $\widehat{\alpha}(\varepsilon\xi)$ decays faster than any negative power of $|\xi|$ as $|\xi| \to \infty$, and therefore, $\widehat{u}_\varepsilon = \widehat{\alpha}(\varepsilon\xi)\widehat{u}(\xi) \in \widetilde{H}_N$ for any N. Let $\chi(x) \in C_0^\infty(\mathbb{R}_n)$, $\chi(x) = 1$ for $|x| \leq 1$. Take $V_\varepsilon(x, \beta) = \chi(\beta x)u_\varepsilon(x) \in C_0^\infty$. Let us show that for any N and $\beta \to 0$, we have $\|u_\varepsilon(x) - V_\varepsilon(x, \beta)\|_N \to 0$. Indeed, taking into account that the norms (2.3.5) and (2.3.1) are equivalent, we see that for $\beta \to 0$,

$$\|u_\varepsilon(x) - V_\varepsilon(x, \beta)\|_N^2 = \sum_{|k| \leq N} \int_{|x| \geq \beta^{-1}} \left| \frac{\partial^k}{\partial x^k}\left[(1 - \chi(\beta x))u_\varepsilon(x)\right] \right|^2 dx \to 0.$$

Consequently, there is $\beta_1(\varepsilon_0)$, $0 < \varepsilon_0 < \varepsilon$, such that for all $0 < \beta < \beta_1$, the following inequality holds:

$$\|u_{\varepsilon_0}(x) - V_{\varepsilon_0}(x, \beta)\|_s < \frac{\delta}{2}. \qquad (2.3.12)$$

From (2.3.11) and (2.3.12), it follows that there is $V_{\varepsilon_0}(x, \beta_0) \in C_0^\infty(\mathbb{R}_n)$ such that

$$\|u - V_{\varepsilon_0}(x, \beta_0)\|_s < \delta.$$

Theorem 2.3.1 is proved.

Because of Theorem 2.3.1, the space H_s can be defined as the completion of $C_0^\infty(\mathbb{R}_n)$ with respect to the norm (2.3.1).

Proposition 2.3.3. *The norm of the space H_λ with $0 < \lambda < 1$ is equivalent to*

$$\|u\|_\lambda'^2 = \int_{\mathbb{R}_n} \int_{\mathbb{R}_n} \frac{|u(x + y) - u(x)|^2}{|y|^{n+2\lambda}} \, dx \, dy + \int_{\mathbb{R}_n} |u(x)|^2 \, dx. \qquad (2.3.13)$$

Proof. For $u(x) \in C_0^\infty(\mathbb{R}_n)$ we have

$$u(x + y) - u(x) = \frac{1}{(2\pi)^n} \int_{\mathbb{R}_n} \widehat{u}(\xi) \left[e^{-i(y,\xi)} - 1 \right] e^{-i(x,\xi)} \, d\xi. \qquad (2.3.14)$$

Writing the Parseval identity for (2.3.14) with y fixed, we obtain

$$\int_{\mathbb{R}_n} |u(x + y) - u(x)|^2 \, dx = \frac{1}{(2\pi)^n} \int_{\mathbb{R}_n} |\widehat{u}(\xi)|^2 \cdot \left| e^{-i(y,\xi)} - 1 \right|^2 d\xi.$$

Therefore,

$$\|u\|_\lambda'^2 = \frac{1}{(2\pi)^n} \int_{\mathbb{R}_n} \int_{\mathbb{R}_n} |u(\xi)|^2 \frac{\left| e^{-i(y,\xi)} - 1 \right|^2}{|y|^{n+2\lambda}} \, dy \, d\xi + \frac{1}{(2\pi)^n} \int_{\mathbb{R}_n} |\widehat{u}(\xi)|^2 \, d\xi. \qquad (2.3.15)$$

Consider the function

$$f_\lambda(\xi) = \int_{\mathbb{R}_n} \frac{\left| e^{-i(y,\xi)} - 1 \right|^2}{|y|^{n+2\lambda}} \, dy. \qquad (2.3.16)$$

The integral in (2.3.16) is convergent, since $\left| e^{-i(y,\xi)} - 1 \right|^2 < C|y|^2$ and $0 < \lambda < 1$. It is easy to see that $f_\lambda(\xi)$ is a spherically symmetrical function, homogeneous of order 2λ (see Example 2.2.6). Hence we find that

$$f(\xi) = C_\lambda |\xi|^{2\lambda}.$$

Now, Proposition 2.3.3 follows from the inequality

$$K_1(\lambda)(1 + |\xi|)^{2\lambda} \le 1 + |\xi|^{2\lambda} \le K_2(\lambda)(1 + |\xi|)^{2\lambda}, \qquad K_i(\lambda) > 0, \quad i = 1, 2,$$

combined with (2.3.15).

Corollary 2.3.1. *For integer m and $0 < \lambda < 1$, the norm in $H_{m+\lambda}$ is equivalent to*

$$\|u\|_{m+\lambda}'^2 = \sum_{|k| \le m} \int_{\mathbb{R}_n} \int_{\mathbb{R}_n} \frac{1}{|y|^{n+2\lambda}} \left| \frac{\partial^k}{\partial x^k} u(x+y) - \frac{\partial^k}{\partial x^k} u(x) \right|^2 dx\, dy + \int_{\mathbb{R}_n} |u(x)|^2 dx. \quad (2.3.17)$$

This statement follows from Propositions 2.3.1 and 2.3.3.

For integer $r \ge 0$, let $C_0^r(\mathbb{R}_n)$ be the space of all functions which are continuous, together with their derivatives up to the order r, and satisfy the conditions

$$\lim_{|x| \to \infty} u(x) = 0, \qquad \lim_{|x| \to \infty} \frac{\partial^k u(x)}{\partial x^k} = 0 \quad \text{for} \quad |k| \le r.$$

The norm in $C_0^r(\mathbb{R}_n)$ is introduced by

$$\|u(x)\| = \sum_{|k|=0}^{r} \sup_{x \in \mathbb{R}_n} \left| \frac{\partial^k u(x)}{\partial x^k} \right|. \quad (2.3.18)$$

Theorem 2.3.2. (**Sobolev imbedding theorem**).

$$H_s(\mathbb{R}_n) \subset C_0^r(\mathbb{R}_n) \quad \text{for} \quad s > \frac{n}{2}, \quad 0 \le r < s - \frac{n}{2},$$

and the imbedding operator is continuous.

Proof. For $u(x) \in C_0^\infty(\mathbb{R}_n)$ and its Fourier transform $\hat{u}(\xi)$ we have

$$\frac{\partial^k u(x)}{\partial x^k} = \frac{1}{(2\pi)^n} \int_{\mathbb{R}_n} (-i\xi)^k \hat{u}(\xi)\, e^{-i(x,\xi)}\, d\xi.$$

By the Cauchy inequality,

$$\left| \frac{\partial^k u(x)}{\partial x^k} \right|^2 \le \frac{1}{(2\pi)^n} \left(\int_{\mathbb{R}_n} |\xi|^{|k|} |\hat{u}(\xi)|\, d\xi \right)^2 \le$$

$$\le \frac{1}{(2\pi)^{2n}} \left(\int_{\mathbb{R}_n} \frac{|\xi|^{2|k|}}{(1+|\xi|)^{2s}}\, d\xi \right) \int_{\mathbb{R}_n} (1+|\xi|)^{2s} |\hat{u}(\xi)|^2\, d\xi. \quad (2.3.19)$$

Since $|k| \le r$ and $s - r > n/2$, we have

$$\int_{\mathbb{R}_n} \frac{|\xi|^{2|k|}}{(1+|\xi|)^{2s}}\, d\xi \le \int_{\mathbb{R}_n} \frac{d\xi}{(1+|\xi|)^{2(s-r)}} < \infty. \quad (2.3.20)$$

From (2.3.18)–(2.3.20), it follows that

$$\|u(x)\|_{C_0^r} \le K \|u(x)\|_s. \quad (2.3.21)$$

By Theorem 2.3.1, for any function $u(x) \in H_s(\mathbb{R}_n)$ there is a sequence $u_m(x) \in C_0^\infty(\mathbb{R}_n)$ converging to $u(x)$ in the norm of $H_s(\mathbb{R}_n)$. Since the space $C_0^r(\mathbb{R}_n)$ is complete, there exists $V(x) \in C_0^r(\mathbb{R}_n)$ such that $\|V(x) - u_m\|_s \to 0$, and thus, $u_m \to V$ in $H_s(\mathbb{R}_n)$. On the other hand, the convergence $\|u_m - u\|_s \to 0$ implies that $u_m \to u$ in $H_s(\mathbb{R}_n)$. Therefore, $u = V$ and $H_s(\mathbb{R}_n) \subset C_0^r(\mathbb{R}_n)$.

The inequality (2.3.21), which holds for u^m, remains valid for u after passing to the limit, and thus Theorem 2.3.2 is proved.

Definition 2.3.2. *The space H_s^+ consists of all functions $u(x) \in H_s(\mathbb{R}_n)$ whose support lies in the closed halfspace $\overline{\mathbb{R}}_n^+ = \{x : x_n \geq 0\}$.*

Definition 2.3.3. *The space H_s^- consists of all functions $u(x) \in H_s(\mathbb{R}_n)$ whose support lies in the closed halfspace $\overline{\mathbb{R}}_n^- = \{x : x_n \leq 0\}$.*

The norm in the spaces $H_s^{\pm}$ is defined by (2.3.1).

If $u(x) \in H_s^+$ with $s \geq 0$, then $u(x) = 0$ almost everywhere for $x_n < 0$. For an arbitrary s, in particular, $s < 0$, the inclusion $u \in H_s^+$ means that $u \in H_s$ and $(u, \varphi) = 0$ for any $\varphi(x) \in C_0^\infty(\mathbb{R}_n^-)$, by the definition of the support of a generalized function.

Proposition 2.3.4. *The subspace H_s^+ is closed in H_s.*

Proof. Let $\{u_n\| \in H_s^+$ and $\|u - u_n\|_s \to 0$. Let us show that $u \in H_s^+$. We have $(u_n, \varphi) = 0$ for all $\varphi \in C_0^\infty(\mathbb{R}_n^-)$. From (2.3.4), we conclude that $(u_n, \varphi) \to (u, \varphi) = 0$. Therefore, $u(x) \in H_s^+$.

Proposition 2.3.5. *The functions in $C_0^\infty(\mathbb{R}_n^+)$ are dense in H_s^+ with respect to the norm (2.3.1).*

Proof. Let $u \in H_s^+$, $\vec{\varepsilon} = (0, \ldots, 0, \varepsilon)$, $\varepsilon > 0$. Denote by $u^{\vec{\varepsilon}}$ the shift of the generalized function u by the vector $\vec{\varepsilon}$ (see Definition 2.1.7). Since $\widehat{u^{\vec{\varepsilon}}} = e^{(\vec{\varepsilon}, \xi)}\widehat{u}(\xi)$ (see (2.2.7)), we have $u^{\vec{\varepsilon}} \in H_s$. Using the Lebesgue theorem on dominated convergence, we find that

$$\left\| u^{\vec{\varepsilon}} - u \right\|_s^2 = \int_{\mathbb{R}_n} |\widehat{u}(\xi)|^2 \left| e^{i\varepsilon\xi_n} - 1 \right|^2 (1 + |\xi|)^{2s}\, d\xi \to 0 \quad \text{as} \quad \varepsilon \to 0. \qquad (2.3.22)$$

It follows that for any $\delta > 0$ there is $\varepsilon_1 > 0$ such that

$$\left\| u - u^{\vec{\varepsilon}_1} \right\|_s < \frac{\delta}{2}. \qquad (2.3.23)$$

Let us show that $\operatorname{supp} u^{\vec{\varepsilon}} \subset \overline{\mathbb{R}_n^\varepsilon} = \{x : x_n \geq \varepsilon\}$. Indeed, take an arbitrary $\varphi(x)$ such that $\operatorname{supp} \varphi \in \mathbb{R}_n \setminus \overline{\mathbb{R}_n^\varepsilon}$. Then $(u^{\vec{\varepsilon}}, \varphi) = (u, \varphi(x', x_n + \varepsilon)) = 0$. Let us approximate the function $u^{\vec{\varepsilon}_1}$ by functions of class $C_0^\infty(\mathbb{R}_n)$, just as we have done in the proof of Theorem 2.3.1. Consider the function $V_{\varepsilon_0} = (u^{\vec{\varepsilon}_1} \times \alpha_{\varepsilon_0})\chi(\beta_0 x)$. Obviously,

$$\operatorname{supp} V_{\varepsilon_0}(x, \beta_0) \subseteq \operatorname{supp}(u^{\vec{\varepsilon}} \times \alpha_{\varepsilon_0}) \subseteq \operatorname{supp} u^{\vec{\varepsilon}} + \operatorname{supp} \alpha_{\varepsilon_0},$$

where $\operatorname{supp} u^{\vec{\varepsilon}} + \operatorname{supp} \alpha_{\varepsilon_0}$ is the vector sum of the two sets (see *Shilov* (1965)).

The above arguments imply that there exist ε_2 and β_2 such that

$$\operatorname{supp} u^{\vec{\varepsilon}_1} + \operatorname{supp} \alpha_{\varepsilon_2} \subset \overline{\mathbb{R}}^+, \qquad \left\| u^{\vec{\varepsilon}_1} - V_{\varepsilon_2}(x, \beta_2) \right\| < \frac{\delta}{2}. \qquad (2.3.24)$$

Now, Proposition 2.3.5 follows from the inequalities (2.3.23) and (2.3.24).

Remark 2.3.1. In a similar way it can be shown that H_s^- is a closed subspace of $H_s(\mathbb{R}_n)$, and also that $C_0^\infty(\mathbb{R}_n)$ is dense in $H_s^-(\mathbb{R}_n)$.

Example 2.3.2. The generalized function $\delta^k = \partial^k \delta / \partial x^k$ belongs to both spaces $H_{-k-\frac{1}{2}-\varepsilon}^+(\mathbb{R}_1)$ and $H_{-k-\frac{1}{2}-\varepsilon}^-(\mathbb{R}_1)$ for any $\varepsilon > 0$.

Let $A(\xi)$ be a locally integrable function satisfying the inequality

$$|A(\xi)| \leq C(1 + |\xi|)^\alpha. \qquad (2.3.25)$$

We define the operator A on $S(\mathbb{R}_n)$ by

$$Au = \frac{1}{(2\pi)^n} \int_{\mathbb{R}_n} A(\xi)\widehat{u}(\xi)\, e^{-i(x,\xi)}\, d\xi = F^{-1}(A(\xi)F(u)), \qquad (2.3.26)$$

where $\widehat{u}(\xi) = F(u)$. The function $A(\xi)$ is called the *symbol of operator A*.

Proposition 2.3.6. *For the operator A given by (2.3.26) with the symbol $A(\xi)$ satisfying the inequality (2.3.25), the following estimate holds*

$$\|Au\|_{s-\alpha} \le C_s \|u\|_s, \qquad \forall\, u \in S(\mathbb{R}_n). \tag{2.3.27}$$

Proof. We have

$$\|Au\|_{s-\alpha}^2 = \int_{\mathbb{R}_n} (1+|\xi|)^{2(s-\alpha)} |A(\xi)|^2 |\widehat{u}(\xi)|^2 \, d\xi \le C \int_{\mathbb{R}_n} (1+|\xi|)^{2s} |\widehat{u}(\xi)|^2 \, d\xi = C\|u\|_s^2.$$

Proposition 2.3.6 is proved.

Corollary 2.3.2. *An operator A for which the assumptions of Proposition 2.3.6 are valid is a bounded operator from H_s to $H_{s-\alpha}$.*

By $\widetilde{H}^{\pm}$ we denote the Fourier images of the spaces $H_s^{\pm}$.

Theorem 2.3.3. *Suppose that the function*

$$A_+(\xi', \xi_n + i\tau), \quad (\text{resp., } A_-(\xi', \xi_n + i\tau)), \quad \xi' = (\xi_1, \dots, \xi_{n-1}),$$

is continuous with respect to all its variables jointly for $\xi' \ne 0$, $\tau \ge 0$ (resp., $\tau \le 0$)), is analytic in $\alpha_n = \xi_n + i\tau$ (resp., $(\tau < 0)$), and satisfies the inequality

$$|A_\pm(\xi', \xi_n + i\tau)| \le C(1 + |\xi'| + |\xi_n| + |\tau|)^\alpha. \tag{2.3.28}$$

Then the operator of multiplication by $A_\pm(\xi', \xi_n)$ is bounded from $\widetilde{H}_s^{\pm}$ to $\widetilde{H}_{s-\alpha}^{\pm}$.

Proof. From Proposition 2.3.6, it follows that the operator of multiplication by $A_+(\xi', \xi_n)$ is bounded from $\widetilde{H}_s^+$ to $\widetilde{H}_{s-\alpha}$. Let us show that $A_+\widehat{u}_+ \in \widetilde{H}_{s-\alpha}^+$ for any $\widehat{u}_+ \in \widetilde{H}_s^+$. First, let $u_+(x) \in C_0^\infty(\mathbb{R}_n)$. Then, by Proposition 2.2.2, the function $\widehat{u}_+(\xi', \xi_n + i\tau)$ is analytic in $\alpha_n = \xi_n + i\tau$ for $\tau > 0$ and satisfies the inequality

$$\left(1 + \xi' + |\xi_n| + \tau\right)^N |\widehat{u}_+(\xi', \xi_n + i\tau)| \le C_N, \qquad \forall N. \tag{2.3.29}$$

Set $\widetilde{V}(\xi', \xi_n) = A_+(\xi', \xi_n)\,\widehat{u}_+(\xi', \xi_n)$. We have

$$V(x', x_n) = \frac{1}{(2\pi)^n} \int_{\mathbb{R}_n} A_+(\xi', \xi_n)\,\widehat{u}(\xi', \xi_n)\, e^{-i(x'.\xi')-ix_n\xi_n} \, d\xi' \, d\xi_n. \tag{2.3.30}$$

Taking into account the estimates (2.3.28), (2.3.29) and the fact that $A_+(\xi', \xi_n + i\tau)$, $\widehat{u}_+(\xi', \xi_n + i\tau)$ are analytic in $\alpha_n = \xi_n + i\tau$ in the halfplane $\tau > 0$ and are continuous in the closed halfplane $\tau \ge 0$ for $\xi' \ne 0$, we can apply the Cauchy theorem (see *Lavrentiev and Shabat* (1973)) to the integral (2.3.30). We get

$$V(x', x_n) = \frac{1}{(2\pi)^n} \int_{\mathbb{R}_n} A_+(\xi', \xi_n + i\tau)\,\widehat{u}(\xi', \xi_n + i\tau)\, e^{-i(x'.\xi')-ix_n(\xi_n+i\tau)} \, d\xi' \, d\xi_n, \tag{2.3.31}$$

where τ is arbitrary. Using (2.3.28) and (2.3.29), we obtain the inequality

$$|V(x', x_n)| \le C\, e^{x_n\tau}, \tag{2.3.32}$$

where C is a constant independent of τ. For $x_n < 0$, since $\tau > 0$ is arbitrary, the inequality (2.3.32) implies that $V(x', x_n) = 0$, which means that $\operatorname{supp} V \subset \overline{\mathbb{R}}_n^+$. Consequently, $V(x) \in H_n^+$, $\widehat{V}(\xi) \in \widetilde{H}_{s-\alpha}^+$. By Proposition 2.3.5, for an arbitrary $u^+ \in H_s^+$, there is a sequence $\varphi_n \in C_0^\infty(\mathbb{R}_n)$ such that $\|u_+ - \varphi_n\|_s \to 0$. It follows that $A_+(\xi)\varphi_n(\xi) \to A_+(\xi)u_+(\xi)$ in $\widetilde{H}_{s-\alpha}$. On the basis of Proposition 2.3.4 we conclude that $A_+(\xi)u_+(\xi) \in \widetilde{H}_{s-\alpha}^+$. In a similar way it can be shown that the operator of multiplication by $A_-(\xi)$ is bounded from $\widetilde{H}_s^-$ to $\widetilde{H}_{s-\alpha}^-$. Theorem 2.3.3 is proved.

Example 2.3.3. For an arbitrary real α, let

$$\Delta_+^\alpha(\xi', \xi_n + i\tau) = (\xi_n + i\tau + i|\xi'|)^\alpha = e^{\alpha \ln(\xi_n + i\tau + i|\xi'|)}, \qquad \tau \geq 0,$$

where the branch of the logarithm has been chosen such that

$$\mathrm{Im} \ln(\xi_n + i\tau + i|\xi'|) = \arg(\xi_n + i\tau + i|\xi'|) \to 0, \quad \text{as} \quad \xi_n \to \infty.$$

The function $\Delta_+^\alpha(\xi', \xi_n + i\tau)$ is analytic with respect to $\alpha = \xi_n + i\tau$ for $\tau > 0$ and any α. Similarly, let

$$\Delta_-^\alpha(\xi', \xi_n + i\tau) = (\xi_n + i\tau - i|\xi'|)^\alpha = e^{\alpha \ln(\xi_n + i\tau - i|\xi'|)}, \qquad \tau \leq 0,$$

where $\arg(\xi_n - i\tau - i|\xi'|) \to 0$ as $\xi_n \to \infty$. The function $\Delta_-^\alpha(\xi', \xi_n + i\tau)$ is analytic with respect to $\alpha_n = \xi_n + i\tau$ on the halfplane $\tau < 0$.

Consider also the functions

$$\hat{\Delta}_+^\alpha(\xi) = \Delta_+^\alpha(\xi', \xi_n + i), \qquad \hat{\Delta}_-^\alpha(\xi) = \Delta_+^\alpha(\xi', \xi_n - i),$$

which allow for analytic extensions $\hat{\Delta}_+^\alpha(\xi', \xi_n + i + i\tau)$ and $\hat{\Delta}_+^\alpha(\xi', \xi_n - i + i\tau)$ to the halfplanes $\tau > 0$ and $\tau < 0$, respectively, and satisfy the inequalities (2.3.28). By Theorem 2.3.2, the operator of multiplication by $\hat{\Delta}_\pm^\alpha(\xi)$ is continuous from $\tilde{H}_s^\pm$ to $\tilde{H}_{s-\alpha}^\pm$. Since the operators of multiplication by $\hat{\Delta}_\pm^\alpha(\xi)$ and $\hat{\Delta}_\pm^{-\alpha}(\xi)$ are mutually inverse, the operator of multiplication by $\hat{\Delta}_\pm^\alpha(\xi)$ establishes an isomorphism between $\tilde{H}_s^\pm$ and $\tilde{H}_{s-\alpha}^\pm$.

Definition 2.3.4. *The space $H_s(\mathbb{R}_+^n)$ consists of all generalized functions $f \in S'(\mathbb{R}_+^n)$ admitting an extension $lf \in H_s(\mathbb{R}_n)$. The norm in $H_s(\mathbb{R}_+^n)$ is given by*

$$\|f\|_s^+ = \inf_l \|lf\|_s, \tag{2.3.33}$$

where the infimum is taken over all extensions of f to $H_s(\mathbb{R}_n)$.

Note that if $l_0 f$ is an extension of f to $H_s(\mathbb{R}_n)$, then $l_0 f + f_-$, with $f_- \in H_s^-$, is also an extension, because $pf_- = 0$, where p is the operator of restriction to $\mathbb{R}_+^n$. Consequently, the space $H_2(\mathbb{R}_+^n)$ is isomorphic to the quotient space $H_s(\mathbb{R}_n)/H_s^-$.

Consider the case $s = 0$. Then $H_0(\mathbb{R}_+^n) = L_2(\mathbb{R}_+^n)$ and the infimum of the norms of extensions of $f \in H_0(\mathbb{R}_+^n)$ is attained on the function f_+ equal to f for $x_n > 0$ and identically vanishing for $x_n < 0$. Thus,

$$\left(\|f\|_0^+\right)^2 = \int_{\mathbb{R}_+^n} |f(x)|^2 \, dx. \tag{2.3.34}$$

For integer $s = m > 0$, consider the norm

$$\left(\|f\|_m'^+\right)^2 = \sum_{|k| \leq m} \int_{\mathbb{R}_+^n} \left| \frac{\partial^k f}{\partial x^k} \right|^2 dx. \tag{2.3.35}$$

Proposition 2.3.7. *The norms (2.3.33) and (2.3.35) are equivalent in $H_m(\mathbb{R}_+^n)$.*

Proof. Since $\|lf\|_m$ is equivalent to the norm $\|lf\|_m'$ defined by (2.3.5) and $\|lf\|_m' \geq \|f\|_m'^+$ for any extension lf, we have

$$\|f\|_m'^+ \leq C \|f\|_m^+. \tag{2.3.36}$$

Let us show that

$$\|f\|_m^+ \leq C_1 \|f\|_m'^+. \tag{2.3.37}$$

For this purpose, we construct an operator of extension form $\mathbb{R}_+^n$ to $\mathbb{R}_n$.

Let $f(x', x_n) \in C^\infty(\overline{\mathbb{R}_+^n})$, $x' = (x_1, \ldots, x_{n-1})$, $f(x', x_n) = 0$ for sufficiently large $|x|$. Denote by L_N the operator

$$L_N f = \begin{cases} f(x', x_n) & \text{for } x_n > 0, \\ \displaystyle\sum_{p=1}^N \lambda_p f(x', -px_n) & \text{for } x_n < 0, \end{cases} \tag{2.3.38}$$

with λ_p chosen such that $L_N f \in C^{N-1}(\mathbb{R}_n)$, i.e., $L_N f$ should be continuous, together with its derivatives up to the order $N-1$, at $x_n = 0$. Therefore, the constants λ_p should satisfy the following equations

$$\sum_{p=1}^N (-p)^k \lambda_p = 1, \qquad 0 \le k \le N-1.$$

Since the determinant of this system (the Vandermonde determinant) is different from zero, the values λ_p, $p = 1, \ldots, N$, are uniquely defined. For $m \le N$, we have

$$\|L_N f\|_m' \le C_2 \|f\|_m^+. \tag{2.3.39}$$

Since functions in $C_0^\infty(\mathbb{R}_n)$ are dense in $H_s(\mathbb{R}_n)$, their restrictions to $\mathbb{R}_+^n$ are dense in $H_s(\mathbb{R}_+^n)$. Hence, by completion, we obtain the estimate (2.3.39) for any $f \in H_m(\mathbb{R}_+^n)$. Thus,

$$\|f\|_m^+ \le \|L_N f\|_m' \le C_2 \|f\|_m'^+,$$

which completes the proof.

Proposition 2.3.8. *Let $A_-(\xi', \xi_n + i\tau)$, $\tau < 0$, be a function for which the assumptions of Theorem 2.3.2 hold, and let A_- be the operator defined by (2.3.26) and p the operator of restriction to $\mathbb{R}_+^n$. Then the operator pA_-lf, where $f \in H_s(\mathbb{R}_+^n)$ and $lf \in H_s$ is an arbitrary extension of f from $\mathbb{R}_+^n$ to $\mathbb{R}_n$, does not depend on the extension lf and is bounded from $H_s(\mathbb{R}_+^n)$ to $H_{s-\alpha}(\mathbb{R}_+^n)$.*

Proof. Let $l_1 f \in H_s$ be another extension of f. Then, $f_- = lf - l_1 f \in H_s^-$. By Theorem 2.3.2, we have $A_- f_- \in H_{s-\alpha}^-$, and therefore, $pA_- f_- = 0$. It follows that $pA_- lf = pA_- l_1 f$. Let us estimate the norm $\|pA_- lf\|_{s-\alpha}^+$. Consider an extension lf such that

$$\|lf\|_s \le 2\|f\|_s^+.$$

Then, by Proposition 2.3.6, we have

$$\|pA_- lf\|_{s-\alpha}^+ \le \|A_- lf\|_{s-\alpha} \le C\|lf\|_s \le C\|f\|_s^+.$$

Proposition 2.3.8 is proved.

For an arbitrary s, consider the space H_s^* dual to H_s, i.e., H_s^* consists of all continuous linear functionals on $H_s(\mathbb{R}_n)$. The space H_s^* is usually defined to within an isomorphism. In particular, since $H_s(\mathbb{R}_n)$ is isomorphic to a Hilbert space with a scalar product, the space H_s^* is isomorphic to H_s. By the Riesz theorem about the representation of continuous linear functionals in Hilbert space, any such functional $\Phi(u)$, $u \in H_s$, is defined by an element $V \in H_s$,

$$\Phi(u) = \int_{\mathbb{R}_n} \widehat{u}(\xi) (1 + |\xi|)^{2s} \overline{\widehat{V}(\xi)} \, d\xi = \langle u, V \rangle,$$

where $\widehat{u}(\xi) = F(u(x))$, $\widehat{V}(\xi) = F(V(x))$, and

$$\|\Phi\| = \sup_{\|u\|_s = 1} |\Phi(u)| = \langle V, V \rangle^{1/2} = \|V\|_s.$$

Let $\widetilde{w}(\xi) = (1 + |\xi|)^{2s} \widehat{V}(\xi)$, $w = F^{-1} \widehat{w}(\xi)$. Then, $w \in H_{-s}$, $\|w\|_{-s} = \|V\|_s$,

$$\Phi(u) = (u, w)_0 = \int_{\mathbb{R}_n} \widehat{u}(\xi) \overline{\widehat{w}(\xi)} \, d\xi, \qquad u \in H_s. \tag{2.3.40}$$

Thus, the relation $\widehat{w}(\xi) = (1 + |\xi|)^{2s} \widehat{V}(\xi)$ establishes an isomorphism between H_s^* and H_{-s}.

Proposition 2.3.9. *Let $v \in H_{-s}$. Then, $v \in H_{-s}^-$ if and only if $(u_+, v) = 0$ for any $u_+ \in H_s^+$.*

This result follows from Proposition 2.3.5 and Remark 2.3.1.

Proposition 2.3.10. *Let $(H_s^+)^*$ be the dual space of H_s^+ with an arbitrary s. The space $(H_s^+)^*$ is isomorphic to $H_{-s}(\mathbb{R}_+^n)$ and the value of a functional $f \in H_{-s}(\mathbb{R}_+^n)$ on an element $u_+ \in H_s^+$ is given by*

$$(u_+, lf)_0 = \int_{\mathbb{R}_n} \widehat{u}^+(\xi) \overline{\widehat{lf}(\xi)} \, d\xi. \tag{2.3.41}$$

Proof. For $f \in H_{-s}(\mathbb{R}_+^n)$, let $lf \in H_{-s}(\mathbb{R}_n)$ be an arbitrary extension of f, and let $u^+ \in H_s^+$. Then the linear functional

$$(u_+, lf)_0 = \int_{\mathbb{R}_n} \widehat{u}^+(\xi) \overline{\widehat{lf}(\xi)} \, d\xi \tag{2.3.42}$$

does not depend on the choice of the extension lf. Indeed, if $l_1 f \in H_{-s}(\mathbb{R}_n)$ is another extension of f, then $lf - l_1 f \in H_{-s}^-(\mathbb{R}_n)$. Therefore, $(u_+, lf)_0 = (u_+, l_1 f)$, since $(u_+, lf - l_1 f) = 0$ by Proposition 2.3.9. From (2.3.42) we obtain the estimate

$$|(u_+, lf)_0| \leq \|u_+\|_s \, \|lf\|_{-s}. \tag{2.3.43}$$

Since the functional $(u_+, lf)_0$ does not depend on the extension lf, we have

$$|(u_+, lf)_0| \leq \|u_+\|_s \, \inf_l \|lf\|_{-s} = \|u_+\|_s \, \|f\|_{-s}^+. \tag{2.3.44}$$

Consequently, any element $f \in H_{-s}(\mathbb{R}_+^n)$ defines a continuous functional on H_s^+ in terms of (2.3.41).

Now, let $\Phi(u_+)$ be an arbitrary continuous linear functional on H_s^+. The space H_s^+ is a Hilbert space with the scalar product (2.3.2). Therefore, by the Riesz theorem, there exists an element $\varphi_+ \in H_s^+$ such that

$$\Phi(u_+) = \int_{\mathbb{R}_n} \widehat{u}_+(\xi)(1 + |\xi|)^{2s} \, \overline{\widehat{\varphi}_+(\xi)} \, d\xi.$$

Let $\widehat{f}_0(\xi) = (1 + |\xi|)^{2s} \widehat{\varphi}_+(\xi)$, $f_0 = F^{-1} \widehat{f}_0(\xi)$. Then, $f_0 \in H_{-s}(\mathbb{R}_n)$, $pf_0 = f \in H_{-s}(\mathbb{R}_+^n)$, where p is the operator of restriction to $\mathbb{R}_+^n$. Moreover,

$$\Phi(u_+) = \int_{\mathbb{R}_n} \widehat{u}_+(\xi) \overline{\widehat{lf}(\xi)} \, d\xi, \qquad \|\Phi\| = \|\varphi_+\|_s = \|f_0\|_{-s} \geq \|f\|_s^+.$$

Consequently, any continuous linear functional on H_s^+ can be represented in the form (2.3.41), so that $\|\Phi\| \geq \|f\|_{-s}^+$. From the inequality (2.3.44) and the last observations, we conclude that $\|\Phi\| = \|f\|_{-s}^+$. Proposition 2.3.10 is proved.

In a similar way, we can prove the following result.

Proposition 2.3.11. *Let $(H_s(\mathbb{R}_+^n))^*$ be the dual space of $H_s(\mathbb{R}_+^n)$ with an arbitrary s. The space $(H_s(\mathbb{R}_+^n))^*$ is isomorphic to H_{-s}^+, and the value of the functional $u_+ \in H_{-s}^+$ on the element $f \in H_s(\mathbb{R}_+^n)$ is given by*

$$(lf, u_+)_0 = \int_{\mathbb{R}_n} \widehat{lf}(\xi) \overline{\widehat{u}_+(\xi)} \, d\xi. \tag{2.3.45}$$

Definition 2.3.5. For an open bounded domain $G \subset \mathbb{R}_n$, *the space $\mathring{H}_s(G)$ consists of all functions $u(x) \in H_s(\mathbb{R}_n)$ compactly supported in $\overline{G}$.*

Similarly to Proposition 2.3.4, it can be shown that $\mathring{H}_s(G)$ is closed in $H_s(\mathbb{R}_n)$.

Definition 2.3.6. For an open bounded domain $G \subset \mathbb{R}_n$, *the space* $H_s(G)$ consists of all generalized functions in G that can be extended to $\mathbb{R}_n$ as elements of $H_s(\mathbb{R}_n)$. The norm in $H_s(G)$ is defined by

$$\|f\|_{H_s(G)} = \inf_l \|lf\|_s,$$

where the infimum is taken over all extensions lf of f.

Definition 2.3.7. An open bounded domain G is said to have $(n-1)$-*dimensional Lipschitz boundary* Γ, if the following conditions hold. For any point $x_0 \in \Gamma$, there are Cartesian coordinates $x = (x_1, \ldots, x_n)$ with the origin at x_0 and there is a rectangle

$$\Delta(x_0) = \{(x_1, \ldots, x_n): \ |x_j| < \xi_j(x_0), \ j = 1, \ldots, n\}$$

such that the intersection $\gamma = \Gamma \cap \Delta$ is given by the equation

$$x_n = \psi(\lambda), \qquad \lambda = (x_1, \ldots, x_{n-1}), \qquad \lambda \in \Delta' = \{|x_j| \le \xi_j(x_0), \ j = 1, \ldots, n-1\}.$$

On Δ', the function $\psi(\lambda)$ satisfies the Lipschitz condition: there is a constant $M > 0$ such that (see *Nikolskii* (1977))

$$|\psi(\lambda_1) - \psi(\lambda_2)| \le M|\lambda_1 - \lambda_2|, \qquad \lambda_1, \lambda_2 \in \Delta'.$$

Remark 2.3.2. We say that *the boundary of the domain G is of class C^r*, if $\psi(\lambda) \in C^r(\overline{\Delta'})$, $r \le \infty$, in Definition 2.3.7.

Definition 2.3.8. Let G be an open bounded domain with a smooth $(n-1)$-dimensional boundary of class C^r. A Cartesian coordinate system $x = (x_1, \ldots, x_n)$ is called a *special local Cartesian coordinate system* (SLCCS) with the origin at the point $x_0 \in \Gamma$ (Γ is the boundary of the domain), if the following conditions hold:

(i) there is a rectangle $\Delta(x_0) = \{|x_j| < \xi_j(x_0), j = 1, \ldots, n\}$ such that the intersection $\gamma = \Gamma \cap \Delta$ is described by the equation $x_n = \psi(\lambda)$, $\lambda \in \Delta'$, $\Delta' = \{|x_j| \le \xi_j(x_0), j = 1, \ldots, n-1\}$, $\psi(\lambda) \in C^r(\Delta')$.

(ii) the plane $x_n = 0$ is tangential to Γ at the point x_0 and the direction of the axis x_n coincides with that of the interior normal to the boundary Γ at x_0;

(iii) the following inclusions hold:

$$A_+ = \{(x_1, \ldots, x_{n-1}) \in \Delta', \ \psi(\lambda) < x_n < \xi_n(x_0)\} \subset G,$$
$$A_- = \{(x_1, \ldots, x_{n-1}) \in \Delta', \ -\xi_n(x_0) < x_n < \psi(\lambda)\} \subset \mathbb{R}_n \setminus \overline{G}. \tag{2.3.46}$$

We introduce the following sets:

$$U_+(x_0) = \{(x_1,, \ldots, x_n) \in U \cap G\},$$
$$U_-(x_0) = \{(x_1,, \ldots, x_n) \in U \cap (\mathbb{R}_n \setminus \overline{G})\},$$
$$U_+(x_0, \varepsilon) = \{(x_1,, \ldots, x_n + \varepsilon): \ (x_1,, \ldots, x_n) \in U_+(x_0)\},$$
$$U_-(x_0, \varepsilon) = \{(x_1,, \ldots, x_n - \varepsilon): \ (x_1,, \ldots, x_n) \in U_-(x_0)\}.$$

Definition 2.3.9. Suppose that for each point $x \in \Gamma$ we have chosen a special local Cartesian coordinate system (SLCCS). *The neighborhood* $U(x_0)$, $x_0 \in \Gamma$, *is called stable with respect to ε-shifts*, if there is $\delta(x_0) > 0$ such that for all ε, $0 < \varepsilon < \delta(x_0)$, we have $U_+(x_0, \varepsilon) \subset G$, $U_-(x_0, \varepsilon) \in \mathbb{R}_n \setminus \overline{G}$, and the following inequalities hold: $C_1^+(x_0, \varepsilon) < \rho_+(\varepsilon)$, $C_1^-(x_0, \varepsilon) < \rho_-(\varepsilon)$, where $\rho_\pm(\varepsilon)$ is the distance from $U_\pm(x_0, \varepsilon)$ to the boundary Γ of the domain G; and $C_1^\pm(x_0, \varepsilon) > 0$ tend to zero as $\varepsilon \to 0$.

Proposition 2.3.12. *Let G be a bounded open domain with a smooth $(n-1)$-dimensional boundary Γ of class C^2. Then the set $C_0^\infty(G)$ is dense in $H_s(G)$ for any real s.*

Proof. From the results obtained in *Nikolskii* (1977), it follows that for each point $x \in \Gamma$ there exist neighborhoods stable with respect to ε-shifts. Let us associate to each $x \in \Gamma$ its neighborhood $U(x)$ which is stable with respect to ε-shifts, and consider its neighborhood $U'(x)$ such that $\overline{U}' \subset U(x)$. Since the boundary Γ is compact, from its covering $\{U'(x)\}$ we can choose a finite covering of Γ, which we denote by $I = \{U_1'(x), \ldots, U_n'(x)\}$. Let us supplement I with a neighborhood U_0', $\overline{U_0'} \subset G$, so as to obtain a covering of the entire $\overline{G}$. Let $\varphi_0, \varphi_1, \ldots, \varphi_N$ be a partition of unity subject to the covering $U_0', U_1', \ldots, U_N'$ (see *Hörmander* (1965)), i.e., $0 \le \varphi_j \le 1$, $\varphi_j \in C_0^\infty(U_j')$, $j = 0, \ldots, N$, and $\psi = \sum_{j=0}^N \varphi_j \equiv 1$ in a neighborhood of $\overline{G}$, $0 \le \psi \le 1$.

Let $f(x) \in \overset{\circ}{H}_s(G)$. Then $f(x) = \sum_{j=0}^N f(x)\varphi_j(x) = \sum_{j=0}^N f_j$, where $f_j = \varphi_j f$. The covering $U_1', \ldots, U_N'$ corresponds to some points $y_1, \ldots, y_N$ of the boundary Γ. For each y_i, let $X^j = (x_1^j, \ldots, x_n^j)$ be its SLCCS (see Definition 2.3.8), and let $\vec{\varepsilon}_j = (0, \ldots, 0, \varepsilon)$, $\varepsilon > 0$, be a vector in the j-th SLCCS. Consider the shifts $f_j^{\vec{\varepsilon}_j}$ of the functions f_j by the vector $\vec{\varepsilon}_j$ (see Definition 2.1.7). From the structure of the covering $U_1', \ldots, U_N'$, we see that $\operatorname{supp} f^{\vec{\varepsilon}_j} \subset G$ for small enough ε, $0 < \varepsilon < \varepsilon_0$. Since the shifts $f_j^{\vec{\varepsilon}_j}$ are continuous with respect to ε in the norm of H_s, Proposition 2.3.12 will be proved if we show that each function $f_0, f_1^{\vec{\varepsilon}_1}, \ldots, f_n^{\vec{\varepsilon}_n}$ can be approximated, with any given accuracy, in the norm of $H_s(\mathbb{R}_n)$ by functions of class $C_0^\infty(G)$. Let $\alpha_\delta(x)$ be the function defined in the proof of Theorem 2.3.1. Consider the families of functions $f_{0\delta} = f_0 \times \alpha_\delta(x)$, $f_{j,\delta}^{\vec{\varepsilon}_j} = f_j^{\vec{\varepsilon}_j} \times \alpha_\delta(x)$. We have $f_{0\delta} \to f_0$, $f_{j,\delta}^{\vec{\varepsilon}_j} \to f_j^{\vec{\varepsilon}_j}$ in the norm of $H_s(\mathbb{R}_n)$ as $\delta \to 0$ (see the proof of Theorem 2.3.1). Since $\operatorname{supp} f_{0\delta} \subseteq \operatorname{supp} f_0 + \operatorname{supp}\alpha_\delta(x)$, $\operatorname{supp} f_{j,\delta}^{\vec{\varepsilon}_j} \subseteq f_j^{\vec{\varepsilon}_j} + \operatorname{supp}\alpha_\delta(x)$, $j = 1, \ldots, n$ (see *Hörmander* (1965)) and the diameter of $\operatorname{supp}\alpha_\delta(x)$ tends to zero as $\delta \to 0$, there is $\delta_0 > 0$ such that for all $0 < \delta < \delta_0$, we have $\operatorname{supp} f_{0\delta} \subset G$, $\operatorname{supp} f_{j,\delta}^{\vec{\varepsilon}_j} \subset G$, $j = 1, \ldots, n$. Moreover, it should be observed that $f_{0,\delta} \in C_0^\infty(G)$, $f_{j,\delta}^{\vec{\varepsilon}_j} \in C_0^\infty(G)$ for $0 < \delta < \delta_0$, $0 < \varepsilon < \varepsilon_0$. The above arguments show that Proposition 2.3.12 indeed holds.

In a similar way, the following result can be proved.

Proposition 2.3.13. *If the assumptions of Proposition 2.3.12 hold for the domain G, then the set $C_0^\infty(CG)$ is dense in $\overset{\circ}{H}_s(CG)$, where $CG = \mathbb{R}_n \setminus \overline{G}$.*

On the basis of Propositions 2.3.12, 2.3.13, and Definition 2.3.6, one establishes statements similar to Propositions 2.3.10 and 2.3.11.

Proposition 2.3.14. *The space $\left(\overset{\circ}{H}_s(G)\right)^*$ dual to $\overset{\circ}{H}_s(G)$ is isomorphic to $H_{-s}(G)$ and the value of the functional $f \in H_{-s}(G)$ on the element $u \in \overset{\circ}{H}_s(G)$ is given by*

$$(u, lf)_0 = \int_{\mathbb{R}_n} \widehat{u}(\xi)\, \overline{\widehat{lf}(\xi)}\, d\xi, \qquad (2.3.47)$$

where $(u, lf)_0$ does not depend on the extension l of the functional f to $\mathbb{R}_n$.

Proposition 2.3.15. *The space $(H_s(G))^*$ dual to $H_s(G)$ is isomorphic to $\overset{\circ}{H}_{-s}(G)$ and the value of the functional $u \in \overset{\circ}{H}_{-s}(G)$ on the element $f \in H_s(G)$ is given by the relation*

$$(lf, u)_0 \int_{\mathbb{R}_n} \widehat{lf}(\xi)\, \overline{\widehat{u(\xi)}}\, d\xi, \qquad (2.3.48)$$

where $(lf, u)_0$ does not depend on the extension l of the functional f to $\mathbb{R}_n$.

2.4. Sobolev–Slobodetskii Spaces on Manifolds

Definition 2.4.1. A compact topological space M is called *an infinitely differentiable n-dimensional manifold* (*without border*), if the following conditions hold: there is a finite covering of M by open sets U_j, $1 \le j \le N$, and for each j, there is a one-to-one continuous mapping $x^{(j)} = S_j(x)$ of U_j to an open ball $V_j \subset \mathbb{R}_n$ such that in each intersection $U_i \cap U_j$, the function $x^{(i)} = S_{ij}(x^{(j)}) = S_i(S_j^{-1}(x^{(j)}))$ is infinitely differentiable and its Jacobian is different from zero, $\left| DS_{ij}(x^{(j)})/Dx^{(j)} \right| \ne 0$. Each set U_j is called *a coordinate neighborhood*, and the coordinates $x^{(j)}$ are called *the j-th local coordinate system*.

Definition 2.4.2. For the manifold M, consider its covering U_k', $1 \le k \le N_1$, together with the mappings $y^{(k)} = S_k'(y)$ of the neighborhood U_k' onto the neighborhood $V_k' \subset \mathbb{R}_n$. The system $\{U_j, S_j\}$, $1 \le j \le N$, is called *equivalent to* the system (U_k', S_k'), $1 \le k \le N_1$, if for $U_k' \cap U_j \ne \varnothing$, the mappings $S_k' S_j^{-1}$ are infinitely differentiable.

By $C^\infty(M)$ we denote *the space of infinitely differentiable functions on M*, i.e., functions which are infinitely differentiable in any local coordinate system.

A sequence $V_n \in C^\infty(M)$ is called *convergent in* $C^\infty(M)$, if it converges uniformly, together with all its derivatives.

Definition 2.4.3. A continuous linear functional on $C^\infty(M)$ is called *a generalized function on M*. The space of generalized functions on M is denoted by $S'(M)$.

Definition 2.4.4. Let $\{\varphi_j(x)\}$ be a partition of unity (see *Hörmander* (1965)) subject to the covering $\{U_j\}$, $1 \le j \le N$, of the manifold M. *The Sobolev–Slobodetskii space* $H_s(M)$ consists of all generalized functions U on M for which the following norm is finite:

$$\|U\|_s^2 = \sum_{j=1}^N \|\varphi_j U\|_s^2, \tag{2.4.1}$$

where $\|\varphi_j U\|_s^2$ is given by (2.3.1) in the j-th local coordinate system (LCS).

Remark 2.4.1. The space $H_s(M)$ can be equivalently defined as the closure of $C^\infty(M)$ with respect to the norm (2.4.1).

The following result is proved in *Shubin* (1978).

Proposition 2.4.1. *The topology in $H_s(M)$ associated to the norm (2.4.1) depends neither on the covering $\{U_j\}$, $1 \le j \le N$, of the manifold M, nor the partition of unity, nor the coordinate diffeomorphisms $S_j(x)$.*

Definition 2.4.5. *An n-dimensional manifold of class C^∞ with border* is a part of a compact n-dimensional manifold M of class C^∞ which coincides with the closure of its interior and has as its boundary a set Γ which is a compact C^∞ manifold of dimension $n-1$ (see *Schwartz* (1967)).

For studying the problems considered in Chapter 4 we do not need abstract manifolds introduced by Definitions 2.4.1 and 2.4.5. Instead, we will use the following definitions.

Definition 2.4.6. A compact set $M \subset \mathbb{R}_n$ is called *an $(n-1)$-dimensional manifold of class C^r*, $r \le \infty$, if the following conditions hold: for any point $x_0 \in M$, there is a Cartesian coordinate system $X = (x_1, \dots, x_n)$ with the origin at x_0 and there is a rectangle $\Delta(x_0) = \left\{ (x_1, \dots, x_n) : |x_j| \le \xi_j(x_0), \ j = 1, \dots, n \right\}$ such that the intersection $m = M \cap \Delta(x_0)$ is described by the equation

$$x_n = \psi(\lambda), \qquad \lambda = (x_1, \dots, x_{n-1}),$$

where

$$\lambda \in \Delta'(x_0) = \left\{(x_1,\ldots,x_{n-1}): \ |x_j| \leq \xi_j(x_0), \ j = 1,\ldots,n-1\right\}, \qquad \psi(\lambda) \in C^r(\Delta').$$

Definition 2.4.7. *An $(n-1)$-dimensional manifold $M \subset \mathbb{R}_n$ of class C^r, $r \leq \infty$, with a border is a part of a compact $(n-1)$-dimensional manifold $\widetilde{M} \subset \mathbb{R}_n$ of class C^r which coincides with the closure of its interior and has as its boundary a compact $(n-2)$-dimensional manifold Γ of class C^r. To be more precise, for any point $x_0 \in \Gamma$, there is a Cartesian coordinate system $X = (x_1,\ldots,x_n)$ with the origin at x_0 and there is a rectangle $\Delta(x_0) = \left\{(x_1,\ldots,x_n): \ |x_j| \leq \xi_j(x_0), \ j = 1,\ldots,n\right\}$ such that the intersection $\gamma = \Gamma \cap \Delta(x_0)$ is described by the equations*

$$x_n = \psi_1(\lambda_1), \qquad x_{n-1} = \psi_2(\lambda_1), \qquad \lambda_1 = (x_1,\ldots,x_{n-2}),$$

where

$$\lambda_1 \in \Delta''(x_0) = \left\{(x_1,\ldots,x_{n-2}): \ |x_j| \leq \xi_j(x_0), \ j = 1,\ldots,n-2\right\},$$
$$\psi_i \in C^r(\Delta''(x_0)), \quad i = 1, 2.$$

The following result is proved in *Shubin* (1978).

Proposition 2.4.2. *Let $M \subset \mathbb{R}_n$ be an $(n-1)$-dimensional manifold of class C^∞. Then any continuous linear functional $l(u)$ on $H_r(M)$ can be represented in the form*

$$l(u) = \int_M u(x)\overline{V(x)}\,dS, \tag{2.4.2}$$

where $u(x) \in H_r(M)$, $V(x) \in H_{-r}(M)$.

In *Shubin* (1978), a more general result is proved.

Definition 2.4.8. *For a compact manifold $M \subset \mathbb{R}_n$ of dimension $(n-1)$ and class C^∞, let G be an open subset of M such that $\overline{G} \neq M$ and G is a manifold of class C^∞ with a border. The space $\mathring{H}_s(G)$ consists of all functions in $H_s(M)$ with support in G.*

Proposition 2.4.3. *The set C_0^∞ is dense in $\mathring{H}_s(G)$.*

This result can be proved on the basis of Definitions 2.4.6, 2.4.7, and Proposition 2.4.1.

Definition 2.4.9. *The space $H_s(G)$ consists of generalized functions on G which can be extended as continuous linear functionals to $C^\infty(M)$. The norm in $H_s(G)$ is defined by*

$$\|f\|_{H_s(G)} = \inf_l \|lf\|_{H_s(M)}, \tag{2.4.3}$$

where $lf \in H_s(M)$ is the extension of the functional f to $C^\infty(M)$.

By analogy with Proposition 2.3.9, the following result can be proved on the basis of Propositions 2.4.2 and 2.4.3.

Proposition 2.4.4. *The space dual to $\mathring{H}_s(G)$ is isomorphic to $H_{-s}(G)$, and the space dual to $H_s(G)$ is isomorphic to $\mathring{H}_{-s}(G)$.*

Chapter 3

Hypersingular Integral Equations

3.1. Pseudodifferential Operators and their Properties

Denote by S_α^0 the class of locally integrable functions $A(\xi)$ satisfying the estimate

$$|A(\xi)| \le C(1 + |\xi|)^\alpha. \tag{3.1.1}$$

Definition 3.1.1. *A pseudodifferential operator* (PDO) *of class* S_α^0 is an operator defined on functions $u(x) \in S(\mathbb{R}_n)$ by the formula

$$Au = \frac{1}{(2\pi)^n} \int_{\mathbb{R}_n} A(\xi)\widehat{u}(\xi)\, e^{-(i,\xi)}\, d\xi, \tag{3.1.2}$$

where $\widehat{u}(\xi)$ is the Fourier transform of $u(\xi) \in S(\mathbb{R}_n)$. The function $A(\xi)$ is called the *symbol* of the operator A.

Example 3.1.1. Let $A(\xi) = \displaystyle\sum_{|k|\le m} a_k(-i\xi)^k$. Then, by the properties of the Fourier transformation, we have

$$Au = \frac{1}{(2\pi)^n} \int_{\mathbb{R}_n} \sum_{|k|\le m} a_k(-i\xi)^k \widehat{u}(\xi)\, e^{-i(x,\xi)}\, d\xi = \sum_{|k|\le m} a_k \frac{\partial^k u(x)}{\partial x^k},$$

which means that A is a differential operator.

Let $\alpha < -n$ for $A(\xi)$ in (3.1.1). Then $A(\xi)$ is an absolutely integrable function, and therefore $a(x) = F^{-1}A(\xi)$ is a bounded continuous function (see *Shilov* (1961)). By the properties of the Fourier transformation, we have

$$Au = F^{-1}(A(\xi) \cdot \widehat{u}(\xi)) = \int_{\mathbb{R}_n} a(x-y)u(y)\, dy,$$

which means that for $\alpha < -n$ the PDO A is an integral operator of convolution type.

If $\alpha > -n$, then there is an integer $m > 0$ such that $\alpha - 2m < -n$. Let $A_1(\xi) = A(\xi)/(1+|\xi|^2)^m$. Then $|A_1(\xi)| \le C(1+|\xi|)^{-\alpha-2m}$. It follows that $A_1(\xi)$ is absolutely integrable. Taking $a_1(x) = F^{-1}(A_1(\xi))$, we get

$$Au = F^{-1}(A(\xi) \cdot \widehat{u}(\xi)) = (-\Delta + 1)^m \int_{\mathbb{R}_n} a_1(x-y)u(y)\, dy = \int_{\mathbb{R}_n} a_1(x-y)(-\Delta + 1)^m u(y)\, dy,$$

where $\Delta = \displaystyle\sum_{k=1}^n \frac{\partial^2}{\partial x_k^2}$ is the Laplace operator. Thus, the PDO A can be represented as an integro-differential operator.

Definition 3.1.2. A function $A_0(\xi) \in C^\infty$ for $|\xi| \neq 0$ is called *homogeneous of the order* $\alpha + i\beta$, if it satisfies the condition $A_0(t\xi) = t^{\alpha+i\beta} A_0(\xi)$ for any $t > 0$. The class of such functions is denoted by $O^\infty_{\alpha+i\beta}$.

Definition 3.1.3. *A pseudodifferential operator of class* $O^\infty_{\alpha+i\beta}$ is an operator defined on functions $u(x) \in S(\mathbb{R}_n)$ by the formula

$$A_0 u = \frac{1}{(2\pi)^n} \int_{\mathbb{R}_n} A_0(\xi)\widehat{u}(\xi)\, e^{-i(x,\xi)}\, d\xi,$$

where $A_0(\xi)$ is a symbol of class $O^\infty_{\alpha+i\beta}$ and $\widehat{u}(\xi)$ is the Fourier transform of $u(x) \in S(\mathbb{R}_n)$.

For $\alpha \leq -n$, the regularization of the integral on the right-hand side of $A_0 u$ is done by analogy with Example 2.2.6 (see *Gelfand and Shilov* (1959), *Eskin* (1973)$_2$).

For $-n < \alpha < 0$, the symbol $A_0(\xi)$ defines a regular functional, so that $a_0(x) = F^{-1}(A_0(\xi)) \in O^\infty_{-\alpha-i\beta-n}$. Hence, using Proposition 2.2.4, we obtain

$$A_0 u = \int_{\mathbb{R}_n} a_0(x-y)u(y)\, dy, \tag{3.1.3}$$

which means that the PDO A_0 is an integral operator with a weak singularity.

If $\alpha \geq 0$, then the PDO A_0 is of class S^0_α.

Example 3.1.2. Denote by $\Lambda^{\alpha+i\beta}$ the PDO with the symbol $|\xi|^{\alpha+i\beta}$, $-n < \alpha < 0$. Then, on the basis of (2.2.31), we get

$$\Lambda^{\alpha+i\beta} u = \frac{2^{\alpha+i\beta}\Gamma\left(\frac{\alpha+i\beta+n}{2}\right)}{\pi^{\frac{n}{2}}\Gamma\left(-\frac{\alpha+i\beta}{2}\right)} \int_{\mathbb{R}_n} \frac{u(y)\, dy}{|x-y|^{\alpha+i\beta+n}}. \tag{3.1.4}$$

For $\alpha = -2$, $\beta = 0$, $n \geq 3$, it follows from (3.1.4) that

$$\Lambda^{-2} u = \frac{\Gamma\left(\frac{n-2}{2}\right)}{4\pi^{\frac{n}{2}}} \int_{\mathbb{R}_n} \frac{u(y)\, dy}{|x-y|^{n-2}}. \tag{3.1.5}$$

Example 3.1.3. Let us find a representation of the PDO P_n with the symbol $A(\xi) = \xi_n/|\xi|^2$ for $n \geq 3$. Since $F^{-1}(\xi_n|\xi|^{-2}) = i(\partial/\partial x_n)\left(F^{-1}(|\xi|^{-2})\right)$, from (3.1.5) we get

$$P_n u = \frac{i(n-2)\Gamma\left(\frac{n-2}{2}\right)}{4\pi^{\frac{n}{2}}} \int_{\mathbb{R}_n} \frac{(x_n - y_n)u(y)\, dy}{|x-y|^{n-2}}. \tag{3.1.6}$$

Example 3.1.4. For $\Lambda^\alpha = \Lambda^{2k}$ with an integer $k > 0$, we have $\Lambda^{2k} u = (-\Delta)^k u$, where Δ is the Laplace operator. If $\alpha > 0$ is not even, then there is an integer $m > 0$ such that $\alpha = 2m + \gamma$, $-2 < \gamma < 0$. Hence, using (3.1.4), we get

$$\Lambda^\alpha u = \Lambda^{2m+\gamma} u = (-\Delta)^m \frac{2^\gamma \Gamma\left(\frac{n+\gamma}{2}\right)}{\pi^{\frac{n}{2}}\Gamma\left(-\frac{\gamma}{2}\right)} \int_{\mathbb{R}_n} \frac{u(y)\, dy}{|x-y|^{n+\gamma}}, \quad n \geq 2. \tag{3.1.7}$$

In particular, for the PDO Λ with the symbol $|\xi|$, we have

$$\Lambda u = -\Delta \frac{\Gamma\left(\frac{n-1}{2}\right)}{2\pi^{\frac{n+1}{2}}} \int_{\mathbb{R}_n} \frac{u(y)\, dy}{|x-y|^{n-1}} = -\frac{\Gamma\left(\frac{n-1}{2}\right)}{2\pi^{\frac{n+1}{2}}} \int_{\mathbb{R}_n} \frac{\Delta_y u(y)\, dy}{|x-y|^{n-1}}. \tag{3.1.8}$$

Definition 3.1.4. *A singular integral of a function* $u(x) \in S(\mathbb{R}_n)$ is an integral of the form

$$V(x) = \int_{\mathbb{R}_n} \frac{f(x, \theta)}{r^n} u(y) \, dy. \tag{3.1.9}$$

where x, y are points of $\mathbb{R}_n$, $r = |x - y| = \left[\sum_{i=1}^{n}(y_i - x_i)^2\right]^{1/2}$, $\theta = r^{-1}(y - x)$. The point x is called a *pole*, $f(x, \theta)$ is called a *characteristic function*, and the value $V(x)$ is defined by the relation

$$V(x) = \lim_{\varepsilon \to 0} \int_{|x-y| > \varepsilon} \frac{f(x, \theta)}{r^n} u(y) \, dy. \tag{3.1.10}$$

As shown by *Mikhlin* (1962), the singular integral (3.1.10) exists if and only if

$$\int_S f(x, \theta) \, dS = 0, \tag{3.1.11}$$

where the vector θ varies over the unit sphere S.

Let $A_0(\xi) \in O_0^{\infty}$, i.e., $A_0(\xi)$ is homogeneous of zero order. Let $A_k(\xi) = \xi_k |\xi|^{-2} A_0(\xi)$. Then (see (3.1.3))

$$A_0(\xi) = \sum_{k=1}^{n} \xi_k A_k(\xi), \qquad A_k(\xi) \in O_{-1}^{\infty}, \qquad a_k(x) = F^{-1}(A_k(\xi)) \in O_{1-n}^{\infty}.$$

Since $F^{-1}(\xi_k A_k(\xi)\hat{u}(\xi)) = i\frac{\partial}{\partial x_k} F^{-1}(A_k(\xi)\hat{u}(\xi))$, it follows from (3.1.3) that

$$A_0 u = i \sum_{k=1}^{n} \frac{\partial}{\partial x_k} \int_{\mathbb{R}_n} a_k(x - y) u(y) \, dy. \tag{3.1.12}$$

Using the formula of differentiation of integrals with a weak singularity (see *Mikhlin* (1962)), we get

$$A_0(u) = \int_{\mathbb{R}_n} \left[i \sum_{k=1}^{n} \frac{\partial}{\partial x_k} a_k(x - y)\right] u(y) \, dy - u(x) \int_S a(\theta) \left(\sum_{k=1}^{n} \frac{iy_k}{|y|}\right) d\theta, \tag{3.1.13}$$

where $\theta = |y|^{-1} y$, $y = (y_1, \ldots, y_n)$, S is the unit sphere in $\mathbb{R}_n$; the first integral in (3.1.13) is understood as a singular integral.

Thus, the pseudodifferential operator A_0 with a symbol $A_0(\xi) \in O_0^{\infty}$ can be represented as the singular integral (3.1.13).

Example 3.1.5. Let Y_k, $1 \le k \le n$, be PDO's with the symbols $h_k = |\xi|^{-1} \xi_k$. Taking into account that

$$F^{-1}\left(\frac{\xi_k}{|\xi|}\right) = i\frac{\partial}{\partial x_k} F^{-1}\left(\frac{1}{|\xi|}\right),$$

and using (3.1.4), (3.1.13), we find that for $n \ge 2$,

$$Y_k u = -\frac{i\Gamma\left(\frac{n-1}{2}\right) \cdot (n-1)}{2\pi^{\frac{n+1}{2}}} \int_{\mathbb{R}_n} \frac{x_k - y_k}{|x - y|^{n+1}} u(y) \, dy - \frac{\Gamma\left(\frac{n-1}{2}\right)}{2\pi^{\frac{n}{2}} \Gamma\left(\frac{1}{2}\right)} iu(x) \int_S y_k \, dS =$$

$$= -\frac{i\Gamma\left(\frac{n-1}{2}\right) \cdot (n-1)}{2\pi^{\frac{n+1}{2}}} \int_{\mathbb{R}_n} \frac{x_k - y_k}{|x - y|^{n+1}} u(y) \, dy. \tag{3.1.14}$$

Singular integral operators Y_k are called *Riesz operators*.

Example 3.1.6. Let Λ be a PDO with the symbol $|\xi|$. Since $|\xi| = \sum_{k=1}^{n} \xi_k(\xi_k|\xi|^{-1})$, this operator can be represented in the form

$$\Lambda u = \sum_{k=1}^{n} Y_k \left(i \frac{\partial u}{\partial x_k} \right) = \frac{\Gamma\left(\frac{n+1}{2}\right)}{\pi^{\frac{n+1}{2}}} \sum_{k=1}^{n} \int_{\mathbb{R}_n} \frac{x_k - y_k}{|x - y|^{n+1}} \frac{\partial u(y)}{\partial y_k} \, dy, \qquad (3.1.15)$$

where we have used the formula $\frac{n-1}{2}\Gamma\left(\frac{n-1}{2}\right) = \Gamma\left(\frac{n+1}{2}\right)$. Another representation of the operator Λ is given by (3.1.8).

Example 3.1.7. Consider the following integrals (see *Hadamard* (1932)) in the sense of Hadamard for functions $u(x) \in S$:

$$I_\lambda = \int_{\mathbb{R}_n} \frac{u(y)\, dy}{|x - y|^{n+\lambda}} = \lim_{\varepsilon \to 0} \left[\int_{\mathbb{R}_n \setminus \Omega_\varepsilon^*(x)} \frac{u(y)\, dy}{|x - y|^{n+\lambda}} - \frac{u(x)\Omega_n}{\lambda \varepsilon^\lambda} \right], \qquad (3.1.16)$$

where Ω_n is the area of the unit sphere in $\mathbb{R}_n$; $\Omega_\varepsilon^*(x)$ is the ball of radius ε with center at x, $0 < \lambda < 2$.

Let us show that the integral in the sense of Hadamard exists for $u(x) \in S$. To that end, we proceed in the same way as in Example 2.2.5. Thus, passing to the spherical coordinates with the origin at x in the integral

$$I_\lambda^*(\varepsilon) = \int_{\mathbb{R}_n \setminus \Omega_\varepsilon^*(x)} \frac{u(y)\, dy}{|x - y|^{n+\lambda}},$$

we get

$$I_\lambda^*(\varepsilon) = \Omega_n \int_\varepsilon^\infty \frac{1}{r^{1+\lambda}} S_u(r)\, dr, \qquad (3.1.17)$$

$$S_u(r) = \frac{1}{\Omega_n} \int_\Omega u(x + r\omega)\, d\omega, \qquad (3.1.18)$$

where $d\omega$ is the area element of the unit sphere Ω with center at $x = (x_1, \ldots, x_n)$; ω is a unit vector. From (3.1.17), we obtain

$$I_\lambda^*(\varepsilon) = \Omega_n \int_\varepsilon^1 \frac{1}{r^{1+\lambda}}[S_u(r) - S_u(0)]\, dr + \Omega_n \int_1^\infty \frac{1}{r^{1+\lambda}} S_u(r)\, dr + \int_\varepsilon^1 S_u(0) \frac{1}{r^{1+\lambda}}\, dr. \qquad (3.1.19)$$

From (3.1.18) it follows that $S_u(0) = u(x)$; moreover, $|S_u(r) - S_u(0)| \le Cr^2$ for $0 \le r \le 1$. Hence,

$$\lim_{\varepsilon \to 0} \Omega_n \int_\varepsilon^1 \frac{1}{r^{1+\lambda}}[S_u(r) - S_u(0)]\, dr = \Omega_n \int_0^1 \frac{1}{r^{1+\lambda}}[S_u - S_u(0)]\, dr, \qquad (3.1.20)$$

$$\Omega_n \int_\varepsilon^1 S_u(0) \frac{1}{r^{1+\lambda}}\, dr = \Omega_n u(x)\left[-\frac{1}{\lambda} + \frac{1}{\lambda \varepsilon^\lambda}\right]. \qquad (3.1.21)$$

From (3.1.16), using (3.1.19)–(3.1.21), we find that

$$I_\lambda = \Omega_n \int_0^1 \frac{1}{r^{1+\lambda}}[S_u(r) - S_u(0)]\, dr + \Omega_n \int_1^\infty \frac{1}{r^{1+\lambda}} S_u(r)\, dr - \frac{\Omega_n u(x)}{\lambda}. \qquad (3.1.22)$$

Relation (3.1.22) shows that the Hadamard integrals I_λ exist not only for $u(x) \in S(\mathbb{R}_n)$ but also for functions $u(x) \in C^2(\mathbb{R}_n)$ such that $|u(x)| \le M$ for all $x \in \mathbb{R}_n$. Moreover, it follows from (3.1.22) that the Hadamard regularization coincides with the regularization (2.2.21) for $k = 1$. Consequently, for x fixed, we have $I_\lambda(x) = (|y - x|^{n+\lambda}, \varphi(y))$ (see Example 2.2.5). Changing the variables, $x - y = t$, in the integral $I_\lambda^*(\varepsilon)$, we get

$$I_\lambda^*(\varepsilon) = \int_{\mathbb{R}_n \setminus \Omega_\varepsilon^*(0)} \frac{u(x - t)}{|t|^{n+\lambda}} \, dt. \tag{3.1.23}$$

From the definition of convolution (2.1.12), the above observations, and formula (3.1.23), it follows that

$$I_\lambda(x) = \frac{1}{|x|^{n+\lambda}} \times u(x). \tag{3.1.24}$$

Proposition 2.2.4 and formula (2.2.31) imply that

$$F(I_\lambda(x)) = \frac{2^{-\lambda} \pi^{\frac{n}{2}} \Gamma\left(-\frac{\lambda}{2}\right)}{\Gamma\left(\frac{n+\lambda}{2}\right)} |\xi|^\lambda \widehat{u}(\xi). \tag{3.1.25}$$

From (3.1.25), it follows that the Hadamard integrals $I_\lambda(x)$ can be represented as pseudodifferential operators,

$$I_\lambda(x) = \frac{1}{(2\pi)^n} \int_{\mathbb{R}_n} A(\xi) \widehat{u}(\xi) \, e^{-i(x,\xi)} \, d\xi \tag{3.1.26}$$

with the symbol

$$A(\xi) = \frac{2^{-\lambda} \pi^{\frac{n}{2}} \Gamma\left(-\frac{\lambda}{2}\right)}{\Gamma\left(\frac{n+\lambda}{2}\right)} |\xi|^\lambda \in S_\lambda^0. \tag{3.1.27}$$

The integrals $(n = 2)$

$$I_1(x, y) = \iint_{\mathbb{R}_2} \frac{u(x_0, y_0) \, dx_0 \, dy_0}{\left((x - x_0)^2 + (y - y_0)^2\right)^{3/2}} \tag{3.1.28}$$

occur in problems of aerodynamics.

The classes of PDO's considered above are too narrow for applications, and therefore, we have to introduce more general classes of PDO's.

For a locally integrable function $\widehat{A}_1(\eta, \xi)$ satisfying the inequalities

$$(1 + |\eta|)^N \left|\widehat{A}_1(\eta, \xi)\right| \leq C_N (1 + |\xi|)^\alpha, \qquad \forall N, \tag{3.1.29}$$

let

$$A_1(x, \xi) = F_\eta^{-1} \widehat{A}_1(\eta, \xi) = \frac{1}{(2\pi)^n} \int_{\mathbb{R}_n} \widehat{A}_1(\eta, \xi) \, e^{-i(x,\xi)} \, d\eta. \tag{3.1.30}$$

Because of (3.1.29), the function $A_1(x, \xi)$ for a fixed ξ belongs to $H_s(\mathbb{R}_n)$ (the Sobolev–Slobodetskii space) for all s. Therefore, $A_1(x, \xi)$ is infinitely differentiable with respect to x and $\partial^k A_1(x, \xi)/\partial x^k \to 0$ as $|x| \to \infty$ for any k.

Definition 3.1.5. *The class* $S_\alpha^0 = S_\alpha^0(\mathbb{R}_n)$ *consists of all functions* $A(x, \xi)$ *such that*

$$A(x, \xi) = A(\infty, \xi) + A_1(x, \xi), \tag{3.1.31}$$

where $A(\infty, \xi)$ is a locally integrable function satisfying the inequality $|A(\infty, \xi)| \leq C(1 + |\xi|)^\alpha$; $F_x A_1(x, \xi) = \widehat{A}_1(\eta, \xi)$ and $\widehat{A}_1(\eta, \xi)$ satisfies the inequality (3.1.29).

Definition 3.1.6. *The class* $S_\alpha^t = S_\alpha^t(\mathbb{R}_n)$, *with* $t = m + \gamma$, *integer* $m \geq 0$, *and* $0 < \gamma \leq 1$, *consists of all functions* $A(x, \xi)$ *for which* $(\partial^k/\partial \xi^k) A(x, \xi)$ *are absolutely integrable and belong to* $S_{\alpha-|k|}^0$ *for* $0 \leq |k| \leq m$, *and* $(\partial^k/\partial \xi^k) A(x, \xi) \in S_{\alpha-m-\gamma}^0$ *for* $|k| = m + 1$.

By S_α^∞ we denote the intersection $\bigcap_{m=0}^{\infty} S_\alpha^m$.

Definition 3.1.7. *A pseudodifferential operator* (PDO) *with symbol* $A(x,\xi) \in S_\alpha^t$, $t \geq 0$, *is the operator defined on functions* $u(x) \in S(\mathbb{R}_n)$ *by the formula*

$$Au = \frac{1}{(2\pi)^n} \int_{\mathbb{R}_n} A(x,\xi)\widehat{u}(\xi)\, e^{-i(x,\xi)}\, d\xi, \tag{3.1.32}$$

where $\widehat{u}(\xi) = F(u(x))$.

Example 3.1.8. Consider the differential operator $Au = \sum_{|k|=0}^{m} a_k(x)\dfrac{\partial^k u}{\partial x^k}$, with the coefficients $a_k(x) \in C^\infty(\mathbb{R}_n)$ being constant outside a ball $|x| > R$. This operator can be represented as the following PDO:

$$Au = \frac{1}{(2\pi)^n} \int_{\mathbb{R}_n} A(x,\xi)\widehat{u}(\xi)\, e^{-i(x,\xi)}\, d\xi, \tag{3.1.33}$$

$$A(x,\xi) = \sum_{|k|=0}^{m} a_k(x)(-i\xi)^k = \sum_{|k|=0}^{m} a_k(\infty)(-i\xi)^k + \sum_{|k|=0}^{m} [a_k(x) - a_k(\infty)](-i\xi)^k.$$

Obviously, $A(x,\xi) \in S_m^\infty$.

Remark 3.1.1. The above example shows that the classes S_α^t impose fairly strong restrictions on the smoothness of the symbols $A(x,\xi)$ with respect to x. The investigation of certain problems requires wider classes of PDO's. The classes of PDO's defined in this book are sufficient for our purposes.

Definition 3.1.8. *The class of symbols* $O_{\alpha+i\beta}^\infty$ *consists of functions* $A_0(x,\xi)$, *differentiable in* x *and* $\xi \neq 0$, *homogeneous of the order* $\alpha + i\beta$ *in* ξ, *and independent of* x *for* $|x| > N$: $A_0(x,\xi) = A_0(\infty,\xi)$ *for* $|x| \geq N$.

Definition 3.1.9. *The class of symbols* $\widehat{O}_{\alpha+i\beta}^\infty$ *consists of functions* $A(x,\xi) \in C^\infty(\mathbb{R}_n \times \mathbb{R}_n)$ *such that* $A(x,\xi) = A(\infty,\xi)$ *for* $|x| \geq N$, *and there is* $A_0(x,\xi) \in O_{\alpha+i\beta}^\infty$ *such that*

$$[A(x,\xi) - A_0(x,\xi)(1-\chi(\xi))] \in S_{\alpha-1}^\infty,$$

where $\chi(\xi) \in C_0^\infty(\mathbb{R}_n)$, $\chi(\xi) = 1$ *for* $|\xi| \leq 1$.

The functions of class $\widehat{O}_{\alpha+i\beta}^\infty$ belong to S_α^∞.

Remark 3.1.2. The pseudodifferential operator with symbol $A(x,\xi)$ is denoted by $A(x,D)$.

Proposition 3.1.1. *Let* $A_1(x,\xi)$ *be a symbol defined by* (3.1.30) *and satisfying conditions* (3.1.29). *Then the Fourier transform of the function*

$$v(x) = A_1(x,D)u = \frac{1}{(2\pi)^n} \int_{\mathbb{R}_n} A_1(x,\xi)\widehat{u}(\xi)\, e^{-i(x,\xi)}\, dx,$$

where $\widehat{u}(\xi) = F(u(x))$, $u(x) \in S(\mathbb{R}_n)$, *has the form*

$$\widehat{v} = \frac{1}{(2\pi)^n} \int_{\mathbb{R}_n} \widehat{A}_1(\eta - \xi, \xi)\widehat{u}(\xi)\, d\xi. \tag{3.1.34}$$

Proof. Consider the generalized function f defined by

$$(f,\varphi) = \frac{1}{(2\pi)^n} \int_{\mathbb{R}_n} \frac{1}{(2\pi)^n} \int_{\mathbb{R}_n} \overline{\widehat{A}_1(\eta - \xi, \xi)\, \widehat{u}(\xi)}\, d\xi\, \widehat{\varphi}(\eta)\, d\eta, \tag{3.1.35}$$

where $\varphi(x) \in S(\mathbb{R}_n)$, $F(\varphi(x)) = \widehat{\varphi}(\eta)$. From Definition 2.2.2 combined with (3.1.35), it follows that

$$Ff = \frac{1}{(2\pi)^n} \int_{\mathbb{R}_n} \widehat{A}_1(\eta - \xi, \xi)\,\widehat{u}(\xi)\,d\xi. \tag{3.1.36}$$

Using the inequalities (3.1.29) and (2.1.4), we obtain

$$|Ff| \leq \frac{1}{(2\pi)^n} \int_{\mathbb{R}_n} \frac{(1 + |\xi|)^\alpha |u(\xi)|\,d\xi}{(1 + |\eta - \xi|)^N} \leq \frac{1}{(2\pi)^n} \int_{\mathbb{R}_n} \frac{(1 + |\xi|)^{\alpha + N} |u(\xi)|\,d\xi}{(1 + |\eta|)^N}. \tag{3.1.37}$$

From (3.1.37), it follows that $Ff \in L_1(\mathbb{R}_n)$, and therefore, the right-hand side of (3.1.35) is a continuous linear functional on S. Let us find the inverse Fourier transform of Ff, taking into account that $A_1(\eta - \xi, \xi)\widehat{u}(\xi) \in L_1(\mathbb{R}_n \times \mathbb{R}_n)$. We have

$$f = \frac{1}{(2\pi)^n} \int_{\mathbb{R}_n} \frac{1}{(2\pi)^n} \left[\int_{\mathbb{R}_n} \widehat{A}_1(\eta - \xi, \xi)\widehat{u}(\xi)\,d\xi \right] e^{-i(x,\eta)}\,d\eta =$$

$$= \frac{1}{(2\pi)^n} \int_{\mathbb{R}_n} \frac{1}{(2\pi)^n} \left[\int_{\mathbb{R}_n} \widehat{A}_1(\eta - \xi, \xi)\,e^{-i(x,\eta)}\,d\eta \right] \widehat{u}(\xi)\,d\xi. \tag{3.1.38}$$

In the interior integral in (3.1.38), let us change the variables, taking $\theta = \eta - \xi$. We get

$$f = \frac{1}{(2\pi)^n} \int_{\mathbb{R}_n} \left[\frac{1}{(2\pi)^n} \int_{\mathbb{R}_n} \widehat{A}_1(\theta, \xi)\,e^{-i(x,\theta)}\,d\theta \right] e^{-i(x,\xi)}\widehat{u}(\xi)\,d\xi. \tag{3.1.39}$$

Definitions 3.1.5 and 3.1.7 show that

$$\frac{1}{(2\pi)^n} \int_{\mathbb{R}_n} \widehat{A}_1(\theta, \xi)\,e^{-i(x,\theta)}\,d\theta = A_1(x, \theta). \tag{3.1.40}$$

From (3.1.39) and (3.1.40), it follows that

$$f = \frac{1}{(2\pi)^n} \int_{\mathbb{R}_n} A_1(x, \xi)\widehat{u}(\xi)\,e^{-i(x,\xi)}\,d\xi = V(x).$$

Hence we obtain the desired result.

Proposition 3.1.2. *Let*

$$f(\eta) = \int_{\mathbb{R}_n} B(\eta - \xi)g(\xi)\,d\xi, \tag{3.1.41}$$

where $g(\xi) \in S(\mathbb{R}_n)$, $\int_{\mathbb{R}_n} |B(\eta)|\,d\eta = B_0 < \infty$. *Then*

$$\|f(\eta)\|_{L_2(\mathbb{R}_n)} \leq B_0 \|g\|_{L_2(\mathbb{R}_n)}. \tag{3.1.42}$$

Proof. Applying the Fourier transformation to (3.1.41), we get $\widehat{f}(x) = \widehat{B}(x)\widehat{g}(x)$, where $\widehat{B}(x) = F(B(\eta))$, $\widehat{g}(x) = F(g(\xi))$. Hence, we obtain the estimate (3.1.42), taking into account the estimate $|\widehat{B}(x)| \leq \int_{\mathbb{R}_n} |B(\eta)|\,d\eta$.

Theorem 3.1.1. *For* $A(x, \xi) \in S^0_\alpha$, *the PDO* $A(x, D)$ *is bounded from* $H_s(\mathbb{R}_n)$ *to* $H_{s-\alpha}(\mathbb{R}_n)$, *for any* s.

Proof. Let $u(x) \in S(\mathbb{R}_n)$. From Definition 3.1.5, it follows that

$$A(x, D) = A(\infty, D)u + A_1(x, D)u. \tag{3.1.43}$$

Proposition 2.3.6 implies the estimate

$$\|A(\infty, D)u\|_{s-\alpha} \le K_s \|u\|_s. \tag{3.1.44}$$

For $\widehat{V}(\eta) = F(A_1(x, D)u)$, Proposition 3.1.1 implies the inequality

$$\left|\widehat{V}(\eta)\right| \le \frac{1}{(2\pi)^n} \int_{\mathbb{R}_n} \frac{C_N (1 + |\xi|)^\alpha |\widehat{u}(\xi)| \, d\xi}{(1 + |\eta - \xi|)^N}, \tag{3.1.45}$$

where N can be taken arbitrary.

From (2.1.4), it follows that

$$(1 + |\eta|)^t \le (1 + |\xi|)^t (1 + |\xi - \eta|)^{|t|}.$$

Using this estimate, we get

$$(1 + |\eta|)^{s-\alpha} \left|\widehat{V}(\eta)\right| \le \frac{1}{(2\pi)^n} \int_{\mathbb{R}_n} \frac{C_N \left(1 + |\xi|\right)^s |\widehat{u}(\xi)| \, d\xi}{\left(1 + |\eta - \xi|\right)^{N - |s-\alpha|}}. \tag{3.1.46}$$

Let us apply Proposition 3.1.2 to the right-hand side of (3.1.46). Taking

$$N = |s - \alpha| + n + 1, \qquad B(\eta - \xi) = \frac{C_N}{(2\pi)^n} \frac{1}{\left(1 + |\eta - \xi|\right)^{n+1}}, \qquad g(\xi) = \left(1 + |\xi|\right)^s |\widehat{u}(\xi)|,$$

and using Proposition 3.1.2, we find

$$\|A_1(x, D)u\|_{s-\alpha} = \|V\|_{s-\alpha} \le C_N' \|u\|_s, \tag{3.1.47}$$

for all $u(x) \in S(\mathbb{R}_n)$. Since $S(\mathbb{R}_n)$ is dense in $H_s(\mathbb{R}_n)$ (see Theorem 2.3.1), the operator $A(x, D)$ can be continuously extended as an operator from $H_s(\mathbb{R}_n)$ to $H_{s-\alpha}(\mathbb{R}_n)$. Theorem 3.1.1 is proved.

Theorem 3.1.2. Let $A(x, \xi) \in S_\gamma^0$, $B(x, \xi) \in S_\beta^0$. Then $C(x, D) = A(x, D)[B(x, D)u]$ is a pseudodifferential operator with the symbol $C(x, \xi) \in S_{\gamma+\beta}^0$ given by

$$C(x, \xi) = A(x, \xi)B(\infty, \xi) + \frac{1}{(2\pi)^n} \int_{\mathbb{R}_n} A(x, \xi + \alpha)\widehat{B}_1(\alpha, \xi) e^{-i(x, \alpha)} \, d\alpha. \tag{3.1.48}$$

Proof. Since $B(x, D) = B(\infty, D) + B_1(x, D)$, we have

$$A(x, D)B(x, D) = A(x, D)B(\infty, D) + A(x, D)B_1(x, D). \tag{3.1.49}$$

For any $u(x) \in S(\mathbb{R}_n)$ with $\widehat{u}(\xi) = F(u(x))$, the following relation holds:

$$A(x, D)B(\infty, D)u(x) = \frac{1}{(2\pi)^n} \int_{\mathbb{R}_n} A(x, \xi)B(\infty, \xi)\widehat{u}(\xi) e^{-i(x, \xi)} \, d\xi. \tag{3.1.50}$$

From (3.1.50), it follows that $A(x, D)B(\infty, D)$ is a PDO with the symbol $A(x.D)B(\infty, \xi) \in S_{\gamma+\beta}^0$.

Proposition 3.1.1 implies that the Fourier transform of $V(x) = B_1(x, D)u(x)$ has the form

$$\widehat{V}(\eta) = \frac{1}{(2\pi)^n} \int_{\mathbb{R}_n} \widehat{B}_1(\eta - \xi, \xi)\widehat{u}(\xi) \, d\xi,$$

where $\widehat{B}_1(\eta, \xi) = F_x B_1(x, \xi)$, $\widehat{u}(\xi) = F(u(x))$. Further,

$$A(x, D)B_1(x, D)u = \frac{1}{(2\pi)^n} \int_{\mathbb{R}_n} A(x, \eta)\widehat{V}(\eta) \, e^{-i(x,\eta)} \, d\eta =$$

$$= \frac{1}{(2\pi)^{2n}} \int_{\mathbb{R}_n} \int_{\mathbb{R}_n} A(x, \eta)\widehat{B}(\eta - \xi, \xi)\widehat{u}(\xi) \, e^{-i(x,\eta)} \, d\xi \, d\eta.$$

Changing the variables, $\alpha = \eta - \xi$, we obtain

$$A(x, D)B_1(x, D)u = \frac{1}{(2\pi)^{2n}} \int_{\mathbb{R}_n} \int_{\mathbb{R}_n} A(x, \xi + \alpha)\widehat{B}_1(\alpha, \xi) \, e^{-i(x,\xi+\alpha)}\widehat{u}(\xi) \, d\alpha \, d\xi. \qquad (3.1.51)$$

Set

$$C_1(x, \xi) = \frac{1}{(2\pi)^n} \int_{\mathbb{R}_n} A(x, \xi + \alpha)\widehat{B}_1(\alpha, \xi) \, e^{-i(x,\alpha)} \, d\alpha. \qquad (3.1.52)$$

Now, from (3.1.51) and (3.1.52), it follows that $A(x, D)B_1(x, D)$ is the PDO with the symbol $C_1(x, \xi)$,

$$A(x, D)B_1(x, D)u = \frac{1}{(2\pi)^{2n}} \int_{\mathbb{R}_n} C_1(x, \xi)\widehat{u}(\xi) \, e^{-i(x,\xi)} \, d\xi. \qquad (3.1.53)$$

Let us show that $C_1(x, \xi) \in S^0_{\gamma+\beta}$. We introduce the notation

$$
\begin{aligned}
C_1(x, \xi) &= C_2(x, \xi) + C_3(x, \xi), \\
C_2(x, \xi) &= \frac{1}{(2\pi)^n} \int_{\mathbb{R}_n} A(\infty, \xi + \alpha)\widehat{B}_1(\alpha, \xi) \, e^{-i(x,\alpha)} \, d\alpha, \\
C_3(x, \xi) &= \frac{1}{(2\pi)^n} \int_{\mathbb{R}_n} A_1(x, \xi + \alpha)\widehat{B}(\alpha, \xi) \, e^{-i(x,\alpha)} \, d\alpha.
\end{aligned}
\qquad (3.1.54)
$$

Since $C_2(x, \xi)$ is the inverse Fourier transform of $f(\alpha, \xi) = A(\infty, \xi + \alpha)B_1(\alpha, \xi)$ with respect to α, we see that for the Fourier transform of $C_2(x, \xi)$, the following relation holds:

$$\int_{\mathbb{R}_n} C_2(x, \xi) \, e^{i(x,\eta)} \, dx = f(\eta, \xi) = A(\infty, \xi + \eta)\widehat{B}_1(\eta, \xi) = \widehat{C}_2(\eta, \xi). \qquad (3.1.55)$$

When calculating the Fourier transform of the function $C_3(x, \xi)$ with respect to x, we can use Proposition 3.1.1, because the inequalities (3.1.29) hold for the function $\widehat{B}_1(\alpha, \xi)$ with respect to α. We get

$$\widehat{C}_3(\eta, \xi) = \int_{\mathbb{R}_n} C_3(x, \xi) \, e^{i(x,\eta)} \, dx = \frac{1}{(2\pi)^n} \int_{\mathbb{R}_n} \widehat{A}_1(\eta - \alpha, \xi + \alpha)\widehat{B}_1(\alpha, \xi) \, d\alpha. \qquad (3.1.56)$$

Since $A(x, \xi) \in S^0_\gamma$ and $B(x, \xi) \in S^0_\beta$, it follows that for any N, the following estimates hold:

$$
\begin{aligned}
\widehat{B}_1(\eta, \xi) &\le C_N \left(1 + |\eta|\right)^{-2N} \left(1 + |\xi|\right)^\beta, \\
\left|\widehat{A}_1(\eta - \xi, \xi + \alpha)\right| &\le C_N \left(1 + |\xi - \eta|\right)^{-N} \left(1 + |\xi + \alpha|\right)^\gamma, \\
\left|A(\infty, \xi + \eta)\right| &\le C \left(1 + |\xi + \eta|\right)^\gamma.
\end{aligned}
\qquad (3.1.57)
$$

The inequality (2.1.4) implies that

$$
\begin{aligned}
\left(1 + |\xi + \eta|\right)^\gamma &\le \left(1 + |\xi|\right)^\gamma \left(1 + |\eta|\right)^{|\gamma|}, \\
\left(1 + |\eta - \xi|\right)^{-N} &\le \left(1 + |\eta|\right)^{-N} \left(1 + |\xi|\right)^N.
\end{aligned}
\qquad (3.1.58)
$$

From (3.1.55), (3.1.57), (3.1.58), we obtain the estimates

$$\left| \widehat{C}_2(\eta, \xi) \right| \leq \frac{CC_N \left(1 + |\xi| \right)^{\gamma + \beta}}{\left(1 + |\eta| \right)^{2N - |\gamma|}}. \tag{3.1.59}$$

Since N is arbitrary, it follows from (3.1.59) that $C_2(x, \xi) \in S^0_{\gamma + \beta}$, and $C_2(\infty, \xi) = 0$. From (3.1.56)–(3.1.58) for $N - |\gamma| \geq n + 1$, we obtain the inequality

$$|C_3(\eta, \xi)| \leq A_N \int_{\mathbb{R}_n} \frac{\left(1 + |\xi + \alpha| \right)^{\gamma} \left(1 + |\xi| \right)^{\beta} d\alpha}{\left(1 + |\eta - \alpha| \right)^N \left(1 + |\alpha| \right)^{2N}} \leq$$

$$\leq A_N \left[\int_{\mathbb{R}_n} \frac{d\alpha}{\left(1 + |\alpha| \right)^{N - |\gamma|}} \right] \frac{1 + |\xi|^{\gamma + \beta}}{\left(1 + |\eta| \right)^N}. \tag{3.1.60}$$

Since N is arbitrary, it follows that $C_3(x, \xi) \in S^0_{\gamma + \beta}$, $C_3(\infty, \xi) = 0$. Theorem 3.1.2 is proved.

Imposing additional restrictions on the smoothness of the symbol $A(x, \xi)$, we can single out the principal part of the symbol of the operator $A(x, D)B(x, D)$ in (3.1.48).

Proposition 3.1.3. *Let* $A_1(x, \xi) \in S^{m + \gamma}_{\alpha}$, $m \geq 0$, $0 < \gamma \leq 1$, $A_1(\infty, \xi) = 0$. *Consider the Taylor expansion for* $\widehat{A}_1(\eta, \xi + \theta)$ *at* $\theta = 0$,

$$\widehat{A}_1(\eta, \xi + \theta) = \sum_{|k|=0}^{m} \frac{1}{|k|!} \frac{\partial^k}{\partial \xi^k} \widehat{A}_1(\eta, \xi) \theta^k + \widehat{R}_{1,m+1}(\eta, \xi, \theta). \tag{3.1.61}$$

The following estimate holds the remainder term $\widehat{R}_{1,m+1}(\eta, \xi, \theta)$:

$$\left(1 + |\eta| \right)^N \left| \widehat{R}_{1,m+1}(\eta, \xi, \theta) \right| \leq C_N \left(1 + |\xi| \right)^{\alpha - m - \gamma} |\theta|^{m + |\alpha| + 1}. \tag{3.1.62}$$

Proof. Consider the case $|\theta| < 2^{-1} \left(1 + |\xi| \right)$. Let us write the remainder $\widehat{R}_{1,m+1}(\eta, \xi, \theta)$ in integral form (see *Shilov* (1972)) as follows:

$$\widehat{R}_{1,m+1}(\eta, \xi, \theta) = \frac{1}{m!} \sum_{|k|=m+1} \left(\int_0^1 \frac{\partial^k \widehat{A}_1(\eta, \xi + t\theta)}{\partial \xi^k} (1 - t)^m dt \right) \theta^k. \tag{3.1.63}$$

Hence, using Proposition 3.1.6, we find that

$$\left(1 + |\eta| \right)^N \left| \widehat{R}_{1,m+1}(\eta, \xi, \theta) \right| \leq C_N |\theta|^{m+1} \int_0^1 \left(1 + |\xi + t\theta| \right)^{\alpha - m - \gamma} dt. \tag{3.1.64}$$

For $|\theta| < 2^{-1} \left(1 + |\xi| \right)$, we have the inequality

$$\left(1 + |\xi + t\theta| \right)^{\alpha - m - \gamma} \leq C \left(1 + |\xi| \right)^{\alpha - m - \gamma}. \tag{3.1.65}$$

The estimate (3.1.62) follows from (3.1.64) and (3.1.65).

Now, let $|\theta| \geq 2^{-1} \left(1 + |\xi| \right)$. Using the representation (3.1.61), we get

$$\left(1 + |\eta| \right)^N \left| \widehat{R}_{1,m+1}(\eta, \xi, \theta) \right| \leq \sum_{|k|=0}^{m} \frac{\left(1 + |\eta| \right)^N}{|k|!} \left| \frac{\partial^k \widehat{A}_1(\eta, \xi)}{\partial \xi^k} \right| |\theta|^{|k|} + \left(1 + |\eta| \right)^N \left| \widehat{A}_1(\eta, \xi + \theta) \right|. \tag{3.1.66}$$

From the inequality (2.1.4), it follows that

$$\left(1 + |\xi + \theta|\right)^{\alpha} \le \left(1 + |\xi|\right)^{\alpha} \left(1 + |\theta|\right)^{|\alpha|}.$$

Further, using the inequality $|\theta| \ge 2^{-1} \left(1 + |\xi|\right)$, we obtain

$$\left(1 + |\eta|\right)^{N} \left|\widehat{R}_{1,m+1}(\eta, \xi, \theta)\right| \le \sum_{|k|=0}^{m} \frac{C_{kN} \left(1 + |\xi|\right)^{\alpha - |k|} |\theta|^{|k|}}{|k|!} + C_N \left(1 + |\xi + \theta|\right)^{\alpha} \le$$

$$\le B_{1N} \left(1 + |\xi|\right)^{\alpha - m - \gamma} |\theta|^{m+\gamma} + C_N \left(1 + |\xi|\right)^{\alpha} \left(1 + |\theta|\right)^{|\alpha|} \le$$

$$\le B_{2N} \left(1 + |\xi|\right)^{\alpha - m - \gamma} |\theta|^{|\alpha| + m + 1}.$$

The last inequality proves Proposition 3.1.3.

In a similar way, we can prove the following result.

Proposition 3.1.4. *Let* $A(\infty, \xi) \in S_{\alpha}^{m+\gamma}$. *Then the function*

$$\widehat{R}_{m+1}(\xi, \theta) = A(\infty, \xi + \theta) - \sum_{|k|=0}^{m} \frac{1}{|k|!} \frac{\partial^k A(\infty, \xi)}{\partial \xi^k} \theta^k \qquad (3.1.67)$$

satisfies the inequality

$$\left|\widehat{R}_{m+1}(\xi, \theta)\right| \le C \left(1 + |\xi|\right)^{\alpha - m - \gamma} |\theta|^{|\alpha| + m + 1}. \qquad (3.1.68)$$

Theorem 3.1.3. *Let* $A(x, \xi) \in S_{\alpha}^{m+\gamma}$, $m \ge 0$, $0 < \gamma \le 1$, $B(x, \xi) \in S_{\beta}^{0}$. *Then the symbol* $C(x, \xi)$ *of the operator* $C(x, D) = A(x, D)B(x, D)$ *can be represented as*

$$C(x, \xi) = \sum_{|k|=0}^{m} \frac{1}{|k|!} \frac{\partial^k A(x, \xi)}{\partial \xi^k} i^{|k|} \frac{\partial^k B(x, \xi)}{\partial x^k} + C_{m+1}(x, \xi), \qquad (3.1.69)$$

where $C_{m+1}(x, \xi) \in S_{\alpha + \beta - m - \gamma}^{0}$, $C_{m+1}(\infty, \xi) = 0$.

Proof. Let us write the Taylor formula for the symbol $A(x, \xi + \theta)$ in (3.1.48) at $\theta = 0$. We have

$$C_1(x, \xi) = \sum_{|k|=0}^{m} \frac{1}{|k|!} \frac{\partial^k A(x, \xi)}{\partial \xi^k} \frac{1}{(2\pi)^n} \int_{\mathbb{R}_n} \theta^k \widehat{B}_1(\theta, \xi) e^{-i(x, \theta)} d\theta + C_{m+1}(x, \xi), \qquad (3.1.70)$$

where

$$C_{m+1}(x, \xi) = \frac{1}{(2\pi)^n} \int_{\mathbb{R}_n} \left(\widehat{R}_{m+1}(\xi, \theta) + R_{1,m+1}(x, \xi, \theta)\right) \widehat{B}_1(\theta, \xi) e^{-i(x, \theta)} d\theta, \qquad (3.1.71)$$

where $\widehat{R}_{m+1}(\xi, \theta)$ is given by (3.1.67), $R_{1,m+1}(x, \xi, \theta) = F_{\eta}^{-1} \widehat{R}_{1,m+1}(\eta, \xi, \theta)$, and the function $\widehat{R}_{1,m+1}(\eta, \xi, \theta)$ is found from (3.1.61).

The Fourier transform of the function $C_{m+1}(x, \xi)$ with respect to x is found by analogy with (3.1.55) and (3.1.56), namely,

$$\widehat{C}_{m+1}(x, \xi) = \int_{\mathbb{R}_n} C_{m+1}(x, \xi) e^{i(x, \eta)} dx = \qquad (3.1.72)$$

$$= \widehat{R}_{m+1}(\xi, \eta) \widehat{B}_1(\eta, \xi) + \frac{1}{(2\pi)^n} \int_{\mathbb{R}_n} \widehat{R}_{1,m+1}(\eta - \theta, \xi, \theta) \widehat{B}_1(\theta, \xi) d\theta.$$

Using the inequalities (3.1.68) and (3.1.62), we obtain, in the same way as (3.1.60), the following estimate:

$$\left(1 + |\eta|\right)^N \left|\widehat{C}_{m+1}(\eta, \xi)\right| \le C_{N,m} \left(1 + |\xi|\right)^{\alpha + \beta - m - \gamma}, \qquad \forall N. \tag{3.1.73}$$

From (3.1.73) it follows that $C_{m+1}(x, \xi) \in S^0_{\alpha + \beta - m - \gamma}$, $C_{m+1}(\infty, \xi) = 0$. Since

$$\frac{1}{(2\pi)^n} \int_{\mathbb{R}_n} \theta^k \widehat{B}_1(\theta, \xi) \, e^{-i(x, \theta)} \, d\theta = i^{|k|} \frac{\partial^k B_1(x, \xi)}{\partial x^k},$$

it follows from (3.1.49) and (3.1.70) that the expansion (3.1.69) is valid. Theorem 3.1.3 is proved.

Proposition 3.1.5. *Let* $A(x, \xi) \in S^{t_1}_\alpha$, $B(x, \xi) \in S^{t_2}_\beta$, $t_1 > 0$, $t_2 > 0$. *Then the commutator*

$$T = A(x, D)B(x, D) - B(x, D)A(x, D)$$

is a pseudodifferential operator with symbol $T(x, \xi) \in S^0_{\alpha + \beta - t_0}$, $t_0 = \min\{1, t_1, t_2\}$.

Proof. Applying Theorem 3.1.3 for $m = 0$, we get

$$\begin{aligned}
A(x, D) \cdot B(x, D) &= C_0(x, D) + T_1(x, D), \\
B(x, D) \cdot A(x, D) &= C_0(x, D) + T_2(x, D),
\end{aligned} \tag{3.1.74}$$

where $C_0(x, D)$ is the PDO with the symbol $A(x, \xi)B(x, \xi)$, $T_1(x, \xi) \in S^0_{\alpha + \beta - t_{10}}$, $T_2(x, \xi) \in S^0_{\alpha + \beta - t_{20}}$, $t_{i0} = \min(1, t_i)$, $i = 1, 2$. It follows that $T(x, \xi) = T_1(x, \xi) - T_2(x, \xi) \in S^0_{\alpha + \beta - t_0}$, and Proposition 3.1.5 is proved.

Example 3.1.9. Let $A(\xi)$ be an infinitely differentiable function of $\xi \ne 0$ such that

$$\left| \frac{\partial^k A(\xi)}{\partial \xi^k} \right| \le C_k \left(1 + |\xi|\right)^{\alpha - k}.$$

Then the PDO $A(D)$ with symbol $A(\xi)$ is of class S^0_α. Consider an operator B with symbol $\psi \in C^\infty_0(\mathbb{R}_n)$ (i.e., $\psi(x) \in S^\infty_0$) associated with the PDO

$$Bu = \frac{1}{(2\pi)^n} \int_{\mathbb{R}_n} \psi(x)\widehat{u}(\xi) \, e^{-i(x, \xi)} \, d\xi.$$

Let us find the class of the commutator $T = BA - AB$. Proposition 3.1.5 cannot be applied in this case. Let us represent the operator A as $A = A_1 + A_2$, where the PDO A_1 has the symbol $A_1(\xi)(1 - \chi(\xi))$, and the PDO A_2 has the symbol $A_2(\xi) = A(\xi)\chi(\xi)$, where $\chi(\xi) \in C^\infty_0(\mathbb{R}_n)$, $\chi(x) = 1$ for $|\xi| \le 1$. Then, $A_1(\xi) \in S^\infty_\alpha$, $A_2(\xi) \in S^0_{-\infty}$. Let us represent the commutator T as

$$T = (BA_1 - A_1 B) + (BA_2) + (-(A_2 B) = T_1 + T_2 + T_3. \tag{3.1.75}$$

By Proposition 3.1.5, we have $T_1 \in S^0_{\alpha - 1}$, and by Theorem 3.1.2, we have $T_2 \in S^0_{-\infty}$. This and (3.1.75) imply that $T \in S^0_{\alpha - 1}$.

3.2. Changing Variables in Pseudodifferential Operators

A pseudodifferential operator $A(x, D)$ can be written in the form

$$A(x, D)u = \frac{1}{(2\pi)^n} \int_{\mathbb{R}_n} \int_{\mathbb{R}_n} A(x, \xi) \, e^{-i(x-y, \xi)} \, u(y) \, dy \, d\xi, \tag{3.2.1}$$

where the integral is understood in the iterated sense, i.e., first, we integrate in y and then in ξ. Let us introduce a class of operators wider than (3.2.1).

Let $\widehat{\widehat{a}}_{11}(\eta, \theta, \xi)$ be a locally integrable function in $\mathbb{R}_{3n}$ subject to the estimate

$$\left(1 + |\eta|\right)^{N_1} \left(1 + |\theta|\right)^{N_2} \left|\widehat{\widehat{a}}_{11}(\eta, \theta, \xi)\right| \le C_{N_1, N_2} \left(1 + |\xi|\right)^{\alpha}, \qquad \forall N_1, N_2. \tag{3.2.2}$$

Definition 3.2.1. *The class* $S_{\alpha}^0(\mathbb{R}_n \times \mathbb{R}_n)$ *consists of functions* $a(x, y, \xi)$ *that can be represented in the form*

$$a(x, y, \xi) = a(x, \infty, \xi) + a_1(x, y, \xi), \tag{3.2.3}$$

where $a(x, \infty, \xi) \in S_{\alpha}^0(\mathbb{R}_n)$ *and for any* $y \in \mathbb{R}_n$, *the function* $a_1(x, y, \xi)$ *belongs to the class* $S_{\alpha}^0(\mathbb{R}_n)$. *Moreover, it is assumed that*

$$a_1(x, y, \xi) = a_1(\infty, y, \xi) + a_{11}(x, y, \xi), \tag{3.2.4}$$

where $a_1(\infty, y, \xi) \in S_{\alpha}^0(\mathbb{R}_n)$, $a_1(\infty, \infty, \xi) = 0$, $F_x F_y a_{11}(x, y, \xi) = \widehat{\widehat{a}}_{11}(\eta, \theta, \xi)$, *with* $\widehat{\widehat{a}}_{11}(\eta, \theta, \xi)$ *satisfying the inequalities (3.2.2)*

Definition 3.2.2. *The class* $S_{\alpha}^t(\mathbb{R}_n \times \mathbb{R}_n)$, *for* $t = m + \gamma$ *with integer* $m \ge 0$ *and* $0 < \gamma \le 1$, *consists of functions* $a(x, y, \xi)$ *absolutely continuous, together with their derivatives in* ξ *up to the order* m, *and satisfying the conditions*

$$\frac{\partial^k a(x, y, \xi)}{\partial \xi^k} \in S_{\alpha-|k|}^0(\mathbb{R}_n \times \mathbb{R}_n) \qquad \text{for} \quad 0 \le |k| \le m,$$

$$\frac{\partial^k a(x, y, \xi)}{\partial \xi^k} \in S_{\alpha-m-\gamma}^0(\mathbb{R}_n \times \mathbb{R}_n) \quad \text{for} \quad |k| = m + 1.$$

Definition 3.2.3. *A pseudodifferential operator* (PDO) *with symbol* $a(x, y, \xi) \in S_{\alpha}^t(\mathbb{R}_n \times \mathbb{R}_n)$, $t \ge 0$, *is an operator defined on functions* $u(y) \in S(\mathbb{R}_n)$ *by*

$$A(u) = \frac{1}{(2\pi)^n} \int_{\mathbb{R}_n} \int_{\mathbb{R}_n} a(x, y, \xi) u(y) \, e^{-i(x-y, \xi)} \, dy \, d\xi. \tag{3.2.5}$$

Pseudodifferential operators (3.1.32) and (3.2.1) are special cases of the operator (3.2.5).

Proposition 3.2.1. *Let* A *be a PDO with symbol* $a(x, y, \xi) \in S_{\alpha}^0(\mathbb{R}_n \times \mathbb{R}_n)$. *Then* A *can be represented as a PDO of the form* (3.1.32) *with symbol* $A(x, \xi) \in S_{\alpha}^0(\mathbb{R}_n)$,

$$A(x, \xi) = a(x, \infty, \xi) + \frac{1}{(2\pi)^n} \int_{\mathbb{R}_n} \widehat{a}_1(x, \theta, \theta + \xi) \, e^{-i(x, \theta)} \, d\theta, \tag{3.2.6}$$

where $a(x, y, \xi) = a(x, \infty, \xi) + a_1(x, y, \xi)$, $\widehat{a}_1(x, \theta, \xi) = F_y(a_1(x, y, \xi))$.

Proof. Let us calculate the integral in y in (3.2.5). By the properties of the Fourier transformation, we have

$$\int_{\mathbb{R}_n} a_1(x, y, \xi) u(y) \, e^{i(y, \xi)} \, dy = \frac{1}{(2\pi)^n} \int_{\mathbb{R}_n} \widehat{a}_1(x, \xi - \eta) \widehat{u}(y) \, d\eta, \tag{3.2.7}$$

where $\widehat{u}(\eta) = F(u(y))$; $\widehat{a}_1(x, \xi - \eta, \xi) = \widehat{a}_1(\infty, \xi - \eta, \xi) + \widehat{a}_{11}(x, \xi - \eta, \xi)$.

Since $\widehat{a}_1(x, \theta, \xi)$ decays faster than any power of $|\theta| \to \infty$, the right-hand side of (3.2.7) rapidly decays with respect to ξ, because of the inequality (2.1.4). Therefore, the integral (3.2.5), regarded as an iterated integral, is convergent. Substituting the right-hand side of (3.2.7) into (3.2.5), we obtain

$$Au = \frac{1}{(2\pi)^n} \int_{\mathbb{R}_n} a(x, \infty, \xi) \widehat{u}(\xi) e^{-i(x,\xi)} d\xi + \frac{1}{(2\pi)^{2n}} \int_{\mathbb{R}_n} \int_{\mathbb{R}_n} \widehat{a}_1(x, \xi - \eta, \xi) \widehat{u}(\eta) e^{-i(x,\xi)} d\xi \, d\eta.$$
(3.2.8)

Changing the variables, $\xi = \theta + \eta$, in the integral (3.2.8), we get

$$\frac{1}{(2\pi)^{2n}} \int_{\mathbb{R}_n} \int_{\mathbb{R}_n} \widehat{a}_1(x, \xi - \eta, \xi) \widehat{u}(\eta) e^{-i(x,\xi)} d\xi \, d\eta = \frac{1}{(2\pi)^n} \int_{\mathbb{R}_n} A_1(x, \eta) \widehat{u}(\eta) e^{-i(x,\eta)} d\eta, \quad (3.2.9)$$

where

$$A_1(x, \eta) = \frac{1}{(2\pi)^n} \int_{\mathbb{R}_n} \widehat{a}_1(x, \theta, \theta + \eta) e^{-i(x,\theta)} d\theta. \tag{3.2.10}$$

From (3.2.8)–(3.2.10), we obtain (3.2.6). It remains to prove that $A(x, \xi) = a(x, \infty, \xi) + A_1(x, \xi) \in S_\alpha^0(\mathbb{R}_n)$. To that end it suffices to show that $A_1(x, \xi) \in S_\alpha^0(\mathbb{R}_n)$, $A_1(\infty, \xi) = 0$. Indeed, the Fourier transform of the function $A_1(x, y)$ has the form

$$\widehat{A}_1(\xi, \eta) = \widehat{a}_1(\infty, \xi, \xi + \eta) + \frac{1}{(2\pi)^n} \int_{\mathbb{R}_n} \widehat{\widehat{a}}_{11}(\xi - \theta, \theta, \theta + \eta) d\theta. \tag{3.2.11}$$

Just as we have obtained (3.1.60), using (3.2.11), (3.1.29), (3.1.58), (3.2.2), we get

$$\left| \widehat{A}_1(\xi, \eta) \right| \leq C_N^1 \frac{(1 + |\xi + \eta|)^\alpha}{(1 + |\xi|)^N} + C_N^2 \int_{\mathbb{R}_n} \frac{(1 + |\theta + \eta|)^\alpha \, d\theta}{(1 + |\xi - \theta|)^N (1 + |\theta|)^{2N}} \leq$$

$$\leq C_N^1 \frac{(1 + |\eta|)^\alpha}{(1 + |\xi|)^{N-|\alpha|}} + C_N^2 \int_{\mathbb{R}_n} \frac{(1 + |\eta|)^\alpha (1 + |\theta|)^{N+|\alpha|}}{(1 + |\xi|)^N (1 + |\theta|)^{2N}} d\theta \leq$$

$$\leq C_N^3 \frac{(1 + |\eta|)^\alpha}{(1 + |\xi|)^{N-|\alpha|}}. \tag{3.2.12}$$

Since N is arbitrary, it follows that $A_1(x, \xi) \in S_\alpha^0$ and $A_1(\infty, \eta) = 0$. Proposition 3.2.1 is proved.

In a similar way, one can prove the following result (see *Eskin* (1973)$_2$).

Proposition 3.2.2. *If $a(x, y, \xi) \in S_\alpha^{m+\varphi}(\mathbb{R}_n \times \mathbb{R}_n)$, then the symbol $A(x, \xi)$ defined by (3.2.6) admits the expansion*

$$A(x, \xi) = a(x, \infty, \xi) + \sum_{|k|=0}^m \frac{1}{|k|!} (i)^{|k|} \frac{\partial^k}{\partial y^k} \frac{\partial^k}{\partial \xi^k} a_1(x, y, \xi) \bigg|_{y=x} + C_{m+1}(x, \xi), \tag{3.2.13}$$

where $C_{m+1}(x, \xi) \in S_{\alpha-m-\gamma}^0$.

Let A be a PDO with symbol $A(x, \xi) \in A_\alpha^0(\mathbb{R}_n)$. Denote by A^* the operator conjugate to A, i.e., the operator for which

$$(Au, v) = (u, A^* v), \qquad \forall \, u, v \in S(\mathbb{R}_n), \tag{3.2.14}$$

where $(u, v) = \int_{\mathbb{R}_n} u(x) \overline{v(x)} \, dx$ is the scalar product in $H_0(\mathbb{R}_n) = L_2(\mathbb{R}_n)$.

By (3.2.1) we have

$$(Au, v) = \int_{\mathbb{R}_n} \frac{1}{(2\pi)^n} \left(\int_{\mathbb{R}_n} \int_{\mathbb{R}_n} A(x, \xi) \, e^{-i(x-y,\xi)} \, u(y) \, dy \, d\xi \overline{v(x)} \right) dx =$$

$$= \int_{\mathbb{R}_n} u(y) \left(\frac{1}{(2\pi)^n} \int_{\mathbb{R}_n} \int_{\mathbb{R}_n} A(x, \xi) \overline{v(x)} \, e^{-i(x-y,\xi)} \, dx \, d\xi \right) dy. \qquad (3.2.15)$$

Interchanging y and x in (3.2.15), we get

$$A^* v = \frac{1}{(2\pi)^n} \int_{\mathbb{R}_n} \overline{A(y, \xi)} \, v(y) \, e^{-i(x-y,\xi)} \, dy \, d\xi, \qquad \forall v \in S(\mathbb{R}_n). \qquad (3.2.16)$$

Hence, it follows that $A^* v$ is a PDO of the form (3.2.5).

Proposition 3.2.3. *Let* $A(x, D)$ *be a PDO with symbol* $A(x, \xi) = A(\infty, \xi) + A_1(x, \xi) \in S_\alpha^0$. *Then the conjugate operator* A^* *is also a PDO whose symbol* $A^*(x, \xi) \in S_\alpha^0(\mathbb{R}_n)$ *is given by*

$$A^*(x, \xi) = \overline{A(\infty, \xi)} + \frac{1}{(2\pi)^n} \int_{\mathbb{R}_n} \overline{\widehat{A}_1(-\theta, \xi + \theta)} \, e^{-i(x,\theta)} \, d\theta, \qquad (3.2.17)$$

where $\widehat{A}_1(\theta, \xi) = F_y A_1(y, \xi)$.

Proof. Formula (3.2.17) follows from Proposition 3.2.1 if we take

$$a(x, y, \xi) = \overline{A(y, \xi)} = \overline{A(\infty, \xi)} + \overline{A_1(y, \xi)}$$

and use the relation $F_y \overline{A_1(y, \xi)} = \overline{\widehat{A}_1(-\theta, \xi)}$.

Proposition 3.2.4. *If* A *is a PDO with symbol* $A(x, \xi) \in S_\alpha^{m+\gamma}$, *where* $m \geq 0$ *is integer and* $0 < \gamma \leq 1$, *then the symbol* $A^*(x, \xi)$ *of the conjugate PDO* A^* *can be represented in the form*

$$A^*(x, \xi) = \overline{A(\infty, \xi)} + \sum_{|k|=0}^{m} \frac{1}{|k|!} i^{|k|} \frac{\partial^k}{\partial x^k} \frac{\partial^k}{\partial \xi^k} \overline{A_1(x, \xi)} + A_{m+1}(x, \xi), \qquad (3.2.18)$$

where $A_{m+1}(x, \xi) \in S_{\alpha-m-\gamma}^0$.

Proposition 3.2.4 follows from Proposition 3.2.2 with $a(x, y, \xi) = \overline{A(y, \xi)} = \overline{A(\infty, \xi)} + \overline{A_1(y, \xi)}$.

Definition 3.2.4. For an operator T which is bounded from $H_s(\mathbb{R}_n)$ to $H_{s-\alpha}(\mathbb{R}_n)$ for any s, we say that T *is of order* α; and we write $\alpha = \text{ord} \, T$.

From Theorem 3.1.1, it follows that for the PDO $A(x, D)$ with symbol $A(x, \xi) \in S_\alpha^0$, we have $\text{ord} \, A(x, D) = \alpha$.

Proposition 3.2.5. *Let* $s_1 < s_2 < s_3$. *Then for any* $\varepsilon > 0$, *the following inequality holds*

$$\|u\|_{s_2} \leq \varepsilon \|u\|_{s_3} + C_\varepsilon \|u\|_{s_1}, \qquad \forall u \in H_{s_3}(\mathbb{R}_n), \qquad (3.2.19)$$

where $\| \cdot \|_s$ *is the norm in the Sobolev–Slobodetskii space* $H_s(\mathbb{R}_n)$; C_ε *is a constant that depends on* ε.

Proof. Clearly, in order to prove this result, it suffices to establish the inequality

$$\left(1 + |\xi|\right)^{2s_2} \leq \varepsilon \left(1 + |\xi|\right)^{2s_3} + C_\varepsilon \left(1 + |\xi|\right)^{2s_1},$$

which is equivalent to

$$\left(1 + |\xi|\right)^{\delta_1} \leq \varepsilon \left(1 + |\xi|\right)^{\delta_2} + C_\varepsilon, \qquad (3.2.20)$$

where $0 < \delta_1 = 2s_2 - 2s_1 < \delta_2 = 2s_3 - 2s_1$.

For any $\varepsilon > 0$, there is $N_\varepsilon > 0$ such that for $|\xi| > N_\varepsilon$, we have $\left(1 + |\xi|\right)^{\delta_1} < \varepsilon \left(1 + |\xi|\right)^{\delta_2}$. Then, taking $C_\varepsilon = \max_{0 < |\xi| \leq N_\varepsilon} \left(1 + |\xi|\right)^{\delta_1}$, we obtain the inequality (3.2.20). Proposition 3.2.5 is proved.

Theorem 3.2.1. *Let $A(x, D)$ be a PDO with symbol $A(x, \xi) \in S_\alpha^t$, $t > 0$. Then, for any $\varepsilon > 0$, the following estimate holds:*

$$\|Au\|_{s-\alpha}^2 \leq (M_0 + \varepsilon)^2 \|u\|_s^2 + C_{\varepsilon,s} \|u\|_{s-\frac{1}{2}}^2, \tag{3.2.21}$$

where $c_{\varepsilon,s}$ is a constant and M_0 is given by

$$M_0 = \overline{\lim_{\xi \to \infty}} \sup_{x \in \mathbb{R}_n} \frac{|A(x, \xi)|}{(1 + |\xi|)^\alpha}. \tag{3.2.22}$$

Proof. By the definition of limit superior, for any $\varepsilon > 0$ we can find N such that

$$\sup_{x \in \mathbb{R}_n} \frac{|A(x, \xi)|}{(1 + |\xi|)^\alpha} < M_0 + \frac{\varepsilon}{2} \qquad \text{for} \qquad |\xi| \geq N.$$

Consider a function $\chi(\xi) \in C_0^\infty(\mathbb{R}_n)$ such that $\chi(\xi) = 1$ for $|\xi| \leq 1$, and $0 \leq \chi(\xi) \leq 1$ for all ξ. Let

$$A_1(x, \xi) = A(x, \xi)(1 - \chi(N^{-1}\xi)), \qquad A_2(x, \xi) = A(x, \xi)\chi(N^{-1}\xi).$$

Since ord $A_2(x, D) = -\infty$, the following estimate holds:

$$\|A_2(x, D)u\|_{s-\alpha}^2 \leq C_s \|u\|_{s-\frac{1}{2}}^2. \tag{3.2.23}$$

Therefore, it suffices to prove the estimate (3.2.21) for the operator $A_1(x, D)$. For all $x \in \mathbb{R}_n$, $\xi \in \mathbb{R}_n$, we have

$$|A_1(x, \xi)| \leq \left(M_0 + \frac{\varepsilon}{2}\right)(1 + |\xi|)^\alpha. \tag{3.2.24}$$

Further,

$$\|A_1 u\|_{s-\alpha}^2 = \left(\Lambda_0^{s-\alpha} A_1 u, \Lambda_0^{s-\alpha} A_1 u\right) = \left(A_1^* \Lambda^{2(s-\alpha)} A_1 u, u\right), \tag{3.2.25}$$

where Λ_0 is the PDO with symbol $(1 + |\xi|) \in S_1^1$. Set

$$B(x, \xi) = \left(\left(M_0 + \frac{\varepsilon}{2}\right)^2 (1 + |\xi|)^{2s} - |A_1(x, \xi)|^2 (1 + |\xi|)^{2(s-\alpha)}\right)^{1/2}. \tag{3.2.26}$$

From (3.2.24), it follows that $B(x, \xi) \in S_s^{t_0}$ with $t_0 = \min\{t, 1\}$. Consider the PDO

$$C(x, D) = \left(M_0 + \frac{\varepsilon}{2}\right)^2 \Lambda_0^{2s} - A_1^* \Lambda_0^{2(s-\alpha)} A_1. \tag{3.2.27}$$

In order to examine the symbol $C(x, \xi)$ of the PDO $C(x, D)$, let us utilize Theorem 3.1.2, Proposition 3.2.3, as well as Theorem 3.1.3 and Proposition 3.3.4 with $m = 0$. We obtain

$$C(x, \xi) = |B(x, \xi)|^2 + T(x, \xi), \tag{3.2.28}$$

where $T(x, \xi) \in S_{2s-t_0}^0$. From (3.2.28) it follows that $C(x, D)$ can be written in the form

$$C(x, D) = B^* B + T, \tag{3.2.29}$$

where T is the operator with symbol $T(x, \xi)$.

Next, consider the expression $(C(x, D)u, u) = (B^* Bu, u) + (Tu, u)$. Using (3.2.25) and (3.2.27), we get

$$\|A_1 u\|_{s-\alpha}^2 = \left(M_0 + \frac{\varepsilon}{2}\right)^2 \|u\|_s^2 - \|Bu\|_0^2 - (Tu, u). \tag{3.2.30}$$

From (3.2.28) we obtain the inequality

$$|(Tu, u)| \leq \|Tu\|_{-s+\frac{t_0}{2}}^2 \|u\|_{s-\frac{t_0}{2}}. \tag{3.2.31}$$

With the help of (3.2.30) and (3.2.31), we get

$$\|A_1 u\|_{s-\alpha}^2 \leq \left(M_0 + \frac{\varepsilon}{2}\right)^2 \|u\|_s^2 + C\|u\|_{s-\frac{t_0}{2}}^2. \tag{3.2.32}$$

In order to estimate the term $C\|u\|_{s-\frac{t_0}{2}}^2$, we apply Proposition 3.2.5 with $s_2 = s - t_0/2$, $s_3 = s$, $s_1 = s - 1/2$, which yields

$$C\|u\|_{s-\frac{t_0}{2}}^2 \leq \frac{\varepsilon}{2}\|u\|_s^2 + C_{\varepsilon,s}\|u\|_{s-\frac{1}{2}}^2. \tag{3.2.33}$$

Theorem 3.2.1 follows from the estimates (3.2.23), (3.2.32), and (3.2.33), and the proof is complete.

Theorem 3.2.2. *Let $A(x, D)$ be a PDO with symbol $A(x, \xi) \in S_\alpha^\infty$. Then, for any $x_0 \in \mathbb{R}_n$ and any $\varepsilon > 0$, there exists a neighborhood U_0 of x_0 such that for any $\varphi(x) \in C_0^\infty(U_0)$ the operator $\varphi(x)A(x, D)$ admits the representation*

$$\varphi(x)A(x, D) = \varphi(x)\big(A(x_0, D) + K(x, D) + T(x, D)\big), \qquad (3.2.34)$$

where $A(x_0, D)$ is the PDO with symbol $A(x_0, \xi)$; $K(x, D)$ is a PDO for which the following estimate holds:

$$\|K(x, D)u\|_{s-\alpha} \le \varepsilon \|u\|_s, \qquad (3.2.35)$$

and $T(x, D)$ is a PDO whose order does not exceed $\alpha - 1$.

Proof. Let us choose a neighborhood U_0 such that for $x \in U_0$, the following inequality holds:

$$\sup_{\xi \in \mathbb{R}_n} |A(x, \xi) - A(x_0, \xi)| \left(1 + |\xi|\right)^{-\alpha} \le \frac{\varepsilon}{2}. \qquad (3.2.36)$$

Let $\varphi(x) \in C_0^\infty(U_0)$. Consider a function $\psi(x) \in C^\infty(U_0)$ such that $\varphi(x)\psi(x) = \varphi(x)$, $|\psi(x)| \le 1$. Taking $B(x, \xi) = \psi(x)[A(x, \xi) - A(x_0, \xi)]$, we get

$$\varphi(x)A(x, \xi) = \varphi(x)A(x_0, \xi) + \varphi(x)B(x, \xi). \qquad (3.2.37)$$

Theorem 3.2.1, combined with (3.2.36), yields

$$\|B(x, D)u\|_{s-\alpha} \le \frac{2}{3}\varepsilon\|u\|_s^2 + C_\varepsilon\|u\|_{s-\frac{1}{2}}^2. \qquad (3.2.38)$$

Let $K(x, \xi) = B(x, \xi)\big(1 - \chi(N^{-1}\xi)\big)$, $T(x, \xi) = B(x, \xi)\chi(N^{-1}\xi)$, where $\chi(\xi)$ is the function defined in the proof of Theorem 3.2.1. From the structure of the symbol $T(x, \xi)$ we see that ord $T(x, D) = -\infty$. Further, from (3.2.38) we obtain the estimate

$$\|K(x, D)\|_{s-\alpha} \le \frac{2}{3}\varepsilon\big\|\big(1 - \chi(N^{-1}D)\big)u\big\|_s + C_\varepsilon\big\|\big(1 - \chi(N^{-1}D)\big)u\big\|_{s-\frac{1}{2}}. \qquad (3.2.39)$$

Since $0 \le 1 - \chi(N^{-1}\xi) \le 1$, we have $\big\|\big(1 - \chi(N^{-1}\xi)\big)u\big\|_s \le \|u\|_s$. Using the definition of the norm in $H_{s-\frac{1}{2}}(\mathbb{R}_n)$, we find that

$$\big\|\big(1 - \chi(N^{-1}D)\big)u\big\|_{s-\frac{1}{2}}^2 \int_{\mathbb{R}_n} \frac{\big(1 - \chi(N^{-1}\xi)\big)^2|\hat{u}(\xi)|^2}{(1 + |\xi|)}\left(1 + |\xi|\right)^{2s} d\xi \le \frac{1}{1 + N}\|u(\xi)\|_s^2. \qquad (3.2.40)$$

For the derivation of (3.2.4) we have used the inequality

$$\frac{\big|1 - \chi(N^{-1}\xi)\big|^2}{1 + |\xi|} \le \frac{1}{1 + N}.$$

Since N does not depend on ε, we can choose it so large that

$$\|K(x, D)\|_{s-\alpha} \le \varepsilon\|u\|_s. \qquad (3.2.41)$$

Theorem 3.2.2 is proved.

Let A be the PDO with symbol $a(x, y, \xi) \in S_\alpha^\infty(\mathbb{R}_n \times \mathbb{R}_n)$ (see Definition 3.2.3),

$$Au = \frac{1}{(2\pi)^n} \int_{\mathbb{R}_n} \int_{\mathbb{R}_n} a(x, y, \xi)\, e^{-i(x-y,\xi)}\, u(y)\, dy\, d\xi. \qquad (3.2.42)$$

We introduce a PDO A_ε by the formula

$$A_\varepsilon u = \frac{1}{(2\pi)^n} \int_{\mathbb{R}_n} \int_{\mathbb{R}_n} a(x, y, \xi) u(y) \chi(\varepsilon\xi) e^{-i(x-y,\xi)} \, dy \, d\xi, \tag{3.2.43}$$

where $\chi(\xi) \in C_0^\infty(\mathbb{R}_n)$, $\chi(\xi) = 1$ for $|\xi| \leq 1$; $0 \leq \chi(\xi) \leq 1$ for all ξ. The operator A_ε is an integral operator with the infinitely differentiable kernel

$$E_\varepsilon(x, y) = \frac{1}{(2\pi)^n} \int_{\mathbb{R}_n} a(x, y, \xi) \chi(\varepsilon\xi) e^{-i(x-y,\xi)} \, d\xi. \tag{3.2.44}$$

Let $u(x, y)$ be an arbitrary function in $S(\mathbb{R}_n \times \mathbb{R}_n)$. Denote by E_0 the linear functional on $S(\mathbb{R}_n \times \mathbb{R}_n)$ given by

$$(E_0, w) = \frac{1}{(2\pi)^n} \int_{\mathbb{R}_n} \int_{\mathbb{R}_n} \int_{\mathbb{R}_n} a(x, y, \xi) e^{-i(x-y,\xi)} w(x, y) \, dx \, dy \, d\xi. \tag{3.2.45}$$

By analogy with Proposition 3.2.1, it can be shown that for the function

$$F(\xi) = \frac{1}{(2\pi)^n} \int_{\mathbb{R}_n} \int_{\mathbb{R}_n} a(x, y, \xi) e^{-i(x-y,\xi)} w(x, y) \, dx \, dy$$

the estimate $|F(\xi)| \leq C_N \left(1 + |\xi|\right)^{-N}$ holds for any $N > 0$. It follows that (3.2.45) exists as an iterated integral and $(E_\varepsilon(x, y), w(x, y)) \to (E_0(x, y), w)$ as $\varepsilon \to 0$, i.e., E_ε converge to E_0 in $S'(\mathbb{R}_n \times \mathbb{R}_n)$ as $\varepsilon \to 0$.

Let us show that outside the diagonal $x = y$ in $\mathbb{R}_n \times \mathbb{R}_n$, the function E_0 is infinitely differentiable.

Theorem 3.2.3. *For any* $\delta > 0$, *the function* $E_0^\delta = \left(1 - \chi(\delta^{-1}(x-y))\right) E_0$ *is in* $C^\infty(\mathbb{R}_n \times \mathbb{R}_n)$.

Proof. Let $E_\varepsilon^\delta = \left(1 - \chi(\delta^{-1}(x-y))\right) E_\varepsilon(x, y)$. Since $(x_k - y_k) e^{-i(x-y,\xi)} = i\partial e^{-i(x-y,\xi)} / \partial\xi_k$, we have

$$E_\varepsilon^\delta(x, y) = \frac{1}{(2\pi)^n} \int_{\mathbb{R}_n} \left(1 - \chi(\delta^{-1}(x-y))\right) \chi(\varepsilon\xi) \frac{a(x, y, \xi)}{|x-y|^{2N}} \Delta_\xi^N e^{-i(x-y,\xi)} \, d\xi, \tag{3.2.46}$$

where $\Delta_\xi = -\sum_{k=1}^n \partial^2 / \partial\xi_k^2$. Note that

$$\frac{\left(1 - \chi(\delta^{-1}(x-y))\right)}{|x-y|^{2N}} \in C^\infty(\mathbb{R}_n \times \mathbb{R}_n),$$

since $1 - \chi(\delta^{-1}(x-y)) = 0$ for $|x-y| \leq \delta$. Integrating by parts in (3.2.46), we get

$$E_\varepsilon^\delta(x, y) = \int_{\mathbb{R}_n} \Delta_\xi^N \left(\chi(\varepsilon\xi) b_N(x, y, \xi)\right) e^{-i(x-y,\xi)} \, d\xi, \qquad \forall N, \tag{3.2.47}$$

$$b_N(x, y, \xi) = \frac{1}{(2\pi)^n} \frac{\left(1 - \chi(\delta^{-1}(x-y))\right)}{|x-y|^{2N}} a(x, y, \xi).$$

Let us show that $E_\varepsilon^\delta(x, y) \to E_0^\delta$ uniformly in x, y, as $\varepsilon \to 0$, where

$$E_0^\delta = \int_{\mathbb{R}_n} \Delta_\xi^N b_N(x, y, \xi) e^{-i(x-y,\xi)} \, d\xi. \tag{3.2.48}$$

Since $a(x, y, \xi) \in S_\alpha^\infty(\mathbb{R}_n \times \mathbb{R}_n)$, for $2M > \alpha + n + 1$ we obtain the inequality

$$\left| \Delta_\xi^N (\chi(\varepsilon\xi) b_N(x, y, \xi)) - \chi \Delta_\xi^N b_N(x, y, \xi) \right| \le$$
$$\le C \sum_{k=1}^{2N} \varepsilon^k \left(1 + |\xi|\right)^{\alpha - 2N + k} \le C \sum_{k=1}^{2N} \varepsilon^k \left(1 + |\xi|\right)^{-n-1+k}. \tag{3.2.49}$$

Since $\partial^k \chi(\varepsilon\xi) / \partial \xi^k \ne 0$ for $|k| > 0$ and $1/\varepsilon \le |\xi| \le C_0/\varepsilon$, it follows from (3.2.49) that

$$\left| \int_{\mathbb{R}_n} \left(\Delta_\xi^N \chi(\varepsilon\xi) b_N(x, y, \xi) - \chi \Delta_\xi^N b_N(x, y, \xi) \right) e^{-i(x-y,\xi)} \, d\xi \right| \le$$
$$\le C \sum_{k=1}^{2N} \varepsilon^k \int_{1/\varepsilon}^{C_0/\varepsilon} \left(1 + |\xi|\right)^{k-2} d|\xi| \to 0 \quad \text{as} \quad \varepsilon \to 0.$$

Therefore, $E_\varepsilon^\delta(x, y) \to \int_{\mathbb{R}_n} \Delta_\xi^N b_N(x, y, \xi) e^{-i(x-y,\xi)} \, d\xi$ uniformly in x and y as $\varepsilon \to 0$. On the other hand, we have $E_\varepsilon^\delta \to E_0^\delta$ in $S'(\mathbb{R}_n \times \mathbb{R}_n)$, and therefore, (3.2.48) holds.

For an arbitrary $M > 0$, take $N > (\alpha + n + 1 + M)/2$. Since $a(x, y, \xi) \in S_\alpha^\infty(\mathbb{R}_n \times \mathbb{R}_n)$, we have

$$\left| \Delta_\xi^N b_N(x, y, \xi) \right| \le C_N \left(1 + |\xi|\right)^{-n-1-M}.$$

Thus, the integral (3.2.48) is an M times continuously differentiable function of x and y. Since M is arbitrary, it follows that $E_0^\delta \in C^\infty(\mathbb{R}_n \times \mathbb{R}_n)$. Theorem 3.2.3 is proved.

Corollary 3.2.1. *Let A be a PDO with symbol $a(x, y, \xi) \in S_\alpha^\infty(\mathbb{R}_n \times \mathbb{R}_n)$, and let $\varphi(x), \psi(x) \in C_0^\infty(\mathbb{R}_n)$ be functions supported on disjoint sets. Then $\psi A \varphi$ is an integral operator with a smooth kernel.*

Proof. Since $\operatorname{supp} \varphi \cap \operatorname{supp} \psi = \varnothing$, there exists $\delta > 0$ such that for the symbol of $\psi A \varphi$ we have $\psi(x) a(x, y, \xi) \varphi(y) = 0$ for $|x - y| < \delta$. Now, Theorem 3.2.3 implies that $\psi A \varphi$ is an integral operator whose kernel is of class $C^\infty(\mathbb{R}_n \times \mathbb{R}_n)$.

Remark 3.2.1. As shown by *Eskin* (1973)$_2$, formula (3.2.13) remains valid for symbols of the form $\chi(\delta^{-1}(x - y)) a(x, y, \xi)$.

Let A be a PDO with symbol $A(x, \xi) \in S_\alpha^\infty(\mathbb{R}_n)$. Denote by $x = S(Z)$ an injective infinitely differentiable mapping of $\mathbb{R}_n$ onto itself and suppose that $S(Z)$ coincides with the identity mapping for $|x| > N$.

Let $V(x) = Au(x)$, $u_1(Z) = u(S(Z))$, $V_1(Z) = V(S(Z))$, and denote by B the operator A written in the new coordinates, i.e., $V_1(Z) = Bu_1(Z)$.

Take $\chi(x) \in C_0^\infty(\mathbb{R}_n)$, $\chi(x) = 1$ for $|x| \le 1$, and $0 \le \chi(x) \le 1$ for all x. Let

$$Au = A_0 u + T_\infty u, \tag{3.2.50}$$

$$A_0 u = \frac{1}{(2\pi)^n} \int_{\mathbb{R}_n} \int_{\mathbb{R}_n} A(x, \xi) \chi(\delta^{-1}(x - y)) e^{-i(x-y,\xi)} u(y) \, dy \, d\xi, \tag{3.2.51}$$

$$T_\infty u = \frac{1}{(2\pi)^n} \int_{\mathbb{R}_n} \int_{\mathbb{R}_n} A(x, \xi)(1 - \chi(\delta^{-1}(x - y))) e^{-i(x-y,\xi)} u(y) \, dy \, d\xi. \tag{3.2.52}$$

Theorem 3.2.3 shows that T_∞ is an integral operator with kernel of class $C^\infty(\mathbb{R}_n \times \mathbb{R}_n)$. Let us change the variables in (3.2.51), (3.2.51) by taking $x = S(Z)$, $y = S(t)$, and let $u_1(t) = u(S(t))$, $B_0 u_1 = A_0 u$, $T_\infty^1 u_1 = T_\infty u$. We see that T_∞^1 is still an integral operator with an infinitely differentiable kernel, and the operator $B_0 u_1$ has the form

$$B_0 u_1 = \frac{1}{(2\pi)^n} \int_{\mathbb{R}_n} \int_{\mathbb{R}_n} A(S(Z), \xi) \chi(\delta^{-1}(S(Z) - S(t))) u_1(y) \left| \frac{DS(t)}{Dt} \right| e^{-i((S(Z) - S(t)),\xi)} \, dt \, d\xi, \tag{3.2.53}$$

where $DS(t)/Dt$ is the Jacobi matrix for the transformation $y = S(t)$ and $|DS(t)/Dt|$ is its determinant. Applying the integral Lagrange formula, we find that

$$S(Z) - S(t) = \int_0^1 \frac{d}{d\lambda} S(t + \lambda(Z - t)) \, d\lambda =$$
$$= \left[\int_0^1 \frac{d}{du} S(t + \lambda(Z - t)) \, d\lambda \right] (Z - t) = H(Z, t)(Z - t), \qquad (3.2.54)$$

where $u = t + \lambda(Z - t)$; $H(Z, t)$ is a matrix of class $C^\infty(\mathbb{R}_n \times \mathbb{R}_n)$. From the Taylor formula

$$S(Z) = S(t) + \frac{DS(Z)}{DZ}(Z - t) + O(Z, t, Z - t), \qquad (3.2.55)$$

with the remainder term $O(Z, t, Z - t)$, it follows that $H(Z, Z) = DS(Z)/DZ$.

Since $|DS(Z)/DZ| \neq 0$, we have $|H(Z, t)| \neq 0$ for small enough $|Z - t|$. Assume that $\delta > 0$ in (3.2.53) has been chosen so small that for $|S(Z) - S(t)|$ we have

$$|H(Z, t)| \neq 0. \qquad (3.2.56)$$

From (3.2.54) we obtain

$$\big(S(Z) - S(t), \xi\big) = \big(H(Z, t)(Z - t), \xi\big) = \big(Z - t, H^*(Z, t)\xi\big), \qquad (3.2.57)$$

where $H^*(Z, t)$ is the conjugate matrix of $H(Z, t)$. Having substituted (3.2.57) into (3.2.53), let us change the variables by

$$\eta = H^*(Z, t)\xi.$$

We further have

$$B_0 u_1 = \frac{1}{(2\pi)^n} \int_{\mathbb{R}_n} \int_{\mathbb{R}_n} A\big(S(Z), (H^*(Z, t))^{-1}\eta\big) \chi\big(\delta^{-1}(S(Z) - S(t))\big) \times$$
$$\times \left| \frac{DS(t)}{Dt} \right| |H(Z, t)|^{-1} u_1(t) \, e^{-i(Z - t, \eta)} \, dt \, d\eta. \qquad (3.2.58)$$

Let

$$b(Z, t, \eta) = A\big(S(Z), (H^*(Z, t))^{-1}\eta\big) \chi\big(\delta^{-1}(S(Z) - S(t))\big) \left| \frac{DS(t)}{Dt} \right| |H(Z, t)|^{-1}. \qquad (3.2.59)$$

The operator B_0 is the PDO with the symbol $b(Z, t, \eta)$. According to Proposition 3.2.1 and Remark 3.2.1, the operator B can be transformed into a PDO of the form (3.1.32) with the symbol

$$B_0(Z, \eta) = \frac{1}{(2\pi)^n} \int_{\mathbb{R}_n} \widehat{b}(Z, \theta, \theta + \eta) \, e^{-i(x, \theta)} \, d\theta, \qquad (3.2.60)$$

where $\widehat{b}(Z, \theta, \eta) = F_t b(Z, t, \eta)$.

Proposition 3.2.2 and Remark 3.2.1 imply that for any N, the following expansion holds:

$$B_0(Z, \eta) = b(Z, Z, \eta) + \sum_{|k|=1}^N \frac{1}{|k|!} i^{|k|} \frac{\partial^k}{\partial t^k} \frac{\partial^k b(Z, t, \eta)}{\partial \eta^k} \bigg|_{t=Z} + C_{N+1}(Z, \eta), \qquad (3.2.61)$$

where $C_{N+1}(Z, \eta) \in S^{\infty}_{\alpha-N-1}(\mathbb{R}_n)$. For $Z = t$, we have $H(Z, Z) = DS(Z)/DZ$, and therefore, the principal part of the symbol $B_0(Z, \eta)$ has the form

$$b(Z, Z, \eta) = A\left(S(Z), \left(\left(\frac{DS(Z)}{DZ}\right)^*\right)^{-1}\eta\right). \tag{3.2.62}$$

Thus, the operator B can be written in the form $B = B_0 + T^1_\infty$, with the PDO $B_0 \in S^\infty_\alpha$, and T^1_∞ being an integral operator with smooth kernel. Let us represent the operator T^1_∞ as a PDO of class $S^\infty_{-\infty}$. The operator $T^1_\infty u_1$ has the form

$$T^1_\infty u_1 = \int_{\mathbb{R}_n} T^1_\infty(Z, t) u_1(t)\, dt, \tag{3.2.63}$$

where (see (3.2.48))

$$T^1_\infty(Z, t) = \int_{\mathbb{R}_n} \Delta^N_\xi \left(\frac{1}{(2\pi)^n} A(S(Z), \xi)\right) \times$$
$$\times \left[\frac{1 - \chi\left(\delta^{-1}(S(Z) - S(t))\right)}{|S(Z) - S(t)|^{2N}}\right] \left|\frac{DS(t)}{Dt}\right| e^{-i(S(Z) - S(T), \xi)}\, d\xi. \tag{3.2.64}$$

Set

$$t_0(Z, Z - t) = \int_{\mathbb{R}_n} \Delta^N_\xi \left(\frac{1}{(2\pi)^n} A(Z), \xi\right) \left[\frac{1 - \chi\left(\delta^{-1}(Z - t)\right)}{|Z - t|^{2N}}\right] e^{-i(Z-t,\xi)}\, d\xi. \tag{3.2.65}$$

By assumption, we have $S(Z) = Z$, $S(t) = t$ for $|Z| \geq N$, $|t| \geq N$. Therefore,

$$t_1(Z, t) = \left[T^1_\infty(Z, t) - t_0(Z, Z - t)\right] \in C^\infty_0(\mathbb{R}_n \times \mathbb{R}_n),$$

and the following estimates hold:

$$\left|\frac{\partial^p}{\partial Z^p} \frac{\partial^k}{\partial t^k} t_1(Z, t)\right| \leq C_{p,k,N}(1 + |Z| + |t|)^{-N}. \tag{3.2.66}$$

Because of the inclusion $A(Z, \xi) \in S^\infty_\alpha(\mathbb{R}_n)$ and the relation (3.2.65), similar estimates hold for the function $t_0(Z, Z - t)$,

$$\left|\frac{\partial^p}{\partial Z^p} \frac{\partial^k}{\partial \lambda^k} t_0(Z, \lambda)\right| \leq C'_{p,k,N}(1 + |Z| + |\lambda|)^{-N}. \tag{3.2.67}$$

Denote by $\widehat{T}^1_\infty(Z, \eta)$ the Fourier transform in t of the complex conjugate of $T^1_\infty(Z, t)$. We have

$$\widehat{T}^1_\infty(Z, \eta) = F_t\left[\overline{t_0(Z, Z - t)} + \overline{t_1(Z, t)}\right] = T_0(Z, \eta)\, e^{i(Z,\eta)} + T_1(Z, \eta). \tag{3.2.68}$$

The estimates (3.2.66) and (3.2.67) imply that $\widehat{T}^1_\infty(Z, \eta) \in S^\infty_{-\infty}$. The Parseval identity yields

$$T^1_\infty u_1 = \int_{\mathbb{R}_n} T^1_\infty(Z, t) u_1(t)\, dt = \frac{1}{(2\pi)^n} \int_{\mathbb{R}_n} \overline{\widehat{T}^1_\infty(Z, \eta)}\, \widehat{u}_1(\eta)\, d\eta =$$
$$= \frac{1}{(2\pi)^n} \int_{\mathbb{R}_n} \left[\overline{T_0(Z, \eta)} + \overline{T_1(Z, \eta)}\, e^{i(Z,\eta)}\right] \widehat{u}_1(\eta)\, e^{-i(Z,\eta)}\, d\eta. \tag{3.2.69}$$

Hence, $T^1_\infty \in S^\infty_{-\infty}$. The above considerations imply the following result.

Theorem 3.2.4. *The operator B obtained from a PDO $A \in S_\alpha^\infty$ by the transformation of the variables is also a PDO of class S_α^∞. The symbol of B has the form $B(Z, \xi) = B_0(Z, \xi) + T_\infty^{(2)}(Z, \xi)$, where $B_0(Z, \xi)$ is given by (3.2.61) and $T_\infty^{(2)} \in S_{-\infty}^\infty$.*

Proposition 3.2.6. *Let $y = f(x) \in C^\infty(\mathbb{R}_n)$ be a diffeomorphism of $\mathbb{R}_n$ such that $y = x$ for $|x| \geq N$. Then the mapping $u(x) \mapsto V(x) = u(f(x))$ of $S(\mathbb{R}_n)$ onto itself can be extended as a continuous injective operator from $H_s(\mathbb{R}_n)$ onto $H_s(\mathbb{R}_n)$ for any s.*

Proof. Let $u(x) \in S(\mathbb{R}_n)$. Then, for $s = 0$, we have

$$\|u\|_0^2 = \int_{\mathbb{R}_n} |u(y)|^2 \, dy = \int_{\mathbb{R}_n} |u(f(x))|^2 \left| \frac{Df}{Dx} \right| dx \leq C \int_{\mathbb{R}_n} |V(x)|^2 \, dx = C\|V\|_0^2,$$

where $|Df(x)/Dx|$ is the Jacobian of the mapping $y = f(x)$. By assumption, the Jacobian is different from zero. Therefore, the opposite inequality holds, $\|V\|_0 \leq C\|u\|_0$.

Consider the case $s \neq 0$. Let Λ_s be a PDO with symbol $\Lambda_s = \left(1 + |\xi|^2\right)^{s/2}$. We have

$$\|u\|_s^2 \leq C\|\Lambda_s u\|_0^2.$$

Let $u_1 = \Lambda_s u$. Passing to the variables $y = f(x)$, we get $u_1(f(x)) = \Lambda_s^{(1)} V$. Theorem 3.2.4 implies that $\Lambda_s^{(1)}$ is an operator of order s. By Theorem 3.1.1, we have $\|u_1\|_0^2 \leq C\|V\|_s^2$, and therefore, $\|u\|_s \leq C\|V\|_s$. If we consider the mapping $x = f^{-1}(y)$, we obtain the inequality $\|V\|_s \leq C\|u\|_s$. To complete the proof, it suffices to note that $S(\mathbb{R}_n)$ is dense in $H_s(\mathbb{R}_n)$.

Remark 3.2.2. Let $x = By + a$ be a linear transformation, where B is a constant $(n \times n)$-matrix, $\det B \neq 0$, and $a \in \mathbb{R}_n$ is a constant vector. Consider the Fourier transform of the function $\psi(y) = u(By + a) \in S(\mathbb{R}_n)$,

$$\widehat{\psi}(\xi) = \int_{\mathbb{R}_n} \psi(y) \, e^{i(y,\xi)} \, dy = \int_{\mathbb{R}_n} u(By + a) \, e^{i(y,\xi)} \, dy = \int_{\mathbb{R}_n} u(x) \, e^{i(B^{-1}x - B^{-1}a, \xi)} \left| \det B^{-1} \right| =$$

$$= \left| \det B^{-1} \right| e^{-i(B^{-1}a, \xi)} \int_{\mathbb{R}_n} u(x) \, e^{i(x, (B^{-1})^* \xi)} \, dx =$$

$$= \left| \det B^{-1} \right| e^{-i(B^{-1}a, \xi)} \widehat{u}\left((B^{-1})^* \xi\right), \tag{3.2.70}$$

where $(B^{-1})^*$ is the transpose of the matrix B^{-1}. Using (3.2.70), we get

$$\|\psi(y)\|_s^2 = \int_{\mathbb{R}_n} \left(1 + |\xi|\right)^{2s} \left| \widehat{\psi}(\xi) \right|^2 d\xi = \left| \det B^{-1} \right|^2 \int_{\mathbb{R}_n} \left(1 + |\xi|\right)^{2s} \left| \widehat{u}((B^{-1})^* \xi) \right|^2 d\xi. \tag{3.2.71}$$

Changing the variables, $(B^{-1})^* \xi = \eta$, in (3.2.71) and using the inequality

$$C_1 \left(1 + |\eta|\right) \leq 1 + |B^* \eta| \leq C_2 \left(1 + |\eta|\right),$$

we get

$$\|\psi(y)\|_s^2 \leq C_3 \|u(x)\|_s^2.$$

In a similar way we establish the inequality $\|\psi(y)\|_s^2 \geq C_4 \|u(x)\|_s^2$. Thus we come to the estimate

$$C_4 \|u(x)\|_s^2 \leq \|\psi(y)\|_s^2 \leq C_3 \|u(x)\|_s^2. \tag{3.2.72}$$

Let $A(x, D)$ be a PDO with symbol $A(x, \xi) \in S_\alpha^\infty$. Changing the variables, $x = By + a$, we obtain

$$A(x, D)u = \frac{1}{(2\pi)^n} \int_{\mathbb{R}_n} A(x, \xi) \widehat{u}(\xi) \, e^{-i(x, \xi)} \, d\xi =$$

$$= \frac{1}{(2\pi)^n} \int_{\mathbb{R}_n} A(By + a) \, e^{-i(z, \xi)} \, e^{-i(By, \xi)} \, \widehat{u}(\xi) \, d\xi. \tag{3.2.73}$$

Taking $B^*\xi = \eta$ in the integral (3.2.73), we get

$$\frac{1}{(2\pi)^n} \int_{\mathbb{R}_n} A\big(By + a, (B^*)^{-1}\eta\big) \, e^{-i(a,(B^*)^{-1}\eta)} \, e^{-i(y,\eta)} \, \widehat{u}\big((B^*)^{-1}\eta\big) |\det(B^*)|^{-1} \, d\eta. \qquad (3.2.74)$$

Comparing (3.2.72) and (3.2.74) and using (3.2.70), we find that $A(x, D)$ transforms into the PDO

$$B(y, D)\psi(y) = \frac{1}{(2\pi)^n} \int_{\mathbb{R}_n} A\big(By + a, (B^*)^{-1}\eta\big) \widehat{\psi}(\eta) \, e^{-i(y,\eta)} \, d\eta. \qquad (3.2.75)$$

3.3. Pseudodifferential and Hypersingular Integral Equations

In this book, we consider pseudodifferential equations of the form

$$pAu = f, \qquad (3.3.1)$$

where A is a PDO, p is the operator of restriction to a domain $G \subset \mathbb{R}_n$, $u \in \mathring{H}_s(G)$, $f \in H_r(G)$ (see Definitions 2.1.9, 2.3.5, and 2.3.6). More general boundary value problems for PDO's are formulated in *Agranovich* (1965), *Dynin* (1961), *Eskin* (1973)$_2$.

For the development of direct numerical methods, such as the method of discrete vortices, it is more convenient to avoid the representation of operators in terms of the Fourier transformation.

Definition 3.3.1. Equations of the form

$$\int_G \frac{f(x, x_0)}{r^\lambda} u(x_0) \, dx_0 = F(x), \qquad (3.3.2)$$

are called *hypersingular integral equations*. Here, $G \subset \mathbb{R}_n$; $x_0, x \in G$, $r = \left[\sum_{i=1}^n (x_i - x_0^i)^2\right]^{1/2}$; $\lambda > n$; $f(x, x_0)$ is a known function, $F(x)$ is a given function, and $u_0(x)$ is unknown.

Remark 3.3.1. The regularization of the hypersingular integral on the left-hand side of equation (3.3.2) will be specified in each particular case. The function $u(x_0)$ is usually sought in $\mathring{H}_s(G)$, and $F(x) \in H_r(G)$.

Let us recall some definitions.

Definition 3.3.2. Let A be a bounded linear operator mapping a Banach space B_1 into another Banach space. The *kernel* of the operator A (denoted by $\ker A$) is the subspace $M \subset B_1$ such that $x \in M \Rightarrow Ax = 0$. The *cokernel* of A (denoted by $\operatorname{coker} A$) is the kernel of the conjugate operator A^*. The *range* of the operator A (denoted by $\mathcal{R}(A)$) is the set of the values of A.

In Banach spaces we consider two types of orthogonal complements.

Definition 3.3.3. Let X be a Banach space with its dual space denoted by X^*. For a linear subspace $M \subset X$, *the complement $M^\perp \subset X^*$* is defined by

$$M^\perp = \big\{ f \in X^* : \ (x, f) = 0, \quad \forall x \in M \big\},$$

where (x, f) is the value of the linear functional f on the element x. For a linear subspace $N \subset X^*$, *the complement $^\perp N$* is defined by

$$^\perp N = \{ x \in X : \ (x, f) = 0 \quad \forall f \in N \}.$$

Proposition 3.3.1. *For Banach spaces B_1, B_2, let $A : B_1 \to B_2$ be a bounded linear operator. Then*

$$\mathcal{R}(A)^\perp = \operatorname{coker} A. \qquad (3.3.3)$$

Proof. Let $f \in \mathcal{R}(A)^\perp$. Then, for all $x \in B_1$, we have $(Ax, f) = 0 = (x, A^* f)$. Since x is arbitrary, it follows that $A^* f = 0$, and therefore, $f \in \operatorname{coker} A$. Proposition 3.2.1 is proved.

Proposition 3.3.2. *For Banach spaces B_1, B_2, let $A : B_1 \to B_2$ be a bounded linear operator. Then*

$$^\perp \operatorname{coker} A = \overline{\mathcal{R}(A)}, \tag{3.3.4}$$

where $\overline{\mathcal{R}(A)}$ is the closure of $\mathcal{R}(A)$.

Proof. Let $y_0 \in \overline{\mathcal{R}(A)}$. Then, there is a sequence $\{x_n\} \in B_1$ such that $\lim_{n \to \infty} A x_n = y_0$. Take any element $f_0 \in \operatorname{coker} A$. Then, for any $x \in B_1$, we have $(Ax, f_0) = (x, A^* f_0) = 0$. It follows that

$$(y_0, f_0) = \left(\lim_{n \to \infty} A x_n, f_0 \right) = \lim_{n \to \infty} (A x_n, f_0) = \lim_{n \to \infty} (x_n, A^* f_0) = 0. \tag{3.3.5}$$

Relations (3.3.5) imply that $\overline{\mathcal{R}(A)} \subseteq {}^\perp \operatorname{coker} A$.

Now, let $y_0 \notin \overline{\mathcal{R}(A)}$. We are going to show that there exists an element of $\operatorname{coker} A$ which is non-orthogonal to y_0. Indeed, the Hahn–Banach theorem claims that there is an element $f_0 \in B_2^*$ such that $(y_0, f_0) \neq 0$, $(Ax, f_0) = (x, A^* f_0) = 0$. The last relation implies that $f_0 \in \operatorname{coker} A$. Therefore, $\overline{\mathcal{R}(A)} \supseteq {}^\perp \operatorname{coker} A$. This, combined with the opposite inclusion established previously, yields (3.3.4).

Definition 3.3.4. A bounded linear operator A from the Banach space B_1 to the Banach space B_2 is called *normally resolvable*, if the following relation holds:

$$^\perp \operatorname{coker} A = \mathcal{R}(A). \tag{3.3.6}$$

Theorem 3.3.1. (Hausdorff). *A bounded linear operator A from the Banach space B_1 to the Banach space B_2 is normally resolvable if and only if $\overline{\mathcal{R}(A)} = \mathcal{R}(A)$.*

Proof. Suppose that $\overline{\mathcal{R}(A)} = \mathcal{R}(A)$. By Proposition 3.3.2, we have $^\perp \operatorname{coker} A = \overline{\mathcal{R}(A)} = \mathcal{R}(A)$, and therefore, the operator A is normally resolvable.

Conversely, suppose that A is a normally resolvable operator. Then

$$^\perp \operatorname{coker} A = \mathcal{R}(A). \tag{3.3.7}$$

Comparing the right-hand sides of (3.3.7) and (3.3.4), we see that $\overline{\mathcal{R}(A)} = \mathcal{R}(A)$, and the Hausdorff theorem is proved.

Definition 3.3.5. A normally resolvable operator A is called *Noetherian*, if both $\ker A$ and $\operatorname{coker} A$ are finite dimensional spaces. The number

$$\chi = \dim \ker A - \dim \operatorname{coker} A$$

is called *the index* of the operator A, where $\dim \ker A$ and $\dim \operatorname{coker} A$ are the dimensions of $\ker A$ and $\operatorname{coker} A$, respectively.

Definition 3.3.6. Let $A : B_1 \to B_2$ be a bounded liner operator in Banach spaces B_1, B_2. A bounded linear operator $R : B_2 \to B_1$ is called *the left regularizer of A*, if

$$RA = I + T_1, \tag{3.3.8}$$

where I is the identity operator and the operator $T_1 : B_1 \to B_1$ is compact. An operator $R : B_2 \to B_1$ is called *the right regularizer of A*, if

$$AR = I + T_2, \tag{3.3.9}$$

where $T : B_2 \to B_2$ is a compact operator.

Proposition 3.3.3. *If R is the right (resp., left) regularizer of operator A, then its conjugate R^* is the left (resp., right) regularizer of A^*.*

Proof. Let R be the right regularizer of A. By the definition of conjugate operator, for $x \in B_2$, $y \in B_2^*$, we get

$$(x, R^*A^*y) = (Rx, A^*y) = (ARx, y) = ((I + T_2)x, y) = \left(x, \left(I + T_2^*\right)y\right). \qquad (3.3.10)$$

It follows that $R^*A^* = I + T_2^*$. Since the conjugate of a compact operator is compact, the last relation implies that R^* is the left regularizer of A^*.

In a similar way it can be shown that if R is a left regularizer of A, then R^* is a right regularizer of A^*. Proposition 3.3.3 is proved.

Proposition 3.3.4. *Let $A = I + T : B \to B$ be a bounded linear operator in a Banach space B, where T is a compact operator. Then $\dim \ker A < \infty$.*

Proof. Consider the unit ball in $\ker A$. Then $S = -TS$, and therefore, the unit ball S in $\ker A$ is a compact set, which implies that $\ker A$ is a finite-dimensional space (see *Kolmogorov and Fomin* (1972)).

Theorem 3.3.2. *Let $A : B_1 \to B_2$ be a bounded linear operator in Banach spaces B_1, B_2. If there exists an operator R which is both the left and the right regularizer of A, then A is Noetherian.*

Proof. It follows from Propositions 3.3.3, 3.3.4, and the definition of the regularizer R that $\ker A$ and $\operatorname{coker} A$ are finite dimensional. Now, in order to prove that A is a Noetherian operator, it suffices to show its normal resolvability. By the Hausdorff theorem, its normal resolvability is equivalent to the relation

$$\overline{\mathcal{R}(A)} = \mathcal{R}(A). \qquad (3.3.11)$$

Let $f \in \mathcal{R}(A)$. Then there is an element $\varphi_0 \in B_1$ such that $A\varphi_0 = f$. If $\varphi_1, \ldots, \varphi_n$ is a basis of the null-space of the operator A, then the general solution of the equation $A\varphi = f$ has the form

$$\varphi = \varphi_0 + \sum_{j=1}^{n} a_j \varphi_j, \qquad (3.3.12)$$

where a_j are arbitrary constants. Since $\dim \ker A < \infty$, among the elements of the form (3.3.12) there is an element $\widetilde{\varphi}$ with the smallest norm. Let us show that for some constant C, we have

$$\|\widetilde{\varphi}\|_{B_1} \leq C \|f\|_{B_2}, \qquad f \in \mathcal{R}(A). \qquad (3.3.13)$$

If R is the left regularizer of A, then

$$RA\widetilde{\varphi} = \widetilde{\varphi} + T_1 \widetilde{\varphi} = Rf, \qquad (3.3.14)$$

where T_1 is a compact operator. Suppose that the ratio $\|\widetilde{\varphi}\|/\|f\|$ is unbounded. Then, there exists a sequence $\widetilde{\varphi}_n$ and a corresponding sequence f_n such that $\|f_n\| \to 0$ as $n \to \infty$ and $\|\widetilde{\varphi}_n\| = 1$. From the bounded sequence $\{\widetilde{\varphi}_n\}$ we can extract a subsequence $\{\widetilde{\varphi}_{n_k}\}$ such that $T_1 \widetilde{\varphi}_{n_k}$ converges to some limit $(-\varphi^0)$. Equation $\widetilde{\varphi}_{n_k} + T_1 \widetilde{\varphi}_{n_k} = Rf_{n_k}$ implies that $\widetilde{\varphi}_{n_k} \to \varphi^0$ as $n_k \to \infty$. On the other hand, $A\widetilde{\varphi}_{n_k} = f_{n_k} \to 0$ as $n_k \to \infty$. Since the operator A is continuous, we have $A\varphi^0 = 0$. Hence, $A\left(\widetilde{\varphi}_{n_k} - \varphi^0\right) = f_{n_k}$. Among the solutions of the equation $A\varphi = f_{n_k}$, the element $\widetilde{\varphi}_{n_k}$ has the smallest norm, and therefore, $\left\|\widetilde{\varphi}_{n_k} - \varphi^0\right\| \geq \|\widetilde{\varphi}_{n_k}\| = 1$, which is in contradiction with the convergence $\widetilde{\varphi}_{n_k} \to \varphi^0$. Thus, the inequality (3.3.13) is proved.

Now, let $f \in \overline{\mathcal{R}(A)}$. Then, there is a sequence $f_n \in \mathcal{R}(A)$, $f_n \to f$ as $n \to \infty$. To each element f_n we associate an element $\widetilde{\varphi}_n$ with the smallest norm and satisfying the equation

$A\widetilde{\varphi}_n = f_n$. Applying the regularizer R to both sides of this equation, we get $\widetilde{\varphi}_n + T_1\widetilde{\varphi}_n = Rf_n$. In view of the inequality (3.3.13), the norms of the elements $\widetilde{\varphi}_n$ are bounded uniformly in n. Therefore, there is a subsequence $\{\widetilde{\varphi}_{n_k}\}$ such that $T_1\widetilde{\varphi}_{n_k}$ converges to some limit. The regularizer R is a bounded operator, and therefore, $Rf_{n_k} \to Rf$ as $n_k \to \infty$. It follows that the sequence $\widetilde{\varphi}_{n_k}$ has a limit φ. Now, $\widetilde{\varphi}_{n_k} \to \varphi$ and $A\widetilde{\varphi}_{n_k} = f_{n_k} \to f$. Because of the continuity of the operator A, we have $A\varphi = f$, and consequently, $f \in \mathcal{R}(A)$. Thus, the set $\mathcal{R}(A)$ is closed. Theorem 3.3.2 is proved.

Definition 3.3.7. Noetherian operators of zero index are called *Fredholm operators*.

Theorem 3.3.3. *The PDO with symbol $T_1(x,\xi) \in S^0_{-\delta}$, $\delta > 0$, such that $T_1(\infty,\xi) = 0$ is a compact operator in $H_s(\mathbb{R}_n)$ for any s.*

Proof. Let $\chi(\xi) \in C_0^\infty(\mathbb{R}_n)$, $\chi(\xi) = 1$ for $|\xi| \le 1$. Consider the function $T_{1\varepsilon} = T_1(x,\xi)\chi(\varepsilon\xi)$. Let us show that the norm of the operator $A_\varepsilon = T_1(x,D) - T_{1\varepsilon}(x,D)$ in $H_s(\mathbb{R}_n)$ tends to zero as $\varepsilon \to 0$. Using (3.1.46), we get

$$\|A_\varepsilon\|_s \le C \sup_{\eta,\xi}\left\{ \left(1+|\eta|\right)^{n+1+|s|} \left|\widetilde{T}_1(\eta,\xi)\right| \left|1-\chi(\varepsilon\xi)\right| \right\}$$

$$\le C_1 \sup_\xi \left(1+|\xi|\right)^{-\delta} \left|1-\chi(\varepsilon\xi)\right| \to 0 \quad \text{as} \quad \varepsilon \to 0. \tag{3.3.15}$$

Now, let us show that $T_{1\varepsilon}(x,D)$ is a compact operator in $H_s(\mathbb{R}_n)$. Then, from (3.3.15) it would follow that the operator $T_1(x,D)$ is compact (see *Kolmogorov and Fomin* (1972)).

Let $v(x) = T_{1\varepsilon}(x,D)u(x)$. Applying the Fourier transformation (see Proposition 3.1.1), we get

$$\widehat{V}(\eta) = \int_{\mathbb{R}_n} \widehat{T}_{1\varepsilon}(\eta-\xi,\xi)\widehat{u}(\xi)\, d\xi, \tag{3.3.16}$$

and for any N the kernel of the integral operator (3.3.16) satisfies the inequality

$$\left|\widehat{T}_{1\varepsilon}(\eta-\xi,\xi)\right| \le C_{\varepsilon,N} \left(1+|\xi|\right)^{-N} \left(1+|\eta|\right)^{-N}. \tag{3.3.17}$$

The integral operator A^1_ε defined by (3.3.16) can be represented in the form

$$A^1_\varepsilon u = \frac{1}{\left(1+|\eta|\right)^s} \int_{\mathbb{R}_n} \left(1+|\eta|\right)^s \widehat{T}_{1\varepsilon}(\eta-\xi,\xi) \left(1+|\xi|\right)^{-s} \frac{\widehat{u}(\xi)}{\left(1+|\xi|\right)^{-s}}\, d\xi. \tag{3.3.18}$$

From (3.3.18), it follows that the operator A_ε can be written as

$$A_\varepsilon = A_3 A_2 A_1, \tag{3.3.19}$$

where $A_1 u = \left(1+|\xi|\right)^s \widehat{u}(\xi) = V_1(\xi)$ is a bounded operator from $\widetilde{H}_2(\mathbb{R}_n)$ to $L_2(\mathbb{R}_n)$. The operator A_2 has the form

$$A_2 V_1 = \int_{\mathbb{R}_n} \left(1+|\eta|\right)^s \widetilde{T}_{1\varepsilon}(\eta-\xi,\xi) \left(1+|\xi|\right)^{-s} \widehat{V}_1(\xi) = V_2(\eta).$$

From the estimates (3.3.17) it follows that the function $A_{2\varepsilon}(\eta,\xi) = \left(1+|\eta|\right)^s \widehat{T}_{1\varepsilon}(\eta-\xi,\xi) \left(1+|\xi|\right)^{-s}$ belongs to $L_2(\mathbb{R}_n \times \mathbb{R}_n)$, and therefore, the operator $A_2 V_1$ in $L_2(\mathbb{R}_n)$ is compact. The operator $A_3 V_2 = \left(1+|\eta|\right)^{-s} V_2(\eta)$ is bounded from $L_2(\mathbb{R}_n)$ to $\widetilde{H}_s(\mathbb{R}_n)$. The operator A_ε is a product of bounded operators and a compact one, and therefore, is compact in $\widetilde{H}_s(\mathbb{R}_n)$, where $\widetilde{H}_s(\mathbb{R}_n)$ is the Fourier image of $H_s(\mathbb{R}_n)$. From the definition of the spaces $H_s(\mathbb{R}_n)$, it follows that $T_{1\varepsilon}(x,D)$ is a compact operator in $H_s(\mathbb{R}_n)$. Theorem 3.3.3 is proved.

Proposition 3.3.5. *For a bounded domain* G *with a smooth boundary, any bounded linear operator* $T_0 : \mathring{H}_s(G) \to \mathring{H}_{s+\varepsilon}$, $\varepsilon > 0$, *is compact in* $\mathring{H}_s(G)$.

Proof. Let Λ_ε be a PDO with symbol $(1 + |\xi|^2)^{\varepsilon/2}$, and $\psi_0(x) \in C_0^\infty(\mathbb{R}_n)$, $\psi_0(x) = 1$ in a neighborhood of $\overline{G}$. For any $u(x) \in \mathring{H}_s(G)$, we have $\psi_0(x)u(x) = u(x)$, and therefore, $T_0 u = \psi_0(x)\Lambda_{-\varepsilon}\Lambda_\varepsilon T_0 u$. The operator $\Lambda_\varepsilon T_0$ is bounded from $\mathring{H}_s(G)$ to $H_s(\mathbb{R}_n)$ and the operator $\psi_0(x)\Lambda_{-\varepsilon}$ is compact in $H_s(\mathbb{R}_n)$ (see Theorem 3.3.3). Consequently, T_0 is also compact in $H_s(G)$. Proposition 3.3.5 is proved.

Proposition 3.3.6. *Any bounded operator* $T : H_s(G) \to H_{s+\varepsilon}$, $\varepsilon > 0$, *is compact in* $H_s(G)$.

Proof. Recall that $H_s(\mathbb{R}_n)$ is a Hilbert space. From the definition of $H_s(G)$, it follows that there is a bounded linear extension operator $l_0 : H_s(G) \to H_s(\mathbb{R}_n)$ satisfying the estimate $\|l_0 g\|_{H_s(\mathbb{R}_n)} \leq C\|g\|_{H_s(G)}$. Representing the operator T in the form $T = p\psi_0(x)\Lambda_{-s}\Lambda_\varepsilon l_0 T$, we see that T is a compact operator in $H_s(G)$, since it is the product of the bounded operator $\Lambda_\varepsilon l_0 T$ and the compact operator $p\psi_0(x)\Lambda_{-\varepsilon}$. Proposition 3.3.6 is proved.

Corollary 3.3.1. *Under the assumptions of Propositions 3.3.5 and 3.3.6, the operator* T_0 *is compact in* $\mathring{H}_{s+\varepsilon}(G)$, *and the operator* T *is compact in* $H_{s+\varepsilon}(G)$.

Proof. Indeed, in this case, the conjugate operators T_0^* and T^* are bounded as operators from $H_{-s-\varepsilon}(G)$ to $H_{-s}(G)$ and $\mathring{H}_{-s-\varepsilon}(G)$ to $\mathring{H}_{-s}(G)$, respectively. Propositions 3.3.5 and 3.3.6 imply that T_0^* and T^* are compact operators in $H_{-s-\varepsilon}$ and $\mathring{H}_{-s-\varepsilon}(G)$, respectively. Hence, we obtain the desired statement. Note that since $\mathring{H}_{s+\varepsilon} \subset \mathring{H}_s(G)$, $H_{s+\varepsilon} \subset H_s(G)$, the operators T_0 and T are a fortiori defined in $\mathring{H}_{s+\varepsilon}(G)$ and $H_{s+\varepsilon}(G)$, respectively.

Proposition 3.3.7. *The unit ball in* $\mathring{H}_s(G)$ *is precompact with respect to the norm of* $\mathring{H}_{s-\varepsilon}(G)$, *for any bounded domain* G *and any* $\varepsilon > 0$.

Proof. Consider the operator $Tu(x) = u(x)$ which is bounded from $\mathring{H}_s(G)$ to $\mathring{H}_{s-\varepsilon}(G)$. Take $\psi \in C_0^\infty(\mathbb{R}_n)$, $\psi(x) \equiv 1$ in a neighborhood of $\overline{G}$. The operator T can be represented as

$$T = \left\{ \psi(x) \left\{ 1 + |\xi|^2 \right\}^{-\varepsilon/2} \right\} \cdot \left[1 + |\xi|^2 \right]^{\varepsilon/2} = T_2(T_1),$$

where $T_1 u = \left[1 + |\xi|^2 \right]^{\varepsilon/2} u$ is bounded as an operator from $\mathring{H}_s(G)$ to $H_{s-\varepsilon}(\mathbb{R}_n)$ and the operator $T_2 v = \psi(x)\left[1 + |\xi|^2 \right]^{-\varepsilon/2} v$ is compact in $H_{s-\varepsilon}(\mathbb{R}_n)$, according to Theorem 3.3.3. Therefore, the operator $T : \mathring{H}_s(G) \to \mathring{H}_{s-\varepsilon}(G)$ is compact, which implies that the image of the unit ball in $\mathring{H}_s(G)$ under the mapping T is a precompact set in $\mathring{H}_{s-\varepsilon}$. Proposition 3.3.7 is proved.

Proposition 3.3.8. *The unit ball in* $H_s(G)$ *is precompact with respect to the norm of* $H_{s-\varepsilon}(G)$, *for any bounded domain* G *and any* $\varepsilon > 0$.

Proof. Consider the operator $Tu(x) = u(x)$ which is bounded from $H_s(G)$ to $H_{s-\varepsilon}(G)$. Let l_0 be the linear extension operator from $H_s(G)$ to $H_s(\mathbb{R}_n)$ specified in Proposition 3.3.6. Representing T in the form

$$T = p\psi_0(x)\left(1 + |\xi|^2 \right)^{-\varepsilon/2} \cdot \left(1 + |\xi|^2 \right)^{\varepsilon/2} l_0 T,$$

where p is the operator of restriction to $H_s(G)$, and arguing as above, we see that T is a compact operator. Hence we obtain the claimed result.

Definition 3.3.8. An operator $A(x, D)$ is called *elliptic*, if outside a ball $|x|^2 + |\xi|^2 \geq N^2$, its symbol $A(x, \xi) \in S_\alpha^t$ admits an inverse $A^{-1}(x, \xi)$, and $A(x, \xi) \geq C \left(1 + |\xi| \right)^{\alpha - \delta}$ for $|x|^2 + |\xi|^2 \geq N^2$, where $0 \leq \delta \leq \min\{t, 1\}$.

Proposition 3.3.9. *Any elliptic PDO* $A(x, D)$ *with symbol* $A(x, t) \in S_\alpha^t$, $t > 0$, *is a Noetherian operator from* $H_s(\mathbb{R}_n)$ *to* $H_{s-\alpha}(\mathbb{R}_n)$ *for any* s.

Proof. Consider a function $\chi(x) \in C_0^\infty(\mathbb{R}_n)$ such that $\chi(x) = 1$ for $|x| \leq 1$, and $0 \leq \chi(x) \leq 1$ for all x. Denote by $R(x, D)$ the PDO with symbol $R(x, \xi) = \left[1 - \chi(N^{-1}x)\chi(N^{-1}\xi)\right]A^{-1}(x, \xi)$. We see that $R(x, \xi) \in S_{-\alpha+\delta}^t$ and $R(\infty, \xi) = A^{-1}(\infty, \xi)$. Theorem 3.1.3 yields the relation

$$A(x, D)R(x, D) = C_0(x, D) + T_1(x, D), \tag{3.3.20}$$

where $C_0(x, D)$ is the PDO with symbol $C_0(x, \xi) = A(x, \xi)R(x, \xi)$, $T_1(x, \xi) \in S_{-t_0}^0$, $t_0 > 0$, $T_1(\infty, \xi) = 0$. Since $[C_0(x, \xi) - 1] \in S_{-N}^t$ with arbitrary N, it follows from (3.3.20) that

$$A(x, D)R(x, D) = I + T_2(x, D), \tag{3.3.21}$$

where I is the identity operator, $T_2(x, \xi) \in S_{-t_0}^0$, $t_0 > 0$, $T_2(\infty, \xi) = 0$. Now, it follows from Theorem 3.3.3 that $T_2(x, D)$ is a compact operator. Therefore, $R(x, D)$ is the right regularizer of the operator $A(x, D)$.

In a similar way, we find that

$$R(x, D)A(x, D) = I + T_3(x, D), \qquad T_3(x, \xi) \in S_{-t_0}^0, \qquad T_3(\infty, \xi) = 0.$$

Therefore, $R(x, D)$ is the left regularizer of $A(x, D)$. By Theorem 3.3.2 , the operator $A(x, D)$ is Noetherian. Proposition 3.3.9 is proved.

Proposition 3.3.10. *For* $u(x) \in \mathring{H}_s(G)$, $s \geq 0$, *the following inequality holds:*

$$\int_{\mathbb{R}_n} |\xi|^{2s} |\widehat{u}(\xi)|^2 \, d\xi \geq C_s(G)\|u\|_s^2, \tag{3.3.22}$$

where G *is a bounded domain,* $C_s(G) = \text{const}$, $\widehat{u}(\xi) = F(u(x))$.

Proof. For any $u(x) \in \mathring{H}_s(G)$, using the Cauchy inequality, we get

$$|\widehat{u}(\xi)| = \left| \int_G u(x) \, e^{i(x,\xi)} \, dx \right| \leq \left(\int_G |u(x)|^2 \, dx \right)^{1/2} \left(\int_G 1 \, dx \right)^{1/2} \leq C(G)\|u(x)\|_{L_2(\mathbb{R}_n)}. \tag{3.3.23}$$

Take r_0 so small that the volume of the ball S_{r_0} of radius r_0 centered at the origin is less than $1/(2C^2(G))$. Then, from (3.3.23) we obtain the inequality

$$\int_{S_{r_0}} |\widehat{u}(\xi)|^2 \, d\xi \leq C_s^2(G)\|u(x)\|_{L_2(\mathbb{R}_n)}^2 \int_{S_{r_0}} d\xi \leq \frac{1}{2}\|u(x)\|_{L_2(\mathbb{R}_n)}^2. \tag{3.3.24}$$

Using the inequality (3.3.24), we find that

$$\int_{\mathbb{R}_n} |\xi|^{2s} |\widehat{u}(\xi)|^2 \, d\xi \geq \int_{\mathbb{R}_n \setminus S_{r_0}} |\xi|_{2s} |\widehat{u}(\xi)|^2 \, d\xi \geq r_0^{2s} \int_{\mathbb{R}_n \setminus S_{r_0}} |\widehat{u}(\xi)|^2 \, d\xi =$$

$$= r_0^{2s} \left[\int_{\mathbb{R}_n} |\widehat{u}(\xi)|^2 \, d\xi - \int_{S_{r_0}} |\widehat{u}(\xi)|^2 \, d\xi \right] \geq$$

$$\geq r_0^{2s} \left[\|u(x)\|_{L_2(\mathbb{R}_n)}^2 - \frac{1}{2}\|u(x)\|_{L_2(\mathbb{R}_n)}^2 \right] = \frac{r_0^{2s}}{2}\|u(x)\|_{L_2(\mathbb{R}_n)}^2. \tag{3.3.25}$$

From (3.3.25), for any $u(x) \in \mathring{H}_s(G)$, we get

$$\int_{\mathbb{R}_n} |\xi|^{2s} |\widehat{u}(\xi)|^2 \, d\xi \geq C_{1,s} \int_{\mathbb{R}_n} \left(1 + |\xi|^{2s}\right) |\widehat{u}(\xi)|^2 \, d\xi. \tag{3.3.26}$$

Since $1 + |\xi|^{2s} \geq C_s \left(1 + |\xi|\right)^{2s}$, we obtain the estimate (3.3.22) from (3.3.26). Proposition 3.3.10 is proved.

Consider the hypersingular integral

$$Au = \int_G r^\lambda u(x_0)\, dx_0, \tag{3.3.27}$$

where $G \subset \mathbb{R}_n$ is a bounded domain, $r = |x - x_0|$, $\lambda < -n$, $\lambda \neq -n - 2k$, $k = 1, 2, \ldots$; $u(x) \in \mathring{H}_s(G)$ is a real-valued function. As shown in Example 2.2.6, this integral can be represented as a pseudodifferential operator

$$Au = \frac{1}{(2\pi)^n} \int_{\mathbb{R}_n} C_\lambda |\xi|^{-\lambda-n} \widehat{u}(\xi)\, e^{-i(x,\xi)}\, d\xi. \tag{3.3.28}$$

Its symbol is $Q(\xi) = C_\lambda |\xi|^{-\lambda-n} \in S_\alpha^0$, $\alpha = -\lambda - n > 0$. Proposition 2.3.6 implies that the PDO A is bounded from $H_s(\mathbb{R}_n)$ to $H_{s-\alpha}(\mathbb{R}_n)$, in particular, it is bounded as an operator from $\mathring{H}_{-\frac{\lambda-n}{2}}(G)$ to $H_{\frac{\lambda+n}{2}}(G)$.

Proposition 3.3.11. *The hypersingular equation*

$$p \int_G r^\lambda u(x_0)\, dx_0 = f(x) \tag{3.3.29}$$

admits a unique solution $u(x_0) \in \mathring{H}_{-\frac{\lambda-n}{2}}(G)$ *for any generalized function* $f \in H_{\frac{\lambda+n}{2}}$, *where* p *is the operator of restriction to* G.

Proof. Let us show that pAu is a self-adjoint operator. Indeed, for any $V(x) \in \mathring{H}_{\frac{\lambda+n}{2}}(G)$, we have (see Propositions 2.3.10, 2.3.14, 2.3.15)

$$(pAu, V) = (Au, V) =$$
$$= \frac{1}{(2\pi)^n} \int_{\mathbb{R}_n} C_\lambda |\xi|^{-\lambda-n} \widehat{u}(\xi) \overline{\widehat{V}(\xi)}\, d\xi = \frac{1}{(2\pi)^n} \int_{\mathbb{R}_n} \overline{C_\lambda |\xi|^{-\lambda-n} \widehat{u}(\xi)}\, \widehat{V}(\xi)\, d\xi =$$
$$= \frac{1}{(2\pi)^n} \int_{\mathbb{R}_n} C_\lambda |\xi|^{-\lambda-n} \overline{\widehat{u}(\xi)}\, \widehat{V}(\xi)\, d\xi = (u, AV) = (u, pAV), \tag{3.3.30}$$

where (pAu, V) is the value of the linear functional $pAu \in H_{\frac{\lambda+n}{2}}(G)$ on the function $V(x) \in \mathring{H}_{-\frac{\lambda-n}{2}}(G)$. From (3.3.30), it follows that $pA = (pA)^*$.

Proposition 3.3.10 yields the estimate

$$|(pAu, u)| = \frac{|C_\lambda|}{(2\pi)^n} \int_{\mathbb{R}_n} |\xi|^{-\lambda-n} |\widehat{u}(\xi)|^2\, d\xi \geq C_{1\lambda}(G) \|u\|^2_{\mathring{H}_{-\frac{\lambda-n}{2}}(G)}. \tag{3.3.31}$$

On the other hand, we have

$$|(pAu, u)| \leq \|pAu\|_{H_{\frac{\lambda+n}{2}}(G)} \|u(x)\|_{\mathring{H}_{-\frac{\lambda-n}{2}}(G)}. \tag{3.3.32}$$

Using (3.3.31) and (3.3.32), we obtain the inequality

$$\|pAu\|_{\mathring{H}_{\frac{\lambda+n}{2}}(G)} \geq C_{1\lambda}(G) \|u(x)\|_{\mathring{H}_{-\frac{\lambda-n}{2}}(G)}. \tag{3.3.33}$$

The estimate (3.3.33) implies that $\overline{\mathcal{R}(pA)} = \mathcal{R}(pA)$ (see *Trenogin* (1980)), and also that $\dim \ker pA = 0$. Since the operator pA is self-adjoint, we also have $\dim \operatorname{coker} pA = 0$. Hence we obtain the desired result. Proposition 3.3.11 is proved.

Remark 3.3.2. The proof of Proposition 3.3.11 can be literally repeated to establish the following statement: *For any $f(x) \in H_{-m}(G)$ with integer $m > 0$, the differential equation*

$$\Delta^m u = f \tag{3.3.34}$$

admits one and only one solution $u(x) \in \mathring{H}_m(G)$, where Δ is the Laplace operator.

Next, we are going to obtain some stronger results than those of Proposition 3.3.11. Let $\omega = |\xi'|^{-1}\xi'$, $\xi' = (\xi_1, \dots, \xi_{n-1})$. For any function $A_0(\xi) \in O_{\alpha+i\beta}^{\infty}$ (see Definition 3.1.2), we introduce the function $\widehat{A}_0(\xi)$ by

$$\widehat{A}_0(\xi', \xi_n) = A_0\big(\left(1 + |\xi'|\right) \omega, \, \xi_n \big). \tag{3.3.35}$$

Proposition 3.3.12. *If $A_0(\xi) \in O_{\alpha+i\beta}^{\infty}$, then the function $A_1(\xi', \xi_n) = \widehat{A}_0(\xi', \xi_n) - A_0(\xi', \xi_n)$ satisfies the inequality*

$$|A_1(\xi', \xi_n)| \leq C\left(|\xi'| + |\xi_n|\right)^{\alpha-1} \quad \text{for} \quad |\xi'| + |\xi_n| \geq 2. \tag{3.3.36}$$

Proof. Using the Lagrange formula, we get

$$|A_1(\xi', \xi_n)| = \left| \sum_{k=1}^{n-1} \frac{\partial A_0(\xi' + \theta\omega, \xi_n)}{\partial \xi_k} \frac{\xi_k}{|\xi'|} \right| \leq \sum_{k=1}^{n-1} \left| \frac{\partial A_0(\xi' + \theta\omega, \xi_n)}{\partial \xi_k} \right| \leq C\left(|\xi' + \theta\omega| + |\xi_n|\right)^{\alpha-1}, \tag{3.3.37}$$

where $0 < \theta < 1$. For $|\xi'| + |\xi_n| \geq 2$, the following inequality holds:

$$\frac{1}{2}\left(|\xi'| + |\xi_n|\right) \leq |\xi'| + |\xi_n| - 1 \leq |\xi'| + |\xi_n| - |\theta\omega| \leq$$
$$\leq |\xi' + \theta\omega| + |\xi_n| \leq |\xi'| + |\theta\omega| + |\xi_n| \leq 2\left(|\xi'| + |\xi_n|\right). \tag{3.3.38}$$

This implies that whatever the sign of $\alpha - 1$, we have the estimate

$$\left(|\xi' + \theta\omega| + |\xi_n|\right)^{\alpha-1} \leq C\left(|\xi'| + |\xi_n|\right)^{\alpha-1}. \tag{3.3.39}$$

Now, (3.3.36) follows from (3.3.37) and (3.3.39). Proposition 3.3.12 is proved.

Definition 3.3.9. A symbol $A(\xi) \in O_{\alpha+i\beta}^{\infty}$ is called *elliptic*, if $A(\xi) \neq 0$ for $|\xi| \neq 0$.

If $A(\xi) \in O_{\alpha+i\beta}^{\infty}$ is an elliptic symbol, then the relation $A(\xi) = |\xi|^{\alpha+C\beta} A(|\xi|^{-1}\xi)$ shows that

$$C_1|\xi|^{\alpha} \leq |A(\xi)| \leq C_2|\xi|^{\alpha}. \tag{3.3.40}$$

Proposition 3.3.13. *Let $A_0(\xi) \in O_{\alpha+i\beta}^{\infty}$ be an elliptic symbol. Then for any s, the pseudodifferential equation*

$$\frac{1}{(2\pi)^n} \int_{\mathbb{R}_n} \widehat{A}_0(\xi', \xi_n)\widehat{u}(\xi)\, e^{-i(x,\xi)}\, d\xi = f \tag{3.3.41}$$

admits one and only one solution in $H_s(\mathbb{R}_n)$ for any $f \in H_{s-\alpha}(\mathbb{R}_n)$.

Proof. Since the inequality (3.3.40) holds for $A_0(\xi'\xi_n)$, we see that $\widehat{A}_0(\xi', \xi_n)$ can be estimated as follows:

$$C_1\left(1 + |\xi'| + |\xi_n|\right)^{\alpha} \leq \left|\widehat{A}_0(\xi', \xi_n)\right| \leq C_2\left|1 + |\xi'| + |\xi_n|\right|^{\alpha}. \tag{3.3.42}$$

After the Fourier transformation with respect to x, equation (3.3.41) becomes

$$\widehat{A}_0(\xi)\widehat{u}(\xi) = \widehat{f}(\xi). \tag{3.3.43}$$

From (3.3.43), we get $\widehat{u}(\xi) = \widehat{f}(\xi)/\widehat{A}_0(\xi)$. Using (3.3.42), we obtain the estimate

$$\|u\|_s^2 = \int_{\mathbb{R}_n} \left(1 + |\xi|\right)^{2s} \frac{\left|\widehat{f}(\xi)\right|^2}{\left|\widehat{A}_0(\xi)\right|^2} \, d\xi \leq C \|f\|_{s-\alpha}^2.$$

It follows that the PDO $\widehat{A}_0^{-1}$ with symbol $\widehat{A}_0^{-1}(\xi)$ is bounded from $H_{s-\alpha}(\mathbb{R}_n)$ to $H_s\mathbb{R}_n)$ and is inverse to $\widehat{A}_0$. Proposition 3.3.13 is proved.

For $\widehat{f}(\xi', \xi_n) \in S(\mathbb{R}_n)$ and $\tau > 0$, consider the integral

$$F_+(\xi', \xi_n + i\tau) = \frac{i}{2\pi} \int_{-\infty}^{\infty} \frac{\widehat{f}(\xi', \eta_n)}{\xi_n + i\tau - \eta_n} \, d\eta_n. \tag{3.3.44}$$

The function $F_+(\xi', \xi_n + i\tau)$ is analytic with respect to the variable $\alpha = \xi_n + i\tau$ in the halfplane $\tau > 0$. The following relation holds (see *Eskin* (1973)$_2$):

$$\lim_{\tau \to +0} F_+(\xi', \xi_n + i\tau) = F_+(\xi', \xi_n + i0) = \frac{1}{2}\widehat{f}(\xi', \xi_n) + \int_{\infty}^{\infty} \frac{\widehat{f}(\xi', \eta_n)}{\xi_n - \eta_n} \, d\eta_n. \tag{3.3.45}$$

In a similar way, for the integral

$$F_-(\xi', \xi_n + i\tau) = -\frac{i}{2\pi} \int_{\infty}^{\infty} \frac{\widehat{f}(\xi', \eta_n)}{\xi_n + i\tau - \eta_n} \, d\eta_n, \qquad \tau < 0, \tag{3.3.46}$$

we have the convergence

$$\lim_{\tau \to -0} F_-(\xi', \xi_n + i\tau) = F_-(\xi', \xi_n - i0) = \frac{1}{2}\widehat{f}(\xi', \xi_n) - \frac{i}{2\pi} \int_{-\infty}^{\infty} \frac{\widehat{f}(\xi', \eta_n)}{\xi_n - \eta_n} \, d\eta_n. \tag{3.3.47}$$

Let us introduce the operators

$$\Pi^0 \widehat{f} = \frac{i}{2\pi} \int_{-\infty}^{\infty} \frac{\widehat{f}(\xi', \eta_n)}{\xi_n - \eta_n} \, d\eta_n, \tag{3.3.48}$$

$$\Pi^+ \widehat{f} = F_+(\xi', \xi_n + i0), \qquad \Pi^- \widehat{f} = F_-(\xi', \xi_n - i0). \tag{3.3.49}$$

From (3.3.45) and (3.3.47), it follows that

$$\Pi^+ \widehat{f} = \frac{1}{2}\widehat{f} + \Pi^0 \widehat{f}, \qquad \Pi^- \widehat{f} = \frac{1}{2}\widehat{f} - \Pi^0 \widehat{f}, \qquad \widehat{f} = \Pi^+ \widehat{f} + \Pi^- \widehat{f}. \tag{3.3.50}$$

Denote by θ^+ the operator of multiplication by a function $\theta(x_n)$ which is equal to 1 for $x_n \geq 0$ and identically vanishes for $x_n < 0$. Let θ^- be the operator of multiplication by the function $(1 - \theta(x_n))$.

Proposition 3.3.14. *For any $f(x', x_n) \in S(\mathbb{R}_n)$, $x' = (x_1, \ldots, x_{n-1})$, we have*

$$F(\theta^+ f) = \Pi^+ \widehat{f}, \tag{3.3.51}$$

$$F(\theta^- f) = \Pi^- \widehat{f}, \tag{3.3.52}$$

where $\widehat{f} = Ff$ is the Fourier transform of $f(x', x_n)$.

Proof. For $\tau > 0$, we have

$$F(\theta(x_n) e^{-x_n \tau}) = \int_0^\infty e^{-x_n \tau + i x_n \xi_n} dx_n = \frac{1}{\xi_n + i\tau}. \tag{3.3.53}$$

For $\varphi(x), \psi(x) \in L_2(\mathbb{R}_n)$, we have (see *Vladimirov (1976)*)

$$F(\varphi(x)\psi(x)) = \frac{1}{(2\pi)^n} \int_{\mathbb{R}_n} \widehat{\varphi}(\xi - \eta) \widehat{\psi}(\eta) \, d\eta. \tag{3.3.54}$$

Passing to the Fourier transform of $\theta(x_n) e^{-x_n \tau} f(x', x_n)$ in x and applying (3.3.54) in x_n, we obtain

$$F\left(\theta(x_n) e^{-x_n \tau} f(x', x_n)\right) = \frac{i}{2\pi} \int_{-\infty}^\infty \frac{\widehat{f}(\xi', \eta_n) \, d\eta_n}{\xi_n + i\tau - \eta_n}. \tag{3.3.55}$$

By the Lebesgue theorem, we get

$$\lim_{\tau \to +0} \int_{\mathbb{R}_n} \theta(x_n) e^{-x_n \tau} f(x) e^{i(x,\xi)} \, dx = \int_{\mathbb{R}_n} \theta(x_n) f(x) e^{i(x,\xi)} \, dx. \tag{3.3.56}$$

Using (3.3.49), (3.3.5) and passing to the limit as $\tau \to +0$, we obtain (3.3.51). Relation (3.3.52) is established in a similar way. Proposition 3.3.14 is proved.

The following result is proved in *Eskin* (1973)$_2$.

Theorem 3.3.4. *The operators $\theta^\pm$ are bounded with respect to the norm in $H_s(\mathbb{R}_n)$ for $|s| < 1/2$.*

Proposition 3.3.15. *For $|s| < 1/2$, any $\widehat{f} \in \widehat{H}_s(\mathbb{R}_n)$ (the Fourier image of $H_s(\mathbb{R}_n)$) can be uniquely represented in the form*

$$\widehat{f} = \widehat{f}_+ + \widehat{f}_-, \quad \text{where} \quad \widehat{f}_+ = \Pi^+ \widehat{f} \in \widehat{H}_s^+, \quad \widehat{f}_- = \Pi^- \widehat{f} \in \widehat{H}_s^- \tag{3.3.57}$$

(see Definitions 2.3.2 and 2.3.3).

Proof. Relation (3.3.57) follows from (3.3.50), combined with Proposition 3.3.14 and Theorem 3.3.4. For any $\varphi(x) \in C_0^\infty(\mathbb{R}_n^\pm)$, we have $\theta^- \theta^+ = 0$, $\theta^+ \theta^- = 0$. Hence, using Propositions 2.3.4 and 2.3.5, we find that if $\widehat{f}_+ + \widehat{f}_- = 0$, then $\Pi^+ \widehat{f}_+ = \widehat{f}_+ = 0$ and $\Pi^- \widehat{f}_- = \widehat{f}_- = 0$. Proposition 3.3.15 is proved.

Definition 3.3.10. *A homogeneous factorization of an elliptic symbol $A(\xi) \in O_{\alpha+i\beta}^\infty$ (see Definition 3.3.3) in the variable ξ_n is the representation of $A(\xi', \xi_n)$ in the form*

$$A(\xi', \xi_n) = A_-(\xi', \xi_n) \cdot A_+(\xi', \xi_n), \tag{3.3.58}$$

with $A_+(\xi', \xi_n)$ and $A_-(\xi', \xi_n)$ satisfying the conditions:

(a) $A_+(\xi', \xi_n)$, for $|\xi'| \neq 0$, can by analytically extended to the upper halfplane $\tau > 0$;

(b) $A_+(\xi', \xi_n + i\tau)$ is continuous with respect to the variables ξ', ξ_n, τ jointly, for $\tau \geq 0$, $|\xi'| + |\xi_n| + |\tau| > 0$;

(c) $A_+(\xi', \xi_n + i\tau)$ is homogeneous of order $\chi = \chi_1 + i\chi_2$, i.e.,

$$A_+(t\xi', t(\xi_n + i\tau)) = t^\chi A_+(\xi', \xi_n + i\tau), \quad \forall t > 0; \tag{3.3.59}$$

(d) $A_+(\xi', \xi_n + i\tau) \neq 0$ for $\tau \geq 0$, $|\xi'| + |\xi_n| + |\tau| > 0$.

Similarly, the function $A_-(\xi', \xi_n)$ admits analytic continuation with respect to ξ_n to the halfplane $\tau < 0$, is continuous and different from zero for $\tau \leq 0$ and $|\xi'| + |\xi_n| + |\tau| > 0$, and $\text{ord}_{\xi, \tau} A_-(\xi', \xi_n + i\tau) = \alpha + i\beta - \chi$.

In order to ensure the uniqueness of the factorization, it is assumed in addition that

$$A_+(0, 1) = 1. \tag{3.3.60}$$

As shown by *Eskin* (1973)$_2$, any elliptic symbol $A(\xi) \in O^\infty_{\alpha + i\beta}$ admits a unique homogeneous factorization. In Chapter 4, we will need the factorization of a particular symbol $A(\xi) \in O^\infty_{\alpha + i\beta}$, which we consider next.

Let $A(\xi) = |D\xi|^\lambda$, $\lambda > 0$, $|D\xi| = (D\xi, D\xi)^{1/2}$, where D is an $(n \times n)$-matrix. Let us represent the function $A(\xi)$ as $A(\xi) = |\xi|^\lambda A_1(\xi)$, $A_1(\xi) = A(\xi)/|\xi|^\lambda$. The factorization of the function $|\xi|^\lambda$ has the form

$$|\xi|^\lambda = (\xi_n + i|\xi'|)^{\frac{\lambda}{2}} (\xi_n - i|\xi'|)^{\frac{\lambda}{2}}, \tag{3.3.61}$$

with $(\xi_n \pm i|\xi'|)^{\lambda/2} = \exp\left\{\frac{\lambda}{2} \ln\left(\xi_n \pm i|\xi'|\right)\right\}$, where the logarithm branch is assumed real on the positive semi-axis. The following relation holds:

$$\lim_{\xi_n \to -\infty} \frac{A(\xi)}{|\xi|^\lambda} = \lim_{\xi_n \to +\infty} \frac{A(\xi)}{|\xi|^\lambda} = A_0 > 0. \tag{3.3.62}$$

Let $A_2(\xi) = A(\xi)/(A_0|\xi|^\lambda)$ and $b_2(\xi) = \ln A_2(\xi)$. Direct verification shows that

$$|b_2(\xi', \xi_n)| \leq \frac{C|\xi'|}{|\xi'| + |\xi_n|}, \quad \left|\frac{\partial b_2(\xi', \xi_n)}{\partial \xi_k}\right| \leq \frac{C}{|\xi'| + |\xi_n|}, \quad 1 \leq k \leq n - 1. \tag{3.3.63}$$

As shown by *Eskin* (1973)$_2$, conditions (3.3.63) ensure that the function

$$B_2^+(\xi', \xi_n + i\tau) = \frac{1}{2\pi} \int_{-\infty}^{\infty} \frac{b_2(\xi', \eta_n)\, d\eta_n}{\xi_n + i\tau - \eta_n}$$

has the following properties: $B_2^+(\xi', \xi_n + i\tau)$ is analytic in $\alpha = \xi_n + i\tau$ for $\tau > 0$; it is continuous for $|\xi'| + |\xi_n| + |\tau| > 0$, $\tau \geq 0$, and homogeneous of zero order; moreover,

$$|B_2^+(\xi', \xi_n + i\tau)| \leq \frac{C_\varepsilon |\xi'|^{1-\varepsilon}}{(|\xi'| + |\xi_n + +\tau)^{1-\varepsilon}}, \quad \forall \varepsilon > 0. \tag{3.3.64}$$

Similarly, the function

$$B_2^-(\xi', \xi_n + i\tau) = -\frac{i}{2\pi} \int_{-\infty}^{\infty} \frac{b_2(\xi', \eta_n)\, d\eta_n}{\xi_n + i\tau - \eta_n}$$

is analytic in $\alpha = \xi_n + i\tau$ for $\tau < 0$, continuous for $|\xi'| + |\xi_n| + |\tau| > 0$, $\tau \leq 0$, and homogeneous of zero order; and it also satisfies the inequality (3.2.64) with τ replaced by $|\tau|$.

We have

$$b_2(\xi', \xi_n) = \ln A_2(\xi', \xi_n) = B_2^+(\xi', \xi_n + i0) + B_2^-(\xi', \xi_n - i0). \tag{3.3.65}$$

From (3.3.65), it follows that

$$A_2(\xi', \xi_n) = \left(\exp \left[B_2^+(\xi', \xi_n + i0) \right] \right) \left(\exp \left[B_2^-(\xi', \xi_n - i0) \right] \right). \qquad (3.3.66)$$

Using (3.3.61) and (3.3.66), we get

$$A(\xi', \xi_n) = A_+(\xi', \xi_n) \cdot A_-(\xi', \xi_n), \qquad (3.3.67)$$

$$A_+(\xi', \xi_n) = (\xi_n + i|\xi'|)^{\frac{\lambda}{2}} \exp \left[B_2^+(\xi', \xi_n + i0) \right], \qquad (3.3.68)$$

$$A_-(\xi', \xi_n) = A_0(\xi_n + i|\xi'|)^{\frac{\lambda}{2}} \exp \left[B_2^-(\xi', \xi_n - i0) \right]. \qquad (3.3.69)$$

For the functions $A_\pm(\xi', \xi_n)$ all conditions of homogeneous factorization hold and the factorization index is $\chi = \lambda/2$.

Remark 3.3.3. Let $A(\xi) = |D\xi|^\lambda$, $\det D \neq 0$, with $A(\xi', \xi_n) = A_+(\xi', \xi_n) \cdot A_-(\xi', \xi_n)$ being the factorization of the symbol $A(\xi)$. Consider the function $\widehat{A}(\xi', \xi_n) = A(1 + |\xi'|\omega, \xi_n)$, $\omega = |\xi'|^{-1}\xi'$. Let us represent $\widehat{A}(\xi', \xi_n)$ in the form

$$\widehat{A}(\xi', \xi_n) = A^+(1 + |\xi'|\omega, \xi_n) \cdot A_-(1 + |\xi'|\omega, \xi_n) = \widehat{A}_+(\xi', \xi_n) \cdot \widehat{A}_-(\xi', \xi_n). \qquad (3.3.70)$$

Then the functions

$$B_+(\xi', \xi_n + i\tau) = \frac{1}{A_+(1 + |\xi'|\omega, \xi_n + i\tau)}, \qquad \text{resp.,} \quad B_-(\xi', \xi_n + i\tau) = \frac{1}{A_-(1 + |\xi'|\omega, \xi_n + i\tau)},$$

are continuous in their arguments jointly, for $|\xi'| \neq 0$, $\tau \geq 0$, resp., $\tau \leq 0$, are analytic in $\alpha = \xi_n + i\tau$ for $\tau > 0$, resp., $\tau < 0$ and satisfy the inequality

$$|B_\pm(\xi', \xi_n + i\tau)| \leq C \ |\xi'| + |\xi_n| + |\tau|)^{-\frac{\lambda}{2}}. \qquad (3.3.71)$$

Let $A(\xi) = |D\xi|^\lambda$. Denote by $\widehat{A}$ the pseudodifferential operator with the symbol $\widehat{A}(\xi', \xi_n) = A (1 + |\xi'|)\omega, \xi_n)$. Consider the following pseudodifferential equation in the halfspace $\mathbb{R}_n^+$:

$$p\widehat{A}u_+ = f, \qquad (3.3.72)$$

where $U_+ \in H_s^+$, $f \in H_{s-\lambda}(\mathbb{R}_n^+)$, p is the operator of restriction to $\mathbb{R}_n^+$.

Theorem 3.3.5. *Suppose that $(\lambda - 1)/2 < s < (\lambda + 1)/2$. Then, for any $f \in H_{s-\lambda}(\mathbb{R}_n^+)$, equation (3.3.72) admits one and only one solution $u_+ \in H_s^+(\mathbb{R}_n)$.*

Proof. Assume first that equation (3.3.72) has a solution $u_+ \in H_s^+$ for $f \in H_{s-\lambda}(\mathbb{R}_n^+)$, and let $lf \in H_{s-\lambda}(\mathbb{R}_n)$ be an arbitrary extension of $f \in H_{s-\lambda}(\mathbb{R}_n^+)$ from $\mathbb{R}_n^+$ to $\mathbb{R}_n$. Set

$$u_- = lf - \widehat{A}u_+. \qquad (3.3.73)$$

Since $lf \in H_{s-\lambda}(\mathbb{R}_n)$ and $\widehat{A}u_+ \in H_{s-\lambda}(\mathbb{R}_n)$, we have $u_- \in H_{s-\lambda}(\mathbb{R}_n)$. Moreover, $u_- = 0$ for $x_n > 0$ because of (3.3.72). Consequently, $u_- \in H_{s-\lambda}^-(\mathbb{R}_n)$.

Applying the Fourier transformation to (3.3.73) with respect to x, we get

$$\widehat{A}(\xi)\widehat{u}_+(\xi) + \widehat{u}_-(\xi) = \widehat{lf}(\xi). \qquad (3.3.74)$$

Substituting the right-hand side of (3.3.70) into (3.3.74), we obtain

$$\widehat{A}_+(\xi)\widehat{u}_+(\xi) + \widehat{A}_-^{-1}(\xi)\widehat{u}_-(\xi) = A_-^{-1}(\xi)\widehat{lf}(\xi). \qquad (3.3.75)$$

Let us introduce the functions

$$\widehat{v}_+(\xi) = \widehat{A}_+(\xi)\widehat{u}_+(\xi), \qquad \widehat{v}_-(\xi) = A_-^{-1}(\xi)\widehat{u}_-(\xi), \qquad \widehat{g}(\xi) = A_-^{-1}(\xi)\widehat{lf}(\xi). \tag{3.3.76}$$

Proposition 2.2.2 and Remark 3.3.3 ensure that

$$\widehat{v}_+(\xi) \in \widehat{H}^+_{s-\frac{\lambda}{2}}(\mathbb{R}_n), \qquad \widehat{v}_-(\xi)\widehat{H}^-_{s-\frac{\lambda}{2}}, \qquad \widehat{g}(\xi) \in \widehat{H}_{s-\frac{\lambda}{2}}(\mathbb{R}_n). \tag{3.3.77}$$

In terms of the new functions, equation (3.3.75) reads

$$\widehat{v}_+(\xi) + \widehat{v}_-(\xi) = \widehat{g}(\xi). \tag{3.3.78}$$

Since $|s - \lambda/2| < 1/2$ by assumption, it follows from Proposition 3.3.15 that the representation (3.3.78) is unique and

$$\widehat{v}_+(\xi) = \Pi^+\widehat{g}(\xi), \qquad \widehat{v}_-(\xi) = \Pi^-\widehat{g}(\xi). \tag{3.3.79}$$

From (3.3.76) and (3.3.79), it follows that

$$\widehat{u}_+(\xi) = \widehat{A}_+^{-1}\Pi^+\widehat{A}_-^{-1}\widehat{lf}(\xi). \tag{3.3.80}$$

Let us show that the right-hand side of (3.3.80) does not depend on the extension lf. Indeed, consider another extension $l_1f \in H_{s-\lambda}(\mathbb{R}_n)$ for $f \in H_{s-\lambda}(\mathbb{R}_n^+)$. Thus, $pl_1f = f$. Since $f_- = lf - f_1 f \in H^-_{s-\lambda}(\mathbb{R}_n)$, we have $A_-^{-1}(\xi)\widehat{f}_-(\xi) \in \widehat{H}_{s-\frac{\lambda}{2}}(\mathbb{R}_n)$ by Proposition 2.2.2. From Proposition 3.3.15, it follows that $\Pi^+ A_-^{-1}(\xi)\widehat{f}_-(\xi) = 0$. Hence, we obtain the relation

$$\Pi^+\widehat{A}_-^{-1}\widehat{lf} = \Pi^+ A_-^{-1}\widehat{l_1 f}.$$

Thus, if equation (3.3.72) has a solution, this solution must be unique, provided that $(\lambda - 1)/2 < s < (\lambda + 1)/2$.

Suppose that we always choose the extension lf such that

$$\|lf\|_{H_{s-\lambda}(\mathbb{R}_n)} \leq 2\|f\|_{H_{s-\lambda}(\mathbb{R}_n^+)}. \tag{3.3.81}$$

Then, Proposition 2.2.2, Remark 3.3.3, and relations (3.3.80), (3.3.81) yield

$$\|u_+\|_s \leq C\|f\|_{H_{s-\lambda}(\mathbb{R}_n^+)}. \tag{3.3.82}$$

Now, let us establish the existence of a solution of equation (3.3.72) with an arbitrary right-hand side $f \in H_{s-\lambda}(\mathbb{R}_n^+)$ by showing that (3.3.72) holds for $u_+ = F^{-1}\left(\widehat{A}_+^{-1}\Pi^+\widehat{A}_-^{-1}\widehat{lf}(\xi)\right)$.

Indeed, since $\Pi^+\widehat{A}_-^{-1}\widehat{lf} = A_-^{-1}\widehat{lf} - \Pi^-\widehat{A}_-^{-1}\widehat{lf}$, we have

$$\widehat{A}(\xi)\widehat{u}_+(\xi) = A_-(\xi)\Pi^+\widehat{A}_-^{-1}\widehat{lf} = \widehat{A}_-(\xi)A_-^{-1}(\xi)\widehat{lf} - \widehat{A}_-(\xi)\Pi^-\widehat{A}_-^{-1}\widehat{lf} =$$
$$= \widehat{lf} - \widehat{A}_-(\xi)\Pi^-\widehat{A}_-^{-1}(\xi)\widehat{lf}. \tag{3.3.83}$$

From Theorem 3.3.4 and Proposition 2.2.2, it follows that $\widehat{A}_-(\xi)\Pi^-\widehat{A}_-^{-1}(\xi)\widehat{lf} \in H^-_{s-\alpha}(\mathbb{R}_n)$. This, combined with (3.3.83), yields

$$p\widehat{A}u_+ = p\left(lf - F^{-1}\left(\widehat{A}_-(\xi)\Pi^-\widehat{A}_-^{-1}(\xi)\widehat{lf}\right)\right) = f. \tag{3.3.84}$$

From (3.3.84), it follows that $u_+ = F^{-1}\left(\widehat{A}_+^{-1}\Pi^+\widehat{A}_-^{-1}\widehat{lf}(\xi)\right)$ is a solution of equation (3.3.72), and $u^+ \in H_s^+$. Theorem 3.3.5 is proved.

Theorem 3.3.6. *For* $(\lambda - 1)/2 < s < (\lambda + 1)/2$, *the equation*

$$pAu = f$$

admits one and only one solution $u \in \overset{\circ}{H}_s(G)$ *for any* $f \in H_{s-\lambda}(G)$, *where* A *is the PDO with symbol* $A(\xi) = |D\xi|^\lambda$, $\det D \neq 0$, p *is the operator of restriction to a bounded domain* G *with the boundary of class* C^∞.

Proof. First, let us show that the operator $pAu : \overset{\circ}{H}_s(G) \to H_{s-\lambda}(G)$ is Noetherian for any $(\lambda - 1)/2 < s < (\lambda + 1)/2$. Let $\chi(\xi) \in C_0^\infty(\mathbb{R}_n)$, $\chi(\xi) = 1$ for $|\xi| \leq 1$. We represent the operator A as $A = A_1 + A_2$, where A_1 is the PDO with symbol $A(\xi)(1 - \chi(\xi)) \in S_\lambda^\infty(\mathbb{R}_n)$, and A_2 is the PDO with symbol $A(\xi)\chi(\xi) \in S_{-\infty}^0(\mathbb{R}_n)$.

For each point x on the boundary Γ, consider its small neighborhood $U(x)$ in which it is possible to introduce a special local Cartesian coordinate system (SLCCS) Y $(y_1, \ldots, y_n)$ (see Definition 2.3.8). The original Cartesian coordinates X $(x_1, \ldots, x_n)$ and those of the SLCCS are related by $x = By + a$, where B is an $(n \times n)$ matrix, $\det B \neq 0$. Then, on the basis of Remark 3.2.2 and formula (3.2.75), the PDO A is transformed into a PDO B such that for any $\varphi(y) \in C_0^\infty(U(x))$, we have

$$C\varphi = \frac{1}{(2\pi)^n} \int_{\mathbb{R}_n} \left| D(B^*)^{-1}\eta \right|^\lambda \widehat{\varphi}(\eta) \, e^{-i(y,\eta)} \, d\eta, \tag{3.3.85}$$

which means that the symbol of the operator C is $\left| D(B^*)^{-1}\eta \right|^\lambda$. Accordingly, the PDO's A_1 and A_2 are transformed into the PDO's C_1 and C_2 with symbols

$$C_1(\eta) = \left| D(B^*)^{-1}\eta \right|^\lambda \left(1 - \chi((B^*)^{-1}\eta)\right), \qquad C_2(\eta) = \left| D(B^*)^{-1}\eta \right| \chi((B^*)^{-1}\eta).$$

By Definition 2.3.8 of the SLCCS, the equation of the boundary Γ in the neighborhood $U(x)$ has the form $y_n = \psi(y_1, \ldots, y_{n-1}) \in C^\infty(\overline{U(x)})$. Extending $\psi(y_1, \ldots, y_{n-1})$ to the entire $\mathbb{R}_{n-1}$ as a function in $C^\infty(\mathbb{R}_{n-1})$ and, for the sake of brevity, denoting the extension again by $\psi(y_1, \ldots, y_{n-1})$, let us pass to the new variables

$$\begin{aligned}
z_i &= y_i, \quad 1 \leq i \leq n-1, \\
z_n &= y_n - \psi(y_1, \ldots, y_{n-1})\alpha(\varepsilon y_n),
\end{aligned} \tag{3.3.86}$$

where $\alpha(\theta) \in C_0^\infty(\mathbb{R}_1)$, $\alpha(\theta) = 1$ for $|\theta| \leq 1$, $0 \leq \alpha(\theta) \leq 1$. The parameter $\varepsilon > 0$ is chosen sufficiently small, such that for the coordinate z_n of any point $M \in G \cap U(x)$ we have $z_n \geq 0$, and for $M \in \Gamma \cap U(x)$, we have $z_n = 0$. In addition, suppose that the following condition holds:

$$\varepsilon \sup_\theta \left| \frac{d\alpha}{d\theta} \right| \leq \frac{1}{2} \left(1 + \sup_{y'} |\psi(y_1, \ldots, y_{n-1})| \right)^{-1}, \tag{3.3.87}$$

where $y' = (y_1, \ldots, y_{n-1})$. This condition ensures the inequality $|D(z)/(Dy)| \geq 1/2$ for the Jacobian of the transformation (3.3.86). Since $\psi(y_1, \ldots, y_{n-1})\alpha(\varepsilon y_n) \in C_0^\infty(\mathbb{R}_n)$, there exists $N > 0$ such that for $|y| \geq N$ the transformation (3.3.86) is identical. Let $y = \beta(z)$ be the inverse transformation. Then, for the PDO C_1 with symbol $C_1(\eta)$ and for the transformation $y = \beta(z)$, all assumptions of Theorem 3.2.4 are valid. Therefore, the PDO F obtained from the PDO C_1 after the transformation $y = \beta(z)$ can be represented as $F = F_1 + F_2$, where the symbol of the operator F_1 is

$$F_1(z, \eta) = C_1(\Xi)\left| D(B^*)^{-1}\Xi \right|^\lambda \left[1 - \chi((B^*)^{-1}\Xi)\right], \qquad \Xi = \left[\left(\frac{D\beta(z)}{Dz}\right)^*\right]^{-1}\eta \tag{3.3.88}$$

(see (3.2.62)). The PDO F_2 is bounded from $H_s(\mathbb{R}_n)$ to $H_{s-\lambda+1}(\mathbb{R}_n)$.

Take a sufficiently small neighborhood $U'(x) \subset U(x)$ of the point x, so as to ensure the result of Theorem 3.2.2 for the PDO F_1. Suppose that the point x has the coordinates z_0 $(z_1^0, \ldots, z_n^0 = 0)$. Then, for any $f_1(z) \in C_0^\infty(U'(x))$, the operator $f_1(z)C_1(z, D)$ can be represented as

$$f_1(z)C_1(z, D) = f_1(z)\big[C_1(z_0, D) + K(z, D) + T(z, D)\big]. \tag{3.3.89}$$

The symbol of the operator $C_1(z_0, D)$ is $F_1(z_0, \eta)$, and for the operator $K(z, D)$ the estimate (3.2.35) holds. The operator $T(z, D)$ is bounded from $H_s(\mathbb{R}_n)$ to $H_{s-\lambda+1}(\mathbb{R}_n)$. The operator $C_1(z_0, D)$ can be written in the form $C_1(z_0, D) = C_{11}(z_0, D) + C_{12}(z_0, D)$. Here, the operator $C_{11}(z_0, D)$ has the symbol $F_{11}(z_0, \eta) = |D_1\eta|^\lambda$, where D_1 is an $(n \times n)$-matrix, $\det D_1 \neq 0$. The operator $C_{12}(z_0, D)$ is bounded from $H_s(\mathbb{R}_n)$ to $H_{s-\lambda+1}(\mathbb{R}_n)$. Consider the PDO $\widehat{C}_{11}(z_0, D)$ with symbol $\widehat{F}_{11}(z_0, \eta) = F_{11}\big((z_0, 1 + |\eta'|)\omega, \eta_n\big)$, where $\eta' = (\eta_1, \ldots, \eta_{n-1})$, $\omega = |\eta'|^{-1}\eta'$. Then, on the basis of Proposition 3.3.12, the PDO $C_{11}(z_0, D)$ can be represented as $C_{11}(z_0, D) = \widehat{C}_{11}(z_0, D) + \widehat{C}_{12}(z_0, D)$, where the PDO $\widehat{C}_{12}(z_0, D)$ is bounded from $H_s(\mathbb{R}_n)$ to $H_{s-\lambda+1}(\mathbb{R}_n)$.

According to Theorem 3.3.5, the operator $p_1\widehat{C}_{11}(z_0, D)$ admits an inverse R_0 acting from $H_{s-\lambda}(\mathbb{R}_n^+)$ to $H_s^+(\mathbb{R}_n)$ for $(\lambda - 1)/2 < s < (\lambda + 1)/2$. Consider a neighborhood $U''(x) \subset U(x)$ of the point x such that

$$\|K(z, D)\| < \frac{1}{2\|R_0\|_s}. \tag{3.3.90}$$

Condition (3.3.90) ensures that the operator $p_1 C_{13}(z_0, D) = p_1\widehat{C}_{11}(z_0, D) + p_1 K(z_0, D)$ admits an inverse operator (see *Trenogin* (1980)) R acting from $H_{s-\lambda}(\mathbb{R}_n^+)$ to $H_s^+(\mathbb{R}_n)$ for $(\lambda - 1)/2 < s < (\lambda + 1)/2$, where p_1 is the operator of restriction to $\mathbb{R}_n^+$. The above considerations show that for any $f_1(z), \varphi(x) \in C_0^\infty(U''(x))$ and $u(x) \in \mathring{H}_s(G)$, the operator value $f_1 p A_1 \varphi u$ can be represented as

$$f_1 p A_1 \varphi u = f_1(z)[p_1 C_{13}(z, D) + T]\varphi u_1 = f_1(z)A_0 \varphi u, \tag{3.3.91}$$

where the operator $T : H_s^+(\mathbb{R}_n) \to H_{s-\lambda+1}(\mathbb{R}_n^+)$ is bounded and the function φu is written in the variables z.

Among the sets $U(x)$ covering Γ, we can choose finitely many still covering Γ and supplement these by an open set U, $\overline{U} \subset G$, so that we obtain a finite covering of $\overline{G}$, denoted by U_i, $1 \leq i \leq k$. Let $\{\varphi_i(z)\}$ be a partition of unity subject to the covering U_i and let $\psi_i \in C_0^\infty(U_i)$ be functions such that $\psi_i\varphi_i = \varphi_i$. It follows from (3.3.91) that for each U_i having common points with Γ, in the i-th local coordinate system $z^{(i)}$, there is an operator $R_i : H_{s-\lambda} \to H_s^+(\mathbb{R}_n)$ which is bounded and

$$R_i A_{0i} = I_1 + T_{i_1}, \qquad A_{0i} R_i = I_2 + T_{i_2}, \tag{3.3.92}$$

where I_1 and I_2 are the identity operators in $H_s^+(\mathbb{R}_n)$ and $H_{s-\lambda}(\mathbb{R}_n^+)$, respectively; $T_{i_1} : H_{s-1}^+(\mathbb{R}_n) \to H_s^+(\mathbb{R}_n)$ and $T_{i_2} : H_{s-\lambda}(\mathbb{R}_n^+) \to H_{s-\lambda+1}(\mathbb{R}_n^+)$ are bounded in their respective spaces; A_{0i} is the operator A_1 in terms of the i-th local coordinate system.

For the domains U_i strictly inside G, we take as the operator R_i the PDO with symbol $A^{-1}(\xi)(1 - \chi(\xi))$, where $\chi(\xi) \in C_0^\infty(\mathbb{R}_n)$, $\chi(\xi) = 1$ for $|\xi| \leq 1$. Then,

$$R_i A_1 = \widehat{I}_1 + \widehat{T}_{i_1}, \qquad A_1 R_i = \widehat{I}_2 + \widehat{T}_{i_2}, \tag{3.3.93}$$

where $\widehat{I}_1$ and $\widehat{I}_2$ are the identity operators in $H_s(\mathbb{R}_n)$ and $H_{s-\lambda}(\mathbb{R}_n)$, respectively; $\widehat{T}_{i_1} : H_{s-1}(\mathbb{R}_n) \to H_s(\mathbb{R}_n)$ and $\widehat{T}_{i_2} : H_{s-\lambda}(\mathbb{R}_n) \to H_{s-\lambda+1}(\mathbb{R}_n)$ are bounded operators in their respective spaces.

We define the operator R on $u(x) \in H_{s-\lambda}(G)$ by

$$Ru = \sum_{i=1}^k \psi_i R_i \varphi_i u. \tag{3.3.94}$$

Let us show that R is bounded as an operator from $H_{s-\lambda}(G)$ to $\mathring{H}_s(G)$. Consider the term $g_i = \psi_i R_i \varphi_i u$. For U_i strictly inside G and some extension $lu \in H_{s-\lambda}(\mathbb{R}_n)$ of $u \in H_{s-\lambda}(G)$, the generalized function $\varphi_i lu$ is supported in G, and therefore, does not depend on the extension operator. In this case, $R_i \varphi_i lu \in H_s(\mathbb{R}_n)$, $\psi_i R_i \varphi_i lu \in \mathring{H}_s(G)$. If U_i crosses the boundary Γ, let us write the function $\varphi_i lu$ in the local coordinate system in terms of the variables $z^{(i)}$. Then the restriction of the generalized function $\varphi_i(z)lu$ to the region $z_n > 0$ does not depend on the extension lu, and therefore, the value $R_i \varphi_i lu \in H_s^+(\mathbb{R}_n)$ is defined uniquely (see Theorem 3.3.5). It follows that $\psi R_i \varphi_i lu \in \mathring{H}_s(G)$. Moreover, Proposition 3.2.6, Remark 3.2.2, and relation (3.2.72) show that R is a bounded operator from $H_{s-\lambda}(G)$ to $\mathring{H}_s(G)$.

Let us show that R is the left regularizer of the operator A. Indeed, for any $u(x) \in \mathring{H}_s(G)$, we have

$$RpAu = RpA_1 u + RpA_2 u = \sum_{i=1}^{k} \psi_i R_i \varphi_i p A_1 \psi_i u + \sum_{i=1}^{k} \psi_i R_i \varphi_i p A_1 (1 - \psi_i) u + RpA_2 u =$$
$$= K_1 u + K_2 u + K_3 u, \qquad (3.3.95)$$

where we have denoted by 1 a function of class $C_0^\infty(\mathbb{R}_n)$ equal to the unity on $\bigcup\limits_{i=1}^{k} \overline{U}_i$.

For the operator A_2 with symbol $A(\xi)\chi(\xi)$, we have $\operatorname{ord} A_2 = -\infty$. Therefore, the operator $K_3 RpA_2$ can be represented as $K_3 = R\mathrm{id}_{s-\lambda}^{s_1} pA_2$, where $\mathrm{id}_{s-\lambda}^{s_1}$ is the operator of the imbedding $H_{s_1}(G) \subset H_{s-\lambda}(G)$ for $s_1 > s - \lambda$. By Proposition 3.3.8, $\mathrm{id}_{s-\lambda}^{s_1}$ is a compact operator for $s_1 > s - \lambda$, and therefore, the operator K_3 is compact, too.

The functions $1 - \psi_i$ and φ_i are supported on disjoint sets, and therefore, by the pseudolocality property (see Corollary 3.2.1), we have $\varphi p A_1 (1 - \psi_i) u \in L_2(G)$. Then, the operator K_2 can be represented as

$$K_2 = \sum_{i=1}^{k} \psi_i R_i \mathrm{id}_{s-\lambda}^{s_1} \varphi_i p A_1 (1 - \psi_i) u. \qquad (3.3.96)$$

On the basis of the above considerations, combined with (3.3.96), we conclude that K_2 is a compact operator.

Taking the commutator of φ_i and A_{0i} (see (3.3.91)) and recalling Example 3.1.9, we obtain

$$K_1 u = \sum_{i=1}^{k} \psi_i R_i [A_{0i} \varphi_i + T_i] \psi_i u = \sum_{i=1}^{k} \psi_i [I_1 + T_{i_1}] \varphi_i \psi_i u + \sum_{i=1}^{k} \psi_i R_i T_i \psi_i u =$$
$$= u + \sum_{i=1}^{k} \psi_i T_{i_1} \varphi_i u + \sum_{i=1}^{k} \psi_i R_i T_i \psi_i u = u + K_1' + K_2' u, \qquad (3.3.97)$$

where T_i is a bounded operator from $\mathring{H}_s(G)$ to $H_{s-\lambda+1}(G)$; the operator T_{i_1} is defined in (3.3.92). On the basis of the properties of the operators T_{i_1} and T_i, taking $u \in \mathring{H}_{s-1}(G)$, we conclude that the operators K_2' and K_3' are bounded from $\mathring{H}_{s-1}(G)$ to $\mathring{H}_s(G)$. Now, taking into account Corollary 3.3.1, we see that K_2' and K_3' are compact operators in $\mathring{H}_s(G)$. Thereby, we have proved that R is the left regularizer of A.

Let us show that R is the right regularizer of the operator A. Consider the functions $\psi_{i_1} \in C_0^\infty(U_i)$ such that $\psi_{i_1} \psi_i = \psi_i$. Then, for any $u(x) \in H_{s-\lambda}(G)$, we have

$$pARu = pA_1 Ru + pA_2 Ru = \sum_{i=1}^{k} \psi_{i_1} p A_1 \psi_i R_i \varphi_i u + \sum_{i=1}^{k} (1 - \psi_{i_1}) p A_1 \psi_i R_i \varphi_i u + pA_2 Ru =$$
$$= K_{11} u + K_{12} u + K_{13} u, \qquad (3.3.98)$$

where p is the operator of restriction to the domain G, and 1 stands for the same function as in (3.3.95).

The operator K_{13} can be represented as $K_{13} = pA_2\, \mathrm{id}_{s_1}^s\, Ru$, where $\mathrm{id}_{s_1}^s$, $s > s_1$, is the imbedding operator: $\overset{\circ}{H}_s(G) \subset \overset{\circ}{H}_{s_1}(G)$. By Proposition 3.3.7, $\mathrm{id}_{s_1}^s$ is a compact operator. Hence we see that K_{13} is compact in $H_{s-\lambda}(G)$. It should be kept in mind that p is the operator of restriction to the domain G.

On the basis of the pseudolocality property, we can at least claim that $(1 - \psi_{i_1})pA_1\psi_i R_i\varphi_i \in L_2(G)$. Therefore, K_{12} can be represented in the form

$$K_{12} = \sum_{k=1}^{k} \mathrm{id}_{s-\lambda}^0 (1 - \psi_{i_1})pA_1\psi_i R_i\varphi_i, \tag{3.3.99}$$

where $\mathrm{id}_{s-\lambda}^0$ is the operator of the imbedding $L_2(G) \subset H_{s-\lambda}(G)$. Therefore, K_{12} is a compact operator in $H_{s-\lambda}(G)$.

Taking the commutator of ψ_i and A_{0i} (see (3.3.91)) and recalling Example 3.1.9, we obtain

$$K_{11}u = \sum_{i=1}^{k} \left(\psi_i A_{0i} + \psi_{i_1} T^{i_1} \right) R_i\varphi_i u =$$

$$= \sum_{i=1}^{k} \varphi_i u + \sum_{i=1}^{k} \psi_i T_{i_2}\varphi_i u + \sum_{i=1}^{k} \psi_{i_1} T^{i_1} R_i\varphi_i u = u + K_{11}' + K_{12}'u, \tag{3.3.100}$$

where T_{i_2} is the operator defined in (3.3.92), and the operator T^{i_1} is bounded from $\overset{\circ}{H}_s(G)$ to $H_{s-\lambda+1}(G)$. Then the operators K_{11}' and K_{12}' can be represented in the form

$$K_{11}'u = \sum_{i=1}^{k} \mathrm{id}_{s-\lambda}^{s-\lambda+1} \psi_i T_{i_2}\varphi_i u, \qquad K_{12}'u = \sum_{i=1}^{k} \mathrm{id}_{s-\lambda}^{s-\lambda+1} \psi_{i_1} T^{i_1} R_i\varphi_i u, \tag{3.3.101}$$

where $\mathrm{id}_{s-\lambda}^{s-\lambda+1}$ is the operator of the imbedding of $H_{s-\lambda+1}(G)$ into $H_{s-\lambda}$. According to Proposition 3.3.8, the operator $\mathrm{id}_{s-\lambda}^{s-\lambda+1}$ is compact. It follows that K_{11}' and K_{12}' are compact operators from $H_{s-\lambda}$ to $H_{s-\lambda}$. Therefore, R is the right regularizer of A. Theorem 3.3.2 implies that the operator A is Noetherian.

Let us show that $\dim \ker A = 0$, $\dim \operatorname{coker} A = 0$ for $(\lambda - 1)/2 < s < (\lambda + 1)/2$. It follows from Proposition 3.3.11 that $\dim \ker A = 0$, $\dim \operatorname{coker} A = 0$ for $s = \lambda/2$. Since $H_s(\mathbb{R}_n) \subset H_{s_1}(\mathbb{R}_n)$ for $s > s_1$, we have $\dim \ker A = 0$ for $\lambda/2 \leq s < (\lambda + 1)/2$. Let $(\lambda - 1)/2 < s < \lambda/2$. For the left regularizer R constructed above, we have

$$RAu = u + Tu. \tag{3.3.102}$$

The operator R_0 (see (3.3.90)) maps $g(x) \in H_{s-\lambda+1}(G)$ to $R_0g \in \overset{\circ}{H}_{\frac{\lambda+1}{2}-\varepsilon}(G)$ for any $\varepsilon > 0$ (see Theorem 3.3.5).

Taking a sufficiently small neighborhood $U''(x)$, we can ensure that the inequality holds uniformly for $a \leq s \leq b$ (see Theorems 3.2.1 and 3.2.2). Hence we see that the function Tu in (3.3.102) belongs to $\overset{\circ}{H}_{\frac{\lambda+1}{2}-\varepsilon}(G)$ for any $\varepsilon > 0$.

Now, let $u_0 \in \ker A$. From (3.3.102), we have $u_0 = (-Tu_0) \in \overset{\circ}{H}_{\frac{\lambda+1}{2}-\varepsilon}(G)$. Consequently, $u_0 = 0$. Thus, $\dim \ker A = 0$ for any s such that $(\lambda - 1)/2 < s < (\lambda + 1)/2$.

Since the symbol $A(\xi)$ of the operator A is real-valued, the symbol of the conjugate operator coincides with $A(\xi)$. It follows that $\dim \operatorname{coker} A = 0$ for $(\lambda - 1)/2 < s < (\lambda + 1)/2$. Theorem 3.3.6 is proved.

Chapter 4

Neumann Problem and Integral Equations with Double Layer Potential

Introduction

In order to find the characteristics of a flow past a body without separation, it is convenient to model the vortex layer next to the body by closed quadrangles and vortex frames. The intensity of these vortex formations coincides with the density of the double layer potential on the surface of the body for which the values of the potential outside the body are the same as in the case of a perturbed flow (see *Sedov* (1973)). Thus we come to the problem of finding the potential outside a body in terms of its normal derivative and the density of the double layer potential. This chapter is dedicated to the theory of the Neumann problem and the corresponding integral equations with the double layer potential.

4.1. Reduction of the Neumann Problem to a Hypersingular Equation

In the Euclidean space $\mathbb{R}_3$, consider closed connected surfaces $S_1^0, \ldots, S_k^0$ of class C^∞ (i.e., compact connected C^∞ manifolds of dimension 2; see Definition 2.4.6).

A manifold $M \subset \mathbb{R}_n$ is called *connected* if any two of its points can be connected by a continuous curve l lying in M.

Suppose also that in $\mathbb{R}_3$ there are non-closed connected surfaces $S_1^1, \ldots, S_n^1$ of class C^∞ (i.e., compact two-dimensional C^∞ manifolds with border; see Definition 2.4.7) such that $\overline{S_i^1} \cap \overline{S_j^1} = \emptyset$ for $i \neq j$.

According to Definition 2.4.7, we assume that there exist two-dimensional compact manifolds (without border) $M_1, \ldots, M_{m_1}$ of class C^∞ such that every surface S_j^1, $j = 1, \ldots, n$, belongs to one of the manifolds $M_1, \ldots, M_{m_1}$, and several surfaces S_i^1 may belong to one and the same manifold M_j. It is also assumed that the manifolds $M_1, \ldots, M_{m_1}$ are connected.

The surfaces S_i^0 coincide with the boundaries of bounded domains Ω_i^0, and the surfaces M_i are the boundaries of bounded domains Ω_i^1. Moreover, $\overline{\Omega_i^0} \cap \overline{\Omega_j^0} = \emptyset$, $\overline{\Omega_i^1} \cap \overline{\Omega_j^1} = \emptyset$ for $i \neq j$, and $\overline{\Omega_i^0} \cap \overline{\Omega_j^1} = \emptyset$ for all i, j.

Consider the following harmonic functions

$$u_i^0(x) = \frac{1}{4\pi} \int_{S_i^0} \nu_i^0(y) \frac{\cos \varphi_{xy}}{|x-y|^2} \, dS_{iy}^0,$$

$$u_i^1(x) = \frac{1}{4\pi} \int_{S_i^1} \nu_i^1(y) \frac{\cos \varphi_{xy}}{|x-y|^2} \, dS_{iy}^1,$$

(4.1.1)

where $x = (x_1, x_2, x_3) \in \mathbb{R}_3$, $y = (y_1, y_2, y_3) \in S_i^0$ or S_i^1; φ_{xy} is the angle between the vector $x - y$ and the normal $\boldsymbol{n}_y$ to the surface S_i^0 or S_i^1 at the point y; $\nu_i^p(y) \in C^\infty(S_i^p)$ is the density of the double layer potential, $p = 0, 1$.

Consider the boundary value problem

$$
\begin{aligned}
\Delta u(x) &= 0, \qquad x \in \mathbb{R}_3 \setminus (\cup S_i^p), \\
\left. \frac{\partial u}{\partial n} \right|_{S_i^p} &= f_i^p(x), \\
p = 0 &: i = 1, \dots k; \qquad p = 1 : i = 1, \dots, n,
\end{aligned}
\tag{4.1.2}
$$

where Δ is the Laplace operator; $(\partial u / \partial n)|_{S_i^p}$ is the derivative along the normal to the surface S_i^p.

We seek the solution of problem (4.1.2) as the double layer potential

$$
u(x) = \frac{1}{4\pi} \sum_p \sum_i \int_{S_i^p} \nu_i^p(y) \frac{\cos \varphi_{xy}}{|x - y|^2} \, dS_{iy}^p,
$$

and for unknown functions $\nu_i^p(y)$ we obtain the following system of equations:

$$
\frac{1}{4\pi} \sum_p \sum_i \frac{\partial}{\partial n_l^m} \int_{S_i^p} \nu_i^p(y) \frac{\cos \varphi_{xy}}{|x - y|^2} \, dS_{iy}^p = f_l^m(x),
\tag{4.1.3}
$$

$$
m = 0 : \ l = 1, \dots, k; \qquad m = 1 : \ l = 1, \dots, n.
$$

In what follows, we will specify more precisely the classes of functions u, f_i^p, and ν_i^p for which problems (4.1.2) and (4.1.3) can be solved.

Obviously, for $x \in \mathbb{R}_3 \setminus \left(\cup_p \cup_i S_i^p \right)$, the left-hand side of (4.1.3), which we denote by $\Pi_l^m(\nu)$, exists. Let us describe the class of functions $\nu_i^p(y)$ for which there is a finite limit

$$
\lim_{x \to S_l^m} \Pi_l^m(\nu).
$$

Clearly, it suffices to consider the case of a single surface S_l^m and a single function $\nu_l^m(y)$, which we denote by σ and $\nu(y)$, respectively. Then $\Pi_l^m(\nu)$ takes the form

$$
\Pi(M, \nu) = \frac{1}{4\pi} \int_\sigma \nu(M_0) \frac{|\overline{M_0 M}|^2 (\boldsymbol{n}_{M_0}, \boldsymbol{n}_M) - 3(\overline{M_0 M}, \boldsymbol{n}_M)(M_0 M, \boldsymbol{n}_M)}{|\overline{M_0 M}|^5} \, d\sigma_{M_0},
\tag{4.1.4}
$$

where $M_0 \in \sigma$, $M \notin \sigma$ belongs to the normal passing through the point $\overline{M} \in \sigma$.

Definition 4.1.1. A *special coordinate system* (SCS) is a coordinate system with the plane $OX_0 Y_0$ parallel to the tangential plane at a fixed point $\overline{M} \in \sigma$.

Consider a neighborhood U of the point $\overline{M}$ such that the equation of σ in the SCS has the form $z_0 = f(x_0, y_0)$. Let $\varphi(P)$, $P \in \sigma$, be an infinitely differentiable function compactly supported in U, $\varphi(P) = 1$, $P \in U_1$, $U_1 \subset \operatorname{supp}\varphi$, $\overline{M} \in U_1$, where U_1 is an open set. Then, $\Pi(\nu) = \Pi(\varphi\nu) + \Pi[(1 - \varphi)\nu]$. In the second term, we can pass to the limit as $M \to \overline{M}$. Therefore, it suffices to calculate the limit of the first term as $M \to \overline{M}$ along the normal. Let us assume that the surface σ is locally convex, i.e., for the tangential plane π to σ at the point $\overline{M} \in \sigma$, there is a neighborhood Q of $\overline{M}$ in $\mathbb{R}_3$ such that $\sigma \cap Q$ lies to one side of the plane π. Assume first that $\nu(M_0) \in C^\infty(\sigma)$. Let $(x, y, z = f(x, y))$ and $(x_0, y_0, z_0 = f(x_0, y_0))$ be the coordinates of the points $\overline{M} \in \sigma$ and $M_0 \in \sigma$, respectively. Then

$$
\boldsymbol{n}_{M_0} = \frac{-f'_{x_0}(x_0, y_0)\, \boldsymbol{i} - f'_{y_0}(x_0, y_0)\, \boldsymbol{j} + \boldsymbol{k}}{\left[1 + f'^2_{x_0}(x_0, y_0) + f'^2_{y_0}(x_0, y_0)\right]^{1/2}}, \qquad \boldsymbol{n}_{\overline{M}} = \boldsymbol{k}.
$$

Let $M(x, y, f(x, y) - t)$ be the current point of the normal to σ passing through $\overline{M}(x, y, f(x, y))$. Then

$$\Pi(M, \varphi\nu) = \frac{1}{4\pi} \iint_{U_{x_0 y_0}} \frac{\varphi(x_0, y_0)\nu(x_0, y_0)}{A^{5/2}} \left\{ \left[(x - x_0)^2 + (y - y_0)^2 + (f(x, y) - f(x_0, y_0) + t)^2 \right] - \right.$$

$$- 3 \left[-f'(x_0, y_0)(x - x_0) - f'_{y_0}(x_0, y_0)(y - y_0) + f(x, y) - f(x_0, y_0) + t \right] \times$$

$$\left. \times [f(x, y) - f(x_0, y_0) + t] \right\} dx_0 \, dy_0 \tag{4.1.5}$$

where $U_{x_0 y_0}$ is the projection of supp φ to the plane $OX_0 Y_0$, $A = |M_0 M|^2$.

Let us choose a system of polar coordinates on the plane $OX_0 Y_0$ with the pole at $O^*(x, y, 0)$ and the polar axis directed parallel to OX_0. Integrating by parts, we get

$$-\frac{r^2}{A^{3/2}} = \int \frac{r \, dr}{A^{3/2}} - 3 \int \frac{r}{A^{5/2}} \left[f(x, y) - f(x + r\cos\varphi_1, \ y + r\sin\varphi_1) + t \right]^2 dr +$$

$$+ 3 \int \frac{r^2}{A^{5/2}} \left[f(x, y) - f(x + r\cos\varphi_1, \ y + r\sin\varphi_1) + t \right] \left[-f'_{x_0}\cos\varphi_1 - f'_{y_0}\sin\varphi_1 \right] dr =$$

$$= K_1 + K_2 + K_3 \tag{4.1.6}$$

In the following integral we can pass to the limit as $t \to 0$ (see *Vladimirov* (1976)),

$$\lim_{t \to 0} \frac{1}{4\pi} \iint_{U_{x_0 y_0}} (-3) \left[f(x, y) - f(x_0, y_0) - f'_{x_0}(x - x_0) + f'_{y_0}(y - y_0) \right] \times$$

$$\times [f(x, y) - f(x_0, y_0)] \varphi\nu \frac{dx_0 \, dy_0}{|M_0 M|^5} = \frac{1}{4\pi} \iint_{U_{x_0 y_0}} B(x, y, x_0, y_0) \frac{dx_0 \, dy_0}{|M_0 \overline{M}|^5}. \tag{4.1.7}$$

Consider the expression corresponding to the terms K_2 and K_3 in (4.1.6).

$$I(t) = -\frac{3}{4\pi} \iint_{U_{x_0 y_0}} \frac{\varphi\nu}{A^{5/2}} \left\{ -r^2 \left[f(x, y) - f(x + r\cos\varphi_1, \ y + r\sin\varphi_1) \right] \left[f'_{x_0}\cos\varphi_1 + f'_{y_0}\sin\varphi_1 \right] - \right.$$

$$\left. - r \left[f(x, y) - f(x + r\cos\varphi_1, \ y + r\sin\varphi_1) \right]^2 \right\} dr \, d\varphi_1 =$$

$$= -\frac{3}{4\pi} \iint_{U_{x_0 y_0}} \frac{\varphi\nu}{A^{5/2}} D(x, y, r, \varphi_1) \, dr d\varphi_1. \tag{4.1.8}$$

Direct calculation shows that

$$\lim_{t \to 0} \int_0^R \frac{t^n \, dr}{(t^2 + r^2)^{n/2}} = 0. \tag{4.1.9}$$

Since $f'_{x_0}(x, y) = 0$, $f'_{y_0}(x, y) = 0$, we have $|f'_{x_0}(x_0, y_0)| \leq Cr$, $|f'_{y_0}(x_0, y_0)| \leq Cr$. The above calculations lead us to

$$\lim_{t \to 0} I(t) = -\frac{3}{4\pi} \iint_{U_{x_0 y_0}} \frac{\varphi\nu D(x, y, r, \varphi_1) \, dr \, d\varphi_1}{\left[r^2 + f(x, y) - (f(x + r\cos\varphi_1, \ y + r\sin\varphi_1))^2 \right]^{5/2}}. \tag{4.1.10}$$

For any bounded continuous function $\Phi(x, y, r, \varphi_1)$, by (4.1.9) and elementary calculations, we find that

$$\lim_{t \to 0} \iint_{U_{x_0 y_0}} \Phi(x, y, r, \varphi_1) r^2 \left[\frac{1}{A^{3/2}} - \frac{1}{(r^2 + t^2)^{3/2}} \right] dr \, d\varphi_1 = \tag{4.1.11}$$

$$= \iint_{U_{x_0 y_0}} \Phi(x, y, r, \varphi_1) r^2 \left[\frac{1}{\left[r^2 + f(x, y) - \left(f(x + r\cos\varphi_1, \ y + r\sin\varphi_1) \right)^2 \right]^{3/2}} - \frac{1}{r^3} \right] dr \, d\varphi_1.$$

Consider the remaining terms in $\Pi(M, \varphi\nu)$ in (4.1.6). Let R be the radius of a circle with center at $O^*(x, y, 0)$ containing the projection of $\operatorname{supp}\varphi$ to the plane $X_0 O Y_0$. Then, integration by parts yields

$$\frac{1}{4\pi} \int_0^{2\pi} d\varphi_1 \int_0^R \frac{\varphi\nu}{A^{5/2}} \left\{ rA - 3r \Big[f(x, y) - f(x + r\cos\varphi_1, \ y + r\sin\varphi_1) + t \Big]^2 + \right.$$

$$\left. + 3r^2 \Big[f(x, y) - f(x + r\cos\varphi_1, \ y + r\sin\varphi_1) + t \Big] \Big[-f'_{x_0}\cos\varphi_1 - f'_{y_0}\sin\varphi_1 \Big] dr = \right.$$

$$= \frac{1}{4\pi} \int_0^{2\pi} d\varphi_1 \int_0^R \Big[(\varphi\nu)'_{x_0} \cos\varphi_1 + (\varphi\nu)'_{y_0} \sin\varphi_1 \Big] \frac{r^2}{A^{3/2}} dr =$$

$$= \frac{1}{4\pi} \int d\varphi_1 \int_0^R \widetilde{\Phi}(x, y, r, \varphi_1) r^2 \left[\frac{1}{A^{3/2}} - \frac{1}{(r^2 + t^2)^{3/2}} \right] dr +$$

$$+ \frac{1}{4\pi} \int_0^{2\pi} d\varphi_1 \int_0^R \frac{\widetilde{\Phi}(x, y, r, \varphi_1) r^2}{(r^2 + t^2)^{3/2}} dr = I_1(t) + I_2(t) = A(t), \qquad (4.1.12)$$

where

$$\widetilde{\Phi}(x, y, r, \varphi_1) = (\varphi\nu)'_{x_0} \cos\varphi_1 + (\varphi\nu)'_{y_0} \sin\varphi_1.$$

For the term $I_1(t)$ we have established relation (4.1.11). We introduce the notation

$$f_1(x + r\cos\varphi_1, \ y + r\sin\varphi_1) = (\varphi\nu)'_{x_0},$$

$$f_2(x + r\cos\varphi_1, \ y + r\sin\varphi_1) = (\varphi\nu)'_{y_0}.$$

Using the Lebesgue theorem, we find that

$$\lim_{t\to 0} I_2(t) = \frac{1}{4\pi} \lim_{t\to 0} \left\{ f_1(x, y) \int_0^{2\pi}\!\!\int_0^R \frac{r^2 \cos\varphi_1}{(r^2+t^2)^{3/2}} \, dr \, d\varphi_1 + f_2(x, y) \int_0^{2\pi}\!\!\int_0^R \frac{r^2 \sin\varphi_1}{(r^2+t^2)^{3/2}} \, dr \, d\varphi_1 \right\} +$$

$$+ \frac{1}{4\pi} \lim_{t\to 0} \int_0^{2\pi}\!\!\int_0^R \frac{r^2}{(r^2+t^2)^{3/2}} \left\{ \Big[f_1(x + r\cos\varphi_1, \ y + r\sin\varphi_1) - f_1(x, y) \Big] \cos\varphi_1 + \right.$$

$$\left. + \Big[f_2(x + r\cos\varphi_1, \ y + r\sin\varphi_1) - f_2(x, y) \Big] \sin\varphi_1 \right\} dr \, d\varphi_1 =$$

$$= \frac{1}{4\pi} \int_0^{2\pi}\!\!\int_0^R \Big[f_1(x + r\cos\varphi_1, \ y + r\sin\varphi_1) - f_1(x, y) \Big] \frac{\cos\varphi_1}{r} \, dr \, d\varphi_1 +$$

$$+ \frac{1}{4\pi} \int_0^{2\pi}\!\!\int_0^R \Big[f_2(x + r\cos\varphi_1, \ y + r\sin\varphi_1) - f_2(x, y) \Big] \frac{\sin\varphi_1}{r} \, dr \, d\varphi_1. \qquad (4.1.13)$$

$$\lim_{t\to 0} A(t) = \frac{1}{4\pi} \int_0^{2\pi}\!\!\int_0^R \frac{\big[(\varphi\nu)'_{x_0}\cos\varphi_1 + (\varphi\nu)'_{y_0}\sin\varphi_1 \big] r^2 \, dr \, d\varphi_1}{\Big[r^2 + \big(f(x,y) - f(x + r\cos\varphi_1, \ y + r\sin\varphi_1) \big)^2 \Big]^{3/2}}, \qquad (4.1.14)$$

where the integral is understood in the singular sense. Using (4.1.14), we get

$$A(0) = \lim_{\varepsilon\to 0} \frac{1}{4\pi} \int_0^{2\pi}\!\!\int_\varepsilon^R \frac{(\varphi\nu)'_r r^2 \, dr \, d\varphi_1}{|M_0\overline{M}|^3} =$$

$$= \frac{1}{4\pi} \lim_{\varepsilon\to 0} \left[-\int_0^{2\pi} \frac{\varepsilon^2 b(\varepsilon, \varphi_1) \, d\varphi_1}{\Big[\varepsilon^2 + \big(f(x,y) - f(x + \varepsilon\cos\varphi_1, \ y + \varepsilon\sin\varphi_1) \big)^2 \Big]^{3/2}} - \right.$$

$$- \int_0^{2\pi}\!\!\int_0^R b(r, \varphi_1) \left\{ 2r|M_0\overline{M}|^3 - \right. \qquad (4.1.15)$$

$$\left. - 3r^2 \Big\{ r + \big[f(x,y) - f(x + r\cos\varphi_1, \ y + r\sin\varphi_1) \big] \big[-f'_{x_0}\cos\varphi_1 - f'_{y_0}\sin\varphi_1 \big] \Big\} \right\} \frac{dr \, d\varphi_1}{|M_0\overline{M}|^5} \right],$$

where $b(r, \varphi_1) = \varphi(x + r\cos\varphi_1, \; y + r\sin\varphi_1)\nu(x + r\cos\varphi_1, \; y + r\sin\varphi_1)$.

From (4.1.8) and (4.1.15), it follows that

$$A(0) + I(0) = \lim_{\varepsilon \to 0} \left[\frac{1}{4\pi} \int_0^{2\pi} \int_\varepsilon^R \frac{b(r, \varphi_1)r\, dr\, d\varphi_1}{\left[r^2 + \left(f(x,y) - f(x + r\cos\varphi_1, \; y + r\sin\varphi_1) \right)^2 \right]^{3/2}} - \right.$$

$$\left. - \frac{1}{4\pi} \int_0^{2\pi} \frac{b(\varepsilon, \varphi)\varepsilon^2\, d\varphi_1}{\left[\varepsilon^2 + \left(f(x,y) - f(x + \varepsilon\cos\varphi_1, \; y + \varepsilon\sin\varphi_1) \right)^2 \right]^{3/2}} \right]. \qquad (4.1.16)$$

In the chosen SCS (see Definition 4.1.1), we have

$$\left[f(x,y) - f(x + \varepsilon\cos\varphi_1, \; y + \varepsilon\sin\varphi_1) \right]^2 \le \alpha\varepsilon^4.$$

Therefore,

$$\frac{\varepsilon^2}{\left[\varepsilon^2 + \left(f(x,y) - f(x + \varepsilon\cos\varphi_1, \; y + \varepsilon\sin\varphi_1) \right)^2 \right]^{3/2}} - \frac{1}{\varepsilon} \to 0 \qquad \text{as} \qquad \varepsilon \to 0.$$

Therefore, instead of the second term in (4.1.16) we can consider

$$\frac{1}{4\pi\varepsilon} \int_0^{2\pi} b(x + \varepsilon\cos\varphi_1, \; y + \varepsilon\sin\varphi_1)\, d\varphi_1 = \frac{\varphi(x,y)\nu(x,y)}{2\varepsilon} + \frac{O(\varepsilon)}{2\varepsilon}, \qquad (4.1.17)$$

where $\lim_{\varepsilon \to 0} O(\varepsilon)/\varepsilon = 0$. From (4.1.7), (4.1.16), 4.1.17), it follows that

$$\lim_{\varepsilon \to 0} \Pi(M, \varphi\nu) = \Pi(\overline{M}, \varphi\nu) =$$

$$= \lim_{\varepsilon \to 0} \left\{ \frac{1}{4\pi} \int_0^{2\pi} \int_\varepsilon^R \frac{\varphi(x + r\cos\varphi_1, \; y + r\sin\varphi_1)\nu(x + r\cos\varphi_1, \; y + r\sin\varphi_1)}{\left[r^2 + \left(f(x,y) - f(x + r\cos\varphi_1, \; y + r\sin\varphi_1) \right)^2 \right]^{3/2}} r\, dr\, d\varphi_1 - \right.$$

$$\left. - \frac{\varphi(x,y)\nu(x,y)}{2\varepsilon} \right\} - \frac{3}{4\pi} \iint_U \varphi\nu \frac{(M_0\overline{M}, n_{M_0})(M_0\overline{M}, n_M)}{|M_0\overline{M}|^5}\, d\sigma. \qquad (4.1.18)$$

Formula (4.1.18) can be taken as a definition of the integral in the sense of Hadamard (see *Hadamard* (1932)).

Definition 4.1.2. For an interior point $M \in \sigma$, let π be the tangential plane to σ at M, and let $K_\varepsilon \subset \pi$ be the circle of radius ε centered at M. Consider the cylinder whose directrix coincides with the border of the circle and generatrix is orthogonal to π. This cylinder cuts the piece σ_ε out of σ. Then *the expression*

$$\lim_{\varepsilon \to 0} \left[\int_{\sigma \setminus \sigma_\varepsilon} \frac{\varphi(M_0)\, d\sigma_{M_0}}{|M_0 M|^3} - \frac{2\pi\varphi(M)}{\varepsilon} \right] = \int_\sigma \frac{\varphi(M_0)\, d\sigma_{M_0}}{|M_0 M|^3}$$

is taken as the value of the integral on the right-hand side in the sense of Hadamard.

Definition 4.1.3. Let $U \subset \sigma$ be an open set which is so small that there is a Cartesian coordinate system $Z_0 X_0 O Y_0$ such that the plane $X_0 O Y_0$ is non-orthogonal to any of the tangential planes to σ at points $M \in U$. Such a coordinate system is called *normal* (NCS).

Suppose that in the NCS, the points of σ belonging to U have the coordinates x_0, y_0, $z_0 = f(x_0, y_0)$. Consider the following integral understood in the sense of Hadamard:

$$A(M, \varphi\nu) = \frac{1}{4\pi} \int_U \frac{\varphi\nu \, d\sigma_{M_0}}{|M_0 M|^3},$$

where $\varphi \in C_0^\infty(U)$, $M_0, M \in U$, and the coordinates of M are $(\hat{x}, y, f(x, y))$.

On the plane $X_0 O Y_0$, we take another coordinate system $X_1 O_1 Y_1$ such that the point O_1 has coordinates $(x, y, 0)$ with respect to $Z_0 O X_0 Y_0$, and the axis $O X_1$ is parallel to the tangential plane to σ at the point $M(x, y, f(x, y))$. In the new coordinates, the local equation of σ is $Z_0 = F(x, y, x_1, y_1)$. Fix a point M with the coordinates $x_1 = 0$, $y_1 = 0$, $Z_0 = F(x, y, 0, 0)$ in the coordinate frame $Z_0 X_1 O Y_1$. Note that $F'_{x_1}(x, y, 0, 0) = 0$.

Consider the family of ellipses U_ε with center at O_1 whose axes coincide with the coordinate axes. The length of the semi-axis along $O_1 X_1$ is $a = \varepsilon$, and that along $O Y_1$ is $b = \varepsilon\left(1 + F'^2_{y_1}(x, y, 0, 0)\right)^{-1/2}$.

We define the functional $A_1(M, \varphi\nu)$ by

$$A_1(M, \varphi\nu) = \qquad\qquad\qquad\qquad\qquad\qquad\qquad\qquad (4.1.19)$$

$$= \lim_{\varepsilon \to 0} \left[\frac{1}{4\pi} \iint_{U_{x_0 y_0} \setminus U_\varepsilon^1} \frac{\varphi(x_1, y_1)\left[1 + F'^2_{x_1}(x, y, x_1, y_1) + F'^2_{y_1}(x, y, x_1, y_1)\right]^{1/2}}{\left\{x_1^2 + y_1^2 + [F(x, y, 0, 0) - F(x, y, x_1, y_1)]^2\right\}^{3/2}} \, dx_1 \, dy_1 - \right.$$

$$\left. - \frac{\varphi(0, 0)\nu(0, 0)}{2\varepsilon} \right],$$

where U_ε^1 is the region bounded by the ellipse U_ε. In the integral term in (4.1.19), denoted by A_ε, we pass to the generalized polar coordinates

$$x_1 = \rho \cos t, \qquad y_1 = \frac{\rho \sin t}{\left[1 + F'^2_{y_1}(x, y, 0, 0)\right]^{1/2}}, \qquad |J| = \frac{1}{\left[1 + F'^2_{y_1}(x, y, 0, 0)\right]^{1/2}}.$$

Then, A_ε can be represented in the form

$$A_\varepsilon = \frac{1}{4\pi} \int_0^{2\pi} \int_\varepsilon^R b(\rho, t) \, K(x, y, \rho, t) \, d\rho \, dt, \qquad\qquad\qquad (4.1.20)$$

where

$$b(\rho, t) = \varphi\left(\rho \cos t, \frac{\rho \sin t}{\left[1 + F'^2_{y_1}(x, y, 0, 0)\right]^{1/2}}\right) \cdot \nu\left(\rho \cos t, \frac{\rho \sin t}{\left[1 + F'^2_{y_1}(x, y, 0, 0)\right]^{1/2}}\right).$$

Moreover, for the sake of convenience, we have taken as the integration region in (4.1.19), (4.1.20) a region U_R^1 such that $U_{x_0 y_0} \subset U_R^1$. The following relations hold:

$$K_1 = \frac{\left[1 + F'^2_{x_1}(x, y, x_1, y_1) + F'^2_{y_1}(x, y, x_1, y_1)\right]^{1/2}}{\left[1 + F'^2_{y_1}(x, y, 0, 0)\right]^{1/2}} = 1 + A_1(x, y)\rho \cos t + A_2(x, y)\rho \sin t + O(\rho^2),$$

$$K_2 = \left\{ 1 - \frac{F'^2_{y_1}(x, y, 0, 0) \sin^2 t}{1 + F'^2_{y_1}(x, y, 0, 0)} + \right.$$

$$\left. + \frac{1}{\rho^2}\left[F(x, y, 0, 0) - F\left(x, y, \rho \cos t, \frac{\rho \sin t}{\left[1 + F'^2_{y_1}(x, y, 0, 0)\right]^{1/2}}\right)\right]^2 \right\}^{-3/2} =$$

$$= 1 + \rho\left[A_3(x, y)\sin^3 t + A_4(x, y)\sin^2 t \cos t + A_5(x, y)\cos^2 t \sin t\right] + O_1(\rho^2),$$

where $|O(\rho^2)| \leq C\rho^2$, $|O_1(\rho^2)| \leq C\rho^2$; the functions $A_i(x,y)$, $i = 1,\ldots,5$, are found from the Taylor expansion. On the basis of the above formulas, we find that

$$K(x,y,\rho,t) = \frac{1}{\rho^2} K_1 K_2 = \frac{1}{\rho^2} + \frac{1}{\rho}\Big[A_1(x,y)\cos t + A_2(x,y)\sin t + B_1(x,y)\sin^3 t +$$
$$+ B_2(x,y)\sin^2 t\cos t + B_3(x,y)\cos^2 t\sin t\Big] + \dot{O}_3(1), \qquad (4.1.21)$$

where $|O_3(1)| \leq C$. From (4.1.20) and (4.1.21) we obtain the following

Proposition 4.1.1. *For $\varphi \in C^\infty(U)$, $\nu \in C^\infty(\sigma)$, $\sigma \in C^\infty$, the functional $A_1(M, \varphi\nu)$ exists.*

Theorem 4.1.1. *For $\varphi, \nu \in C^\infty(U)$, there exists*

$$\lim_{\varepsilon \to 0}\left[A_\varepsilon - \frac{\varphi(0,0)\nu(0,0)}{2\varepsilon}\right] = A(M, \nu\varphi), \qquad (4.1.22)$$

where $A(M, \nu\varphi)$ is the value of the integral in the sense of Hadamard.

Proof. By Proposition 4.1.1, the left-hand side of (4.1.22) exists. On the tangential plane to the surface σ at the point M, consider the circle K_ε of radius ε with center at M. Consider a cylinder passing through the border of the circle with generatrix orthogonal to the plane, and let σ_ε be the piece of σ inside the cylinder. Denote by Γ the boundary of σ_ε and let Γ^* be the projection of Γ to the plane $X_0 O Y_0$. Note that the projection of the circle coincides with the ellipse U_ε. On the tangential plane π, take two mutually orthogonal straight lines l_1 and l_2 passing through M, $l_1 \| O X_1$. Let A be a point on the border of the circle K_ε, and let $B \in \Gamma$, $AB \perp \pi$. Then, the following relations hold:

$$|A_1 B_1| = O_1(\varepsilon^2), \qquad |A_1 B_1| = |C_1 D_1| = O_2(\varepsilon^3), \qquad (4.1.23)$$

where $A_1 B_1$ and $C_1 D_1$ are the projections of the segments AB and CD to the plane $X_0 O Y_0$; the points A and C are symmetrical with respect to the center of the circle K_ε; $D \in \Gamma$, $CD \perp \pi$.

On the border of the ellipse U_ε, consider two arcs cut out by the straight lines forming the angle β_0 with the axis $O X_1$. Let l be the straight line passing through the center of the ellipse at an angle β to $O X_1$, $\beta_0 \leq \beta \leq \pi - \beta_0$. Let M, N and M_1, N_1 be the points at which l crosses the curve Γ^* and the border of the ellipse U_ε, respectively. Then,

$$|MN| = |M_1 N_1| + O_3(\varepsilon^3). \qquad (4.1.24)$$

For the line l forming the angle β with $O X_1$, $-\beta_0 \leq \beta \leq \beta_0$, we have

$$|MN| = O_4(\varepsilon^2), \qquad |M_1 N_1| = O_5(\varepsilon^2). \qquad (4.1.25)$$

Let $\tilde{\sigma}_\varepsilon$ be the part of σ inside the cylinder whose directrix coincides with the border of U_ε and generatrix is orthogonal to the plane $X_0 O Y_0$. Consider the sets $\sigma'_\varepsilon = \sigma \setminus \sigma_\varepsilon$, $\sigma''_\varepsilon = \sigma \setminus \tilde{\sigma}_\varepsilon$, $\sigma'''_\varepsilon = \sigma'_\varepsilon \Delta \sigma''_\varepsilon$, and let D_ε be the projection of σ'''_ε to the plane $X_0 O Y_0$, $D_\varepsilon = \alpha_1 \cup \alpha_2$, where α_1 is inside the ellipse U_ε and α_2 is outside.

The plane $X_0 O Y_0$ is divided into the regions D_1, D_2, D_3, D_4 by the half-lines issuing from the center of the ellipse at the angles $(-\beta_0)$, β_0, $\pi - \beta_0$, $-\pi + \beta_0$, respectively. Let $D'_1 = D_1 \cap \alpha_1$, $D'_2 = D_2 \cap \alpha_1$, $D'_3 = D_3 \cap \alpha_1$, $D'_4 = D_4 \cap \alpha_1$, $D''_1 = D_1 \cap \alpha_2$, $D''_2 = D_2 \cap \alpha_2$, $D''_3 = D_3 \cap \alpha_2$, $D''_4 = D_4 \cap \alpha_2$.

For a fixed β_0 and sufficiently small ε, we have either $D''_2 = \emptyset$, $D'_4 = \emptyset$ or $D'_2 = \emptyset$, $D''_4 = \emptyset$, according to the angle between the tangential plane π and the plane $X_0 O Y_0$.

Consider the integral

$$\iint_{\sigma'_\varepsilon} \frac{\varphi\nu\, d\sigma}{|M_0 M|^3} = \iint_{U^1_R \setminus U^1_\varepsilon} \frac{\varphi\nu\left[1 + F'^2_{x_1} + F'^2_{y_1}\right]^{1/2} dx_1\, dy_1}{|M_0 M|^3} +$$

$$+ \left\{ \iint_{\alpha_1} \frac{\varphi\nu\left[1 + F'^2_{x_1} + F'^2_{y_1}\right]^{1/2} dx_1\, dy_1}{|M_0 M|^3} - \iint_{\alpha_2} \frac{\varphi\nu\left[1 + F'^2_{x_1} + F'^2_{y_1}\right]^{1/2} dx_1\, dy_1}{|M_0 M|^3} \right\} =$$

$$= \gamma^\varepsilon_1 + \gamma^\varepsilon_2. \tag{4.1.26}$$

Let us show that

$$\lim_{\varepsilon \to 0} \gamma^\varepsilon_2 = 0. \tag{4.1.27}$$

Relation (4.1.21) shows that it suffices to prove (4.1.27) for the term with $1/\rho^2$.

Fix $\delta > 0$. Relations (4.1.23)–(4.1.25) imply that there exists an angle $\beta_0(\delta)$ such that in the region D'_1 we have

$$\left| \iint_{D'_1} \frac{\varphi\nu}{\rho^2}\, d\rho\, d\varphi \right| \le C \int_{-\beta_0(\delta)}^{\beta_0(\delta)} d\varphi \int_{\rho_1(\varphi)}^{\rho_2(\varphi)} \frac{d\rho}{\rho^2} \le C_1 \int_{-\beta_0(\delta)}^{\beta_0(\delta)} \le \frac{\delta}{8}. \tag{4.1.28}$$

Similar inequalities hold for D''_1, D'_3, D''_3.

Consider the case of $D''_2 = \varnothing$, $D'_4 = \varnothing$. Then

$$\left| \iint_{D'_2} \frac{\varphi\nu}{\rho^2}\, d\rho\, d\varphi - \iint_{D''_4} \frac{\varphi\nu}{\rho^2}\, d\rho\, d\varphi \right| \le$$

$$\le \left| \iint_{D'_2} \frac{\varphi\nu - \varphi(0,0)\nu(0,0)}{\rho^2}\, d\rho\, d\varphi - \iint_{D''_4} \frac{\varphi\nu - \varphi(0,0)\nu(0,0)}{\rho^2}\, d\rho\, d\varphi \right| +$$

$$+ |\varphi(0,0)\nu(0,0)| \left| \iint_{D'_2} \frac{1}{\rho^2}\, d\rho\, d\varphi - \iint_{D''_4} \frac{1}{\rho^2}\, d\rho\, d\varphi \right| = I'_\varepsilon + I''_\varepsilon. \tag{4.1.29}$$

It follows from (4.1.24) that $I'_\varepsilon \to 0$ as $\varepsilon \to 0$. Let us estimate I''_ε. We have

$$|I''_\varepsilon| \le C \left\{ \left| \int_{\beta_0}^{\pi - \beta_0} d\varphi \left[\int_{\varepsilon - \varepsilon(\varphi)}^\varepsilon \frac{d\rho}{\rho^2} - \int_\varepsilon^{\varepsilon + \varepsilon(\varphi)} \frac{d\rho}{\rho^2} \right] \right| + \left| \int_{\beta_0}^{\pi - \beta_0} d\varphi \int_{\Delta I} \frac{d\rho}{\rho^2} \right| \right\} = C\left\{ \widetilde{I}'_\varepsilon + \widetilde{I}''_\varepsilon \right\},$$

$$\Delta I = \left[\varepsilon_1,\, \varepsilon_1 + O_6(\varepsilon^3) \right], \quad \varepsilon_1 = \varepsilon + O_7(\varepsilon^2), \quad \varepsilon(\varphi) = O_7(\varepsilon^2),$$

$$\left| O_6(\varepsilon^3) \right| \le C_1 \varepsilon^3, \left| O_7(\varepsilon^2) \right| \le C_1 \varepsilon^2.$$

Direct calculations show that $\widetilde{I}'_\varepsilon \to 0$ and $\widetilde{I}''_\varepsilon \to 0$, and therefore $I''_\varepsilon \to 0$ as $\varepsilon \to 0$.

From (4.1.28) and (4.1.29) we obtain (4.1.27). Now, the statement of Theorem 4.1.1 follows from (4.1.27) and the definition of the integral in the sense of Hadamard. The proof is complete.

Consider the polar coordinates with the origin at $O_1(x, y)$ and the polar axis being parallel to OX_0 and having the same direction. Thus, $x_0 - x = r\cos\theta$, $y_0 - y = r\sin\theta$. Using the Taylor formula, we obtain the relation

$$K(x, y, x_0, y_0) = \frac{\left[1 + f'^2_{x_0}(x, y) + f'^2_{y_0}(x_0, y_0)\right]^{1/2}}{\left[(x - x_0)^2 + (y - y_0)^2 + [f(x, y) - f(x_0, y_0)]^2\right]^{3/2}} =$$

$$= \frac{1}{r^3} \frac{\left[1 + f_{x_0}'^2(x, y) + f_{y_0}'^2(x, y)\right]^{1/2}}{\left[1 + \left[f_{x_0}'(x, y) \cos \theta + f_{y_0}'(x_0, y_0) \sin \theta\right]^2\right]^{3/2}} +$$

$$+ \frac{1}{r^2} \frac{1}{\left[1 + \left[f_{x_0}'(x, y) \cos \theta + f_{y_0}'(x_0, y_0) \sin \theta\right]^2\right]^{3/2}} \left\{A_1(x, y) \cos \theta + A_2(x, y) \sin \theta + \right.$$

$$\left. + A_3(x, y) \cos^3 \theta + A_4(x, y) \cos^2 \theta \sin \theta + A_5(x, y) \cos \theta \sin^2 \theta + A_6(x, y) \sin^3 \theta\right\} +$$

$$+ \frac{1}{r} K_3(x, y, r, \theta) = \frac{1}{r^3} K_1(x, y, \theta) + \frac{1}{r^2} K_2(x, y, \theta) + \frac{1}{r} K_3(x, y, r, \theta), \tag{4.1.30}$$

where the kernel $K_3(x, y, r, \theta)$ admits the estimate

$$|K_3(x, y, r, \theta)| \le C. \tag{4.1.31}$$

Direct calculations show that

$$\int_{-\pi}^{\pi} \frac{\sin \theta \, d\theta}{[L(x, y, \theta)]^{3/2}} = 0, \qquad \int_{-\pi}^{\pi} \frac{\cos \theta \, d\theta}{[L(x, y, \theta)]^{3/2}} = 0, \qquad \int_{-\pi}^{\pi} \frac{\sin^m \theta \cos^n \theta \, d\theta}{[L(x, y, \theta)]^{3/2}} = 0, \quad m + n = 3, \tag{4.1.32}$$

where $L(x, y, \theta) = 1 + \left[f_{x_0}'(x, y) \cos \theta + f_{y_0}'(x_0, y_0) \sin \theta\right]^2$.

Consider the following integral in the sense of Hadamard:

$$A(M, \varphi \nu) = \lim_{\varepsilon \to 0} \left[\frac{1}{4\pi} \iint_{U_R^1 \setminus U_\varepsilon^1} \varphi \nu K(x, y, x_0, y_0) \, dx_0 \, dy_0 - \frac{\varphi(x, y) \nu(x, y)}{2\varepsilon}\right] =$$

$$= \lim_{\varepsilon \to 0} \left[\frac{1}{4\pi} \iint_{U_R^1 \setminus U_\varepsilon^1} \left[\frac{1}{r^2} K_1(x, y, \theta) + \frac{1}{r} K_2(x, y, \theta) + K_3(x, y, r, \theta)\right] dr \, d\theta - \frac{\varphi(x, y) \nu(x, y)}{2\varepsilon}\right] =$$

$$= A_1 + A_2 + A_3. \tag{4.1.33}$$

Because of the inequality (4.1.31), the integral A_3 is absolutely convergent. Let us show that A_2 exists as a singular integral.

Proposition 4.1.2. *For $\varphi, \nu \in C^1(\mathbb{R}_2)$, we have*

$$A_2 = \lim_{\varepsilon \to 0} \iint_{U_R^1 \setminus U_\varepsilon^*} \frac{1}{r} K_2(x, y, \theta) \nu \varphi \, dr \, d\theta,$$

where U_ε^* is the circle of radius ε with center at $M(x, y)$.

Proof. It suffices to show that

$$\lim_{\varepsilon \to 0} \iint_{U_\varepsilon^* \setminus U_\varepsilon^1} \frac{1}{r} K_2(x, y, \theta) b(r, \theta) \, dr \, d\theta = 0,$$

where $b(r, \theta) = \varphi(x + r \cos \theta, y + r \sin \theta) \nu(x + r \cos \theta, y + r \sin \theta)$.

Let us change the variables by $\theta = \psi + \alpha$, where α is such that $f_{y_0}'(x, y) \cos \alpha - f_{x_0}'(x, y) \sin \alpha = 0$. Then, the equation of the ellipse becomes

$$r = \frac{\varepsilon}{\left[1 + \left[f_{x_0}'^2(x, y) + f_{y_0}'^2(x, y)\right] \cos^2 \psi\right]^{1/2}} = r_0(\psi)$$

Further, we find that

$$\iint_{U_\varepsilon^* \backslash U_\varepsilon^1} \frac{1}{r} K_2(x, y, \theta) b(r, \theta)\, dr\, d\theta = \iint_{U_\varepsilon^* \backslash U_\varepsilon^1} \frac{b(r, \psi + \alpha) - b(x, y)}{r} K_2(x, y, \psi + \alpha)\, dr\, d\psi +$$

$$+ b(x, y) \int_0^{2\pi} d\psi \int_{r_0(\psi)}^{\varepsilon} \frac{1}{r} K_2(x, y, \psi + \alpha)\, dr = I_1(\varepsilon) + I_2(\varepsilon).$$

The integral $I_1(\varepsilon)$ is absolutely convergent, and therefore, $\lim_{\varepsilon \to 0} I_1(\varepsilon) = 0$.

By analogy with (4.1.32), we verify the relation

$$I_2(\varepsilon) = \int_0^{2\pi} K_2(x, y, \psi + \alpha) \ln \left[1 + \left[f_{x_0}'^2(x, y) + f_{y_0}'^2(x, y) \right] \cos^2 \psi \right]^{1/2} d\psi = 0.$$

Proposition 3.1.2 is proved.

Consider the term A_1,

$$A_1 = \lim_{\varepsilon \to 0} \frac{1}{4\pi} \iint_{U_R^1 \backslash U_\varepsilon^1} \frac{1}{r^2} K_1(x, y, \psi + \alpha) b(r, \psi + \alpha)\, dr\, d\psi - \frac{\varphi(x, y)\nu(x, y)}{\varepsilon} =$$

$$= \frac{1}{4\pi} \iint_{U_R^1 \backslash U_\varepsilon^*} \frac{b(r, \psi + \alpha) - b(0, 0)}{r^2} K_1(x, y, \psi + \alpha)\, dr\, d\psi +$$

$$+ \iint_{U_\varepsilon^* \backslash U_\varepsilon^1} \frac{b(r, \psi + \alpha) - b(0, 0)}{r^2} K_1(x, y, \psi + \alpha)\, dr\, d\psi + \qquad (4.1.34)$$

$$+ \left[\varphi(x, y)\nu(x, y) \iint_{U_R^1 \backslash U_\varepsilon^1} \frac{K_1(x, y, \psi + \alpha)}{r^2}\, dr\, d\psi - \frac{\varphi(x, y)\nu(x, y)}{\varepsilon} \right] = B_1 + B_2 + B_3,$$

where $b(0, 0) = \varphi(x, y)\nu(x, y)$. Direct calculation shows that

$$B_3 = -\frac{\varphi(x, y)\nu(x, y)}{2R}. \qquad (4.1.35)$$

By analogy with Proposition 4.1.2, it can be shown that $B_2 = 0$.

Let us transform the term B_1. Integrating by parts with respect to r, for $\varphi, \nu \in C^2(\mathbb{R}_2)$, we obtain

$$B_1 = \lim_{\varepsilon \to 0} \left[\iint_{U_R^1 \backslash U_\varepsilon^*} \frac{1}{4\pi} \frac{K_1(x, y, \theta)}{r} \frac{\partial}{\partial r} b(r, \theta)\, dr\, d\theta + \frac{\varphi(x, y)\nu(x, y)}{2R} + \alpha(\varepsilon) \right], \qquad (4.1.36)$$

where

$$\lim_{\varepsilon \to 0} \alpha(\varepsilon) = \lim_{\varepsilon \to 0} \int_0^{2\pi} K_1(x, y, \theta) \frac{b(\varepsilon, \theta) - b(0, 0)}{\varepsilon}\, d\theta = 0.$$

From (4.1.34)–(4.1.36), it follows that

$$A_1 = \lim_{\varepsilon \to 0} \iint_{U_R^1 \backslash U_\varepsilon^*} \frac{K_1(x, y, \theta)}{r} \frac{\partial}{\partial r} b(r, \theta)\, dr\, d\theta. \qquad (4.1.37)$$

Combining (4.1.37) with Proposition 4.1.2, we obtain the following result.

Proposition 4.1.3. *For $\varphi, \nu \in C^2(\sigma)$, in any normal coordinate system we have*

$$\frac{1}{4\pi} \int_U \frac{\varphi \nu \, d\sigma_{M_0}}{|M_0 M|^3} = \frac{1}{4\pi} \iint_{U_R^1} \frac{K_1(x, y, \theta)}{r} \frac{\partial}{\partial r} b(r, \theta) \, dr \, d\theta + \frac{1}{4\pi} \iint_{U_R^1} \frac{K_2(x, y, \theta)}{r} b(r, \theta) \, dr \, d\theta +$$

$$+ \frac{1}{4\pi} \iint_{U_R^1} K_3(x, y, r, \theta) b(r, \theta) \, dr \, d\theta, \tag{4.1.38}$$

where the first two integrals on the right-hand side are understood in the singular sense.

Note that according to Definition 4.1.2, the Neumann problem (4.1.2) for σ is reduced to the hypersingular integral equation

$$\frac{1}{4\pi} \int_\sigma \nu(M_0) \left[\frac{|M_0 M|^2 (n_{M_0}, n_M) - 3(\overline{M_0 M}, n_M)(\overline{M_0 M}, n_{M_0})}{|M_0 M|^5} \, d\sigma_{M_0} \right] = f. \tag{4.1.39}$$

The integral with the double layer density $\nu(M_0)$ is understood in the sense of Hadamard; $M_0, M \in \sigma$.

Definition 4.1.4. The operator defined by the left-hand sides of relations (4.1.3) is called the *Prandtl operator.*

The Prandtl operator for a single surface can be represented in the form

$$\Pi(\nu) = A(\nu) + A_4(\nu), \tag{4.1.40}$$

where

$$A(\nu) = \frac{1}{4\pi} \int_\sigma \frac{\nu(M_0)}{|M_0 M|^3} \, d\sigma_{M_0}, \tag{4.1.41}$$

$$A_4(\nu) = \frac{1}{4\pi} \int_\sigma \nu(M_0) \left[\frac{|M_0 M|^2 [(n_{M_0}, n_M) - 1] - 3(\overline{M_0 M}, n_M)(\overline{M_0 M}, n_{M_0})}{|M_0 M|^5} \, d\sigma_{M_0} \right], \tag{4.1.42}$$

and the following estimate holds:

$$|K_4(M_0, M)| = \left| \frac{|M_0 M|^2 [(n_{M_0}, n_M) - 1] - 3(\overline{M_0 M}, n_M)(\overline{M_0 M}, n_{M_0})}{|M_0 M|^5} \, d\sigma_{M_0} \right| \leq \frac{C}{|M_0 M|}. \tag{4.1.43}$$

4.2. The Noetherian Property of the Prandtl Operator

Consider the generalized function $f = |Dx|^\lambda$, $\lambda \neq -n-2, -n-4, \ldots$, $\lambda < -n$, where D is a constant $(n \times n)$-matrix, $\det D \neq 0$. Let us find the Fourier transform of f. Using (2.2.31) and (3.2.7), for any $\varphi \in S(\mathbb{R}_n)$ we find that

$$(f, \varphi) = \frac{1}{|\det D|} \left(|y|^\lambda, \varphi(D^{-1} y) \right) = \frac{1}{|\det D|} \left(|y|^\lambda, \varphi_1(y) \right) = \frac{1}{(2\pi)^n |\det D|} \left(C_\lambda |\xi|^{-\lambda-n}, \widehat{\varphi}_1(\xi) \right) =$$

$$= \frac{1}{(2\pi)^n |\det D|} |\det D| \left(C_\lambda |\xi|^{-\lambda-n}, \widehat{\varphi}(D^* \xi) \right) =$$

$$= \frac{1}{(2\pi)^n |\det D|} \left(C_\lambda |(D^*)^{-1} \eta|^{-\lambda-n}, \widehat{\varphi}(\eta) \right). \tag{4.2.1}$$

It follows that

$$\widehat{f} = \frac{1}{|\det D|} C_\lambda |(D^*)^{-1}\eta|^{-\lambda-n}. \tag{4.2.2}$$

Consider the generalized function $|D(x)(x - x_0)|^\lambda = g(x, x_0)$ depending on x as a parameter, where $D(x)$ is an $(n \times n)$-matrix depending on x, $\det D(x) \neq 0$ for all x. Let us find the Fourier transform of $g(x, x_0)$ with respect to x_0. Using (4.2.2), we get

$$(g, \varphi) = \left(|D(x)y|^\lambda, \varphi(x - y)\right) = \frac{1}{(2\pi)^n} \left(\frac{1}{|\det D(x)|} C_\lambda |D^*(x)^{-1}\eta|^{-\lambda-n}, e^{i(x,\eta)} \widehat{\varphi}(-\eta) \right) =$$

$$= \frac{1}{(2\pi)^n} \left(\frac{1}{|\det D(x)|} C_\lambda |D^*(x)^{-1}\eta|^{-\lambda-n} e^{i(x,\eta)}, \widehat{\varphi}(\eta) \right). \tag{4.2.3}$$

Hence,

$$\widehat{g} = \frac{1}{|\det D(x)|} C_\lambda |D^*(x)^{-1}\eta|^{-\lambda-n} e^{i(x,\eta)}. \tag{4.2.4}$$

Since $\widehat{g}(\eta)$ is a regular functional, relation (4.2.3) with a real-valued $\varphi(x_0)$ can be written as

$$\int_{\mathbb{R}_n} |D(x)(x - x_0)|^\lambda \varphi(x_0)\, dx_0 = \frac{1}{(2\pi)^n} \int_{\mathbb{R}_n} \frac{1}{|\det D(x)|} C_\lambda |D^*(x)^{-1}\eta|^{-\lambda-n} e^{i(x,\eta)} \overline{\widehat{\varphi}(\eta)}\, d\eta =$$

$$= \frac{1}{(2\pi)^n} \int_{\mathbb{R}_n} \frac{1}{|\det D(x)|} C_\lambda |D^*(x)^{-1}\eta|^{-\lambda-n} e^{-i(x,\eta)} \widehat{\varphi}(\eta)\, d\eta, \tag{4.2.5}$$

where the value of the hypersingular integral on the left-hand side of (4.2.5) is understood in the sense of regularization, just as in Example 2.2.5.

It should be observed that (4.2.5) can be interpreted as a representation of the hypersingular integral in terms of a pseudodifferential operator with the symbol

$$A(\eta) = \frac{1}{|\det D(x)|} C_\lambda |D^*(x)^{-1}\eta|^{-\lambda-n}. \tag{4.2.6}$$

Let us utilize relations (4.2.5) and (4.2.6) to represent

$$A_1(\varphi\nu) = \iint_{\mathbb{R}_2} \frac{1}{4\pi} \frac{1}{r^2} K_1(x, y, \theta) b(r, \theta)\, dr\, d\theta$$

as a pseudodifferential operator. From (4.1.30), we obtain

$$A_1(\varphi\nu) = \iint_{\mathbb{R}_2} \frac{1}{4\pi} \frac{\left[1 + f_{x_0}'^2(x, y) + f_{y_0}'^2(x, y)\right]^{1/2} \nu(x_0, y_0)\varphi(x_0, y_0)\, dx_0\, dy_0}{\left[(x - x_0)^2 + (y - y_0)^2 + \left[f_{x_0}'(x - x_0) + f_{y_0}'(y - y_0)\right]^2\right]^{3/2}}. \tag{4.2.7}$$

The quadratic form

$$B = (x - x_0)^2 + (y - y_0)^2 + f_{x_0}'^2(x - x_0)^2 + 2f_{x_0}'(x, y)f_{y_0}'(x, y)(x - x_0)(y - y_0) + f_{y_0}'^2(x, y)(y - y_0)^2$$

is positive definite. Denote the matrix of this quadratic form by

$$D_1 = \begin{pmatrix} 1 + A^2 & AB \\ AB & 1 + B^2 \end{pmatrix}. \tag{4.2.8}$$

The eigenvalues of the matrix D_1 are $\lambda_1 = 1$, $\lambda_2 = 1 + A^2 + B^2$, where $A = f_{x_0}'(x, y)$, $B = f_{y_0}'(x, y)$. The matrix D_1 can be represented in the form $D_1 = C^* D_{1\lambda} C$, where C is an orthogonal matrix,

$D_{1\lambda} = \begin{pmatrix} \lambda_1 & 0 \\ 0 & \lambda_2 \end{pmatrix}$. Therefore, $D_1 = D^* D$, where $D = \begin{pmatrix} \lambda_1^{1/2} & 0 \\ 0 & \lambda_1^{1/2} \end{pmatrix} C$, and D^* is the transpose matrix of D. Thus, (4.2.7) reduces to

$$A_1(\varphi\nu) = \frac{1}{4\pi} \iint_{\mathbb{R}_2} \left| D(x,y) \begin{pmatrix} x - x_0 \\ y - y_0 \end{pmatrix} \right|^{-3} |\det D| \, \nu(x_0, y_0) \varphi(x_0, y_0) \, dx_0 \, dy_0. \qquad (4.2.9)$$

Let us consider the right-hand side of (4.2.9) as the value of the generalized function

$$g(x, y, x_0, y_0) = \frac{1}{4\pi} \left| D(x,y) \begin{pmatrix} x - x_0 \\ y - y_0 \end{pmatrix} \right|^{-3} |\det D|$$

on $\nu(x, y_0)\varphi(x_0, y_0)$. Then, taking into account Example 3.1.7, we find

$$\left(g(x, y, x_0, y_0), \ \nu(x_0, y_0)\varphi(x_0, y_0) \right) = \frac{1}{4\pi} \left((x_1^2 + y_1^2)^{-3/2}, \nu(x, y, x_1, y_1)\varphi(x, y, x_1, y_1) \right) =$$

$$= \iint_{\mathbb{R}_2} \frac{1}{4\pi} \frac{\nu_1(x, y, x_1, y_1)\varphi_1(x, y, x_1, y_1)}{(x_1^2 + y_1^2)^{3/2}} \, dx_1 \, dy_1 =$$

$$= \lim_{\varepsilon \to 0} \left[\frac{1}{4\pi} \iint_{\mathbb{R}_2 \setminus K_\varepsilon} \frac{\nu_1(x, y, x_1, y_1)\varphi_1(x, y, x_1, y_1)}{(x_1^2 + y_1^2)^{3/2}} \, dx_1 \, dy_1 - \frac{\nu(x, y)\varphi(x, y)}{2\varepsilon} \right], \quad (4.2.10)$$

where K_ε is the circle of radius ε with center at $(x_1, y_1) = (0, 0)$; $\nu_1(x, y, x_1, y_1)$, $\varphi_1(x, y, x_1, y_1)$ are the functions obtained from $\nu(x_0, y_0)$, $\varphi(x_0, y_0)$ after the transformation

$$D(x, y) \begin{bmatrix} x - x_0 \\ y - y_0 \end{bmatrix} = \begin{bmatrix} x_1 \\ y_1 \end{bmatrix}. \qquad (4.2.11)$$

On the right-hand side of (4.2.10), we go back to the variables x_0 and y_0, which yields

$$\left(g(x, y, x_0, y_0), \ \nu(x_0, y_0)\varphi(x_0, y_0) \right) = \qquad (4.2.12)$$

$$= \lim_{\varepsilon \to 0} \left[\frac{1}{4\pi} \iint_{\mathbb{R}_2 \setminus U_\varepsilon^1} |\det D(x,y)| \left| D(x,y) \begin{pmatrix} x - x_0 \\ y - y_0 \end{pmatrix} \right|^{-3} \nu(x_0, y_0)\varphi(x_0, y_0) \, dx_0 \, dy_0 - \right.$$

$$\left. - \frac{\nu(x, y)\varphi(x, y)}{2\varepsilon} \right],$$

where the ellipse U_ε^1 is the image of the circle K_ε. It follows that the definition of the integral $A_1(\varphi\nu)$ in the sense of Hadamard (see (4.1.33)) coincides with the definition of the generalized function $g(x, y, x_0, y_0)$. Now, using (4.2.5), we can represent the hypersingular integral $A_1(\varphi\nu)$ as the pseudodifferential operator

$$A_1(\varphi\nu) = \frac{1}{4\pi^2} \iint_{\mathbb{R}_2} \left(-\frac{1}{2} \right) \left| (D^*(x, y))^{-1}\xi \right| \, \widehat{\nu}_1(\xi) \, e^{-i(x,\xi)} \, d\xi, \qquad (4.2.13)$$

where $\widehat{\nu}_1(\xi) = F\varphi\nu$; the matrix D is determined from the relation $D_1 = D^* D$, and the matrix D_1 is defined by (4.2.8).

Let S_i^0, $1 \le i \le k$, and S_j^1, $1 \le j \le n$ be the surfaces introduced in the beginning of Section 4.1.

Consider the Sobolev–Slobodetskii spaces (see Definitions 2.4.4, 2.4.7, and 2.4.9) $H_r(S_i^0)$ containing the functions ν_i^0 defined on S_i^0, and $\mathring{H}_r(S_j^1)$ containing the functions ν_j^1 defined on S_j^1.

Denote by $H_r(\nu)$ the direct sum of all spaces $H_r(S_i^0)$ and $\mathring{H}_r(S_j^1)$, and by $\widehat{H}_r(\nu)$ the direct sum of all spaces $H_r(S_i^0)$ and $H_r(S_j^1)$.

Let $\mathbf{\Pi}(\nu) = \left\{\Pi_l^m(\nu)\right\}$, $m = 0 : l = 1,\ldots,k$; $m = 1 : l = 1,\ldots,n$, where

$$\Pi_l^m = \frac{1}{4\pi}\sum_i\sum_p \frac{\partial}{\partial n_l^m}\int_{S_i^p}\nu_i^p(y)\frac{\cos\varphi_{xy}}{|x-y|^2}\,dS_{iy}^p = \frac{1}{4\pi}\frac{\partial}{\partial n_l^m}\int_{S^m}\nu_l^m(y)\frac{\cos\varphi_{xy}}{|x-y|^2}\,dS_{ly}^m +$$

$$+ \sum_i\sum_p{}' \frac{\partial}{\partial n_l^m}\frac{1}{4\pi}\int_{S_i^p}\nu_i^p(y)\frac{\cos\varphi_{xy}}{|x-y|^2}\,dS_{iy}^p = L_l^m(\nu) + N_l^m(\nu), \qquad (4.2.14)$$

where $'$ indicates that in the sum the term corresponding to the surface S_l^m is omitted. The operator $L_l^m(\nu)$ is defined by (4.1.40)–(3.1.43), $p = 0$: $i = 1,\ldots,k$; $p = 1$: $i = 1,\ldots,n$.

Proposition 4.2.1. *The operator* $\mathbf{\Pi}(\nu) : H_r(\nu) \to \widehat{H}_{r-1}(\nu)$ *is bounded for* $0 \le r \le 1$.

Proof. Obviously, the operator $N = \left\{N_l^m(\nu)\right\}$ in (4.2.14) acting from $H_r(\nu)$ to $\widehat{H}_{r-1}(\nu)$ is bounded for $0 \le r \le 1$. The operator L can be represented in the form $L = L_1 + L_2$, where $L_1 = \{A(\nu,m,l)\}$, $L_2 = \{A_4(\nu,m,l)\}$ (see (4.1.40)–(4.1.43); the indices m, l refer to the surface S_l^m). The estimate (4.1.43) implies that the operators $A_4(\nu,m,l)$ are bounded from $L_2(S_l^m)$ to $L_2(S_l^m)$. Since $H_r(S_l^m) \subset L_2(S_l^m) \subset H_{r-1}(S_l^m)$, $0 \le r \le 1$, it follows that $A_4(\nu,m,l)$ is a bounded operator from $H_r(\nu)$ to $\widehat{H}_{r-1}(\nu)$. Let $\{U_i(m,l)\}$, $0 \le i \le N(m,l)$ be a covering of the surface $\overline{S_l^m}$. For $m = 1$, S_l^1 is a manifold with border and belongs to a manifold M_j (without border). In this case, for $l_1 \ne l_2$, the covering $\{U_i(1,l_1)\}$ of $S_{l_1}^1 \subset M_j$ and the covering $\{U_i(1,l_2)\}$ of $S_{l_2}^1 \subset M_j$ are assumed to have no points in common. Consider functions $\varphi_i(m,l) \in C_0^\infty(U_i(m,l))$ forming a partition of unity subject to these coverings, $\sum_i\varphi_i(1,l) = 1$ on $\overline{S_l^1}$. Moreover, the open sets of the coverings are assumed sufficiently small, so that in each U_i there is a normal coordinate system (see Definition 4.1.3) in which $S_l^m \subset U_i(m,l)$ is given by the equation $z_i(m,l) = f_i(x_i,y_i,m,l)$. Let $\psi_i(m,l) \in C_0^\infty(U_i(m,l))$ be functions such that $\psi_i(m,l) = 1$ on the open set U_i', $\overline{U_i'} \subset U_i$, $\operatorname{supp}\varphi_i \subset U_i'$. Then the operator $A(\nu,m,l)$ can be represented in the form

$$A(\nu,m,l) = \sum_i \psi_i A(\varphi_i\nu,m,l) + \sum_i (1-\psi_i)A(\varphi_i\nu,m,l) = B_1^{m,l}(\nu) + B_2^{m,l}(\nu). \qquad (4.2.15)$$

Since the functions $(1-\psi_i)$ and φ_i are supported on mutually disjoint sets, we have $B_2^{m,l} \in L_2(S_l^m)$. From (4.1.33), it follows that the term $\psi_i A(\varphi_i\nu,m,l)$ can be represented as

$$\psi_i A(\varphi_i\nu,m,l) = \psi_i A_1(\varphi_i\nu,m,l) + \psi_i A_2(\varphi_i\nu,m,l) + \psi_i A_3(\varphi_i\nu,m,l). \qquad (4.2.16)$$

The kernel of the integral operator $\psi_i A_3(\varphi_i\nu,m,l)$ satisfies the estimate (4.1.31), and therefore, $\psi_i A_3(\varphi_i\nu,m,l) \in L_2(S_l^m)$.

Each surface S_l^m is of class C^∞, and therefore, each $f_i(x_i,y_i,m,l)$ can be extended as a function in C^∞ with compact support. Consequently, the singular operator $\psi_i A_2(\varphi_i\nu,m,l)$ is bounded from $L_2(S_l^m)$ to $L_2(S_l^m)$ (see Mikhlin (1962)). The integral $\psi_i A_1(\varphi_i\nu,m,l)$ in the sense of Hadamard can be represented as a pseudodifferential operator of the form (4.2.13) with symbol $A_1(x,\xi) \in S_1^0(\mathbb{R}_2)$. Therefore, this operator is bounded from $H_r(S_l^m)$ $(\mathring{H}_m(S_l^m))$ to $H_{r-1}(S_l^m)$.

Thus, $L_1 = \{A(\nu,m,l)\}$ is a bounded operator from $H_r(\nu)$ to $\widehat{H}_{r-1}(\nu)$. Proposition 4.2.1 is proved.

Definition 4.2.1. The set $U = \left\{U_l^m\right\}$ is called *open* if U_l^0 are open sets on the surfaces S_l^0, and the sets U_l^1 are open sets on the manifolds M_j (without border), $S_l^1 \subset M_j$.

Let us establish a theorem about commutators.

Theorem 4.2.1. *Let* $U = \{U_l^m\}$ *be an open set, and let* U_l^m *be coordinate neighborhoods of the surfaces* S_l^m. *Then, for* $0 \le r \le 1$, *the commutator* $K = \varphi \Pi(\nu) - \Pi(\varphi\nu)$ *is a bounded operator from* $H_r(\nu)$ *to* $H_0(\nu)$ $(= L_2)$, *where* $\varphi = \{\varphi_l^m\}$, $\varphi_l^m \in C_0^\infty(U_l^m)$.

Proof. Consider functions $\psi_l^m \in C_0^\infty(U_l^m)$ such that $\psi_l^m = 1$ on an open set $U_l'^m \subset U_l^m$ such that $\operatorname{supp}\psi_l^m \subset U_l'^m$. We represent the commutator K in the form

$$K = \psi K + (1 - \psi)K = \psi[\varphi \Pi(\nu) - \Pi(\varphi\nu)] - (1 - \psi)\Pi(\varphi\nu). \tag{4.2.17}$$

The functions φ and $(1 - \psi)$ are supported on mutually disjoint sets, and therefore, $(1-\psi)\Pi(\varphi\nu) \in H_0(\nu)$. It can be seen from the proof of Proposition 4.2.1 that in order to establish the desired result, it suffices to show that $\psi_l^m[\varphi_l^m A_1(\nu_l^m) - A_1(\varphi_l^m \nu_l^m)] \in L_2(S_l^m)$. Indeed, let us extend the function $f(x, y, m, l)$ to the plane $\mathbb{R}_2$, so that $lf \in C_0^\infty(\mathbb{R}_2)$. Then, $A_1(\varphi_l^m \nu_l^m)$ can be represented as a PDO $\hat{A}_1$ with symbol $A_1(x, \xi) \in S_1^0$. Let us write $\hat{A}_1 = \hat{B}_1 + \hat{B}_2$, where $\hat{B}_1$ is the PDO with symbol $B_1(x, \xi) = A_1(x, \xi)\chi(\xi)$, and the symbol of $\hat{B}_2$ is $B_2(x, \xi) = A_1(x, \xi)(1 - \chi(\xi))$, $\chi(\xi) \in C_0^\infty(\mathbb{R}_2)$, $\chi(x) = 1$ for $|\xi| \le 1$. We have $B_1(x, \xi) \in S_{-\infty}^0$ and $B_2(x, \xi) \in S_1^\infty$ (see Definition 3.1.5). Now, it follows from Theorem 3.1.1 and Proposition 3.1.5 that $\psi_l^m[\varphi_l^m A_1(\nu_l^m) - A_1(\varphi_l^m \nu_l^m)] \in L_2(S_l^m)$. Theorem 4.2.1 is proved.

Remark 4.2.1. Let p be the operator of restriction to the domain G. Proposition 4.2.1 implies that

$$p\varphi\Pi(\nu) = p\Pi(\varphi\nu) + pK. \tag{4.2.18}$$

Theorem 4.2.2. *For* $0 < r < 1$, *the Prandtl operator* $\Pi(\nu)$, *which maps* $H_r(\nu)$ *into* $\hat{H}_{r-1}(\nu)$, *is Noetherian.*

Proof. For each interior point $M \in S_l^m$, consider its neighborhood $U_l^m(M)$ whose projection to the plane $X_0 O Y_0$ of some normal coordinate system is a circle (the normal coordinate system depends on M). We denote this projection by $U_l^m(M, x_0, y_0)$. Then, for functions $\varphi, \psi \in C_0^\infty(U_l^m)$, we have the representation of the form (4.2.16). Let $z_0 = f(x_0, y_0)$ be the equation of the piece $U_l^m \cap S_l^m$. We assume that $f(x_0, y_0)$ is extended to $\mathbb{R}_2$ as a function in C_0^∞. Then, according to (4.2.13), the operator $A(\varphi\nu, m, l)$ can be represented as a pseudodifferential operator $\hat{A}_l^m$ with symbol $\hat{A}_l^m \in S_1^0$. We represent this operator as a sum of two pseudodifferential operators, $\hat{A}(m, l) = \hat{A}_1(m, l) + \hat{A}_2(m, l)$, with symbols $\hat{A}_{1l}^m(x, y, \xi) = \hat{A}_l^m(x, y, \xi)(1 - \chi(\xi))$ and $\hat{A}_{2l}^m(x, y, \xi) = \hat{A}_l^m(x, y, \xi)\chi(\xi) \in S_{-\infty}^0$, where $\chi(\xi) \in C_0^\infty(\mathbb{R}_2)$, $\chi(\xi) = 1$ for $|\xi| \le 1$. Since $\hat{A}_{l1}^m \in S_1^\infty$, we can use Theorem 3.2.2, which ensures that there exists a neighborhood $U_l'^m \subset U_l^m$ of the point M such that for any $\psi \in C_0^\infty(U_l'^m)$, we have

$$\psi\hat{A}_{1l}^m = \psi\left(\hat{A}_{1l}^m(x_0, y_0, D) + K_l(x, y, D) + T_{1l}(x, y, D)\right) =$$
$$= \psi\left(\hat{A}_l^m(x_0, y_0, D) + K_l(x, y, D) + T_{2l}(x, y, D)\right), \tag{4.2.19}$$

where $\operatorname{ord} T_{il}(x, y, D) \le 0$, $i = 1, 2$, and for the operator $K_l(x, y, D)$ and any fixed $\varepsilon \in (0, 1)$, the estimate

$$\|K_l(x, y, D)\| \le \frac{1}{2\|R_{0l}^m\|_r} \tag{4.2.20}$$

holds uniformly in r for $\varepsilon < r \le 1-\varepsilon$, where $R_{0l}^m : H_{r-1}(U_l^m) \to \hat{H}_r(U_l^m)$, $R_{0l}^m \hat{A}_l^m(x_0, y_0, D) = I_1$, $p\hat{A}_l^m(x_0, y_0, D)R_{0l}^m = I_2$; I_1 and I_2 are identity operators, p is the operator of restriction to U_l^m. The existence of the operator R_{0l}^m follows from Theorem 3.3.6. By (4.2.20), there is an operator R_l^m such that $R_l^m p\tilde{A}_l(x, y, D) = I_1$, where $\tilde{A}_l^m(x, y, D) = A_l^m(x_0, y_0, D) + K(x, y, D)$; (x_0, y_0) are the coordinates of the point M.

By definition, each surface S_l^1 belongs to one of the closed surfaces $M_1, \ldots, M_m$ (see the beginning of Section 4.1). Let $S_l^1 \subset M_j(l)$. For each point of the boundary Γ_l of S_l^1, we take a neighborhood U_l^1 for which the following condition holds: the projection of the domain $D_l^1 = S_l^1 \cap U_l^1$ to the plane $X_0 O Y_0$ of some normal coordinate system is a domain with C^∞ boundary; such a neighborhood can always be found, because the boundary Γ_l^1 of the surface S_l^1 is of class C^∞. Now, arguing as above, we find a neighborhood U_l'' of the point $M \in \Gamma_l^1$ for which (4.2.19) and (4.2.20) hold, where $R_{0l}^1 : H_{r-1}(D_l^1) \to \hat{H}_r(D_l^1)$. It follows that there is an operator R_l^1 such that

$$R_l^1 p \widetilde{A}_l^1(x, y, D) = I_1, \qquad p \widetilde{A}_l^1(x, y, D) R_l^1 = I_2, \tag{4.2.21}$$

where I_1, I_2 are identity operators, $\widetilde{A}_l^1(x, y, D) = \hat{A}_l^1(x_0, y_0, D) + K_l^1(x, y, D)$.

From the covering constructed above for each surface S_l^m, we choose a finite covering U_{il}^{1m}, $1 \le i \le N(m, l)$ (for each surface S_l^1 we have a covering of $\overline{S_l^1}$), with the sets U_{il}^{1m} sufficiently small, as specified in Proposition 4.2.1.

Let $\varphi_{il}^m \in C_0^\infty(U_{il}^{1m})$ be a partition of unity subjected to the covering $\{U_{il}^{1m}\}$, and $\sum_i \varphi_{il}^1 \equiv 1$ on $\overline{S_l^1}$. Consider functions $\psi_{il}^m \in C_0^\infty(U_{il}^{1m})$, $\psi_{il}^m = 1$ on a set G_{il}^m such that $\overline{G_{il}^m} \subset U_{il}^{1m}$, $\operatorname{supp} \varphi_{il}^m \subset G_{il}^m$. We define the operator

$$R_l^m = \sum_i \psi_{il}^m R_l^{im} \varphi_{il}^m, \tag{4.2.22}$$

where the operators R_l^{im} are specified by (4.2.20), (4.2.21). Let us introduce the operator

$$R = \{R_l^m\}, \qquad m = 0: l = 1, \ldots, k; \qquad m = 1: l = 1, 2, \ldots, n. \tag{4.2.23}$$

Using Proposition 2.4.1, as it has been done in Theorem 3.3.6, we can prove that R is a bounded operator from $\hat{H}_{r-1}(\nu)$ to $H_r(\nu)$ for $0 < r < 1$.

Let us show that R is the left regularizer of the Prandtl operator $\Pi(\nu)$. Indeed, for any function $\nu(x) \in H_r(\nu)$ we find, with the help of (4.2.14), that

$$R\Pi(\nu) = RL(\nu) + RN(\nu). \tag{4.2.24}$$

Since $N(\nu)$ always belongs to $\hat{H}_0(\nu)$, we can represent the operator $RN(\nu)$ as

$$RN(\nu) = R \operatorname{id}_{r-1}^0 N(\nu), \tag{4.2.25}$$

where $\operatorname{id}_{r-1}^0$ is the imbedding operator. As shown by *Shubin* (1978), the imbedding operator $\operatorname{id}_{s_2}^{s_1}$ is compact for $s_1 > s_2$. Hence we deduce that $RN(\nu)$ is a compact operator in $H_r(\nu)$. From (4.2.24), we obtain $RL(\nu) = \{R_l^m L_l^m\}$, $m = 0: l = 1, \ldots, k; m = 1: l = 1, \ldots, n$. Further, from (4.1.40)–(4.1.43) and (4.2.22), we find that

$$R_l^m L_l^m = \sum_i \psi_{il}^m R_l^{im} \varphi_{il}^m A_l^m(\psi_{il}^m) + \sum_i \psi_{il}^m R_l^{im} \varphi_{il}^m A_l^m\big((1 - \psi_{il}^m)\nu\big) + R_l^m A_{4l}^m(\nu). \tag{4.2.26}$$

Since $A_{4l}^m(\nu) \in \hat{H}_0(\nu)$, we can use the above reasoning to show that the operator $R_l^m A_{4l}^m(\nu)$ is compact in $H_r(\nu)$.

The functions φ_{il}^m and $(1 - \psi_{il}^m)$ are supported on mutually disjoint sets, and therefore, $\varphi_{il}^m A_l^m((1 - \psi_{il}^m)\nu)$ is a bounded operator from $H_r(\nu)$ to $H_0(\nu)$. Consequently, the operator $\sum_i \psi_{il}^m R_l^{im} \varphi_{il}^m A_l^m\big((1 - \psi_{il}^m)\nu\big)$ is compact in $\hat{H}_r(S_i^1)$ $(H_r(S_i^0))$.

Using the commutation formula (4.2.18), where p_{il}^m is the operator of restriction either to U_{il}^m or D_{il}^1, and taking into account relations (4.1.33), (4.2.21), we get

$$\sum_i \psi_{il}^m R_l^{im} \varphi_{il}^m A_l^m(\psi_{il}^m \nu) = \nu_l^m + \sum_i \psi_{il}^m R_l^{im} p_{il}^m T_{il}^m(\nu_l^m) = \nu_l^m + T_l^m \nu, \tag{4.2.27}$$

where $T_{il}^m(\nu_l^m)$ are bounded operators from $\mathring{H}_r(S_i^1)$ $(H_r(S_i^0))$ to $L_2(S_l^m)$.

Representing the operator T_l^m in terms of imbedding operators id_{r-1}^0, we see that T_l^m is a compact operator in $H_r(\nu)$. It follows from (4.2.27) and the above reasoning that R is the left regularizer. Note that we have

$$R\Pi(\nu) = \nu + T_1\nu, \qquad (4.2.28)$$

where $T_1\nu \in H_{\frac{1}{2}+\varepsilon}(\nu)$, $0 < \varepsilon < 1/2$.

Consider functions $\psi_{il}^{1m} \in C_0^\infty(U_{il}^{1m})$ such that $\psi_{il}^{1m} = 1$ on an open set G_{il}^{1m} for which $\overline{G_{il}^{1m}} \subset U_{il}^{1m}$, $\mathrm{supp}\,\psi_{il}^m \subset G_{il}^{1m}$. Then, for any $\nu(x) \in \widehat{H}_{r-1}(\nu)$ we obtain

$$\Pi R(\nu) = LR(\nu) + NR(\nu). \qquad (4.2.29)$$

The operator $R(\nu)$ is bounded from $\widehat{H}_{r-1}(\nu)$ to $H_r(\nu)$, and the operator N is bounded from $H_r(\nu)$ to $\widehat{H}_0(\nu)$. Representing the operator $NR(\nu)$ as $\mathrm{id}_{r-1}^0 NR(\nu)$, we see that $NR(\nu)$ is a compact operator in $\widehat{H}_{r-1}(\nu)$.

The operator $LR(\nu)$ has the form $LR(\nu) = \{L_l^m R_l^m\}$. Therefore, by analogy with (4.2.26), we find that

$$L_l^m R_l^m(\nu_l^m) = \sum \psi_{il}^{1m} A_l^m \psi_{il}^m R_l^{im} \varphi_{il}^m \nu_l^m + \sum_i \left(1 - \psi_{il}^{1m}\right) A_l^m \psi_{il}^m R_l^{im} \varphi_{il}^m \nu_l^m + A_{4l}^m R_l^m \nu_l^m =$$

$$= B_{1l}^m \nu_l^m + B_{2m}^l \nu_l^m + B_{2m}^l \nu_l^m. \qquad (4.2.30)$$

Since $(1 - \psi_{il}^{1m})$ and ψ_{il}^m are supported on mutually disjoint sets, the operator $(1 - \psi_{il}^{1m}) A_l^m \psi_{il}^m$ is bounded from $\mathring{H}_r(S_i^1)$ $(H_r(S_i^0))$ to $\widehat{H}_0(S_l^m)$. Hence, arguing as above, we conclude that B_{2m}^l is a compact operator in $H_{r-1}(S_l^m)$. The operator A_l^{4m} is bounded from $\mathring{H}_r(S_l^1)$ $(H_r(S_l^0))$ to $\widehat{H}_0(S_l^0)$, which implies that B_{3m} is a compact operator in $H_{r-1}(S_l^m)$.

On the basis of Remark 4.2.1, by analogy with (4.2.27), we obtain

$$B_{1l}^m \nu_l^m = \sum_i \psi_{il}^{1m} A_l^m \psi_{il}^m R_l^{im} \varphi_{il}^m \nu_l^m = \nu_l^m + \sum_i \psi_{il}^{1m} p_{il}^m T_{il}^{1m} R_l^{im} \varphi_{il}^m \nu_l^m = \nu_l^m + B_{4m}^l \nu_l^m, \quad (4.2.31)$$

where T_{il}^{1m} are bounded operators from $\mathring{H}_r(S_i^1)$ $(H_r(S_i^0))$ to $L_2(S_i^m)$. It follows that the operator B_{4m}^l is compact in $H_{r-1}(S_l^m)$. From (4.2.29)–(4.2.31), it follows that R is the right regularizer of the Prandtl operator. Theorem 3.3.2 shows that $\Pi(\nu)$ is a Noetherian operator. Theorem 4.2.2 is proved.

4.3. Index of the Prandtl Operator

In this section, we are going to find the index of the Prandtl operator and study the Neumann problem.

Proposition 4.3.1. *For $\nu \in H_r(\nu)$, $r \geq 1/2$, the function*

$$u(x) = \sum_p \sum_i \int_{S_i^p} \nu_i^p(y) \frac{\cos\varphi_{xy}}{|x-y|^2}\, dS_{iy}^p \qquad (4.3.1)$$

belongs to $H_1(\Omega)$, where $\Omega = \mathbb{R}_3 \setminus \left(\bigcup_p \bigcup_i S_i^p\right)$.

Proof. Assume first that $\nu_i^0 \in C^\infty(S_i^0)$, $\nu_i^1 \in C_0^\infty(S_i^1)$. Clearly, it suffices to prove our statement for a single surface. Consider the expression

$$\int_\Omega \left[u_i^p(x)\right]^2 dx = \int_{\Omega_1} \left[\iint_{S_i^p} \nu_i^p(y) \frac{\cos\varphi_{xy}}{|x-y|^2}\, dS_{iy}^p\right]^2 dx + \int_{\Omega_2} \left[u_i^p(x)\right]^2 dx = I_1 + I_2,$$

where $\Omega = \Omega_1 \cup \Omega_2$, Ω_1 is the exterior of a sufficiently large ball containing all surfaces S_i^p. The integral I_1 obviously exists. Let us estimate I_2. We have

$$I_2 \leq \int_{\Omega_2} \left[\int_{S_i^p} |\nu_i^p(y)| \frac{|\cos \varphi_{xy}|}{|x-y|^2} \, dS_{iy}^p \right]^2 dx \leq$$

$$\leq \int_{\Omega_2} \left[\int_{S_i^p} \frac{|\cos \varphi_{xy}|}{|x-y|^2} \, dS_{iy}^p \right] \left[\int_{S_i^p} |\nu_i^p(y)|^2 \frac{|\cos \varphi_{xy}|}{|x-y|^2} \, dS_{iy}^p \right] dx \leq$$

$$\leq K \int_{S_i^p} |\nu_i^p(y)|^2 \left\{ \int_{\Omega_2} \frac{|\cos \varphi_{xy}|}{|x-y|^2} \, dx \right\} dS_{iy}^p \leq$$

$$\leq K K_1 \int_{S_i^p} |\nu_i^p(y)|^2 \, dS_{iy}^p, \tag{4.3.2}$$

where we have used the inequalities (see *Vladimirov* (1976))

$$\int_{S_i^p} \frac{|\cos \varphi_{xy}|}{|x-y|^2} \, dS_{iy}^p \leq K, \quad \forall x \in \Omega_2; \qquad \int_{\Omega_2} \frac{|\cos \varphi_{xy}|}{|x-y|^2} \, dx \leq K_1, \quad \forall y \in S_i^p.$$

By assumption, the sets S_i^1 belong to smooth surfaces M_j. We consider normals to S_i^1 which coincide with exterior normals to M_j. Let $S_{i\delta\pm}^p$ be the surfaces formed by the points x' on the normals to S_i^p at the points $x \in S_i^p$, the distance between x' and x being equal to δ. By $\Omega_{i\delta\pm}^p$ we denote the domains bounded by the surfaces $S_{i\delta\pm}^p$, respectively. Let Ω_R be a ball of a sufficiently large radius such that its surface, the sphere S_R, embraces all domains $\Omega_{i\delta\pm}^p$. Consider the set

$$D_{\delta R} = \Omega_R \setminus \left\{ \bigcup_p \bigcup_i \left(\Omega_{i\delta+}^p \setminus \Omega_{i\delta-}^p \right) \right\}$$

Using the Green formula, we find that

$$\lim_{\delta \to 0} \frac{1}{4\pi} \int_{D_{\delta R}} |\text{grad } u|^2 \, dx = \lim_{\delta \to 0} \left[\frac{1}{4\pi} \sum_i \sum_p \left(\int_{S_{i\delta-}^p} \frac{\partial u}{\partial n} \, dS - \int_{S_{i\delta+}^p} \frac{\partial u}{\partial n} \, dS \right) \right] +$$

$$+ \frac{1}{4\pi} \int_{S_R} \frac{\partial u}{\partial n} u \, dS = -\frac{1}{4\pi} \sum_i \sum_p \int_{S_i^p} \frac{\partial u}{\partial n} \nu_i^p(y) \, dS_{iy}^p + \frac{1}{4\pi} \int_{S_R} \frac{\partial u}{\partial n} u \, dS_R. \tag{4.3.3}$$

For large R, the following estimates hold

$$\left| \frac{\partial u}{\partial n} \right| \leq \frac{C}{R^3}, \qquad |u| \leq \frac{C}{R^2},$$

and therefore,

$$\lim_{R \to \infty} \frac{1}{4\pi} \int_{S_R} \frac{\partial u}{\partial n} u \, dS_R = 0.$$

Now, from (4.3.3), it follows that

$$\frac{1}{4\pi} \int_\Omega |\text{grad } u|^2 \, dx = -\sum_i \sum_p \int_{S_i^p} \frac{\partial u}{\partial n} \nu_i^p(y) \, dS. \tag{4.3.4}$$

Since $\nu_i^p(y) \in \mathring{H}_r(S_i^1)$ $(H_r(S_i^0))$, Proposition 4.2.1 yields $\partial u / \partial n \in H_{r-1}(S_i^p)$. Then, for $r \geq 1/2$, we have $\partial u / \partial n \in H_{-r}(S_i^p)$, and therefore, the right hand side of (4.3.4) exists for $\nu \in H_r(\nu)$ with

$r \geq 1/2$. Hence, using the density of the functions $\nu_i^p \in C_0^\infty$ in $H_r(\nu)$, we obtain relations (4.3.2) and (4.3.4) for $\nu \in H_r(\nu)$. Proposition 4.3.1 is proved.

Next, we examine the jump of the double layer potential for $\nu \in H_r(\nu)$. Let us introduce the function (see *Vladimirov* (1976))

$$W_i^p(\delta, x) = \int_{S_i^p} \left[\nu_i^p(y) - \nu_i^p(x)\right] \frac{\cos \varphi_{x'y}}{|x' - y|^2} \, dS_{iy}^p, \qquad x' \in S_{i\delta\pm}^p. \tag{4.3.5}$$

Proposition 4.3.2. *If* $\nu_i^p(y) \in \mathring{H}_r(S_i^1)$ $(H_r(S_i^0))$, $r > 0$, *then* $W_i^p(\delta, x) \to W_i^p(0, x)$ *in the norm of* $L_2(S_i^p)$ *as* $\delta \to 0$.

Proof. Assume first that $\nu_i^p(x) \in C_0^\infty(S_i^p)$. For any $x \in S_i^p$, consider its neighborhood U_x consisting of all y such that $|x - y| < \varepsilon_1$ and for $\varepsilon > 0$, we have

$$\int_{U_x} \frac{|\cos \varphi_{xy}|}{|x - y|^2} \, dS_{iy}^p < \varepsilon. \tag{4.3.6}$$

Since the surface is smooth, for any $\varepsilon > 0$ we can find $\delta > 0$ such that if x, x', y satisfy the conditions $|x - x'| < \delta$, $y \notin U_x$, then (see *Vladimirov* (1976)) the following uniform inequality holds

$$\frac{|\cos \varphi_{x'y}|}{|x' - y|^2} - \frac{|\cos \varphi_{xy}|}{|x - y|^2} < \varepsilon. \tag{4.3.7}$$

We can write

$$W_i^p(\delta, x) - W_i^p(0, x) = \int_{U_x} \left[\nu_i^p(y) - \nu_i^p(x)\right] \frac{\cos \varphi_{x'y}}{|x' - y|^2} \, dS_{iy}^p +$$

$$+ \int_{U_x} \left[\nu_i^p(y) - \nu_i^p(x)\right] \left[-\frac{\cos \varphi_{xy}}{|x - y|^2}\right] dS_{iy}^p +$$

$$+ \int_{S_i^p \setminus U_x} \left[\nu_i^p(y) - \nu_i^p(x)\right] \left[\frac{\cos \varphi_{x'y}}{|x' - y|^2} - \frac{\cos \varphi_{xy}}{|x - y|^2}\right] dS_{iy}^p = I_1 + I_2 + I_3. \tag{4.3.8}$$

Using (4.3.6), (4.3.7), and the Jensen inequality (see *Zygmund* (1965)), we find

$$\int_{S_i^p} I_3^2 \, dS_{ix}^p \leq \int_{S_i^p} \left[\int_{S_i^p \setminus U_s} \left|\frac{\cos \varphi_{x'y}}{|x' - y|^2} - \frac{\cos \varphi_{xy}}{|x - y|^2}\right| dS_{iy}^p\right] \times$$

$$\times \left[\int_{S_i^p \setminus U_x} |\nu_i^p(y) - \nu_i^p(x)|^2 \left|\frac{\cos \varphi_{x'y}}{|x' - y|^2} - \frac{\cos \varphi_{xy}}{|x - y|^2}\right| dS_{iy}^p\right] dS_{ix}^p \leq C\varepsilon \|\nu_i^p\|_{L_2(S_i^p)}^2. \tag{4.3.9}$$

$$\int_{S_i^p} |I_2| \, dS_{ix}^p \leq \int_{S_i^p} \left[\int_{U_x} \frac{|\cos \varphi_{xy}|}{|x - y|^2} \, dS_{iy}^-\right] \cdot \left[\int_{U_x} |\nu_i^p(y) - \nu_i^p(x)|^2 \frac{|\cos \varphi_{xy}|}{|x - y|^2} \, dS_{iy}^p\right] dS_{ix}^p \leq$$

$$\leq C_1 \varepsilon \int_{S_i^p} \left\{\int_{U_x} |x - y|^{1+2r} \frac{|\nu_i^p(y) - \nu_i^p(x)|}{|x - y|^{2+2r}}\right\} dS_{ix}^p \leq C_2 \varepsilon \|\nu_i^p\|_{\mathring{H}_r(S_i^p)}^2. \tag{4.3.10}$$

Since the surfaces S_i^p are of class C^∞, there is a small ε_1 such that for $|x - y| \leq \varepsilon_1$, the inequality

$$|x' - y| \geq C_0 |x - y|, \qquad C_0 > 0,$$

holds uniformly in x. Hence we obtain the estimate

$$\int_{S_i^p} |I_1|^2 \, dS_{ix}^p \leq \int_{S_i^p} \left[\iint_{U_x} \frac{|\cos \varphi_{x'y}|}{|x'-y|^2} \, dS_{iy}^p \right] \cdot \left[\iint_{U_x} |\nu_i^p(y) - \nu_i^p(x)|^2 \frac{|\cos \varphi_{x'y}|}{|x'-y|^2} \, dS_{iy}^p \right] dS_{ix}^p \leq$$

$$\leq \frac{K_1}{C_0} \int_{S_i^p} \left[\iint_{U_x} |x-y|^{2r} \frac{|\nu_i^p(y) - \nu_i^p(x)|}{|x-y|^{2+2r}} \, dS_{iy}^p \right] dS_{ix}^p \leq$$

$$\leq \frac{K_1 \varepsilon_1^{2r}}{C_0} \int_{S_i^p} \left[\iint_{U_x} \frac{|\nu_i^p(y) - \nu_i^p(x)|^2}{|x-y|^{2+2r}} \, dS_{iy}^p \right] dS_{ix}^p \leq A \varepsilon_1^{2r} \|\nu_i^p(x)\|_{\dot{H}_r(S_i^p)}^2. \tag{4.3.11}$$

Since functions $\nu_i^p \in C^\infty(S_i^p)$ are dense in $H_r(S_i^p)$ ($\dot{H}_r(S_i^p)$), the estimates (4.3.9)–(4.3.11) prove Proposition 4.3.2.

Proposition 4.3.2 implies the usual limit relations (the limit is understood in the sense of Proposition 4.3.2)

$$u_{i+}^p = 2\pi \nu_i^p + \int_{S_i^p} \nu_i^p(y) \frac{\cos \varphi_{xy}}{|x-y|^2} \, dS_{iy}^p, \quad x \in S_i^p,$$

$$u_{i-}^p = -2\pi \nu_i^p + \int_{S_i^p} \nu_i^p(y) \frac{\cos \varphi_{xy}}{|x-y|^2} \, dS_{iy}^p, \quad x \in S_i^p, \tag{4.3.12}$$

Theorem 4.3.1. *Suppose that the operator* $\Pi(\nu)$ *maps* $H_r(\nu)$ *into* $\hat{H}_{r-1}(\nu)$, $0 < r < 1$. *Then,* $\dim \ker \Pi = k$, $\dim \operatorname{coker} \Pi = k$, *and equations (4.1.3) have solutions if*

$$\int_{S_i^p} f_i^0(y) \, dS_{iy}^0 = 0, \quad i = 1, 2, \ldots, k. \tag{4.3.13}$$

Proof. Suppose that for some $\nu \in H_r(\nu)$, $r \geq 1/2$, we have $\Pi(\nu) = 0$. Then, by (4.3.4), $u = \text{const}$ in any connected region. Since $u(\infty) = 0$, we see from (4.3.12) that $\nu_i^1 = 0$, $i = 1, \ldots, n$, whereas $\nu_i^0 = C_i$, $i = 1, \ldots, k$, on any of the closed surfaces. It follows that $\dim \ker \Pi = k$, where k is the number of given closed surfaces.

Let us find the kernel of the conjugate operator $\Pi^*(\nu)$ from $H_{1-r}(\nu)$ to $\hat{H}_{-r}(\nu)$. Note that the conjugate operator $\Pi^*(\nu)$ has the same form as $\Pi(\nu)$. As shown in the proof of Theorem 4.2.2, the application of the left regularizer to equations (4.1.3) corresponds to the original equation being transformed to (see (4.2.28))

$$\nu + T\nu = Rf,$$

where the operator $T\nu$ maps $H_r(\nu)$, $0 < r < 1$, to $H_l(\nu)$, $l > 1/2$. Therefore, if $f = 0$, then $\nu \in H_l(\nu)$. Thus, as we have shown above, $\nu_i^1 = 0$, $i = 1, \ldots, n$; $\nu_i^0 = C_i$, $i = 1, \ldots, k$. It follows that $\dim \operatorname{coker} \Pi = k$. In a similar way, the statement of the theorem is established for $0 \leq r \leq 1/2$. Thus, the index of the operator $\Pi(\nu)$ (see Definition 3.3.5) is equal to zero, $\chi = 0$, which means that $\Pi(\nu)$ is an operator of Fredholm type (see Definition 3.3.7). The solvability conditions for equations (4.1.3) have the form

$$\int_{S_i^0} f_i^0(y) \, dS_{iy}^0 = 0.$$

Theorem 4.3.1 is proved.

Consider the Neumann problem

$$\Delta u = 0, \quad x \in \Omega = \mathbb{R}_3 \setminus \left(\bigcup_p \bigcup_i S_i^p \right),$$

$$\left. \frac{\partial u}{\partial n} \right|_{S_i^p} = f_i^p, \quad \int_{S_i^0} f_i^0 \, dS_{iy}^0 = 0, \tag{4.3.14}$$

$$p = 0: i = 1, 2, \ldots, k; \quad p = 1: i = 1, 2, \ldots, n,$$

where Δ is the Laplace operator, $f_i^p \in H_{r-1}(S_i^p)$, $1/2 \leq r < 1$.

Definition 4.3.1. A function $u(x) \in H_1(\Omega)$ is called a *generalized solution* of problem (4.3.14) if there exists a sequence $u_m(x) \in H_1(\Omega)$ with the following properties:

(i) $u_m(x) \to u(x)$ as $m \to \infty$ in $H_1(\Omega)$, $\Delta u_m(x) = 0$ for $x \in \Omega$;

(ii) $u_m(x)$ are twice continuously differentiable in Ω;

(iii) $\dfrac{\partial u_m}{\partial n}\Big|_{S_i^p} = f_{im}^p \in C^1(S_i^p)$, f_{im}^p to f_i^p in the norm of $H_{r-1}(S_i^p)$, $1/2 \le r < 1$;

(iv) $\displaystyle\int_{S_i^0} f_{im}^0 \, dS_{iy}^0 = 0$.

Definition 4.3.2. *A solution of the exterior Neumann problem* is defined as the restriction of the generalized solution of problem (4.3.14) to the set $\mathbb{R}_3 \setminus \left[\bigcup_{i,j} \left(\overline{\Omega_i^0} \cup S_j^1 \right) \right]$, where Ω_i^0 are the domains whose boundaries coincide with the surfaces S_i^0.

Theorem 4.3.2. *For $1/2 \le r < 1$, the exterior Neumann problem admits one and only one generalized solution.*

Proof. The existence of the solution is ensured by the representation of the solution of problem (4.3.14) as a double layer potential and the solvability condition for system (4.1.3) (see Theorem 4.3.1).

Let us show that the solution is unique. It suffices to establish the uniqueness in the case of zero boundary values. Let $u(x) \in H_1(\Omega)$ be a generalized solution of problem (4.3.14) with zero boundary conditions. Then, there is a sequence $\{u_m(x)\} \in H_1(\Omega)$ for which the assumptions of Definition 4.3.1 are valid. Denote by $D(R_0, R)$ the domain between two balls of radii R_0, R centered at the origin, with all surfaces S_i^p inside the ball of radius $R_0 < R$. We have

$$I_m(R_0, R) = \int_{D(R_0,R)} \left[|\mathrm{grad}\, u_m|^2 + u_m^2 \right] dx = I_{1m}(R_0, R) + I_{2m}(R_0, R) = \int_{R_0}^{R} G_m \, dr, \quad (4.3.15)$$

where (S is the unit sphere)

$$G_m(r) = r^2 \int_S \left[|\mathrm{grad}\, u_m|^2 + u_m^2 \right] dS = G_{1m}(r) + G_{2m}(r).$$

Since the integral (4.3.15) is absolutely convergent as $R \to \infty$, there is a sequence R_q such that $G_{1m}(R_q) \to 0$, $G_{2m}(R_q) \to 0$. Then, using the Cauchy inequality, we find that

$$\left| \int_{S_{R_q}} \frac{\partial u_m}{\partial n} u_m \, dS_{R_q} \right| \le \left(\int_{S_{R_q}} |\mathrm{grad}\, u_m|^2 \, dS_{R_q} \right)^{1/2} \left(\int_{S_{R_q}} |u_m|^2 \, dS_{R_q} \right)^{1/2} \to 0, \quad (4.3.16)$$

as $R_q \to \infty$, where S_{R_q} is the sphere of radius R_q. By the Green formula, from (4.3.16) we get

$$\lim_{R_q \to \infty} I_{1m}(R_0, R_q) = \lim_{R_q \to \infty} \left[-\int_{S_{R_0}} \frac{\partial u_m}{\partial n} u_m \, dS_{R_0} + \int_{S_{R_q}} \frac{\partial u_m}{\partial n} u_m \, dS_{R_q} \right] =$$

$$= -\int_{S_{R_0}} \frac{\partial u_m}{\partial n} u_m \, dS_{R_0}. \quad (4.3.17)$$

From (4.3.17), it follows that

$$\int_{\mathbb{R}_3} |\mathrm{grad}\, u_m|^2 \, dx = \sum_p \sum_i \int_{S_i^p} \frac{\partial u_m}{\partial n} \nu_{im}^p \, dS_{iy}^0 = I_{1m}. \quad (4.3.18)$$

Since $\partial u_m / \partial n \to 0$ in $H_{r-1}(S_i^p)$ for $r - 1 \ge -1/2$, and $\{\nu_{im}^p\} \in H_r(\nu)$, we have $I_{1m} \to 0$ as $m \to \infty$. Now, it follows from (4.3.17) that $u(x) = C$ for $x \in \mathbb{R}_3 \setminus \left[\bigcup_{i,j} \left(\overline{\Omega_i^0} \cup S_j^1 \right) \right]$. Since $u \in H_1(\Omega)$, we have $u = 0$, which proves the uniqueness of the solution of the exterior Neumann problem with zero boundary conditions. Theorem 4.3.2 is proved.

4.4. Equation of the Double Layer Potential in the Plane Case

In $\mathbb{R}_2$, consider closed connected curves $S_1^0, \ldots, S_k^0$ of class C^∞ (compact connected C^∞ manifolds of dimension 1; see Definition 2.4.6).

Suppose also that in $\mathbb{R}_2$ there are non-closed connected curves $S_1^1, \ldots, S_n^1$ of class C^∞ (i.e., compact one-dimensional C^∞ manifolds with border) such that $\overline{S_i^1} \cap \overline{S_j^1} = \varnothing$ for $i \neq j$.

According to Definition 2.4.7, we assume that there exist one-dimensional compact manifolds (without border) $M_1, \ldots, M_{m_1}$ of class C^∞ such that every surface S_j^1, $j = 1, \ldots, n$, belongs to one of the manifolds $M_1, \ldots, M_{m_1}$, and several surfaces S_i^1 may belong to one and the same manifold M_j.

The surfaces S_i^0 coincide with the boundaries of bounded domains Ω_i^0, and the surfaces M_i are the boundaries of bounded domains Ω_i^1. Moreover, $\overline{\Omega_i^0} \cap \overline{\Omega_j^0} = \varnothing$, $\overline{\Omega_i^1} \cap \overline{\Omega_j^1} = \varnothing$ for $i \neq j$, and $\overline{\Omega_i^0} \cap \overline{\Omega_j^1} = \varnothing$ for all i, j.

As the normal to the curves we choose the normal exterior to the domains Ω_i^k.

Consider the boundary value problem

$$
\begin{aligned}
\Delta u &= 0, \qquad x \in \mathbb{R}_2 \setminus (\cup S_i^p), \\
\frac{\partial u}{\partial n}\Big|_{S_i^p} &= f_i^p, \qquad p = 0: i = 1, \ldots, k; \quad p = 1: i = 1, \ldots, n,
\end{aligned}
\tag{4.4.1}
$$

where Δ is the Laplace operator, $\partial u/\partial n\big|_{S_i^p}$ is the derivative along the normal to the curve S_i^p.

Let us seek a solution of problem (4.4.1) as the double layer potential

$$
u(x) = \sum_p \sum_i \frac{1}{2\pi} \int_{S_i^p} \nu_i^p(y) \frac{\cos \varphi_{xy}}{|x - y|} \, dS_{iy}^p,
\tag{4.4.2}
$$

where $x = (x_1, x_2) \in \mathbb{R}_2$ $y = (y_1, y_1) \in S_i^p$, φ_{xy} is the angle between the vector $x - y$ and the normal $\boldsymbol{n}_y$ to the curve S_i^p at the point y; $\nu_i^p(y)$ is the density of the double layer potential. For the unknown function ν_i^p we obtain the following equations

$$
\sum_i \sum_p \frac{1}{2\pi} \frac{\partial}{\partial n_l^m} \int_{S_i^p} \nu_i^p(y) \frac{\cos \varphi_{xy}}{|x - y|} \, dS_{iy}^p = f_l^m,
\tag{4.4.3}
$$

$$
m = 0: l = 1, \ldots, k; \quad m = 1: l = 1, \ldots, n.
$$

The investigation of this problem mainly follows Sections 4.1–4.3, but is much simpler, and thus, our exposition will be brief. For $x \in \mathbb{R}_2 \setminus \left(\cup_p \cup_i S_i^p \right)$, the left-hand side of (4.4.3), which we denote by $\Pi_l^m(\nu)$, always exists. Just as in Sect 4.1, let us find

$$
\lim_{x \to S_l^m} \Pi_l^m(\nu).
$$

It suffices to consider a single curve S_l^m and a single function $\nu_l^m(y)$, which we denote by σ and $\nu(y)$, respectively. Then $\Pi(\nu)$ becomes

$$
\Pi(\nu) = \frac{1}{2\pi} \int_\sigma \nu(y) \frac{|\overline{M_0 M}|^2 \left(n_{M_0}, n_{\overline{M}}\right) - 2\left(\overline{M_0 M}, n_{\overline{M}}\right)\left(\overline{M_0 M}, n_{M_0}\right)}{|M_0 M|^4} \, d\sigma_{M_0},
\tag{4.4.4}
$$

where $M_0 \in \sigma$, $\overline{M} \in \sigma$, $M \notin \sigma$, M lies on the normal passing through $\overline{M}$.

Definition 4.4.1. $Y_0 O X_0$ is said to be *a special coordinate system* (SCS) if the axis $O X_0$ is parallel to the tangent to the curve σ at a fixed point $\overline{M} \in \sigma$.

For the point $\overline{M}$, consider its neighborhood U such that the piece $\sigma \cap U$ is given by the equation $y_0 = f(x_0)$ in the SCS. Let $\varphi(P)$, $P \in \sigma$, be an infinitely differentiable function compactly supported in U, $\varphi(P) = 1$ in a neighborhood U_1 such that $\overline{M} \in U_1$. Then, $\Pi(\nu) \doteq \Pi(\varphi \nu) + \Pi[(1 - \varphi)\nu]$. In the second term, we can pass to the limit as $M \to \overline{M}$, and it remains to pass to the limit in the first term. Assume first that $\nu(M_0) \in C_0^\infty(\sigma)$. Let $(x, f(x))$ and $(x_0, f(x_0))$ be the coordinates of the points $\overline{M}$, $M_0 \in \sigma$, respectively. Then

$$
n_{M_0} = \frac{-f'(x_0)\,i + j}{\sqrt{1 + f'^2(x_0)}}, \qquad n_{\overline{M}} = j.
$$

Let $M(x, f(x){-}t))$ be the current point of the normal to σ passing through the point $\overline{M}(x, f(x))$. Then, by analogy with (4.1.18) we obtain the relation

$$
\lim_{t \to 0} \Pi(\varphi \nu) = \frac{1}{2\pi} \lim_{\varepsilon \to 0} \left[\int_{I \setminus U_\varepsilon^*} \frac{\varphi \nu \, dx_0}{(x - x_0)^2 + [f(x) - f(x_0)]^2} - \frac{2\varphi(x)\nu(x)}{\varepsilon} \right] -
$$
$$
- \frac{1}{\pi} \int_U \frac{\nu(y)}{|M_0 \overline{M}|^4} \left(\overline{M_0 \overline{M}}, n_{\overline{M}} \right) \left(\overline{M_0 \overline{M}}, n_{M_0} \right) d\sigma_{M_0}, \tag{4.4.5}
$$

where $I = [a, b]$ is the segment containing the projection of U to OX_0, $U_\varepsilon^* = [x - \varepsilon, x + \varepsilon]$. On the basis of the first term on the right-hand side of (4.4.5), we can give the following definition.

Definition 4.4.2. The relation

$$
\int_\sigma \frac{\nu \, d\sigma_{M_0}}{|M_0 \overline{M}|^2} = \lim_{\varepsilon \to 0} \left[\int_{\sigma \setminus U_\varepsilon} \frac{\nu \, d\sigma_{M_0}}{|M_0 \overline{M}|^2} - \frac{2\nu(\overline{M})}{\varepsilon} \right] \tag{4.4.6}
$$

is taken as a definition of the *integral in the sense of Hadamard*, where U_ε is a neighborhood of the point $\overline{M}$ whose projection to the tangent line at $\overline{M}$ coincides with the interval of length 2ε having $\overline{M}$ as its midpoint.

Consider a coordinate system $X_0 O Y_0$ such that the tangent line to σ at $\overline{M} \in \sigma$ forms the angle α with the axis $O X_0$, $0 \le \alpha < \pi/2$. Let U_ε be the neighborhood of $\overline{M} \in \sigma$ whose projection to $O X_0$ is $(x - \varepsilon, x + \varepsilon)$. Let us prove the relation

$$
\int_U \frac{\varphi \nu \, d\sigma_{M_0}}{|M_0 \overline{M}|^2} = \lim_{\varepsilon \to 0} \left[\int_{I \setminus U_\varepsilon} \frac{\varphi \nu \sqrt{1 + f'^2(x_0)} \, dx_0}{(x - x_0)^2 + [f(x) - f(x_0)]^2} - \frac{2\varphi(x)\nu(x)}{\varepsilon \sqrt{1 + f'^2(x)}} \right]. \tag{4.4.7}
$$

Denote by $M_1 \in \sigma$, $M_2 \in \sigma$ two points whose projections to the axis OX_0 are $\widetilde{M}_1(x - \varepsilon, 0)$, $\widetilde{M}_2(x + \varepsilon, 0)$; and let K and D, respectively, be the crossing points of the lines $M_1\widetilde{M}_1$ and $M_2\widetilde{M}_2$ and the tangent line l to σ at $\overline{M}$. We have $K\overline{M} = D\overline{M} = \varepsilon / \cos \alpha$. Denote by $M_1^* \in \sigma$, $M_2^* \in \sigma$ the points whose projections to the tangent l coincide with K and D, respectively. The points $N_1(x_1, 0)$, $N_2(x_2, 0)$ are the projections of M_1^* and M_2^* to the axis OX_0.

According to the above definition of the integral in the sense of Hadamard, we have

$$
\int_\sigma \frac{\varphi \nu \, d\sigma_{M_0}}{|M_0 \overline{M}|^2} = \tag{4.4.8}
$$
$$
= \lim_{\varepsilon \to 0} \left[\int_a^{x_1} \frac{\varphi \nu \sqrt{1 + f'^2(x_0)} \, dx_0}{(x - x_0)^2 + [f(x) - f(x_0)]^2} + \int_{x_2}^b \frac{\varphi \nu \sqrt{1 + f'^2(x_0)} \, dx_0}{(x - x_0)^2 + [f(x) - f(x_0)]^2} - \frac{2\nu(x)\varphi(x)}{\varepsilon \sqrt{1 + f'^2(x)}} \right].
$$

Suppose that in the SCS associated to the tangent line l, the origin coincides with $\overline{M}$ and the equation of $U \subset \sigma$ is $y_1 = f_1(x_1)$. Then the ordinates of the points M_1^* and M_2^* are $f_1(-\varepsilon/\cos\alpha)$ and $f_1(\varepsilon/\cos\alpha)$. By Taylor's formula, we get

$$
\begin{aligned}
x_1 - (x - \varepsilon) = \sin\alpha f_1\left(\frac{-\varepsilon}{\cos\alpha}\right) = \frac{\sin\alpha}{2} f_1''(0)\frac{\varepsilon^2}{\cos^2\alpha} + O_1(\varepsilon^2), \\
x_2 - (x - \varepsilon) = \sin\alpha f_1\left(\frac{\varepsilon}{\cos\alpha}\right) = \frac{\sin\alpha}{2} f_1''(0)\frac{\varepsilon^2}{\cos^2\alpha} + O_2(\varepsilon^2),
\end{aligned}
\tag{4.4.9}
$$

where $\lim_{\varepsilon\to 0} O_i(\varepsilon^2)/\varepsilon^2 = 0$, $i = 1, 2$. From (4.4.9) it follows that for sufficiently small ε, we have either

$$
\text{(i)} \quad \begin{cases} x_1 \geq x - \varepsilon, \\ x_2 \geq x + \varepsilon, \end{cases} \quad \text{or} \quad \text{(ii)} \quad \begin{cases} x_1 \leq x - \varepsilon, \\ x_2 \leq x + \varepsilon. \end{cases}
\tag{4.4.10}
$$

Consider case (i) (case (ii) can be considered similarly). From (4.4.9), it follows that

$$
\begin{aligned}
\lim_{\varepsilon\to 0}\int_{x-\varepsilon}^{x_1} \frac{\varphi(x_0)\nu(x_0)\sqrt{1+f'^2(x_0)} - \varphi(x)\nu(x)\sqrt{1+f'^2(x)}}{(x-x_0)^2 + [f(x)-f(x_0)]^2}\,dx_0 = 0, \\
\lim_{\varepsilon\to 0}\int_{x+\varepsilon}^{x_2} \frac{\varphi(x_0)\nu(x_0)\sqrt{1+f'^2(x_0)} - \varphi(x)\nu(x)\sqrt{1+f'^2(x)}}{(x-x_0)^2 + [f(x)-f(x_0)]^2}\,dx_0 = 0,
\end{aligned}
\tag{4.4.11}
$$

Using (4.4.9), we obtain the inequalities

$$
\frac{1}{\varepsilon^2 + \tilde{O}_1(\varepsilon^2)}\left[\frac{\sin\alpha}{2}f_1''(0)\frac{\varepsilon^2}{\cos^2\alpha} + O_1(\varepsilon^2)\right] \leq \int_{x-\varepsilon}^{x_1}\frac{dx_0}{(x-x_0)^2 + [f(x)-f(x_0)]^2} \leq
$$
$$
\leq \frac{1}{\varepsilon^2}\left[\sin\alpha\frac{f_1''(0)}{2}\frac{\varepsilon^2}{\cos^2\alpha} + O_1(\varepsilon^2)\right]
$$

where $\tilde{O}_1(\varepsilon^2) > 0$ and $\lim_{\varepsilon\to 0}\tilde{O}_1(\varepsilon^2)/\varepsilon^2 = 0$. The last estimate implies that

$$
\lim_{\varepsilon\to 0}\int_{x-\varepsilon}^{x_1}\frac{dx_0}{(x-x_0)^2 + [f(x)-f(x_0)]^2} = f_1''(0)\frac{\sin\alpha}{\cos^2\alpha}.
\tag{4.4.12}
$$

In a similar way we find that

$$
\lim_{\varepsilon\to 0}\int_{x+\varepsilon}^{x_2}\frac{dx_0}{(x-x_0)^2 + [f(x)-f(x_0)]^2} = f_1''(0)\frac{\sin\alpha}{\cos^2\alpha}.
\tag{4.4.13}
$$

From (4.4.8), (4.4.12), and (4.4.13) we obtain (4.4.7).

Consider the function

$$
F(x, x_0) = \sqrt{1+f'^2(x_0)}\left[1 + \left(\frac{f(x)-f(x_0)}{x-x_0}\right)^2\right]^{-1}.
$$

Let us change the variables by setting $x_0 = x + (x_0 - x) = x + t$. Since $f(x) \in C^\infty(\overline{U})$, we have $\psi(x, t) = [f(x+t) - f(x)]/t \in C^\infty$. Taylor's formula for the function $F(x, x+t)$ at $t = 0$ yields

$$
\psi(x, t) = F(x, x+t) = \frac{1}{\sqrt{1+f'^2(x)}} + \psi_t'(x, 0)t + O(x, t),
\tag{4.4.14}
$$

where $|O(x, t)| \leq Ct^2$. Hence, we obtain the relation

$$\frac{\sqrt{1 + f'^2(x_0)}}{(x - x_0)^2 + [f(x) - f(x_0)]^2} = \frac{1}{\sqrt{1 + f'^2(x)}} \frac{1}{(x - x_0)^2} + \frac{A(x)}{(x - x_0)} + B(x, x_0), \qquad (4.4.15)$$

where $|B(x, x_0)| \leq C$.

Let $\psi(M) \in C_0^\infty(\sigma)$ be a function such that $\psi(M) = 1$ for $M \in \operatorname{supp} \varphi$, $\operatorname{supp} \psi \subset U$, where U is a neighborhood given by the equation $y_0 = f(x_0)$ in some Cartesian coordinate system. Then, it follows from (4.4.15) that the operator $\psi \Pi(\varphi \nu)$ can be represented in the form

$$\psi \Pi(\varphi \nu) = \frac{\psi(x)}{2\pi \sqrt{1 + f'^2(x)}} \int_a^b \frac{\varphi \nu \, dx_0}{(x - x_0)^2} + \frac{A(x)\psi(x)}{2\pi} \int_a^b \frac{\varphi \nu \, dx_0}{x - x_0} + \frac{\psi(x)}{2\pi} \int_a^b B(x, x_0)\varphi \nu \, dx_0 +$$

$$+ \frac{\psi(x)}{2\pi} \int_a^b C(x, x_0)\sqrt{1 + f'^2(x_0)}\varphi \nu \, dx_0 = I_1(\varphi \nu) + I_2(\varphi \nu) + I_3(\varphi \nu) + I_4(\varphi \nu), \qquad (4.4.16)$$

where $C(x, x_0)$ is determined by the kernel

$$K(\overline{M}, M_0) = \frac{[(\boldsymbol{n}_{M_0}, \, \boldsymbol{n}_M) - 1] - 2(\overline{M_0 M}, \, \boldsymbol{n}_{\overline{M}})(M_0 \overline{M}, \, \boldsymbol{n}_{M_0})}{|M_0 \overline{M}|^4}, \qquad (4.4.17)$$

with $|K(\overline{M}, M_0)| \leq C$ (see *Vladimirov* (1976)). Thus, I_3 and I_4 are operators with bounded kernels, and therefore, they are bounded in $L_2(\sigma)$, whereas I_2 is a singular operator, and therefore, it is also bounded in $L_2(\sigma)$. The operator I_1 can be represented as a pseudodifferential operator (see *Brychkov and Prudnikov* (1977)), namely,

$$I_1(\varphi \nu) = \frac{1}{2} \int_{-\infty}^{\infty} \left(-\frac{\psi(x)}{\sqrt{1 + f'^2(x)}} |\xi| \right) \widehat{L}(\xi) \, e^{-i(\xi, x)} \, dx, \qquad (4.4.18)$$

where $\widehat{L}(\xi) = F(\varphi \nu)$ is the Fourier transform of the function $\varphi \nu$.

Next, let we turn to the general operator $\Pi(\nu)$ associated with the left hand side of system (4.4.3). We introduce the spaces

$$H_r(S) = H_r(S_1^0) \times \cdots \times H_r(S_k^0) \times \hat{H}_r(S_1^1) \times \cdots \times \hat{H}_r(S_n^1),$$

$$\widehat{H}_r(S) = H_r(S_1^0) \times \cdots \times H_r(S_k^0) \times H_r(S_1^1) \times \cdots \times H_r(S_n^1).$$

By analogy with Proposition 4.2.1 and Theorem 4.2.2, we can prove the following results.

Proposition 4.4.1. *The operator* $\Pi(\nu)$ *is bounded from* $H_r(S)$ *to* $\widehat{H}_{r-1}(S)$ *for* $0 \leq r \leq 1$.

Theorem 4.4.1. *The operator* $\Pi(\nu)$ *is a Noetherian operator from* $H_r(S)$ *to* $\widehat{H}_{r-1}(S)$ *for* $0 < r < 1$.

Just as we have obtained (4.3.12), it can be shown that

$$u_{i+}^p = \pi \nu_i^p + \int_{S_i^p} \nu_i^p(y) \frac{\cos \varphi_{xy}}{|x - y|} \, dS_{iy}^p,$$

$$u_{i-}^p = -\pi \nu_i^p + \int_{S_i^p} \nu_i^p(y) \frac{\cos \varphi_{xy}}{|x - y|} \, dS_{iy}^p, \qquad (4.4.19)$$

where $\nu_i^p(y) \in H_r(S)$, $r > 0$, and the limits are understood in the sense of $L_2(S)$, just as in Proposition 4.3.2.

By means to (4.4.19), we can establish the following analogue of Theorem 4.2.1.

Theorem 4.4.2. *If $\nu \in H_r(S)$, $0 < r < 1$, then $\dim \ker \Pi(\nu) = k$, $\dim \operatorname{coker} \Pi(\nu) = k$, and equations* (4.4.3) *admit a solution if*

$$\int_{S_i^0} f_i^0(y)\, dS_{iy}^0 = 0.$$

For $\Omega = \mathbb{R}_2 \setminus \left(\cup_p \cup_i S_i^p \right)$ and $p(x) > 0$ in Ω, by $H_{1,p(x)}(\Omega)$ we denote the space of functions $u(x)$ with finite norm

$$\|u\|^2 = \int_\Omega \left[|\operatorname{grad} u|^2 + p(x) u^2(x) \right] dx.$$

Consider the Neumann problem (4.3.14) on the plane $\mathbb{R}_2$. We seek its solution in the space $H_{1,p(x)}$ with $p(x) = 1/\ln^2(|x|+1)$, where $|x|$ is the distance from $x \in \mathbb{R}_2$ to a fixed point $O \in \mathbb{R}_2$. The the generalized solution of this problem and that of the exterior Neumann problem are defined by analogy with Section 4.3.

Theorem 4.4.3. *For $\Omega_1 = \mathbb{R}_2 \setminus \left[\bigcup_{i,j} \left(\overline{\Omega_i^0} \cup S_j^1 \right) \right]$, the exterior Neumann problem admits one and only one solution $u(x) \in H_{1,p(x)}(\Omega_1)$, $1/2 \le r < 1$.*

Proof. The existence of the solution follows from its representation as a double layer potential (see Theorem 4.4.2). By analogy with Proposition 4.3.1 it can be shown that if $\nu(x) \in H_r(S)$ with $r \ge 1/2$, then the solution belongs to $H_{1,p(x)}(\Omega_1)$.

It suffices to establish the uniqueness of the solution in the case of zero boundary conditions. Let $u(x) \in H_{1,p(x)}(\Omega_1)$ be a solution of the exterior Neumann problem with zero boundary conditions. Then, by definition, there is a function $lu(x) \in H_{1,p(x)}(\Omega)$ which is a solution of problem (4.4.1) with zero boundary conditions. In view of Definition 4.3.1, there exists a sequence $\{u_m(x)\} \in H_{1,p(x)}(\Omega)$ for which conditions (i)–(iv) of that definition hold. Denote by $D(R_0, R)$ the region between two circles of radii R_0 and R centered at the origin, and suppose that all curves S_i^p belong to the circle of radius $R_0 < R$. Then, we have

$$I_m(R_0, R) = \int_{D(R_0,R)} \left[|\operatorname{grad} u_m|^2 + p(x) u_m^2(x) \right] dx = \int_{R_0}^R G_m(r)\, dr = I_{1m} + I_{2m},$$

$$G_m(r) = \int_{S_r} \left[|\operatorname{grad} u_m|^2 + p(x) u_m^2(x) \right] dS_r,$$

where S_r is the circular line of radius r. Since $G_m(r)$ is an absolutely integrable function, there is a sequence $R_g \to \infty$ as $g \to \infty$ such that

$$\lim_{g \to \infty} G_m(R_g) = 0. \tag{4.4.20}$$

Using (4.4.20), we find that

$$\lim_{g \to \infty} I_{1m}(R_0, R_g) = \lim_{g \to \infty} \left[-\int_{S_{R_0}} \frac{\partial u_m}{\partial n} u_m\, dS_{R_0} + \int_{S_{R_g}} \frac{\partial u_m}{\partial n} u_m\, dS_{R_g} \right] =$$

$$= -\int_{S_{R_0}} \frac{\partial u_m}{\partial n} u_m\, dS_{R_0}, \tag{4.4.21}$$

and therefore,

$$I_{1m} = \int_\Omega |\operatorname{grad} u_m|^2\, dx = -\sum_p \sum_i \int_{S_i^p} \frac{\partial u_m}{\partial n} \nu_{im}^p\, dS_{iy}^p. \tag{4.4.22}$$

Since $\partial u_m/\partial n \to 0$ in $H_{r-1}(S)$ with $r - 1 \ge -1/2$, and $\nu_m \in \widehat{H}_r(S)$, it follows that $\lim_{m \to \infty} I_{1m} = 0$. Consequently, $u(x) = C$. Since $u(x) \in H_{1,p(x)}(\Omega_1)$, we have $u(x) = 0$. Theorem 4.4.3 is proved.

Chapter 5

Spaces of Fractional Quotients and Their Properties

5.1. Discrete Fourier Transformation and Pseudo-Difference Operators

Let $OX_1, \ldots, X_n$ be a Cartesian coordinate system in the Euclidean space $\mathbb{R}_n$. We partition each X_k-axis with step h_k by the points $x_k^i = h_k \left(i - \frac{1}{2}\right)$, $i \in \mathbb{Z}$, where $\mathbb{Z}$ is the set of all integer numbers. Then the entire $\mathbb{R}_n$ is split into the cells

$$D(k_1, \ldots, k_n, h_1, \ldots, h_n) = \prod_{j=1}^{n} \left(h_j \left(k_j - \frac{1}{2}\right), \, h_j \left(k_j + \frac{1}{2}\right) \right), \qquad k_j \in \mathbb{Z}.$$

Here $\prod$ denotes the Cartesian product of the intervals $h_j \left(k_j - \frac{1}{2}\right) < x_j < h_j \left(k_j + \frac{1}{2}\right)$. In what follows, we use the notation

$$D(k_1, \ldots, k_n, h_1, \ldots, h_n) = D(k, h), \qquad k = (k_1, \ldots, k_n), \quad h = (h_1, \ldots, h_n).$$

Definition 5.1.1. The partition of $\mathbb{R}_n$ into the cells $D(k, h)$ is called the *canonical partition*, and the set of the points $M(k_1 h_1, \ldots, k_n h_n)$ is called the *canonical grid* in $\mathbb{R}_n$. Functions $a(k_1, \ldots, k_n) = a(k)$ defined at the nodes $M(k_1 h_1, \ldots, k_n h_n)$ of the canonical grid are called *grid functions*.

Let M_h be the set of complex-valued step functions $a(x, h)$ taking constant values in each cell of the canonical partition $D(k, h)$. The set of complex valued grid functions $a(k)$ is denoted by M_k.

Let $\mathbb{Z}_n$ be the Cartesian product of n copies of the set of integers $\mathbb{Z}$, and let $\mathbb{Z}_n(N)$ be the Cartesian product of n copies of the set of integers k such that $-N \le k \le N$.

Definition 5.1.2. The *standard map* of the set M_h onto the set M_k is the operator π defined by the relation $\pi a(x, h) = a(k)$, where $a(x, h) = a(k)$ for $x \in D(k, h)$.

Let $a(x, h) \in M_h$ be a function that also belongs to $L_p(\mathbb{R}_n)$, $1 \le p \le 2$, which means that $\int_{\mathbb{R}_n} |a(x, h)|^p \, dx \le C$, or equivalently, $\sum_{k \in \mathbb{Z}_n} |\pi a(x, h)|^p \le C/\Omega_h$, where $\Omega_h = h_1 \cdots h_n$.

Definition 5.1.3. The (*discrete*) *step Fourier transformation* $F_h a(x, h)$ of a function $a(x, h) \in M_h \cap L_p(\mathbb{R}_n)$, $1 \le p \le 2$, is the function defined by the relation

$$F_h a(x, h) = \widehat{a}_h(\varphi) = \Omega_h \sum_{z \in \mathbb{Z}_n} a(k) \, e^{i(k\varphi, h)}, \qquad (5.1.1)$$

where the right-hand side is understood as follows:

$$\widehat{a}_h(\varphi) = \lim_{N \to +\infty} \Omega_h \sum_{k \in \mathbb{Z}_n(N)} a(k)\, e^{i(ik\varphi, h)}, \qquad (k\varphi, h) = \sum_{l=1}^{n} k_l \varphi_l h_l,$$

$$a(k) = \pi a(x, h), \qquad k_l \in \mathbb{Z}, \qquad -\infty < \varphi_l < \infty. \tag{5.1.2}$$

The function $\widehat{a}(\varphi)$ defined by (5.1.1) is periodic in each variable φ_l with period $2\pi/h_l$. Moreover, $\widehat{a}_h(\varphi) \in L_{p_1}[D_h]$, where $D_h = \prod_{l-1}^{n} \left[-\pi/h_l \le \varphi_l \le \pi/h_l\right]$, and $1/p + 1/p_1 = 1$ (see *Lifanov and Poltavskii* (1999)$_{1,2}$).

The value $a(k)$ of the step function $a(x, h)$ on the cell $D(k, h)$ can be recovered by the formula

$$a(k) = \frac{1}{(2\pi)^n} \int_{D_h} \widehat{a}(\varphi)\, e^{-i(k\varphi, h)}\, d\varphi, \tag{5.1.3}$$

where $d\varphi = d\varphi_1 \cdot d\varphi_n$.

Definition 5.1.4. The *(discrete) grid Fourier transform* $F_c a(k)$ of a grid function $a(k) \in l_p$, $1 \le p \le 2$ (i.e., $\|a(k)\|_p^p = \sum_{k \in \mathbb{Z}_n} |a(k)|^p < \infty$) is defined by the relation

$$F_c a(k) = \widehat{a}_c(\varphi) = \lim_{N \to \infty} \sum_{k \in \mathbb{Z}_n} a(k)\, e^{i(k\varphi)}, \tag{5.1.4}$$

where $(k\varphi) = \sum_{l=1}^{n} k_l \varphi_l$, $k_l \in \mathbb{Z}$.

The values of the grid function $a(k)$ can be recovered by the formula

$$a(k) = \frac{1}{(2\pi)^n} \int_U \widehat{a}_c(\varphi)\, e^{-i(k\varphi)}\, d\varphi, \tag{5.1.5}$$

where $U = \prod_{l=1}^{n} [-\pi \le \varphi_l \le \pi]$.

The discrete Fourier transform of the step function $a(x, h)$ and that of the corresponding grid function $a(k) = \pi a(x, h)$ are related by

$$\widehat{a}_h(\varphi_1, \ldots, \varphi_n) = \Omega_h \cdot \widehat{a}_c(\varphi_1 h_1, \ldots, \varphi_n h_n) = \Omega_h \widehat{a}_c(\varphi, h). \tag{5.1.6}$$

Definition 5.1.5. The *convolution of grid functions* $a(k)$ and $b(k)$ is the grid function

$$c(k_1, \ldots, k_n) = \sum_{i_1=-\infty}^{\infty} \cdots \sum_{i_n=-\infty}^{\infty} a(i_1, \ldots, i_n) b(k_1 - i_1, \ldots, k_n - i_n) = a(k) \times b(k). \tag{5.1.7}$$

We introduce the norms

$$\|a(k)\|_p = \left(\sum_{k \in \mathbb{Z}_n} |a(k)|^p \right)^{1/p}, \qquad 1 \le p < \infty; \qquad \|a(k)\|_\infty = \sup_{k \in \mathbb{Z}_n} |a(k)|.$$

The following result holds (see *Lifanov and Poltavskii* (1999)$_{1,2}$).

Proposition 5.1.1. *If $a(k) \in l_1$, $b(k) \in l_p$ with $1 \le p < \infty$, then*

$$\|a(k) \times b(k)\|_p \le \|a(k)\|_1 \cdot \|b(k)\|_p.$$

Let $a(k) \in l_p$, $1 \le p \le 2$. We define a linear operator T from the space l_2 of grid functions $x(k)$ to the space of grid functions l_∞ by letting

$$T(x) = a(k) \times x(k). \tag{5.1.8}$$

The function $a(k)$ is called *the characteristic function of the operator* T (see *Lifanov and Poltavskii* (1999)[1,2]), and T is called the *convolution operator*.

Definition 5.1.6. The discrete Fourier transform of the function $a(k)$ is called *the symbol of the convolution operator* T.

Let $\widehat{A}(\varphi) = \widehat{a}_c(\varphi)$ be the grid Fourier transform of $a(k)$. Then (see *Lifanov and Poltavskii* (1999)[1,2]), relation (5.1.9) can be written as follows:

$$T(x) = \frac{1}{(2\pi)^n} \int_U \widehat{A}(\varphi) \widehat{x}_c(\varphi) \, e^{-i(k\varphi)} \, d\varphi. \tag{5.1.9}$$

Proposition 5.1.1. implies the following result.

Proposition 5.1.2. *Let $T_0 : l_2 \to l_2$ be an operator in the space of grid functions $x(k)$. If T_0 is a composition of convolution operators $T_1, \ldots, T_p$ with the characteristic functions $a_m(k) \in l_1$, $m = 1, \ldots, p$, then*

$$\widehat{A}_0(\varphi) = \widehat{A}_1(\varphi) \cdots \widehat{A}_p(\varphi), \tag{5.1.10}$$

where $\widehat{A}_m(\varphi)$, $m = 0, 1, \ldots, p$ are the symbols of the operators T_m.

Examples: one-dimensional case

Example 5.1.1. The shift operator T on grid functions $x(k)$: $Tx(k) = x(k+n)$, where n is a fixed integer. Consider the grid function $a(k)$ such that $a(-n) = 1$, $a(k) = 0$ of all other k. Then

$$c(k) = a(k) \times x(k) = \sum_{m=-\infty}^{\infty} a(m)x(k-m) = x(k+n),$$

which shows that the shift operator is of convolution type and its symbol is $\widehat{A}(\varphi) = e^{-in\varphi}$.

Example 5.1.2. The first order right difference operator

$$T_r x = \frac{x(k+1) - x(k)}{h}.$$

Consider the grid function $a(k)$ such that $a(0) = -1/h$, $a(-1) = 1/h$, and $a(k) = 0$ of the other k. We have

$$c(k) = a(k) \times x(k) = \sum_{m=-\infty}^{\infty} a_m x(k-m) = \frac{x(k+1) - x(k)}{h}.$$

Thus, T_r is a convolution operator with the symbol

$$\widehat{A}(\varphi) = \frac{1}{h} \sum_{k=-\infty}^{\infty} a(k) e^{ik\varphi} = \frac{e^{-i\varphi} - 1}{h}.$$

In a similar way, the symbol of the left difference operator $T_l x(k) = h^{-1}(x(k) - x(k-1))$ is $\widehat{A}(\varphi) = h^{-1}(1 - e^{i\varphi})$.

By Proposition 5.1.2, the k-th order right difference operator $T_r^k = (T_r)^k$ has the symbol $\widehat{A}(\varphi) = h^{-k}(e^{-i\varphi} - 1)^k$, whereas the k-th order left difference operator $T_l^k = (T_l)^k$ has the symbol $\widehat{A}(\varphi) = h^{-k}(1 - e^{i\varphi})^k$.

Example 5.1.3. The second order symmetric difference operator

$$T_2^* x(k) = \frac{x(k+1) - 2x(k) + x(k-1)}{h^2}.$$

Consider the grid function $a(k)$ such that $a(0) = -2/h^2$, $a(1) = a(-1) = 1/h^2$, $a(k) = 0$ for the rest of k. Then

$$c(k) = a(k) \times x(k) = \sum_{m=-\infty}^{\infty} a(m)x(k-m) = \frac{x(k+1) - 2x(k) + x(k-1)}{h^2}.$$

Hence we see that T_2^* is a convolution operator whose symbol is

$$\widehat{A}(\varphi) = \frac{e^{-i\varphi} - 2 + e^{i\varphi}}{h^2} = -\frac{4}{h^2}\sin^2\frac{\varphi}{2}.$$

On the other hand, we have $\widehat{A}(\varphi) = e^{i\varphi}\left(\dfrac{e^{-i\varphi} - 1}{h}\right)^2$. It follows that $T_2^* = T(-1)T_r^2$, where $T(-1)$ is the shift operator: $T(-1)x(k) = x(k-1)$.

The symbol of the symmetric difference operator $T_{2m}^* = (T_2^*)^m$ of order $2m$ is

$$\widehat{A}(\varphi) = \left[-\frac{4}{h^2}\sin^2\frac{\varphi}{2}\right]^m = (-1)^m\frac{4^m}{h^{2m}}\sin^{2m}\frac{\varphi}{2}.$$

Multidimensional case

Example 5.1.4. Consider the first order right difference operator $T_r(i)$ with respect to the i-th variable:

$$T_r(i)x(k) = \frac{x(k_1, \ldots, k_{i-1}, k_i + 1, k_{i+1}, \ldots, k_n) - x(k_1, \ldots, k_n)}{h_i}.$$

The right difference operator $T_r^l(i)$ of the l-th order in the i-th variable is defined by the relation $T_r^l(i) = (T_r(i))^l$. The mixed right difference operator $T_r^m(j_1, \ldots, j_p)$ of the order $m = m_1 + \cdots + m_p$ in the variables $j_1, \ldots, j_p$, with m_k being the order in j_k, $k = 1, \ldots, p$, is defined by

$$T_r^m(j_1, \ldots, j_p) = T_r^{m_1}(j_1) \cdots T_r^{m_p}(j_p), \qquad m_1 + \cdots + m_p = m, \quad m_k \geq 0, \quad m_k \in \mathbb{Z}.$$

Proposition 5.1.2 shows that the symbol $\widehat{A}(\varphi)$ of the operator $T_r^m(j_1, \ldots, j_p)$ has the form

$$\widehat{A}(\varphi) = \prod_{k=1}^{p}\left(\frac{e^{-i\varphi_{j_k}} - 1}{h_{j_k}}\right)^{m_k}.$$

Similarly, we define the m-th order mixed left difference operator $T_l^m(j_1, \ldots, j_p)$ with the symbol

$$\widehat{A}(\varphi) = \prod_{k=1}^{p}\left(\frac{1 - e^{i\varphi_{j_k}}}{h_{j_k}}\right)^{m_k}.$$

Example 5.1.5. The second order symmetric difference operator $T_2^*(i)$ in the variable i is the operator defined by the relation

$$T_2^*(i)x(k) = \frac{1}{h_i^2}\Big[x(x_1, \ldots, k_i + 1, \ldots, k_n) - 2x(k_1, \ldots, k_i, \ldots, k_n) + x(k_1, \ldots, k_i - 1, \ldots, k_n)\Big].$$

The Laplace difference operator Δ_p is defined by

$$\Delta_p = \sum_{i=1}^{n} T_2^*(i),$$

and its symbol has the form

$$\widehat{A}(\varphi) = -4 \sum_{i=1}^{n} \frac{1}{h_i^2} \sin^2 \frac{\varphi_i}{2}.$$

The m-th order Laplace difference operator is $T = (\Delta_p)^m = \Delta_p^m$, and its symbol is

$$\widehat{A}(\varphi) = (-1)^m 4^m \left[\sum_{i=1}^{n} \frac{1}{h_i^2} \sin^2 \frac{\varphi_i}{2} \right]^m.$$

Since the shifts of the space $\mathbb{R}_n$ by vectors $x^0 = \left\{ k_1^0 h_1, \ldots, k_n^0 h_n \right\}$, $k_i^0 \in \mathbb{Z}$, map the cells $D(k, h)$ to similar cells, it follows that the shifts $a(x - x_0, h)$ of step functions $a(x, h) \in M_h$ also belong to M_h. Therefore, all difference operators defined on grid functions are naturally defined on step functions of class M_h and map M_h into M_h.

The above examples show that it would be natural to introduce the following notion.

Definition 5.1.7. *A Pseudodifference operator* A mapping a class of grid functions $x(k)$ to the set of grid functions M_k is an operator of the form

$$Ax = \frac{1}{(2\pi)^n} \int_U \widehat{A}(\varphi, k, h) \widehat{x}_c(\varphi) \, e^{-i(k,\varphi)} \, d\varphi,$$

where the function $\widehat{A}(\varphi, k, h)$, called the *symbol of the operator* A, is periodic in each variable φ_i, $1 \le i \le n$; $\widehat{x}_c(\varphi)$ us the discrete Fourier transform of the function $x(k)$.

Remark 5.1.1. The classes of the functions $\widehat{A}(\varphi, k, h)$ and $x(k)$ depend on a particular problem under consideration.

Example 5.1.6. The method of discrete vortices (one-dimensional case). Consider the partition of the real axis with step $h > 0$ by the points $t_i = ih$, $i \in \mathbb{Z}$. For $0 < \alpha < 1$, consider another system of points $\{ t_{0j} \}$, $t_{0j} = (j + \alpha)h$. The grid function x is defined at the nodes t_i, $x(t_i) = x_i$. The method of discrete vortices deals with the operator

$$Tx = \sum_{i=-\infty}^{\infty} \frac{x_i h}{t_{0j} - t_i} = \sum_{i=-\infty}^{\infty} \frac{x_i}{(j - i) + \alpha}. \tag{5.1.11}$$

Suppose that $\sum_{i=-\infty}^{\infty} |x_i| < \infty$. From (5.1.11), it follows that Tx is a convolution operator with the symbol

$$\widehat{A}(\varphi) = \sum_{i=-\infty}^{\infty} \frac{1}{k + \alpha} e^{ik\varphi} = \frac{\pi}{\sin \pi \alpha} e^{-i(\pi - \varphi)\alpha}. \tag{5.1.12}$$

For $\alpha = 1/2$ (which is the case in problems of aerodynamics), we have $\widehat{A}(\varphi) = i\pi \, e^{-i\varphi/2}$. The above observation shows that the values $f(k)$ of the grid function Tx are found from the relation

$$f_k = \frac{1}{2\pi} \int_0^{2\pi} i\pi \, e^{-i\varphi/2} x_c(\varphi) \, e^{-ik\varphi} \, d\varphi,$$

which means that the operator of the method of discrete vortices can be represented as a pseudodifference operator.

Example 5.1.7. The discrete Prandtl operator in the one-dimensional case is defined by the relation (see *Lifanov* (1996))

$$Tx = \sum_{i=-\infty}^{\infty} \left(\frac{1}{t_{0k} - t_i} - \frac{1}{t_{0k} - t_{i+1}} \right) x_i. \tag{5.1.13}$$

The operator Tx can be represented as the following pseudodifference operator:

$$f_k = \frac{1}{2\pi} \int_0^{2\pi} \widehat{A}(\varphi) \widehat{x}_c \, e^{-ik\varphi} \, d\varphi,$$

$$\widehat{A}(\varphi) = \frac{1}{h} \sum_{p=-\infty}^{\infty} \left[\frac{1}{p+\alpha} - \frac{1}{p+\alpha-1} \right] e^{ip\varphi} =$$

$$= \frac{1}{h} \left[\frac{\pi}{\sin \pi\alpha} e^{i(\pi-\varphi)\alpha} - \frac{\pi}{\sin \pi(\alpha-1)} e^{i(\pi-\varphi)(\alpha-1)} \right].$$

In problems of aerodynamics, $\alpha = 1/2$, $\widehat{A}(\varphi) = (2\pi/h) \sin(\varphi/2)$.

Example 5.1.8. Consider the difference operator

$$Tx = \sum_{p=0}^{m} a_p(k) \, \mathrm{T}_r^p x(k),$$

where $a_p(k)$ are given grid functions, $a_p(k) \in l_\infty$; T_r^p are mixed right difference operators of order p. In this case, Example 5.1.4 shows that Tx can be written as

$$Tx = \frac{1}{(2\pi)^n} \int_{-\pi}^{\pi} \sum_{p=0}^{m} a_p(k) \prod_{k=1}^{n} \left(\frac{e^{-i\varphi_k} - 1}{h_k} \right)^{m_{k_p}} \widehat{x}_c(\varphi) \, e^{-ik\varphi} \, d\varphi,$$

where $\sum_{k=1}^{n} m_{k_p} = p$ and some m_{k_p} may be equal to zero.

5.2. Special Trigonometric Series

Consider the canonical partition of $\mathbb{R}_n$ (see Definition 5.1.1) with the vector-valued step $h = (1, 1, \ldots, 1)$ and the cells $D(k, h)$ denoted by $D(k)$, $|k| = \sqrt{\sum_{j=1}^{n} k_j^2}$.

Proposition 5.2.1. *If $|k| \neq 0$, then for all $D(k)$ and λ, the following inequality holds:*

$$C_1(\lambda)|k|^\lambda \leq \int_{D(k)} r^\lambda \, dx \leq C_2(\lambda)|k|^\lambda,$$

where $C_1(\lambda) \geq 0$, $C_2(\lambda) \geq 0$ are constants, and $r = \sqrt{\sum_{j=1}^{n} x_j^2} = |x|$.

Proof. Let $x \in D(k)$. Then $\dfrac{|x|}{|k|} \leq \dfrac{|k| + \sqrt{n}}{|k|} \leq 1 + \sqrt{n}$. On the other hand,

$$\frac{|k|}{|x|} \leq \frac{|x| + \sqrt{n}}{|x|} \leq 1 + 2\sqrt{n}.$$

Therefore, for $x \in D(k)$, $|k| \neq 0$, we have

$$\frac{1}{1 + 2\sqrt{n}} \frac{|x|}{|k|} \leq 1 + \sqrt{n} .$$

(5.2.1)

From (5.2.1), using the mean value theorem, we get

$$\left(\frac{1}{1 + 2\sqrt{n}}\right)^{\lambda} |k|^{\lambda} \leq \int_{D(k)} r^{\lambda} \, dx \leq \left(1 + \sqrt{n}\right)^{\lambda} |k|^{\lambda} \quad \text{for} \quad \lambda \geq 0,$$

$$\left(\frac{1}{1 + \sqrt{n}}\right)^{|\lambda|} |k|^{\lambda} \leq \int_{D(k)} r^{\lambda} \, dx \leq \left(1 + 2\sqrt{n}\right)^{|\lambda|} |k|^{\lambda} \quad \text{for} \quad \lambda < 0,$$

as required. Proposition 5.2.1 is proved.

Consider the functions

$$F(\lambda, \varphi) = \sum_{|k| \neq 0} \frac{\sin^2 \frac{(k\varphi)}{2}}{|k|^{n+\lambda}} , \qquad 0 < \lambda < 2,$$

(5.2.2)

where $(k\varphi) = \sum_{i=1}^{n} k_i \varphi_i$, $k = (k_1, \ldots, k_n) \in \mathbb{Z}_n$

By Proposition 5.2.1, the series on the right-hand side of (5.2.2) is absolutely convergent for any φ. Therefore, the functions $F(\lambda, \varphi)$ are defined for all φ. Moreover, the functions $F(\lambda, \varphi)$ are periodic in each φ_i with period 2π.

Proposition 5.2.2. *On the cube $U = \prod_{j=1}^{n} \left[-\pi \leq \varphi_j \leq \pi\right]$, the following estimate holds:*

$$F(\lambda, \varphi) \leq C_1(\lambda) |\varphi|^{\lambda} .$$

Proof. For each $N = 1, 2, \ldots$ consider the set $1/(2N) \leq |\varphi| \leq 1/N$ and let ε be the set of integer n-tuples $k = (k_1, \ldots, k_n)$ with $|k| \geq N$. Then, using Proposition 5.2.1 and the inequality (5.2.1), we find that

$$\sum_{k \notin \varepsilon} \frac{\sin^2 \frac{(k\varphi)}{2}}{|k|^{n+\lambda}} \leq C(\lambda) \int_{\Omega_N^*} \frac{dx}{r_{n+\lambda}} = C(\lambda) \int_S ds \int_{N(1+2\sqrt{n})}^{\infty} \frac{dr}{r^{1+\lambda}} \leq \frac{A(\lambda)}{N^{\lambda}} \leq A_1(\lambda) |\varphi|^{\lambda}, \quad (5.2.3)$$

where Ω_N^* is the exterior of the ball of radius $R/(1 + 2\sqrt{n})$, S is the surface of the unit sphere.

Next, consider the inequality

$$\sum_{k \notin \varepsilon} \frac{\sin^2 \frac{(k\varphi)}{2}}{|k|^{n+\lambda}} \leq \frac{1}{4} \sum_{k \notin \varepsilon} \frac{|k|^2 |\varphi|^2}{|k|^{n+\lambda}} \leq C_1(\lambda) |\varphi|^2 \int_S ds \int_{1/2}^{N(1+\sqrt{n})} r^{1-\lambda} \, dr \leq$$

$$\leq B(\lambda) |\varphi|^2 \left(\frac{1}{[(1 + \sqrt{n})N]^{\lambda-2}} + \frac{1}{2^{2-\lambda}}\right) \leq B_1(\lambda) |\varphi|^{\lambda} .$$

(5.2.4)

From (5.2.3) and (5.2.4), it follows that $F(\lambda, \varphi) \leq C_2(\lambda) |\varphi|^{\lambda}$ for $1/(2N) \leq |\varphi| \leq 1/N$. For all positive integers q and p such that $q \geq p$, we have

$$N = \left[\frac{q}{p}\right] \leq \frac{q}{p} \leq \left[\frac{q}{p}\right] + 1 \leq 2N,$$

where $[q/p]$ is the integer part of q/p. Thus, the intervals $1/(2N) \leq |\varphi| \leq 1/N$, $N = 1, 2, \ldots$, cover the set $(0, 1]$. Hence we obtain the desired result. Proposition 5.2.2 is proved.

Proposition 5.2.3. *On the cube* $U = \prod_{j=1}^{n} \left[-\pi \le \varphi_j \le \pi \right]$, *the following estimate holds:*

$$F(\lambda, \varphi) \ge C_2(\lambda)|\varphi|^{\lambda}, \qquad C_2(\lambda) > 0.$$

Proof. Let ε be the set of all n-tuples $k = (k_1, \ldots, k_n)$ with $k_1 \ne 0$. Since the function $|k|$ is even with respect to each k_i, we have

$$F(\lambda, \varphi, \varepsilon) = \sum_{k \in \varepsilon} \frac{\sin^2 \frac{(k\varphi)}{2}}{|k|^{n+\lambda}} = \sum_{k \in \varepsilon_1} \frac{1 - (\cos k_1 \varphi_1) \cos \sum_{i=2}^{n} k_i \varphi_i}{|k|^{n+\lambda}}, \tag{5.2.5}$$

where ε_1 is the set of all $k = (k_1, \ldots, k_n)$ with $k_1 > 0$.

Consider the sets $\pi/(16N) \le \varphi_1 \le \pi/(8N)$, $N = 1, 2, \ldots$. For $N \le k_1 \le 2N$, we have $\pi/16 \le k_1 \varphi_1 \le \pi/4$. Therefore,

$$\cos \frac{\pi}{4} \le \cos k_1 \varphi_1 \le \cos \frac{\pi}{16}, \qquad 1 - \cos k_1 \varphi_1 \cos \left(\sum_{i=2}^{n} k_i \varphi_i \right) \ge \delta.$$

Now, from (5.2.5), we get

$$F(\lambda, \varphi, \varepsilon) \ge \delta \sum_{k_1=N}^{2N} \sum_{k \in \varepsilon_2} \frac{1}{|k|^{n+\lambda}}, \tag{5.2.6}$$

where $\sum_{k \in \varepsilon_2}$ denotes summation over all $k_2, \ldots, k_n$.

The cubes $D(k)$ corresponding to $k = (k_1, \ldots, k_n)$ in (5.2.6) make up the set

$$M = \left\{ x : \ N - \frac{1}{2} \le x_1 \le 2N + \frac{1}{2}, \quad -\infty < x_i < +\infty, \ i = 2, 3, \ldots, n \right\}.$$

Then, it follows from (5.2.6) and Proposition 5.2.1 that

$$F(\lambda, \varphi, \varepsilon) \ge \delta A(\lambda) \int_{N-1/2}^{2N+1/2} dx_1 \int_{\mathbb{R}_{n-1}} \frac{1}{r^{n+\lambda}} \, dx_2 \cdots dx_n.$$

Consider the spherical coordinates: r is the radius vector of the point $x = (x_1, \ldots, x_n)$, Θ_1 is the angle between r and the axis Ox_1, $0 \le \Theta_1 \le \pi$; the vector r_1 is the projection of r to $\mathbb{R}_{n-1}$; Θ_2 is the angle between r_1 and the axis OX_2, $0 \le \Theta_2 \le \pi$, etc; r_{n-2} is the projection of r_{n-3} to $\mathbb{R}_2$, Θ_{n-1} is the angle between Ox_{n-1} and r_{n-2}, $-\pi < \Theta_{n-1} \le \pi$. Thus,

$$x_1 = r \cos \Theta_1, \quad x_2 = r \sin \Theta_1 \cos \Theta_2, \quad \ldots,$$
$$x_{n-1} = r \sin \Theta_1 \cdots \sin \Theta_{n-2} \cos \Theta_{n-1}, \quad x_n = r \sin \Theta_1 \sin \Theta_2 \cdots \sin \Theta_{n-1},$$
$$|I| = r^{n-1} \sin^{n-2} \Theta_1 \sin_{n-3} \Theta_2 \cdots \sin \Theta_{n-2},$$

where $|I|$ is the Jacobian of the transformation. Denote by S_1 the range of the variables $\Theta_2, \ldots, \Theta_{n-1}$, namely,

$$0 \le \Theta_i \le \pi, \quad i = 1, 2, \ldots, n-2; \qquad -\pi \le \Theta_{n-1} \le \pi;$$

and let $dS_1 = \sin^{n-3} \Theta_2 \cdots \sin \Theta_{n-2} \, d\Theta_2 \cdots d\Theta_{n-2} \, d\Theta_{n-1}$. Then (5.2.6) can be written as

$$F(\lambda, \varphi, \varepsilon) \ge \delta A(\lambda) \int_{S_1} dS_1 \int_0^{\pi/2} \sin^{n-2} \Theta_1 \left\{ \int_{\frac{N-1/2}{\cos \Theta_1}}^{\frac{2N+1/2}{\cos \Theta_1}} \frac{1}{r^{1+\lambda}} \, dr \right\} d\Theta_1 =$$

$$= \delta A_1(\lambda) \left[\left(N - \frac{1}{2} \right)^{-\lambda} - \left(2N + \frac{1}{2} \right)^{-\lambda} \right] \ge A_2(\lambda)|\varphi_1|^{\lambda}. \tag{5.2.7}$$

Here, we have used the condition $\pi/(16N) \le \varphi_1 \le \pi/(8N)$. Since $\bigcup_{N=1}^{\infty} \left[\frac{\pi}{16N}, \frac{\pi}{8N}\right] = \left(0, \frac{\pi}{8}\right]$, it follows that (5.2.7) holds for all $\varphi_2, \ldots, \varphi_n$ and $0 \le \varphi_1 \le \pi/8$. Note that $F(\lambda, \varphi, \varepsilon)$ is an even function of φ_1, and $F(\lambda, \varphi) \ge \sin^2(\varphi_1/2)$. Therefore, $F(\lambda, \varphi) \ge A_3(\lambda)|\varphi_1|^\lambda$ for all $\varphi \in U$. In a similar way, we obtain the inequalities $F(\lambda, \varphi) \ge A_3(\lambda)|\varphi_2|^\lambda, \ldots, F(\lambda, \varphi) \ge A_3(\lambda)|\varphi_n|^\lambda$. Hence,

$$F(\lambda, \varphi) \ge \frac{A_3(\lambda)}{n} \sum_{j=1}^{n} |\varphi_j|^\lambda \ge C_2(\lambda)|\varphi|^\lambda.$$

Proposition 5.2.3 is proved.

Consider the function

$$J(\lambda, \varphi, h) = \sum_k \gamma(\lambda, k) e^{i(k\varphi)} \quad \text{with} \quad \gamma(\lambda, k) = \int_{D(k,h)} \frac{1}{r^{n+\lambda}} dx_1 \cdots dx_n, \quad 0 < \lambda < 2, \quad (5.2.8)$$

where $D(k, h)$ are the cells of the canonical partition of $\mathbb{R}_n$ with the vector-valued step $h = (h_1, \ldots, h_n)$ (see Definition 5.1.1). For $k = (0, \ldots, 0)$ the integral us understood in the sense of Hadamard (see *Lifanov and Poltavskii* (1999)$_{1,2}$), namely,

$$\gamma(\lambda, 0) = \lim_{\varepsilon \to 0} \left[\int_{D(0,h) \setminus \Omega_\varepsilon^*} \frac{dx}{r^{n+\lambda}} - \frac{\Omega_n}{\lambda \varepsilon^\lambda} \right], \tag{5.2.9}$$

where Ω_n is the area of the unit sphere in $\mathbb{R}_n$; Ω_ε^* is the ball of radius $\varepsilon > 0$. We have (see (5.2.9))

$$\sum_k \gamma(\lambda, k) = \int_{\mathbb{R}_n} \frac{dx}{r^{n+\lambda}} \lim_{\varepsilon \to 0} \left[\Omega_n \int_\varepsilon^\infty \frac{dr}{r^{\lambda+1}} - \frac{\Omega_n}{\lambda \varepsilon^\lambda} \right] = 0. \tag{5.2.10}$$

From (5.2.10), it follows that $\gamma(\lambda, 0) < 0$, $\gamma(\lambda, k) > 0$ for $|k| \ne 0$, and $\gamma(\lambda, 0) = -\sum_{|k| \ne 0} \gamma(\lambda, k)$. Therefore, the function $J(\lambda, \varphi, h)$ can be represented in the form

$$J(\lambda, \varphi, h) = \sum_{|k| \ne 0} \gamma(\lambda, k) \left[e^{i(k\varphi)} - 1 \right]. \tag{5.2.11}$$

We have $\gamma(\lambda, -k) = \gamma(\lambda, k)$, since the function $r^{n+\lambda}$ is symmetric with respect to the origin and the partition of $\mathbb{R}_n$ into cells $D(k, h)$ is symmetric. It follows that $\operatorname{Im} J(\lambda, \varphi, h) = 0$. The above observations and (5.2.11) show that

$$J(\lambda, \varphi, h) = -2 \sum_{|k| \ne 0} \gamma(\lambda, k) \sin^2 \frac{(k\varphi)}{2}. \tag{5.2.12}$$

Changing the variables, $x_j = h_j y_j$, $j = 1, \ldots, n$, we get

$$\gamma(\lambda, k) = \frac{\Omega_h}{h_1^{n+\lambda}} \int_{k_1-1/2}^{k_1+1/2} \cdots \int_{k_n-1/2}^{k_n+1/2} \frac{dy_1 \cdots dy_n}{\left[\sum_{i=1}^n \left(\frac{h_j}{h_1} \right) y_j^2 \right]^{\frac{n+\lambda}{2}}}. \tag{5.2.13}$$

Hence, using Proposition 5.2.1, we obtain the inequality

$$\frac{C_1(\lambda)}{h_1^\lambda |k|^{\lambda+n}} \le \gamma(\lambda, k) \le \frac{C_2(\lambda)}{h_1^\lambda |k|^{\lambda+n}}. \tag{5.2.14}$$

Propositions 5.2.2, 5.2.3, combined with the inequality (5.2.14), imply that for $\varphi \in U$,

$$\frac{A_1(\lambda)}{h_1^\lambda} |\varphi|^\lambda \leq |J(\lambda, \varphi, h)| \leq \frac{A_2(\lambda)}{h_1^\lambda} |\varphi|^\lambda. \tag{5.2.15}$$

Consider the functions

$$B_N(\lambda, \varphi, h) = - \sum_{k \in M_N} \gamma(\lambda, k) e^{i(\varphi k)}, \tag{5.2.16}$$

where M_N is the set of the indices $k = (k_1, \ldots, k_n)$ such that $-N \leq k_j \leq N$, $j = 1, \ldots, n$, $N > 0$ is integer. Because if the symmetry of the set M_N, we have $\operatorname{Im} B(\lambda, \varphi, h) = 0$. Hence, using (5.2.10), we obtain

$$B_N(\lambda, \varphi, h) = 2 \sum_{\substack{k \in M_N \\ |k| \neq 0}} \gamma(\lambda, k) \sin^2 \frac{(k\varphi)}{2} + \sum_{k \notin M_N} \gamma(\lambda, k) = B_{1N}(\lambda, \varphi, h) + d_N(\lambda, h). \tag{5.2.17}$$

We estimate $d_N(\lambda, h)$ as follows:

$$d_N(\lambda, h) \leq \int_{\mathbb{R}_n \setminus \Omega^*((N+\frac{1}{2})h_0)} \frac{1}{r^{n+\lambda}} \, dx \leq \frac{C_1}{N^\lambda h_1^\lambda}, \tag{5.2.18}$$

where $\Omega^*((N + 1/2)h_0)$ is the ball of radius $R = (N + 1/2)h_0$ with center at the origin, $h_0 = \min(h_1, \ldots, h_n)$. On the other hand,

$$d_N(\lambda, h) \geq \int_{\mathbb{R}_n \setminus \Omega^*(\sqrt{n}(N+\frac{1}{2})h_{10})} \frac{1}{r^{n+\lambda}} \, dx \geq \frac{C_2}{N^\lambda h_1^\lambda}, \tag{5.2.19}$$

where $\Omega^*((N + 1/2)\sqrt{n}h_{10})$ is the ball of radius $R = (N + 1/2)\sqrt{n}h_{10}$ with center at the origin, $h_{10} = \max(h_1, \ldots, h_n)$.

From (5.2.15), (5.2.17), and (5.2.8), we obtain the inequality

$$0 \leq B_N(\lambda, \varphi, h) \leq \frac{C_3(\lambda)}{h_1^\lambda} \left[|\varphi|^\lambda + \frac{1}{N^\lambda} \right]. \tag{5.2.20}$$

Next, from (5.2.15), (5.2.17), we get

$$\frac{|\varphi|^\lambda A_1(\lambda)}{h_1^\lambda} \leq -J(\lambda, \varphi, h) = 2 \sum_{k \in M_N, \, |k| \neq 0} \gamma(\lambda, k) \sin^2 \frac{(k\varphi)}{2} + 2 \sum_{k \notin M_N} \gamma(\lambda, k) \sin^2 \frac{(k\varphi)}{2} \leq$$

$$\leq 2 B_N(\lambda, \varphi, h). \tag{5.2.21}$$

From (5.2.17), (5.2.19), and (5.2.21), we obtain the inequality

$$B_N(\lambda, \varphi, h) \geq \frac{C_4(\lambda)}{h_1^\lambda} \left[|\varphi|^\lambda + \frac{1}{N^\lambda} \right]. \tag{5.2.22}$$

Finally, (5.2.20) and (5.2.22) imply that

$$\frac{C_4(\lambda)}{h_1^\lambda} \left[|\varphi|^\lambda + \frac{1}{N^\lambda} \right] \leq B_N(\lambda, \varphi, h) \leq \frac{C_3(\lambda)}{h_1^\lambda} \left[|\varphi|^\lambda + \frac{1}{N^\lambda} \right]. \tag{5.2.23}$$

5.3. Spaces of Fractional Quotients $M(r, h)$

Definition 5.3.1. For an arbitrary real r, the *space of fractional quotients* $M(r, h)$ consists of step functions $a(x, h) \in M_h$ for which

$$\|a(x, h)\|^2_{M(r,h)} = \int_{D_h} \left(1 + |\varphi|\right)^{2r} |\widehat{a}_h(\varphi)|^2 \, d\varphi < \infty, \tag{5.3.1}$$

where $\widehat{a}_h(\varphi) = F_h a(x, h)$ is the discrete step Fourier transform of the function $a(x, h)$ (see Definition 5.1.3); $|\varphi| = \left[\sum_{i=1}^{n} \varphi_j^2\right]^{1/2}$.

It should be observed that $a(x, h) \in M(r, h)$ implies that $a(x, h) \in L_2(\mathbb{R}_n)$, and the converse is also true. However, the constants in the estimate of the norms $\|a(x, h)\|_{M(r,h)}$ by the norms $\|a(x, h)\|_{L_2(\mathbb{R}_n)}$ depend on $h_1, \ldots, h_n$, in general.

Let us pass to new variables, $t_i = \varphi_i h_i$, $1 \le i \le n$, in the right-hand side of (5.1.7). We get

$$\|a(x, h)\|^2_{M(r,h)} = \Omega_n \int_{-\pi}^{\pi} \cdots \int_{-\pi}^{\pi} \left(\left[\sum_{i=1}^{n} \left(\frac{t_i}{h_i}\right)^2 \right]^{1/2} + 1 \right)^{2r} |\widehat{a}_c(t_1, \ldots, t_n)|^2 \, dt_1 \cdots dt_n, \tag{5.3.2}$$

where $\widehat{a}_c(t)$ is the discrete Fourier transform of the step function $\pi a(x, h)$ (see Definition 5.1.4).

In the sequel, it will be assumed that

$$0 < \Theta_0 \le \frac{h_j}{h_1} \le \Theta_j \le k, \qquad j = 1, \ldots, n. \tag{5.3.3}$$

Recall that the norm in the Sobolev–Slobodetskii spaces is defined by

$$\|u(x)\|^2_{H_r(\mathbb{R}_n)} = \int_{\mathbb{R}_n} \left(1 + |\varphi|\right)^{2r} |\widetilde{u}(\varphi)|^2 \, d\varphi$$

where $\widetilde{u}(\varphi)$ is the Fourier transform of $u(x)$.

Proposition 5.3.1. *For any $a(x, h) \in M(r, h)$ the following estimate holds:*

$$\|a(x, h)\|_{M(r,h)} \le C_1(r) \|a(x, h)\|_{H_r(\mathbb{R}_n)},$$

where $C_1(r)$ does not depend on $h = (h_1, \ldots, h_n)$.

Proof. The Fourier transform $F(a(x, h)) = \int_{\mathbb{R}_n} a(x, h) \, e^{i(x,\varphi)} \, dx$ of the step function $a(x, h)$ has the form

$$F(a(x, h)) = \prod_{p=1}^{n} \frac{e^{ih_p\varphi_p/2} - e^{-ih_p\varphi_p/2}}{i\varphi_p} \sum_{k \in \mathbb{Z}_n} a(k) \, e^{i(k\varphi, h)} =$$

$$= \frac{1}{\Omega_h} \prod_{p=1}^{n} \frac{e^{ih_p\varphi_p/2} - e^{-ih_p\varphi_p/2}}{i\varphi_p} \widehat{a}_h(\varphi),$$

Hence,

$$|F(a(x, h))|^2 = \frac{2^{2n}}{\Omega_h^2} \prod_{p=1}^{n} \frac{\sin^2(\varphi_p h_p/2)}{\varphi_p^2} |\widehat{a}_h(\varphi)|^2.$$

On the parallelepiped $D_h = \prod_{j=1}^{n} \left[-\pi/h_j \le \varphi_j \le \pi/h_j\right]$, the following estimate holds

$$C_1 \left|\frac{\varphi_j h_j}{2}\right| \le \left|\sin \frac{\varphi_j h_j}{2}\right| \le \left|\frac{\varphi_j h_j}{2}\right|,$$

and therefore,

$$C_2 |F(a(x, h))|^2 \le |\widehat{a}_h(\varphi)|^2 \le C_3 |F(a(x, h))|^2 \quad \text{on} \quad D_h. \tag{5.3.4}$$

From (5.3.1) and (5.3.4), it follows that

$$\|a(x, h)\|^2_{M(r,h)} \le C_3 \int_{D_h} \left(1 + |\varphi|\right)^{2r} |F(a(x, h))|^2 \, d\varphi \le C_3 \|a(x, h)\|^2_{H_r(\mathbb{R}_n)},$$

where C_3 does not depend on $h = (h_1, \ldots, h_n)$. Proposition 5.3.1 is proved.

Theorem 5.3.1. *For any $a(x, h) \in M(r, h)$ with $r < 1/2$, the estimate*

$$\|a(x, h)\|_{H_r(\mathbb{R}_n)}^2 \leq C_2(r) \|a(x, h)\|_{M(r,h)}$$

with a constant $C_2(r)$ independent of $h = (h_1, \ldots, h_n)$.

Proof. Let us split the space $\mathbb{R}_n$ into cells $D_h(k) = D_h(k_1, \ldots, k_n)$ of the form

$$D_h(k_1, \ldots, k_n) = \prod_{p=1}^{n} \left[-\frac{\pi}{h_p} + \frac{2\pi k_p}{h_p} \leq \varphi_p \leq \frac{\pi}{h_p} + \frac{2\pi k_p}{h_p} \right], \qquad \forall k \in \mathbb{Z}_n.$$

Note that (5.3.4) implies the inequality

$$\int_{D_h(0)} \left(1 + |\varphi|\right)^{2r} |F(a(x, h))|^2 \, d\varphi \leq C \|a(x, h)\|_{M(r,h)}, \qquad (5.3.5)$$

where $D_h(0) = D_h(0, \ldots, 0)$. Let us estimate the integral

$$I(k) = \int_{D_h(k)} \left(1 + |\varphi|\right)^{2r} |F(a(x, h))|^2 \, d\varphi = 2^{2n} \int_{D_h(k)} \left(1 + |\varphi|\right)^{2r} \prod_{p=1}^{n} \frac{\sin^2(\varphi_p h_p/2)}{\varphi_p^2} |\widehat{a}_c(\varphi h)|^2 \, d\varphi.$$

Observe that the function $\widehat{a}_x(\varphi_1 h_1, \ldots, \varphi_n h_n)$ is periodic in each φ_j with period $T_j = 2\pi/h_j$. Passing to the variables $z_p = \varphi_p - 2\pi k_p/h_p$, $p = 1, \ldots, h$, we get

$$I(k) = \int \cdots \int_{D_h[0]} \left(\left[\sum_{p=1}^{n} \left(z_p + \frac{2\pi k_p}{h_p} \right)^2 \right]^{1/2} + 1 \right)^{2r} \prod_{p=1}^{n} \frac{\sin^2(z_p h_p/2)}{\left(z_p + 2\pi k_p/h_p \right)^2} |\widehat{a}_c(zh)|^2 \, dz_1 \cdots dz_n =$$

$$= 2^{2n} \int_{D_h[0]} A(k, z) \left(1 + |z|\right)^{2r} \prod_{p=1}^{n} \frac{\sin^2(z_p h_p/2)}{z_p^2} |\widehat{a}_c(zh)|^2 \, dz, \qquad (5.3.6)$$

where

$$A(k, z) = \frac{\left(\left[\sum_{p=1}^{n} \left(z_p + \frac{2\pi k_p}{h_p} \right)^2 \right]^{1/2} + 1 \right)^{2r} \prod_{p=1}^{n} z_p^2}{\left(\left[\sum_{p=1}^{n} z_p^2 \right]^{1/2} + 1 \right)^{2r} \prod_{p=1}^{n} \left(z_p + \frac{2\pi k_p}{h_p} \right)^2}. \qquad (5.3.7)$$

Let us estimate $A(k, z)$ for $z \in D_h(0)$. Changing the variables $z_p = \pi y_p/h_p$, $-1 < y_0 < 1$, in (5.3.7), we get

$$A\left(k, \frac{\pi}{h_1} y_1, \ldots, \frac{\pi}{h_n} y_n \right) = \frac{\left(h_1 + \pi \left[\sum_{p=1}^{n} \Theta_p^2 (y_p + 2k_p)^2 \right]^{1/2} \right)^{2r} \prod_{p=1}^{n} y_p^2}{\left(h_1 + \pi \left[\sum_{p=1}^{n} \Theta_p^2 y_p^2 \right]^{1/2} \right)^{2r} \prod_{p=1}^{n} (y_p + 2k_p)^2},$$

where $0 < \Theta_0 \leq \Theta_p \leq k$ with Θ_0 and k independent of h, because of (5.3.3). Consider the case $r < 0$. It should be pointed out that $\sum_{p=1}^{n} \Theta_p^2 (y_p + 2k_p)^2 \geq \sum_{p=1}^{n} \Theta_p^2 y_p^2$, which follows from the condition $-1 \leq y_p \leq 1$, $k_p \in \mathbb{Z}$. Hence,

$$A\left(k, \frac{\pi}{h} y\right) \leq \frac{C}{\displaystyle\prod_{j \in O_1} k_j^2}, \tag{5.3.8}$$

where O_1 is the subset of the indices $j = 1, \ldots, n$ such that for $j \in O_1$, we have $k_j \neq 0$, while O_2 is the set of the remaining indices, i.e., if $j \in O_2$, then $k_j = 0$; the constant C is independent of $h = (h_1, \ldots, h_n)$.

From (5.3.5), (5.3.6), and (5.3.8), it follows that

$$I(k) \leq \frac{C_1}{\displaystyle\prod_{j \in O_1} k_j^2} 2^{2n} \int_{D_h(0)} \left(1 + |z|\right)^{2r} \prod_{p=1}^{n} \frac{\sin^2 \frac{z_p h_p}{2}}{z_p^2} |\widehat{a}_c(zh)|^2 \, dz =$$

$$= \frac{C_1}{\displaystyle\prod_{j \in O_1} k_j^2} \int_{D_h(0)} \left(1 + |z|\right)^{2r} |F(a(x, h))|^2 \, dz \leq \frac{C_2}{\displaystyle\prod_{j \in O_1} k_j^2} \|a(x, h)\|_{M(r,h)}^2.$$

Hence, we find that

$$\|a(x, h)\|_{H_r(\mathbb{R}_n)}^2 \leq C_2 \|a(x, h)\|_{M(r,h)}^2 \sum_{O_2} \left(\sum_{j \in O_1} \frac{1}{\displaystyle\prod_{j \in O_1} k_j^2} \right) \leq C_3 \|a(x, h)\|_{M(r,h)}^2, \tag{5.3.9}$$

where $\sum_{j \in O_1}$ stands for $\sum_{k_{i_1}=-\infty}^{\infty} \cdots \sum_{k_{i_l}=-\infty}^{\infty}$; $i_1, \ldots, i_l$ is the set of the indices in O_1; $\sum_{O_2}$ is the sum over all sets O_2; this sum involves finitely many terms, because the set of all subsets of a finite set of indices $(1, \ldots, n)$ is finite. From (5.3.9) we obtain the assertion of our theorem for $r \leq 0$.

Consider the case $0 < r < 1/2$.

In our further transformations, we require two obvious inequalities:

1. If $a \geq 1$, $b > 1$, then $a + b \leq 2ab$; This inequality will be used for $|y_p + 2k_p| \geq 1$.
2. If $0 \leq a \leq 1$, $0 \leq b \leq 1$, then $a + b \geq ab$. This inequality will be applied to $0 \leq y_p^2 \leq 1$, $p = 1, 2, \ldots, p$.

We have

$$A\left(k, \frac{\pi}{h} y\right) = \left(\frac{h_1}{h_1 + \pi \left[\displaystyle\sum_{p=1}^{n} \Theta_p^2 y_p^2 \right]^{1/2}} + \frac{\pi \left[\displaystyle\sum_{p=1}^{n} \Theta_p^2 (y_p + 2k_p)^2 \right]^{1/2}}{h_1 + \pi \left[\displaystyle\sum_{p=1}^{n} \Theta_p^2 y_p^2 \right]^{1/2}} \right)^{2r} \frac{\displaystyle\prod_{p \in O_1} y_p^2}{\displaystyle\prod_{p \in O_1} (y_p + 2k_p)^2} \leq$$

$$\leq \left(\frac{\displaystyle\sum_{p=1}^{n} \Theta_p^2 (y_p + 2k_p)^2}{\displaystyle\sum_{p=1}^{n} \Theta_p^2 y_p^2} \right)^{r} \left(\frac{\left[\displaystyle\sum_{p=1}^{n} \Theta_p^2 y_p^2 \right]^{1/2}}{\left[\displaystyle\sum_{p=1}^{n} \Theta_p^2 (y_p + 2k_p)^2 \right]^{1/2}} + 1 \right)^{2r} \frac{\displaystyle\prod_{p \in O_1} y_p^2}{\displaystyle\prod_{p \in O_1} (y_p + 2k_p)^2} \leq$$

$$\leq \left(\frac{\lambda_1}{\lambda_0}2\right)^{2r} 2^{(n-1)r} \frac{\prod_{p=1}^{n} |y_p + 2k_p|^{2r} \sum_{p\in O_1} y_p^2}{\left(\sum_{p=1}^{n} y_p^2\right)^r \prod_{p\in O_1}(y_p + 2k_p)^2} \leq \frac{C}{\prod_{p\in O_1} |k_p|^{2(1-r)}}, \qquad (5.3.10)$$

where $\lambda_1 = \max_p \Theta_p$, $\lambda_0 = \min_p \Theta_p$.

Similarly to (5.3.9), we find that

$$\|a(x,h)\|_{H_r(\mathbb{R}_n)}^2 \leq C_1 \|a(x,h)\|_{M(r,h)}^2 \sum_{O_2}\left(\sum_{j\in O_1} \frac{1}{\prod_{j\in O_1} k_j^{2(1-r)}}\right) \leq C_3 \|a(x,h)\|_{M(r,h)}^2.$$

This proves the theorem for $r < 1/2$.

An Analogue of the Sobolev Imbedding Theorem

Definition 5.3.2. *The space* $K_0^r(h_1,\ldots,h_n) = K_0^r(h)$ *consists of step functions* $a(x,h) \in M_h$ satisfying the conditions:

(i) $a(x,h)$ are bounded on $\mathbb{R}_n$;
(ii) $\lim_{|x|\to\infty} a(x,h) = 0$; $\lim_{|x|\to\infty} T_r^p a(x,h) = 0$, $p = 1, 2, \ldots, r$.

The space $K_0^r(h)$ is endowed with the norm

$$\|a(x,h)\| = \sup_{x\in\mathbb{R}_n} \sum_{|p|=0}^{r} |T_r^p(a(x,h)|, \qquad (5.3.11)$$

where T_r^p is the right difference operator of order p defined by

$$T_r^p a(x,h) = T_r^{p_1}(x_1)\cdots T_r^{p_n} a(x,h), \qquad \forall\, p_i \in \mathbb{Z}, \quad p_i \geq 0, \quad p_1 + \cdots + p_n = p = |p|,$$

$$T_r(x_i)a(x,h) = \frac{1}{h_i}\Big[a(x_1,\ldots,x_{i-1}, x_i + h_i, x_{i+1},\ldots,x_n,h) - a(x_1,\ldots,x_n,h)\Big];$$

where $M(x_1,\ldots,x_n) \in D(k,h)$.

Proposition 5.3.2. *The space* $K_0^r(h)$ *is complete.*

Proof. Suppose that $\{a_n(x,h)\}_{n=1}^{\infty} \in K_0^r(h)$ is a Cauchy sequence with respect to the norm (5.3.11). Then, for any $x \in D(k,h)$, the numerical sequence $\{a_n(x,h)\}_{n=1}^{\infty}$ is a Cauchy sequence, and therefore, it converges pointwise to some function $a_0(x,h)$. Let us show that $\lim_{|x|\to\infty} a_0(x,h) = 0$. Indeed, assuming the contrary, we can find a sequence of points M_k, $k = 1, 2, \ldots$, such that $\lim_{k\to\infty} |OM_k| \to \infty$ and $|a_0(M_k,h)| \geq \varepsilon_0 > 0$ for any k, where O is the origin. Since for any fixed $a_n(x,h)$, we have $\lim_{|x|\to\infty} a_n(x,h) = 0$, it follows that for any n there exists $k_0(n)$ such that $|a_n(M_k,h)| \leq \varepsilon_0/10$ for all $k > k_0(n)$. Therefore, $|a_0(M_k,h) - a_n(M_k,h)| \geq 9\varepsilon_0/10$ for $k > k_0(n)$, which means that $\{a_n(x,h)\}_{n=1}^{\infty}$ cannot be convergent to $a_0(x,h)$ in the norm of $K_0^r(h)$. This contradiction shows that $\lim_{|x|\to\infty} a_0(x,h) = 0$.

In a similar way, it can be shown that there exist $\lim_{|x|\to\infty} T_r^p a_n(x,h) = b_p(x,h)$, $p = 1,\ldots,r$, and $\lim_{|x|\to\infty} b_p(x,h) = 0$. For any point $x \in D(k,h)$, we have $b_p(x,h) = \lim_{n\to\infty} T_r^p a_n(x,h) = T_r^p a_0(x,h)$, $1 \leq p \leq r$, and therefore, Proposition 5.3.2 is valid.

Proposition 5.3.3. *Let $s > r + n/2$ for a positive integer $r \in \mathbb{N}$. Then, for any $a(x, h) \in M(s, h)$, the estimate*

$$\|a(x, h)\|_{K_0^r(h)} \leq C \|a(x, h)\|_{M(s,h)}$$

holds with a constant C independent of $h = (h_1, \ldots, h_n)$.

Proof. Note that any function $a(x, h) \in M(s, h)$, with $s \geq 0$ and $h = (h_1, \ldots, h_n)$ fixed, belongs to $K_0^r(h)$ for any r. Therefore, it suffices to show that the constant C does not depend on h.

Consider an arbitrary function $a(x, h) \in M(s, h)$ and let $T_r^m(v_1, \ldots, v_n)$ by an arbitrary difference operator of order $m \leq r$. The symbol of this operator has the form (see Example 5.1.4)

$$\widehat{A}(\varphi) = \prod_{k=1}^{n} \left(\frac{e^{-i\varphi} - 1}{h_k} \right)^{m_k}, \qquad m = m_1 + \cdots + m_n.$$

Then, for $x \in D(k, h)$, by virtue of (5.1.4) and (5.1.10), we have

$$T_r^m(a(x, h)) = \frac{1}{(2\pi)^n} \int_U \widehat{A}(\varphi) \widehat{a}_c(\varphi) e^{-i(k\varphi)} \, d\varphi = \frac{\Omega_n}{(2\pi)^n} \int_{D_h} \widehat{A}(ht) \widehat{a}_c(ht) e^{-i(kt,h)} \, dt =$$

$$= \frac{1}{(2\pi)^n} \int_{D_h} \widehat{A}(ht) \widehat{a}_h(t) e^{-i(kt,h)} \, dt, \tag{5.3.12}$$

where $\widehat{a}_c(\varphi)$ is the discrete grid Fourier transform of the function $\pi a(x, h)$ (see Definition 5.1.4), and $\widehat{a}_h(t)$ is the discrete step Fourier transform of $a(x, h)$. From (5.3.12) we obtain

$$|T_r^m a(x, h)| \leq C \int_{D_h} \frac{|\widehat{A}(ht)|}{(1 + |t|)^s} \left[(1 + |t|)^s |\widehat{a}_h(t)| \right] dt \leq$$

$$\leq C \left[\int_{D_h} \frac{|\widehat{A}(ht)|^2}{(1 + |t|)^{2s}} \, dt \right]^{1/2} \|a(x, h)\|_{M(s,h)}. \tag{5.3.13}$$

Let us estimate the expression

$$I = \int_{D_h} \frac{|\widehat{A}(ht)|^2}{(1 + |t|)^{2s}} \, dt = \int_{D_h} \frac{\prod_{k=1}^{n} \left(\dfrac{4 \sin^2 \frac{h_k t_k}{2}}{h_k^2} \right)^{m_k}}{(1 + |t|)^{2s}} \, dt \leq$$

$$\leq C_1 \int_{D_h} \frac{\prod_{k=1}^{n} |t_k|^{2m_k}}{(1 + |t|)^{2s}} \, dt \leq C_1 \int_{K(\sqrt{n}\,\pi/h_0)} \frac{\prod_{k=1}^{n} |t_k|^{2m_k}}{(1 + |t|)^{2s}} \, dt \tag{5.3.14}$$

where $K(\sqrt{n}\,\pi/h_0)$ is the ball of radius $R = \sqrt{n}\,\pi/h_0$, $h_0 = \min(h_1, \ldots, h_n)$. Passing to spherical coordinates in (5.3.14), we obtain

$$I \leq C_2 \int_S f(\omega) \, d\omega \int_0^{\sqrt{n}\,\pi/h_0} \frac{\rho^{2m+n+1}}{(1 + \rho)^{2s}} \, d\rho \leq C_3 \int_S f(\omega) \int_0^{\infty} \frac{d\rho}{(1 + \rho)^{1+\varepsilon}} \leq C_4, \tag{5.3.15}$$

where S is the unit sphere, $f(\omega) \, d\omega$ is the area element of the unit sphere; the constants C_i, $i = 1, \ldots, 4$, do not depend on $h = (h_1, \ldots, h_n)$. Now, Proposition 5.3.3 follows from (5.3.13) and (5.3.15).

Definition 5.3.3. *The space of step functions* $M_1(r, h)$ *with* $0 < r < 1$ *consists of all* $a(x, h) \in M_h$ with the finite norm

$$\|a(x, h)\|^2_{M_1(r,h)} = \Omega_h \sum_{|l| \neq 0} \frac{1}{|l|^{n+2r}(\Omega_h)^{1+2r/n}} \Omega_h \sum_{k \in \mathbb{Z}_n} |a(k+l) - a(k)|^2 + \Omega_h \sum_{k \in \mathbb{Z}_n} |a(k)|^2 =$$

$$= I_1(r, h) + I_2(r, h), \tag{5.3.16}$$

where $\Omega_h = \prod_{j=1}^n h_j$, $a(k) = \pi a(x, h)$ (see Definition 5.3.2); $\sum_{|l| \neq 0}$ is the sum over all $l \in \mathbb{Z}_n$ such that $|l| \neq 0$.

Proposition 5.3.4. *For any* $a(x, h) \in M_1(r, h)$, $0 < r < 1$, *the inequality*

$$C_1(r)\|a(x, h)\|_{M(r,h)} \leq \|a(x, h)\|_{M_1(r,h)} \leq C_2(r)\|a(x, h)\|_{M(r,h)}$$

holds with constants $C_i(r) > 0$, $i = 1, 2$, *independent of* $h = (h_1, \ldots, h_n)$.

Proof. Using Example 5.1.1 and Parseval's identity, we get

$$\Omega_h \sum_{k \in \mathbb{Z}_n} |a(k+l) - a(k)|^2 = \frac{\Omega_h}{(2\pi)^n} \int_U \widehat{a}_c(\varphi)\left[e^{-i(l\varphi)} - 1\right] \cdot \overline{\widehat{a}_c(\varphi)\left[e^{-i(l\varphi)} - 1\right]} \, d\varphi.$$

Hence, by (5.3.16), it follows that

$$I_1(r, h) = \frac{\Omega_h}{(2\pi)^n} \int_U |\widehat{a}_c(\varphi)|^2 \sum_{|l| \neq 0} \frac{1}{\Omega_h^{2r/n}} \frac{4\sin^2 \frac{(l\varphi)}{2}}{|l|^{n+2r}} \, d\varphi = \frac{\Omega_h}{(2\pi)^n} \int_U |\widehat{a}_c(\varphi)|^2 \frac{4F(2r, \varphi)}{\Omega_h^{2r/n}} \, d\varphi,$$

where $F(\lambda, \varphi)$ is the function defined by (5.2.2).

According to Propositions 5.2.2 and 5.2.3, the functions $F(2r, \varphi)$ satisfy the inequalities

$$C_1(r)|\varphi|^{2r} \leq F(2r, \varphi) \leq C_2(r)|\varphi|^{2r}. \tag{5.3.17}$$

From (5.3.3) we deduce

$$K_1(r)\frac{|\varphi|^{2r}}{\Omega_h^{2r/n}} \leq \left(\left[\sum_{i=1}^n \left|\frac{\varphi_i}{h_i}\right|^2\right]^{1/2}\right)^{2r} \leq K_2(r)\frac{|\varphi|^{2r}}{\Omega_h^{2r/n}}. \tag{5.3.18}$$

From (5.3.17) and (5.3.18), it follows that

$$K_3(r)\left[\frac{4F(2r, \varphi)}{\Omega_h^{2r/n}} + 1\right] \leq \left(\left[\sum_{i=1}^n \left|\frac{\varphi_i}{h_i}\right|^2\right]^{1/2} + 1\right)^{2r} \leq K_4(r)\left[\frac{4F(2r, \varphi)}{\Omega_h^{2r/n}} + 1\right].$$

This inequality, together with (5.3.2), proves Proposition 5.3.4.

Definition 5.3.4. *The space of step functions* $M_1(r, h)$ *with integer* $r > 0$ *consists of all* $a(x, h) \in M_h$ *with finite norm*

$$\|a(x, h)\|^2_{M_1(r,h)} = \Omega_h \sum_{|m| \leq r} \sum_{k \in \mathbb{Z}_n} |T_r^m a(k)|^2, \tag{5.3.19}$$

where T_r^m is the right difference operator of order m; $a(k) = \pi a(x, h)$.

Proposition 5.3.5. *For any $a(x, h) \in M_1(r, h)$ with integer $r > 0$, the inequality*

$$C_1(r)\|a(x, h)\|_{M_1(r,h)} \leq \|a(x, h)\|_{M(r,h)} \leq C_2(r)\|a(x, h)\|_{M_1(r,h)}$$

holds with constants $C_i(r) > 0$, $i = 1, 2$, independent of $h = (h_1, \ldots, h_n)$.

Proof. According to Example 5.1.4, for the symbol $\widehat{A}(\varphi)$ of the difference operator T_r^m ($m = m_1 + \cdots + m_n$), we have

$$\left|\widehat{A}(\varphi)\right| = \prod_{k=1}^{n} \frac{2^{m_k}\left|\sin\frac{\varphi_k}{2}\right|^{m_k}}{h_k^{m_k}}.$$

Using Parseval's identity for (5.3.19), we find that

$$\|a(x, h)\|_{M_1(r,h)} = \frac{\Omega_h}{(2\pi)^n} \sum_{|m|\leq r} 2^{2m} \int_{-\pi}^{\pi} \cdots \int_{-\pi}^{\pi} \prod_{k=1}^{n} \frac{\left|\sin\frac{\varphi_k}{2}\right|^{2m_k}}{h_k^{2m_k}} |\widehat{a}_c(\varphi)|^2 \, d\varphi_1 \cdots d\varphi_n. \quad (5.3.20)$$

Further, we have

$$\prod_{k=1}^{n} \frac{\left|\sin\frac{\varphi_k}{2}\right|^{2m_k}}{h_k^{2m_k}} \leq C\left(\sum_{k=1}^{n} \frac{|\varphi_k|}{h_k}\right)^{2m} \leq C_1\left(\left[\sum_{i=1}^{n} \left|\frac{\varphi_i}{h_i}\right|^2\right]^{1/2}\right)^{2m} \leq$$

$$\leq C_1\left(\left[\sum_{i=1}^{n} \left|\frac{\varphi_i}{h_i}\right|^2\right]^{1/2} + 1\right)^{2m}. \quad (5.3.21)$$

From (5.3.2), (5.3.20), and (5.3.21), it follows that $\|a(x, h)\|_{M_1(r,h)} \leq C_2(r)\|a(x, h)\|_{M(r,h)}$. On the other hand, by the Hölder inequality, we have

$$\sum_{|m|\leq r} \prod_{k=1}^{n} \frac{\left|\sin\frac{\varphi_k}{2}\right|^{2m_k}}{h_k^{2m_k}} \geq 1 + \sum_{i=1}^{n} \frac{\left|\sin\frac{\varphi_i}{2}\right|^{2r}}{h_i^{2r}} \geq C_2\left(\sum_{i=1}^{n} \frac{|\varphi_i|}{h_i} + 1\right)^{2r} \geq$$

$$\geq C_2\left(\left[\sum_{i=1}^{n} \left|\frac{\varphi_i}{h_i}\right|^2\right]^{1/2} + 1\right)^{2r}, \quad (5.3.22)$$

where $\varphi = (\varphi_1, \ldots, \varphi_n) \in U = \prod_{i=1}^{n} [-\pi \leq \varphi_i \leq \pi]$. Now, (5.3.2), (5.3.20), and (5.3.22) imply that $\|a(x, h)\|_{M_1(r,h)} \geq C_2(r)\|a(x, h)\|_{M(r,h)}$, and thus Proposition 5.3.5 is proved.

Definition 5.3.5. *The space of step functions $M_1(r, h)$ with $r = r_0 + \lambda$, integer $r_0 \geq 0$ and $0 < \lambda < 1$ consists of all $a(x, h) \in M_h$ with the finite norm*

$$\|a(x, h)\|_{M_1(r,h)}^2 = \quad\quad\quad\quad\quad\quad\quad\quad\quad\quad\quad\quad\quad\quad\quad\quad\quad (5.3.23)$$

$$= \Omega_h \sum_{|l|\neq 0} \frac{1}{|l|^{n+2\lambda}\Omega_h^{1+2\lambda/n}} \sum_{|m|\leq r_0} \Omega_h \sum_{k\in\mathbb{Z}_n} |T_r^m a(k+l) - T_r^m a(k)|^2 + \Omega_h \sum_{k\in\mathbb{Z}_n} |a(k)|^2,$$

where $a(k) = \pi a(x, h)$, T_r^m is the right difference operator of order m.

Proposition 5.3.6. *For any $a(x, h) \in M_1(r, h)$, $r = r_0 + \lambda$, $0 < \lambda < 1$ and integer $r_0 \geq 0$, the inequality*

$$C_1(r)\|a(x, h)\|_{M(r,h)} \leq \|a(x, h)\|_{M_1(r,h)} \leq C_2(r)\|a(x, h)\|_{M(r,h)}$$

holds with constants $C_i(r) > 0$, $i = 1, 2$, that do not depend on $h = (h_1, \ldots, h_n)$.

Proof. Using Proposition 5.3.4, we obtain

$$C_1(\lambda)\Omega_h \int_U \left(\left[\sum_{i=1}^n \left| \frac{\varphi_i}{h_i} \right|^2 \right]^{1/2} + 1 \right)^{2\lambda} |B(\varphi)|^2 \, d\varphi \le \|a(x,h)\|^2_{M(r,h)} \le$$

$$\le C_2(\lambda)\Omega_h \int_U \left(\left[\sum_{i=1}^n \left| \frac{\varphi_i}{h_i} \right|^2 \right]^{1/2} + 1 \right)^{2\lambda} |B(\varphi)|^2 \, d\varphi, \qquad (5.3.24)$$

where $B(\varphi)$ is the discrete grid Fourier transform (see Definition 5.1.4) of the function

$$\sum_{|m|\le r_0} \mathbf{T}_r^m a(k), \qquad a(k) = \pi a(x,h).$$

As shown in Proposition 5.3.5, for $B(\varphi)$ the following estimate is valid:

$$A_1(r_0) \left(\left[\sum_{i=1}^n \left| \frac{\varphi_i}{h_i} \right|^2 \right]^{1/2} + 1 \right)^{2r_0} |\widehat{a}_c(\varphi)|^2 \le |B(\varphi)|^2 \le$$

$$\le A_2(r_0) \left(\left[\sum_{i=1}^n \left| \frac{\varphi_i}{h_i} \right|^2 \right]^{1/2} + 1 \right)^{2r_0} |\widehat{a}_c(\varphi)|^2, \qquad (5.3.25)$$

where $\widehat{a}_c(\varphi)$ is the discrete grid Fourier transform of $a(k) = \pi a(x,h)$. The inequalities (5.3.24) and (5.3.25) prove Proposition 5.3.6.

Definition 5.3.6. *The space of step functions $M_2(r,h)$ for integer $r > 0$ consists of all $a(x,h) \in M_h$ with the finite norm*

$$\|a(x,h)\|^2_{M_2(r,h)} = \Omega_h \sum_{i=1}^n \sum_{k\in\mathbb{Z}_n} |T_r^r(i)a(k)|^2 + \Omega_h \sum_{z\in\mathbb{Z}_n} |a(k)|^2; \qquad (5.3.26)$$

and for $r = r_0 + \lambda$, with integer $r_0 \ge 0$ and $1 < \lambda < 1$, the space $M_2(r,h)$ consists of all $a(x,h) \in M_h$ with the finite norm

$$\|a(x,h)\|^2_{M_2(r,h)} = \qquad (5.3.27)$$

$$= \Omega_h \sum_{|l|\ne 0} \frac{1}{|l|^{n+2\lambda}\Omega_h^{1+2\lambda/n}} \Omega_h \sum_{k\in\mathbb{Z}_n} \sum_{i=1}^n |\mathbf{T}_r^{r_0}[a(k+l) - a(k)]|^2 + \Omega_h \sum_{k\in\mathbb{Z}_n} |a(k)|^2,$$

where $\mathbf{T}_r^{r_0}(i)$ is the right difference operator with respect to the index i, $a(k) = \pi a(x,h)$.

Proposition 5.3.7. *For any $a(x,h) \in M_2(r,h)$, $r > 0$, the inequality*

$$C_1(r)\|a(x,h)\|_{M(r,h)} \le \|a(x,h)\|_{M_2(r,h)} \le C_2(r)\|a(x,h)\|_{M(r,h)}, \qquad (5.3.28)$$

holds with constants $C_i(r) > 0$, $i = 1, 2$, independent of $h = (h_1, \ldots, h_n)$.

Proof. First, consider the case of integer $r > 0$. Using the Parseval identity for (5.3.26), we find

$$\|a(x,h)\|_{M_2(r,h)} = \frac{\Omega_h}{(2\pi)^n} \left[2^{2r} \sum_{i=1}^n \frac{|\sin \frac{\varphi_i}{2}|^{2r}}{h_i^{2r}} + 1 \right] |\widehat{a}_c(\varphi)|^2 \, d\varphi. \qquad (5.3.29)$$

It follows from the proof of Proposition 5.3.5 that for $\varphi \in U = \prod_{i=1}^{n} [-\pi \leq \varphi_i \leq \pi]$, we have

$$C_1(r) \left(\left[\sum_{i=1}^{n} \left| \frac{\varphi_i}{h_i} \right|^2 \right]^{1/2} + 1 \right)^{2r} \leq \left[2^{2r} \sum_{i=1}^{n} \frac{\left| \sin \frac{\varphi_i}{2} \right|^{2r}}{h_i^{2r}} + 1 \right] \leq$$

$$\leq C_2(r) \left(\left[\sum_{i=1}^{n} \left| \frac{\varphi_i}{h_i} \right|^2 \right]^{1/2} + 1 \right)^{2r}. \tag{5.3.30}$$

From (5.3.2), (5.3.29), and (5.3.3), we obtain the desired inequality (5.3.28).

Now, let $r = r_0 + \lambda$ with integer $r_0 \geq 0$ and $0 < \lambda < 1$. Using the Parseval identity for (5.3.27), in combination with Proposition 5.3.4, we find

$$C_1(r) \int_U \left(\left[\sum_{i=1}^{n} \left| \frac{\varphi_i}{h_i} \right|^2 \right]^{1/2} + 1 \right)^{2\lambda} \left[2^{2r_0} \sum_{i=1}^{n} \frac{\left| \sin \frac{\varphi_i}{2} \right|^{2r_0}}{h_i^{2r_0}} + 1 \right] |\hat{a}_c(\varphi)|^2 \, d\varphi \leq \|a(x,h)\|_{M_2(r,h)} \leq$$

$$\leq C_2(r) \int_U \left(\left[\sum_{i=1}^{n} \left| \frac{\varphi_i}{h_i} \right|^2 \right]^{1/2} + 1 \right)^{2\lambda} \left[2^{2r_0} \sum_{i=1}^{n} \frac{\left| \sin \frac{\varphi_i}{2} \right|^{2r_0}}{h_i^{2r_0}} + 1 \right] |\hat{a}_c(\varphi)|^2 \, d\varphi. \tag{5.3.31}$$

The inequalities (5.3.30) and (5.3.31) imply (5.3.28). Proposition 5.3.7 is proved.

5.4. Integral Projector

Consider the canonical partition of the space $\mathbb{R}_n$ into cells $D(k,h)$ with vector-valued step $h = (h_1, \ldots, h_n)$ (see Definition 5.1.1). To each locally integrable function $f(x)$ we assign a grid function $a(k)$ defined by

$$a(k) = \frac{1}{\Omega_h} \int_{D(k,h)} f(x) \, dx.$$

And to the grid function $a(k)$ we associate the step function $a(x,h) \in M_h$ such that $a(k) = \pi a(x,h)$ (see Definition 5.1.2).

Definition 5.4.1. The above operator that maps $f(x) \in L(\mathbb{R}_n)$ to $a(x,h) \in M_h$ is called the *integral projector* and is denoted by Π_h; $\Pi_h f(x) = a(x,h)$ (see *Lifanov and Poltavskii* (1999)[1,2]).

Proposition 5.4.1. *The integral projector* Π_h *is a bounded operator from the Sobolev–Slobodetskii space* $H_r(\mathbb{R}_n)$, $0 \leq r < 1$, *to the space of fractional quotients* $M(r,h)$.

Proof. Consider the case of $r = 0$, i.e., $H_0(\mathbb{R}_n) = L_2(\mathbb{R}_n)$. For $\nu(x) \in L_2(\mathbb{R}_n)$, by the definition of the integral projector and the Jensen inequality, we have

$$\|\Pi_h(\nu(x))\|_{L_2(\mathbb{R}_n)}^2 = \Omega_h \sum_{k \in \mathbb{Z}_n} |a(k)|^2 = \Omega_h \sum_{k \in \mathbb{Z}_n} \left| \int_{D(k,h)} \frac{1}{\Omega_h} \nu(x) \, dx \right|^2 \leq$$

$$\leq \Omega_h \sum_{k \in \mathbb{Z}_n} \int_{D(k,h)} |\nu(x)|^2 \frac{1}{\Omega_h} \, dx \leq \int_{\mathbb{R}_n} |\nu(x)|^2 \, dx. \tag{5.4.1}$$

Hence, we obtain the desired statement for $r = 0$.

Let $0 < r < 1$. For $\nu(x) \in H_r(\mathbb{R}_n)$, $a(x,h) = \Pi_h \nu(x)$ we have

$$\int_{\mathbb{R}_n} |a(x,h) - a(x)lh,h)|^2 \, dx \leq \int_{\mathbb{R}_n} |\nu(x) - \nu(x+lh)|^2 \, dx, \tag{5.4.2}$$

where $l = (l_1, \ldots, l_n)$, $l_i \in \mathbb{Z}$.

Applying the Parseval identity for ordinary Fourier transforms, from (5.4.2) we get

$$\int_{\mathbb{R}_n} |a(x,h) - a(x+lh,h)|^2 \, dx \le \frac{1}{(2\pi)^n} \int_{\mathbb{R}_n} 4\sin^2 \frac{(lh,\xi)}{2} |\hat{\nu}(\xi)|^2 \, d\xi, \tag{5.4.3}$$

where $\hat{\nu}(\xi) = F(\nu(x))$ is the Fourier transform of $\nu(x)$. In view of Proposition 5.3.4, it suffices to establish our statement for the space of step functions $M_1(r,h)$.

Using (5.3.16), from (5.4.1) and (5.4.3) we find

$$\|\Pi_h \nu(x)\|^2_{M_1(r,h)} \le \frac{1}{(2\pi)^n} \int_{\mathbb{R}_n} |\hat{\nu}(\xi)|^2 \left[\frac{1}{\Omega_h^{2r/n}} \sum_{|l| \ne 0} \frac{4\sin^2 \frac{(lh,\xi)}{2}}{|l|^{n+2r}} + 1 \right] d\xi. \tag{5.4.4}$$

According to Proposition 5.2.2, the function $F(2r,(h,\xi)) = \sum_{|l| \ne 0} \frac{4\sin^2 \frac{(lh,\xi)}{2}}{|l|^{n+2r}}$ satisfies the inequality

$$0 \le F(2r,(h,\xi)) \le C_1(r)|(h,\xi)|^{2r} \le C_1(r)|h|^{2r}|\xi|^{2r}.$$

Condition (5.3.3) implies that $C_2(r) \le |h|^{2r}/\Omega_h^{2r/n} \le C_3(r)$ with $C_i(r) > 0$, $i = 2,3$, independent of $h = (h_1, \ldots, h_n)$. Therefore, it follows from (5.4.4) that

$$\|\Pi_h \nu(x)\|^2_{M_1(r,h)} \le C_3(r) \int_{\mathbb{R}_n} \left(1 + |\xi|\right)^{2r} |\hat{\nu}(\xi)|^2 \, d\xi \le C_4(r) \|\nu(x)\|^2_{H_r(\mathbb{R}_n)},$$

which completes the proof of Proposition 5.4.1.

Proposition 5.4.2. *For integer $r > 0$, the integral projector Π_h is bounded from $H_r(\mathbb{R}_n)$ to $M(r,h)$.*

Proof. Because of Proposition 5.3.7, it suffices to establish this result for the space of step functions $M_2(r,h)$ (see Definition 5.3.6). Let $\nu(x) \in H_r(\mathbb{R}_n)$, $a(k) = \pi \Pi_h \nu(x)$. Then

$$T_r^r(i)a(k) = \frac{1}{\Omega_h} T_r^r(x_i)\nu(x) \, dx. \tag{5.4.5}$$

Using the Parseval identity combined with (5.3.26) and (5.4.5), we find

$$\|\Pi_h \nu(x)\|^2_{M_2(r,h)} = \Omega_h \sum_{i=1}^n \sum_{k \in \mathbb{Z}_n} |T_r^r(i)a(k)|^2 + \Omega_h \sum_{k \in \mathbb{Z}_n} |a(k)|^2 =$$

$$= \Omega_h \sum_{i=1}^n \sum_{k \in \mathbb{Z}_n} \left| \int_{D(k,h)} T_r^r(x_i) \cdot \frac{\nu(x)}{\Omega_h} \, dx \right|^2 + \Omega_h \sum_{k \in \mathbb{Z}_n} |a(k)|^2 \le$$

$$\le \sum_{i=1}^n \int_{\mathbb{R}_n} |T_r^r(x_i)\nu(x)|^2 \, dx + \int_{\mathbb{R}_n} |\nu(x)|^2 \, dx =$$

$$= \frac{1}{(2\pi)^n} \int_{\mathbb{R}_n} |\hat{\nu}(\xi)|^2 \left(\sum_{j=1}^n \frac{|1 - e^{-ih_j \xi_j}|^{2r}}{h_j^{2r}} + 1 \right) d\xi \le$$

$$\le C(r) \int_{\mathbb{R}_n} \left(|\xi|^{2r} + 1\right) |\hat{\nu}(\xi)|^2 \, d\xi \le C_1(r) \|\nu(x)\|^2_{H_r(\mathbb{R}_n)}.$$

The last inequality yields the desired statement.

Theorem 5.4.1. *For any* $\nu(x) \in H_r(\mathbb{R}_n)$, $r \geq 0$, *the following inequality holds:*

$$\|\Pi_h \nu(x)\|_{M(r,h)} \leq C(r)\|\nu(x)\|_{H_r(\mathbb{R}_n)},$$

where $C(r) > 0$ *is a constant independent of* $h = (h_1, \dots, h_n)$.

Proof. This result has already been established for $0 \leq r \leq 1$ and integer $r \geq 1$. Consider the case of $r = r_0 + \lambda$, with $0 < \lambda < 1$ and integer $r_0 \geq 1$.

In view of Proposition 5.3.7, it suffices to prove the theorem for the space $M_2(r, h)$. Let $\nu(x) \in H_r(\mathbb{R}_n)$ and $a(k) = \pi \Pi_h \nu(x)$ (see Definition 5.1.2). Then, using the Parseval identity, together with (5.4.27) and (5.4.5), we find

$$\|\Pi_h \nu(x)\|_{M_2(r,h)} =$$

$$= \Omega_h \sum_{|l| \neq 0} \frac{1}{|l|^{n+2\lambda} \Omega_h^{1+2\lambda/n}} \Omega_h \sum_{k \in \mathbb{Z}_n} \sum_{i=1}^{n} |\mathrm{T}_r^{r_0}(i)[a(k+l) - a(k)]|^2 + \Omega_h \sum_{k \in \mathbb{Z}_n} |a(k)|^2 \leq$$

$$\leq \sum_{|l| \neq 0} \frac{1}{|l|^{n+2\lambda} \Omega_h^{1+2\lambda/n}} \sum_{i=1}^{n} \int_{\mathbb{R}_n} |\mathrm{T}_r^{r_0}(x_i)[\nu(x+lh) - \nu(x)]|^2 \, dx + \int_{\mathbb{R}_n} |\nu(x)|^2 \, dx =$$

$$= \frac{1}{(2\pi)^n} \int_{\mathbb{R}_n} |\hat{\nu}(\xi)|^2 \left[2^{2r_0} \sum_{i=1}^{n} \frac{\sin^{2r_0} \frac{h_i \xi_i}{2}}{h_i^{2r_0}} \sum_{|l| \neq 0} \frac{4\sin^2 \frac{(lh, \xi)}{2}}{|l|^{n+2\lambda} \Omega_h^{2\lambda/n}} + 1 \right] d\xi \leq$$

$$\leq C(r) \int_{\mathbb{R}_n} |\hat{\nu}(\xi)|^2 \left(|\xi|^{2r_0}|\xi|^{2\lambda} + 1\right) d\xi \leq C_1(r)\|\nu(x)\|_{H_r(\mathbb{R}_n)}^2. \tag{5.4.6}$$

Theorem 5.4.1 is proved.

The space of fractional quotients $M(r, h)$ is a Hilbert space with the scalar product

$$(a(x, h), b(x, h)) = \int_{D_h} \left(1 + |\varphi|\right)^{2r} \hat{a}_h(\varphi)\overline{\hat{b}_h(\varphi)} \, d\varphi. \tag{5.4.7}$$

Therefore, every continuous linear functional $f(a(x, h)) = f(a)$ can be represented in the form

$$f(a) = \int_{D_h} \left(1 + |\varphi|\right)^{2r} \hat{a}_h(\varphi)\overline{\hat{b}_h(\varphi)} \, d\varphi, \tag{5.4.8}$$

where $\hat{b}_h(\varphi) \in \widetilde{M}(r, h)$; $\widetilde{M}(r, h)$ is the Fourier image of the space $M(r, h)$.

Consider the function $\hat{w}_h(\varphi)$ defined as follows:

$$\hat{w}_h(\varphi) = \left(1 + |\varphi|\right)^{2r} \hat{b}_h(\varphi) \quad \text{for} \quad \varphi \in D_h = \prod_{j=1}^{n} \left[-\frac{\pi}{h_j} \leq \varphi_j \leq \frac{\pi}{h_j}\right],$$

and $\hat{w}_h(\varphi)$ is periodic in $\mathbb{R}_n$ with respect to every φ_j with period $T_j = 2\pi/h_j$. Since $\hat{w}_h \in \widetilde{M}(-r, h)$, any linear functional on $M(r, h)$ can be represented as

$$f(a) = \int_{D_h} \hat{a}_h(\varphi)\overline{\hat{w}_h(\varphi)} \, d\varphi. \tag{5.4.9}$$

According to Theorem 5.4.1, the integral projector Π_h is a bounded operator from $H_r(\mathbb{R}_n)$ to $M(r, h)$, $r \geq 0$. Therefore, its kernel (i.e., the set of $\nu(x)$ such that $\Pi_h \nu(x) = 0$) is a closed linear subspace.

Denote by L_h^* the kernel of Π_h and by L_h the orthogonal complement of L_h^*. Then, $H_r(\mathbb{R}_n) = L_h \oplus L_h^*$. Theorem 5.3.1 shows that the inclusion $a(x, h) \in M(-r, h)$ implies that $a(x, h) \in H_{-r}(\mathbb{R}_n)$. Therefore, it would be correct to introduce the following notation: $M(-r, h) = Q_h \subset H_{-r}(\mathbb{R}_n)$. It follows from Proposition 5.3.1 and Theorem 5.3.1 that the norms of $M(-r, h)$ and $H_{-r}(\mathbb{R}_n)$ are equivalent, and $M(-r, h)$ is a complete space. Therefore, Q_h is a closed subspace of $H_{-r}(\mathbb{R}_n)$. Let Q_h^* be the orthogonal complement of Q_h. Then, $H_{-r} = Q_h \oplus Q_h^*$. Denote by $\widetilde{H}_r(\mathbb{R}_n)$, $\widetilde{H}_{-r}(\mathbb{R}_n)$, $\widetilde{L}_h$, $\widetilde{L}_h^*$, $\widetilde{Q}_h$, and $\widetilde{Q}_h^*$ the Fourier images of the respective spaces.

Note that the operator $\lambda \widehat{u}(t) = \left(1 + |t|\right)^{2r} \widehat{u}(t)$, $\widehat{u}(t) \in \widetilde{H}_r(\mathbb{R}_n)$, is an injective isometric mapping of $\widetilde{H}_r(\mathbb{R}_n)$ to $\widetilde{H}_{-r}(\mathbb{R}_n)$.

Proposition 5.4.3. *The operator λ^{-1} maps $\widetilde{Q}_h$ into $\widetilde{L}_h$ and $\widetilde{Q}_h^*$ into $\widetilde{L}_h^*$.*

Proof. First let us show that λ^{-1} maps $\widetilde{Q}_h$ into $\widetilde{L}_h$. Consider

$$\widehat{u}(t) \in \widetilde{Q}_h, \qquad \lambda^{-1}\widehat{u}(t) = \frac{\widehat{u}(t)}{\left(1 + |t|\right)^{2r}} = \nu_1(t),$$

and take any $\widehat{\nu}(t) \in \widetilde{L}_h^*$. Then

$$(\widehat{\nu}_1(t), \widehat{\nu}(t)) = \int_{\mathbb{R}_n} \left(1 + |t|\right)^{2r} \frac{\widehat{u}(t)}{\left(1 + |t|\right)^{2r}} \overline{\widehat{\nu}(t)} \, dt = \int_{\mathbb{R}_n} \widehat{u}(t) \overline{\widehat{\nu}(t)} \, dt. \qquad (5.4.10)$$

Since $\widehat{u}(t) \in L_2(\mathbb{R}_n)$ for any fixed $h = (h_1, \ldots, h_n)$, and the function $\widehat{\nu}(t) \in \widetilde{L}_h^* \subset \widetilde{H}_r(\mathbb{R}_n)$ always belongs to $L_2(\mathbb{R}_n)$, we can apply the Parseval identity in (5.4.10), which yields

$$(\widehat{\nu}_1(t), \widehat{\nu}(t)) = (2\pi)^n \int_{\mathbb{R}_n} u(x) \overline{\nu(x)} \, dx = (2\pi)^n \int_{\mathbb{R}_n} u(x) \overline{\Pi_h \nu(x)} \, dx = 0. \qquad (5.4.11)$$

To obtain the last relation in (5.4.11), we have used the inclusion $u(x) \in M(-r, h)$. From (5.4.11) it follows that $\lambda^{-1} \widetilde{Q}_h \subset \widetilde{L}_h$.

Let us show that λ^{-1} maps $\widetilde{Q}_h^*$ into $\widetilde{L}_h^*$. For $\widehat{u}(t) \in \widetilde{Q}_h^*$ and any $\widehat{\nu}(t) \in \widetilde{Q}_h$, we have

$$0 = (\widehat{\nu}(t), \widehat{u}(t)) = \int_{\mathbb{R}_n} \widehat{\nu}(t) \frac{\overline{\widehat{u}(t)}}{\left(1 + |t|\right)^{2r}} \, dt = (2\pi)^n \int_{\mathbb{R}_n} \nu(x) \overline{u_1(x)} \, dx = (2\pi) \int_{\mathbb{R}_n} \nu(x) \overline{\Pi_h u_1(x)} \, dx,$$
$$(5.4.12)$$

where $u_1(x) = F^{-1}\left(\widehat{u}(t) / \left(1 + |t|\right)^{2r}\right)$, $\nu(x) = F^{-1}(\widehat{\nu}(t)) \in Q_h$.

Since the step function $\nu(x)$ can be chosen arbitrary, it follows from (5.4.12) that $\Pi_h u_1(x) = 0$, and therefore, $\widehat{u}(t) / \left(1 + |t|\right)^{2r} \in \widetilde{L}_h^*$. Proposition 5.4.3 is proved.

Proposition 5.4.4. *The operator λ^{-1} is an injective mapping from $\widetilde{Q}_h$ onto $\widetilde{L}_h$, and from $\widetilde{Q}_h^*$ onto $\widetilde{L}_h^*$.*

Proof. Let us show that λ^{-1} is an injective mapping from $\widetilde{Q}_h$ onto $\widetilde{L}_h$. Suppose the contrary. Then, there is a function $\widehat{\nu}_0(t) \in \widetilde{L}_h$ such that $\lambda \widehat{\nu}_0(t) \notin \widetilde{Q}_h$. Therefore, there exist functions $\widetilde{u}_1(t) \in \widetilde{Q}_h$ and $u_2(t) \in \widetilde{Q}_h^*$ such that $\widehat{u}_2(t) \neq 0$ and $\lambda \widehat{\nu}_0(t) = \widehat{u}_1(t) + \widehat{u}_2(t)$. Then, $\lambda^{-1}(\lambda \widehat{\nu}_0(t)) = \lambda^{-1} \widehat{u}_1(t) + \lambda^{-1} \widehat{u}_2(t) = \widehat{\nu}_0(t)$, but it follows from Proposition 5.4.3 that $\lambda^{-1} \widehat{u}_1(t) \in \widetilde{L}_h$ and $\lambda^{-1} \widehat{u}_2(t) \in \widetilde{u}_2(t) \in \widetilde{L}_h^*$ $(\lambda^{-1} \widehat{u}_2(t) \neq 0)$. Hence, $\widehat{\nu}_0(t) \notin \widetilde{L}_h$, which is impossible. Thus, the operator λ^{-1} is an injective mapping from $\widetilde{Q}_h$ to $\widetilde{L}_h$. It follows, by virtue of Proposition 5.4.3 and the injectivity of the mapping $\lambda^{-1} : H_{-r} \to H_r(\mathbb{R}_n)$, that λ^{-1} is an injective mapping from $\widetilde{Q}_h^*$ onto $\widetilde{L}_h^*$.

Proposition 5.4.5. *For any $\nu(x) \in L_h$, the following estimate holds:*

$$\|\nu(x)\|_{H_r(\mathbb{R}_n)} \leq C \|\Pi_h \nu(x)\|_{M(r,h)},$$

where $C > 0$ is a constant independent of $h = (h_1, \ldots, h_n)$, $\nu(x)$, $r \geq 0$.

Proof. Let $\hat{\nu}_0(t) \in \tilde{L}_h$. Consider the continuous linear functional $l(u)$ on $H_{-r}(\mathbb{R}_n)$ given by

$$l(u) = \int_{\mathbb{R}_n} \hat{u}(t) \overline{\hat{\nu}_0(t)} \, dt.$$

For any $\hat{u}(t) \in \tilde{Q}_h^*$, we have $l(u) = 0$. Indeed, in view of Proposition 5.4.4, any element $\hat{\nu}_0(t) \in \tilde{L}_h$ can be represented in the form $\hat{\nu}_0(t) = \hat{\nu}_1(t) / \left(1 + |t|\right)^{2r}$ with $\hat{\nu}_1(t) \in \tilde{Q}_h$. But the subspaces $\tilde{Q}_h^*$ and $\tilde{Q}_h$ are orthogonal in $H_{-r}(\mathbb{R}_n)$, and therefore, $l(u) = 0$ for $\hat{u}(t) \in \tilde{Q}_h^*$. Hence,

$$\|l(u)\| \leq \|l(u)\|_{H_{-r}(\tilde{Q}_h)}, \tag{5.4.13}$$

where $\|l(u)\|_{H_{-r}(\tilde{Q}_h)}$ is the norm of the functional $l(u)$ on the space $\tilde{Q}_h$ (Q_h).

On the subspace Q_h, the Parseval identity yields the representation

$$l(u) = (2\pi)^n \int_{\mathbb{R}_n} u(x) \overline{\Pi_h \nu_0(x)} \, dx.$$

By Theorem 5.3.1, for any $u(x) \in Q_h$, we have

$$C_1 \|u(x)\|_{M(-r,h)} \leq \|u(x)\|_{H_{-r}(\mathbb{R}_n)} \leq C_2 \|u(x)\|_{M(-r,h)},$$

with C_1 and C_2 independent of $h = (h_1, \ldots, h_n)$, $u(x)$. Hence we obtain

$$\|\nu_0(x)\|_{H_r(\mathbb{R}_n)} \leq \|l(u)\| \leq \|l(u)\|_{H_{-r}(\tilde{Q}_h)} \leq C \sup_{u(x) \in Q_h} \frac{\|u(x)\|_{M(-r,h)} \|\Pi_h \nu_0(x)\|_{M(r,h)}}{\|u\|_{H_{-r}(\mathbb{R}_n)}} \leq$$

$$\leq K \sup_{u(x) \in Q_h} \frac{\|u(x)\|_{M(-r,h)} \|\Pi_h \nu_0(x)\|_{M(r,h)}}{\|u\|_{M(-r,h)}} = K \|\Pi_h \nu_0(x)\|_{M(r,h)},$$

where K does not depend on h, $\nu_0(x)$. The last inequality proves Proposition 5.4.5.

Theorem 5.4.2. *The integral projector Π_h is an injective operator from $L_h \subset H_r(\mathbb{R}_n)$ onto the space $M(r, h)$, $r \geq 0$, and for any $\nu(x) \in L_h$ the inequality*

$$K_1 \|\nu(x)\|_{H_r(\mathbb{R}_n)} \leq \|\Pi_h \nu(x)\|_{M(r,h)} \leq K_2 \|\nu(x)\|_{H_r(\mathbb{R}_n)}, \tag{5.4.14}$$

holds with constants K_1 and K_2 independent of h and $\nu(x)$.

Proof. The inequality (5.4.14) follows from Theorem 5.4.1 and Proposition 5.4.5. Let us show that Π_h is a mapping of L_h onto $M(r, h)$. Consider an arbitrary $a(x, h) \in M(r, h)$. This function defines a linear functional on the subspace Q_h and its $H_{-r}(\mathbb{R}_n)$ norm is bounded. Let us extend this functional to the entire $H_{-r}(\mathbb{R}_n)$ by setting it equal to zero on all elements $u(x) \in Q_h^*$ and denote the resulting functional by $l(u)$. It follows that there exists $\nu_0(x) \in H_r(\mathbb{R}_n)$ such that $l(u) = \int_{\mathbb{R}_n} \hat{u}(t) \overline{\hat{\nu}_0(t)} \, dt$. Let us show that $\hat{\nu}_0(t) \in \tilde{L}_h$. Indeed, let $\hat{\nu}_0(t) = \hat{\nu}_1(t) + \hat{\nu}_2(t)$, where $\hat{\nu}_1(t) \in \tilde{L}_h$, $\hat{\nu}_2(t) \in \tilde{L}_h^*$. By Proposition 5.4.4, there exist $\hat{u}_1(t) \in \tilde{Q}_h$ and $\hat{u}_2(t) \in \tilde{Q}_h^*$ such that

$\hat{\nu}_1(t) = \hat{u}_1(t)/\left(1 + |t|\right)^{2r}$ and $\hat{\nu}_2(t) = \hat{u}_2(t)/\left(1 + |t|\right)^{2r}$. Take $\hat{u}(t) = \hat{u}_2(t) \in \tilde{Q}_h^*$. By the definition of the functional $l(u)$, we have $l(u_2(x)) = 0$. Further, we get

$$0 = l(u_2(x)) = \int_{\mathbb{R}_n} \hat{u}_2(t) \left[\frac{\overline{\hat{u}_1(t)}}{\left(1 + |t|\right)^{2r}} + \frac{\overline{\hat{u}_2(t)}}{\left(1 + |t|\right)^{2r}} \right] dt = \|u_2(x)\|_{H_{-r}(\mathbb{R}_n)}^2.$$

Hence we conclude that $\nu_0(x) \in L_h$.

Now, let $\hat{u}(t) \in \tilde{Q}_h$. Using the Parseval identity and the inclusion $u(x) \in M(-r, h)$, we can represent the functional $l(u)$ as

$$l(u) = (2\pi)^n \int_{\mathbb{R}_n} u(x) \overline{\Pi_h \nu_0(x)} \, dx.$$

On the other hand, according to the definition of $l(u)$, we have $l(u) = (2\pi)^n \int_{\mathbb{R}_n} u(x) \overline{a(x, h)} \, dx$. Since the step function $u(x)$ is arbitrary, it follows that $a(x, h) = \Pi_h \nu_0(x)$. The injectivity of the mapping Π_h follows, for instance, from the inequality (5.4.14) established above. Theorem 5.4.2 is proved.

Proposition 5.4.6. *For any $a(x) \in H_r(\mathbb{R}_n)$, $r \geq 0$, the following convergence takes place:*

$$\lim_{\Omega_h \to 0} \|\Pi_h a(x)\|_{M(r,h)} = \|a(x)\|_{H_r(\mathbb{R}_n)}.$$

Proof. Assume first that $a(x) \in C_0^\infty(\mathbb{R}_n)$, i.e., $a(x)$ is an infinitely differentiable function with a compact support. In this case, for $r = 0$, the mean value theorem implies that

$$\lim_{\Omega_h \to 0} \|a(x) - \Pi_h a(x)\|_{L_2(\mathbb{R}_n)} = 0,$$

and our statement is proved.

Let $r > 0$. The function $a(x) \in C_0^\infty()$ belongs to $H_r(\mathbb{R}_n)$ for any r, and Theorem 5.4.1 ensures that there is a constant $C(r)$, independent of $h = (h_1, \ldots, h_n)$ and $a(x)$, such that

$$\|\Pi_h a(x)\|_{M(r,h)} \leq C(r) \|a(x)\|_{H_r(\mathbb{R}_n)}.$$

Note that

$$\|a(x)\|_{H_r(\mathbb{R}_n)}^2 = \lim_{\Omega_h \to 0} \int_{D_h} \left(1 + |\varphi|\right)^{2r} |\hat{a}(\varphi)|^2 \, d\varphi,$$

where $\hat{a}(\varphi) = F(a(x))$ is the Fourier transform of $a(x)$.

Taking into account the above observations, we obtain

$$\left| \|\Pi_h a(x)\|_{M(r,h)}^2 - \int_{D_h} \left(1 + |\varphi|\right)^{2r} |\hat{a}(\varphi)|^2 \, d\varphi \right| = \left| \int_{D_h} \left(1 + |\varphi|\right)^{2r} \left(|\hat{a}_h(\varphi)|^2 - |\hat{a}(\varphi)|^2\right) d\varphi \right| \leq$$

$$\leq \left[\int_{D_h} |\hat{a}_h(\varphi) - \hat{a}(\varphi)|^2 \, d\varphi \right]^{1/2} \left[\int_{D_h} \left(1 + |\varphi|\right)^{4r} \left(|\hat{a}_h(\varphi)| + |\hat{a}(\varphi)|\right)^2 d\varphi \right]^{1/2} \leq$$

$$\leq C \left[\int_{D_h} |\hat{a}_h(\varphi) - \hat{a}(\varphi)|^2 \, d\varphi \right]^{1/2} \left(\|\Pi_h a(x)\|_{M(2r,h)} + \|a(x)\|_{H_{2r}(\mathbb{R}_n)} \right), \qquad (5.4.15)$$

where $\hat{a}_h(\varphi)$ is the discrete step Fourier transform (see Definition 5.1.3) of the function $a(x, h) = \Pi_h a(x)$.

Let us show that $\lim_{\Omega_h \to 0} \int_{D_h} |\widehat{a}_h(\varphi) - \widehat{a}(\varphi)|^2 \, d\varphi \to 0$. Take $\varepsilon > 0$ and choose $N > 1$ such that

$$\frac{2}{N^2} \left[\|\widehat{a}_h(\varphi)\|^2_{\widetilde{M}(1,h)} + \|\widehat{a}(\varphi)\|^2_{\widetilde{H}_1(\mathbb{R}_n)} \right] \leq \varepsilon.$$

Then, we can find $\delta > 0$ such that for all $h = (h_1, \ldots, h_n)$ with $\Omega_h < \delta$, we have $|\widehat{a}_h(\varphi) - \widehat{a}(\varphi)| < \varepsilon/V(N)$ for all $|\varphi| < N$, where $V(N)$ is the volume of the ball of radius N. Further, we get

$$\int_{D_h} |\widehat{a}_h(\varphi) - \widehat{a}(\varphi)|^2 \, d\varphi = \int_{K(N)} |\widehat{a}_h(\varphi) - \widehat{a}(\varphi)|^2 \, d\varphi + \int_{D_h \setminus K(N)} (1 + |\varphi|)^2 \frac{|\widehat{a}_h(\varphi) - \widehat{a}(\varphi)|^2}{(1 + |\varphi|)^2} \, d\varphi \leq$$

$$\leq \varepsilon^2 + \frac{2}{N^2} \left[\|\widehat{a}_h(\varphi)\|^2_{\widetilde{M}(1,h)} + \|\widehat{a}(\varphi)\|^2_{\widetilde{H}_1(\mathbb{R}_n)} \right] \leq \varepsilon^2 + \varepsilon, \tag{5.4.16}$$

where $K(N)$ is the ball of radius N with center at the origin, $D_h = \prod_{j=1}^n \left[-\pi/h_j \leq \varphi_j \leq \pi/h_j \right]$. Thus, the integral on the left-hand side of (5.4.16) tends to zero as $\Omega_h \to 0$. Hence, in view of (5.4.15), we obtain Proposition 5.4.6 for any $a(x) \in C_0^\infty(\mathbb{R}_n)$.

Now, suppose that $a(x) \in H_r(\mathbb{R}_n)$, $r > 0$. Since $C_0^\infty(\mathbb{R}_n)$ is dense in $H_r(\mathbb{R}_n)$, it follows that for any $\varepsilon > 0$, we can find $\varphi_\varepsilon(x) \in C_0^\infty(\mathbb{R}_n)$ such that $\|a(x) - \varphi_\varepsilon(x)\|_{H_r(\mathbb{R}_n)} < \varepsilon$. Applying the result just proved for $\varphi_\varepsilon(x) \in C_0^\infty(\mathbb{R}_n)$, we can find $\delta > 0$ such that $\left| \|\varphi_\varepsilon(x)\|_{H_r(\mathbb{R}_n)} - \|\Pi_h \varphi_\varepsilon(x)\|_{M(r,h)} \right| < \varepsilon$ for all $\Omega_h < \delta$. Hence, by Theorem 5.4.1, we get

$$\left| \|a(x)\|_{H_r(\mathbb{R}_n)} - \|\Pi_h a(x)\|_{M(r,h)} \right| =$$

$$= \left| \|[a(x) - \varphi_\varepsilon(x)] + \varphi_\varepsilon(x)\|_{H_r(\mathbb{R}_n)} - \|\Pi_h[(a(x) - \varphi_\varepsilon(x)) + \varphi_\varepsilon(x)]\|_{M(r,h)} \right| \leq$$

$$\leq \|a(x) - \varphi_\varepsilon(x)\|_{H_r(\mathbb{R}_n)} + \|\Pi_h(a(x) - \varphi_\varepsilon(x))\|_{M(r,h)} + \left| \|\varphi_\varepsilon(x)\|_{H_r(\mathbb{R}_n)} - \|\Pi_h \varphi_\varepsilon(x)\|_{M(r,h)} \right| \leq$$

$$\leq \varepsilon(2 + C), \tag{5.4.17}$$

where $C > 0$ is a constant independent of $h = (h_1, \ldots, h_n)$, $a(x)$, $\varphi_\varepsilon(x)$. Since $\varepsilon > 0$ is arbitrary, Proposition 5.4.6 follows from (5.4.17).

Proposition 5.4.7. *For $\varphi(x) \in C_0^\infty(\mathbb{R}_n)$ the following inequality holds:*

$$\left| \frac{d^m \varphi(x)}{dx_1^{m_1} dx_2^{m_2} \cdots dx_n^{m_n}} - T_r^m(x_1, \ldots, x_n) \varphi(x) \right| \leq K(m) \sqrt[n]{\Omega_h}, \tag{5.4.18}$$

where $K(m) > 0$ does not depend on $h = (h_1, \ldots, h_n)$; $T_r^m(x_1, \ldots, x_n)$ is the right difference operator of order $m = m_1 + \cdots + m_n$ (see Example 5.1.4).

Proof. The ordinary Fourier transform of the function $T_r^m(x_1, \ldots, x_n) \varphi(x_1, \ldots, x_n)$ is

$$F \, T_r^m(x_1, \ldots, x_n) \varphi(x_1, \ldots, x_n) = \widehat{\varphi}(t_1, \ldots, t_n) \prod_{j=1}^n \left(\frac{e^{-it_j h_j} - 1}{h_j} \right)^{m_j}, \tag{5.4.19}$$

where $\widehat{\varphi}(t_1, \ldots, t_n) = F(\varphi(x_1, \ldots, x_n))$. The Taylor formula yields

$$\prod_{j=1}^n \left(\frac{e^{-it_j h_j} - 1}{h_j} \right)^{m_j} = \prod_{j=1}^n (-it_j)^{m_j} + \sqrt[n]{\Omega_h} \, O(t, h). \tag{5.4.20}$$

Without loss of generality, we can consider the case $|h| < 1$, because for $|h| \geq 1$, the inequality (5.4.18) is obvious. In this case, taking into account (5.3.3), we have the estimate

$$|O(t, h)| \leq K_1 \left(|t|^{2m} + 1\right), \tag{5.4.21}$$

with K_1 independent of h. From (5.4.19)–(5.4.21), it follows that

$$\Delta = \left| T_r^m(x_1, \ldots, x_n) \varphi(x) - \frac{d^m \varphi(x)}{dx_1^m \cdots dx_n^{m_n}} \right| =$$

$$= \left| \frac{1}{(2\pi)^n} \int_{\mathbb{R}_n} \left[\prod_{j=1}^n \left(\frac{e^{-it_j h_j} - 1}{h_j} \right)^{m_j} \widehat{\varphi}(t) - \prod_{j=1}^n (-it_j)^{m_j} \widehat{\varphi}(t) \right] e^{-i(t,x)} \, dt \right| \leq$$

$$\leq K_2 \Omega_h^{1/n} \int_{\mathbb{R}_n} \left(|t|^{2m} + 1\right) |\widehat{\varphi}(t)| \, dt. \tag{5.4.22}$$

Since $\varphi(x) \in C_0^\infty(\mathbb{R}_n)$, its Fourier transform $F(\varphi(x)) = \widehat{\varphi}(t)$ satisfies the inequality $|\widehat{\varphi}(t)| \leq C_N \left(1 + |t|\right)^{-N}$ for any $N > 0$. Taking $N = 2m + n + 1$, from (5.4.22) we get

$$\Delta \leq K_2 C_N \Omega_h^{1/n} \int_{\mathbb{R}_n} \frac{dt}{\left(1 + |t|\right)^{n+1}} \leq K_3(m) \Omega_h^{1/n}, \tag{5.4.23}$$

with $K_3(m)$ independent of h. Now, Proposition 5.4.7 follows from (5.4.23).

Proposition 5.4.8. *For any* $\varphi(x) \in C_0^\infty(\mathbb{R}_n)$ *the following inequality holds:*

$$\left| \Delta^m \varphi(x) - \Delta_p^m \varphi(x) \right| \leq K(m) \Omega_h^{2/n}, \tag{5.4.24}$$

where Δ^m *is the Laplace operator of order* m, *and* Δ_p^m *is the difference Laplace operator of order* m *(see Example 5.1.5).*

Proof. For the Fourier transforms (F denotes the Fourier transformation) we have

$$\begin{cases} F\left[\Delta_p^m \varphi(x)\right] = \widehat{\varphi}(t)(-4)^m \left[\sum_{k=1}^n \frac{\sin^2 \frac{t_k h_k}{2}}{h_k^2} \right]^m, \\[4mm] F\left[\Delta^m \varphi(x)\right] = \widehat{\varphi}(t)(-1)^m \left(\sum_{k=1}^n t_m^2 \right). \end{cases} \tag{5.4.25}$$

Using the Taylor formula, we find

$$4^m \left[\sum_{k=1}^n \frac{\sin^2 \frac{t_k h_k}{2}}{h_k^2} \right]^m = \left(\sum_{k=1}^n t_k^2 \right)^m + \Omega_h^{2/n} O(t, h). \tag{5.4.26}$$

Without the loss of generality, we assume that $|h| < 1$, and therefore,

$$|O(t, h)| \leq K(m) \left(|t|^{4m} + 1\right). \tag{5.4.27}$$

Using (5.4.25)–(5.4.27), we obtain

$$\left| \Delta_p^m \varphi(x) - \Delta^m \varphi(x) \right| =$$

$$= \frac{1}{(2\pi)^n} \left| \int_{\mathbb{R}_n} \left\{ (-1)^m \left[\left(\sum_{k=1}^n t_k^2 \right)^m + \Omega_h^{2/n} O(t, h) \right] - (-1)^m \sum_{k=1}^n t_k^2 \right\} \widehat{\varphi}(t) \, e^{-i(t,x)} \, dt \right| \leq$$

$$\leq K_1(m) \Omega_h^{2/n} \int_{\mathbb{R}_n} \frac{dt}{\left(1 + |t|\right)^{n+1}} \leq K_2(m) \Omega_h^{2/n}. \tag{5.4.28}$$

Proposition 5.4.8 follows from (5.4.28).

To any continuous function $f(x) \in C(\mathbb{R}_n)$ we can associate a grid function $a(k)$ by setting $a(k_1, \ldots, k_n) = f(k_1 h_1, \ldots, k_n h_n)$. To the grid function $a(k)$ we associate the step function $a(x, h)$ such that $\pi a(x, h) = a(k)$ (see Definition 5.1.2).

Definition 5.4.2. The operator that maps $f(x) \in C(\mathbb{R}_n)$ to its associated step function $a(x, h)$ is denoted by Π_h^*; thus, $\Pi_h^* f(x) = a(x, h)$.

Proposition 5.4.9. *For any $\varphi(x) \in C_0^\infty(\mathbb{R}_n)$ and any $r \geq 0$, the following convergence takes place:*

$$\lim_{\Omega_h \to 0} \|\Pi_h^* \varphi(x) - \Pi_h \varphi(x)\|_{M(r,h)} = 0.$$

Proof. Since $\varphi(x) \in C_0^\infty(\mathbb{R}_n)$, it follows from Proposition 5.4.7 that $\|\Pi_h^* \varphi(x)\|_{M_2(r,h)} \leq C(r, \varphi(x))$, where the constant $C(r, \varphi(x))$ does not depend on $h = (h_1, \dots, h_n)$ and is determined by $\varphi(x)$ and r (the space $M_2(r, h)$ is introduced by Definition 5.3.4). Hence, by Proposition 5.4.3, we obtain

$$\|\Pi_h^* \varphi(x)\|_{M(r,h)} \leq C_1(r, \varphi(x)). \tag{5.4.29}$$

For the discrete Fourier transforms $\widehat{a}_{1h}(t)$ and $\widehat{a}_h(t)$ of the functions $\Pi_h^* \varphi(x)$ and $\Pi_h \varphi(x)$, respectively, Definition 5.1.3 of the discrete Fourier transformation of step functions and the mean value theorem yield the convergence

$$\lim_{\Omega_h \to 0} |\widehat{a}_{1h}(t) - \widehat{a}_h(t)| = 0, \tag{5.4.30}$$

which holds uniformly with respect to t.

Take $\varepsilon > 0$ and $N > 1$ such that $1/N < \varepsilon$. Then, in view of (5.4.30), we can find $\delta > 0$ such that for all $\Omega_h < \delta$ and $|t| < N$, the following inequality holds:

$$|\widehat{a}_{1h}(t) - \widehat{a}_h(t)| \leq \frac{\varepsilon}{V(N)}, \qquad V(N) = \int_{K(N)} \left(1 + |t|\right)^{2r} dt,$$

where $K(N)$ is the ball of radius N with center at the origin. Using the above observations, in combination with (5.4.29) and Theorem 5.4.1, we obtain

$$\|\Pi_h^* \varphi(x) - \Pi_h \varphi(x)\|_{M(r,h)}^2 = \int_{D_h} \left(1 + |t|\right)^{2r} |\widehat{a}_{1h}(t) - \widehat{a}_h(t)|^2 dt =$$

$$= \int_{K(N)} \left(1 + |t|\right)^{2r} |\widehat{a}_{1h}(t) - \widehat{a}_h(t)|^2 dt + \int_{D_h \setminus K(N)} \left(1 + |t|\right)^{2r+1} \frac{|\widehat{a}_{1h}(t) - \widehat{a}_h(t)|^2}{|t| + 1} dt \leq$$

$$\leq \varepsilon + \frac{2}{N} \left(\|\Pi_h^* \varphi(x)\|_{M(r+\frac{1}{2}, h)}^2 + \|\Pi_h \varphi(x)\|_{M(r+\frac{1}{2}, h)} \right) \leq$$

$$\leq \varepsilon + \varepsilon C_2 \left(r + \frac{1}{2}, \varphi(x)\right), \tag{5.4.31}$$

where $D_h = \prod_{i=1}^n \left[-\pi/h_i \leq \varphi_i \leq \pi/h_i \right]$. This inequality proves Proposition 5.4.9, since $\varepsilon > 0$ is arbitrary.

Proposition 5.4.10. *For $u(x) \in L_2(\mathbb{R}_n)$, we have $\Pi_h u(x) \to u(x)$ in the norm of $L_2(\mathbb{R}_n)$ as $|h| \to 0$.*

Proof. According to Proposition 5.4.1, we have $\|\Pi_h u(x)\|_{L_2(\mathbb{R}_n)} \leq C \|u(x)\|_{L_2}$ with a constant C independent of $h = (h_1, \dots, h_n)$ and $u(x)$. Take an arbitrary $\varepsilon > 0$. Since $C_0^\infty(\mathbb{R}_n)$ is dense $L_2(\mathbb{R}_n)$, there exists $\varphi_\varepsilon \in C_0^\infty(\mathbb{R}_n)$ such that

$$\|u(x) - \varphi_\varepsilon(x)\|_{L_2(\mathbb{R}_n)} \leq \min \left\{ \frac{\varepsilon}{3C}, \frac{\varepsilon}{3} \right\}. \tag{5.4.32}$$

By the mean value theorem and the Lagrange theorem, there exists $\delta(\varepsilon) > 0$ such that for $|h| < \delta(\varepsilon)$,

$$\|\Pi_h \varphi_\varepsilon(x) - \varphi_\varepsilon(x)\|_{L_2(\mathbb{R}_n)} < \frac{\varepsilon}{3}. \tag{5.4.33}$$

Hence, for $|h| < \delta(\varepsilon)$, we obtain the inequality

$$\|\Pi_h u(x) - u(x)\|_{L_2(\mathbb{R}_n)} \leq \|\Pi_h (u(x) - \varphi_\varepsilon(x))\|_{L_2(\mathbb{R}_n)} + \|u(x) - \varphi_\varepsilon(x)\|_{L_2(\mathbb{R}_n)} +$$

$$+ \|\Pi_h \varphi_\varepsilon(x) - \varphi_\varepsilon(x)\|_{L_2(\mathbb{R}_n)} < \varepsilon.$$

Proposition 5.4.10 is proved.

Proposition 5.4.11. *Let $a(x, h) \in M(r, h)$, $r > 0$, be a family of functions such that*

$$\|a(x, h)\|_{M(r,h)} \leq C$$

with C independent of $h = (h_1, \ldots, h_n)$, and $a(x, h)$ converge to $u(x)$ in the norm of $L_2(\mathbb{R}_n)$ as $|h| \to 0$. Then, $u(x) \in H_r(\mathbb{R}_n)$.

Proof. Let $K(N)$ be an arbitrary ball $|\varphi| < N$, and $\varepsilon = 1/(N + 1)^{2r}$. Then, there is $\delta(\varepsilon) > 0$ such that for h with $|h| < \delta(\varepsilon)$, the inequality

$$\int_{K(N)} |\widehat{u}(\varphi) - \widehat{a}(\varphi, h)|^2 \, d\varphi < \frac{1}{(N + 1)^{2r}} \tag{5.4.34}$$

holds for the Fourier transforms $\widehat{u}(\varphi)$ and $\widehat{a}(\varphi, h)$ of the functions $u(x)$ and $a(x, h)$. Hence we find that

$$I(N) = \int_{K(N)} \left(1 + |\varphi|\right)^{2r} |\widehat{u}(\varphi)|^2 \, d\varphi \leq$$

$$\leq L\left(\int_{K(N)} \left(1 + |\varphi|\right)^{2r} |\widehat{u}(\varphi) - \widehat{a}(\varphi, h)|^2 \, d\varphi + \int_{K(N)} \left(1 + |\varphi|\right)^{2r} |\widehat{a}(\varphi, h)|^2 \, d\varphi\right) \leq$$

$$\leq L\left(1 + \int_{K(N)} \left(1 + |\varphi|\right)^{2r} |\widehat{a}(\varphi, h)|^2 \, d\varphi\right). \tag{5.4.35}$$

Suppose that we have chosen $\delta(\varepsilon)$ so small that $K(N) \subset \prod_{j=1}^{n} \left[-\pi/h_j \leq \varphi_j \leq \pi/h_j\right] = D_h$. According to (5.4.4), for $\varphi \in D_h$, the following inequality is valid:

$$|\widehat{a}(\varphi, h)| \leq L_1 |\widehat{a}_h(\varphi)|, \tag{5.4.36}$$

where L and L_1 do not depend on h; $\widehat{a}_h(\varphi)$ is the discrete Fourier transform of $a(x, h)$ (see Definition 5.1.3). It follows from (5.4.36) that

$$\int_{K(N)} \left(1 + |\varphi|\right)^{2r} |\widehat{a}(\varphi, h)|^2 \, d\varphi \leq L_1 \|a(x, h)\|_{M(r,h)}^2. \tag{5.4.37}$$

The inequalities (5.4.35) and (5.4.37) imply that

$$I(N) \leq L_2, \tag{5.4.38}$$

with L_2 independent of N. Therefore, there exists a finite norm $\|u(x)\|_{H_r(\mathbb{R}_n)}^2 = \lim_{N \to \infty} I(N)$, which proves Proposition 5.4.11.

Definition 5.4.3. *The space of step functions M_h^∞ consists of all functions of the form $\Pi_h \varphi(x)$, where Π_h is the integral projector and $\varphi(x) \in C_0^\infty(\mathbb{R}_n)$.*

Remark 5.4.1. The space M_h^∞ is dense in $M(r, h)$ for any $r \geq 0$. Indeed, according to Theorem 5.4.2, for any $a(x, h) \in M(r, h)$ with $r > 0$, there is $u(x) \in H_r(\mathbb{R}_n)$ such that $\Pi_h u(x) = a(x, h)$. Since $C_0^\infty(\mathbb{R}_n)$ is dense in $H_r(\mathbb{R}_n)$, it follows that there is a sequence $\{u_k(x)\} \in C_0^\infty(\mathbb{R}_n)$ which converges to $u(x)$ in $H_r(\mathbb{R}_n)$ as $k \to \infty$. Therefore,

$$\|a(x, h) - \Pi_h u_k(x)\|_{M(r,h)} = \|\Pi_h(u(x) - u_k(x))\|_{M(r,h)} \leq C\|u(x) - u_k(x)\|_{H_r(\mathbb{R}_n)} \to 0$$

as $k \to \infty$.

5.5. Spaces $\mathring{M}(r, h, \Omega_h)$ and $M(r, h, \Omega_h)$

Let us recall the definition of a domain with Lipschitz boundary.

Definition 5.5.1. A bounded domain Ω with the boundary Γ is said to have *Lipschitz boundary*, if for any $x_0 \in \Gamma$ there is an orthogonal coordinate system $\xi = (\xi_1, \ldots, \xi_n)$ with the origin at x_0 and there is a rectangular parallelepiped

$$\Delta = \left\{ |\xi_j| < h_j, \ \ j = 1, \ldots, n \right\}$$

such that the piece $\gamma = \Gamma \cap \Delta$ is described by the equation $\xi_n = \varphi(\lambda)$, $\lambda = (\xi_1, \ldots, \xi_{n-1})$, $\lambda \in \Delta^1 = \left\{ |\xi_j| < h_j, \ j = 1, \ldots, n-1 \right\}$, where $\varphi(\lambda)$ satisfies the Lipschitz condition on Δ^1, i.e., for some constant $M > 0$, the inequality $|\varphi(\lambda_1) - \varphi(\lambda)| \leq M|\lambda_1 - \lambda|$ holds for all $\lambda^1, \lambda \in \Delta^1$.

In the sequel, we consider (open) bounded domains Ω with Lipschitz boundaries.

Consider the canonical partition of the space $\mathbb{R}_n$ into cells $D(k, h)$ with the vector-valued step $h = (h_1, \ldots, h_n)$ (see Definition 5.1.1). By $\overline{\Omega}_{1h}$ we denote the closed domain formed by closed cells $\overline{D(k, h)}$ belonging to $\overline{\Omega}$. Let Γ_{1h} be the boundary of $\overline{\Omega}_{1h}$, and let $\Omega_{1h} = \overline{\Omega}_{1h} \setminus \overline{\Gamma}_{1h}$. By $\overline{\Omega}_{2h}$ we denote the closed domain that consists of closed cells $\overline{D(k, h)}$ having common points with $\overline{\Omega}$. The boundary of $\overline{\Omega}_{2h}$ is denoted by Γ_{2h}, and $\overline{\Omega}_{2h} \setminus \overline{\Gamma}_{2h} = \Omega_{2h}$.

Definition 5.5.2. *The space* $\mathring{M}(r, h, \Omega_{1h})$ *consists of all functions* $a(x, h) \in M(r, h)$ *that vanish identically outside* $\overline{\Omega}_{1h}$. *The norm in* $\mathring{M}(r, h, \Omega_{1h})$ *is taken the same as in* $M(r, h)$.

The above definition shows that $\mathring{M}(r, h, \Omega_{1h})$ is a closed subspace of $M(r, h)$.

Definition 5.5.3. *The space* $\mathring{M}(r, h, \Omega_{1h}^*)$ *consists of all* $a(x, h) \in M(r, h)$ *that vanish identically on* Ω_{1h}. *The norm in* $\mathring{M}(r, h, \Omega_{1h}^*)$ *is taken the same as in* $M(r, h)$.

The space $\mathring{M}(r, h, \Omega_{1h}^*)$, too, is closed in $M(r, h)$.

Definition 5.5.4. *The space* $M(r, h, \Omega_{1h})$ *consists of all complex-valued step functions* $a(x, h)$ *such that* $a(x, h)$ *is constant on every cell* $D(k, h) \subset \Omega_{1h}$ *and can be extended to* $\mathbb{R}_n$ *as a step function belonging to* $M(r, h)$ *and constant on every cell* $D(k, h)$. *The norm in* $M(r, h, \Omega_{1h})$ *is introduced by*

$$\|a(x, h)\|_{M(r, h, \Omega_{1h})} = \inf_l \|la(x, h)\|_{M(r, h)}, \tag{5.5.1}$$

where $la(x, h)$ is the extension to $\mathbb{R}_n$ described above.

Let $l_1 a(x, h) \in M(r, h)$ and $l_2 a(x, h) \in M(r, h)$ be two extensions of $a(x, h) \in M(r, h, \Omega_{1h})$. Then, $b(x, h) = l_1 a(x, h) - l_2 a(x, h) \in \mathring{M}(r, h, \Omega_{1h}^*)$. It follows that

$$\|a(x, h)\|_{M(r, h, \Omega_{1h})} = \inf_{b(x, h) \in \mathring{M}(r, h, \Omega_{1h}^*)} \|la(x, h) + b(x, h)\|_{M(r, h)}. \tag{5.5.2}$$

Hence we conclude that $M(r, h, \Omega_{1h})$ is isometrically isomorphic to the orthogonal complement of $\mathring{M}(r, h, \Omega_{1h}^*)$ in $M(r, h)$.

Proposition 5.5.1. *The space dual to* $\mathring{M}(r, h, \Omega_{1h})$ *is isometrically isomorphic to the space* $M(-r, h, \Omega_{1h})$.

Proof. Let us denote by $\widetilde{M}(r, h)$ and $\widetilde{M}(-r, h)$ the Fourier images of the spaces $M(r, h)$ and $M(-r, h)$, respectively. Consider the linear operator λ from $\widetilde{M}(r, h)$ into $\widetilde{M}(-r, h)$ defined as follows: for any $\widehat{V}_h(\varphi) \in \widetilde{M}(r, h)$, let $\lambda \widehat{V}_h(\varphi) = \left(1 + |\varphi|\right)^{2r} \widehat{V}_h(\varphi)$ for $\varphi \in D_h = \prod_{j=1}^{n} \left[-\pi/h_j \leq \varphi_j \leq \pi/h_j \right]$, while for other $\varphi \in \mathbb{R}_n$, the function $\left(1 + |\varphi|\right)^{2r} \widehat{V}_h(\varphi)$ is periodically extended in every φ_j with period $2\pi/hj$. By the definition of the operator λ, we have $\lambda \widehat{V}_h(\varphi) \in M(-r, h)$ and

$$\left\| \lambda \widehat{V}_h(\varphi) \right\|_{\widetilde{M}(-r, h)} = \left\| \widehat{V}_h(\varphi) \right\|_{\widetilde{M}(r, h)}.$$

It should be observed that λ is an isometrical isomorphism from $\widetilde{M}(r,h)$ onto $\widetilde{M}(-r,h)$. Let us show that λ is an isometrical isomorphism from $\widetilde{M}(r,h,\Omega_h^*)$ onto $\overset{\circ}{\widetilde{M}}(-r,h,\Omega_{1h}^*)$, and from $\overset{\circ}{\widetilde{M}}(r,h,\Omega_{1h})$ onto $\widetilde{M}(-r,h,\Omega_{1h})$, where $\overset{\circ}{\widetilde{M}}(-r,h,\Omega_{1h}^*)$ and $\overset{\circ}{\widetilde{M}}(r,h,\Omega_{1h})$ are the Fourier images of the spaces $\overset{\circ}{M}(-r,h,\Omega_{1h}^*)$ and $\overset{\circ}{M}(r,h,\Omega_{1h})$, whereas $\widetilde{M}(r,h,\Omega_{1h}^*)$ and $\widetilde{M}(-r,h,\Omega_{1h})$ are the Fourier images of their orthogonal complements in $M(r,h)$ and $M(-r,h)$, respectively. For any $\widehat{V}_h(\varphi) \in \widetilde{M}(r,h,\Omega_{1h}^*)$ and any $\widehat{u}_h(\varphi) \in \overset{\circ}{\widetilde{M}}(r,h,\Omega_{1h})$, we have

$$\int_{D_h} \left(1+|\varphi|\right)^{2r} \widehat{u}_h(\varphi)\overline{\widehat{V}_h(\varphi)}\, d\varphi = \int_{D_h} \widehat{u}_h(\varphi)\lambda\overline{\widehat{V}_h(\varphi)}\, d\varphi = 0. \tag{5.5.3}$$

Hence, using the Parseval identity, we obtain

$$\int_{D_h} \widehat{u}_h(\varphi)\lambda\overline{\widehat{V}_h(\varphi)}\, d\varphi = (2\pi)^n \int_{\Omega_{1h}} u(x,h)\overline{w(x,h)}\, dx = 0, \tag{5.5.4}$$

where $u(x,h)$ and $w(x,h)$ are the inverse discrete Fourier transforms of the functions $\widehat{u}_h(\varphi)$ and $\lambda\widehat{V}_h(\varphi)$.

Since $u(x,h) \in \overset{\circ}{M}(r,h,\Omega_{1h})$ is arbitrary, it follows from (5.5.4) that $w(x,h) \in \overset{\circ}{M}(-r,h,\Omega_{1h}^*)$, and therefore, $\lambda\widehat{V}_h(\varphi) \in \overset{\circ}{\widetilde{M}}(-r,h,\Omega_{1h}^*)$. Since (5.5.4) holds for any $w(x,h) \in \overset{\circ}{M}(-r,h,\Omega_{1h}^*)$, the operator λ^{-1} maps $\overset{\circ}{\widetilde{M}}(-r,h,\Omega_{1h}^*)$ into $\widetilde{M}(r,h,\Omega_{1h}^*)$. Consequently, λ maps $\widetilde{M}(r,h,\Omega_{1h}^*)$ onto $\overset{\circ}{\widetilde{M}}(-r,h,\Omega_{1h}^*)$. The operator λ transforms orthogonal elements of $\widetilde{M}(r,h)$ into orthogonal elements of $\widetilde{M}(-r,h)$, and therefore, it maps $\overset{\circ}{\widetilde{M}}(r,h\Omega_{1h})$ into $\widetilde{M}(-r,h\Omega_{1h})$. It follows that the operator $(2\pi)^{-1}F_{\text{discr}}^{-1}\lambda$ (with the inverse discrete Fourier transformation denoted by F_{discr}^{-1}) is an isometric isomorphism from $\overset{\circ}{M}(r,h,\Omega_{1h})$ onto $\widetilde{M}(-r,h,\Omega_{1h})$, and since $\overset{\circ}{M}(r,h,\Omega_{1h})$ is a Hilbert space, it coincides with its dual. Proposition 5.5.1 follows from these observations.

Corollary 5.5.1. *Any bounded linear functional on $\overset{\circ}{M}(r,h,\Omega_{1h})$ can be represented in the form*

$$l(u) = \int_{D_g} \widehat{u}_h(\varphi)\overline{\widehat{w}_h(\varphi)}\, d\varphi = \int_{\Omega_{1h}} u(x,h)\overline{w(x,h)}\, dx, \tag{5.5.5}$$

where $\widehat{u}_h(\varphi) \in \overset{\circ}{\widetilde{M}}(r,h,\Omega_{1h})$, $\widehat{w}_h(\varphi) \in \widetilde{M}(-r,h,\Omega_{1h})$.

This result follows from Proposition 5.5.1.

Proposition 5.5.2. *Let $a(x,h) \in \overset{\circ}{M}(r,h,\Omega_{1h})$, $r>0$, be a family of functions with the parameter $h = (h_1,\ldots,h_n)$. Suppose that the estimate $\|a(x,h)\|_{\overset{\circ}{M}(r,h,\Omega_{1h})} \leq C$ holds uniformly in h. Then, there is a sequence $h^k = (h_1^k,\ldots,h_n^k)$, $|h^k| \to 0$ as $k \to \infty$, such that the sequence $a(x,h^k)$ converges to a function $a_0(x)$ in the norm of $\overset{\circ}{H}_{r_0}(\Omega)$, $0 < r_0 < 1/2$, $r_0 < r$; moreover, $a_0(x) \in \overset{\circ}{H}_r(\Omega)$ (here $\overset{\circ}{H}_{r_0}(\Omega)$, $\overset{\circ}{H}_r(\Omega)$ are the Sobolev–Slobodetskii spaces).*

Proof. Take r_1 such that $0 < r_1 < 1/2$, $r_1 < r$. By Theorem 5.3.1, $a(x,h) \in \overset{\circ}{H}_{r_1}(\Omega)$ and the estimate $\|a(x,h)\|_{\overset{\circ}{H}_{r_1}(\Omega)} \leq C_1$ holds uniformly in h, where $\overset{\circ}{H}_{r_1}(\Omega)$ is the Sobolev space of functions $V(x)$ with $\operatorname{supp} V \in \Omega$. For r_0 such that $0 < r_0 < r_1$, the functions $a(x,h)$ form a precompact set in $\overset{\circ}{H}_{r_0}(\Omega)$. Consequently, there is a sequence $a(x,h^k)$ which converges to some $a_0(x)$ in the norm of $\overset{\circ}{H}_{r_0}(\Omega)$, as $|h^k| \to 0$ $(k \to \infty)$. Let us show that $a_0(x) \in \overset{\circ}{H}_r(\Omega)$.

Consider the ball $K(N) = \{|\varphi| \leq N\}$ and let $\varepsilon = 1/(N+1)^{2r}$. Then, there is k_0 such that for all $k > k_0$, the following inequality holds for the Fourier transforms $\widehat{a}_0(\varphi)$ and $\widehat{a}(\varphi,h^k)$ of the functions $a_0(x)$ and $a(x,h^k)$:

$$\int_{K(N)} \left(1+|\varphi|\right)^{2r_0} \left|\widehat{a}_0(\varphi) - \widehat{a}(\varphi,h^k)\right|^2 d\varphi < \frac{1}{(N+1)^{2r}}. \tag{5.5.6}$$

Hence, using the assumptions of Proposition 5.5.2, we get

$$
\begin{aligned}
I(N) &= \int_{K(N)} \left(1 + |\varphi|\right)^{2r} |\widehat{a}_0(\varphi)|^2 \, d\varphi = \\
&= \int_{K(N)} \left(1 + |\varphi|\right)^{2r} \left| \left(\widehat{a}_0(\varphi) - \widehat{a}(\varphi, h^k)\right) + \widehat{a}(\varphi, h^k)\right|^2 d\varphi \leq \\
&\leq L \left(\int_{K(N)} \left(1 + |\varphi|\right)^{2r} |\widehat{a}_0(\varphi) - \widehat{a}(\varphi, h^k)|^2 + \int_{K(N)} \left(1 + |\varphi|\right)^{2r} |\widehat{a}(\varphi, h^k)|^2 \, d\varphi \right),
\end{aligned}
\tag{5.5.7}
$$

where $L > 0$ does not depend of $h = (h_1, \ldots, h_n)$. Further, we find that

$$
\int_{K(N)} \left(1 + |\varphi|\right)^{2r} |\widehat{a}_0(\varphi) - \widehat{a}(\varphi, h^k)|^2 \, d\varphi \leq (N+1)^{2(r - r_0)} \frac{1}{(N+1)^{2r}} = \frac{1}{(N+1)^{2r_0}}. \tag{5.5.8}
$$

Assume that k_0 is so large that for $k > k_0$, we have $K(N) \subset \prod_{j=1}^{n} \left[-\pi/h_j^k \leq \varphi \leq \pi/h_j^k \right] = D_h$. By virtue of (5.3.4), for $\varphi \in D_h$ and the discrete Fourier transform $\widehat{a}_{h^k}(\varphi)$ of $a(x, h^k)$ the following inequality holds

$$
\left|\widehat{a}(\varphi, h^k)\right|^2 \leq L_1 |\widehat{a}_{h^k}(\varphi)|^2, \tag{5.5.9}
$$

where the constant L_1 does not depend on $h = (h_1, \ldots, h_n)$. From (5.5.9), it follows that

$$
\begin{aligned}
\int_{K(N)} \left(1 + |\varphi|\right)^{2r} |\widehat{a}(\varphi, h^k)|^2 \, d\varphi &\leq L_1 \int_{K(N)} \left(1 + |\varphi|\right)^{2r} |\widehat{a}_{h^k}|^2 \, d\varphi \leq \\
&\leq L_1 \left\| a(x, h^k) \right\|^2_{\mathring{M}(r, h, \Omega_{1h})} \leq C_1.
\end{aligned}
\tag{5.5.10}
$$

Combining (5.5.7)–(5.5.9) with (5.5.11), we obtain

$$
I(N) = \int_{K(N)} \left(1 + |\varphi|\right)^{2r} |\widehat{a}_0(\varphi)|^2 \, d\varphi \leq L_2, \tag{5.5.11}
$$

where the constant L_2 does not depend on N. It follows that there exists $\lim_{N \to \infty} I(N) = \|a_0(x)\|^2_{\mathring{H}_r(\Omega)}$ and $a_0(x) \in \mathring{H}_r(\Omega)$. Proposition 5.5.2 is proved.

Consider the differences

$$
\begin{aligned}
\Delta_\lambda f(x) &= f(x) - f(\lambda x), \\
\Delta_\lambda^{(2)} f(x) &= \Delta_\lambda f(x) - \Delta_\lambda f(\lambda x), \\
&\ldots\ldots\ldots\ldots\ldots\ldots\ldots\ldots\ldots\ldots \\
\Delta_\lambda^{(n)} f(x) &= \Delta_\lambda^{(n-1)} f(x) - \Delta_\lambda^{(n-1)} f(\lambda x).
\end{aligned}
\tag{5.5.12}
$$

In Let us calculate the Fourier transform of $\Delta_\lambda^{(n)} f(x)$. We have

$$
\begin{aligned}
\widehat{\Delta}_\lambda \widehat{f}(\sigma) &= F(\Delta_\lambda f(x)) = \widehat{f}(\sigma) - \frac{1}{\lambda} \widehat{f}\left(\frac{\sigma}{\lambda}\right), \\
F(\Delta_\lambda f(\lambda x)) &= \frac{1}{\lambda} \widehat{f}\left(\frac{\sigma}{\lambda}\right) - \frac{1}{\lambda^2} \widehat{f}\left(\frac{\sigma}{\lambda^2}\right) = \frac{1}{\lambda} \widehat{\Delta}_\lambda \widehat{f}\left(\frac{\sigma}{\lambda}\right), \\
\widehat{\Delta}_\lambda^{(2)} \widehat{f}(\sigma) &= F\left(\Delta_\lambda^{(2)} f(x)\right) = \widehat{\Delta}_\lambda \widehat{f}(\sigma) - \frac{1}{\lambda} \widehat{f}\left(\frac{\sigma}{\lambda}\right).
\end{aligned}
$$

Let us prove the relation

$$
\widehat{\Delta}_\lambda^{(n)} \widehat{f}(\sigma) = F\left(\Delta_\lambda^{(n)} f(x)\right) = \widehat{\Delta}_\lambda^{(n-1)} \widehat{f}(\sigma) - \frac{1}{\lambda} \widehat{\Delta}_\lambda^{(n-1)} \widehat{f}\left(\frac{\sigma}{\lambda}\right). \tag{5.5.13}
$$

Indeed,

$$\widehat{\Delta}_\lambda^{(n)} \widehat{f}(\sigma) = F\left(\Delta_\lambda^{(n)} f(x)\right) = F\left(\Delta_\lambda^{(n-1)} f(x) - \Delta_\lambda^{(n-1)} f(\lambda x)\right) =$$

$$= F\left(\Delta_\lambda^{(n-1)} f(x)\right) - \frac{1}{\lambda} F\left(\Delta_\lambda^{(n-1)} \lambda f(\lambda x)\right) = \widehat{\Delta}_\lambda^{(n-1)} \widehat{f}(\sigma) - \frac{1}{\lambda}\widehat{\Delta}_\lambda^{(n-1)} \widehat{f}\left(\frac{\sigma}{\lambda}\right).$$

Next, we establish the formula

$$\widehat{\Delta}_\lambda^{(n)} \widehat{f}(\sigma) = \sum_{k=0}^{n} (-1)^k C_n^k \widehat{f}\left(\frac{\sigma}{\lambda^k}\right) \frac{1}{\lambda^k}, \qquad C_n^k = \frac{n!}{(n-k)!k!}. \tag{5.5.14}$$

Clearly, this relation is valid for $n = 1$. Assuming that it holds for $n = p$, let us prove it for $n = p+1$. From (5.5.14), using the induction hypothesis, we get

$$\widehat{\Delta}_\lambda^{(p+1)} \widehat{f}(\sigma) = \widehat{\Delta}_\lambda^{(p)} \widehat{f}(\sigma) - \frac{1}{\lambda}\widehat{\Delta}_\lambda^{(p)} \widehat{f}\left(\frac{\sigma}{\lambda}\right) =$$

$$= \sum_{k=0}^{p} (-1)^k C_p^k \frac{1}{\lambda^k} \widehat{f}\left(\frac{\sigma}{\lambda^k}\right) - \frac{1}{\lambda} \sum_{k=0}^{p} (-1)^k C_p^k \frac{1}{\lambda^k} \widehat{f}\left(\frac{\sigma}{\lambda^{k+1}}\right) =$$

$$= \widehat{f}(\sigma) + \sum_{k=1}^{p} \left[(-1)^k C_p^k \frac{1}{\lambda^k} - C_p^{k-1}(-1)^{k-1} \frac{1}{\lambda^k}\right] \widehat{f}\left(\frac{\sigma}{\lambda^k}\right) + \frac{1}{\lambda^{p+1}}(-1)^{p+1} \widehat{f}\left(\frac{\sigma}{\lambda^{p+1}}\right) =$$

$$= \sum_{k=0}^{p+1} (-1)^k C_{p+1}^k \frac{1}{\lambda^k} \widehat{f}\left(\frac{\sigma}{\lambda^k}\right),$$

where we have used the fact that $C_p^k + C_p^{k-1} = C_{p+1}^k$. Thereby relation (5.5.14) is proved.

For $\widehat{f}(\sigma) \in C^\infty(\mathbb{R}_n)$, consider the Taylor expansion for each function $\widehat{f}(t^k) = \widehat{f}(\sigma/\lambda^k)$, $t^k = (t_1^k, \ldots, t_n^k)$, at the point $M(\sigma_1, \ldots, \sigma_n)$. We have

$$\widehat{f}\left(\frac{\sigma}{\lambda^k}\right) = \widehat{f}(\sigma) + \sum_{k=1}^{m} \frac{1}{k!} d^k \widehat{f}(\sigma) + \frac{1}{(m+1)!} d^{m+1} \widehat{f}\left(\sigma + \theta^k\left(\frac{\sigma}{\lambda^k} - \sigma\right)\right), \tag{5.5.15}$$

where $0 < \theta_i^k < 1$, $\theta^k = (\theta_1^k, \ldots, \theta_n^k)$,

$$d^i \widehat{f}(\sigma) = \left.\left(\frac{d}{dt_1^k} dt_1^k + \cdots + \frac{d}{dt_n^k} dt_n^k\right)^i \widehat{f}(t)\right|_{t=\sigma},$$

$$dt_1^1 = \sigma_1 \left(\frac{1-\lambda^k}{\lambda^k}\right), \quad \cdots, \quad dt_n^k = \sigma_n \left(\frac{1-\lambda^l}{\lambda^k}\right). \tag{5.5.16}$$

Substituting the right-hand side of (5.5.15) into (5.5.14) for $n = p$ and $m > p$, we obtain

$$\widehat{\Delta}_\lambda^{(p)} \widehat{f}(\sigma) = \sum_{k=0}^{p} (-1)^k C_p^k \frac{1}{\lambda^k} \sum_{i=0}^{p} \frac{d^i g_k(\sigma)}{i!} +$$

$$+ \frac{1}{(m+1)!} \sum_{k=0}^{p} (-1)^k C_p^k \frac{1}{\lambda^k} d^{m+1} g_k\left(\sigma + \theta^k\left(\frac{\sigma}{\lambda^k} - \sigma\right)\right) =$$

$$= \sum_{k=0}^{p} (-1)^k C_p^k \frac{1}{\lambda^k} \sum_{i=0}^{p} \frac{d^i g_k(\sigma)}{i!} + \sum_{k=0}^{p} (-1)^k C_p^k \frac{1}{\lambda^k} \sum_{i=p+1}^{m} \frac{d^i g_k(\sigma)}{i!} +$$

$$+ \frac{1}{(m+1)!} \sum_{k=0}^{p} (-1)^k C_p^k \frac{1}{\lambda^k} d^{m+1} g_k\left(\sigma + \theta^k\left(\frac{\sigma}{\lambda^k} - \sigma\right)\right), \tag{5.5.17}$$

where $g_k(t^k) = \widehat{f}(t^k)$.

Next, for $i \geq 1$, we have

$$d^i \widehat{\Delta}_\lambda^{(p)} \widehat{f}(\sigma) = \sum_{k=1}^p (-1)^k C_p^k \frac{1}{\lambda^k} d^i g_k(\sigma) = \sum_{k_1 + \cdots + k_n = i} \frac{\partial^i \widehat{f}(\sigma)}{\partial \sigma_1^{k_1} \cdots \partial \sigma_n^{k_n}} \sigma_1^{k_1} \cdots \sigma_n^{k_n} A_{k_1, \ldots, k_n}^p(\lambda). \quad (5.5.18)$$

The summation on the right-hand side of (5.5.18) starts at $k = 1$, because in view of (5.5.16) we have $d^i g_0(\sigma) = d^i \widehat{f}(t_0)\big|_{t_0 = \sigma} = 0$ for $i \geq 1$.

For $i = 0$, we get

$$d^0 \widehat{\Delta}_\lambda^{(p)} \widehat{f}(\sigma) = \sum_{k=0}^p (-1)^k C_p^k \frac{1}{\lambda^k} \widehat{f}(\sigma) = \widehat{f}(\sigma) \frac{(1 - \lambda)^p}{\lambda^p}. \quad (5.5.19)$$

Formulas (5.5.16)–(5.5.18) show that $A_{k_1, \ldots, k_n}^p(\lambda)$ are rational functions of λ with a pole at $\lambda = 0$. Let us show that $\lambda = 1$, for any integer i, $0 \leq i \leq p$, is a root of $A_{k_1, \ldots, k_n}^p(\lambda)$, $k_1 + \cdots + k_n = i$, and its multiplicity is equal to p. Relation (5.5.19) ensures our statement for $i = 0$. Let us prove it for $p = 1$. Since the result holds for $i = 0$, it suffices to consider the case of $i = 1$. We have

$$d\widehat{\Delta}_\lambda \widehat{f}(\sigma) = -\frac{1}{\lambda} d\widehat{f}(t^1)\Big|_{t^1 = \sigma} = \left(-\frac{1}{\lambda}\right) \left(\frac{\partial \widehat{f}(\sigma)}{\partial \sigma_1} \sigma_1 \frac{1 - \lambda}{\lambda} + \cdots + \frac{\partial \widehat{f}(\sigma)}{\partial \sigma_n} \sigma_n \frac{1 - \lambda}{\lambda}\right). \quad (5.5.20)$$

From (5.5.20) we deduce that $\lambda = 1$ is a root of multiplicity 1 for the function $A_1^1(\lambda) = -(1 - \lambda)/\lambda^2$.

Assuming that our statement is true for $p = l$, let us prove it for $p = l + 1$. Consider the quantity

$$\widehat{\Delta}_\lambda^{(l)} \widehat{f}\left(\frac{\sigma}{\lambda}\right) = \sum_{k=0}^l (-1)^k C_l^k \frac{1}{\lambda^k} \widehat{f}\left(\frac{\sigma}{\lambda^{k+1}}\right). \quad (5.5.21)$$

Passing to the variables $z = \sigma/\lambda$ in (5.5.21), let us represent each function $g_k(z^k) = \widehat{f}(z^k) = \widehat{f}(z/\lambda^k)$ by Taylor's formula at the point $M(z_1, \ldots, z_n)$ (see (5.5.17)). We obtain

$$\widehat{\Delta}_\lambda^{(l)} \widehat{f}(z) = \sum_{k=0}^l (-1)^k C_l^k \frac{1}{\lambda^k} \sum_{i=0}^l \frac{d^i g_k(z)}{i!} + \sum_{k=0}^l (-1)^k C_l^k \frac{1}{\lambda^k} \sum_{i=l+1}^m \frac{d^i g_k(z)}{i!} +$$
$$+ \frac{1}{(m+1)!} \sum_{k=0}^l (-1)^k C_l^k \frac{1}{\lambda^k} d^{m+1} g_k\left(z + \widehat{\theta}^k \left(\frac{z}{\lambda^k} - z\right)\right), \quad (5.5.22)$$

where $\widehat{\theta}^k = (\widehat{\theta}_1^k, \ldots, \widehat{\theta}_n^k)$, $0 < \widehat{\theta}_i^k \leq 1$, $i = 1, 2, \ldots, n$. Using (5.5.18), we find that

$$d^i \widehat{\Delta}_\lambda^{(l)} \widehat{f}(z) = \sum_{k=0}^l (-1)^k C_l^k \frac{1}{\lambda^k} d^i g_k(z) = \sum_{k_1 + \cdots + k_n = i} \frac{\partial^i \widehat{f}(z)}{\partial z_1^{k_1} \cdots \partial z_n^{k_n}} z_1^{k_1} \cdots z_n^{k_n} A_{k_1, \ldots, k_n}^l(\lambda). \quad (5.5.23)$$

It follows from the induction assumption that $\lambda = 1$ is a root of multiplicity l for all $A_{k_1, \ldots, k_n}^l(\lambda)$ with $k_1 + \cdots + k_n = i$, $0 \leq i \leq l$. We have to show that $\lambda = 1$ is a root of multiplicity λ of the coefficients $A_{k_1, \ldots, k_n}^{l+1}(\lambda)$ of the functions $\dfrac{\partial^i \widehat{f}(\sigma)}{\partial \sigma_1^{k_1} \cdots \partial \sigma_n^{k_n}}$ with $0 \leq k_1 + \cdots + k_n = i \leq l + 1$. It follows from (5.5.16) that this result holds for $k_1 + \cdots + k_n = l + 1$. Therefore, it suffices to prove it for $0 \leq i \leq l$ (for $i = 0$, it follows from (5.5.19)).

Let us transform the right-hand side of (5.5.23). For this purpose, we represent the functions

$$Q^i_{k_1,\ldots,k_n}\left(\frac{\sigma}{\lambda}\right) = \frac{\partial^i \widehat{f}(z)}{\partial z_1^{k_1}\cdots\partial z_n^{k_n}}\bigg|_{z=\sigma/\lambda}$$

by the Taylor formula at the point $M(\sigma_1,\ldots,\sigma_n)$ as follows:

$$Q^i_{k_1,\ldots,k_n}\left(\frac{\sigma}{\lambda}\right) = \frac{\partial^i \widehat{f}(\sigma)}{\partial \sigma_1^{k_1}\cdots\partial \sigma_n^{k_n}} + \sum_{q=1}^{j}\frac{d^q Q^i_{k_1,\ldots,k_n}(\sigma)}{q!} + \frac{d^{j+1}Q^i_{k_1,\ldots,k_n}\left(\sigma + \theta^{i,k_1,\ldots,k_n}\left(\frac{\sigma}{\lambda}-\sigma\right)\right)}{(j+1)!},$$

(5.5.24)

where $\theta^{i,k_1,\ldots,k_n} = \left(\theta_1^{i,k_1,\ldots,k_n},\ldots,\theta_n^{i,k_1,\ldots,k_n}\right),\ 0 < \theta_j^{i,k_1,\ldots,k_n} < 1$.

Substituting the right-hand side of (5.5.24) into (5.5.23), we obtain

$$
\begin{aligned}
d^i\widehat{\Delta}_\lambda^{(l)}\widehat{f}\left(\frac{\sigma}{\lambda}\right) &= \sum_{k_1+\cdots+k_n=i}\left[\frac{\partial^i \widehat{f}(\sigma)}{\partial \sigma_1^{k_1}\cdots\partial \sigma_n^{k_n}} + \sum_{q=1}^{j}\frac{d^q Q^i_{k_1,\ldots,k_n}(\sigma)}{q!}+\right.\\
&\quad \left.+\frac{d^{j+1}Q^i_{k_1,\ldots,k_n}\left(\sigma + \theta^{i,k_1,\ldots,k_n}\left(\frac{\sigma}{\lambda}-\sigma\right)\right)}{(j+1)!}\right]\frac{\sigma_1^{k_1}\cdots\sigma_n^{k_n}}{\lambda^i}A^l_{k_1,\ldots,k_n}(\lambda) =\\
&= \sum_{k_1+\cdots+k_n=i}\frac{\partial^i \widehat{f}(\sigma)}{\partial \sigma_1^{k_1}\cdots\partial \sigma_n^{k_n}}\frac{\sigma_1^{k_1}\cdots\sigma_n^{k_n}}{\lambda^i}A^l_{k_1,\ldots,k_n}(\lambda) + B_i(\sigma,\lambda). \qquad (5.5.25)
\end{aligned}
$$

In (5.5.24) we take j such that $i+j\geq l+1$. From (5.5.16), (5.5.24), (5.5.25), and the induction assumption, it follows that the coefficients $B^p_{k_1,\ldots,k_n}(\lambda)$ of $\dfrac{\partial^i \widehat{f}(\sigma)}{\partial \sigma_1^{k_1}\cdots\partial \sigma_n^{k_n}}$, with $k_1+\cdots+k_n=p>i$, involved in $B_i(\sigma,\lambda)$ are equal to zero for $\lambda=1$, and the multiplicity of the root $\lambda=1$ is not less than $l+1$.

For $0 < i \leq l$, using (5.5.13), (5.5.23), (5.5.25), we obtain

$$d^i\widehat{\Delta}_\lambda^{(l+1)}\widehat{f}(\sigma) = \qquad\qquad\qquad\qquad\qquad\qquad\qquad\qquad\quad (5.5.26)$$

$$= \sum_{k_1+\cdots+k_n}\sigma_1^{k_1}\cdots\sigma_n^{k_n}\frac{\partial^i \widehat{f}(\sigma)}{\partial \sigma_1^{k_1}\cdots\partial \sigma_n^{k_n}}\left[A^l_{k_1,\ldots,k_l}(\lambda) - \frac{1}{\lambda^{i+1}}A^l_{k_1,\ldots,k_l}(\lambda) + C^l_{k_1,\ldots,k_l}(\lambda)\right],$$

where $C^l_{k_1,\ldots,k_l}(\lambda)$ stands for the sum of the coefficients of $\sigma_1^{k_1}\cdots\sigma_n^{k_n}\dfrac{\partial^i \widehat{f}(\sigma)}{\partial \sigma_1^{k_1}\cdots\partial \sigma_n^{k_n}}$ involved in the expansion (5.5.25) of $\left(-\dfrac{1}{\lambda}d^j\widehat{\Delta}_\lambda^{(l)}\widehat{f}\left(\dfrac{\sigma}{\lambda}\right)\right)$ with $j<i$. As we have just mentioned, $\lambda=1$ is the root of multiplicity $l+1$ of the functions $C^l_{k_1,\ldots,k_l}(\lambda)$. Hence, by virtue of the induction hypothesis, it follows that $\lambda=1$ is a root of multiplicity $l+1$ of the functions

$$A^{l+1}_{k_1,\ldots,k_l}(\lambda) = A^l_{k_1,\ldots,k_l}(\lambda) - \frac{1}{\lambda^{i+1}}A^l_{k_1,\ldots,k_l}(\lambda) + C^l_{k_1,\ldots,k_l}(\lambda),$$

with $0 < k+1+\cdots+k_n \leq l$.

Thus, we have shown that $\lambda=1$ is a root of multiplicity p of the functions $A^p_{k_1,\ldots,k_l}(\lambda)$ with $0 < k+1+\cdots+k_n \leq p$ involved in (5.5.18). The above arguments prove the following result.

Proposition 5.5.3. Let $f(x) \in \overset{\circ}{H}_r(\Omega)$, where $r > 0$ and Ω is a bounded domain in $\mathbb{R}_n$. Then, for the Fourier transform $\widehat{f}(\sigma) = F(f(x))$ the following relations hold:

$$\widehat{\Delta}_\lambda^{(p)}\widehat{f}(\sigma) = \widehat{f}(\sigma)\frac{(1-\lambda)^p}{\lambda^p} + \sum_{i=1}^p \frac{1}{i!} \sum_{k_1+\cdots+k_n=i} \frac{\partial^i \widehat{f}(z)}{\partial z_1^{k_1}\cdots\partial z_n^{k_n}}\sigma_1^{k_1}\cdots\sigma_n^{k_n}(1-\lambda)^p\widehat{A}_{k_1,\ldots,k_n}^p(\lambda)+$$

$$+ \sum_{i=p+1}^m \frac{1}{i!}\sum_{k=1}^p (-1)^k C_p^k \frac{1}{\lambda^k}d^i g_k(\sigma) + \frac{1}{(m+1)!}\sum_{k=1}^p (-1)^k C_p^k \frac{1}{\lambda^k}d^{m+1}g_k\left(\sigma + \theta^k\left(\frac{\sigma}{\lambda^k}\right)\right),$$

$$(5.5.27)$$

where $0 < \theta_j^k < 1$, $\theta^k = (\theta_1^k, \ldots, \theta_n^k)$, $(1-\lambda)^p\widehat{A}_{k_1,\ldots,k_n}^p(\lambda) = A_{k_1,\ldots,k_n}^p(\lambda)$, $g_k(t^k) = \widehat{f}(t^k)$, $t^k = \sigma/\lambda^k$; and $d^i g_k(\sigma)$ are the differentials defined by (5.5.16).

Proposition 5.5.4. Let $u(x) \in \overset{\circ}{H}_r(\Omega)$, where Ω is a bounded domain in $\mathbb{R}_n$, $r > 0$. Then, for $1/2 < \lambda < 3/2$ and integer p such that $0 \le p - 1 < r \le p$, the following inequality holds:

$$\left\|\Delta_\lambda^{(p)}u(x)\right\|_{L_2(\mathbb{R}_n)} \le C_r|\lambda - 1|^r \|u(x)\|_{\overset{\circ}{H}_r(\Omega)}, \tag{5.5.28}$$

where C_r is a constant independent of $u(x)$.

Proof. Using the Parseval identity, we find that

$$\left\|\Delta_\lambda^{(p)}u(x)\right\|_{L_2(\mathbb{R}_n)} = \frac{1}{(2\pi)^{n/2}}\left[\int_{\mathbb{R}_n}\left|\widehat{\Delta}_\lambda^{(p)}\widehat{u}(\sigma)\right|^2 d\sigma\right]^{1/2} \le \tag{5.5.29}$$

$$\le C\left(\left[\int_{\Omega^*}\left|\widehat{\Delta}_\lambda^{(p)}\widehat{u}(\sigma)\right|^2 d\sigma\right]^{1/2} + \left[\int_{\mathbb{R}_n\setminus\Omega^*}\left|\widehat{\Delta}_\lambda^{(p)}\widehat{u}(\sigma)\right|^2 d\sigma\right]^{1/2}\right) = I_1 + I_2,$$

where $\Omega^* = \left\{\sigma : |\lambda - 1||\sigma| < 1/2\right\}$. Using (5.5.14), we get

$$I_2 \le C_1 \sum_{k=0}^p C_p^k \frac{1}{\lambda^k}\left[\int_{\mathbb{R}_n\setminus\Omega^*}\left|\widehat{u}\left(\frac{\sigma}{\lambda^k}\right)\right|^2 d\sigma\right]^{1/2} \le$$

$$\le C_1 \sum_{k=0}^p C_p^k \frac{1}{\lambda^k}\left[\int_{\mathbb{R}_n\setminus\Omega^*}(1+|\sigma|)^{2r}\frac{\left|\widehat{u}\left(\frac{\sigma}{\lambda^k}\right)\right|^2}{(1+|\sigma|)^{2r}}d\sigma\right]^{1/2} \le$$

$$\le C_2 \sum_{k=0}^p C_p^k \frac{1}{\lambda^k}\frac{1}{\left(1 + \frac{1}{2|\lambda-1|}\right)^r}\left[\int_{\mathbb{R}_n\setminus\Omega^*}(1+|\sigma|)^{2r}\left|\widehat{u}\left(\frac{\sigma}{\lambda^k}\right)\right|^2 d\sigma\right]^{1/2} \le$$

$$\le A_r|\lambda - 1|^r \|u(x)\|_{\overset{\circ}{H}_r(\Omega)}. \tag{5.5.30}$$

In order to estimate the term I_1, we utilize (5.5.27),

$$I_1 \le C\frac{|1-\lambda|^p}{\lambda^p}\left[\int_{\Omega^*}|\widehat{u}(\sigma)|^2 d\sigma\right]^{1/2} +$$

$$+ C|1-\lambda|^p \sum_{i=1}^p \frac{1}{i!}\sum_{k_1+\cdots+k_n=i}\left|\widehat{A}_{k_1,\ldots,k_n}^p(\lambda)\right|\left[\int_{\Omega^*}|\sigma|^{2i}\left|\frac{\partial^i \widehat{u}(\sigma)}{\partial\sigma_1^{k_1}\cdots\partial\sigma_n^{k_n}}\right|^2 d\sigma\right]^{1/2} +$$

$$+ C\sum_{i=p+1}^m \frac{1}{i!}\sum_{k=1}^p C_p^k \frac{1}{\lambda^k}\left[\int_{\Omega^*}|d^i g_k(\sigma)|^2 d\sigma\right]^{1/2} +$$

$$+ \frac{1}{(m+1)!}\sum_{k=1}^p C_p^k \frac{1}{\lambda^k}\left[\int_{\Omega^*}\left|d^{m+1}g_k\left(\sigma + \theta^k\left(\frac{\sigma}{\lambda^k} - \sigma\right)\right)\right|^2 d\sigma\right]^{1/2} =$$

$$= A_1 + A_2 + A_3 + A_4. \tag{5.5.31}$$

The term A_1 obviously satisfies the inequality (5.5.28). We have (F denotes the Fourier transformation)

$$\frac{\partial^p \widehat{u}(\sigma)}{\partial \sigma_1^{k_1} \cdots \partial \sigma_n^{k_n}} = F\big((ix_1)^{k_1} \cdots (ix_1)^{k_n} u(x)\big),$$

and the operator $T_j u(x) = (ix_j) u(x)$ is bounded from $\mathring{H}_r(\Omega)$ to $\mathring{H}_r(\Omega)$ for any r. Therefore,

$$\left[\int_{\Omega^*} |\sigma|^{2r} \left| \frac{\partial^i \widehat{u}(\sigma)}{\partial \sigma_1^{k_1} \cdots \partial \sigma_n^{k_n}} \right|^2 d\sigma \right]^{1/2} \le C_1^i(r) \|u(x)\|_{\mathring{H}_r(\Omega)}. \tag{5.5.32}$$

It follows that for the term A_2 the inequality (5.5.28) is valid. From (5.5.16) and (5.5.32), for $i > p$ we have

$$\left[\int_{\Omega^*} |d^i g_k(\sigma)|^2 d\sigma \right]^{1/2} \le \sum_{k_1 + \cdots + k_n = i} \frac{1}{\lambda^{ki}} \left[\int_{\Omega^*} |\sigma|^{2i} |1 - \lambda|^{2i} \left| \frac{\partial^i \widehat{u}(\sigma)}{\partial \sigma_1^{k_1} \cdots \partial \sigma_n^{k_n}} \right|^2 d\sigma \right]^{1/2} \le$$

$$\le C_2^i(r) \sum_{k_1 + \cdots + k_n = i} \frac{1}{\lambda^{ki}} \left[\int_{\Omega^*} |\sigma|^{2r} |1 - \lambda|^{2r} (|\sigma||1 - \lambda|)^{2i - 2r} \left| \frac{\partial^i \widehat{u}(\sigma)}{\partial \sigma_1^{k_1} \cdots \partial \sigma_n^{k_n}} \right|^2 d\sigma \right]^{1/2} \le$$

$$\le C_3^i(r) |1 - \lambda|^r \|u(x)\|_{\mathring{H}_r(\Omega)}, \tag{5.5.33}$$

where we have used the inequalities $1/2 < \lambda < 3/2$, $|\sigma||1 - \lambda| < 1/2$ for $\sigma \in \Omega^*$.

By virtue of (5.5.33), we find

$$A_3 \le C \sum_{i = p+1}^{m} \frac{1}{i!} \sum_{k=1}^{p} C_p^k \frac{1}{\lambda^k} C_3^i(r) |1 - \lambda|^r \|u(x)\|_{\mathring{H}_r(\Omega)} \le$$

$$\le C_1 \sum_{i = p+1}^{m} \frac{C_3^i(r)}{i!} |1 - \lambda|^r \|u(x)\|_{\mathring{H}_r(\Omega)} \le C_4(r) |1 - \lambda|^r \|u(x)\|_{\mathring{H}_r(\Omega)}. \tag{5.5.34}$$

In order to estimate the term A_4, we note that

$$\left| \frac{\partial^{m+1} \widehat{u}(\sigma)}{\partial \sigma_1^{k_1} \cdots \partial \sigma_n^{k_n}} \right| \le \int_{\Omega} \left| (ix_1)^{k_1} \cdots (ix_1)^{k_n} u(x) e^{i(x,\sigma)} \right| dx \le K^{m+1} \sqrt{V_\Omega} \|u(x)\|_{L_2(\Omega)}, \tag{5.5.35}$$

where $K = \max_\Omega \{|ix_1|, \dots |ix_n|\}$, V_Ω is the volume of Ω. Using (5.5.35), we obtain

$$\left[\int_{\Omega^*} \left| d^{m+1} g_k \left(\sigma + \theta^k \left(\frac{\sigma}{\lambda^k} - \sigma \right) \right) \right|^2 d\sigma \right]^{1/2} \le$$

$$\le C_1^{m+1} \sum_{k_1 + \cdots + k_n = m+1} \left[\int_{\Omega^*} |\sigma|^{2(m+1)} |1 - \lambda|^{2(m+1)} \left| \frac{\partial^{m+1} \widehat{u}\left(\sigma + \theta^k\left(\frac{\sigma}{\lambda^k} - \sigma\right)\right)}{\partial \sigma_1^{k_1} \cdots \partial \sigma_n^{k_n}} \right|^2 d\sigma \right]^{1/2} \le$$

$$\le C_1^{m+1} K^{m+1} n^{m+1} [V_\Omega]^{1/2} \left[\int_{\Omega^*} \left(|\sigma|(1 - \lambda) \right)^{2(m+1)} d\sigma \right]^{1/2} \|u(x)\|_{L_2(\Omega)} \le$$

$$\le K_1 C_2^{m+1} \|u(x)\|_{L_2(\Omega)}, \tag{5.5.36}$$

where K_1 is a constant independent of m. Hence,

$$A_4 \le \frac{K_2}{(m+1)!} \sum_{k=1}^{p} C_p^k \frac{1}{\lambda^k} C_2^{m+1} \|u(x)\|_{L_2(\Omega)} \le K_3 \frac{C_2^{m+1}}{(m+1)!} \|u(x)\|_{L_2(\Omega)} = A_5(m) \|u(x)\|_{L_2(\Omega)}.$$

Since $\lim_{m \to \infty} A_5(m) = 0$, there exists m_0 such that for $m > m_0$, the following inequality holds:

$$A_4 \le A_5(x) \|u(x)\|_{L_2(\Omega)} \le C_r |\lambda - 1|^r \|u(x)\|_{\mathring{H}_r(\Omega)}. \tag{5.5.37}$$

Since the inequality (5.5.28) holds for A_1 and A_2, Proposition 5.5.4 follows from (5.5.31), (5.5.34), (5.5.37).

Definition 5.5.5. A discrete operator Π_{1h} is called the *restriction of the integral projector* Π_h defined on $\mathring{H}_r(\Omega)$, if for any $u \in \mathring{H}_r(\Omega)$, $r \geq 0$, we have $\Pi_{1h}u(x) = \Pi_h u(x)$ for $x \in \Omega_{1h}$, and $\Pi_{1h}u(x) = 0$ for $x \in \mathbb{R}_n \setminus \overline{\Omega}_{1h}$.

Definition 5.5.6. We say that a bounded domain Ω with $(n-1)$-dimensional Lipschitz boundary *admits a partition*, if there exists $h_0 > 0$ such that for any $h = (h_1, \ldots, h_n)$ with $|h| < h_0$ and any $u(x) \in \mathring{H}_r(\Omega)$, the following inequality holds:

$$\|\Pi_{1h}u(x)\|_{\mathring{M}(r,h,\Omega_{1h})} \leq C_r \|u(x)\|_{\mathring{H}_r(\Omega)}, \tag{5.5.38}$$

where the constant C_r depends neither on $h = (h_1, \ldots, h_n)$ nor $u(x)$.

Proposition 5.5.5. *Suppose that the domain Ω admits a partition, and for any $a(x, h) \in M(-r, h, \Omega_{1h})$, $r \geq 0$, let $\widetilde{a}(x, h) = a(x, h)$ for $x \in \Omega_{1h}$, and $\widetilde{a}(x, h) = 0$ for $x \in \Omega \setminus \Omega_{1h}$. Then*

$$\|\widetilde{a}(x, h)\|_{H_{-r}(\Omega)} \leq C_1(r) \|a(x, h)\|_{M(-r,h,\Omega_{1h})}, \tag{5.5.39}$$

for $|h| < h_0$, where the constant $C_1(r)$ depends neither on h nor $a(x, h)$.

Proof. Extending $a(x, h) \in M(-r, h, \Omega_{1h})$ to the entire $\mathbb{R}_n$ as zero outside $\mathbb{R}_n \setminus \Omega_{1h}$, we obtain a function in $L_2(\mathbb{R}_n)$. Therefore, $\widetilde{a}(x, h) \in H_{-r}(\Omega)$. Thus, in order to prove Proposition 5.5.5, it suffices to establish the inequality (5.5.39). The properties of Sobolev spaces ensure that

$$\|\widetilde{a}(x, h)\|_{H_{-r}(\Omega)} = \sup_{\|u(x)\|=1} |(l\widetilde{a}(x, h), u(x))|, \tag{5.5.40}$$

where $(l\widetilde{a}(x, h), u(x))$ is the linear functional on $\mathring{H}_r(\Omega)$ associated with the element $\widetilde{a}(x, h) \in H_{-r}(\Omega)$; $u(x) \in \mathring{H}_r(\Omega)$; $l\widetilde{a}(x, h) \in H_{-r}(\mathbb{R}_n)$ is an extension of $\widetilde{a}(x, h)$. Since the linear functional $(l\widetilde{a}(x, h), u(x))$ does not depend on the extension $l\widetilde{a}(x, h)$, we can take as $l\widetilde{a}(x, h)$ the extension described above, $l\widetilde{a}(x, h) \in L_2(\mathbb{R}_n)$. Using the Parseval identity and taking into account that Ω admits a partition, we obtain

$$\|\widetilde{a}(x, h)\|_{H_{-r}(\mathbb{R}_n)} = (2\pi)^n \sup_{\|u(x)\|=1} \left| \int_{\Omega_{1h}} u(x)\overline{a(x, h)} \, dx \right| =$$

$$= (2\pi)^n \sup_{\|u(x)\|=1} \left| \int_{\Omega_{1h}} \Pi_{1h}u(x)\overline{a(x, h)} \, dx \right| \leq$$

$$\leq C_1(r) \|a(x, h)\|_{M(-r,h,\Omega_{1h})} \|\Pi_{1h}u(x)\|_{\mathring{M}(r,h,\Omega_{1h})} \leq$$

$$\leq C_2(r) \|a(x, h)\|_{M(-r,h,\Omega_{1h})}.$$

Proposition 5.5.5 is proved.

Definition 5.5.7. Consider a bounded domain Ω which is star-shaped with respect to some point $P \in \Omega$ and has $(n-1)$-dimensional Lipschitz boundary Γ. Let $\Omega(\lambda)$ be the domain obtained from Ω by its homothetic transformation with ratio λ with respect to the point P. The domain Ω is called *normally starshaped*, if the following inequality holds:

$$C_1|\lambda - 1| \leq \rho \leq C_2|\lambda - 1|, \tag{5.5.41}$$

where ρ is the distance between the boundaries of Ω and $\Omega(\lambda)$, and the constants $C_i > 0$, $i = 1, 2$ do not depend on λ, $1/2 < \lambda < 3/2$.

Proposition 5.5.6. *Any normally star-shaped domain Ω admits a partition.*

Proof. Without loss of generality, we can assume that the domain Ω is star-shaped with respect to the origin. Let $\lambda = 1 - Mh_0$, where $M > 0$ and $h_0 = \max_j h_j$, $h = (h_1, \ldots, h_n)$ is the vector-valued step of the canonical partition of $\mathbb{R}_n$ into cells $D(k, h)$ (see Definition 5.1.1). We assume that $|h| < \delta$ and δ is such that $1/2 < \lambda < 1$. Then any $u(x) \in \mathring{H}_r(\Omega)$ can be represented in the form

$$u(x) = u_1(x) + u_2(x), \tag{5.5.42}$$

where $u_1(x) \in \mathring{H}_r(\Omega(\lambda))$, $u_2(x) = \Delta_\lambda^{r_0} u(x)$ (see (5.5.12)), $r_0 - 1 < r \le r_0$, with integer $r_0 \ge 1$; moreover,

$$\|u_1(x)\|_{H_r(\mathbb{R}_n)} \le C(r)\|u(x)\|_{\mathring{H}_r(\Omega)}. \tag{5.5.43}$$

Let us show that the distance ρ_1 from the domain $\overline{\Omega}_{1h}$ to the boundary of Ω satisfies the inequality $\rho_1 \le nh_0$, where n is the dimension of $\mathbb{R}_n$. Denote by D_h the subset of Ω consisting of the points whose distance from the boundary Γ is not less than nh_0. For $M_1 \in D_h$ there is a cell $D(k, h)$ such that $M_1 \in \overline{D(k, h)}$. Let us show that $D(k, h) \subset \Omega$. Assuming the contrary, we can find a point $M_2 \in \Gamma \cap D(k, h)$. From the definition of D_h, it follows that $\rho(M_2, M_1) \ge nh_0$, which is impossible, since the distance between any two points in $D(k, h)$ does not exceed $\sqrt{n}h_0$. Therefore, $D(k, h) \subset \Omega$.

Now, from (5.5.41) we see that if $C_1 M \ge n$ and $1 - Mh_0 = \lambda$, then $\Omega(\lambda) \subseteq \Omega_{1h}$. Therefore, $\operatorname{supp} u_1(x) \subseteq \overline{\Omega}_{1h}$, which implies that $\Pi_h u_1(x) = \Pi_{1h} u_1(x)$, where Π_h is the integral projector and Π_{1h} is its restriction (see Definitions 5.4.1 and 5.5.5).

Using Theorem 5.3.1, we get

$$\|\Pi_{1h} u_1(x)\|_{\mathring{M}(r, h, \Omega_{1h})} \le C(r)\|u_1(x)\|_{H_r(\mathbb{R}_n)} \le C_1(r)\|u(x)\|_{\mathring{H}_r(\Omega)}. \tag{5.5.44}$$

Proposition 5.5.4 implies the inequality

$$\left\|\Delta_\lambda^{(r_0)} u(x)\right\|_{L_2(\mathbb{R}_n)} \le C_2(r)|\lambda - 1|^r \|u(x)\|_{\mathring{H}_r(\Omega)}. \tag{5.5.45}$$

From (5.5.45) and Theorem 5.4.1 we obtain the inequalities

$$\|\Pi_{1h} u_2(x)\|_{L_2(\mathbb{R}_n)} \le \|\Pi_h u_2(x)\|_{L_2(\mathbb{R}_n)} \le C\|u(x)\|_{L_2(\mathbb{R}_n)} \le C_3(r)h_0^r\|u(x)\|_{\mathring{H}_r(\Omega)}. \tag{5.5.46}$$

Let us estimate $\Pi_{1h} u_2(x)$ in the norm of $\mathring{M}(r, h, \Omega_{1h})$ To that and, we utilize (5.3.2) and (5.5.46) as follows:

$$\|\Pi_{1h} u_2(x)\|_{\mathring{M}(r, h, \Omega_{1h})}^2 = \|\Pi_{1h} u_2(x)\|_{M(r, h)}^2 =$$

$$= \Omega_h \int_{-\pi}^{\pi} \cdots \int_{-\pi}^{\pi} \left(\left[\sum_{i=1}^n \left(\frac{t_i}{h_i}\right)^2\right]^{1/2} + 1\right)^{2r} |\widehat{a}_c(t_1, \ldots, t_n)|^2 \, dt_1 \cdots dt_n \le$$

$$\le \frac{K}{\delta_1^{2r}} \Omega_h \int_U |\widehat{a}_c(t)|^2 \, dt \le C_3(r) \frac{h_0^{2r}}{\delta_1^{2r}} \|u(x)\|_{\mathring{H}_r(\Omega)}^2 \le C_4(r)\|u(x)\|_{\mathring{H}_r(\Omega)}^2, \tag{5.5.47}$$

where $\delta_1 = \min_j h_j$; $\widehat{a}_c(t)$ is the discrete Fourier transform (see Definition 5.1.4) of the gird function $a(k) = \pi\Pi_{1h} u_2(x)$ (the operator π is introduced by Definition 5.1.4), $\Omega_h = h_1 \cdots h_n$. While deriving (5.5.47), we have used (5.3.3) and the relation

$$\Omega_h \int_U |\widehat{a}_c(t)|^2 \, dt = (2\pi)^n \int_{\mathbb{R}_n} |\Pi_{1h} u_2(x)|^2 \, dx.$$

The inequalities (5.5.44) and (5.5.47) imply that for $|h| < \delta$ the following inequality holds:

$$\|\Pi_{1h} u_1(x)\|_{\mathring{M}(r, h, \Omega_{1h})} \le C_3(r)\|u(x)\|_{\mathring{H}_r(\Omega)}.$$

Therefore, Ω admits a partition. Proposition 5.5.6 is proved.

Remark 5.5.1. A ball is a normally star-shaped domain with respect to its center, and a rectangle is also a star-shaped domain with respect to the crossing point of its diagonals.

Definition 5.5.8. A *star-shaped neighborhood* of a point $x_0 \in \Gamma$ on the boundary Γ of an open domain Ω is a ball U with center at x_0 and the following properties:

(i) there is a point $M_1 \in \Omega$ and a ball U_1 with center at x_0 such that $(K \setminus \Gamma U_1) \subset \Omega$, $U \cap \Omega \subset K$, where K is the domain bounded by the cone with vertex at M_1 and base $\gamma = \overline{\Gamma U_1}$; $\Gamma U_1 = \Gamma \cap U_1$;

(ii) let $K(\lambda)$ be the homothetic transformation of K with ratio $\lambda \in (0, 1)$ and center at M_1; then
    (a) $K(\lambda) \subset \Omega$;
    (b) $\rho \geq C(x_0)|\lambda - 1|$, where ρ is the distance from $K(\lambda)$ to the boundary Γ of the domain Ω; $C(x_0) > 0$.

Proposition 5.5.7. *Any bounded domain Ω with $(n-1)$-dimensional boundary Γ of class C^2 admits a partition.*

Proof. From the results described in *Nikolskii* (1977), it follows that under our assumptions, any point $x_0 \in \Gamma$ has a star-shaped neighborhood $U(x_0)$. Since Γ is a compact set, from these neighborhoods we can select finitely many, say $U_1, \ldots, U_m$, covering Γ. We can also find an open domain U_0, $\overline{U}_0 \subset \Omega$, such that U_j, $j = 0, \ldots, m$, form a finite covering of $\overline{\Omega}$. Then, there exist functions $\varphi_j(x) \in C_0^\infty(\mathbb{R}_n)$ such that $\operatorname{supp}\varphi_j(x) \subset U_j$ and $\psi(x) = \sum_{j=0}^n \varphi_j(x) = 1$ in a neighborhood of $\overline{\Omega}$.

Let $\lambda = 1 - Mh_0$, where $M > 0$, $h_0 = \max_j h_j$, $h = (h_1, \ldots, h_n)$ is the vector-valued step of the canonical partition of $\mathbb{R}_n$ into cells $D(k, h)$ (see Definition 5.1.1). We also assume that $|h| < \delta$ and $\delta > 0$ is so small that $1/2 < \lambda < 1$; and $nh_0 < \rho$, where ρ is the distance from $\overline{U}_0$ to the boundary of Ω.

Any function $u(x) \in \mathring{H}_r(\Omega)$, $r \geq 0$, can be represented in the form

$$u(x) = \sum_{j=0}^n u_j(x) = \sum_{j=0}^n u(x)\varphi_j(x). \tag{5.5.48}$$

Since $U_0 \subset \Omega_{1h}$, Theorem 5.4.1 yields

$$\|\Pi_{1h}u_0(x)\|_{\mathring{M}(r,h,\Omega_{1h})} = \|\Pi_h u_0\|_{M(r,h)} \leq C(r)\|u_0(x)\|_{\mathring{H}_r(\Omega)} \leq C_1(r)\|u(x)\|_{\mathring{H}_r(\Omega)}. \tag{5.5.49}$$

For $0 \leq r_0 - 1 < r \leq r_0$ with integer r_0, any function $u_j(x) \in \mathring{H}_r(U_j)$ admits the representation

$$u_j(x) = z_{ij}(x) + z_{2j}(x), \tag{5.5.50}$$

where $z_{1j} \in \mathring{H}_r(K_j(\lambda))$; $K_j(\lambda)$ is the domain specified in Definition 5.5.8; $z_{2j} = \Delta_\lambda^{(r_0)}u_j(x)$ is defined by (5.5.12) with $u_j(\lambda x)$ replaced by $u_j(M_j + \lambda \cdot \overline{M_j M})$, M_j being the vertex of the cone from Definition 5.5.8.

The inequalities similar to (5.5.43) and (5.5.45) hold for z_{ij}, $i = 1, 2$, namely

$$\begin{aligned}
\|z_{1j}(x)\|_{\mathring{H}_r(\Omega)} &\leq C(r)\|u(x)\|_{\mathring{H}_r(\Omega)}, \\
\|z_{2j}(x)\|_{L_2(\mathbb{R}_n)} &\leq C_1(r)|\lambda - 1|^r\|u(x)\|_{\mathring{H}_r(\Omega)}.
\end{aligned} \tag{5.5.51}$$

Since the neighborhoods $U_1, \ldots, U_m$ are star-shaped, we can choose (see the proof of Proposition 5.5.6) $M > 0$ independent of h_0 such that $1/2 < \lambda = 1 - M_0 h_0 < 1$ and all $K_j(\lambda)$ belong to Ω_{1h}. Thus, from (5.5.51) and Theorem 5.4.1, we obtain

$$\|\Pi_{1h}z_{1j}\|_{\mathring{M}(r,h,\Omega_{1h})} \leq C_2(r)\|u(x)\|_{\mathring{H}_r(\Omega)}, \tag{5.5.52}$$

where Π_{1h} is the restriction of the integral projector to Ω_{1h} (see Definition 5.5.5). Using (5.5.51), we obtain the inequality similar to (5.5.47), namely,

$$\|\Pi_{1h} z_{2j}\|_{\dot{M}(r,h,\Omega_{1h})} \leq C_3(r)\|u(x)\|_{\dot{H}_r(\Omega)}, \tag{5.5.53}$$

From (5.5.48)–(5.5.50), (5.5.53) we get

$$\|\Pi_{1h} u(x)\|_{\dot{M}(r,h,\Omega_{1h})} \leq C_4(r)\|u(x)\|_{\dot{H}_r(\Omega)},$$

which implies that Ω admits a partition.

Proposition 5.5.8. *For any $u(x) \in H_r(\mathbb{R}_n)$ with $r < 0$, there is a family of functions $b(x, h) \in M(r, h)$ such that $b(x, h) \to u(x)$ in the norm of $H_r(\mathbb{R}_n)$ as $|h| \to 0$, and $b(x, h) = 0$ for $x \in D(k, h) \subset \Omega_{2h} \setminus \overline{\Omega}_{1h}$, where Ω is a bounded domain.*

Proof. Let $0 < \varepsilon_n \to 0$ be a monotonically decreasing sequence, $n \to \infty$. Since $C_0^\infty(\mathbb{R}_n)$ is dense in $H_r(\mathbb{R}_n)$, there is a sequence of $\varphi_n(x) \in C_0^\infty(\mathbb{R}_n)$ such that

$$\|u(x) - \varphi_n(x)\|_{H_r(\mathbb{R}_n)} < \varepsilon_n. \tag{5.5.54}$$

For the sequence ε_n we can find a sequence $\delta_n > 0$, $\delta_n \searrow 0$ such that for any $h = (h_1, \ldots, h_n)$ with $|h| < \delta_n$ and any fixed $\varphi_n(x)$, we have

$$\|\Pi_{1h}^* \varphi_n(x) - \varphi_n(x)\|_{L_2(\mathbb{R}_n)} < \varepsilon_n. \tag{5.5.55}$$

Here, Π_{1h}^* is the operator defined as follows:

(i) $\Pi_{1h}^* \varphi_n(x) = \Pi_h^* \varphi_n(x)$ for $x \in D(k, h) \subset \Omega_{1h}$ or $x \in D(k, h) \subset \mathbb{R}_n \setminus \overline{\Omega}_{2h}$, with Π_h^* being the operator from Definition 5.4.2;

(ii) $\Pi_{1h}^* \varphi_n(x) = 0$ for $x \in D(k, h) \subset \Omega_{2h} \setminus \Omega_{1h}$.

The inequality (5.5.55) holds since $\varphi_n(x) \in C_0^\infty(\mathbb{R}_n)$ and Ω is a measurable set.

For any $h = (h_1, \ldots, h_n)$ such that $\delta_{m+1} \leq |h| < \delta_m$, let $b(x, h) = \Pi_{1h}^* \varphi_m(x)$.

Taking any $\varepsilon > 0$, we find $N > 0$ such that for all $m \geq N$ we have $\varepsilon_m \leq \varepsilon/2$. For all $h = (h_1, \ldots, h_n)$ with $\delta_{N+1} \leq |h| < \delta_N$, from (5.5.54) and (5.5.55) we obtain the inequality

$$\|u(x) - b(x, h)\|_{H_r(\mathbb{R}_n)} \leq \|u(x) - \varphi_N(x)\|_{H_r(\mathbb{R}_n)} + \|\varphi_N(x) - b(x, h)\|_{H_r(\mathbb{R}_n)} < \varepsilon,$$

which proves Proposition 5.5.8.

Corollary 5.5.2. *Let Ω be a bounded domain and $u(x) \in H_r(\Omega)$, $r > 0$. Then, there is a family of functions $b(x, h) \in M(r, h, \Omega_{1h})$ such that $\tilde{b}(x, h) \to u(x)$ in the norm of $H_r(\Omega)$ as $|h| \to 0$, where $\tilde{b}(x, h) \in H_r(\Omega)$, $\tilde{b}(x, h) = b(x, h)$ for $x \in D(k, h) \subset \Omega_{1h}$ and $\tilde{b}(x, h) = 0$ for $x \in \Omega \setminus \overline{\Omega}_{1h}$.*

This result follows from Proposition 5.5.8 and the definition of $H_r(\Omega)$ and $M(r, h, \Omega_{1h})$ (see *Eskin* (1973)$_2$ and Definition 5.5.4).

Proposition 5.5.9. *Let Ω be a bounded domain that admits a partition, and let $0 < r_0 < 1/2$, $r_0 < r$. Then, for any $u(x) \in \dot{H}_r(\Omega)$, we have $\Pi_{1h} u(x) \to u(x)$ in the norm of $\dot{H}_{r_0}(\Omega)$ as $|h| \to 0$.*

Proof. Assume the contrary. Then, there exist $\varepsilon_0 > 0$ and $u_0(x) \in \dot{H}_r(\Omega)$ such that for some sequences $\delta_m \searrow 0$ and $|h^m| < \delta_m$, $h^m = (h_1, \ldots, h_n^m)$, we have

$$\|u_0(x) - \Pi_{1h^m} u_0(x)\|_{\dot{H}_{r_0}(\varepsilon)} \geq \varepsilon_0 \quad \text{as} \quad m \to \infty. \tag{5.5.56}$$

Since the domain Ω admits a partition, we have $\|\Pi_{1h} u_0(x)\|_{\dot{M}(r,h,\Omega_{1h})} \leq C\|u_0(x)\|_{\dot{H}_r(\Omega)}$. Therefore, $\Pi_{1h} u_0(x) \in \dot{H}_{r_0}(\Omega)$ by Theorem 5.3.1. This observation, together with Proposition 5.5.2, implies that there is a subsequence h^{m_p} such that $\Pi_{1h^{m_p}} u_0(x) \to u_0(x)$ in the norm of $L_2(\mathbb{R}_n)$ as $p \to \infty$. Therefore, $b(x) = u_0(x)$ and $\|u_0(x) - \Pi_{1h^{m_p}} u_0(x)\| \to 0$ as $p \to \infty$, which is in contradiction with (5.5.56). Proposition 5.5.9 is proved.

Proposition 5.5.10. *Let* Ω *be a bounded domain and* $0 < r_0 < r$, $r_0 < 1/2$. *Then for any* $u(x) \in \mathring{H}_r(\Omega)$, *the convergence* $\Pi_h u(x) \to u(x)$ *takes place in the norm of* $H_{r_0}(\Omega)$ *as* $|h| \to 0$.

Proof. Let Ω_1 be a bounded domain with $(n-1)$-dimensional boundary of class C^2, $\overline{\Omega} \subset \Omega_1$. Then, there is $\delta_0 > 0$ such that $\Pi_h u(x) \in M(r, h, \Omega_{1h})$ for all h with $|h| < \delta_0$. By Proposition 5.5.7, the domain Ω_1 admits a partition, and therefore, the desired result follows from Proposition 5.5.9.

Proposition 5.5.11. *Let* $0 < r_0 < r$, $r_0 < 1/2$. *Then for any* $u(x) \in H_r(\mathbb{R}_n)$, *the convergence* $\Pi_h u(x) \to u(x)$ *takes place in the norm of* $H_{r_0}(\mathbb{R}_n)$ *as* $|h| \to 0$.

Proof. From Theorems 5.3.1 and 5.4.1 it follows that

$$\|\Pi_h u(x)\|_{H_{r_0}(\mathbb{R}_n)} \le C \|u(x)\|_{H_{r_0}(\mathbb{R}_n)}$$

with a constant C independent of $u(x)$, $h = (h_1, \ldots, h_n)$. For an arbitrary $\varepsilon > 0$ we can find a function $\varphi_\varepsilon \in C_0^\infty(\mathbb{R}_n)$ such that

$$\|u(x) - \varphi_\varepsilon(x)\|_{H_{r_0}(\mathbb{R}_n)} < \min\left\{ \frac{\varepsilon}{3C}, \frac{\varepsilon}{3} \right\}. \tag{5.5.57}$$

By Proposition 5.5.10, there exists $\delta(\varepsilon) > 0$ such that for $|h| < \delta(\varepsilon)$ the following inequality holds:

$$\|\Pi_h \varphi_\varepsilon(x) - \varphi_\varepsilon(x)\|_{H_{r_0}(\mathbb{R}_n)} < \frac{\varepsilon}{3}. \tag{5.5.58}$$

Then, for $|h| < \delta(\varepsilon)$, we obtain the inequality

$$\|\Pi_h u(x) - u(x)\|_{H_{r_0}(\mathbb{R}_n)} \le$$
$$\le \|\Pi_h(u(x) - \varphi_\varepsilon(x))\|_{H_{r_0}(\mathbb{R}_n)} + \|u(x) - \varphi_\varepsilon(x)\|_{H_{r_0}(\mathbb{R}_n)} + \|\Pi_h \varphi_\varepsilon(x) - \varphi_\varepsilon(x)\|_{H_{r_0}(\mathbb{R}_n)} < \varepsilon.$$

Proposition 5.5.11 is proved.

Definition 5.5.9. A bounded domain $\Omega \subset \mathbb{R}_n$ is called *normal*, if each function $u(x) \in H_r(\mathbb{R}_n)$ with support in $\overline{\Omega}$ (resp., $\mathbb{R}_n \setminus \overline{\Omega}$) is the limit of functions $\varphi \in C_0^\infty(\mathbb{R}_n)$ (resp., $C_0^\infty(\mathbb{R}_n \setminus \overline{\Omega})$) in the norm of $H_r(\mathbb{R}_n)$.

Remark 5.5.2. The class of normal domains includes bounded domains with $(n-1)$-dimensional smooth boundaries, as well as normally star-shaped domains with Lipschitz boundaries (see *Eskin* (1973)$_2$, *Nikolskii* (1977)).

Proposition 5.5.12. *Let* $a(x, h), b(x, h) \in M(r, h)$, $r \le 0$ *be two families of functions such that*

$$\lim_{|h| \to 0} \|a(x, h) - l_1 u(x)\|_{H_r(\mathbb{R}_n)} = 0, \qquad \lim_{|h| \to 0} \|b(x, h) - l_2 u(x)\|_{H_r(\mathbb{R}_n)} = 0,$$

where $l_1 u(x), l_2 u(x) \in H_r(\mathbb{R}_n)$ *are bounded extensions of a function* $u(x) \in H_r(\Omega)$ *in a normal domain* Ω. *Then*

$$\lim_{|h| \to 0} \|P_h a(x, h) - P_h b(x, h)\|_{M(r, h, \Omega_{1h})} = 0, \tag{5.5.59}$$

where $P_h a(x, h)$ *and* $P_h b(x, h)$ *are the restrictions of the functions* $a(x, h)$ *and* $b(x, h)$ *to* Ω_{1h}.

Proof. Take any $\varepsilon > 0$. Since $l_1 u(x) - l_2 u(x) = u_0(x) \in \mathring{H}_r(\mathbb{R}_n \setminus \overline{\Omega})$ and the domain Ω is normal, there is a function $\varphi_\varepsilon(x) \in C_0^\infty(\mathbb{R}_n \setminus \overline{\Omega})$ such that

$$\|u_0(x) - \varphi_\varepsilon(x)\|_{H_r(\mathbb{R}_n)} < \varepsilon. \tag{5.5.60}$$

Proposition 5.4.9 implies that there is $\delta_1(\varepsilon, \varphi_\varepsilon(x))$ such that for all h with $|h| < \delta_1(\varepsilon, \varphi_\varepsilon(x))$, the following inequality holds:

$$\|\Pi_h \varphi_\varepsilon(x) - \varphi_\varepsilon(x)\|_{H_r(\mathbb{R}_n)} < \varepsilon. \tag{5.5.61}$$

The assumptions of Proposition 5.5.12 ensure the existence of $\delta_2(\varepsilon)$ such that for $|h| < \delta_2(\varepsilon)$, we have

$$\|a(x, h) - l_1 u(x)\|_{H_r(\mathbb{R}_n)} < \varepsilon, \qquad \|b(x, h) - l_2 u(x)\|_{H_r(\mathbb{R}_n)} < \varepsilon. \tag{5.5.62}$$

From Definition 5.5.4 of the spaces $M(r, h, \Omega_{1h})$ we obtain the estimate

$$\|P_h a(x, h) - P_h b(x, h)\|_{M(r,h,\Omega_{1h})} \leq \|a(x, h) - b(x, h) - \Pi_h \varphi_\varepsilon(x)\|_{M(r,h)}. \tag{5.5.63}$$

Using the inequalities (5.5.60)–(5.5.63), together with Theorem 5.3.1, we find that for $|h| < \delta_0(\varepsilon, \varphi_\varepsilon(x)) = \min\{\delta_1(\varepsilon, \varphi(x)), \delta_2(\varepsilon)\}$, the following inequalities are valid:

$$\|P_h a(x, h) - P_h b(x, h)\|_{M(r,h,\Omega_{1h})} \leq C(r)\|a(x, h) - b(x, h) - \Pi_h \varphi_\varepsilon(x)\|_{H_r(\mathbb{R}_n)} \leq$$

$$\leq C(r)\Big\|[a(x, h) - l_1 u(x)] + [l_2 u(x) - b(x, h)] + [\varphi_\varepsilon(x) - \Pi_h \varphi_\varepsilon(x)] +$$

$$+ [l_1 u(x) - l_2 u(x) - \varphi_\varepsilon(x)]\Big\|_{H_r(\mathbb{R}_n)} < C(r) 4\varepsilon.$$

Since the constant $C(r)$ is independent of h and ε is arbitrary, the last inequality implies Proposition 5.5.12.

Proposition 5.5.13. *For a domain Ω admitting a partition, let $a(x, h) \in M(r, h, \Omega_{1h})$, $r < 0$, be a family of functions depending on the parameter h and satisfying the conditions:*

   (i) $\|a(x, h)\|_{M(r,h,\Omega_{1h})} \leq C$;

   (ii) $\lim\limits_{|h| \to 0} \|a(x, h)\|_{M(r_0,h,\Omega_{r_0})} = 0$ *for some $r_0 < r$.*

Then, the functions

$$\tilde{a}(x, h) = \begin{cases} a(x, h) & \text{for} & x \in \Omega_{1h}, \\ 0 & \text{for} & x \in \Omega \setminus \overline{\Omega}_{1h}, \end{cases}$$

are weakly convergent to zero in $H_r(\Omega)$ as $|h| \to 0$.

Proof. By Proposition 5.5.5, we have

$$\|\tilde{a}(x, h)\|_{H_r(\Omega)} \leq C \tag{5.5.64}$$

with a constant C independent of h. For any $u(x) \in \mathring{H}_{-r}(\Omega)$, $r < 0$, $u(x) \neq 0$, and any $\varepsilon > 0$, we can find $\varphi_\varepsilon(x) \in C_0^\infty(\Omega)$ such that

$$\|u(x) - \varphi_\varepsilon(x)\|_{\mathring{H}_{-r}(\Omega)} \leq \frac{\varepsilon}{C}. \tag{5.5.65}$$

Since $\varphi_\varepsilon(x) \in \mathring{H}_m(\Omega)$ for any $m > 0$, Theorem 5.3.1 on the boundedness of the integral projector yields

$$\|\Pi_h \varphi_\varepsilon(x)\|_{M(-r_0,h)} \leq C_1 \|\varphi_\varepsilon(x)\|_{\mathring{H}_{-r_0}(\Omega)}. \tag{5.5.66}$$

The condition $\lim_{|h| \to 0} \|a(x, h)\|_{M(r_0,h,\Omega_{1h})} = 0$ ensures that there is $\delta(\varepsilon, \varphi_\varepsilon(x)) > 0$ such that for $|h| < \delta(\varepsilon, \varphi_\varepsilon(x))$, the following inequality holds:

$$\|a(x, h)\|_{M(r_0,h,\Omega_{1h})} < \frac{\varepsilon}{C_1 \|\varphi_\varepsilon(x)\|_{\mathring{H}_{r_0}(\Omega)}}. \tag{5.5.67}$$

From the inequalities (5.5.65)–(5.5.67) we obtain

$$|(\tilde{a}(x, h), u(x))| \leq |(\tilde{a}(x, h), u(x) - \varphi_\varepsilon(x)) + (\tilde{a}(x, h), \varphi_\varepsilon(x))| \leq \tag{5.5.68}$$

$$\leq \|\tilde{a}(x, h)\|_{H_r(\Omega)} \|u(x) - \varphi_\varepsilon(x)\|_{\mathring{H}_{-r}(\Omega)} + \|a(x, h)\|_{M(r_0,h,\Omega_{1h})} \|\Pi_h \varphi_\varepsilon(x)\|_{M(-r_0,h)} < 2\varepsilon,$$

where $(a(x, h), u(x))$ stands for the value of the linear functional on $H_r(\Omega)$ associated with $u(x) \in \mathring{H}_{-r}(\Omega)$. The inequality (5.5.68) implies Proposition 5.5.13.

Proposition 5.5.14. *For $r > 0$, let $a(x,h)$ and $b(x,h)$ be two families of functions such that*

$$\|a(x,h)\|_{M(r,h)} \le C_1, \qquad \|b(x,h)\|_{M(r,h)} \le C_2, \qquad \lim_{|h|\to 0} \|a(x,h) - b(x,h)\|_{L_2(\mathbb{R}_n)} = 0,$$

where the constants C_1 and C_2 do not depend on $h = (h_1, \ldots, h_n)$. Then,

$$\lim_{|h|\to 0} \|a(x,h) - b(x,h)\|_{M(r_0,h)} = 0, \qquad \forall \, r_0 < r. \qquad (5.5.69)$$

Proof. Let $\widehat{a}_h(\varphi)$ and $\widehat{b}_h(\varphi)$ be the discrete Fourier transforms of the step functions $a(x,h)$ and $b(x,h)$ (see Definition 5.5.3). The Parseval identity yields

$$\lim_{|h|\to 0} \int_{D_h} \left| \widehat{a}_h(\varphi) - \widehat{b}_h(\varphi) \right|^2 d\varphi = 0. \qquad (5.5.70)$$

For $r_0 < r$ and an arbitrary $\varepsilon > 0$, consider the ball $K(N) = \{\varphi : |\varphi| < N\}$ with N such that

$$\frac{1}{(N+1)^{2(r-r_0)}} < \varepsilon. \qquad (5.5.71)$$

In view of (5.5.70), there is $\delta(\varepsilon, N, r_0)$ such that for all h with $|h| < \delta(\varepsilon, N, r_0)$, the following estimate holds:

$$\int_{K(N)} \left(1 + |\varphi|\right)^{2r_0} \left| \widehat{a}_h(\varphi) - \widehat{b}_h(\varphi) \right|^2 d\varphi < \varepsilon. \qquad (5.5.72)$$

Now, using the assumptions of Proposition 5.5.14, together with (5.5.71) and (5.5.72), we obtain

$$\|a(x,h) - b(x,h)\|_{M(r_0,h)}^2 =$$

$$= \int_{K(N)} \left(1 + |\varphi|\right)^{2r_0} \left| \widehat{a}_h(\varphi) - \widehat{b}_h(\varphi) \right|^2 d\varphi + \int_{D_h \backslash K(N)} \left(1 + |\varphi|\right)^{2r_0} \left| \widehat{a}_h(\varphi) - \widehat{b}_h(\varphi) \right|^2 d\varphi <$$

$$< \varepsilon + \frac{1}{(N+1)^{2(r-r_0)}} \int_{D_h \backslash K(N)} \left(1 + |\varphi|\right)^{2r} \left| \widehat{a}_h(\varphi) - \widehat{b}_h(\varphi) \right|^2 d\varphi < \varepsilon(L+1),$$

where L is a constant independent of h, N; $D_h = \prod_{j=1}^n \left[-\pi/h_j \le \varphi_j \le \pi/h_j \right]$. Since ε is arbitrary, the last inequality implies Proposition 5.5.14.

Proposition 5.5.15. *Let Ω be a domain that admits a partition. For a given $r < 0$, let $a(x,h) \in M(r,h,\Omega_{1h})$ be a family of functions satisfying the conditions*

$$\|a(x,h)\|_{M(r,h,\Omega_{1h})} \le C, \qquad \lim_{|h|\to 0} \|\widetilde{a}(x,h)\|_{H_r(\Omega)} = 0,$$

where $\widetilde{a}(x,h)$ is the extension of $a(x,h)$ as zero to Ω. Then, for any $r_0 < r$, we have

$$\lim_{|h|\to 0} \|a(x,h)\|_{M(r_0,h,\Omega_{1h})} = 0. \qquad (5.5.73)$$

Proof. Suppose the contrary. Then, there exist $r_0 < 0$, $m_0 > 0$, and a sequence of vector-valued steps $h^k = (h_1, \ldots, h_n^k)$ (see Definition 5.1.1) such that $\lim_{k\to\infty} |h^k| = 0$ and

$$\|a(x,h^k)\|_{M(r_0,h^k,\Omega_{1h^k})} \ge m_k > m_0 \qquad (5.5.74)$$

for all $k = 1, 2, \ldots$. The inequality $\|a(x, h)\|_{M(r_0, h^k, \Omega_{1k})} \leq C_1 \|a(x, h)\|_{M(r, h^k, \Omega_{1k})}$, with C_1 independent of h and $a(x, h)$, implies that this sequence of step functions determines a sequence of continuous functionals on $\mathring{M}(-r_0, h^k, \Omega_{1h^k})$. By Proposition 5.5.1, it follows that there exists a sequence of functions $u(x, h^k) \in \mathring{M}(-r_0, h^k, \Omega_{1h^k})$ such that

$$\|u(x, h^k)\|_{\mathring{M}(-r_0, h^k, \Omega_{1h^k})} = 1, \qquad (a(x, h^k), u(x, h^k)) = m_k,$$

where $(a(x, h^k), u(x, h^k))$ denotes the corresponding linear functional on $\mathring{M}(-r_0, h^k, \Omega_{1h^k})$.

Proposition 5.5.2 ensures that there is a subsequence $u(x, h^{k_l})$ such that

$$\lim_{|h^{k_l}| \to 0} u(x, h^{k_l}) = u_0(x), \qquad u_0(x) \in \mathring{H}_{-r_0}(\Omega),$$

where the limit is with respect to the norm of $\mathring{H}_{r_1}(\Omega)$ for $0 < r_1 < 1/2$, $r_1 < -r_0$.

Since $-r < -r_0$, we have $u_0(x) \in H_{-r}(\Omega)$. Then, the assumptions of Proposition 5.5.15 yield

$$\lim_{|h^{k_l}| \to 0} (\widetilde{a}(x, h^{k_l}), u_0(x)) = 0, \tag{5.5.75}$$

where $(\widetilde{a}(x, h^{k_l}), u_0(x))$ is the value of the linear functional from $\mathring{H}_{-r_0}(\Omega)$ associated with $\widetilde{a}(x, h^{k_l})$ on the element $u_0(x)$.

Since $\widetilde{a}(x, h^{k_l}) \in L_2(\Omega)$ and $u_0(x) \in L_2(\Omega)$, we have

$$(\widetilde{a}(x, h^{k_l}), u_0(x_0)) = (a(x, h^{k_l}) m \Pi_{1k^{k_l}} u_0(x)), \tag{5.5.76}$$

where $\Pi_{1h^{k_l}} u_0(x)$ is from Definition 5.5.5; $(a(x, h^{k_l}), \Pi_{1k^{k_l}} u_0(x))$ is the linear functional on $\mathring{M}(-r, h^{k_l}, \Omega_{1h^{k_l}})$ associated with $a(x, h^{k_l})$. From (5.5.76), it follows that

$$(\widetilde{a}(x, h^{k_l}), u_0(x)) = (a(x, h^{k_l}), u(x, h^{k_l})) + (a(x, h^{k_l}), \Pi_{1h^{k_l}} u_0(x) - u(x, h^{k_l})) \geq$$
$$\geq m_0 - |(a(x, h^{k_l}), \Pi_{1h^{k_l}} u_0(x) - u(x, h^{k_l}))|. \tag{5.5.77}$$

According to Proposition 5.5.9, we have $\Pi_{1h^{k_l}} u_0(x) \to u_0(x)$ in the norm of $L_2(\Omega)$ as $|h^{k_l}| \to 0$. Moreover, $\|\Pi_{1h^{k_l}} u_0(x)\|_{\mathring{M}(-r_0, h^{k_l}, \Omega_{1h^{k_l}})} \leq C \|u_0(x)\|_{\mathring{H}_{-r_0}(\Omega)}$, since the domain Ω admits a partition. Hence, using Proposition 5.5.14, we see that for any r_2 such that $-r < r_2 < -r_0$, the following convergence takes place:

$$\lim_{|h^{k_l}| \to 0} \|\Pi_{1h^{k_l}} u_0(x) - u(x, h^{k_l})\|_{\mathring{M}(-r_2, h^{k_l}, \Omega_{1h^{k_l}})} = 0, \tag{5.5.78}$$

and therefore,

$$\lim_{|h^{k_l}| \to 0} |(a(x, h^{k_l}), \Pi_{1h^{k_l}} u_0(x) - u(x, h^{k_l}))| = 0. \tag{5.5.79}$$

But the relations (5.5.77) and (5.5.79) are in contradiction with (5.5.75). Proposition 5.5.15 is proved.

5.6. Weighted Spaces

Let $\omega^+(x)$ be a non-decreasing function on $[0, +\infty)$ such that $w^+(0) = 1$, $\lim_{x \to \infty} w^+(x) = \infty$; and let $w^-(x)$ be a non-increasing function on $[0, +\infty)$ such that $w^-(x) > 0$, $w^-(0) = 1$, $\lim_{x \to \infty} w^-(x) = 0$.

Definition 5.6.1. *The space* $H_r(\mathbb{R}_n, w^+)$ *consists of all functions* $u(x)$ *with finite norm*

$$\|u(x)\|_{r,w^+}^2 = \int_{\mathbb{R}_n} \left(1 + |t|\right)^{2r} w^+(|t|)|\widehat{u}(t)|^2 \, dt < \infty, \tag{5.6.1}$$

where $w^+(t)$ is fixed, $\widehat{u}(t)$ is the Fourier transform of $u(x)$, r is an arbitrary real number.

Definition 5.6.2. *The space* $H_r(\mathbb{R}_n, \Omega^+)$ *consists of all functions* $u(x)$ *with finite norm*

$$\|u(x)\|_{r,w^-}^2 = \int_{\mathbb{R}_n} \left(1 + |t|\right)^{2r} w^-(|t|)|\widehat{u}(t)|^2 \, dt < \infty, \tag{5.6.2}$$

where $w^-(t)$ is fixed, $\widehat{u}(t)$ is the Fourier transform of $u(x)$, r is an arbitrary real number.

Definition 5.6.3. *The spaces* $M(r, h, w^+)$ *and* $M(r, h, w^-)$, *respectively, consist of all step functions* $a(x, h)$ *with finite norms*

$$\|a(x, h)\|_{M(r,h,w^+)}^2 = \int_{D_h} \left(1 + |t|\right)^{2r} w^+(|t|)|\widehat{a}_h(t)|^2 \, dt, \tag{5.6.3}$$

$$\|a(x, h)\|_{M(r,h,w^-)}^2 = \int_{D_h} \left(1 + |t|\right)^{2r} w^-(|t|)|\widehat{a}_h(t)|^2 \, dt, \tag{5.6.4}$$

where $\widehat{a}_h(t)$ is the discrete Fourier transform of the step function $a(x, h)$; $h = (h_1, \ldots, h_n)$ is the vector-valued step of the canonical partition of the space $\mathbb{R}_n$ into cells $D(k, h)$ (see Definition 5.1.1); r is an arbitrary real number.

Proposition 5.6.1. *Let* $a(x, h)$, $b(x, h)$ *be two families of functions such that*

$$\|a(x, h)\|_{M(r,h)} \leq C, \qquad \|b(x, h)\|_{M(r,h)} \leq C, \qquad \lim_{|h| \to \infty} \|a(x, h) - b(x, h)\|_{L_2(\mathbb{R}_n)} = 0,$$

where $r > 0$ *and the constant* C *does not depend on* h. *Then, for any* $w^-(x)$, *the following convergence takes place:*

$$\lim_{|h| \to 0} \|a(x, h) - b(x, h)\|_{M(r,h,w^-)} = 0.$$

This result is proved in the same way as Proposition 5.5.14.

Proposition 5.6.2. *For any* $u(x) \in H_r(\mathbb{R}_n)$, *there exists* $w^+(x)$ *(depending on* $u(x)$*) such that* $u(x) \in H_r(\mathbb{R}_n, w^+)$.

Proof. Let $K(N)$ be the ball of radius N with center at the origin, and let $\|u\|_{H_r(\mathbb{R}_n)}^2 = B_0$. Then, there is $N_1 > 0$ such that

$$\int_{\mathbb{R}_n \setminus K(N_1)} \left(1 + |t|\right)^{2r} |\widehat{u}(t)|^2 \, dt < \frac{B_0}{4},$$

and we can find $N_2 > N_1 + 1$ such that

$$\int_{\mathbb{R}_n \setminus K(N_2)} \left(1 + |t|\right)^{2r} |\widehat{u}(t)|^2 \, dt < \frac{B_0}{4^2}, \quad \text{etc.}$$

Thus, we obtain a sequence $N_{k+1} > N_k + 1$ such that

$$\int_{\mathbb{R}_n \setminus K(N_{k+1})} \left(1 + |t|\right)^{2r} |\widehat{u}(t)|^2 \, dt < \frac{B_0}{4^{k+1}}.$$

We construct the function $w^+(x)$ by setting $w^+(x) = 2^i$ for $x \in [N_i, N_{i+1})$, $i = 0, 1, \ldots$, where $N_0 = 0$. We have

$$\|u(x)\|_{H_r(\mathbb{R}_n, w^+)}^2 = \int_{\mathbb{R}_n} \left(1 + |t|\right)^{2r} w^+(|t|)|\widehat{u}(t)|^2 \, dt =$$

$$= \int_{K(N_1)} \lambda(t) \, dt + \sum_{i=1}^{\infty} \int_{K(N_{i+1}) \setminus K(N_i)} \lambda(t) \, dt < B_0 + \sum_{i=1}^{\infty} \frac{2^i B_0}{4^i} = 2B_0. \tag{5.6.5}$$

Hence follows Proposition 5.6.2.

Proposition 5.6.3. *Let $b_n(x) \in H_r(\mathbb{R}_n)$ be a sequence such that $b_n(x) \to b_0(x)$ in $H_r(\mathbb{R}_n)$ as $n \to \infty$. Then there exists $w^+(x)$ such that the following estimate holds uniformly in n:*

$$\int_{\mathbb{R}_n} \left(1 + |t|\right)^{2r} w^+(|t|) \left|\widehat{b}_h(t)\right|^2 dt < C. \tag{5.6.6}$$

Proof. Let us show that

$$\lim_{N \to \infty} A(N, n) = \lim_{N \to \infty} \int_{\mathbb{R}_n \setminus K(N)} \left(1 + |t|\right)^{2r} \left|\widehat{b}_0(t) - \widehat{b}_n(t)\right|^2 dt = 0, \tag{5.6.7}$$

where the convergence is uniform with respect to n. Assume the contrary. Then there exists $\varepsilon > 0$ and there are sequences $N_k \to \infty$ and $n_k \to \infty$ as $k \to \infty$ such that $A(N_k, n_k) \geq \varepsilon$. Therefore,

$$\|b_0(x) - b_{n_k}(x)\|^2_{H_r(\mathbb{R}_n)} =$$

$$= \int_{K(N_k)} \left(1 + |t|\right)^{2r} \left|\widehat{b}_0(t) - \widehat{b}_{n_k}(t)\right|^2 dt + \int_{\mathbb{R}_n \setminus K(N_k)} \left(1 + |t|\right)^{2r} \left|\widehat{b}_0(t) - \widehat{b}_{n_k}(t)\right|^2 dt \geq \varepsilon.$$

The last inequality is incompatible with the convergence $b_n(x) \to b_0(x)$. Now, proceeding as in the proof of Proposition 5.6.2, we construct a function $\nu^+(x)$ such that

$$\int_{\mathbb{R}_n} \left(1 + |t|\right)^{2r} \nu^+(|t|) \left|\widehat{b}_0(t) - \widehat{b}_{n_k}(t)\right|^2 dt < C_1$$

with a constant C_1 independent of n.

Let ν_1^+ be the function whose existence has been established in Proposition 5.6.2 for $b_0(x)$. Set $w^+(x) = \min\left\{\nu^+(x), \nu_1^+(x)\right\}$. It is easy to verify that $w^+(0) = 1$, $\lim_{x \to \infty} w^+(x) = +\infty$, and $w^+(x)$ is non-decreasing on $[0, +\infty)$.

Next, we consider the inequality

$$\|b_n(x)\|^2_{H_r(\mathbb{R}_n, w^+)} \leq$$

$$\leq K \left(\int_{\mathbb{R}_n} \left(1 + |t|\right)^{2r} w^+(|t|) \left|\widehat{b}_0(t) - \widehat{b}_n(t)\right|^2 dt + \int_{\mathbb{R}_n} \left(1 + |t|\right)^{2r} w^+(|t|) \left|\widehat{b}_0(t)\right|^2 dt\right) \leq C,$$

with the constant C independent of n. Hence follows Proposition 5.6.3.

Proposition 5.6.4. *Let $a(x, h^m) \in H_r(\mathbb{R}_n)$, $r < 0$, $h^m = (h_1^m, \ldots, h_n^m)$, be a sequence of step functions which converges to $a_0(x)$ in the norm of $H_r(\mathbb{R}_n)$ as $|h^m| \to 0$, $m \to \infty$. Then, there exists $w^+(x)$ such that*

$$\|a(x, h^m)\|_{M(r, h^m, w^+)} \leq C. \tag{5.6.8}$$

Proof. According to Proposition 5.6.3, there exists $w^+(x)$ for which

$$\int_{\mathbb{R}_n} \left(1 + |t|\right)^{2r} w^+(|t|) |\widehat{a}(t, h^m)| dt < C_1 \tag{5.6.9}$$

with C_1 independent of m. Here $\widehat{a}(t, h^m)$ is the Fourier transform of $a(x, h^m)$.

By virtue of (5.3.4), for the discrete Fourier transform $\widehat{a}_{h^m}(t)$ of the step function $\widehat{a}(t, h^m)$ (see Definition 5.1.3) the estimate

$$|\widehat{a}_{h^m}(t)| \leq C_2 |\widehat{a}(t, h^m)|$$

holds on the set $D_h = \prod_{j=1}^n \left[-\pi/h_j \leq t_j \leq \pi/h_j\right]$ with a constant C_2 independent of h^m.

Further, it can be shown that

$$\|a(x, h^m)\|^2_{M(r, h^m, w^+)} \leq C_2 \int_{D_h} \left(1 + |t|\right)^{2r} w^+(|t|) |\widehat{a}(t, h^m)|^2 dt \leq C_2 C_1,$$

which implies Proposition 5.6.4.

Proposition 5.6.5. *Any element* $a(x, h) \in M(-r, h, 1/w^+)$ *can be associated with the continuous linear functional on* $M(r, hw^+)$ *defined by*

$$f(u) = \int_{D_h} \widehat{u}_h(t) \overline{\widehat{a}_h(t)} \, dt, \qquad u(x, h) \in M(r, h, w^+). \qquad (5.6.10)$$

This statement is true if w^+ *is replaced by* w^-.

Proof. For $a(x, h) \in M(-r, h, 1/w^+)$, we have

$$f(u) = \left| \int_{D_h} \left(1 + |t| \right)^r \left[w^+(|t|) \right]^{1/2} \widehat{u}_h(t) \frac{\overline{\widehat{a}_h(t)}}{\left(1 + |t| \right)^r \left[w^+(|t|) \right]^{1/2}} \, dt \right| \leq$$

$$\leq \left[\left(1 + |t| \right)^{2r} w^+(|t|) |\widehat{u}_h(t)|^2 \, dt \right]^{1/2} \left[\int_{D_h} \frac{\overline{\widehat{a}_h(t)}}{\left(1 + |t| \right)^r \left[w^+(|t|) \right]^{1/2}} \, dt \right]^{1/2} \leq$$

$$\leq \| a(x, h) \|_{M(-r, h, 1/w^+)} \| u(x, h) \|_{M(r, h, w^+)}.$$

Therefore, the functional $f(u)$ is bounded on $M(r, h, w^+)$

In exactly the same way, this result is proved if w^+ is replaced by w^-, for $a(x, h) \in M(-r, h, 1/w^-)$.

Chapter 6

Discrete Operators in Quotient Spaces

6.1. Bounded Operator Families in Quotient Spaces $M(r, h)$

6.1.1. One-Dimensional Discrete Singular Operators

Let $D(k, h) = \big(h(k - 1/2), h(k + 1/2)\big)$ be a partition of the real axis with step h and integer $z \in Z$. Consider a family of operators T_h defined on functions $u(x, h) \in M(r, h)$ by the relation

$$\pi \mathrm{T}_h u(x, h) = \lim_{N \to \infty} \sum_{i=-N}^{N} \gamma(i) u(k - i) = \sum_{i=-\infty}^{\infty} \gamma(i) u(k - i), \qquad (6.1.1)$$

where π is the operator introduced in Definition 5.1.1; $u(k) = \pi u(x, h)$; the numbers $\gamma(i)$ have the form

$$\gamma(i) = \int_{(i-1/2)h}^{(i+1/2)h} \frac{dx}{x}. \qquad (6.1.2)$$

It follows from (6.1.2) that $\gamma(0) = 0$, $\gamma(-i) = -\gamma(i)$. Using (5.1.9), we get

$$\pi \mathrm{T} u(x, h) = \frac{1}{2\pi} \int_{-\pi}^{\pi} S_1(\varphi) \widehat{u}_c(\varphi) e^{-ik\varphi} \, d\varphi, \qquad (6.1.3)$$

where $\widehat{u}_c(\varphi)$ is the discrete Fourier transform (see Definition 5.1.4) of the grid function $u(k) = \pi a(x, h)$, and the function $S_1(\varphi)$ is given by

$$S_1(\varphi) = \lim_{N \to \infty} \sum_{k=-N}^{N} \gamma(k) e^{ik\varphi}. \qquad (6.1.4)$$

Since $\gamma(k)$ is odd, it follows from (6.1.4) that

$$S_1(\varphi) = 2i \sum_{k=1}^{\infty} \gamma(k) \sin k\varphi.$$

Note that the definition of $\gamma(k)$ implies the inequalities

$$|\gamma(k)| < \frac{C}{k}, \qquad |\gamma(k) - \gamma(k + 1)| < \frac{C}{k^2}. \qquad (6.1.5)$$

Moreover, for the kernel

$$D_n(x) = \sum_{\nu=1}^{n} \sin \nu x = \left(\cos \frac{x}{2} - \cos \left(n + \frac{1}{2} \right) x \right) \left(2 \sin \frac{x}{2} \right)^{-1}$$

the following estimates hold

$$|D_n(x)| \le n, \qquad |D_n(x)| \le \left(\sin \left| \frac{x}{2} \right| \right)^{-1}. \tag{6.1.6}$$

Let us show that the function $S_1(\varphi)$ is bounded on the interval $[\pi, \pi]$.

Take $\delta > 0$, say $\delta = 100^{-1}$, and consider an arbitrary φ such that $|\varphi| \ge \delta$. Then, using the Abel transformation, we find that

$$|S_1(N, \varphi)| = \left| 2 \sum_{k=1}^{N} \gamma(k) \sin k\varphi \right| \le$$

$$\le 2 \sum_{k=1}^{N-1} |\gamma(k) - \gamma(k+1)| \left| \sum_{i=1}^{k} \sin i\varphi \right| + 2|\gamma(N)| \left| \sum_{i=1}^{N} \sin i\varphi \right| \le$$

$$\le C \sum_{k=1}^{N-1} \left(k^2 \sin \frac{\delta}{2} \right)^{-1} + C \left(N \sin \frac{\delta}{2} \right)^{-1} \le C \left(\sin \frac{\delta}{2} \right)^{-1}. \tag{6.1.7}$$

Now, let $|\varphi| < \delta$. For $\varphi = 0$, we have $S_1(\varphi) = 0$, and therefore, we may assume that $\varphi \ne 0$. The sets $K(N) = \{ \varphi : (2N)^{-1} < |\varphi| \le N^{-1} \}$ cover the set $K = \{ \varphi : 0 < |\varphi| \le 1 \}$, and thus, for any fixed $\varphi \in K$ there exists N such that $\varphi \in K(N)$. Let $\varphi \in K(N)$.

Consider the function

$$S_1(p, \varphi) = 2i \sum_{k=1}^{p} \gamma(k) \sin k\varphi = 2i \sum_{k=1}^{N-1} \gamma(k) \sin k\varphi + 2i \sum_{k=N}^{p} \gamma(k) \sin k\varphi =$$

$$= S_1(N-1, \varphi) + S_2(N, p, \varphi).$$

For the term $S_1(N-1, \varphi)$ we obtain the estimate

$$S_1(N-1, \varphi) \le 2 \sum_{k=1}^{N-1} |\gamma(k)| k |\varphi| \le 2CN\varphi \le 2C. \tag{6.1.8}$$

Using the Abel transformation, we get

$$|S_2(N, p, \varphi)| = 2 \left| \sum_{k=N}^{p} \gamma(k) \sin k\varphi + \gamma(N) \sum_{k=1}^{N-1} \sin k\varphi - \gamma(N) \sum_{k=1}^{N-1} \sin k\varphi \right| \le$$

$$\le 2 \sum_{k=N}^{p-1} |\gamma(k) - \gamma(k+1)| \left| \sum_{l=1}^{k} \sin l\varphi \right| + 2|\gamma(p)| \left| \sum_{k=1}^{p} \sin k\varphi \right| + 2|\gamma(N)| \left| \sum_{k=1}^{N-1} \sin k\varphi \right| \le$$

$$\le C_1 \sum_{k=N}^{p-1} \left(k^2 \sin \frac{|\varphi|}{2} \right)^{-1} + C_1 \left(p \sin \frac{|\varphi|}{2} \right)^{-1} + C_1 \left(N \sin \frac{|\varphi|}{2} \right)^{-1}. \tag{6.1.9}$$

Since $\varphi \in K(N)$, it follows from (6.1.9) that

$$|S_2(N, p, \varphi)| \le C_2. \tag{6.1.10}$$

From (6.1.9), it follows that the series $2i \sum_{k=1}^{\infty} \gamma(k) \sin k\varphi$ is convergent for any $\varphi \ne 0$. Its convergence for $\varphi = 0$ is obvious. Therefore, it is convergent for any $\varphi \in [-\pi, \pi]$. The estimates (6.1.7), (6.1.8), (6.1.10) imply that $S_1(\varphi)$ is bounded on $[-\pi, \pi]$.

Definition 6.1.1. The operators T_h defined by (6.1.1) are said to form a *family of discrete one-dimensional singular operators*.

Proposition 6.1.1. *The family of discrete one-dimensional singular operators T_h from $M(r,h)$ to $M(r,h)$ is uniformly bounded for any h.*

Proof. The symbol $S_1(\varphi)$ of the operators T_h does not depend on h and is bounded on $[-\pi, \pi]$. For any fixed h, any function $u(x,h) \in M(r,h)$ is an element of $L_2(R_1)$. Therefore, $S_1(\varphi)\widehat{u}_c(\varphi) \in L_2[-\pi, \pi]$. Now, (6.1.3) implies that $T_h u(x,h) \in L_2(R_1)$. It follows that there is a function $\widehat{a}_c(\varphi)$ which is the discrete Fourier transform of the grid function $a(k) = \pi T_h u(x,h)$ (see Definition 5.1.4)) and $\widehat{a}_c(\varphi) = S_1(\varphi)\widehat{u}_c(\varphi)$. This observation, together with the definition of the norm in $M(r,h)$, implies Proposition 6.1.1.

6.1.2. Multidimensional Discrete Singular Operators

Consider a singular integral in $\mathbb{R}_n$

$$\int \frac{f(\vec{\theta})\nu(x)}{r^n}\, dx,$$

where $r = r_{MM_0}$ is the distance between the points $M(x_1, \ldots, x_n)$ and $M_0(x_1^0, \ldots, x_n^0)$; $f(\vec{\theta})$ is the characteristic function of the singular integral. It is a function of the vector $\vec{\theta} = |r|^{-1}r$, satisfies the Lipschitz condition, and

$$\int_{S_n} f(\vec{\theta})\, dS_n = 0, \tag{6.1.11}$$

where S_n is the unit sphere in $\mathbb{R}_n$.

Consider a canonical partition of the space $\mathbb{R}_n$ with vector-valued step $h = (h, \ldots, h)$ (i.e., the step is the same in the direction of each axis) and let $D(k,h)$ be a cell of this partition (see Definition 5.1.1). Let T_h be a family of operators defined on functions $u(x,h) \in M(r,h)$ and given by the relation

$$\pi T_h u(x,h) = \lim_{N \to \infty} \sum_{i_1=-N}^{N} \cdots \sum_{i_n=-N}^{N} \gamma(i_1, \ldots, i_n) u(k_1 - i_1, \ldots, k_n - i_n) =$$

$$= \sum_{i \in \mathbb{Z}_n} \gamma(i) u(k-i), \tag{6.1.12}$$

where π is the operator described in Definition 5.1.2; $u(k) = \pi T_h u(x,h)$; the numbers $\gamma(i)$ are given by the formulas

$$\gamma(i) = \int\limits_{(i_1-1/2)h}^{(i_1+1/2)h} \cdots \int\limits_{(i_n-1/2)h}^{(i_n+1/2)h} \frac{f(\vec{\theta})}{r^n}\, dx, \tag{6.1.13}$$

where $r = \left[\sum_{i=1}^{n} x_i^2\right]^{1/2}$.

Let us change the variables in (6.1.13) by setting $x_j = hy_j$, $j = 1, \ldots, n$. Since this transformation does not change the vector $\vec{\theta} = |r|^{-1}r$, we obtain the following expression from (6.1.13):

$$\gamma(i) = \int_{D(i,1)} \frac{f(\vec{\theta})}{r^n}\, dx. \tag{6.1.14}$$

It follows from (6.1.14) that $\gamma(i)$ does not depend on h and the grid function $\gamma(i)$ belongs to l_2. Therefore, the symbol $S_n(\varphi_1, \ldots, \varphi_n)$ of the operators T_h belongs to $L_2(U)$, $U = \prod_{j=1}^{n}[-\pi \le \varphi_j \le \pi]$. The symbol $S_n(\varphi)$ is defined by (see Definition 5.1.6)

$$S_n(\varphi) = \lim_{N \to \infty} \sum_{k \in \mathbb{Z}_n(N)} a(k) e^{i(k\varphi)}. \tag{6.1.15}$$

Hence, similarly to (6.1.3), we obtain

$$\pi T_h u(x, h) = \frac{1}{(2\pi)^n} \int_U S_n(\varphi) \widehat{u}_c(\varphi) e^{-(k\varphi)} \, d\varphi, \tag{6.1.16}$$

where $(k\varphi) = \sum_{l=1}^{n} \varphi_l k_l$.

By analogy with (6.1.14), we get

$$\Gamma(N) = \sum_{k \in \mathbb{Z}_n(N)} \gamma(k) = \int_{D(0,1)} \frac{f(\vec{\theta})}{r^n} \, dx = \gamma(0). \tag{6.1.17}$$

Let us introduce the quantities

$$\Gamma_l(m) = \sum_{i=1, \, i \neq l} \sum_{k_i = -\infty}^{\infty} \left| \gamma(k_1, \ldots, k_{l-1}, k_l = m, k_{l+1}, \ldots, k_n) \right|. \tag{6.1.18}$$

We estimate $\Gamma_1(0)$ as follows:

$$\Gamma_1(0) \leq |\gamma(0)| + \sum_{k_2 = -\infty}^{\infty} \cdots \sum_{k_n = -\infty}^{\infty} |\gamma(0, k_2, \ldots, k_n)| \leq$$

$$\leq C + \int_{-1/2}^{1/2} dx_1 \int_{\Omega_{n-1}(1/2)} \frac{\left| f(\vec{\theta}) \right|}{r^n} \, dx_2 \cdots dx_n \leq$$

$$\leq C + C_1 \int_{-1/2}^{1/2} dx_1 \int_{S_{n-1}} \left\{ \int_{1/2}^{\infty} \frac{dr_1}{r_1^2} \right\} dS_{n-1} \leq C_2, \tag{6.1.19}$$

where $\Omega_{n-1}(1/2)$ is the exterior of the ball of radius $R = 1/2$ in the $(n-1)$-dimensional space, S_{n-1} is the area of the unit sphere in that space, $r_1^2 = \sum_{i=2}^{n} x_i^2$.

In order to estimate the quantities $\Gamma_1(m)$, $m > 0$, we introduce spherical coordinates, just as we have done for the derivation of (5.2.7). We obtain

$$\Gamma_1(m) \leq C \int_{S_{n-1}} dS_{n-1} \int_{-\pi/2}^{\pi/2} \sin^{n-2} \theta_1 \int_{\frac{m-1/2}{\cos \theta_1}}^{\frac{m+1/2}{\cos \theta_1}} \frac{dr}{r} \leq$$

$$\leq C_1 \ln \left(1 + \frac{2}{2m - 1} \right) \leq \frac{C_2}{|m| + 1}. \tag{6.1.20}$$

A similar estimate is obtained in the same way for $\Gamma_l(m)$ with arbitrary l and m,

$$\Gamma_l(m) \leq \frac{C}{|m| + 1} \tag{6.1.21}$$

Next, we estimate the quantities

$$\Delta\Gamma_l(m) = \sum_{i=1, \, i \neq l} \sum_{k_i = -\infty}^{\infty} \left| \gamma(k_1, \ldots, k_l = m, \ldots, k_n) - \gamma(k_1, \ldots, k_l = m + 1, \ldots, k_n) \right| \tag{6.1.22}$$

Using the Lipschitz condition for the function $f(\vec{\theta})$ and performing simple transformations, we obtain the estimate

$$\left|\gamma(m, k_2, \ldots, k_n) - \gamma(m+1, k_2, \ldots, k_n)\right| \leq C \int\limits_{D(m, k_2, \ldots, k_n, 1)} \frac{dx}{r^{n+1}} \tag{6.1.23}$$

for $m \neq 0$, $m + 1 \neq 0$. Hence, introducing spherical coordinates as above, we obtain

$$\Delta\Gamma_1(m) \leq C_1 \int\limits_{S_{n-1}} dS_{n-1} \int\limits_{-\pi/2}^{\pi/2} \sin^{n-2}\theta_1 \, d\theta_1 \int\limits_{\frac{m-1/2}{\cos\theta_1}}^{\frac{m+1/2}{\cos\theta_1}} \frac{dr}{r^2} \leq \frac{C_2}{m^2 + 1} \tag{6.1.24}$$

for $m > 0$. In the same way, we obtain an estimate for any $l = 1, \ldots, n$ and any $m \in Z$,

$$\Delta\Gamma_l(m) \leq \frac{C_2}{m^2 + 1}. \tag{6.1.25}$$

Next, we estimate the quantities

$$\Delta(l)\Gamma(k_j = N) =$$
$$= \sum_{k_j = N+1}^{+\infty} \sum_{p=1, \, p \neq j}^{n} \sum_{k_p = -\infty}^{\infty} \left|\gamma(k_1, \ldots, k_l, \ldots, k_n) - \gamma(k_1, \ldots, k_l + 1, \ldots, k_n)\right|,$$

$$\Delta(l)\Gamma(k_j = -N) = \tag{6.1.26}$$
$$= \sum_{j = -\infty}^{-N-1} \sum_{p=1, \, p \neq j}^{n} \sum_{k_p = -\infty}^{\infty} \left|\gamma(k_1, \ldots, k_l, \ldots, k_n) - \gamma(k_1, \ldots, k_l + 1, \ldots, k_n)\right|,$$

where $N > 0$ is integer and $l \neq j$.

Using the estimate (6.1.23), we get

$$\Delta(l)\Gamma(k_1 = N) \leq C \int\limits_{S_{n-1}} dS_{n-1} \int\limits_{-\pi/2}^{\pi/2} \sin^{n-2}\Theta_1 \left(\int\limits_{\frac{N+1}{\cos\Theta_1}} \frac{dr}{r^2} \right) d\Theta_1 \leq \frac{C}{N + 1/2}.$$

In a similar way, we obtain estimates for all $l = 1, \ldots, n$ and all $k_j = N$, $k_j = -N$, $l \neq j$,

$$\Delta(l)\Gamma(k_j = N) \leq \frac{C}{N+1}, \qquad \Delta(l)\Gamma(k_j = -N) \leq \frac{C}{N+1}. \tag{6.1.27}$$

Proposition 6.1.2. *If the characteristic function of the singular integral satisfies the Lipschitz condition, then*
$$S_n(\varphi) = \lim_{N \to \infty} \sum_{k \in \mathbb{Z}_n(N)} \gamma(k) \, e^{i(k\varphi)}$$

is a bounded function of $\varphi \in \prod_{j=1}^{n} [-\pi \leq \varphi_j \leq \pi]$.

Proof. Let $M(\varphi_1, \ldots, \varphi_n)$ be an arbitrary point with $|\varphi| \geq \delta > 0$. Then, there is a coordinate φ_j with $|\varphi_j| \geq \delta/\sqrt{n}$. Because of the symmetry of the estimates (6.1.21), (6.1.25), and (6.1.27), we may assume that $|\varphi_1| \geq \delta/\sqrt{n}$.

Consider the function

$$S_n(N, \varphi) = \sum_{k \in \mathbb{Z}_n(N)} \gamma(k) \, e^{i(k,\varphi)} = \tag{6.1.28}$$

$$= \sum_{k_1=-N}^{-1} \sum_{p=2}^{n} \sum_{k_p=-N}^{N} \gamma(k) \, e^{i(k,\varphi)} + \sum_{k_1=0}^{N} \sum_{p=2}^{n} \sum_{k_p=-N}^{N} \gamma(k) \, e^{i(k,\varphi)} = S_n^1(N, \varphi) + S_n^2(N, \varphi),$$

where $(k, \varphi) = \sum_{j=1}^{n} k_j \varphi_j$, $k = (k_1, \ldots, k_n)$, $\varphi = (\varphi_1, \ldots, \varphi_n)$.

Applying the Abel transformation with respect to the variable k_1, for each fixed $k' = (k_2, \ldots, k_n)$, we obtain

$$S_n^2(N, \varphi) = \sum_{k_1=0}^{N-1} \sum_{p=2}^{n} \sum_{k_p=-N}^{N} \left[\gamma(k_1, k') - \gamma(k_1 + 1, k') \right] e^{i(k',\varphi')} \sum_{l=0}^{k_1} e^{il\varphi_1} +$$

$$+ \sum_{p=2}^{n} \sum_{k_p=-N}^{N} \gamma(N, k') \, e^{i(k',\varphi')} \sum_{l=0}^{N} e^{il\varphi_1} = \Delta_1 + \Delta_2. \tag{6.1.29}$$

Using the estimates (6.1.21) and (6.1.25), we get

$$|\Delta_1| \le \sum_{m=0}^{N-1} \Delta\Gamma_1(m) 2 \left(\sin \frac{|\varphi_1|}{2} \right)^{-1} \le C_1 \left(\sin \frac{\delta}{2} \right)^{-1} \sum_{m=0}^{N-1} \frac{1}{m^2 + 1} \le C_2 \left(\sin \frac{\delta}{2} \right)^{-1},$$

$$|\Delta_2| \le \Gamma_1(N) 2 \left(\sin \frac{\delta}{2} \right)^{-1} \le C \left(N \sin \frac{\delta}{2} \right)^{-1}. \tag{6.1.30}$$

We proceed in the same way with the function $S_n^1(N, \varphi)$. To that end, it is convenient to change the summation index k_1 by letting $k_1 = -l$, $1 \le l \le N$. We get

$$S_n^1(N, \varphi) = \sum_{l=1}^{N-1} \sum_{p=2}^{n} \sum_{k_p=-N}^{N} \left[\gamma(-l, k') - \gamma(-l - 1, k') \right] e^{i(k',\varphi')} \sum_{l_1=1}^{l} e^{-\varphi_1 l_1} +$$

$$+ \sum_{p=2}^{n} \sum_{k_p=-N}^{N} \gamma(-N, k') \sum_{l_1=1}^{N} e^{-i\varphi_1 l_1} = \Delta_1' + \Delta_2'. \tag{6.1.31}$$

The form of the right-hand side of (6.1.31) is similar to (6.1.29), and therefore,

$$\left| S_n^1(N, \varphi) \right| \le C \left(\sin \frac{\delta}{2} \right)^{-1}. \tag{6.1.32}$$

From (6.1.28)–(6.1.32), we obtain the following estimate:

$$S_n(N, \varphi) \le C_3 \left(\sin \frac{\delta}{2} \right)^{-1} \tag{6.1.33}$$

for $|\varphi| \ge \delta$, $\varphi \in U = \prod_{j=1}^{n} [-\pi \le \varphi_j \le \pi]$.

Now, let $|\varphi| < \delta < 1$. For $\varphi = 0$, we have $S_n(N, 0) = \Gamma(N) = \gamma(0)$. Therefore,

$$S_n(0) = \lim_{N \to \infty} S_n(N, 0) = \gamma(0).$$

Thus, we may assume that $|\varphi| \neq 0$. The sets $K(N) = \{\varphi : (2N)^{-1} < |\varphi| \leq N^{-1}\}$ cover the set $K = \{\varphi : 0 < |\varphi| \leq 1\}$. Therefore, for any fixed $\varphi \in K$, there exists N such that $\varphi \in N(N)$. Then, for at least one coordinate φ_j, the following inequality holds:

$$\frac{1}{2N\sqrt{n}} < |\varphi_j| \leq \frac{1}{N}.$$

Without loss of generality, just as above, we may assume that this inequality holds for φ_1.

Let us represent the function $S_n(q, \varphi)$, $q > N$, in the form

$$S_n(q, \varphi) = S_n(N, \varphi) + \widetilde{S}_n(q, \varphi), \tag{6.1.34}$$

where $\widetilde{S}_n(q, \varphi)$ contains all the terms not included in $S_n(N, \varphi)$.

For $S_n(N, \varphi)$, taking into account (6.1.17), we obtain

$$|S_n(N, \varphi)| \leq \left| \sum_{k \in \mathbb{Z}_n(N)}' \gamma(k)\left(e^{i(k,\varphi)} - 1\right) \right| + |\gamma(0)| \leq \sum_{k \in \mathbb{Z}_n(N)}' |\gamma(k)| |k| \, |\varphi| + C \leq$$

$$\leq C \left(|\varphi| \int_{S_n} \int_{1/2}^{\sqrt{n}\,(N+1/2)} dr + 1 \right) \leq C(C_1 N |\varphi| + 1) \leq C_2. \tag{6.1.35}$$

Here S_n is the unit sphere in $\mathbb{R}_n$ and Σ' indicates that there is no term with $|k| = 0$.

For the index k_j, let us divide the interval $-q \leq k_j \leq q$ into the intervals

$$I_1 = \{k_j : -q \leq k_j \leq -N-1\}, \quad I_2 = \{k_j : -N \leq k_j \leq N\}, \quad I_1 = \{k_j : N+1 \leq k_j \leq q\}.$$

Let α_j take three values 1, 2, 3. Then $\sum_{k_j \in I_{\alpha_j}}$ denotes the sum over all $k_j \in I_{\alpha_j}$.

We introduce the notation

$$S_{\alpha_1, \ldots, \alpha_n}(\varphi) = \sum_{k_1 \in I_{\alpha_1}} \cdots \sum_{k_n \in I_{\alpha_n}} \gamma(k_1, \ldots, k_n) \exp\left(i \sum_{l=1}^{n} k_l \varphi_l\right).$$

In particular, $S_{2,\ldots,2} = S_n(N, \varphi)$. Then, the function $\widetilde{S}_n(q, \varphi)$ from (6.1.34) can be written in the form

$$\widetilde{S}_n(q, \varphi) = \sum_{\alpha}' S_{\alpha_1, \ldots, \alpha_n}(\varphi), \tag{6.1.36}$$

where $\sum_{\alpha}'$ is the sum over all $(\alpha_1, \ldots, \alpha_n)$ except $(2, \ldots, 2)$.

Relation (6.1.36) holds, since for any multi-index $k = (k_1, \ldots, k_n)$, every k_j belongs to one of the intervals I_1, I_2, I_3. Therefore, the term $\gamma(k_1, \ldots, k_n) \exp\left(i \sum_{l=1}^{n} k_l \varphi_l\right)$ is always present in at least one function $S_{\alpha_1, \ldots, \alpha_n}(\varphi)$; and the same term cannot be present in two different functions, since the intervals I_1, I_2, I_3 are mutually disjoint.

Let us estimate the function $S_{3,\alpha_2,\ldots,\alpha_n}(\varphi)$, using the Abel transformation with respect to k_1 for each fixed $k' = (k_2 \ldots, k_n)$. We have

$$|S_{3,\alpha_2,\ldots,\alpha_n}(\varphi)| \leq \left| \sum_{k_1=N+1} \sum_{k' \in (\alpha_2,\ldots,\alpha_n)} \left[\gamma(k_1, k') - \gamma(k_1+1, k')\right] e^{i(k',\varphi')} \sum_{l=N+1}^{k_1} e^{il\varphi_1} \right| +$$

$$+ \left| \sum_{k' \in (\alpha_2,\ldots,\alpha_n)} \gamma(q, k') e^{i(k',\varphi')} \sum_{l=N+1}^{q} e^{il\varphi_1} \right| \leq$$

$$\leq C \left(\sin\frac{|\varphi_1|}{2}\right)^{-1} \left(\frac{1}{N+1} + \frac{1}{q+1}\right). \tag{6.1.37}$$

To obtain (6.1.37), we have used the estimates (6.1.21) and (6.1.25).

The estimate (6.1.37) for the function $S_{1,\alpha_2,\ldots,\alpha_n}(\varphi)$ is obtained in a similar manner. Using the Abel transformation, we obtain the following relation for the function $S_{2,\alpha_2,\ldots,\alpha_n}(\varphi)$ involved in $\widetilde{S}_n(q,\varphi)$:

$$S_{2,\alpha_2,\ldots,\alpha_n}(\varphi) = \sum_{k_1=-N}^{N-1} \sum_{k'\in(\alpha_2,\ldots,\alpha_n)} \left[\gamma(k_1,k') - \gamma(k_1+1,k')\right] e^{i(k',\varphi')} \sum_{l=-N}^{k_1} e^{il\varphi_1} +$$

$$+ \sum_{k'\in(\alpha_2,\ldots,\alpha_n)} \gamma(N,k') e^{i(k',\varphi')} \sum_{i=-N}^{N} e^{il\varphi_1}. \tag{6.1.38}$$

In this case, among α_j, $2 \leq j \leq n$, there is α_j such that either $\alpha_j = 1$ or $\alpha_j = 3$. Therefore, from (6.1.27) we get

$$\sum_{k_1=-N}^{N-1} \sum_{k'\in(\alpha_1,\ldots,\alpha_n)} |\gamma(k_1,k') - \gamma(k_1+1,k')| \leq \frac{C}{N+1}. \tag{6.1.39}$$

From (6.1.21), (6.1.38), and (6.1.39), it follows that

$$|S_{2,\alpha_2,\ldots,\alpha_n}(\varphi)| \leq C \left(\sin \frac{|\varphi_1|}{2}\right)^{-1} \frac{1}{N+1}. \tag{6.1.40}$$

From the inequalities (6.1.37) and (6.1.40), we obtain

$$\left|\widetilde{S}_n(q,\varphi)\right| \leq \frac{C_2}{|\varphi|}\left(\frac{1}{N} + \frac{1}{q}\right). \tag{6.1.41}$$

It follows from (6.1.41) that $\lim_{q\to\infty} S_n(q,\varphi)$ exists for any $\varphi \in U = \prod_{j=1}^{n}[-\pi \leq \varphi_j \leq \pi]$. Since the estimate (6.1.41) holds uniformly in $|\varphi| > \delta > 0$, the function $S_n(\varphi) = \lim_{q\to\infty} S_n(q,\varphi)$ is continuous for $|\varphi| \neq 0$. Moreover, from (6.1.41) we obtain estimate

$$\left|\widetilde{S}_n(q,\varphi)\right| \leq C_0 \tag{6.1.42}$$

for $(2N)^{-1} < |\varphi| \leq N^{-1}$.

It follows from (6.1.35) and (6.1.42) that $|S_n(q,\varphi)| \leq B_0$ for $0 \leq |\varphi| < 1$. This inequality and (6.1.33) imply Proposition 6.1.2.

Definition 6.1.2. The operators T_h defined by (6.1.12) are said to form a *family of discrete singular operators*.

Proposition 6.1.3. *The family of discrete singular operators* T_h, $h = (h_1,\ldots,h_n)$, *from* $M(r,h)$ *to* $M(r,h)$ *is uniformly bounded for any* h.

This result is proved in the same way as Proposition 6.1.1.

6.1.3. Discrete Vortex Operators

Consider the family of operators $T_{h,\lambda}$ defined on functions $u(x,h) \in M(r,h)$, $h = (h_1,\ldots,h_n)$, and given by

$$\pi T_{h,\lambda} u(x,h) = \tag{6.1.43}$$

$$= \lim_{N\to\infty} \sum_{i_1=-N}^{N} \cdots \sum_{i_n=-N}^{N} \gamma(i_1,\ldots,i_n,\lambda) u(k_1-i_1,\ldots,k_n-i_n) = \sum_{i\in\mathbb{Z}_n} \gamma(i,\lambda)u(k-i),$$

where $u(k) = \pi u(x,h)$ and the numbers $\gamma(i,\lambda)$ are defined by

$$\gamma(i,\lambda) = \int_{D(i,h)} \frac{dx}{r^{n+\lambda}}, \tag{6.1.44}$$

$0 < \lambda < 2$. For $|i| = 0$ this integral is understood in the sense of Hadamard (see (5.2.9)).

Definition 6.1.3. The operators $T_{h,\lambda}$ defined by (6.1.43) and (6.1.44) are said to form a family of discrete vortex operators.

Proposition 6.1.4. *The family of discrete vortex operators $T_{h,\lambda}$ from $M(r,h)$ to $M(r-\lambda,h)$ is uniformly bounded for any h.*

Proof. Since the series $\sum_{i\in\mathbb{Z}_n}\gamma(i,\lambda)$ is absolutely convergent for any fixed $h=(h_1,\dots,h_n)$, it follows from (5.1.9) that the grid function $\pi T_{h,\lambda}u(x,h)$ can be represented in the form

$$\pi T_{h,\lambda}u(x,h) = \frac{1}{(2\pi)^n}\int_U J(\lambda,\varphi,h)\widehat{u}_c(\varphi)\,e^{-i(k\varphi)}\,d\varphi, \tag{6.1.45}$$

where $\widehat{u}_c(\varphi)$ is the discrete Fourier transform of the grid function $u(k)=\pi u(x,h)$ (see Definition 5.1.4); the function $J(\lambda,\varphi,h)$ is defined by

$$J(\lambda,\varphi,h) = \sum_{k\in\mathbb{Z}_n}\gamma(k,\lambda)\,e^{i(k\varphi)}. \tag{6.1.46}$$

Denote by $\widehat{a}_c(\varphi)$ the discrete Fourier transform of the grid function $\pi T_{h,\lambda}u(x,h)$. The above results imply that $\widehat{a}_c(\varphi)=J(\lambda,\varphi,h)\widehat{u}_c(\varphi)$. Then, using (5.3.2), (5.3.3), and (5.2.15), we get

$$\left\|\pi T_{h,\lambda}u(x,h)\right\|^2_{M(r-\lambda,h)} =$$

$$= \Omega_h\int_{-\pi}^{\pi}\cdots\int_{-\pi}^{\pi}\left(\left[\sum_{i=1}^{n}\left(\frac{\varphi_i}{h_i}\right)^2\right]^{1/2}+1\right)^{2r-2\lambda}|\widehat{a}_c(\varphi_1,\dots,\varphi_n)|^2\,d\varphi_1\cdots d\varphi_n \le$$

$$\le C(\lambda)\Omega_h\int_{-\pi}^{\pi}\cdots\int_{-\pi}^{\pi}\left(\left[\sum_{i=1}^{n}\left(\frac{\varphi_i}{h_i}\right)^2\right]^{1/2}+1\right)^{2r}h_1^{2\lambda}\left(\left[\varphi_1^2+\sum_{i=2}^{n}\left(\frac{h_1}{h_i}\varphi_i\right)^2\right]^{1/2}\right)^{-2\lambda}\times$$

$$\times\frac{|\varphi|^{2\lambda}}{h_1^{2\lambda}}|\widehat{u}_c(\varphi)|^2\,d\varphi_1\cdots d\varphi_n \le C_1(\lambda)\|u(x,h)\|^2_{M(r,h)},$$

where $\Omega_h=\prod_{j=1}^{n}h_j$. The last inequality implies Proposition 6.1.4.

6.1.4. Difference Operators

(a) Consider the right difference operator $T^m_{h_{\text{right}}}$ of order m,

$$\pi T^m_{h_{\text{right}}}u(x,h) = \frac{1}{(2\pi)^n}\int_{-\pi}^{\pi}\cdots\int_{-\pi}^{\pi}\prod_{j=1}^{n}\left(\frac{e^{-i\varphi_j}-1}{h_j}\right)^{m_j}\widehat{u}_c(\varphi)\,e^{-i(k,\varphi)}\,d\varphi_1\dots d\varphi_n, \tag{6.1.47}$$

where $m_1+\dots+m_n=m$, $\widehat{u}_c(\varphi)$ is the discrete Fourier transform of the grid function $u(k)=\pi u(x,h)$; π is the operator specified in Definition 5.1.2; $u(x,h)\in M(r,h)$. The symbol $\widehat{A}_h(\varphi)$ of the operator $T^m_{h_{\text{right}}}$ has the form

$$\widehat{A}_h(\varphi) = \prod_{k=1}^{n}\left(\frac{e^{-i\varphi_k}-1}{h_k}\right)^{m_k}.$$

Let us show that the family of operators $T^m_{h_{\text{right}}}$ from $M(r,h)$ to $M(r-m,h)$ is uniformly bounded. Indeed, using (5.3.2) and (5.3.3), we obtain

$$\left\|T^m_{h_{\text{right}}}u(x,h)\right\|^2_{M(r-m,h)} =$$

$$= \Omega_h \int\limits_{-\pi}^{\pi} \cdots \int\limits_{-\pi}^{\pi} \left(\left[\sum_{i=1}^{n} \left(\frac{\varphi_i}{h_i} \right)^2 \right]^{1/2} + 1 \right)^{2r-2m} \left| \prod_{k=1}^{n} \left(\frac{e^{-i\varphi_k} - 1}{h_k} \right)^{m_k} \widehat{u}_c(\varphi) \right|^2 d\varphi_1 \cdots d\varphi_n \leq$$

$$\leq C\Omega_h \int\limits_{-\pi}^{\pi} \cdots \int\limits_{-\pi}^{\pi} \left(\left[\sum_{i=1}^{n} \left(\frac{\varphi_i}{h_i} \right)^2 \right]^{1/2} + 1 \right)^{2r} \times$$

$$\times h_1^{2m} \left(\left[\varphi_1^2 + \sum_{i=2}^{n} \left(\frac{\varphi_i h_1}{h_i} \right)^2 \right]^{1/2} + h_1 \right)^{-2m} \frac{|\varphi|^{2m}}{h_1^{2m}} |\widehat{u}_c|^2 \, d\varphi_1 \cdots d\varphi_n \leq$$

$$\leq C_1 \|u(x,h)\|_{M(r,h)}^2. \tag{6.1.48}$$

The last inequality implies the desired uniform boundedness. A similar result can be established for a mixed difference operator of order m.

(b) Iterated Laplace difference operator Δ_{hp}^m. The symbol of the iterated Laplace operator has the form (see Example 5.1.5)

$$\widehat{A}_c(\varphi) = (-1)^m 4^m \left[\sum_{k=1}^{n} \frac{1}{h_k^2} \sin^2 \frac{\varphi_k}{2} \right]^m. \tag{6.1.49}$$

By analogy with (6.1.48), using (6.1.49), we can show that the family of operators Δ_{hp}^m from $M(r, h)$ to $M(r - 2m, h)$ is uniformly bounded.

(c) Hypersingular discrete operators.

Definition 6.1.4. Operators $T_n(m, \lambda)$ of the form

$$T_h(m, \lambda)u(x, h) = T_{h,\lambda} \left(\Delta_{hp}^m u(x, h) \right), \tag{6.1.50}$$

where $u(x, h) \in M(r, h)$ and $T_{h,\lambda}$ is a discrete vortex operator, are called discrete hypersingular operators.

Proposition 6.1.4 and the result of Section 6.1.4 (b) imply that the family of discrete hypersingular operators $T_h(m, \lambda)$ from $M(r, h)$ to $M(r - 2m - \lambda, h)$ is uniformly bounded.

6.2. Approximation of Operators

Definition 6.2.1. We say that a *family of operators* T_h *from* $M(r, h)$ *to* $M(q, h)$ *is bounded* if $\|T_h\| \leq C$ with C independent of h.

Definition 6.2.2. A bounded family of operators T_h from $M(r, h)$, $r > 0$, to $M(q, h)$, $q < 0$, is *a weak approximation of a bounded operator* $T: H_r(\mathbb{R}_n) \to H_q(\mathbb{R}_n)$, if $T_h \Pi_h^* a(x) \to Ta(x)$ in the norm of $H_q(\mathbb{R}_n)$ as $|h| \to 0$, for any $a(x) \in C_0^\infty(\mathbb{R}_n)$ (see Definition 5.4.2).

Proposition 6.2.1. *Let* $T_h: M(r, h) \to M(q, h)$, $r > 0$, $q < 0$, *be a family of operators weakly approximating a bounded operator* $T: H_r(\mathbb{R}_n) \to H_q(\mathbb{R}_n)$. *Then, for any* $a(x) \in H_r(\mathbb{R}_n)$, *we have*

$$\lim_{|h| \to 0} \|Tu(x) - T_h \Pi_h u(x)\|_{H_q(\mathbb{R}_n)} = 0, \tag{6.2.1}$$

where Π_h *is an integral projector* (see Definition 5.4.1).

Proof. Theorem 5.4.1 implies that $\Pi_h u(x) \in M(r, h)$, and therefore, $T_h \Pi_h u(x) \in M(q, h)$. From Theorem 5.3.1, it follows that $T_h \Pi_h u(x) \in H_q(\mathbb{R}_n)$. Therefore, the left-hand side of (6.2.1) makes sense. For $\varphi(x) \in C_0^\infty(\mathbb{R}_n)$, using Proposition 5.4.9 and the inequality

$$\|T_h \Pi_h \varphi(x) - T\varphi(x)\|_{H_q(\mathbb{R}_n)} \le$$
$$\le \|T_h \Pi_h^* \varphi(x) - T\varphi(x)\|_{H_q(\mathbb{R}_n)} + \|T_h \left[\Pi_h \varphi(x) - \Pi_h^* \varphi(x)\right]\|_{H_q(\mathbb{R}_n)},$$

we obtain

$$\lim_{|h| \to 0} \|T_h \Pi_h \varphi(x) - T\varphi(x)\|_{H_q(\mathbb{R}_n)} = 0. \tag{6.2.2}$$

Let $u(x) \in H_r(\mathbb{R}_n)$. Since $C_0^\infty(\mathbb{R}_n)(\mathbb{R}_n)$ is dense in $H_r(\mathbb{R}_n)$, for any $\varepsilon > 0$ there is $\varphi_\varepsilon(x) \in C_0^\infty(\mathbb{R}_n)$ such that $\|u(x) - \varphi_\varepsilon(x)\| < \varepsilon$. From (6.2.2), it follows that there is $\delta = \delta(\varepsilon, \varphi)$ such that $\|T_h \Pi_h \varphi_\varepsilon(x) - T\varphi_\varepsilon(x)\|_{H_q(\mathbb{R}_n)} < \varepsilon$ for all h with $|h| < \delta$. Using Theorem 5.4.1 on the boundedness of integral projectors, we find that

$$\|T_h \Pi_h u(x) - Tu(x)\|_{H_q(\mathbb{R}_n)} =$$
$$= \|T_h \Pi_h (u(x) - \varphi_\varepsilon(x)) + T_h \Pi_h \varphi_\varepsilon(x) - T\varphi_\varepsilon(x) + T(\varphi_\varepsilon(x) - u(x))\|_{H_q(\mathbb{R}_n)} \le$$
$$\le C\|T_h\| \|\Pi_h(u(x) - \varphi_\varepsilon(x))\|_{M(r,h)} + \|T_h \Pi_h \varphi_\varepsilon(x) - T\varphi_\varepsilon(x)\|_{H_q(\mathbb{R}_n)} +$$
$$+ \|T\| \|\varphi_\varepsilon(x) - u(x)\|_{H_q(\mathbb{R}_n)} \le C_1 \varepsilon + \varepsilon + C_2 \varepsilon,$$

where the constants C_1 and C_2 do not depend on h. The last inequality implies Proposition 6.2.1.

Proposition 6.2.2. *Let* $T_h : M(r, h) \to M(q, h)$, $r > 0$, $q < 0$, *be a family of operators that weakly approximate a bounded operator* $T : H_r(\mathbb{R}_n) \to H_q(\mathbb{R}_n)$, *and let* Ω *be a normal domain that admits a partition. Then, for any* $u(x) \in \mathring{H}_r(\Omega)$, *we have*

$$\lim_{|h| \to 0} \|T_h \Pi_{1h} u(x) - Tu(x)\|_{H_q(\mathbb{R}_n)} = 0, \tag{6.2.3}$$

where the operator Π_{1h} *is the restriction of the integral projector* Π_h *to the domain* Ω_h.

Proof. Let $u(x) \in \mathring{H}(\Omega)$. Since $C_0^\infty(\Omega)$ is dense in $\mathring{H}_r(\Omega)$, for any $\varepsilon > 0$ there is $\varphi_\varepsilon(x) \in C_0^\infty(\Omega)$ such that $\|u(x) - \varphi_\varepsilon(x)\|_{\mathring{H}_r(\Omega)} < \varepsilon$. Since $\text{supp}\,\varphi_\varepsilon(x) \subset \Omega$, there exists $\delta > 0$ such that $\Omega_{1h} \supset \text{supp}\,\varphi_\varepsilon(x)$ for $|h| < \delta$. Therefore, $\Pi_h \varphi_\varepsilon(x) = \Pi_{1h} \varphi_\varepsilon(x)$ for $|h| < \delta$. By Proposition 6.2.1, there exists $\delta_1 = \delta_1(\varepsilon, \varphi_\varepsilon)$ such that $\|T_h \Pi_h \varphi_\varepsilon(x) - T\varphi_\varepsilon(x)\|_{H_q(\mathbb{R}_n)} < \varepsilon$ for $|h| < \delta_1$. Now, taking $\delta_2 = \min(\delta, \delta_1)$, we obtain the inequality $\|T_h \Pi_{1h} \varphi_\varepsilon(x) - T\varphi_\varepsilon(x)\| < \varepsilon$ for $|h| < \delta_2$. Since the domain Ω admits a partition, the operators Π_{1h} from $\mathring{H}_r(\Omega)$ to $\mathring{M}(r, h, \Omega_{1h})$ are bounded uniformly in $h = (h_1, \ldots, h_n)$. Using this fact, we obtain

$$\|T_h \Pi_{1h} u(x) - Tu(x)\|_{H_q(\mathbb{R}_n)} =$$
$$= \|T_h \Pi_{1h}(u(x) - \varphi_\varepsilon(x)) + T_h \Pi_{1h} \varphi_\varepsilon(x) - T\varphi_\varepsilon(x) + T(\varphi_\varepsilon(x) - u(x))\|_{H_q(\mathbb{R}_n)} \le$$
$$\le C\|T_h\| \|\Pi_{1h}\| \|u(x) - \varphi_\varepsilon(x)\|_{\mathring{H}_r(\Omega)} + \|T_h \Pi_{1h} \varphi_\varepsilon(x) - T\varphi_\varepsilon(x)\|_{H_q(\mathbb{R}_n)} +$$
$$+ \|T\| \|\varphi_\varepsilon(x) - u(x)\|_{\mathring{H}_r(\Omega)} \le C_1 \varepsilon + \varepsilon + C_2 \varepsilon \quad \text{for} \quad |h| < \delta_2,$$

where the constants C_1 and C_2 do not depend on ε. This inequality implies Proposition 6.2.2.

Definition 6.2.3. A bounded operator $T : H_r(\mathbb{R}_n) \to H_q(\mathbb{R}_n)$, $r > 0$, $q < 0$, is called an *operator of finite order*, if $T\varphi(x) \in L_2(\mathbb{R}_n)$ for any $\varphi(x) \in C_0^\infty(\mathbb{R}_n)$.

Definition 6.2.4. A bounded family of linear operators $T_h : M(r, h) \to M(q, h)$, $r > 0$, $q < 0$, *weakly approximates a bounded linear operator* $T : H_r(\mathbb{R}_n) \to H_q(\mathbb{R}_n)$ of finite order, if

$$\lim_{|h| \to 0} \|T_h \Pi_h \varphi(x) - T\varphi(x)\|_{L_2(\mathbb{R}_n)}, \qquad \forall\, \varphi(x) \in C_0^\infty(\mathbb{R}_n). \tag{6.2.4}$$

Definition 6.2.5. A bounded family of operators T_h : $M(r, h) \to M(q, h)$, $r > 0$, $q < 0$, *approximates a bounded finite order operator* T: $H_r(\mathbb{R}_n) \to H_q(\mathbb{R}_n)$ *in the domain* Ω, *if*

$$\lim_{|h| \to 0} \|P_h T_h \Pi_h \varphi(x) - P_h T \varphi(x)\|_{L_2(\Omega_{1h})} = 0, \qquad \forall \, \varphi(x) \in C_0^\infty(\mathbb{R}_n), \qquad (6.2.5)$$

where P_h is the operator of restriction to the domain Ω_{1h}.

Proposition 6.2.3. *Let* T: $H_r(\mathbb{R}_n) \to H_q(\mathbb{R}_n)$ *be a bounded operator of finite order, and let* T_h: $M(r, h) \to M(q, h)$, $r > 0$, $q < 0$, *be a bounded family of operators that approximate* T *in a bounded domain* Ω. *Then, for any* $\varphi(x) \in C_0^\infty(\Omega)$, *there exists an extension* l_h *of the function* $P_h T_h \Pi_{1h}(\varphi(x)$ *from the domain* Ω_{1h} *to the entire* $\mathbb{R}_n$, *and this extension satisfies the conditions:*

(i) $l_h P_h T_h \Pi_{1h} \varphi(x)$ *is a step function which is constant on the cells* $D(k, h)$,

$$a(y, h) = l_h P_h T_h \Pi_{1h} \varphi(x) = 0 \quad \text{for} \quad y \in D(k, h) \subset \Omega_{2h} \setminus \overline{\Omega}_{1h}; \qquad (6.2.6)$$

(ii) $\lim_{|h| \to 0} \|l_h P_h T_h \Pi_{1h} \varphi(x) - T\varphi(x)\|_{L_2(\mathbb{R}_n)} = 0$.

Proof. Consider a sequence $\{\varepsilon_n > 0\}_{n=1}^\infty$, $\varepsilon_n \searrow 0$, and an arbitrary function $\varphi_0(x) \in C_0^\infty(\Omega)$. Since $T\varphi(x) \in L_2(\mathbb{R}_n)$ for any $\varphi(x) \in C_0^\infty(\mathbb{R}_n)$ and $C_0^\infty(\mathbb{R}_n)$ is dense in $L_2(\mathbb{R}_n)$, there is a sequence $\{\varphi_n(y)\}_{n=1}^\infty$, $\varphi_n(y) \in C_0^\infty(\mathbb{R}_n)$, such that $\|T\varphi_0(x) - \varphi_n(y)\|_{L_2(\mathbb{R}_n)} < \varepsilon_n$. For the given sequence $\{\varepsilon_n\}_{n=1}^\infty$, we can find a sequence $\{\delta_n > 0\}_{n=1}^\infty$, $\delta_n \searrow 0$, such that

$$\|\Pi_h^* \varphi_n(y) - \varphi_n(y)\|_{L_2(\mathbb{R}_n)} < \varepsilon_n, \qquad (6.2.7)$$

for $|h| < \delta_n$ and any fixed $\varphi_n(y)$.

Take any h satisfying the inequality $\delta_{n+1} \le |h| < \delta_n$ and define the extension $l_h P_h T_h \Pi_{1h} \varphi_0(x)$ as follows:

$$l_h P_h T_h \Pi_{1h} \varphi_0(x) = \begin{cases} P_h T_h \Pi_{1h} \varphi_0(x), & y \in D(k, h) \subset \Omega_{1h}; \\ \Pi_h^* \varphi_n(x), & y \in D(k, h) \subset \mathbb{R}_n \setminus \overline{\Omega}_{2h}; \\ 0, & y \in D(k, h) \subset \Omega_{2h} \setminus \overline{\Omega}_{1h}. \end{cases}$$

Now, the property (i) claimed in Proposition 6.2.3 obviously follows from the definition of l_h. We further have

$$\|l_h P_h T_h \Pi_{1h} \varphi_0(x) - T\varphi_0(x)\|_{L_2(\mathbb{R}_n)}^2 = \|l_h P_h T_h \Pi_{1h} \varphi_0(x) - T\varphi_0(x)\|_{L_2(\Omega_{1h})}^2 +$$

$$+ \|l_h P_h T_h \Pi_{1h} \varphi_0(x) - T\varphi_0(x)\|_{L_2(\Omega_{2h} \setminus \overline{\Omega}_{1h})}^2 +$$

$$+ \|l_h P_h T_h \Pi_{1h} \varphi_0(x) - T\varphi_0(x)\|_{L_2(\mathbb{R}_n \setminus \overline{\Omega}_{2h})}^2 = I_1(h) + I_2(h) + I_3(h). \qquad (6.2.8)$$

Since $\Omega_{1h} \supset \overline{\operatorname{supp} \varphi_0(x)}$ for $|h| \to 0$, and $\varphi_0(x) \in C_0^\infty(\Omega)$, it follows from the condition of approximation that

$$\lim_{|h| \to 0} I_1(h) = 0. \qquad (6.2.9)$$

Since $\mu(\Omega_{2h} \setminus \overline{\Omega}_{1h}) \to 0$ as $|h| \to 0$, where $\mu(\Omega_{2h} \setminus \overline{\Omega}_{1h})$ is the measure of the set $\Omega_{2h} \setminus \overline{\Omega}_{1h}$, we have

$$\lim_{|h| \to 0} I_2(h) = 0. \qquad (6.2.10)$$

For $\delta_{n+1} \le |h| < \delta_n$, it follows from (6.2.7) that

$$I_3^{1/2}(h) \le \|l_h P_h T_h \Pi_{1h} \varphi_0(x) - \varphi_n(y)\|_{L_2(\mathbb{R}_n \setminus \overline{\Omega}_{2h})} + \|\varphi_n(y) - T\varphi_0(x)\|_{L_2(\mathbb{R}_n)} \le$$

$$\le \|\Pi_h^* \varphi_n(y) - \varphi_n(y)\|_{L_2(\mathbb{R}_n)} + \varepsilon_n < 2\varepsilon_n. \qquad (6.2.11)$$

Since $n \to \infty$, $\varepsilon_n \to \infty$ for $|h| \to 0$, it follows from (6.2.11) that

$$\lim_{|h| \to 0} I_3(h) = 0. \qquad (6.2.12)$$

Relations (6.2.9), (6.2.10), (6.2.12) imply (6.2.6). This completes the proof of Proposition 6.2.3.

Definition 6.2.6. The operator P_{1h} acting on step-functions $a(x, h)$ and defined by

$$P_{1h}a(x, h) = \begin{cases} a(x, h), & x \in D(k, h) \subset \Omega_{1h}, \\ 0, & x \in \Omega \setminus \overline{\Omega}_{1h} \end{cases}$$

is called *the operator of restriction to the bounded domain* Ω.

Proposition 6.2.4. *Let* $T : H_r(\mathbb{R}_n) \to H_q(\mathbb{R}_n)$ *be a bounded operator of finite order, and let* $T_h : M(r, h) \to M(q, h)$, $r > 0$, $q < 0$, *be a bounded family of operators approximating* T *in a normal bounded domain* Ω *that admits a partition. Then for any* $u(x) \in \mathring{H}_r(\Omega)$,

$$\lim_{|h| \to 0} \|P_{1h}T_h\Pi_{1h}u(x) - PTu(x)\|_{H_q(\Omega)} = 0, \tag{6.2.13}$$

where $PTu(x)$ *is the restriction of* $Tu(x)$ *to the domain* Ω, *i.e.,* $PTu(x)$ *is the restriction of the linear functional* $Tu(x)$ *defined on* $H_{-q}(\mathbb{R}_n)$ *to the subspace* $\mathring{H}_{-q}(\Omega)$.

Proof. Let $u(x) \in \mathring{H}_r(\Omega)$. For any $\varepsilon > 0$, there is a function $\varphi_\varepsilon(x) \in C_0^\infty(\Omega)$ such that

$$\|u(x) - \varphi_\varepsilon(x)\|_{\mathring{H}_r(\Omega)} < \varepsilon. \tag{6.2.14}$$

By Proposition 6.2.3, there exist $\delta = \delta(\varepsilon, \varphi_\varepsilon)$ and a function $l_hP_hT_h\Pi_{1h}\varphi_\varepsilon(x)$ such that

$$\|T\varphi_\varepsilon(x) - l_hP_hT_h\Pi_{1h}\varphi_\varepsilon(x)\|_{L_2(\mathbb{R}_n)} < \varepsilon, \tag{6.2.15}$$

for $|h| < \delta$. Further, for $|h| < \delta$, we have

$$\|P_{1h}T_h\Pi_{1h}u(x) - PTu(x)\|_{H_q(\Omega)} \le \|P_{1h}T_h\Pi_{1h}u(x) - P_{1h}T_h\Pi_{1h}\varphi_\varepsilon(x)\|_{H_q(\Omega)} +$$
$$+ \|P_{1h}T_h\Pi_{1h}\varphi_\varepsilon(x) - PT\varphi_\varepsilon(x)\|_{H_q(\Omega)} + \|PT\varphi_\varepsilon(x) - PTu(x)\|_{H_q(\Omega)} =$$
$$= I_1(h) + I_2(h) + I_3(h). \tag{6.2.16}$$

From the definition of the norm in $H_q(\Omega)$, it follows that

$$I_3(h) \le \|T\|\varepsilon. \tag{6.2.17}$$

Since the domain Ω admits a partition (see Definition 6.2.6), the operators Π_{1h} from $\mathring{H}_r(\Omega)$ to $\mathring{M}(r, h, \Omega_{1h})$ are uniformly bounded. Proposition 5.5.5 implies that the operators P_{1h} from $M(q, h, \Omega_{1h})$ to $H_q(\Omega)$ are uniformly bounded. And the operators T_h from $\mathring{M}(r, h, \Omega_{1h})$ to $M(q, h, \Omega_{1h})$ are uniformly bounded because of the assumptions of Proposition 6.2.4 and the definition of the norm in $M(q, h, \Omega_{1h})$ (see Definition 5.5.4). Therefore,

$$I_1(h) \le \|P_{1h}\| \|T_h\Pi_{1h}(u(x) - \varphi_\varepsilon(x))\|_{M(q,h,\Omega_{1h})} \le$$
$$\le \|P_{1h}\| \|T_h\| \|\Pi_{1h}\| \|u(x) - \varphi_\varepsilon(x)\|_{\mathring{H}_r(\Omega)} \le k\varepsilon, \tag{6.2.18}$$

where k does not depend on $h = (h_1, \dots, h_n)$.

Consider the term $I_2(h)$. As an extension of the function $P_{1h}T_h\Pi_{1h}$ we take $l_hP_{1h}T_h\Pi_{1h}\varphi(x)$, which is admissible because of the properties of the extension operator l_h established in Proposition 6.2.3. An an extension of $PT\varphi_\varepsilon(x)$ we take $T\varphi_\varepsilon(x)$. Then, using the definition of the norm in $H_q(\Omega)$ and (6.2.15), we obtain

$$I_2 \le \|T\varphi_\varepsilon(x) - l_hP_hT_h\Pi_{1h}\varphi_\varepsilon(x)\|_{H_q(\mathbb{R}_n)} \le$$
$$\le C\|T\varphi_\varepsilon(x) - l_hP_hT_h\Pi_{1h}\varphi_\varepsilon(x)\|_{L_2(\mathbb{R}_n)} \le C\varepsilon, \tag{6.2.19}$$

where C is independent of h.

Now, Proposition 6.2.4 follows from (6.2.16)–(6.2.19).

Proposition 6.2.5. *Let* T *be a bounded operator from* $H_r(\mathbb{R}_n)$ *to* $H_q(\mathbb{R}_n)$, *and let* $T_h :$ $M(r, h) \rightarrow M(q, h)$, $r > 0$, $q < 0$ *be a bounded family of operators approximating* T *in a normal bounded domain that admits a partition. Then, for any* $u(x) \in \mathring{H}_r(\Omega)$, *there is an extension* $l_h P_h T_h \Pi_{1h} u(x) \in M(q, h)$ *of the function* $P_h T_h \Pi_{1h} u(x) \in M(q, h, \Omega_{1h})$ *such that*

$$\lim_{|h| \to 0} \| l_h P_h T_h \Pi_{1h} u(x) - T u(x) \|_{H_q(\mathbb{R}_n)} = 0. \tag{6.2.20}$$

Proof. Let $u(x) \in \mathring{H}_r(\Omega)$. Consider a sequence $\{\varepsilon > 0\}_{n=1}^\infty$, $\varepsilon \searrow 0$. Since the operator T is bounded and $C_0^\infty(\Omega)$ is dense in $\mathring{H}_r(\Omega)$, there is a sequence $\{\varphi_n(x) > 0\}_{n=1}^\infty$, $\varphi_n(x) \in C_0^\infty(\Omega)$, such that $\| T u(x) - T \varphi_n(x) \|_{H_q(\mathbb{R}_n)} \leq \varepsilon_n$. Proposition 6.2.3 implies that there is a sequence $\delta_n = \delta_n(\varepsilon_n, \varphi_n)$, $\delta_n \searrow 0$, and there exist extensions $l_h P_h T_h \Pi_{1h} \varphi_n(x)$ such that for all h, $|h| < \delta_n$ and any $\varphi_n(x)$, the following inequality holds:

$$\| l_h P_h T_h \varphi_n(x) - T \varphi_n(x) \|_{H_q(\mathbb{R}_n)} < \varepsilon_n, \tag{6.2.21}$$

and we have $\Omega_{1h} \supset \operatorname{supp} \varphi_n(x)$.

Take any h such that $\delta_{n+1} \leq |h| < \delta_n$ and define the extension $l_h P_h T_h \Pi_{1h} u(x)$ by the relation

$$l_h P_h T_h \Pi_{1h} u(x) = T_h \Pi_{1h} u(x) - T_h \Pi_{1h} \varphi_n(x) + l_h P_h T_h \Pi_{1h} \varphi_n(x). \tag{6.2.22}$$

The structure of $l_h P_h T_h \Pi_{1h} \varphi_n(x)$ specified in Proposition 6.2.3 implies that

$$P_h l_h T_h \Pi_{1h} u(x) = P_h T_h \Pi_{1h} u(x),$$

and therefore, the extension (6.2.22) is well-defined.

Let $|h| < \delta_{n_0} = \delta_{n_0}(\varepsilon_{n_0}, \varphi_{n_0})$. By Theorem 5.3.1, for $\delta_{m+1} < |h| < \delta_m < \delta_{n_0}$, we obtain

$$\| l_h P_h T_h \Pi_{1h} u(x) - T u(x) \|_{H_q(\mathbb{R}_n)} =$$
$$= \| T_h \Pi_{1h} u(x) - T_h \Pi_{1h} \varphi_m(x) + l_h P_h T_h \Pi_{1h} \varphi_m(x) - T \varphi_m(x) + T \varphi_m(x) - T u(x) \|_{H_q(\mathbb{R}_n)} \leq$$
$$\leq C \| T_h \Pi_{1h}(u(x) - \varphi_m(x)) \|_{M(q,h)} + \| l_h P_h T_h \Pi_{1h} \varphi_m(x) - T \varphi_m(x) \|_{H_q(\mathbb{R}_n)} +$$
$$+ \| T(u(x) - \varphi_m(x)) \|_{H_q(\mathbb{R}_n)} < C_1 \varepsilon_m + 2\varepsilon_m = k \varepsilon_m, \tag{6.2.23}$$

where the constant k does not depend on h. The inequality (6.2.23) implies Proposition 6.2.5.

Definition 6.2.7. Let $T \colon H_r(\mathbb{R}_n) \rightarrow H_q(\mathbb{R}_n)$ be a bounded operator of finite order. A bounded family of operators $T_h : M(r, h) \rightarrow M(q, h)$, $r > 0$, $q < 0$, is said to be *a* Π_h^*-*approximation of the operator* T *in a bounded domain* Ω, if

$$\lim_{|h| \to 0} \| P_h T_h \Pi_h^* \varphi(x) - P_h T \varphi(x) \|_{L_2(\Omega_{1h})} = 0, \tag{6.2.24}$$

for any $\varphi \in C_0^\infty(\mathbb{R}_n)$, where Π_h^* is the operator from Definition 5.4.2.

Remark 6.2.1. Let the family of operators $T_h : M(r, h) \rightarrow M(q, h)$, $r > 0$, $q < 0$ be a Π_h^*-approximation of an operator $T \colon H_r(\mathbb{R}_n) \rightarrow H_q(\mathbb{R}_n)$ in the domain Ω. Suppose also that for some r_0, the family the operators T_h from $M(r_0, h)$ to $M(0, h)$ is bounded uniformly in h. Then, it follows from Proposition 5.4.9 that the operators T_h approximate T in Ω (see Definition 6.2.5).

Proposition 6.2.6. *Let* $T \colon H_r(\mathbb{R}_n) \rightarrow H_q(\mathbb{R}_n)$ *be a bounded operator of finite order, and let* $T_h : M(r, h) \rightarrow M(q, h)$, $r > 0$, $q < 0$ *be its* Π_h^*-*approximation in a bounded domain* Ω. *Then, for any* $\varphi(x) C_0^\infty(\mathbb{R}_n)$, *there exists an extension* l_h *of the function* $P_h T_h \Pi_{1h}^* \varphi(x)$ *which satisfies the conditions:*

(i) $l_h P_h T_h \Pi_{1h}^* \varphi(x)$ *is a step-function which is constant on the cells* $D(k, h)$, *and* $a(y, h) = l_h P_h T_h \Pi_{1h}^* \varphi(x) = 0$ *for* $y \in D(k, h) \subset \Omega_{2h} \setminus \overline{\Omega}_{1h}$;

(ii) $\lim\limits_{|h| \to 0} \| l_h P_h T_h \Pi_{1h}^* \varphi(x) - T \varphi(x) \|_{L_2(\mathbb{R}_n)} = 0$, *where* Π_{1h}^* *is the restriction of the operator* Π_h^* *to the domain* Ω_{1h}.

This result is proved similarly to Proposition 6.2.3

Proposition 6.2.7. *Let* $T \colon H_r(\mathbb{R}_n) \to H_q(\mathbb{R}_n)$ *be a bounded operator, and let* $T_h \colon M(r,h) \to M(q,h)$, $r > 0$, $q < 0$ *be a bounded family of operators yielding its* Π_h^**-approximation in a normal bounded domain* Ω *admitting a partition. Then*

$$\lim_{|h| \to 0} \|P_{1h} T_h \Pi_{1h} u(x) - PTu(x)\|_{H_q(\Omega)} = 0, \tag{6.2.25}$$

for any $u(x) \in \mathring{H}_r(\Omega)$.

Proof. Let $u(x) \in \mathring{H}_r(\Omega)$. For any $\varepsilon > 0$, there is a function $\varphi_\varepsilon(x) \in C_0^\infty(\Omega)$ such that

$$\|u(x) - \varphi_\varepsilon(x)\|_{\mathring{H}_r(\Omega)} < \varepsilon. \tag{6.2.26}$$

Proposition 6.2.6 ensures the existence of $\delta = \delta(\varepsilon, \varphi_\varepsilon)$ and a function $l_h P_h T_h \Pi_{1h}^* \varphi_\varepsilon(x)$ such that

$$\|T\varphi_\varepsilon(x) - l_h P_h T_h \Pi_{1h}^* \varphi_\varepsilon(x)\|_{L_2(\mathbb{R}_n)} < \varepsilon,$$

for all h such that $|h| < \delta$.

Proposition 5.4.9 implies that there is $\delta_1 = \delta_1(\varepsilon, \varphi_\varepsilon(x))$ such that for all h, $|h| < \delta_1$, we have $\Omega_{1h} \supset \operatorname{supp} \varphi_\varepsilon(x)$ and

$$\|\Pi_{1h}^* \varphi_\varepsilon(x) - \Pi_{1h} \varphi_\varepsilon(x)\|_{M(r,h)} < \varepsilon. \tag{6.2.27}$$

Further, for all h such that $|h| < \delta_2 = \min\{\delta, \delta_1\}$, we get

$$\|P_{1h} T_h \Pi_{1h} u(x) - PTu(x)\|_{H_q(\Omega)} \leq \|P_{1h} T_h \Pi_{1h} u(x) - P_{1h} T_h \Pi_{1h} \varphi_\varepsilon(x)\|_{H_q(\Omega)} +$$
$$+ \|P_{1h} T_h \Pi_{1h} \varphi_\varepsilon(x) - P_{1h} T_h \Pi_{1h}^* \varphi_\varepsilon(x)\|_{H_q(\Omega)} + \|P_{1h} T_h \Pi_{1h}^* \varphi_\varepsilon(x) - PT\varphi_\varepsilon(x)\|_{H_q(\Omega)} +$$
$$+ \|PT\varphi_\varepsilon(x) - PTu(x)\|_{H_q(\Omega)} = \sum_{k=1}^{4} I_k(h). \tag{6.2.28}$$

Similarly to Proposition 6.2.4, it can be shown that

$$I_i(h) < k\varepsilon, \qquad i = 1, 3, 4, \tag{6.2.29}$$

where k does not depend on $h = (h_1, \ldots, h_n)$.

In order to estimate $I_2(u)$, we utilize Proposition 6.2.5 and the inequality (6.2.7). We have

$$I_2(h) = C \|P_h T_h \left(\Pi_{1h} \varphi_\varepsilon(x) - \Pi_{1h}^* \varphi_\varepsilon(x) \right)\|_{M(q,h,\Omega_{1h})} \leq$$
$$\leq C_1 \|\Pi_{1h} \varphi_\varepsilon(x) - \Pi_{1h}^* \varphi_\varepsilon(x)\|_{M(r,h)} = C_1 \varepsilon. \tag{6.2.30}$$

The inequalities (6.2.29) and (6.2.30) imply Proposition 6.2.7.

Proposition 6.2.8. *Let* $T \colon H_r(\mathbb{R}_n) \to H_q(\mathbb{R}_n)$ *be bounded operator of finite order, and let* $T_h \colon M(r,h) \to M(q,h)$, $r > 0$, $q < 0$, *be a bounded family of operators yielding its* Π_h^**-approximation in a normal bounded domain* Ω *that admits a partition. Then, for any* $u(x) \in \mathring{H}_r(\Omega)$, *there exists an extension* $l_h P_h T_h \Pi_{1h} u(x) \in M(q,h)$ *of the function* $P_h T_h \Pi_{1h} u(x) \in M(q,h,\Omega_{1h})$ *such that*

$$\lim_{|h| \to 0} \|l_h P_h T_h u(x) - Tu(x)\|_{H_q(\mathbb{R}_n)} = 0. \tag{6.2.31}$$

Proof. Let $u(x) \in \mathring{H}_r(\Omega)$. Consider a sequence $\{\varepsilon_n > 0\}_{n=1}^{\infty}$, $\varepsilon_n \searrow 0$. Then there is a sequence $\{\varphi_n(x)\}_{n=1}^{\infty}$, $\varphi_n(x) \in C_0^\infty(\Omega)$, such that

$$\|u(x) - \varphi_n(x)\|_{\mathring{H}_r(\Omega)} < \frac{\varepsilon_n}{\|T\|}. \tag{6.2.32}$$

Proposition 6.2.6 implies that there is a sequence $\delta_n = \delta_n(\varepsilon_n, \varphi_n) > 0$, $\delta_n \searrow 0$, and there are extensions $l_h P_h T_h \Pi_{1h}^* \varphi_n(x)$ such that $\Omega_{1h} \subset \operatorname{supp} \varphi_n(x)$ and

$$\left\| l_h P_h T_h \Pi_{1h}^* \varphi_n(x) - T \varphi_n(x) \right\|_{H_q(\mathbb{R}_n)} < \varepsilon_n . \tag{6.2.33}$$

For h such that $\delta_{n+1} \le |h| < \delta_n$, we define the extension $l_h P_h T_h \Pi_{1h} u(x)$ by

$$l_h P_h T_h \Pi_{1h} u(x) = T_h \Pi_{ih} u(x) - T_h \Pi_{1h}^* \varphi_n(x) + l_h P_h T_h \Pi_{1h}^* \varphi_n(x). \tag{6.2.34}$$

The structure of $l_h P_h T_h \Pi_{1h}^*$ is the same as that of the extension in Proposition 6.2.3. Thus, $P_h l_h P_h T_h \Pi_{1h} u(x) = P_h T_h \Pi_{1h} u(x)$, and therefore, the extension given by (6.2.34) is well-defined. Let $|h| < \delta_{n_0}$. Then, applying Theorem 5.3.1, we obtain for $\delta_{m+1} \le |h| \le \delta_m \le \delta_{n_0}$,

$$\left\| l_h P_h T_h \Pi_{1h} u(x) - T u(x) \right\|_{H_q(\mathbb{R}_n)} =$$
$$= \left\| T_h \Pi_{1h} u(x) - T_h \Pi_{1h}^* \varphi_m(x) + l_h P_h T_h \Pi_{1h}^* \varphi_m(x) - T \varphi_m(x) + T \varphi_m(x) - T u(x) \right\|_{H_q(\mathbb{R}_n)} \le$$
$$\le C \left\| T_h \Pi_{1h}(u(x) - \varphi_m(x)) \right\|_{M(q,h)} + C \left\| T_h \left(\Pi_{1h} \varphi_m(x) - \Pi_{1h}^* \varphi_m(x) \right) \right\|_{M(q,h)} +$$
$$+ \left\| l_h P_h T_h \Pi_{1h}^* \varphi_m(x) - T \varphi_m(x) \right\|_{H_q(\mathbb{R}_n)} + \|T\| \, \|\varphi_m(x) - u(x)\|_{\mathring{H}_r(\mathbb{R}_n)} < k \varepsilon_{n_0}, \tag{6.2.35}$$

where k does not depend on h. The inequality (6.2.35) implies Proposition 6.2.8.

Definition 6.2.8. Let $T : \mathring{H}_r(\Omega) \to H_q(\mathbb{R}_n)$ be a bounded operator of finite order. A bounded family of operators $T_h : \mathring{M}(r, h, \Omega_{1h}) \to M(q, h)$, $r > 0$, $q < 0$, is said to be *a Π_h^*-approximation of the operator T in a bounded domain Ω*, if

$$\lim_{|h| \to 0} \left\| P_h T_h \Pi_h \varphi(x) - P_h T \varphi(x) \right\|_{L_2(\Omega_{1h})} = 0, \tag{6.2.36}$$

for any $\varphi(x) \in C_0^\infty(\Omega)$.

Remark 6.2.2. Let $T : \mathring{H}_r(\Omega) \to H_q(\mathbb{R}_n)$ be a bounded finite order operator and let $T_h : \mathring{M}(r, h, \Omega_{1h}) \to M(q, h)$, $r > 0$, $q < 0$ be a bounded family of operators Π_h^*-approximating T in a bounded normal domain Ω admitting a partition. Then, Propositions 6.2.3–6.2.8 remain valid, which can be seen from their proof.

6.3. Quadrature Formulas in Sobolev–Slobodetskii Spaces

6.3.1. Quadrature Formulas for Integrals

$$u(x_0) = \int_{\mathbb{R}_n} \frac{\varphi(x)}{r^{n+\lambda}} \, dx , \qquad 0 < \lambda < 2, \tag{6.3.1}$$

where $r = |x^0 - x|$ and the integral is understood in the sense of Hadamard.

Consider a canonical partition of $\mathbb{R}_n$ with a vector-valued step $h = (h_1, \ldots, h_n)$ and cells $D(k, h)$ (see Definition 5.1.1). In each cell $D(k, h)$, we take a point $M_{0k}(h_1 k_1, \ldots, h_n k_n)$. Let $r_{0k} = |M M_{0k}|$, where M is the point with the coordinates $x = (x_1, \ldots, x_n)$. In what follows, it is assumed that $\varphi(x) \in C_0^\infty(\mathbb{R}_n)$. In order to calculate the integral (6.3.1) at the point M_{0m}, we use the quadrature formula

$$S(M_{0m}) = \sum_k \varphi(M_k) \int_{D(k,h)} \frac{dx}{r_{0m}^{n+\lambda}} . \tag{6.3.2}$$

where the sum is taken over all $D(k, h)$.

Let us estimate the quadrature error $\Delta u(M_{0m}) = u(M_{0m}) - S(M_{0m})$. The sum (6.3.2) contains only finitely many terms, since $\varphi(x) \in C_0^\infty(\mathbb{R}_n)$. Denote by $\varepsilon(\varphi)$ the set of all indices $k = (k_1, \ldots, k_n)$ corresponding to the cells $D(k, h)$ having a non-empty intersection with the support of $\varphi(x)$. Then

$$\Delta u(M_{0m}) = \sum_{k \in \varepsilon(\varphi)} \int_{D(k,h)} \frac{\varphi(x) - \varphi(M_{0k})}{r_{0m}^{n+\lambda}} \, dx. \tag{6.3.3}$$

Case 1. Let $M_{0m} \in \operatorname{supp} \varphi(x)$ and consider the expression

$$L = \sum_{i=1}^n \varphi'_{x_i}(M_{0m})\left(x_i - x_i^{0k}\right) = \sum_{i=1}^n \varphi'_{x_i}\left(x_i - x_i^{0m}\right) + \sum_{i=1}^n \varphi'_{x_i}\left(x_i^{0m} - x_i^{0k}\right) = L_1 + L_2,$$

where x_i^{0m} and x_i^{0k}, $i = 1, \ldots, n$, are the coordinates of the points M_{0m} and M_{0k}, respectively. Consider a ball $K_R(M_{0m})$ with center at M_{0m} and a sufficiently large radius R, such that the cells $D(k, h)$ completely lying in K_R cover the support of $\varphi(x)$. The set of the indices k corresponding to such cells we denote by ε_1. Let us prove the relation

$$A = \sum_{k \in \varepsilon_1} \int_{D(k,h)} \frac{L(s)\, dx}{r_{0m}^{n+\lambda}} = \sum_{k \in \varepsilon_1} \int_{D(k,h)} \frac{L_1(x)\, dx}{r_{0m}^{n+\lambda}} + \sum_{k \in \varepsilon_1} \int_{D(k,h)} \frac{L_2(x)\, dx}{r_{0m}^{n+\lambda}} = A_1 + A_2 = 0. \tag{6.3.4}$$

Indeed, introducing new variables $y_j = x_j - x_j^{0m}$, $j = 1, \ldots, n$, we see that $A_1 = 0$, since A_1 is an integral of an odd function over a symmetric domain (the integral in the sense of Hadamard has been defined in (5.2.9)). Since the cells $D(k, h)$ are symmetric with respect to the new coordinate axes and the origin, for each $D(k, h)$ there is $D(\tilde{k}, h)$ such that $\left(x_j^{0m} - x_j^{0k}\right) = -\left(x_j^{0m} - x_j^{0\tilde{k}}\right)$. Therefore, $A_2 = 0$.

Representing $\varphi(x)$ by the Taylor formula at the point M_{0k}, we obtain

$$\varphi(x) - \varphi(M_{0k}) - L(x) = \sum_{i=1}^n \left(x_i - x_i^{0k}\right)\left(\varphi'_{x_i}(M_{0k}) - \varphi'_{x_i}(M_{0m})\right) + O(x, M_{0k}) =$$
$$= B(x) + O(x, M_{0k}), \tag{6.3.5}$$

where $O(x, M_{0k})$ is the residual of the Taylor formula. It follows that $O(M_{0k}, M_{0k}) = 0$ for $x \in D(k, h)$, $|O(x, M_{0k})| \leq C|h|^2$.

In view of (6.3.4) and (6.3.5), relation (6.3.3) can be written in the form

$$\Delta u(M_{0m}) = \sum_{k \in \varepsilon_1} \int_{D(k,h)} \frac{B(x)}{r_{0m}^{n+\lambda}} \, dx + \sum_{k \in \varepsilon_1} \int_{D(k,h)} \frac{O(x, M_{0k})}{r_{0m}^{n+\lambda}} \, dx = I_1 + I_2. \tag{6.3.6}$$

Let us estimate the two terms in (6.3.6). First, consider I_2. For $k = m$, we have

$$|O(x, M_{0m})| \leq C r_{0m}^2, \qquad O(M_{0m}, M_{0m}) = 0.$$

Then, by (5.2.9), we obtain

$$\left| \int_{D(m,h)} \frac{O(x, M_{0m})}{r_{0m}^{n+\lambda}} \, dx \right| = \lim_{\varepsilon \to 0} \left| \int_{D(m,h) \setminus D(\varepsilon)} \frac{O(x, M_{0m})}{r_{0m}^{n+\lambda}} \, dx - \frac{S_n O(M_{0m}, M_{0m})}{\lambda \varepsilon^\lambda} \right| \leq$$

$$\leq C \lim_{\varepsilon \to 0} \int_\varepsilon^{|h|\sqrt{n}} \frac{r_{0m}^2}{r_{0m}^{n+\lambda}} \, r_{0m}^{n-1} \, dr_{0m} \leq C_1(\lambda)|h|^{2-\lambda}, \tag{6.3.7}$$

where $D(\varepsilon)$ is the ball of radius $\varepsilon > 0$ with center at M_{0m}; S_n is the area of the unit sphere.

We further have

$$\left| \sum_{k \in \varepsilon_1, k \neq m} \int_{D(k,h)} \frac{O(x, M_{0k})}{r_{0m}^{n+\lambda}} \, dx \right| \le C|h|^2 \int_{|h|}^{\infty} \frac{r_{0m}^{n-1}}{r_{0m}^{n+\lambda}} \, dr_{0m} \le C(\lambda)|h|^{2-\lambda} \tag{6.3.8}$$

From (6.3.7), (6.3.8), it follows that

$$|I_2(h)| \le C_2(\lambda)|h|^{2-\lambda}, \tag{6.3.9}$$

where $C_2(\lambda) \to \infty$ as $\lambda \to 2$ or $\lambda \to 0$.

Consider the term I_1. Let $k = m$. Then,

$$B(M_{0m}) = 0, \qquad |B(x)| \le Cr_{0m}^2 \quad \text{for} \quad x \in D(k, h).$$

Hence, using (5.2.9), we obtain

$$\left| \int_{D(m,h)} \frac{B(x)}{r_{0m}^{n+\lambda}} \, dx \right| = \left| \lim_{\varepsilon \to 0} \int_{D(m,h) \setminus D(\varepsilon)} \frac{B(x)}{r_{0m}^{n+\lambda}} \, dx - \frac{S_n B(M_{0m})}{\lambda \varepsilon^\lambda} \right| \le$$

$$\le C(\lambda) \lim_{\varepsilon \to 0} \int_{\varepsilon}^{|h|\sqrt{n}} r_{0m}^{1-\lambda} \, dr_{0m} \le C_1(\lambda)|h|^{2-\lambda}. \tag{6.3.10}$$

Next, we have

$$I_3(h) = \left| \sum_{k \in \varepsilon_1, \, k \neq m} \int_{D(k,h)} \frac{B(x)}{r_{0m}^{n+\lambda}} \, dx \right| \le C(\lambda)|h| \sum_{k \in \varepsilon_1, \, k \neq m} \int_{D(k,h)} \frac{dx}{r_{0m}^{n+\lambda-1}} \le$$

$$\le C_1(\lambda)|h| \int_{|h|}^{R_{0m}} r_{0m}^{-\lambda} \, dr_{0m} = \begin{cases} C_2(\lambda)|h| \left| \ln(|h|^{-1} R_{0m}) \right| & \text{for} \quad \lambda = 1, \\ C_2(\lambda)|h| \left| R_{0m}^{1-\lambda} - |h|^{1-\lambda} \right| & \text{for} \quad \lambda \neq 1. \end{cases} \tag{6.3.11}$$

For a fixed function $\varphi(x)$, we have $\max R_{0m} < D_0$, where D_0 is a constant depending on $\text{supp}\,\varphi(x)$. Therefore, from (6.3.11) we obtain the inequality

$$I_3(h) \le \begin{cases} C_3(\lambda)|h| & \text{for} \quad 0 < \lambda < 1, \\ C_3(\lambda)|h| \ln|h| & \text{for} \quad \lambda = 1, \\ C_3(\lambda)|h|^{2-\lambda} & \text{for} \quad 1 < \lambda < 2. \end{cases} \tag{6.3.12}$$

From (6.3.9), (6.3.10), and (6.3.11), it follows that for $|\Delta u(M_{0m})|$ the estimate (6.3.12) holds.

Case 2. Fix $\delta > 0$. Obviously, for all points M_{0m} whose distance from $\text{supp}\,\varphi(x)$ does not exceed δ (i.e., $\rho_{0m} \le \delta$)), the estimate (6.3.12) takes place.

Case 3. Consider the points M_{0m} whose distance from $\text{supp}\,\varphi(x)$ is greater than δ (i.e., $\rho_{0m} > \delta$). In this case, we have the following estimate for the quadrature error:

$$|\Delta u(M_{0m})| \le \sum_{k \in \varepsilon(\varphi)} \int_{D(k,h)} \frac{|\varphi(x) - \varphi(M_{0k})|}{r_{0m}^{n+\lambda}} \, dx \le C(\lambda) \frac{|h|}{r_{0m}^{n+\lambda}}. \tag{6.3.13}$$

Let $u(x, h)$ be the step-function defined by the relations $u(x, h) = |\Delta u(M_{0m})|$ for $x \in D(m, h)$. Let us estimate $\Delta = \int_{\mathbb{R}_n} |u(x, h)|^2 \, dx$.

We have

$$\Delta = \int_{J(h)} |h(x, h)|^2 \, dx + \int_{\mathbb{R}_n \setminus J(h)} |u(x, h)|^2 \, dx = \Delta_1 + \Delta_2, \tag{6.3.14}$$

where $J(h)$ is the domain formed by the cells $D(m, h)$ whose centers M_{0m} satisfy the condition: either $M_{0m} \in \operatorname{supp} \varphi(x)$ or the distance from M_{0m} to $\operatorname{supp} \varphi(x)$ does not exceed δ. This definition of $J(h)$, together with (6.3.12), implies that

$$\Delta_1 \leq C(\lambda) O(|h|), \tag{6.3.15}$$

where $O(|h|) \to 0$ as $|h| \to 0$.

When estimating the term Δ_2, we can assume, without loss of generality, that the origin O belongs to $\operatorname{supp} \varphi(x)$. Let r be the current radius-vector of the point $M \in D(m, h) \subset \mathbb{R}_n \setminus \overline{J(h)}$. Since $\rho_{0m \geq \delta}$, we have $1/\rho_{0m} \leq C/r$, where C is a constant that does not depend on $D(m, h)$. Hence, we obtain

$$\Delta_2 \leq C|h|^2 \int_\delta^\infty \frac{dr}{r^{n+2\lambda+1}} \leq C_1 |h|^2. \tag{6.3.16}$$

It follows from (6.3.14)–(6.3.16) that

$$\Delta \leq C_1(\lambda) O_1(|h|), \tag{6.3.17}$$

where $O_1(|h|) \to 0$ as $|h| \to 0$.

Proposition 6.3.1. *Let $\varphi(x) \in C_0^\infty(\mathbb{R}_n)$. Then the functions $S(x, h)$, $\pi S(x, h) = S(M_{0m})$, converge to $u(x_0) = \int_{\mathbb{R}_n} \varphi(x) r^{-n-\lambda} \, dx$ in the norm of $L_2(\mathbb{R}_n)$ as $|h| \to 0$.*

Proof. Since $\varphi(x) \in C_0^\infty(\mathbb{R}_n)$, we have $u(x) \in H_r(\mathbb{R}_n)$ for any r. Therefore, $u(x) \in C^\infty(\mathbb{R}_n)$, $u(x) \in L_2(\mathbb{R}_n)$. Consider a family of functions $A(x, h)$ such that $\pi A(x, h) = u(M_{0m})$ (the operator π is that of Definition 5.1.2). Let us show that

$$\lim_{|h| \to 0} \|u(x) - A(x, h)\|_{L_2(\mathbb{R}_n)} = 0. \tag{6.3.18}$$

For an arbitrary $\varepsilon > 0$, let $K(R)$ be a ball of a sufficiently large radius R such that

$$\int_{\mathbb{R}_n \setminus K(R)} |u(x)|^2 \, dx < \varepsilon. \tag{6.3.19}$$

Consider the set $M_h = \cup_m D(m, h)$, where each cell $D(m, h)$ has a nonempty intersection with the ball $K(R)$. Since $u(x) \in C^\infty(\mathbb{R}_n)$, there exists $\delta(\varepsilon) > 0$ such that for $|h| < \delta(\varepsilon)$, we have

$$\mu(M_h) < 2\mu(K(R)), \qquad |A(x, h) - u(x)|^2 < \frac{\varepsilon}{2\mu(K(R))} \quad \text{for} \quad x \in D(m, h) \subset M_h,$$

where $\mu(M_h)$ and $\mu(K(R))$ are the volumes of M_h and $K(R)$. For $|h| < \delta(\varepsilon)$, we have

$$\int_{M_h} |u(x) - A(x, h)|^2 \, dx < \varepsilon. \tag{6.3.20}$$

Without loss of generality, we may assume that the center of the ball $K(R)$ is at the origin and its radius R has been chosen so large that the boundary of $K(R)$ is separated from $\operatorname{supp} \varphi(x)$ by a positive distance δ_0. This allows us to claim that

$$|A(M_{0m}, h| = |u(M_{0m})| \leq \int_{\mathbb{R}_n} \frac{|\varphi(x)|}{r_{0m}^{n+\lambda}} \, dx \leq \frac{C}{R_{0m}^{n+\lambda}} \quad \text{for} \quad M_{0m} \in \mathbb{R}_n \setminus K(R), \tag{6.3.21}$$

where R_{0m} is the distance from M_{0m} to the origin. This inequality implies the estimate

$$\int_{\mathbb{R}_n \setminus K(R)} |A(x, h)|^2 \, dx \leq C_1 \int_R^\infty \frac{dr}{r^{n+2\lambda+1}} = O(R), \qquad (6.3.22)$$

where $O(R) \to 0$ as $R \to \infty$.

Note that the parameters h and R can be chosen independently, and thus, the estimates (6.3.19), (6.3.20), and (6.3.22) imply (6.3.18).

Now, since

$$\|u(x) - S(x, h)\|_{L_2(\mathbb{R}_n)} \leq \|u(x) - A(x, h)\|_{L_2(\mathbb{R}_n)} + \|A(x, h) - S(x, h)\|_{L_2(\mathbb{R}_n)},$$

Proposition 6.3.1 follows from (6.3.17) and (6.3.18).

6.3.2. Quadrature Formulas for Singular Integrals

Consider the singular integral

$$u(x_0) = \int_{\mathbb{R}_n} \frac{f(\vec{\theta})\varphi(x)}{r^n} \, dx, \qquad (6.3.23)$$

where $r = |x^0 - x|$, $f(\vec{\theta})$ is the characteristic function of the singular integral, $\vec{\theta} = r^{-1} r$, $f(\vec{\theta})$ satisfies the Lipschitz condition and

$$\int_{S_n} f(\vec{\theta}) \, dS_n = 0,$$

where S_n is the unit sphere.

Consider a partition of $\mathbb{R}_n$ into cells $D(k, h)$, the same as in Section 6.3.1. In what follows, we assume that $\varphi(x) \in C_0^\infty(\mathbb{R}_n)$. For the approximation of the integral (6.3.23) at the points $M_{0m} = (m_1 h_1, \ldots, m_n h_n)$, we use the formula

$$S(M_{0m}) = \sum_k \varphi(M_{0k}) \int_{D(k,h)} \frac{f(\vec{\theta})}{r_{0m}^n} \, dx. \qquad (6.3.24)$$

Let us estimate the quadrature error $\Delta u(M_{0m}) = u(M_{0m}) - S(M_{0m})$. We have

$$\Delta u(M_{0m}) = \sum_{k \in \varepsilon(\varphi), \, k \neq m} \int_{D(k,h)} \frac{(\varphi(x) - \varphi(M_{0k}))f(\vec{\theta})}{r_{0m}^n} \, dx +$$

$$+ \int_{D(m,h)} \frac{(\varphi(x) - \varphi(M_{0m}))f(\vec{\theta})}{r_{0m}^n} \, dx = I_1(h) + I_2(h), \qquad (6.3.25)$$

where $r_{0m} = |x^{0m} - x|$, x^{0m} are the coordinates of the point M_{0m}; $\varepsilon(\varphi)$ is the set of indices k corresponding to all $D(k, h)$ having a nonempty intersection with $\operatorname{supp} \varphi(x)$.

Let us examine the case, when either $M_{0m} \in \operatorname{supp} \varphi(x)$ or the distance from M_{0m} to $\operatorname{supp} \varphi(x)$ does not exceed a fixed δ. Assume that $h = (h_1, \ldots, h_1)$, i.e., the partition step is the same for each axis, and $|h| = \sqrt{\sum_{i=1}^n h_1^2} = h_1 \sqrt{n} < \delta_0$ for some constant $\delta_0 > 0$. Let $K_R(M_{0m})$ be a ball with center at M_{0m} and a sufficiently large radius R, such that the cells $D(k, h)$ lying in $K_R(M_{0m})$ cover $\operatorname{supp} \varphi(x)$. The radius R is chosen the same for all M_{0m}, which is possible, since $\varphi(x) \in C_0^\infty(\mathbb{R}_n)$. We further have

$$|I_1(h)| \leq \sum_{k \in \varepsilon(\varphi), \, k \neq m} \int_{D(k,h)} \frac{|\varphi(x) - \varphi(M_{0k})| |f(\vec{\theta})|}{r_{0m}^n} \, dx \leq C_1 |h| \int_{|h|}^R \frac{dr}{r} \leq C_1 |h| \, |\ln |h||. \quad (6.3.26)$$

Let us estimate the term $I_2(h)$, using the definition of a singular integral. We have

$$|I_2(h)| \leq \left| \lim_{\varepsilon \to 0} \int_{D(m,h) \setminus D(\varepsilon)} \frac{f(\vec{\theta})(\varphi(x) - \varphi(M_{0m}))}{r^n} \, dx \right| \leq C \lim_{\varepsilon \to 0} \int_{\varepsilon}^{|h|\sqrt{n}} dr \leq C_1 |h|, \qquad (6.3.27)$$

where $D(\varepsilon)$ is the ball of radius ε with center at M_{0m}.

From (6.3.26) and (6.3.27), it follows that

$$|\Delta u(M_{0m})| \leq C|h| \big| \ln |h| \big|. \qquad (6.3.28)$$

Now, consider the case of the points M_{0m} separated from $\operatorname{supp} \varphi(x)$ by the distance $\rho_{0m} > \delta$. At these points, we obtain the estimate

$$|\Delta u(M_{0m})| \leq \sum_{k \in \varepsilon(\varphi)} \int_{D(k,h)} \frac{\left| f(\vec{\theta}) \right| |\varphi(x) - \varphi(M_{0k})|}{r_{0m}^n} \, dx \leq C \frac{|h|}{\rho_{0m}^n}. \qquad (6.3.29)$$

Denote by $u(x, h)$ the step function defined by the relations $u(x, h) = |\Delta u(M_{0m})|$ for $x \in D(m, h)$. By analogy with (6.3.17), we obtain

$$\Delta = \int_{\mathbb{R}_n} |u(x, h)|^2 \, dx = O(|h|),$$

where $O(|h|) \to 0$ as $|h| \to 0$.

Proposition 6.3.2. *For* $\varphi(x) \in C_0^\infty(\mathbb{R}_n)$, *the functions* $S(x, h)$, $\pi S(x, h) = S(M_{0m})$ *converge to* $u(x_0) = \int_{\mathbb{R}_n} f(\vec{\theta}) \varphi(x) r^{-n} \, dx$ *in the norm of* $L_2(\mathbb{R}_n)$ *as* $|h| \to 0$.

This result is proved in the same way as Proposition 6.3.1.

6.3.3. Quadrature Formulas for Hypersingular Integrals

$$u(x_0) = \int_{\mathbb{R}_n} \frac{\varphi(x)}{r^{n+2p+\lambda}} \, dx, \qquad (6.3.30)$$

where $0 < \lambda < 2$, $p > 0$, $p \in \mathbb{N}$ is a positive integer.

The Fourier transforms of the generalized functions $f_1(x) = r^{-n-2p-\lambda}$ and $f_2(x) = r^{-n-\lambda}$ have the form (see *Brychkov and Prudnikov (1977)*)

$$F(r^{-n-2p-\lambda}) = \frac{\Gamma\left(-p - \frac{\lambda}{2}\right)}{\Gamma\left(\frac{n+2p+\lambda}{2}\right)} 2^{-2p-\lambda} \pi^{n/2} |\sigma|^{2p+\lambda},$$

$$F(r^{-n-\lambda}) = \frac{\Gamma\left(-\frac{\lambda}{2}\right)}{\Gamma\left(\frac{n+\lambda}{2}\right)} 2^{-\lambda} \pi^{n/2} |\sigma|^\lambda, \qquad (6.3.31)$$

where $\Gamma(t)$ is the gamma function. On account of the above relations, the symbol (6.3.30) will be understood in the following sense:

$$u(x_0) = C(\lambda, p) \int_{\mathbb{R}_n} \frac{\Delta^p \varphi(x)}{r^{n+\lambda}} \, dx, \qquad (6.3.32)$$

where $C(\lambda, p)$ is the constant defined by (6.3.31); Δ is the Laplace operator, and the integral in (6.3.32) is understood in the sense of Hadamard.

In what follows, we use the notation from Sections 6.3.1 and 6.3.2 and assume that $\varphi(x) \in C_0^\infty(\mathbb{R}_n)$. For the approximation of the integral (6.3.32) at the points M_{0m}, we use the quadrature formula

$$S(M_{0m}) = C(\lambda, p) \sum_{k \in \varepsilon(\varphi)} \int_{D(k,h)} \frac{\Delta_{hp}^p \varphi(M_{0k})}{r_{0m}^{n+\lambda}} \, dx, \tag{6.3.33}$$

where Δ_{hp}^p is the iterated Laplace difference operator of order p; $\varepsilon(\varphi)$ is the set of all indices k corresponding to the cells $D(k, h)$ having a nonempty intersection with $\operatorname{supp}\Delta_{hp}^p \varphi(M_{0k})$.

Let us estimate the quadrature error $\Delta u(M_{0m}) = u(M_{0m}) - S(M_{0m})$. For this purpose, we transform $\Delta u(M_{0m})$ as follows:

$$\Delta u(M_{0m}) = C(\lambda, p) \sum_{k \in D(k,h)} \int_{D(k,h)} \frac{\Delta^p \varphi(x) - \Delta^p \varphi(M_{0k})}{r_{0m}^{n+\lambda}} \, dx +$$

$$+ C(\lambda, p) \sum_{k \in \varepsilon(\varphi)} \int_{D(k,h)} \frac{\Delta^p(M_{0k}) - \Delta_{hp}^p \varphi(M_{0k})}{r_{0m}^{n+\lambda}} \, dx =$$

$$= \Delta_1 u(M_{0m}) + \Delta_2(M_{0m}). \tag{6.3.34}$$

For the error $\Delta_1 u(M_{0m})$, the results of Section 6.3.1 remain valid. Let us estimate $\Delta_2 u(M_{0m})$. To that end, we use the transformations

$$|\Delta_2 u(M_{0m})| \leq |C(\lambda, p)| \left| \Delta^p \varphi(M_{0m}) - \Delta_{hp}^p \varphi(M_{0m}) \right| \left| \int_{D(m,h)} \frac{dx}{r_{0m}^{n+\lambda}} \right| +$$

$$+ \sum_{k \in \varepsilon(\varphi),\, k \neq m} |C(\lambda, p)| \left| \Delta^p \varphi(M_{0k}) - \Delta_{hp}^p \varphi(M_{0k}) \right| \int_{D(k,h)} \frac{dx}{r_{0m}^{n+\lambda}} =$$

$$= B_1(h) + B_2(h). \tag{6.3.35}$$

Since $\varphi(x) \in C_0^\infty(\mathbb{R}_n)$, Proposition 5.4.8 implies the uniform estimate

$$\left| \Delta^p \varphi(M_{0k}) - \Delta_{hp}^p \varphi(M_{0k}) \right| \leq k(\Omega_h)^{2/n},$$

with k independent of h, $\Omega_h = \prod_{j=1}^n h_j$. Using (5.3.3), we obtain the inequality

$$\left| \Delta^p \varphi(M_{0k}) - \Delta_{hp}^p \varphi(M_{0k}) \right| \leq k_1 |h|^2. \tag{6.3.36}$$

Let $|h| < \delta_0$. Consider the case of the points M_{0m} such that either $M_{0m} \in \operatorname{supp}\Delta_{hp}^p \varphi(M_{0k})$ or the distance from M_{0m} to $\operatorname{supp}\Delta_{hp}^p \varphi(M_{0k})$ does not exceed $\delta > 0$.

From the definition of the integral in the sense of Hadamard (5.2.9), it follows that

$$\sum_{k \neq m} \int_{D(k,h)} \frac{dx}{r_{0m}^{n+\lambda}} \leq \frac{C}{|h|^\lambda}, \qquad \left| \int_{D(m,h)} \frac{dx}{r_{0m}^{n+\lambda}} \right| \leq \frac{C}{|h|^\lambda}. \tag{6.3.37}$$

The estimates (6.3.35)–(6.3.37) imply that

$$|\Delta_2 u(M_{0m}))| \leq C_1 |h|^{2-\lambda}. \tag{6.3.38}$$

Consider the case of the points M_{0m} being at a distance $\rho_{0m} \geq \delta$ from $\operatorname{supp} \Delta_{hp}^p \varphi(M_{0k})$. For these points, we have

$$|\Delta_2 u(M_{0m})| \leq |C(\lambda, p)| \sum_{k \in \varepsilon(\varphi)} \int_{D(k,h)} \frac{\left| \Delta^p \varphi(M_{0k}) - \Delta_{hp}^p \varphi(M_{0k}) \right|}{r_{0m}^{n+\lambda}} \, dx \leq$$

$$\leq C_1(\lambda, p) \frac{|h|^2}{r_{0m}^{n+\lambda}} . \tag{6.3.39}$$

Since the results of Section 6.3.1 are valid for $\Delta_1 u(M_{0m})$, it follows from (6.3.39) that the function $u(x, h)$, $\pi u(x, h) = \Delta u(M_{0m})$, satisfies the relation

$$\int_{\mathbb{R}_n} |u(x, h)|^2 \, dx = \Theta(|h|),$$

where $\Theta(h) \to 0$ as $|h| \to 0$.

Proposition 6.3.3. *For $\varphi(x) \in C_0^\infty(\mathbb{R}_n)$, the functions $S(x, h)$, $\pi S(x, h) = S(M_{0m})$, converge to $u(x_0) = C(\lambda, p) \int_{\mathbb{R}_n} r^{-n-\lambda} \Delta^p \varphi(x) \, dx$ in the norm of $L_2(\mathbb{R}_n)$ as $|h| \to 0$.*

The proof of this result is similar to that of Proposition 6.3.1.

Remark 6.3.1. Since $H_r(\mathbb{R}_n) \supset L_2(\mathbb{R}_n)$ for all $r \leq 0$, all quadrature formulas considered in Section 6.3 converge in the norm of $H_r(\mathbb{R}_n)$ for any $r \leq 0$.

Chapter 7

Stability of Discrete Operators in Quotient Spaces

7.1. Convergence of Approximate Solutions and the Existence of Solutions of Operator Equations

Definition 7.1.1. A bounded family of operators T_h from $\overset{\circ}{M}(r, h, \Omega_{1h})$ to $M(-r, h)$, $r > 0$, is called *stable*, if there exist $\delta > 0$ and $C > 0$ such that for any $a(x, h) \in \overset{\circ}{M}(r, h, \Omega_{1h})$ and all h, $|h| < \delta_0$, the following estimate holds:

$$\|P_h T_h a(x, h)\|_{M(-r, h, \Omega_{1h})} \geq C \|a(x, h)\|_{\overset{\circ}{M}(r, h, \Omega_{1h})}, \tag{7.1.1}$$

where P_h is the operator of restriction to Ω_{1h}, and C is a constant independent of $h = (h_1, \ldots, h_n)$.

Definition 7.1.2. *Functions* $a(x, h) \in \overset{\circ}{M}(r, h, \Omega_{1h})$, $r \geq 0$, *are said to converge in* $\overset{\circ}{M}(r, h, \Omega_{1h})$ *to a function* $u(x) \in \overset{\circ}{H}_r(\Omega)$, *if*

$$\lim_{|h| \to 0} \|\Pi_{1h} u(x) - a(x, h)\|_{\overset{\circ}{M}(r, h, \Omega_{1h})} = 0.$$

Definition 7.1.3. *Functions* $a(x, h) \in \overset{\circ}{M}(r, h, \Omega_{1h})$, $r \geq 0$, *are said to converge weakly in* $\overset{\circ}{M}(r, h, \Omega_{1h})$ *to a function* $u(x) \in \overset{\circ}{H}_r(\Omega)$, *if*

$$\lim_{|h| \to 0} \|\Pi_{1h} u(x) - a(x, h)\|_{\overset{\circ}{M}(r_0, h, \Omega_{1h})} = 0, \qquad \forall r_0 < r.$$

Proposition 7.1.1. *Suppose that a stable family of operators* $T_h : M(r, h) \to M(-r, h)$ *weakly approximates a bounded operator* $T : H_r(\mathbb{R}_n) \to H_{-r}(\mathbb{R}_n)$, $r > 0$, *and the equation* $Tu = f$ *admits a solution for any* $f(x) \in H_{-r}(\mathbb{R}_n)$. *Then*

(i) *the equation* $Tu = f$ *can have only one solution;*

(ii) *the solutions* $a(x, h) \in M(r, h)$ *of the equation* $T_h a(x, h) = f(x, h)$, *where*

$$f(x, h) \in M(-r, h), \qquad \lim_{|h| \to 0} \|f(x) - f(x, h)\|_{H_{-r}(\mathbb{R}_n)} = 0,$$

converge in $M(r, h)$ *to the solution of the equation* $Tu = f$.

Proof. Let $u_1(x), u_2(x) \in H_r(\mathbb{R}_n)$ be two solutions of equation $Tu = f$. Then, by Proposition 6.2.1, we have

$$T_h \Pi_h u_1(x) \xrightarrow{H_{-r}(\mathbb{R}_n)} f(x), \qquad T_h \Pi_h u_2(x) \xrightarrow{H_{-r}(\mathbb{R}_n)} f(x) \qquad \text{as} \qquad |h| \to 0.$$

Hence, by Theorem 5.3.1, we have $T_h \Pi_h (u_1(x) - u_2(x)) \xrightarrow{M(-r,h)} 0$ as $|h| \to 0$. Since the family of operators T_h is stable, it follows that $\Pi_h (u_1(x) - u_2(x)) \xrightarrow{M(r,h)} 0$ as $|h| \to 0$. Using Proposition 5.4.11, we obtain $u_1(x) = u_2(x)$ and thereby, the uniqueness of the solution of equation $Tu = f$ is proved.

The operators T_h map $M(r, h)$ onto $M(-r, h)$, and therefore, equation $T_h a(x, h) = f(x, h)$ admits a solution. Since the family of operators T_h is stable, there is $\delta > 0$ such that for $|h| < \delta$, the equation can have only one solution. From Theorem 5.3.1 and Proposition 6.2.1, we have $T_h(a(x, h) - \Pi_h u(x)) \xrightarrow{M(-r,h)} 0$ as $|h| \to 0$. Hence, taking into account the stability of T_h, we deduce that

$$\lim_{|h| \to 0} \|\Pi_h u(x) - a(x, h)\|_{M(r,h)} = 0.$$

Proposition 7.1.1 is proved.

Proposition 7.1.2. *Let* $T : \mathring{H}_r(\Omega) \to H_{-r}(\mathbb{R}_n)$ *be a bounded finite order operator and let* $T_h :$ $\mathring{M}(r, h, \Omega_{1h}) \to M(-r, h)$, $r > 0$, *be a stable family of the operators yielding a* Π_h^*-*approximation of* T *in a normal domain* Ω *admitting a partition. Suppose that for any* $f(x) \in H_{-r}$, *the equation* $PTu = f$ *admits a solution* (P *is the operator of restriction to* Ω). *Then*

(i) *the equation* $PTu = f$ *can have only one solution;*

(ii) *the solutions* $a(x, h) \in \mathring{M}(r, h, \Omega_{1h})$ *of the equations* $P_h T_h a(x, h) = P_h f(x, h)$, *where*

$$f(x, h) \in M(-r, h), \qquad \lim_{|h| \to 0} \|f(x) - f(x, h)\|_{H_{-r}(\mathbb{R}_n)} = 0,$$

converge in $\mathring{M}(r, h, \Omega_{1h})$ *to the solution* $u(x)$ *of the equation* $PTu = f$, *where* P_h *is the operator of restriction to* Ω_{1h}; $lf(x)$ *is an extension of* $f(x) \in H_{-r}(\Omega)$, $lf(x) \in H_{-r}(\mathbb{R}_n)$.

Proof. Let $u_1(x), u_2(x) \in \mathring{H}_r(\Omega)$ be two solutions of the equation $PTu = f$, $f(x) \in H_{-r}(\Omega)$. Then, according to Remark 6.2.1, there exist extensions $l_{1h} P_h T_h \Pi_{1h} u_1(x)$ and $l_{2h} P_h T_h \Pi_{1h} u_2(x)$ of the functions $P_h T_h \Pi_{1h} u_1(x)$ and $P_h T_h \Pi_{1h} u_2(x)$ such that

$$l_{1h} P_h T_h \Pi_{1h} u_1(x) \xrightarrow{H_{-r}(\mathbb{R}_n)} Tu_1(x), \qquad l_{2h} P_h T_h \Pi_{1h} u_2(x) \xrightarrow{H_{-r}(\mathbb{R}_n)} Tu_2(x) \qquad \text{as} \quad |h| \to 0.$$

From Proposition 5.5.12, it follows that $P_h T_h \Pi_{1h} (u_1(x) - u_2(x)) \xrightarrow{M(-r,h,\Omega_{1h})} 0$ as $|h| \to 0$. Hence, using the stability of the operators T_h, we conclude that $\Pi_{1h}(u_1(x) - u_2(x)) \xrightarrow{\mathring{M}(-r,h,\Omega_{1h})} 0$ as $|h| \to 0$. By Proposition 5.5.9, we get $u_1(x) = u_2(x)$, which proves the uniqueness of the solution of the equation $PTu = f(x)$.

The spaces $\mathring{M}(r, h, \Omega_{1h})$ and $M(-r, h, \Omega_{1h})$ have the same finite dimension. The family of operators T_h is stable, and therefore, there is $\delta > 0$ such that for $|h| < \delta$ there exists a unique solution of the equation $P_h T_h a(x, h) = P_h f(x, h)$.

According to Remark 6.2.2, there is an extension $l_h P_h T_h \Pi_{1h} u(x)$ of $P_h T_h \Pi_{1h} u(x)$ such that $l_h P_h T_h \Pi_{1h} u(x) \xrightarrow{H_{-r}(\mathbb{R}_n)} Tu(x)$ as $|h| \to 0$.

Proposition 5.5.12 implies that $P_h T_h(\Pi_{1h} u(x) - a(x, h)) \xrightarrow{M(-r,h,\Omega_{1h})} 0$ as $|h| \to 0$. This, together with the stability of the operators T_h, ensures that $\|\Pi_{1h} u(x) - a(x, h)\|_{\mathring{M}(r,h,\Omega_{1h})} \to 0$ as $|h| \to 0$. Proposition 7.1.2 is proved.

Proposition 7.1.3. *Let* $T : \mathring{H}_r(\Omega) \to H_{-r}(\mathbb{R}_n)$ *be a bounded finite order operator, and let* $T_h : \mathring{M}(r, h, \Omega_{1h}) \to M(-r, h)$, $r > 0$, *be a stable family of operators* Π_h^*-*approximating* T *in a normal domain* Ω *admitting a partition. Suppose for some* $r_0 < r$, *the family of operators* T_h *from* $\mathring{M}(r_0, h, \Omega_{1h})$ *to* $M(q, h)$ *is bounded. Then, for any* $f \in H_{-r}(\Omega)$, *there exists one and only one solution of the equation* $PTu = f$ (*here* $u(x) \in \mathring{H}_r(\Omega)$, P *is the operator of restriction to* Ω).

Proof. The uniqueness of the solution follows from Proposition 7.1.2. The relation $PTu = f$ is understood as the equality of two linear functionals on $\mathring{H}_r(\Omega)$, i.e.,

$$(\nu, PTu) = (\nu, f), \qquad \forall \, \nu(x) \in \mathring{H}_r(\Omega),$$

where (ν, PTu) and (ν, f) denote the linear functionals associated with the functions $PTu(x)$ and $f(x)$. It should be observed that these functionals do not depend on the extensions of PT and f (see *Eskin* (1973)$_2$) as linear functionals on $H_r(\mathbb{R}_n)$. As an extension of PTu we take Tu, and as an extension lf of f we take that with the smallest norm in $H_{-r}(\mathbb{R}_n)$ (this extension exists, because $H_{-r}(\mathbb{R}_n)$ is a Hilbert space). Thus, we define a solution $u(x)$ of the equation $PTu = f$ as a function $u(x) \in \mathring{H}_r(\Omega)$ such that

$$(\nu, Tu) = (\nu, lf), \qquad \forall \, \nu(x) \in \mathring{H}_r(\Omega). \tag{7.1.2}$$

According to Proposition 5.5.8, there is a family of functions $f(x, h) \in M(-r, h)$ such that

$$f(x, h) \xrightarrow{H_{-r}(\mathbb{R}_n)} lf(x), \qquad P_{1h} f(x, h) \xrightarrow{H_{-r}(\Omega)} f(x) \quad \text{as} \quad |h| \to 0,$$

where P_{1h} is the operator of restriction to Ω (see Definition 6.2.6). Consider the family of equations

$$P_h T_h a(x, h) = P_h f(x, h), \tag{7.1.3}$$

where P_h is the operator of restriction to the domain Ω_{1h}.

Since the family of operators T_h is stable, there exists $\delta > 0$ such that for each h, $|h| < \delta$, the solution $a(x, h)$ of equation (7.1.3) exists and is unique. Moreover,

$$\|a(x, h)\|_{\mathring{M}(r, h, \Omega_{1h})} \leq C \|f(x)\|_{H_{-r}(\Omega)}.$$

Therefore, according to Proposition 5.5.2, there is a sequence $\{h^n\}_{n=1}^{\infty}$, $|h^n| \to 0$ as $n \to \infty$, and there exists r_0, $0 < r_0 < 1/2$, $r_0 < r$, and a function $a(x) \in \mathring{H}_r(\Omega)$ such that $a(x, h^n) \xrightarrow{\mathring{H}_{r_0}(\Omega)} a(x)$ as $|h^n| \to 0$. Let us show that (7.1.2) holds for the function $a(x)$. According to Remark 6.2.2 (see Proposition 6.2.7), the restriction $PTa(x)$ of the linear functional $Ta(x)$ on $\mathring{H}_r(\Omega)$ can be represented in the form

$$PTa(x) = P_{1h^n} T_{h^n} \Pi_{1h^n} a(x) + \varepsilon(x, h^n), \tag{7.1.4}$$

where $\varepsilon(x, h^n) \xrightarrow{H_{-r}(\Omega)} 0$ as $|h^n| \to 0$. It follows that

$$(\nu, Ta(x)) = \lim_{|h^n| \to 0} \left(\nu(x), P_{1h^n} T_{h^n} \Pi_{1h^n} a(x) \right). \tag{7.1.5}$$

We further have

$$\begin{aligned}
(\nu, Ta(x)) &= \lim_{|h^n| \to 0} (\nu, P_{1h^n} T_{h^n} a(x, h^n)) + \lim_{|h^n| \to 0} (\nu(x), P_{1h^n} T_{h^n} (\Pi_{1h^n} a(x) - a(x, h^n))) = \\
&= \lim_{|h^n| \to 0} (\nu, P_{1h^n} f(x, h^n) + \lim_{|h^n| \to 0} (\Pi_{1h^n} \nu(x), P_{1h^n} T_{h^n} (\Pi_{1h^n} a(x) - a(x, h^n))) = \\
&= (\nu, lf(x))) + \lim_{|h^n| \to 0} I(h^n). \tag{7.1.6}
\end{aligned}$$

Proposition 5.5.14 implies that $\|\Pi_{1h^n}a(x) - a(x,h^n)\|_{\mathring{M}(r_1,h,\Omega_{1h})} \to 0$ for any $r_1 < r$. The assumptions of Proposition 7.1.3 ensure the existence of $r_0 < r$ such that the family of operators T_{h^n} is bounded from $\mathring{M}(r_0, h^n, \Omega_{1h^n})$ to $M(q, h^n)$. It follows that for $|h^n| \to 0$, we have

$$\|P_{h^n}T_{h^n}(\Pi_{1h^n}a(x) - a(x,h^n))\|_{M(q,h^n,\Omega_{1h^n})} \le \|T_{h^n}(\Pi_{1h^n}a(x) - a(x,h^n))\|_{M(q,h^n)} \le$$
$$\le \|T_{h^n}\|_{r_0,q}\|\Pi_{1h^n}a(x) - a(x,h^n)\|_{\mathring{M}(r_0,h\Omega_{1h})} \to 0, \tag{7.1.7}$$

where $\|T_{h^n}\|_{r_0,q}$ is the norm of the operator $T_{h^n}\colon \mathring{M}(r_0,h^n,\Omega_{1h}) \to M(q,h^n)$. If $q \ge -r$, then for $I(h^n)$ from (7.1.6) we have

$$|I(h^n)| \le \|\Pi_{1h^n}\nu(x)\|_{\mathring{M}(r,h^n,\Omega_{1h^n})}\|P_{h^n}T_{h^n}(\Pi_{1h^n}a(x) - a(x,h^n))\|_{M(-r,h^n,\Omega_{1h^n})} \le$$
$$\le C\|\nu(x)\|_{\mathring{H}_r(\Omega)}\|P_{h^n}T_{h^n}(\Pi_{1h^n}a(x) - a(x,h^n))\|_{M(q,h^n,\Omega_{1h^n})}. \tag{7.1.8}$$

From (7.1.7) and (7.1.8), it follows that $I(h^n) \to 0$ as $|h^n| \to 0$.

If $q < -r$, it follows from Proposition 5.5.13 that $\lim_{|h^n| \to 0} = 0$.

The above considerations and (7.1.6) show that the function $a(x)$ is a solution of equation $PTu = f(x)$. Proposition 7.1.3 is proved.

Definition 7.1.4. *A bounded family of operators* $T_h\colon \mathring{M}(r,h,\Omega_{1h}) \to M(-r,h)$, $r > 0$, *is called weakly convergent, if for any* $\nu(x) \in \mathring{H}_r(\Omega)$, *there is a function* $u(x) \in H_{-r}(\mathbb{R}_n)$ *and there are extensions* $l_h(P_hT_h)^*\Pi_{1h}\nu(x) \in M(-r,h)$ *of the function* $(P_hT_h)^*\Pi_{1h}\nu(x) \in M(-r,h,\Omega_{1h})$ *such that*

$$l_h(P_hT_h)^*\Pi_{1h}\nu(x) \xrightarrow{H_{-r}(\mathbb{R}_n)} u(x) \quad \text{as} \quad |h| \to 0,$$

where $(P_hT_h)^*$ *are the operators conjugate to* P_hT_h.

Proposition 7.1.4. *Let* T *be a bounded finite order operator from* $\mathring{H}_r(\Omega)$ *to* $H_{-r}(\mathbb{R}_n)$, *where* Ω *is a normal bounded domain that admits a partition. Let* $T_h\colon \mathring{M}(r,h,\Omega_{1h}) \to M(-r,h)$, $r > 0$, *be a weakly convergent stable family of operators* Π_h^*-*approximating the operator* T. *Then, for any* $f \in H_{-r}(\Omega)$, *the equation* $PTu = f$ *admits one and only one solution* $u(x) \in \mathring{H}_r(\Omega)$.

Proof. Just as in the proof of Proposition 7.1.3 (see (7.1.6)), we obtain the relation

$$(\nu, Ta(x)) = (\nu, lf) + \lim_{|h| \to 0} I(h), \tag{7.1.9}$$

where we have replaced $a(x,h^n)$ by $a(x,h)$, and h^n by h, to simplify notation.

According to the assumptions of Proposition 7.1.4, there is an extension

$$l_h(P_hT_h)^*\Pi_{1h}\nu(x) \in M(-r,h)$$

of the function

$$(P_hT_h)^*\Pi_{1h}\nu \in M(-r,h,\Omega_{1h}),$$

and we have $(\Pi_{1h}a(x) - a(x,h)) \in \mathring{M}(r,h,\Omega_{1h})$. Therefore,

$$I(h) = \overline{\left(\Pi_{1h}a(x) - a(x,h),\, l_h(P_hT_h)^*\Pi_{1h}\nu(x)\right)}, \tag{7.1.10}$$

where the bar indicates complex conjugation.

Proposition 5.6.4 implies that there is a function $w^+(x)$ such that

$$\|l_h(P_hT_h)\Pi_{1h}\nu(x)\|_{M(-r,h,w^+)} \le C, \tag{7.1.11}$$

where C does not depend on h.

By Proposition 5.6.1, we obtain

$$\lim_{|h| \to 0} \|\Pi_{1h}a(x) - a(x,h)\|_{M(r,h,1/w^+)} = 0. \tag{7.1.12}$$

From Proposition 5.6.5, combined with (7.1.10)–(7.1.12), we find that $\lim_{|h| \to 0} I(h) = 0$. This, together with (7.1.8), implies the desired result. Proposition 7.1.4 is proved.

Corollary 7.1.1. *Suppose that the assumptions of Proposition 7.1.3 or 7.1.4 hold for the operators* T *and* T_h. *Consider the family of equations*

$$P_h T_h a(x, h) = f(x, h), \qquad (7.1.13)$$

where $f(x, h) \in M(-r, h, \Omega_{1h})$, $\|f(x, h)\|_{M(-r, h, \Omega_{1h})} \leq C$ *with* C *independent of* h. *Moreover, consider the functions*

$$\widetilde{f}(x, h) = \begin{cases} f(x, h) & \text{for } x \in D(k, h) \subset \Omega_{1h} \\ 0 & \text{for } x \in \Omega \setminus \overline{\Omega}_{1h}, \end{cases}$$

and assume that $\widetilde{f}(x, h) \to f(x) \in H_{-r}(\Omega)$ *in the norm of* $H_{-r}(\Omega)$ *as* $|h| \to 0$. *Then, there exists* $\delta > 0$ *such that for* $|h| < \delta$, *each equation (7.1.13) admits a unique solution* $a(x, h) \in \overset{\circ}{M}(r, h, \Omega_{1h})$, *and these solutions weakly converge to the solution* $a(x)$ *of the equation* $PTa(x) = f(x)$, *i.e., for any* $r_0 < r$, *we have*

$$\lim_{|h| \to 0} \|\Pi_{1h} a(x) - a(x, h)\|_{\overset{\circ}{M}(r_0, h, \Omega_{1h})} = 0.$$

This statement is immediately obtained by the arguments used in the proof of Propositions 7.1.3 and 7.1.4, combined with Proposition 7.1.1.

Proposition 7.1.5. *Let* T, T_h *be operators satisfying the assumptions of Propositions 7.1.3 and 7.1.4. Let* $F : \overset{\circ}{H}_s(\Omega) \to H_{-r+\varepsilon}(\mathbb{R}_n)$ *be a bounded operator of finite order, where* Ω *is a normal domain admitting a partition. Let* $F_h : \overset{\circ}{M}(s, h, \Omega_{1h}) \to M(-r+\varepsilon, h)$, $0 < r_0 < s \leq r$, $\varepsilon > 0$, *be a bounded family of operators* Π_h^*-*approximating* F *in* Ω, *and suppose that the kernel of the restriction* PA *of the operator* $A = T + F$ *to* $\overset{\circ}{H}_r(\Omega)$ *is trivial. Then, there is* $\delta > 0$ *such that for* $|h| < \delta$, *the family of operators* $P_h A_h = P_h T_h + P_h F_h$ *from* $\overset{\circ}{M}(r, h, \Omega_{1h})$ *to* $M(-r, h, \Omega_{1h})$ *is stable.*

Proof. Suppose the contrary. Then, there exists a sequence $a(x, h^n) \in \overset{\circ}{M}(r, h^n, \Omega_{1h^n})$ such that $\|a(x, h^m)\|_{\overset{\circ}{M}(r, h^n, \Omega_{1h^n})} = 1$, $|h^n| \to 0$ as $n \to \infty$,

$$\lim_{|h^n| \to 0} \|P_{h^n} A_{h^n} a(x, h^n)\|_{M(-r, h^n, \Omega_{1h^n})} = 0. \qquad (7.1.14)$$

Let $b(x, h^n) = P_{1h^n} F_{h^n} a(x, h^n)$, where P_{1h^n} is the operator of restriction to the domain Ω (see Definition 6.2.6). The properties of the operators F_h and Proposition 5.5.5 ensure the inequalities

$$\|b(x, h^n)\|_{M(-r+\varepsilon, h^n, \Omega_{1h^n})} \leq C, \qquad \|b(x, h^n)\|_{H_{-r+\varepsilon}(\Omega)} \leq C_1,$$

where the constants C and C_1 do not depend on h. Since $H_{-r+\varepsilon}(\Omega)$ is compactly imbedded into $H_m(\Omega)$, (see *Eskin* (1973)$_1$), $-r < m < -r+\varepsilon$, we can find a subsequence of $b(x, h^n)$ that converges to some $b(x) \in H_m(\Omega)$. This subsequence is still denoted by $b(x, h^n)$. Let us show that $b(x) \neq 0$. Indeed, if $b(x) = 0$ as an element of $H_m(\Omega)$, then Proposition 5.5.15 implies that $P_{1h^n} F_{1h^n} a(x, h^n) \to 0$ in the norm of $M(-r, h^n, \Omega_{1h^n})$ as $|h^n| \to 0$. This fact is in contradiction with the stability of the operator family T_h and the relation (7.1.14). Thus, $b(x) \neq 0$.

From (7.1.14), it follows that

$$P_{h^n} T_{h^n} a(x, h^n) = -P_{h^n} F_{h^n} a(x, h^n) + \varepsilon(x, h^n), \qquad (7.1.15)$$

where $\varepsilon(x, h^n) \to 0$ in the norm of $M(-r, h^n, \Omega_{1h^n})$ as $|h^n| \to 0$.

Let us regard (7.1.15) as equations with the unknown $a(x, h^n)$ in the left-hand side of (7.1.15) and assume its right-hand side known. Proposition 5.5.5 implies that

$$P_{1h^n} \varepsilon(x, h^n) \xrightarrow{H_{-r}(\Omega)} 0, \qquad b(x, h^n) \xrightarrow{H_{-r}} b(x).$$

According to Corollary 7.1.1, the functions $a(x, h)$ converge weakly to $u(x) \in \mathring{H}_r(\Omega)$ which is a solution of the equation $PTu(x) = -b(x)$, $u(x) \neq 0$. Let us show that $PAu(x) = 0$. To that end, it suffices to show that $PFu(x) = b(x)$ or, equivalently, $(\nu, Fu(x)) = (\nu, b(x))$ for any $\nu(x) \in \mathring{H}_r(\Omega)$.

According to Remark 6.2.2 (see Proposition 6.2.7), the restriction $PFu(x)$ of the linear functional $Fu(x)$ to $\mathring{H}_r(\Omega)$ can be represented in the form

$$PFu(x) = P_{1h^n} F_{h^n} \Pi_{1h^n} u(x) + \varepsilon_1(x, h^n), \qquad (7.1.16)$$

where $\varepsilon_1(x, h^n) \to 0$ in the norm of $H_{-r}(\Omega)$ as $|h^n| \to 0$.

Using (7.1.16), we obtain the relation

$$
\begin{aligned}
(\nu, Fu(x)) &= \lim_{|h^n| \to 0} (\nu(x), P_{1h^n} F_{h^n} \Pi_{1h^n} u(x)) = \\
&= \lim_{|h^n| \to 0} (\nu(x), P_{1h^n} F_{h^n} a(x, h^n)) + \lim_{|h^n| \to 0} (\nu(x), P_{1h^n} F_{h^n} (\Pi_{1h^n} u(x) - a(x, h^n))) = \\
&= (\nu(x), b(x)) + \lim_{|h^n| \to 0} (\Pi_{1h^n} \nu(x), P_{1h^n} F_{h^n} (\Pi_{1h^n} u(x) - a(x, h^n))) = \\
&= (\nu(x), b(x)) + I(h^n). \qquad (7.1.17)
\end{aligned}
$$

Since the family of operators $F_h : \mathring{M}(s, h, \Omega_{1h}) \to M(-r, +\varepsilon, h)$, $\varepsilon > 0$, is bounded for any s satisfying the inequality $0 < r_0 \leq s < r$, we have

$$
\begin{aligned}
|I(h^n)| &\leq \|\Pi_{1h^n} \nu(x)\|_{\mathring{M}(r, h^n, \Omega_{1h^n})} \|P_{1h^n} F_{h^n} (\Pi_{1h^n} u(x) - a(x, h^n))\|_{M(-r, h^n, \Omega_{1h^n})} \leq \\
&\leq \|\Pi_{1h^n} \nu(x)\|_{\mathring{M}(r, h^n, \Omega_{1h^n})} \|F_{h^n} (\Pi_{1h^n} u(x) - a(x, h^n))\|_{M(-r, h^n)} \leq \\
&\leq C \|\Pi_{1h^n} \nu(x)\|_{\mathring{M}(r, h^n, \Omega_{1h^n})} \|F_{h^n} (\Pi_{1h^n} u(x) - a(x, h^n))\|_{M(-r+\varepsilon, h^n)} \leq \\
&\leq C_1 \|F_{h^n}\|_{s_0, -r+\varepsilon} \|\Pi_{1h^n} u(x) - a(x, h^n)\|_{\mathring{M}(s_0, h^n, \Omega_{1h^n})} \|\nu(x)\|_{\mathring{H}_r(\Omega)}, \qquad (7.1.18)
\end{aligned}
$$

where $r_0 \leq s_0 < r$, and $\|F_{h^n}\|_{s_0, -r+\varepsilon}$ is the norm of the operator F_h from $\mathring{M}(s_0, h^n, \Omega_{1h^n})$ to $M(-r + \varepsilon, h^n)$. The functions $a(x, h^n)$ converge weakly to $u(x) \in \mathring{H}_r(\Omega)$ (see Definition 7.1.3), and therefore, it follows from (7.1.18) that $\lim_{|h^n| \to 0} I(h^n) = 0$. This, together with (7.1.17), implies that $PAu(x) = 0$, $u(x) \neq 0$, which is in contradiction with the assumptions of Proposition 7.1.5. Proposition 7.1.5 is proved.

7.2. Stability of Discrete Operators in Quotient Spaces

In what follows, it is assumed that Ω is an open bounded domain with Lipschitz boundary and that the domain Ω is normal and admits a partition (see Definitions 5.5.1–5.5.4, 5.5.6, 5.5.9).

7.2.1. Discrete Vortex Operators

Discrete vortex operators are defined by relations (6.1.43) and (6.2.44). It is convenient to represent such operators in an equivalent form as follows ($0 < \lambda < 2$):

$$\pi T_{h,\lambda} u(x, h) = \sum_k u(k) \gamma(k, m), \qquad \gamma(k, m) = \int_{D(kh)} \frac{dx}{r_{0m}^{n+\lambda}}, \qquad (7.2.1)$$

where $r_{0m} = |MM_{0m}|$ (see Section 6.3.1); the operator π is that of Definition 5.1.2; $u(x, h) \in \mathring{M}(r, h, \Omega_{1h})$ (see Definition 5.5.2); the sum is taken over all k such that $D(k, h) \subset \Omega_{1h}$.

Proposition 6.1.4 implies that the family of operators $T_{h,\lambda} : M(r, h) \to M(r - \lambda, h)$ is bounded. From the definition of the spaces $\mathring{M}(r, h, \Omega_{1h})$ and $M(r, h, \Omega_{1h})$, it follows that the family of operators $P_h T_{h,\lambda} : \overset{\circ}{M}(\lambda/2, h, \Omega_{1h}) \to M(-\lambda/2, h, \Omega_{1h})$ is bounded, where P_h is the operator of restriction to Ω_{1h}.

Let us examine the stability of the operator family $P_h T_{h,\lambda}$ acting from $\overset{\circ}{M}(\lambda/2, h, \Omega_{1h})$ to $M(-\lambda/2, h, \Omega_{1h})$.

It is easy to check that $\gamma(k, m) = \gamma(m, k)$ for any k, m. Consider the linear functional $I = (2\pi)^{-n} l(\nu)$ associated with an element $u(x) \in \overset{\circ}{M}(\lambda/2, h, \Omega_{1h})$. Using (5.5.5), we get

$$I = \frac{1}{(2\pi)^n} l(P_h T_h u(x, h)) = \sum_{m \in \Lambda} \Omega_h \overline{u(m)} \sum_{k \in \Lambda} \gamma(k, m) =$$

$$= \Omega_h \sum_{m \in \Lambda} \overline{u(m)} \left[\sum_{k \in \Lambda,\, k \neq m} u(k)\gamma(k, m) + \gamma(m, m)u(m) \right], \qquad (7.2.2)$$

where Λ is the set of all k such that $D(k, h) \subset \Omega_{1h}$; $\pi u(x, h) = u(k)$; $\Omega_h = \prod_{j=1}^{n} h_j$. From the definition of the integral in the sense of Hadamard (see (5.2.9)), it follows that

$$\gamma(k, m) > 0 \quad \forall\, k, m, \quad k \neq m; \qquad \sum_k \gamma(k, m) = 0, \qquad \gamma(m) = \sum_{k \in \Lambda} \gamma(k, m) < 0.$$

Therefore, we can transform (7.2.2) as follows

$$I = \Omega_h \sum_{m \in \Lambda} \overline{u(m)} \left[\sum_{k \in \Lambda,\, k \neq m} u(k)\gamma(k, m) - \sum_{k \in \Lambda,\, k \neq m} \gamma(k, m)u(m) + \gamma(m)u(m) \right] =$$

$$= \Omega_h \sum_{m \in \Lambda} \overline{u(m)}\Omega_h \sum_{m \in \Lambda} [u(k) - u(m)]\gamma(k, m) + \Omega_h \sum_{m \in \Lambda} \gamma(m)|u(m)|^2. \qquad (7.2.3)$$

From (7.2.3), interchanging summation variables k, m and taking into account that $\gamma(k, m) = \gamma(m, k)$, we get

$$I = \Omega_h \sum_{k \in \Lambda} \overline{u(k)} \sum_{m \in \Lambda} \gamma(k, m)[u(m) - u(k)] + \Omega_h \sum_{m \in \Lambda} \gamma(m)|u(m)|^2. \qquad (7.2.4)$$

From (7.2.3) and (7.2.4), it follows that

$$I = -\frac{\Omega_h}{2} \sum_{m \in \Lambda} \sum_{k \in \Lambda} \gamma(k, m)|u(k) - u(m)|^2 + \Omega_h \sum_{m \in \Lambda} \gamma(m)|u(m)|^2. \qquad (7.2.5)$$

Let us show that $|\gamma(m)| > C > 0$. By assumption, the domain Ω is bounded, and therefore, there exists $R_0 > 0$ such that for all $M_{0m} \in \Omega$ the ball $K(R_0, M_{0m})$ of radius R_0 and center at M_{0m} contains Ω. It follows that

$$|\gamma(m)| = \left| \sum_{k \in \Lambda} \gamma(k, m) \right| \geq \left| \int\!\!\!\int_{K(R_0, M_{0m})} \frac{dx}{r_{0m}^{n+\lambda}} \right| \geq C. \qquad (7.2.6)$$

From (7.2.5) and (7.2.6), we get

$$|I| \geq C\Omega_h \sum_{m \in \Lambda} |u(m)|^2. \qquad (7.2.7)$$

On the other hand, taking into account that $u(k) = 0$, $u(m) = 0$ for $k, m \notin \Lambda$, we obtain

$$I = \Omega_h \sum_m \overline{u(m)} \sum_k u(k)\gamma(k, m) = \Omega_h \sum_m \overline{u(m)} \sum_k \gamma(k, m)[u(k) - u(m)]. \qquad (7.2.8)$$

Interchanging the summation variables in (7.2.8), we get

$$I = -\Omega_h \sum_k \overline{u(k)} \sum_m \gamma(k,m)[u(k) - u(m)]. \tag{7.2.9}$$

Using (7.2.8) and (7.2.9), we find that

$$I = -\frac{\Omega_h}{2} \sum_k \sum_m \gamma(k,m)|u(k) - u(m)|^2. \tag{7.2.10}$$

Note that for $\gamma(k,m)$ with $k \neq m$, the following inequality holds:

$$C_1 \frac{\Omega_h}{(|k-m|)|h|)^{n+\lambda}} \leq \gamma(k,m) \leq C_2 \frac{\Omega_h}{(|k-m|)|h|)^{n+\lambda}}. \tag{7.2.11}$$

From (7.2.10) and (7.2.11), we obtain the estimate

$$|I| \geq C_3 \Omega_h \sum_{k \neq m} \frac{\Omega_h}{(|k-m|)|h|)^{n+\lambda}} |u(k) - u(m)|^2. \tag{7.2.12}$$

The inequalities (7.2.7) and (7.2.12), combined with Proposition 5.3.4, yield

$$|I| \geq C_4 \|u(x,h)\|^2_{\mathring{M}(\lambda/2,h,\Omega_{1h})}, \tag{7.2.13}$$

where C_4 does not depend on h.

Using (7.2.13), we get

$$C_4 \|u(x,h)\|^2_{\mathring{M}(\lambda/2,h,\Omega_{1h})} \leq |I| \leq$$
$$\leq \frac{1}{(2\pi)^n} \|u(x,h)\|_{\mathring{M}(\lambda/2,h,\Omega_{1h})} \|P_h T_{h,\lambda} u(x,h)\|_{\mathring{M}(-\lambda/2,h,\Omega_{1h})}. \tag{7.2.14}$$

Now, from (7.2.14) and Definition 7.1.1 we obtain the following result.

Proposition 7.2.1. *The bounded family of discrete vortex operators* $P_h T_{h,\lambda}$ *from the space* $\mathring{M}(\lambda/2, h, \Omega_{1h})$ *to* $M(-\lambda/2, h, \Omega_{1h})$ *is stable.*

7.2.2. Iterated Difference Laplace Operator of Order m

We start with a preliminary statement.

Proposition 7.2.2. *For any* $u(x,h) \in \mathring{M}(r, h, \Omega_{1h})$, $r \geq 0$, *the following inequality holds:*

$$\Omega_h \int_{-\pi}^{\pi} \cdots \int_{-\pi}^{\pi} \left(\sum_{i=1}^{n} \left(\frac{t_i}{h_i} \right)^2 \right)^r |\hat{u}_c(t)|^2 \, dt \geq C \|u(x,h)\|^2_{\mathring{M}(r,h,\Omega_{1h})}, \tag{7.2.15}$$

where $\hat{u}_c(t)$ *is the discrete Fourier transform of the grid function* $\pi u(x,h) = u(k)$; $C > 0$ *does not depend on* $h = (h_1, \ldots, h_n)$.

Proof. In view of (5.3.2), it suffices to show that

$$\Omega_h \int_{-\pi}^{\pi} \cdots \int_{-\pi}^{\pi} \left(\sum_{i=1}^{n} \left(\frac{t_i}{h_i} \right)^2 \right)^r |\hat{u}_c(t)|^2 \, dt \geq C_1 \Omega_h \int_U |\hat{u}_c(t)|^2 \, dt, \tag{7.2.16}$$

where C_1 does not depend on h. We have

$$|\widehat{u}_c(t)| = \left|\sum_k u(k)\, e^{i(k,t)}\right| \le \frac{1}{\Omega_h}\left[\sum_k |u(k)|^2 \Omega_h\right]^{1/2}\left[\sum_k 1^2 \Omega_h\right]^{1/2} \le$$

$$\le \frac{\sqrt{V(\Omega)}}{\Omega_h}\|u(x,h)\|_{L_2(\Omega_{1h})}, \tag{7.2.17}$$

where the sum is taken over all k such that $D(k,h) \subset \Omega_{1h}$; $V(\Omega)$ is the volume of the domain Ω; $\Omega_h = \prod_{j=1}^n h_j$.

Since the domain Ω is bounded, there exists $\delta > 0$ such that for $|h| < \delta$, the spaces $\mathring{M}(r,h,\Omega_{1h})$ make sense. Therefore, for some $A(\Omega) > 2$, we have

$$U_A \subset U = \prod_{j=1}^n [-\pi \le t_j \le \pi], \qquad U_A = \prod_{j=1}^n I_j(A), \quad I_j(A) = \left\{t_j : |t_j| \le \frac{h_j}{A\sqrt[n]{V(\Omega)}}\right\}.$$

From (7.2.17), we get

$$\Omega_h \int_{U_A} |\widehat{u}_c(t)|^2\, dt \le \Omega_h \frac{V(\Omega)}{\Omega_h^2}\|u(x,h)\|_{L_2(\Omega_{1h})}^2 \frac{\Omega_h}{V(\Omega)A^n}. \tag{7.2.18}$$

From the definition of U_A and (7.2.18), it follows that

$$\Omega_h \int_U \left(\sum_{i=1}^n \left(\frac{t_i}{h_i}\right)^2\right)^r |\widehat{u}_c(t)|^2\, dt \ge C_1\Omega_h \int_{U\backslash U_A} |\widehat{u}_c(t)|^2\, dt \ge C_2\Omega_h \int_U |\widehat{u}_c(t)|^2\, dt,$$

where $C_2 > 0$ does not depend on h. The above inequality implies Proposition 7.2.2.

Proposition 7.2.3. *The family of operators* $P_hT_h = P_h\Delta_{hp}^m : \mathring{M}(m,h,\Omega_{1h}) \to M(-m,h,\Omega_{1h})$, *where* Δ_{hp}^m *is the iterated difference Laplace operator of order* m, *is stable.*

Proof. From Section 6.1.4, it follows that $P_hT_h : \mathring{M}(m,h,\Omega_{1h}) \to M(-m,h,\Omega_{1h})$ is a bounded family of operators. Consider the linear functional $I = (2\pi)^{-1}l(V)$ on $M(-m,h,\Omega_{1h})$ associated with the element $u(x,h) \in \mathring{M}(m,h,\Omega_{1h})$. Then, by virtue of (5.5.5), we obtain

$$|I| = \left|\frac{1}{(2\pi)^n}l(P_hT_h u(x,h))\right| = \left|\frac{1}{(2\pi)^n}\Omega_h \int_U \widehat{A}_c(t)|\widehat{u}_c(t)|^2\, dt\right| =$$

$$= \left|\frac{1}{(2\pi)^n}\Omega_h \int_U (-1)^m 4^m\left[\sum_{k=1}^n \frac{1}{h_k^2}\sin^2\frac{t_k}{2}\right]^m |\widehat{u}_c(t)|^2\, dt\right| \ge$$

$$\ge C_1\Omega_h \int_U \left(\left[\sum_{i=1}^n \left(\frac{t_i}{h_i}\right)^2\right]^{1/2}\right)^{2m} |\widehat{u}_c(t)|^2\, dt \ge C_2\|u(x,h)\|_{\mathring{M}(m,h,\Omega_{1h})}^2, \tag{7.2.19}$$

where

$$\widehat{A}_c(t) = (-1)^m 4^m\left[\sum_{k=1}^n \frac{1}{h_k^2}\sin^2\frac{t_k}{2}\right]^m$$

is the symbol of the operator Δ_{hp}^m; the constants C_1 and C_2 do not depend on h. In (7.2.19), we have used Proposition 7.2.2. The inequality (7.2.19) implies the stability of the operator family P_hT_h. Proposition 7.2.3 is proved.

7.2.3 Hypersingular Operators

Discrete hypersingular operators $T_h(m, \lambda)$ have been defined by (6.1.50).

Proposition 7.2.4. *The family of operators $P_h T_h(m, \lambda)$ from the space $\mathring{M}(m + \lambda/2, h, \Omega_{1h})$ to $M(-m - \lambda/2, h, \Omega_{1h})$ is stable.*

Proof. From Section 6.1.4, it follows that the family of operators under consideration is bounded. Consider the linear functional $I = (2\pi)^{-n} l(V)$ on $M(-m - \lambda/2, h, \Omega_{1h})$ associated with the element $u(x, h) \in \mathring{M}(m + \lambda/2, h, \Omega_{1h})$. Using (5.5.5), we find that

$$|I| = \left| \frac{1}{(2\pi)^n} l(P_h T_h(m, \lambda) u(x, h)) \right| = \left| \frac{1}{(2\pi)^n} \Omega_h \int_U \widehat{A}_c(t) |\widehat{u}_c(t)|^2 \, dt \right| =$$

$$= \left| \frac{1}{(2\pi)^n} \Omega_h \int_U (-1)^m 4^m \left[\sum_{k=1}^n \sin^2 \frac{t_k}{2} \right]^m J(\lambda, t, h) |\widehat{u}_c(t)|^2 \, dt \right| \geq$$

$$\geq C_1 \Omega_h \int_U \left(\left[\sum_{i=1}^n \left(\frac{t_i}{h_i} \right)^2 \right]^{1/2} \right)^{2m} \frac{|t|^\lambda}{h_1^\lambda} |\widehat{u}_c(t)|^2 \, dt \geq$$

$$\geq C_2 \Omega_h \int_U \left(\sum_{i=1}^n \left(\frac{t_i}{h_i} \right)^2 \right)^{m+\lambda/2} |\widehat{u}_c(t)|^2 \, dt \geq C_3 \|u(x, h)\|_{\mathring{M}(m+\lambda/2, h, \Omega_{1h})}, \qquad (7.2.20)$$

where

$$\widehat{A}_c(t) = (-1)^m 4^m \left[\sum_{k=1}^n \sin^2 \frac{t_k}{2} \right]^m J(\lambda, t, h);$$

$\widehat{A}_c(t)$ is the symbol of a hypersingular discrete operator; $J(\lambda, t, h)$ is the symbol of a discrete vortex operator and satisfies the inequality (see (5.2.15))

$$|J(\lambda, t, h)| \geq \frac{C_1 |t|^\lambda}{h_1^\lambda}.$$

The inequality (7.2.20) implies the stability of the operator family $T_h(m, \lambda)$. Proposition 7.2.4 is proved.

7.2.4. Second Order Elliptic Difference Operator with Variable Coefficients

Consider the operator

$$L(u) = \sum_{i=1}^n \frac{\partial}{\partial x_i} \sum_{j=1}^n a_{ij}(x) \frac{\partial u}{\partial x_j}, \qquad (7.2.21)$$

satisfying the condition of uniform ellipticity in a bounded domain $\Omega \subset \mathbb{R}_n$,

$$\nu \sum_{i=1}^n \xi_i^2 \leq \sum_{i,j=1}^n a_{ij}(x) \xi_i \xi_j \leq \mu \sum_{i=1}^n \xi_i^2, \qquad (7.2.22)$$

where $\nu, \mu > 0$ are constants, $a_{ij}(x) = a_{ji}(x)$, $a_{ij}(x) \in C_1(\overline{\Omega})$.

We define a discrete operator $L_h u(x, h)$ by the formula

$$L_h u(x, h) = \sum_{i=1}^n T_{\text{left}}(x_i) \sum_{j=1}^n a_{ij}(x, h) T_{\text{right}}(x_j) u(x, h), \qquad (7.2.23)$$

where $u(x, h) \in \overset{\circ}{M}(r, h, \Omega_{1h})$, $T_{\text{right}}(x_i)$ and $T_{\text{left}}(x_i)$ are the right and the left difference ratios with respect to the variable x_i (see Example 5.1.2);

$$a_{ij}(x, h) = a_{ij}(k_1 h_1, \ldots, k_n h_n) \quad \text{for} \quad x \in D(k_1, \ldots, k_n, h_1, \ldots, h_n) \subset \Omega_{1h}.$$

Proposition 7.2.5. *The family of operators* $P_h L_h : \overset{\circ}{M}(1, h, \Omega_{1h}) \to M(-1, h, \Omega_{1h})$ *is stable.*

Proof. The condition $a_{ij} \in C_1(\overline{\Omega}_{1h})$ and the results of Section 6.1.4 imply that this family of operators from $\overset{\circ}{M}(1, h, \Omega_{1h})$ to $M(-1, h, \Omega_{1h})$ is bounded. Let

$$\nu_i(x, h) = \sum_{j=1}^{n} a_{ij}(xh) T_{\text{right}}(x_j) u(x, h),$$

and denote by $\widehat{\nu}_i(\varphi)$ the discrete Fourier transform of the grid function $\nu_i(k) = \pi \nu_i(x, h)$ (see Definition 5.1.2 regarding π).

Consider the linear functional $I = (2\pi)^{-1} l(V(x))$ on $M(-1, h, \Omega_{1h})$ associated with $u(x, h) \in \overset{\circ}{M}(1, h, \Omega_{1h})$. Then, using (5.5.5), we get

$$
\begin{aligned}
I &= \frac{1}{(2\pi)^n} l(P_h L_h u(x, h)) = \frac{\Omega_h}{(2\pi)^n} \sum_{k=1}^{n} \int_U \widehat{\nu}_k(\varphi) \frac{1 - e^{i\varphi_k}}{h_k} \overline{\widehat{u}_c(\varphi)} \, d\varphi = \\
&= -\frac{\Omega_h}{(2\pi)^n} \sum_{k=1}^{n} \int_U \widehat{\nu}_k(\varphi) \frac{\overline{-1 + e^{-i\varphi_k}}}{h_k} \overline{\widehat{u}_c(\varphi)} \, d\varphi = \\
&= -\int_{\mathbb{R}^n} \sum_{k=1}^{n} \sum_{j=1}^{n} a_{kj}(x, h) \big(T_{\text{right}}(x_k) u(x, h)\big)^2 \, dx.
\end{aligned}
$$
(7.2.24)

From (7.2.22) and (7.2.24), we obtain the inequality

$$|I| \geq \nu \int_{\mathbb{R}^n} \sum_{k=1}^{n} \big[T_{\text{right}}(x_k) u(x, h)\big]^2 \, dx \geq C_0 \Omega_h \int_U \sum_{i=1}^{n} \left(\frac{t_i}{h_i}\right)^2 |\widehat{u}_c(t)| \, dt.$$
(7.2.25)

This, together with Proposition 7.2.2, implies that

$$\|P_h L_h u(x, h)\|_{M(-1, h, \Omega_{1h})} \geq C_1 \|u(x, h)\|_{\overset{\circ}{M}(1, h, \Omega_{1h})},$$

with C_1 independent of h. The last inequality implies the desired stability. Proposition 7.2.5 is proved.

7.3. Some Equations

Definition 7.3.1. We say that *step-functions* $u(x, h)$ *form a Lipschitz family*, if the following conditions hold:

(i) $|u(x, h) \leq C_1$, where C_1 does not depend on h;

(ii) $|u(x_1, h) - u(x_2, h)| \leq C_2 |m - k| |h|$ for any $x_1 \in D(k, h)$, $x_2 \in D(m, h)$, where C_2 does not depend on h.

Proposition 7.3.1. *Let* $u(x, h) \in M(r, h)$, $|r| < 1$, *and let* $u_0(x, h)$ *be a Lipschitz family of step-functions. Then, for* $\nu(x, h) = u(x, h) u_0(x, h)$, *the inequality*

$$\|\nu(x, h)\|_{M(r, h)} \leq C(r) \|u(x, h)\|_{M(r, h)}$$
(7.3.1)

holds with a constant $C(r)$ independent of h.

Proof. Consider the case $r \geq 0$. For $r = 0$, our statement is obviously true. By Proposition 5.3.4, the norm in $M_1(r, h)$ is equivalent to the norm in $M(r, h)$ for $0 \leq r < 1$. We have

$$\|\nu(x,h)\|_{M_1(r,h)}^2 = \Omega_h \sum_{|l| \neq 0} \frac{1}{|l|^{n+2r}\Omega_h^{1+2r/n}} \, \Omega_h \sum_{k \in \mathbb{Z}_n} |\nu(k+l) - \nu(k)|^2 + \Omega_h \sum_{k \in \mathbb{Z}_n} |\nu(k)|^2 =$$
$$= I_1(r,h) + I_2(r,h), \tag{7.3.2}$$

where $\Omega_h = \prod_{j=1}^{n} h_j$, $\nu(k) = \pi\nu(x, h)$.

Since the functions $u_0(x, h)$ form a Lipschitz family, the following estimate holds:

$$I_2(r,h) \leq C_1 \|u(x,h)\|_{M(r,h)}^2, \tag{7.3.3}$$

where the constant C_1 does not depend on $h = (h_1, \ldots, h_n)$. Using the obvious inequality

$$|\nu(k+l) - \nu(k)|^2 \leq$$
$$\leq A\Big\{|u_0(k+l)|^2|u(k+l) - u(k)|^2 + |u(k)|^2|u_0(k+l) - u_0(k)|^2\Big\}, \tag{7.3.4}$$

together with (7.3.4), we find that

$$I_1(r,h) \leq A\Omega_h \sum_{|l| \neq 0} \frac{1}{|l|^{n+2r}\Omega_h^{1+2r/n}} \, \Omega_h \sum_{k \in \mathbb{Z}_n} |u_0(k+l)|^2|u(k+l) - u(k)|^2 +$$
$$+ A\Omega_h \sum_{|l| \neq 0} \frac{1}{|l|^{n+2r}\Omega_h^{1+2r/n}} \, \Omega_h \sum_{k \in \mathbb{Z}_n} |u(k)|^2|u_0(k+l) - u_0(k)|^2 =$$
$$= B_1(r,h) + B_2(r,h). \tag{7.3.5}$$

Since the functions $u_0(x, h)$ form a Lipschitz family, we have

$$B_1(r,h) \leq C_2(r)\|u(x,h)\|_{M(r,h)}^2, \tag{7.3.6}$$

where $C_2(r)$ does not depend on h.

Let $h_0 = \min h_j$. For $h_0 \geq 1$, the series in $I_1(r, h)$ is absolutely convergent, and therefore, (7.3.1) holds.

Now, let $0 < h_0 < 1$. Then, it follows from (5.3.3) that for some δ independent of h we have $|h| < \delta$. Let $L = 10\delta$. Denote by $\Lambda_1(h)$ the set of all multi-indices l, $|l| \neq 0$, for which $\sum_{i=1}^{n}(l_i h_i)^2 < L^2$, and let $\Lambda_2(h)$ be the set of the remaining multi-indices l, $|l| \neq 0$. The term $B_2(r, h)$ can be written in the form

$$B_2(r,h) = \sum_{i=1}^{2} A\Omega_h \sum_{l \in \Lambda_i(h)} \frac{1}{|l|^{n+2r}\Omega_h^{1+2r/n}} \, \Omega_h \sum_{k \in \mathbb{Z}_n} |u(k)|^2|u_0(k+l) - u_0(k)|^2 =$$
$$= C_1(r,h) + C_2(r,h). \tag{7.3.7}$$

From (7.3.7), using the Lipschitz property of $u_0(x, h)$, we get

$$C_2(r,h) \leq A_1(r)\Omega_h \sum_{l \in \Lambda_2(h)} \frac{\|u(x,h)\|_{L_2(\mathbb{R}_n)}^2}{(|l||h|)^{n+2r}} \leq A_2(r) \int_{9\delta}^{\infty} \frac{d\rho}{\rho^{1+2r}} \|u(x,h)\|_{L_2(\mathbb{R}_n)}^2 \leq$$
$$\leq A_3(r)\|u(x,h)\|_{L_2(\mathbb{R}_n)}^2, \tag{7.3.8}$$

$$C_1(r,h) \leq A_4(r)\Omega_h \sum_{l \in \Lambda_1(h)} \frac{\|u(x,h)\|_{L_2(\mathbb{R}_n)}^2|l|^2|h|^2}{(|l||h|)^{n+2r}} \leq A_5(r) \int_0^{11\delta} \frac{d\rho}{\rho^{-1+2r}} \|u(x,h)\|_{L_2(\mathbb{R}_n)}^2 \leq$$
$$\leq A_6(r)\|u(x,h)\|_{L_2(\mathbb{R}_n)}^2. \tag{7.3.9}$$

From (7.3.3), (7.3.5)–(7.3.9), we obtain (7.3.1) for $r > 0$.

Let us prove (7.3.1) for $-1 < r < 0$. To that end, consider the set of linear functionals I_λ on $M(r, h)$ associated with the functions $\lambda(x, h) \in M(-r, h)$, $0 < -r < 1$, $\|\lambda(x, h)\|_{M(-r,h)} = 1$. In order to prove (7.3.1), it suffices to show that

$$|I_\lambda(\nu(x, h)| \le C(r)\|u(x, h)\|_{M(r,h)}, \tag{7.3.10}$$

with $C(r)$ independent of h, $\lambda(x, h)$. We have

$$I_\lambda(\nu(x, h)| = \int_{\mathbb{R}_n} u_0(x, h)u(x, h)\,\overline{\lambda(x, h)}\,dx. \tag{7.3.11}$$

Since $0 < -r < 1$, we can apply the inequality (7.3.1) to the family of functions $b(x, h) = u_0(x, h)\overline{\lambda(x, h)}$. This, together with (7.3.11), implies (7.3.10). Proposition 7.3.1 is proved.

7.3.1. The Dirichlet Problem for the General Second Order Elliptic Equation

Let Ω be an open bounded domain with Lipschitz boundary. It is assumed that Ω is a normal domain, admits a partition, and any $a(x) \in C_1(\overline{\Omega})$ can be extended as $\tilde{a}(x) \in C_1(\mathbb{R}_n)$ (see Definitions 7.3.1–7.3.4, 7.3.6, 7.3.9).

Consider a uniformly elliptic equation

$$\sum_{i=1}^n \frac{\partial}{\partial x_i} \sum_{j=1}^n a_{ij}(x)\frac{\partial u}{\partial x_j} = f(x), \tag{7.3.12}$$

where $a_{ij}(x) \in C_1(\overline{\Omega})$, $u(x) \in \mathring{H}_1(\Omega)$, $f(x) \in H_{-1}(\Omega)$; $a_{ij}(x) = a_{ji}(x)$; and the inequality (7.2.22) holds.

In order to find an approximate solution $u(x, h)$ of equation (7.3.12), consider the difference equation

$$L_h u(x, h) = f(x, h), \tag{7.3.13}$$

where L_h as the family of operators defined by (7.3.23); $u(x, h) \in \mathring{M}(1, h, \Omega_{1h})$, $f(x, h) \in M(-1, h, \Omega_{1h})$; and there is an extension $\tilde{f}(x, h) \in M(-1, h)$ of the function $f(x, h)$ which converges in the $H_{-1}(\mathbb{R}_n)$ norm to an extension $lf(x) \in H_{-1}(\mathbb{R}_n)$ as $|h| \to 0$.

It follows from Proposition 7.2.5 that $P_h L_h$ form a stable family of operators from $\mathring{M}(1, h\Omega_{1h})$ to $M(-1, h\Omega_{1h})$, where P_h is the operator of restriction to Ω_{1h}. Proposition 7.3.1 and the assumptions about Ω endure that L_h is a bounded family of operators from $\mathring{M}(r, h, \Omega_{1h})$, $0 < r \le 1$ to $M(r - 2, h)$. Therefore, in order to apply Proposition 7.1.3, it suffices to show that the condition of Π_h^*-approximation holds (see Definition 6.2.7). It should be kept in mind that the operator

$$L(u) = \sum_{i=1}^n \frac{\partial}{\partial x_i} \sum_{j=1}^n a_{ij}(x)\frac{\partial u}{\partial x_j}$$

from $\mathring{H}_1(\Omega)$ to $H_{-1}(\Omega)$ is bounded.

Let $\varphi(x) \in C_0^\infty(\Omega)$. Then $L(\varphi) \in C(\overline{\Omega})$, which means that the operator L is of finite order (see Definition 6.2.3). Since $\varphi(x) \in C_0^\infty(\Omega)$, there exists $\delta > 0$ such that $\operatorname{supp} L_h \Pi_h^* \varphi(x) \subset \Omega_{1h}$ for $|h| < \delta$, where Π_h^* is the operator from Definition 5.4.2. We introduce the notation

$$\psi_j(k_1 h_1, \ldots, k_n h_n) = \pi T_{\text{right}}(x_j)\Pi_h^*\varphi(x),$$

where π is the operator from Definition 5.1.2. We have

$$\pi T_{\text{left}}(x_i)[a_{ij(x,h)}\psi_j(x,h)] = \tag{7.3.14}$$

$$= \frac{a_{ij}(k_1h_1,\ldots,k_ih_i,\ldots,k_nh_n) - a_{ij}(k_1h_1,\ldots,(k_i-1)h_i,\ldots,k_nh_n)}{h_i}\,\psi_j(k_1h_1,\ldots,k_nh_n)+$$

$$+\,\frac{\psi_j(k_1h_1,\ldots,k_ih_i,\ldots,k_nh_n) - \psi_j(k_1h_1,\ldots,(k_i-1)h_i,\ldots,k_nh_n)}{h_i}\times$$

$$\times\, a_{ij}(k_1h_1,\ldots,(k_i-1)h_i,\ldots,k_nh_n).$$

Hence, taking into account that $a_{ij}(x) \in C_1(\overline{\Omega})$, we obtain

$$|L\varphi(k_1h_1,\ldots,k_nh_n) - \pi L_h\Pi_h^*\varphi(x)| = O(|h|), \tag{7.3.15}$$

where $O(|h|) \to 0$ as $|h| \to 0$ uniformly in k such that $D(k,h) \in \Omega_{1h}$. Since $L(\varphi) \in C(\overline{\Omega})$, it follows from (7.3.15) that the operators L_h yield a Π_h^*-approximation for the operator L in the domain Ω.

We see that all assumptions of Proposition 7.1.3 hold, and therefore, there is one and only one solution of equation (7.3.12) for any $f(x) \in H_{-1}(\Omega)$. The solutions $u(x,h)$ of equation (7.2.13) converge strongly in $\mathring{H}_r(\Omega)$ for any r, $0 < r < 1/2$, and their limit is the actual solution $u(x) \in \mathring{H}_1(\Omega)$; moreover,

$$\|u(x,h) - \Pi_{1h}u(x)\|_{\mathring{M}(1,h,\Omega_{1h})} \to 0 \quad \text{as} \quad |h| \to 0,$$

where P_{1h} is the restriction of the integral projector Π_h to the domain Ω_{1h}.

Consider the operator

$$M(u) = L(u) + F(u) = \sum_{i=1}^{n}\frac{\partial}{\partial x_i}\sum_{j=1}^{n}a_{ij}(x)\frac{\partial u}{\partial x_j} + \sum_{i=1}^{n}b_i(x)\frac{\partial u}{\partial x_i} + \lambda(x)u(x),$$

where $b_i(x), \lambda(x) \in C(\overline{\Omega})$; $a_{ij}(x) = a_{ji}(x) \in C_1(\overline{\Omega})$; and condition (7.2.22) holds for the operator $L(u)$.

Consider the equation

$$M(u) = f, \tag{7.3.16}$$

where $u(x) \in \mathring{H}_1(\Omega)$, $f(x) \in H_{-1}(\Omega)$, and suppose that the kernel of the operator M consists of the null-element only.

In order to find an approximate solution $u(x,h)$ of equation (7.3.16), we introduce the difference equation

$$M_h u(x,h) = f(x,h), \tag{7.3.17}$$

where

$$M_h = L_h + F_h, \qquad F_h = \sum_{i=1}^{n}b_i(x,h)T_{\text{right}}(x_i)u(x,h) + \lambda(x,h)u(x,h),$$

$$\pi b_i(x,h) = b_i(k_1h_1,\ldots,k_nh_n), \qquad \pi\lambda(x,h) = \lambda(k_1h_1,\ldots,k_nh_n),$$

$$u(x,h) \in \mathring{M}(1,h,\Omega_{1h}), \qquad f(x,h) \in M(-1,h\Omega_{1h}),$$

and there exists an extension $\widetilde{f}(x,h) \in M(-1,h)$ of the function $f(x,h)$ which converges in the $H_{-1}(\mathbb{R}_n)$ norm to an extension $lf(x) \in H_{-1}(\mathbb{R}_n)$ of $f(x)$ as $|h| \to 0$.

Just as above, one can verify all assumptions of Proposition 7.1.5 in this case, and therefore, there is one and only one solution $u(x) \in \mathring{H}_1(\Omega)$ of equation (7.3.16) for any $f(x) \in H_{-1}(\Omega)$. For some $\delta > 0$ and $|h| < \delta$, each equation (7.3.17) admits a solution $u(x,h)$ and these solutions converge strongly in $\mathring{H}_r(\Omega)$ to the actual solution $u(x) \in \mathring{H}_1(\Omega)$ for any r, $0 < r < 1/2$. Moreover,

$$\|u(x,h) - \Pi_{1h}u(x)\|_{\mathring{M}(1,h,\Omega_{1h})} \to 0 \quad \text{as} \quad |h| \to 0.$$

7.3.2. Characteristic Hypersingular Equation

$$\int_\Omega \frac{u(x)}{r^{n+2p+\lambda}} \, dx = f(x_0), \qquad\qquad (7.3.18)$$

where $0 < \lambda < 1$; $p = 0, 1, 2, \ldots$; $u(x) \in \mathring{H}_{p+\lambda/2}(\Omega)$, $f(x_0) \in H_{-p-\lambda/2}(\Omega)$.

The symbol in the left-hand side of (7.3.18) is defined in (6.3.32). The operator

$$Tu = \int_\Omega \frac{u(x)}{r^{n+2p+\lambda}} \, dx$$

from $\mathring{H}_{p+\lambda/2}(\Omega)$ to $H_{-p-\lambda/2}(\Omega)$ is bounded.

In order to find an approximate solution $u(x, h)$ of equation (7.3.18), consider the equation

$$C(\lambda, p)T_h(\lambda, p)u(x, h) = f(x, h), \qquad\qquad (7.3.19)$$

where the operators $T_h(\lambda, p)$ are defined by (7.2.1) and (6.1.50); $C(\lambda, 0) = 1$,

$$u(x, h) \in \mathring{M}\left(p + \frac{\lambda}{2}, \Omega_{1h}\right), \qquad f(x, h) \in M\left(-p - \frac{\lambda}{2}, h, \Omega_{1h}\right),$$

and there are extensions $\widetilde{f}(x, h) \in M(-p - \lambda/2, h)$ of $f(x, h)$ which converge in the $H_{-p-\lambda/2}(\mathbb{R}_n)$ norm to an extension $lf(x) \in H_{-p-\lambda/2}(\mathbb{R}_n)$ of $f(x)$ as $|h| \to 0$.

It follows from Propositions 6.1.4, 6.3.1, and 6.3.3, and the results of Sections 6.1.4, 7.2.2, and 7.2.3 that all assumptions of Theorem 7.1.3 hold for the operator T and the operator family $C(\lambda, p)T_h(\lambda, p)$. Therefore, for any $f \in H_{-p-\lambda/2}(\Omega)$, equation (7.3.18) admits a unique solution $u(x) \in \mathring{H}_{p+\lambda/2}$. For $p = 0$, $0 < \lambda < 1$, the solutions $u(x, h)$ of equation (7.3.9) converge to $u(x) \in \mathring{H}_{\lambda/2}$ in the $\mathring{H}_r(\Omega)$-norm for any r, $0 < r < \lambda/2$. For $p = 1, 2, \ldots$, and $0 < \lambda < 2$, the solutions $u(x, h)$ converge to $u(x) \in \mathring{H}_{p+\lambda/2}(\Omega)$ in the $\mathring{H}_r(\Omega)$-norm for any r, $0 < r < 1/2$, as $|h| \to 0$. In all these cases, we have

$$\|u(x, h) - \Pi_{1h}u(x)\|_{\dot{M}(p+\lambda/2, h, \Omega_{1h})} \to 0 \quad \text{as} \quad |h| \to 0.$$

7.3.3. Iterated Laplace Operator of Order m (the Dirichlet Problem)

Consider the equation

$$\Delta^m u(x) = f(x), \qquad\qquad (7.3.20)$$

where Δ^m is the m-th order iterated Laplace operator; $u(x) \in \mathring{H}_m(\Omega)$, $f(x) \in H_{-m}(\Omega)$.

To find an approximate solution $u(x, h)$ of equation (7.3.20), we use the difference equation

$$\Delta_{hp}^m u(x, h) = f(x, h), \qquad\qquad (7.3.21)$$

where Δ_{hp}^m is the difference Laplace operator of order m; $u(x, h) \in \mathring{M}(m, h, \Omega_{1h})$, $f(x, h) \in M(-m, h, \Omega_{1h})$; and there are extensions $\widetilde{f}(x, h) \in M(-m, h)$ of $f(x, h)$ which converge in the $H_{-m}(\mathbb{R}_n)$-norm to an extension $lf(x) \in H_{-m}(\mathbb{R}_n)$ of $f(x)$ as $|h| \to 0$.

It follows from Propositions 7.2.3 and 7.2.8, and the results of Section 6.1.4 that all assumptions of Proposition 7.1.3 hold in this case. Therefore, for any $f(x) \in H_{-m}(\Omega)$, equation (7.3.20) admits a unique solution $u(x) \in \mathring{H}_m(\Omega)$. The approximate solutions $u(x, h)$ of equations (7.3.21) converge to $u(x) \in \mathring{H}_m(\Omega)$ in the $\mathring{H}_m(\Omega)$-norm for any r, $0 < r < 1/2$, as $|h| \to 0$. Moreover,

$$\|u(x, h) - \Pi_{1h}u(x)\|_{\dot{M}(m, h, \Omega_{1h})} \to 0 \quad \text{as} \quad |h| \to 0.$$

7.3.4. An Equation Related to Diffraction Problems

Let $u(x)$ be a solution of the Helmholtz equation

$$\Delta u + \lambda^2 u = 0, \quad x \in \mathbb{R}_3 \setminus \Omega,$$
$$\left. \frac{\partial u}{\partial n} \right|_\Omega = f, \tag{7.3.22}$$

where Δ is the Laplace operator; Ω is a bounded piece of the plane α; n is the normal to Ω; f is a given function in Ω. It is assumed that the domain Ω is normal and admits a partition (see Definitions 5.5.6 and 5.5.9).

Suppose that Sommerfeld's radiation conditions hold,

$$u(x) = O\big(|x|^{-1}\big), \qquad \frac{\partial u(x)}{\partial |x|} = i\lambda u(x) = o\big(|x|^{-1}\big) \quad \text{as} \quad |x| \to \infty. \tag{7.3.23}$$

Moreover, we assume that

$$\operatorname{Re} \lambda > 0, \qquad \operatorname{Im} \lambda \leq 0. \tag{7.3.24}$$

The fundamental solution of the equation $\Delta u + \lambda^2 u = 0$ with the conditions (7.3.23) in $\mathbb{R}_3$ has the form

$$\varepsilon = -\frac{1}{4\pi r} - \frac{1}{4\pi} \frac{e^{-i\lambda r} - 1}{r}. \tag{7.3.25}$$

We seek a solution of problem (7.3.22), (7.3.23) in terms of the double layer potential $u(x, y)$ distributed over Ω. Thus, for the function $u(x, y)$ we have the equation

$$T(u) + F(u) = g(M_0), \tag{7.3.26}$$

where

$$T(u) = \int_\Omega \frac{u(x, y)}{r_{MM_0}^3} \, dx \, dy, \qquad F(u) = \int_\Omega \frac{J(r_{MM_0})}{r_{MM_0}^3} \, dx \, dy,$$

$$g(M_0) = -4\pi f(x_0, y_0), \quad r_{MM_0} = \big[(x_0 - x)^2 + (y_0 - y)^2 \big]^{1/2},$$

$$J(r_{MM_0}) = e^{-i\lambda r_{MM_0}} + i\lambda e^{-i\lambda r_{MM_0}} r_{MM_0} - 1,,$$

$$u(x, y) \in \mathring{H}_{1/2}(\Omega), \qquad g(x_0, y_0) \in H_{-1/2}(\Omega).$$

Let λ be such that the kernel of the operator $M(u) = T(u) + F(u)$ is trivial. Since Ω is bounded, there exist a bounded domain $\Omega_1 \supset \Omega$ and a function $\alpha(OM_0) \in C_0^\infty(\mathbb{R}_2)$ (O is the origin), $\alpha(x_0, y_0) = 1$ for $M_0(x_0, y_0) \in \Omega_1$, such that the equation

$$\int_\Omega \frac{u(x, y)}{r_{MM_0}^3} \, dx \, dy + \int_\Omega \frac{J_1(r_{MM_0}) u(x, y)}{r_{MM_0}^3} \, dx \, dy = g(M_0), \tag{7.3.27}$$

is equivalent to (7.3.26). Here

$$J_1(r_{MM_0}) = J(r_{MM_0}) \alpha(r_{MM_0}) \quad \text{for} \quad \alpha(r_{MM_0}) = 1, \quad M, M_0 \in \Omega.$$

The following inequality holds

$$|J_2(r_{MM_0})| = \left| \frac{J_1(r_{MM_0})}{r_{MM_0}^3} \right| \leq \frac{C}{r_{MM_0}},$$

and therefore, $J_2(r_{MM_0}) \in L_2(\mathbb{R}_2)$ for each fixed M_0. Let us write equation (7.3.27) in operator form

$$T(u) + F_1(u) = g(M_0). \qquad (7.3.28)$$

In order to find an approximate solution $u(x, y, h)$ of equation (7.3.28), we consider a partition of $\mathbb{R}_2$ with vector-valued step $h = (h_1, h_2)$ and cells $D(k, h)$ (see Definition 5.1.1) such that condition 5.3.3 is satisfied. The grid function $u(k) = \pi u(x, y, h)$ (see Definition 5.1.2 regarding π) is found from the equation

$$\sum_k u(k)\gamma_1(k, m) + \sum_k u(k)\gamma_2(k, m) = g(M_{0m}), \qquad M_{0m} \in \Omega_{1h}, \qquad (7.3.29)$$

where the sum is taken over all $k = (k_1, k_2)$ such that $D(k, h) \subset \Omega_{1h}$, $M_{0m} = (m_1 h_1, m_2 h_2)$; $\pi g(x, h) = g(M_{0m}, h)$, and there are extensions $\widetilde{g}(x, y, h) \in M(-1/2, h)$ of $g(x, y, h)$ which converge to an extension $lg(x, y) \in H_{-1/2}(\mathbb{R}_n)$ of $g(x, y)$. The coefficients of equations (7.3.29) are found from the relations

$$\gamma_1(k, m) = \int_{D(k,h)} \frac{dx\, dy}{r_{MM_{0m}}^3}, \qquad \gamma_2(k, m) = \int_{D(k,h)} J_2(r_{MM_{0m}})\, dx\, dy. \qquad (7.3.30)$$

The operator $T_{h1} = \sum_k u(k)\gamma_1(k, m)$ is of vortex type and all assumptions of Proposition 7.1.3 hold for this operator, as shown above. The operator $F_{1h}u(k) = \sum_k u(k)\gamma_2(k, m)$ is a discrete convolution operator and its symbol $\widehat{F}_{1h}(\varphi)$ satisfies the inequality (see (5.1.9))

$$\left| \widehat{F}_{1h}(\varphi) \right| = \left| \sum_k \gamma_2(k, 0)\, e^{i(k, \varphi)} \right| \leq \int_{\mathbb{R}_2} |J_2(r_{MM_{0m}})|\, dx\, dy \leq C. \qquad (7.3.31)$$

It follows that the family of operators F_{1h} from $M(r, h)$ to $M(r, h)$ is bounded for any r. Since

$$\sum_k |\gamma_2(k, m)| \leq C_1$$

with C_1 independent of m, h, we see that for the operator F_1 and the family F_{1h} the Π_h^*-approximation condition holds in the domain Ω. Thus, all assumptions of Proposition 7.1.5 hold for the operators T, F_1 and the families T_{h1}, F_{h1}. It follows that equations (7.3.26), (7.3.28) admit one and only one solution $u(x, y) \in \mathring{H}_{1/2}(\Omega)$. For some δ and $|h| < \delta$, the system of equations (7.3.29) admits a solution. The solutions $u(x, y, h)$ of system (7.3.29) converge to $u(x, y) \in \mathring{H}_{1/2}(\Omega)$ in the $\mathring{H}_r(\Omega)$-norm for any r, $0 < r < 1/2$, as $|h| \to 0$. Moreover,

$$\|u(x, y, h) - \Pi_{1h}u(x, y)\|_{\mathring{M}(1/2, h, \Omega_{1h})} \to 0 \quad \text{as} \quad |h| \to 0.$$

Chapter 8

Asymptotic Estimates of the Discrete Green Function

8.1. Restriction Problems for Pseudodifference Operators

Definition 8.1.1. Let ε be a set of indices $k = (k_1, \ldots, k_n)$ and let A be a pseudodifference operator (see Definition 7.1.7). The operator pA defined by

$$pA(x) = \frac{1}{(2\pi)^n} \int_{-\pi}^{\pi} \widehat{A}(\varphi, k, h) \widehat{x}_c(\varphi) e^{-ik\varphi} d\varphi, \quad k \in \varepsilon,$$

is called *the restriction of A to the set ε*.

Definition 8.1.2. *The problem of restriction of a pseudodifference operator* is formulated as follows: given a grid function $f(k_1, \ldots, k_n) = f(k)$ for $k \in \varepsilon$, find a grid function $x(k)$ defined on the same set of nodes and satisfying the relations

$$pA(x) = f, \quad k \in \varepsilon. \tag{8.1.1}$$

Let us examine some restriction problems. Here, we assume that the symbol $\widehat{A}(\varphi, k, h)$ of the pseudodifference operator A does not depend on k, and the symbol $\widehat{A}(\varphi, h)$ is denoted by $\widehat{A}(\varphi)$.

Definition 8.1.3. Consider the sets of integer numbers

$$\varepsilon_1 = \{k \in \mathbb{Z} : k \geq 0\}, \qquad \varepsilon_2 = \{k \in \mathbb{Z} : k \leq 0\}.$$

Problems of restriction to ε_1 or ε_2 are called problems of restriction to L_+ or L_-, respectively. For $k = (k_1, \ldots, k_n)$ and $\varepsilon_{1i} = \{k \in \mathbb{Z}_n : k_i \geq 0\}$, $\varepsilon_{2i} = \{k \in \mathbb{Z}_n : k_i \leq 0\}$, problems of restriction to ε_{1i} or ε_{2i} are called problems of restriction to L_{i+} or L_{i-}, respectively.

Consider the problem of restriction to L_+ for a pseudodifference operator with the symbol $\widehat{A}(\varphi) \in H(\alpha)$. Thus, $\widehat{A}(\varphi)$ is a Hölder continuous function on the curve L described by the equation $|z| = 1$ (or $|e^{i\varphi}| = 1$); $\widehat{A}(\varphi) \neq 0$ on L; the index of the function $A(\varphi)$ is equal to zero (see *Gakhov* (1966), *Muskhelishvili* (1968)). Then the function $\widehat{A}(\varphi)$ admits a unique factorization (see *ibid.*), i.e., it can be represented in the form

$$\widehat{A}(\varphi) = A_+(\varphi) A_-(\varphi), \qquad A_+(\varphi) = \lim_{z \to L+0} A_+(z), \qquad A_-(\varphi) = \lim_{z \in L-0} A_-(z),$$

where the functions $A_+(z)$ and $A_-(z)$ are analytic for $|z| < 1$ and $|z| > 1$, respectively; $A_{\pm}(\varphi)$ are Hölder continuous functions, $A_{\pm}(\varphi) \neq 0$ on L, $\lim_{z \to \infty} A_-(z) = 0$.

Proposition 8.1.1. *Suppose that $\widehat{f}_c(\varphi)$ is a Hölder continuous function and the symbol $\widehat{A}(\varphi)$ satisfies the above conditions. Then, the problem of restriction to L_+ of the form (8.1.1) admits a unique solution $\widehat{x}_c(\varphi)$ in the class of Hölder continuous functions ($\widehat{x}_c(\varphi)$ and $\widehat{f}_c(\varphi)$ are the discrete Fourier transforms of the functions $f(k)$ and $x(k)$; see Definition 5.1.4.).*

Proof. Any function $b(\varphi)$, $0 \le \varphi \le 2\pi$, can be regarded as a function of the complex variable $\tau = e^{i\varphi}$, $b(\varphi) = b^*(\tau)$. In what follows, we write $b(\tau)$ instead of $b^*(\tau)$.

Suppose that problem (8.1.1) has a solution $\widehat{x}_c(\tau) = \sum_{k=0}^{\infty} x_k \tau^k$. Then

$$\widehat{A}(\tau)\widehat{x}_c(\tau) = \widehat{f}_c(\tau) + y_-(\tau), \tag{8.1.2}$$

where $y_-(e^{i\varphi}) = \sum_{k=-\infty}^{-1} y_k e^{ik\varphi}$ is the unknown "tail". Relation (8.1.2), combined with the factorization formula $\widehat{A}(\tau) = A_+(\tau)A_-(\tau)$, implies that

$$A_+(\tau)\widehat{x}_c(\tau) = \frac{\widehat{f}_c(\tau)}{A_-(\tau)} + \frac{y_-(\tau)}{A_-(\tau)}. \tag{8.1.3}$$

The function $g(\tau) = \widehat{f}_c(\tau)/A_-(\tau)$ can be represented in the form (see *Gakhov* (1966), *Muskhelishvili* (1968))

$$\frac{\widehat{f}_c(\tau)}{A_-(\tau)} = \Phi^+(\tau) - \Phi^-(\tau). \tag{8.1.4}$$

The functions $\Phi^+(\tau)$ and $\Phi^-(\tau)$ are given by the Sokhotski–Plemelj formulas

$$
\begin{aligned}
\Phi^+(\tau) &= \frac{1}{2}\frac{\widehat{f}_c(\tau)}{A_-(\tau)} + \frac{1}{2\pi i}\int_L \frac{\widehat{f}_c(\tau_1)}{A_-(\tau_1)}\frac{d\tau_1}{\tau_1 - \tau}, \\
\Phi^-(\tau) &= -\frac{1}{2}\frac{\widehat{f}_c(\tau)}{A_-(\tau)} + \frac{1}{2\pi i}\int_L \frac{\widehat{f}_c(\tau_1)}{A_-(\tau_1)}\frac{d\tau_1}{\tau_1 - \tau}.
\end{aligned}
\tag{8.1.5}
$$

From (8.1.3) and (8.1.4), it follows that

$$A_+(\tau)\widehat{x}_c(\tau) = \Phi^+(\tau), \qquad \frac{y_-(\tau)}{A_-(\tau)} - \Phi^-(\tau) = 0. \tag{8.1.6}$$

From (8.1.5) and (8.1.6), we find that

$$\widehat{x}_c(\tau) = \frac{1}{A_+(\tau)}\left[\frac{1}{2}\frac{\widehat{f}_c(\tau)}{A_-(\tau)} + \frac{1}{2\pi i}\int_L \frac{\widehat{f}_c(\tau_1)}{A_-(\tau_1)}\frac{d\tau_1}{\tau_1 - \tau}\right]. \tag{8.1.7}$$

Relation (8.1.7) and the properties of singular integrals (see *Gakhov* (1966), *Muskhelishvili* (1968)) allow us to claim that $x(\tau)$ is a Hölder continuous function.

Let us show that $x(\tau)$ given by (8.1.7) is a solution of the restriction problem (8.1.1).

Problem (8.1.1) is equivalent to the equation

$$\widehat{f}_x(z) = \frac{1}{2\pi i}\int_L \frac{\widehat{A}(\tau)\widehat{x}_c(\tau)}{\tau - z}\,d\tau,$$

where $z \in D_+ = \{z : |z| < 1\}$. We further have

$$
\frac{1}{2\pi i} \int_L \frac{\widehat{A}(\tau)\widehat{x}_c(\tau)\,d\tau}{\tau - z} =
$$

$$
= \frac{1}{2\pi i} \int_L \frac{\widehat{A}(\tau)}{\tau - z} \left[\frac{1}{2} \frac{\widehat{f}_c(\tau)}{\widehat{A}(\tau)} + \frac{1}{2\pi i} \frac{1}{A^+(\tau)} \int_L \frac{\widehat{f}_c(\tau_1)}{A_-(\tau_1)} \frac{d\tau_1}{\tau_1 - \tau} \right] d\tau =
$$

$$
= \frac{1}{2} \frac{1}{2\pi i} \int_L \frac{f_c(\tau)}{\tau - z}\,d\tau +
$$

$$
+ \frac{1}{2\pi i} \int_L \frac{f_c(\tau_1)}{A_-(\tau_1)(\tau_1 - z)} \left[\frac{1}{2\pi i} \int_L \frac{A_-(\tau)\,d\tau}{\tau - z} + \frac{1}{2\pi i} \int_L \frac{A_-(\tau)\,d\tau}{\tau_1 - \tau} \right] d\tau_1 =
$$

$$
= \frac{1}{2} f_c(z) + \frac{1}{2} f_c(z) = f_c(z), \tag{8.1.8}
$$

where we have used the relations

$$
\frac{1}{2\pi i} \int_L \frac{A_-(\tau)}{\tau - z}\,d\tau = 0, \quad z \in D_+; \qquad \frac{1}{2\pi i} \int_L \frac{A_-(\tau)}{\tau - \tau_1}\,d\tau = \frac{1}{2} A_-(\tau_1).
$$

The uniqueness of the solution follows from the first part of the proof. Proposition 8.1.1 is proved.

Consider the two-dimensional problem of restriction to L_{1+} for a pseudodifference operator A with symbol $\widehat{A}(\varphi_1, \varphi_2)$. In this case, the restriction set is $\varepsilon_1 = \{k = (k_1, k_2) \in \mathbb{Z}_2 : k_1 \geq 0\}$.

Suppose that the symbol $\widehat{(}\varphi_1, \varphi_2)$ is a Hölder continuous function non-vanishing on the edge $L = L_1 \times L_2$, where L_1 and L_2 are respectively given by the equations $|z_1| = 1$ and $|z_2| = 1$ for complex z_1 and z_2. Functions $b(\varphi_1, \varphi_2)$ defined on the edge $L = L_1 \times L_2$ will be regarded as functions of complex variables $\tau_1 = e^{i\varphi_1}$, $\tau_2 = e^{i\varphi_2}$, i.e., $b(\varphi_1, \varphi_2) = b^*(\tau_1, \tau_2)$; and we will write $b(\tau_1, \tau_2)$ instead of $b^*(\tau_1, \tau_2)$. It is assumed that the partial indexes of the function $\widehat{A}(\tau_1, \tau_2)$ are equal to zero (see *Kakichev* (1959), *Kakichev* (1967)). Then, the symbol $\widehat{A}(\tau_1, \tau_2)$ admits factorization (see *Kakichev* (1959), *Kakichev* (1967)),

$$
\widehat{A}(\tau_1, \tau_2) = A^{++}(\tau_1, \tau_2)A^{+-}(\tau_1, \tau_2)A^{-+}(\tau_1, \tau_2)A^{--}(\tau_1, \tau_2), \tag{8.1.9}
$$

where $A^{++}(\tau_1, \tau_2)$, $A^{+-}(\tau_1, \tau_2)$, $A^{-+}(\tau_1, \tau_2)$, $A^{--}(\tau_1, \tau_2)$ are the boundary values (values on the edge L) of the respective functions $A^{++}(z_1, z_2)$, $A^{+-}(z_1, z_2)$, $A^{-+}(z_1, z_2)$, $A^{--}(z_1, z_2)$ which are analytic in the respective domains

$$
D^{++} = D_1^+ \times D_2^+, \quad D^{+-} = D_1^+ \times D_2^-, \quad D^{-+} = D_1^- \times D_2^+, \quad D^{--} = D_1^- \times D_2^-,
$$

where $D_1^+ = \{z : |z_1| < 1\}$, $D_2^+ = \{z : |z_2| < 1\}$, $D_1^- = \{z : |z_1| > 1\}$, $D_2^- = \{z : |z_2| > 1\}$.

Factorization (8.1.9) is unique if

$$
A^{\pm=}(z_1, \infty) = 1, \qquad A^{=\pm}(\infty, z_2) = 1.
$$

In what follows, we use the Sokhotski formulas for boundary values of a piecewise analytic function $\Phi(z_1, z_2)$ defined by the Cauchy integral

$$
\Phi(z_1, z_2) = -\frac{1}{4\pi^2} \int_L \frac{x(\tau_1, \tau_2)\,d\tau}{(\tau_1 - z_1)(\tau_2 - z_2)},
$$

where $L = L_1 \times L_2$, $\tau_1 \in L_1$, $\tau_2 \in L_2$. The boundary values satisfy the following relations:

$$\left.\begin{array}{c} \Phi^{++}(\tau_1, \tau_2) \\ \Phi^{--}(\tau_1, \tau_2) \end{array}\right\} = \frac{1}{4}\left(x \pm S_1(x) \pm S_2(x) + S_{12}(x)\right),$$

$$\left.\begin{array}{c} \Phi^{+-}(\tau_1, \tau_2) \\ \Phi^{-+}(\tau_1, \tau_2) \end{array}\right\} = \frac{1}{4}\left(-x \mp S_1(x) \pm S_2(x) + S_{12}(x)\right),$$

where

$$S_1(x) = \frac{1}{i\pi} \int_{L_1} \frac{x(t_1, \tau_2)\, dt_1}{(t_1 - \tau_1)}, \qquad S_2(x) = \frac{1}{i\pi} \int_{L_1} \frac{x(\tau_1, t_2)\, dt_2}{(t_2 - \tau_2)},$$

$$S_{12} = -\frac{1}{\pi^2} \int_L \frac{x(t_1, t_2)\, dt_1\, dt_2}{(t_1 - \tau_1)(t_2 - \tau_2)}. \tag{8.1.10}$$

Proposition 8.1.2. *Let $\widehat{f}_c(\varphi_1, \varphi_2)$ be a Hölder continuous function and the above assumptions hold for the symbol $\widehat{A}(\varphi_1, \varphi_2)$. Then, problem (8.1.1) of restriction to L_{1+} admits a unique solution $\widehat{x}_c(\varphi_1, \varphi_2)$ in the Hölder class (here $\widehat{f}_c(\varphi_1, \varphi_2)$) and $\widehat{x}_c(\varphi_1, \varphi_2)$ are the discrete Fourier transforms of the grid functions $f(k)$ and $x(k)$; see Definition 5.1.4).*

Proof. Let

$$\widehat{x}_c(\tau_1, \tau_2) = \sum_{k_2=-\infty}^{\infty} \sum_{k_1=0}^{\infty} x(k_1, k_2)\tau_1^{k_1}\tau_2^{k_2}$$

be a solution of problem (8.1.1) (it should be kept in mind that we retain the notation of the functions, when passing to the variables $\tau_1 = e^{i\varphi_1}$, $\tau_2 = e^{i\varphi_2}$). Then, the following relation holds:

$$\widehat{A}(\tau_1, \tau_2)\widehat{x}_c(\tau_1, \tau_2) = \widehat{f}_c(\tau_1, \tau_2) - y^{-+}(\tau_1, \tau_2) + y^{--}(\tau_1, \tau_2), \tag{8.1.11}$$

where $y^{-+}(\tau_1, \tau_2)$, $y^{--}(\tau_1, \tau_2)$ are the unknown "tails". Hence, using the factorization formula (8.1.9) for the symbol $\widehat{A}(\tau_1, \tau_2)$, we obtain

$$A^{++}(\tau_1, \tau_2)A^{+-}(\tau_1, \tau_2)\widehat{x}_c(\tau_1, \tau_2) =$$

$$= \frac{\widehat{f}_c(\tau_1, \tau_2)}{A^{-+}(\tau_1, \tau_2)A^{--}(\tau_1, \tau_2)} + \frac{y^{--}(\tau_1, \tau_2) - y^{-+}(\tau_1, \tau_2)}{A^{-+}(\tau_1, \tau_2)A^{--}(\tau_1, \tau_2)}. \tag{8.1.12}$$

The function

$$g(\tau_1, \tau_2) = \frac{\widehat{f}_c(\tau_1, \tau_2)}{A^{-+}(\tau_1, \tau_2)A^{--}(\tau_1, \tau_2)}$$

can be represented in the form

$$g(\tau_1, \tau_2) = \Phi^{++}(\tau_1, \tau_2) - \Phi^{+-}(\tau_1, \tau_2) - \Phi^{-+}(\tau_1, \tau_2) + \Phi^{--}(\tau_1, \tau_2).$$

By the Sokhotski–Plemelj formulas we get

$$\Phi^{++}(\tau_1, \tau_2) - \Phi^{+-}(\tau_1, \tau_2) = \tag{8.1.13}$$

$$= \frac{1}{2} \frac{\widehat{f}_c(\tau_1, \tau_2)}{A^{-+}(\tau_1, \tau_2)A^{--}(\tau_1, \tau_2)} + \frac{1}{2\pi i} \int_{L_1} \frac{\widehat{f}_c(t_1, \tau_2)}{A^{-+}(t_1, \tau_2)A^{--}(t_1, \tau_2)} \frac{dt_1}{(t_1 - \tau_1)}.$$

From (8.1.12) and (8.1.13), it follows that

$$\widehat{x}_c(\tau_1, \tau_2) = \frac{1}{A^{++}A^{+-}}\left[\frac{1}{2}\frac{\widehat{f}_c(\tau_1, \tau_2)}{A^{-+}A^{--}} + \frac{1}{2\pi i} \int_{L_1} \frac{\widehat{f}_c(t_1, \tau_2)}{A^{-+}(t_1, \tau_2)A^{--}(t_1, \tau_2)} \frac{dt_1}{(t_1 - \tau_1)}\right]. \tag{8.1.14}$$

This relation and the known properties of singular integrals (see *Gakhov* (1966)) allow us to claim that $x(\tau_1, \tau_2)$ is a Hölder continuous function. Let us show that $x(\tau_1, \tau_2)$ given by (8.1.14) is a solution of problem (8.1.1) on L_{1+}.

The function $\widehat{f}_c(\tau_1, \tau_2)$ can be represented in the form

$$\widehat{f}_c(\tau_1, \tau_2) = f^{++}(\tau_1, \tau_2) - f^{+-}(\tau_1, \tau_2),$$

where $f^{++}(\tau_1, \tau_2)$ and $f^{+-}(\tau_1, \tau_2)$ are the boundary values of the analytic functions $f^{++}(z_1, z_2)$ and $f^{+-}(z_1, z_2)$ on L. Therefore, problem (8.1.1) of the restriction to L_{1+} is equivalent to the system of equations

$$-\frac{1}{4\pi^2} \int_L \frac{\widehat{A}(\tau_1, \tau_2)\widehat{x}_c(\tau_1, \tau_2)\, d\tau_1\, d\tau_2}{(\tau_1 - z_1)(\tau_2 - z_2)} = f^{++}(z_1, z_2), \qquad (z_1, z_2) \in D^{++},$$

$$-\frac{1}{4\pi^2} \int_L \frac{\widehat{A}(\tau_1, \tau_2)\widehat{x}_c(\tau_1, \tau_2)\, d\tau_1\, d\tau_2}{(\tau_1 - z_1)(\tau_2 - z_2)} = f^{+-}(z_1, z_2), \qquad (z_1, z_2) \in D^{+-},$$

(8.1.15)

Substituting (8.1.14) into the left-hand side of (8.1.15), we obtain

$$-\frac{1}{4\pi^2} \int_L \frac{\widehat{A}(\tau_1, \tau_2)}{(\tau_1 - z_1)(\tau_2 - z_2)} \left[\frac{1}{2}\frac{\widehat{f}_c(\tau_1, \tau_2)}{\widehat{A}(\tau_1, \tau_2)} + \right.$$

$$\left. + \frac{1}{2\pi i}\frac{1}{A^{+-}(\tau_1, \tau_2)A^{++}(\tau_1, \tau_2)} \int_{L_1} \frac{\widehat{f}_c(t_1, \tau_2)}{A^{-+}(t_1, \tau_2)A^{--}(t_1, \tau_2)} \frac{dt}{(t_1 - \tau_1)} \right] d\tau_1\, d\tau_2 =$$

$$= -\frac{1}{4\pi^2} \int_L \frac{2^{-1}\widehat{f}_c(\tau_1, \tau_2)\, d\tau_1\, d\tau_2}{(\tau_1 - z_1)(\tau_2 - z_2)} - \frac{1}{4\pi^2} \int_L \frac{\widehat{f}_c(t_1, \tau_2)}{A^{-+}(t_1, \tau_2)A^{--}(t_1, \tau_2)} \frac{1}{(\tau_1 - z_1)(\tau_2 - z_2)} \times$$

$$\times \left[\frac{1}{2\pi i} \int_{L_1} \frac{A^{--}(\tau_1, \tau_2)A^{-+}(\tau_1, \tau_2)\, d\tau_1}{\tau_1 - z_1} + \frac{1}{2\pi i} \int_{L_1} \frac{A^{--}(\tau_1, \tau_2)A^{-+}(\tau_1, \tau_2)\, d\tau_1}{t_1 - \tau_1} \right] dt_1\, d\tau_2 =$$

$$= -\frac{1}{4\pi^2} \int_L \frac{\widehat{f}_c(\tau_1, \tau_2)\, d\tau_1\, d\tau_2}{(\tau_1 - z_1)(\tau_2 - z_2)}.$$

(8.1.16)

For the derivation of (8.1.16) we have used the relations (see *Gakhov* (1966))

$$\frac{1}{2\pi i} \int_{L_1} \frac{A^{--}(\tau_1, \tau_2)A^{-+}(\tau_1, \tau_2)\, d\tau_1}{\tau_1 - z_1} = 0, \qquad z_1 \in D_1^+,$$

$$\frac{1}{2\pi i} \int_{L_1} \frac{A^{--}(\tau_1, \tau_2)A^{-+}(\tau_1, \tau_2)\, d\tau_1}{t_1 - \tau_1} = \frac{1}{2}A^{--}(t_1, \tau_2)A^{-+}(t_1, \tau_2).$$

Relation (8.1.16) implies (8.1.15) for $(z_1, z_2) \in D^{++}$, $(z_1, z_2) \in D^{+-}$. The uniqueness of $x(\tau_1, \tau_2)$ follows from the first part of the proof. Proposition 8.1.2 is proved.

8.2. Estimates of the Discrete Green Function for the Discrete Prandtl Operator in the Halfplane

Consider a canonical partition of the plane $\mathbb{R}_2$ with vector-valued step $h = (h, h)$, $h > 0$, and cells $D(k, h)$ (see Definition 5.1.1). For a given domain $G \subset \mathbb{R}_2$, let ε be the set of all indices $k = (k_1, k_2)$ for which $D(k, h) \subset G$. Let R be the linear space of step-functions taking constant values on $D(k, h)$. As a basis in R we take functions $e_{k_1 k_2} = e_k$ equal to 1 on $D(k, h)$ and identically vanishing elsewhere. Any element $x \in R$ can be represented in the form $x = \sum_{k \in \varepsilon} x(k) e_k$. Consider an operator $A : R \to R$ for which

$$A(e_k) = \sum_{k \in \varepsilon} f(k_1, k_2, m_1, m_2) e_{m_1 m_2} = \sum_{k \in \varepsilon} f(k, m) e_m.$$

Thus, we have

$$A(x) = A\left(\sum_{k \in \varepsilon} x(k) e_k \right) =$$

$$= \sum_{k \in \varepsilon} x(k) A(e_k) = \sum_{k \in \varepsilon} x(k) \sum_{m \in \varepsilon} f(k, m) e_m = \sum_{m \in \varepsilon} e_m \sum_{k \in \varepsilon} f(k, m) x(k).$$

The function $f(k, m)$ will be called *the characteristic function of the operator* A. Let us find the characteristic function of the conjugate operator A^*. In R, we introduce the scalar product

$$(x, y) = \sum_{k \in \varepsilon} x(k) y(k) \quad \text{for} \quad x = \sum_{k \in \varepsilon} x(k) e_k, \quad y = \sum_{k \in \varepsilon} y(k) e_k.$$

Let

$$A^*(x) = \sum_{m \in \varepsilon} e_m \sum_{k \in \varepsilon} f^*(k, m) x(k).$$

By the definition of the conjugate operator, $(Ax, y) = (x, A^*, y)$, we have

$$(Ax, y) = \sum_{m \in \varepsilon} y(m) \sum_{k \in \varepsilon} f(k, m) x(k) = \sum_{k \in \varepsilon} x(k) \sum_{m \in \varepsilon} f(k, m) y(m),$$

$$(x.A^* y) = \sum_{k \in \varepsilon} x(k) \sum_{m \in \varepsilon} f^*(m, k) y(m).$$

Therefore,

$$f^*(m, k) = f(k, m), \qquad f(k, m) = f(m, k). \tag{8.2.1}$$

Let $f(k, m)$ and $g(k, m)$ be the characteristic functions of operators A and B, respectively. Let us find the characteristic function of the product AB. We have

$$A(B(x)) = A\left(\sum_{m \in \varepsilon} e_m \sum_{k \in \varepsilon} g(k, m) x(k) \right) = \sum_{p \in \varepsilon} e_p \sum_{m \in \varepsilon} f(m, p) \sum_{k \in \varepsilon} g(k, m) x(k) =$$

$$= \sum_{p \in \varepsilon} e_p \sum_{k \in \varepsilon} x(k) \sum_{m \in \varepsilon} g(k, m) f(m, p),$$

where $p = (p_1, p_2)$. Hence, we obtain the relation

$$F(k, p) = \sum_{m \in \varepsilon} g(k, m) f(m, p). \tag{8.2.2}$$

It is assumed here that all numerical series are absolutely convergent.

Let $\delta(k, m)$ be the characteristic function of the identity operator I. We have

$$I(x) = \sum_{m \in \varepsilon} e_m \sum_{k \in \varepsilon} \delta(k, m) x(k) = \sum_{m \in \varepsilon} e_m x(m). \tag{8.2.3}$$

Therefore,

$$\delta(k, m) = \begin{cases} 1 & \text{for} \quad k = m, \\ 0 & \text{for} \quad k \neq m. \end{cases}$$

Suppose that the operator A admits an inverse A^{-1} whose characteristic function is $g(k, m)$. Then, for the characteristic of the operator $I = AA^{-1}$ we have

$$\delta(k, p) = \sum_{m \in \varepsilon} g(k, m) f(m, p), \tag{8.2.4}$$

and for $I = A^{-1}A$ we have

$$\delta(k, p) = \sum_{m \in \varepsilon} f(k, m) g(m, p). \tag{8.2.5}$$

Now, let $\widetilde{g}^*(k, m)$ be the characteristic function of the operator $(A^*)^{-1}$. Then, from (8.2.4) we obtain the following relation for the characteristic function of $A^*(A^*)^{-1}$:

$$\delta(k, p) = \sum_{m \in \varepsilon} \widetilde{g}^*(k, m) f^*(m, p) = \sum_{m \in \varepsilon} \widetilde{g}^*(k, m) f(p, m).$$

We also have

$$\delta(k, p) = \delta(p, k) = \sum_{m \in \varepsilon} f(k, m) \widetilde{g}^*(p, m).$$

Hence, taking into account the uniqueness of a solution of system (8.2.5), we find that

$$\widetilde{g}^*(p, m) = g(m, p). \tag{8.2.6}$$

The characteristic function $F^*(k, m)$ of the operator $(A^{-1})^*$ is expressed through the characteristic function of A^{-1} as follows:

$$F^*(k, m) = g(m, k). \tag{8.2.7}$$

Relations (8.2.6) and (8.2.7) show that the operators $(A^{-1})^*$ and $(A^*)^{-1}$ have the same characteristic function.

Consider the vector

$$A^{-1}(x) = \sum_{k \in \varepsilon} e_k \sum_{m \in \varepsilon} g(k, m) e(m) = \sum_{k \in \varepsilon} e_k \sum_{m \in \varepsilon} \widetilde{g}^*(k, m) x(m).$$

In order to find $\widetilde{g}^*(k, m)$ we should solve the system

$$\sum_{m \in \varepsilon} f^*(m, p) \widetilde{g}^*(k, m) = \delta(k, p), \qquad p \in \varepsilon.$$

for each $k \in \varepsilon$ fixed.

Definition 8.2.1. The characteristic function of the operator A^{-1} is called the *discrete Green function*.

Let A be a convolution operator and $f(k, m)$ its characteristic function. For the characteristic function of a convolution operator, the following property holds:

$$f(k - p, m - p) = f(k, m).$$

Let us find the discrete Fourier transform of the characteristic function of the operator A^* (see Definition 5.1.4). We have

$$\Phi^*(\varphi_1, \varphi_2) = \sum_{m_1=-\infty}^{+\infty} \sum_{m_2=-\infty}^{+\infty} f^*(0, 0, m_1, m_2)\, e^{i(m_1\varphi_1 + m_2\varphi_2)} =$$

$$= \sum_{m_1=-\infty}^{+\infty} \sum_{m_2=-\infty}^{+\infty} f(m_1, m_2, 0, 0)\, e^{i(m_1\varphi_1 + m_2\varphi_2)} =$$

$$= \sum_{m_1=-\infty}^{+\infty} \sum_{m_2=-\infty}^{+\infty} f(0, 0, -m_1, -m_2)\, e^{i(m_1\varphi_1 + m_2\varphi_2)} = \overline{\Phi(\varphi_1, \varphi_2)}, \qquad (8.2.8)$$

where $\Phi(\varphi_1, \varphi_2)$ is the discrete Fourier transform of the characteristic function of A.

Definition 8.2.2. Let $\{A_n\}$ be a family of pseudodifference operators. We say that *operators A_n are of Prandtl type*, if their symbols $\widehat{A}_n(\varphi_1, \varphi_2)$ are Hölder continuous, 2π-periodic in each variable, and satisfy the following conditions:

$$D_1|\varphi_1| + D_2|\varphi_2| + C_1 n^{-1} < \widehat{A}_n(\varphi_1, \varphi_2) \le C_2,$$
$$C_i > 0, \quad D_i > 0, \quad i = 1, 2, ; -\pi \le \varphi_1, \varphi_2 \le \pi; \qquad (8.2.9_1)$$

$$\int_{-\pi}^{\pi}\int_{-\pi}^{\pi} \frac{1}{\widehat{A}_n^2(\varphi_1, \varphi_2)}\, d\varphi_1\, d\varphi_2 \le K \ln n, \qquad \int_{-\pi}^{\pi}\int_{-\pi}^{\pi} \frac{d\varphi_1\, d\varphi_2}{\widehat{A}_n(\varphi_1, \varphi_2)} \le C_1';$$

$$\int_{-\pi}^{\pi} \left| \ln \widehat{A}_n(\theta, \varphi_2) \right|\, d\theta \le C_1'; \qquad (8.2.9_2)$$

$$\left| \left[\widehat{A}_n(\varphi_1, \varphi_2) \right]'_{\varphi_i} \right| \le C_1' \ln n, \qquad i = 1, 2. \qquad (8.2.9_3)$$

The symbols $\widehat{A}_n(\varphi_1, \varphi_2)$ can be represented in the form (8.1.9). Consider the function (see *Gakhov* (1966))

$$G_{1n}^*(\varphi_1, \varphi_2) = A_n^{++}(\varphi_1, \varphi_2) A^{+-}(\varphi_1, \varphi_2) = \qquad (8.2.10)$$

$$= \exp\left(\frac{1}{2} \ln A_n(\varphi_1, \varphi_2) + \frac{1}{4\pi i} \int_0^{2\pi} \cot \frac{\theta - \varphi_1}{2} \ln A_n(\theta, \varphi_2)\, d\theta + \frac{1}{4\pi} \int_0^{2\pi} \ln A_n(\theta, \varphi_2)\, d\theta \right) =$$

$$= \sqrt{A_n(\varphi_1, \varphi_2)} \exp\left(\frac{1}{4\pi i} \int_0^{2\pi} \cot \frac{\theta - \varphi_1}{2} \ln A_n(\theta, \varphi_2)\, d\theta + \frac{1}{4\pi} \int_0^{2\pi} \ln A_n(\theta, \varphi_2)\, d\theta \right).$$

Let us estimate the Fourier coefficients of the function

$$G_n(\varphi_1, \varphi_2) = \frac{1}{G_{1n}^*(\varphi_1, \varphi_2)}. \qquad (8.2.11)$$

First, we calculate $G'_{n\varphi_i}(\varphi_1, \varphi_2)$, $i = 1, 2$ (see *Gakhov* (1966)),

$$
G'_{n\varphi_1}(\varphi_1, \varphi_2) = -\frac{1}{2} \frac{A'_{n\varphi_1}(\varphi_1, \varphi_2)}{\sqrt{A_n^3(\varphi_1, \varphi_2)}} \times
$$

$$
\times \exp\left(-\frac{1}{4\pi i}\int_0^{2\pi} \cot\frac{\theta - \varphi_1}{2}\ln A_n(\theta, \varphi_2)\,d\theta - \frac{1}{4\pi}\int_0^{2\pi}\ln A_n(\theta, \varphi_2)\,d\theta\right) +
$$

$$
+ \frac{1}{\sqrt{A_n(\varphi_1, \varphi_2)}}\exp\left(-\frac{1}{4\pi i}\int_0^{2\pi}\cot\frac{\theta - \varphi_1}{2}\ln A_n(\theta, \varphi_2)\,d\theta - \frac{1}{4\pi}\int_0^{2\pi}\ln A_n(\theta, \varphi_2)\,d\theta\right) \times
$$

$$
\times \left\{-\frac{1}{4\pi i}\int_0^{2\pi}\cot\frac{\theta - \varphi_1}{2}\frac{A'_{n\theta}(\theta, \varphi_2}{A_n(\theta, \varphi_2)}\,d\theta\right\} = J_1 + J_2. \tag{8.2.12}
$$

$$
G'_{n\varphi_1}(\varphi_1, \varphi_2) = -\frac{1}{2} \frac{A'_{n\varphi_2}(\varphi_1, \varphi_2)}{\sqrt{A_n^3(\varphi_1, \varphi_2)}} \times
$$

$$
\times \exp\left(-\frac{1}{4\pi i}\int_0^{2\pi} \cot\frac{\theta - \varphi_1}{2}\ln A_n(\theta, \varphi_2)\,d\theta - \frac{1}{4\pi}\int_0^{2\pi}\ln A_n(\theta, \varphi_2)\,d\theta\right) +
$$

$$
+ \frac{1}{\sqrt{A_n(\varphi_1, \varphi_2)}}\exp\left(-\frac{1}{4\pi i}\int_0^{2\pi}\cot\frac{\theta - \varphi_1}{2}\ln A_n(\theta, \varphi_2)\,d\theta - \frac{1}{4\pi}\int_0^{2\pi}\ln A_n(\theta, \varphi_2)\,d\theta\right) \times
$$

$$
\times \left\{-\frac{1}{4\pi i}\int_0^{2\pi}\cot\frac{\theta - \varphi_1}{2}\frac{A'_{n\varphi_2}(\theta, \varphi_2)}{A_n(\theta, \varphi_2)}\,d\theta - \frac{1}{4\pi}\int_0^{2\pi}\frac{A'_{n\varphi_2}(\theta, \varphi_2)}{A_n(\theta, \varphi_2)\,d\theta}\right\} =
$$

$$
= J'_1 + J'_2. \tag{8.2.13}
$$

In view of (8.2.9), the term J_1 is estimated as follows:

$$
\int_{-\pi}^{\pi} |J_1|^{4/3}\,d\varphi_1\,d\varphi_2 \le C_1(\ln n)(\ln n)^{4/3} = C(\ln n)^{7/3}. \tag{8.2.14}
$$

From (8.2.9), we get

$$
\int_{-\pi}^{\pi}\int_{-\pi}^{\pi} \frac{|A'_n(\varphi_1, \varphi_2)|^2}{A_n^2(\varphi_1, \varphi_2)}\,d\varphi_1\,d\varphi_2 \le C_1\ln^3 n, \tag{8.2.15}
$$

and since singular integrals with Hilbert kernel are bounded in L_2, an estimate similar to (8.2.15) holds for the functions

$$
\rho_n(\varphi_1, \varphi_2) = -\frac{1}{4\pi i}\int_0^{2\pi}\cot\frac{\theta - \varphi_1}{2}\frac{A'_{n\theta}(\theta, \varphi_2)}{A_n(\theta, \varphi_2)}\,d\theta.
$$

Therefore, for the term J_2 in (8.2.12), the following estimate holds:

$$
\int_{-\pi}^{\pi}\int_{-\pi}^{\pi} |J_2|^{4/3}\,d\varphi_1\,d\varphi_2 \le \tag{8.2.16}
$$

$$
\le C_2 \int_{-\pi}^{\pi}\int_{-\pi}^{\pi}\frac{1}{|A_n((\varphi_1, \varphi_2))|^{2/3}}|\rho_n(\varphi_1, \varphi_2)|^{4/3}\,d\varphi_1\,d\varphi_2 \le
$$

$$
\le C_2\left(\int_{-\pi}^{\pi}\int_{-\pi}^{\pi}\frac{d\varphi_1\,d\varphi_2}{A_n^2(\varphi_1, \varphi_2)}\right)^{1/3}\left(\int_{-\pi}^{\pi}\int_{-\pi}^{\pi}|\rho_n(\varphi_1, \varphi_2)|^2\,d\varphi_1\,d\varphi_2\right)^{2/3} \le (\ln n)^{7/3}.
$$

From (8.2.14), (8.2.16), it follows that $G'_{n\varphi_1}(\varphi_1, \varphi_2) \in L_{4/3}(D)$, where $D = [-\pi, \pi] \times [-\pi, \pi]$, and the estimate of the form (8.1.16) holds for $G'_{n\varphi_1}(\varphi_1, \varphi_2)$. Therefore, by the generalized Riesz–Fisher theorem, the Fourier coefficients $c_n(k_1, k_2)$ of the function $G'_{n\varphi_1}(\varphi_1, \varphi_2)$ satisfy the inequality

$$\left(\sum_k |c_n(k)|^4 \right)^{1/4} \leq M \left(\int_{-\pi}^{\pi} \int_{-\pi}^{\pi} \left| G'_{n\varphi_1}(\varphi_1, \varphi_2) \right|^{4/3} d\varphi_1 \, d\varphi_2 \right)^{3/4} \leq M_1 (\ln n)^{7/4}. \qquad (8.2.17)$$

In a similar way, it can be shown that for the Fourier coefficients $c'_n(k_1, k_2)$ of the function $G'_{n\varphi_2}(\varphi_1, \varphi_2)$ defined by (8.2.13), the estimate (8.2.17) holds. Therefore, the Fourier coefficients $g_n(k)$ of the function $G_n(\varphi_1, \varphi_2)$ satisfy the estimates

$$|g_n(k)| \leq \frac{|g'_n(k)|}{|k_1| + |k_2| + 1}, \qquad \left(\sum_k |g'_n(k)|^4 \right)^{1/4} \leq M_1 (\ln n)^{7/4}. \qquad (8.2.18)$$

Now, consider the function

$$F_n^*(\varphi_1, \varphi_2) = A_n^{-+}(\varphi_1, \varphi_2) A_n^{--}(\varphi_1, \varphi_2) =$$
$$= \sqrt{A_n(\varphi_1, \varphi_2)} \exp \left(-\frac{1}{4\pi i} \int_0^{2\pi} \cot \frac{\theta - \varphi_1}{2} \ln A_n(\theta, \varphi_2) \, d\theta - \frac{1}{4\pi} \int_0^{2\pi} \ln A_n(\theta, \varphi_2) \, d\theta \right).$$

By the arguments similar to those used for $G_n(\varphi_1, \varphi_2)$, it can be shown that for the Fourier coefficients $b_n(k)$ of the function $F_n(\varphi_1, \varphi_2) = 1/F_n^*(\varphi_1, \varphi_2)$ estimates of the form (8.2.18) hold.

Consider the problem of restriction to L_{1+} for pseudodifference operators A_n of Prandtl type. The problem of restriction to L_{1+} is the problem of restriction to the halfplane. In the sequel, we will need some estimates of the Green function for the operators A_n. Since the symbols $\widehat{A}_n(\varphi_1, \varphi_2)$ are positive real functions and (8.2.8) holds, the values of the Green function coincide with the Fourier coefficients of the function $\widehat{x}_{nc}(\varphi_1, \varphi_2)$ such that $x_n(k)$ is the solution of the restriction problem (8.1.1) with $f(k) = 1$ for $k_1 = k_1^0 > 0$, $k_2 = k_2^0$, and $f(k) = 0$ for the remaining k. The solution of this problem of restriction to L_{1+} given by (8.1.14) can be written in the form

$$\widehat{x}_{nc}(k_1^0, k_2^0, \varphi_1, \varphi_2) = G_n(\varphi_1, \varphi_2) e^{i(k_1^0 \varphi_1 + k_2^0 \varphi_2)} \frac{1}{2\pi} \int_{-\pi}^{\pi} F_n(\psi_1, \varphi_2) D_{k_1^0}(\psi_1 - \varphi_1) \, d\psi_1,$$

where

$$D_{k_1^0}(\varphi_1) = \sum_{k=0}^{k_1^0} e^{ik\varphi_1}.$$

Consider the function

$$y_n(k_1^0, k_2^0, \varphi_1, \varphi_2) = \frac{\widehat{x}_{nc}(k_1^0, k_2^0, \varphi_1, \varphi_2)}{e^{i(k_1^0 \varphi_1 + k_2^0 \varphi_2)}} =$$
$$= \left\{ \frac{1}{2\pi} \int_{-\pi}^{\pi} F_n(\psi_1, \varphi_2) D_{k_1^0}(\psi_1 - \varphi_1) \, d\psi_1 \right\} G_n(\varphi_1, \varphi_2). \qquad (8.2.19)$$

We have $G_n(\varphi_1, \varphi_2) \in L_4(D)$, $F_n(\varphi_1, \varphi_2) \in L_4(D)$, $G'_{n\varphi} \in L_{4/3}(D)$, $F'_{n\varphi_i} \in L_{4/5}(D)$, $i = 1, 2$,

$D = [-\pi, \pi] \times [-\pi, \pi]$, and in view of (8.2.9), (8.2.14), (8.2.16), the following estimates hold:

$$\int\limits_{-\pi}^{\pi}\int\limits_{-\pi}^{\pi} |G_n(\varphi_1, \varphi_2)|^4 \, d\varphi_1 \, d\varphi_2 \leq L \ln n,$$

$$\int\limits_{-\pi}^{\pi}\int\limits_{-\pi}^{\pi} |G'_{n\varphi_i}(\varphi_1, \varphi_2)|^{4/3} \, d\varphi_1 \, d\varphi_2 \leq C_1(\ln n)^{7/3}, \quad i = 1, 2,$$

$$\int\limits_{-\pi}^{\pi}\int\limits_{-\pi}^{\pi} |F_n(\varphi_1, \varphi_2)|^4 \, d\varphi_1 \, d\varphi_2 \geq L \ln n, \tag{8.2.20}$$

$$\int\limits_{-\pi}^{\pi}\int\limits_{-\pi}^{\pi} |F'_{n\varphi_i}(\varphi_1, \varphi_2)|^{4/3} \, d\varphi_1 \, d\varphi_2 \leq C_1(\ln n)^{7/3}, \quad i = 1, 2.$$

It follows that the functions

$$b'_{n\varphi_1}(\varphi_1, \varphi_2) = \frac{1}{2\pi} \int\limits_{-\pi}^{\pi} F'_{n\varphi_1}(\psi_1, \varphi_2) D_{k_1^0}(\psi_1 - \varphi_1) \, d\psi_1, \quad i = 1, 2,$$

belong to $L_{4/3}$ and satisfy an inequality similar to (8.2.20).

Let us obtain estimates for the derivatives of the function $y_n(k_1^0, k_2^0, \varphi_1, \varphi_2)$,

$$y'_{n\varphi_i}(k_1^0, k_2^0, \varphi_1, \varphi_2) = G'_{n\varphi_1}(\varphi_1, \varphi_2)b_n(\varphi_1, \varphi_2) + G_n(\varphi_1, \varphi_2)b'_{n\varphi_1}(\varphi_1, \varphi_2), \quad i = 1, 2.$$

Using the Hölder inequality, we find that

$$\int\limits_{-\pi}^{\pi}\int\limits_{-\pi}^{\pi} |y'_{n\varphi_i}(k_1^0, k_2^0, \varphi_1, \varphi_2)| \, d\varphi_1 \, d\varphi_2 \leq \int\limits_{-\pi}^{\pi}\int\limits_{-\pi}^{\pi} |G'_{n\varphi_i}(\varphi_1, \varphi_2)b_n(\varphi_1, \varphi_2)| \, d\varphi_1 \, d\varphi_2 +$$

$$+ \int\limits_{-\pi}^{\pi}\int\limits_{-\pi}^{\pi} |G_n(\varphi_1, \varphi_2)b'_{n\varphi_i}(\varphi_1, \varphi_2)| \, d\varphi_1 \, d\varphi_2 \leq$$

$$\leq \left(\int\limits_{-\pi}^{\pi}\int\limits_{-\pi}^{\pi} |G'_{\varphi_i}(\varphi_1, \varphi_2)|^{4/3} \, d\varphi_1 \, d\varphi_2\right)^{3/4} \left(\int\limits_{-\pi}^{\pi}\int\limits_{-\pi}^{\pi} |b_n(\varphi_1, \varphi_2)|^4 \, d\varphi_1 \, d\varphi_2\right)^{1/4} +$$

$$+ \left(\int\limits_{-\pi}^{\pi}\int\limits_{-\pi}^{\pi} |b'_{n\varphi_i}(\varphi_1, \varphi_2)|^{4/3} \, d\varphi_1 \, d\varphi_2\right)^{3/4} \left(\int\limits_{-\pi}^{\pi}\int\limits_{-\pi}^{\pi} |G_n(\varphi_1, \varphi_2)|^4 \, d\varphi_1 \, d\varphi_2\right)^{1/4} \leq$$

$$\leq C\left((\ln n)(\ln n)^{4/3}\right)^{3/4} (\ln n)^{1/4} = C(\ln n)^2. \tag{8.2.21}$$

Using the inequality (8.2.21), we obtain an estimate for the Fourier coefficients $y_n(k_1^0, k_2^0, p_1, p_2)$ of the function $y_n(k_1^0, k_2^0, \varphi_1, \varphi_2)$,

$$|y_n(k_1^0, k_2^0, p_1, p_2)| \leq \frac{C \ln^2 n}{|p_1| + |p_2| + 1}. \tag{8.2.22}$$

8.3. Estimates of the Discrete Green Function for the Prandtl Operators in Bounded Domains

Consider the restriction problem for a pseudodifference operator A with symbol $\widehat{A}(\varphi_1, \varphi_2)$.

Definition 8.3.1. *The problem of restriction* (8.1.1) *for a pseudodifference operator A is called finite*, if the restriction set ε is finite.

A finite restriction problem is equivalent to the equation

$$\int_{-\pi}^{\pi} \int_{-\pi}^{\pi} \widehat{A}(\varphi_1, \varphi_2)\widehat{x}_c(\varphi_1, \varphi_2)D_\varepsilon(\alpha - \varphi_1, \beta - \varphi_2)\, d\varphi_1\, d\varphi_2 = \widehat{f}_c(\alpha, \beta), \tag{8.3.1}$$

where

$$\widehat{f}_c(\alpha, \beta) = \sum_{k \in \varepsilon} f(k)\, e^{i(k_1\alpha + k_2\beta)}, \qquad D_\varepsilon = \frac{1}{4\pi^2} \sum_{k \in \varepsilon} e^{i(k_1\varphi_1 + k_2\varphi_2)}.$$

Definition 8.3.2. *The discrete Green function* for a finite problem (8.1.1) is the grid function $y(m, k)$ defined on the set $Q = \varepsilon \times \varepsilon$ (ε is the restriction set) and such that for any grid function $f(k)$, $k \in \varepsilon$, the finite restriction problem admits a unique solution $x(k)$ of the form

$$x(k) = \sum_{m \in \varepsilon} f(m)y(m, k).$$

Consider a function $G(\varphi_1, \varphi_2) \geq 0$ which is 2π-periodic in each argument and has a finite integral $\int_{-\pi}^{\pi} \int_{-\pi}^{\pi} G^2(\varphi_1, \varphi_2)\, d\varphi_1\, d\varphi_2$. Let us introduce the linear space E_ε which consists of linear combinations of functions $\exp(i(k_1\varphi_1 + k_2\varphi_2))$, $k = (k_1, k_2) \in \varepsilon$, where ε is a finite set. Consider the following two functionals on E_ε:

$$\|x\|_1 = \left(\int_{-\pi}^{\pi} \int_{-\pi}^{\pi} G(\varphi_1, \varphi_2)|x(\varphi_1, \varphi_2)|^2\, d\varphi_1\, d\varphi_2 \right)^{1/2}, \tag{8.3.2}$$

$$\|x\|_2 = \max_{\|y\|_1 = 0} \left| \int_{-\pi}^{\pi} \int_{-\pi}^{\pi} x(\varphi_1, \varphi_2)\overline{y(\varphi_1, \varphi_2)}\, d\varphi_1\, d\varphi_2 \right|, \tag{8.3.3}$$

$$x = x(\varphi_1, \varphi_2) = \sum_{k \in \varepsilon} x_{k_1 k_2}\, e^{i(k_1\varphi_1 + k_2\varphi_2)}.$$

Let us show that the functional $\|x\|_2$ specifies a norm in E_ε. Indeed, $\|x\|_2 \geq 0$ and $\|x\|_2 > 0$ for $x \neq 0$. Obviously, $\|\alpha x\|_2 = |\alpha| \|x\|_2$. We further have

$$\|x_1 + x_2\|_2 = \max_{\|y\|_1 = 1} \left| \int_{-\pi}^{\pi} \int_{-\pi}^{\pi} [x_1(\varphi_1, \varphi_2) + x_2(\varphi_1, \varphi_2)]\overline{y(\varphi_1, \varphi_2)}\, d\varphi_1\, d\varphi_2 \right| \leq$$

$$\leq \max_{\|y\|_1 = 1} \left[\left| \int_{-\pi}^{\pi} \int_{-\pi}^{\pi} x_1(\varphi_1, \varphi_2)\overline{y(\varphi_1, \varphi_2)}\, d\varphi_1\, d\varphi_2 \right| + \left| \int_{-\pi}^{\pi} \int_{-\pi}^{\pi} x_2(\varphi_1, \varphi_2)\overline{y(\varphi_1, \varphi_2)}\, d\varphi_1\, d\varphi_2 \right| \right] \leq$$

$$\leq \max_{\|y\|_1 = 1} \left| \int_{-\pi}^{\pi} \int_{-\pi}^{\pi} x_1(\varphi_1, \varphi_2)\overline{y(\varphi_1, \varphi_2)}\, d\varphi_1\, d\varphi_2 \right| +$$

$$+ \max_{\|y\|_1 = 1} \left| \int_{-\pi}^{\pi} \int_{-\pi}^{\pi} x_2(\varphi_1, \varphi_2)\overline{y(\varphi_1, \varphi_2)}\, d\varphi_1\, d\varphi_2 \right| = \|x_1\|_2 + \|x_2\|_2.$$

Thus, all conditions in the definition of the norm are fulfilled. Denote by $E(\| \ \|_1)$ and $E(\| \ \|_2)$ the linear space E_ϵ equipped with the norm $\| \ \|_1$ or $\| \ \|_2$, respectively.

Consider the operator

$$T(x) = \int\limits_{-\pi}^{\pi}\int\limits_{-\pi}^{\pi} G(\varphi_1, \varphi_2) x(\varphi_1, \varphi_2) D_\epsilon(\alpha - \varphi_1, \beta - \varphi_2)\, d\varphi_1\, d\varphi_2, \qquad (8.3.4)$$

where

$$D_\epsilon(\varphi_1, \varphi_2) = \frac{1}{4\pi^2} \sum_{k \in \epsilon} e^{i(k_1\varphi_1 + k_2\varphi_2)}. \qquad (8.3.5)$$

Proposition 8.3.1. *The operator $T(x)$ is an isomorphism from $E(\| \ \|_1)$ to $E(\| \ \|_2)$.*

Proof. By the definition of $\| \ \|_2$, we have

$$\|T(x)\|_2 = \max_{\|y\|_1=1} \left| \int\limits_{-\pi}^{\pi}\int\limits_{-\pi}^{\pi} T(x)\overline{y(\alpha, \beta)} \right| =$$

$$= \max_{\|y\|_1=1} \left| \int\limits_{-\pi}^{\pi}\int\limits_{-\pi}^{\pi} \overline{y(\alpha, \beta)} \left\{ \int\limits_{-\pi}^{\pi}\int\limits_{-\pi}^{\pi} G(\varphi_1, \varphi_2) x(\varphi_1, \varphi_2) D_\epsilon(\alpha - \varphi_1, \beta - \varphi_2)\, d\varphi_1\, d\varphi_2 \right\} d\alpha\, d\beta \right| =$$

$$= \max_{\|y\|_1=1} \left| \int\limits_{-\pi}^{\pi}\int\limits_{-\pi}^{\pi} G(\varphi_1, \varphi_2) x(\varphi_1, \varphi_2) \left\{ \int\limits_{-\pi}^{\pi}\int\limits_{-\pi}^{\pi} \overline{y(\alpha, \beta) D_\epsilon(\varphi_1 - \alpha, \varphi_2 - \beta)}\, d\alpha\, d\beta \right\} d\varphi_1\, d\varphi_2 \right| =$$

$$= \max_{\|y\|_1=1} \left| \int\limits_{-\pi}^{\pi}\int\limits_{-\pi}^{\pi} G(\varphi_1, \varphi_2) x(\varphi_1, \varphi_2) \overline{y(\varphi_1, \varphi_2)}\, d\varphi_1\, d\varphi_2 \right| \leq$$

$$\leq \max_{\|y\|_1=1} \left[\int\limits_{-\pi}^{\pi}\int\limits_{-\pi}^{\pi} G(\varphi_1, \varphi_2) |x(\varphi_1, \varphi_2)|^2\, d\varphi_1\, d\varphi_2 \right]^{1/2} \left[\int\limits_{-\pi}^{\pi}\int\limits_{-\pi}^{\pi} G(\varphi_1, \varphi_2) |y(\varphi_1, \varphi_2)|^2\, d\varphi_1\, d\varphi_2 \right]^{1/2} =$$

$$= \|x\|_1. \qquad (8.3.6)$$

On the other hand, taking $y(\alpha, \beta) = \|x(\alpha, \beta)\|^{-1} x(\alpha, \beta)$, we get

$$\left| \int\limits_{-\pi}^{\pi}\int\limits_{-\pi}^{\pi} T(x)\overline{y(\alpha, \beta)}\, d\alpha\, d\beta \right| = \frac{1}{\|x\|_1} \int\limits_{-\pi}^{\pi}\int\limits_{-\pi}^{\pi} G(\varphi_1, \varphi_2) |x(\varphi_1, \varphi_2)|^2\, d\varphi_1\, d\varphi_2 = \|x\|_1. \qquad (8.3.7)$$

Now, Proposition 8.3.1 is a consequence of (8.3.6) and (8.3.7).

Proposition 8.3.2. *If*

$$\int\limits_{-\pi}^{\pi}\int\limits_{-\pi}^{\pi} (G(\varphi_1, \varphi_2))^{-2}\, d\varphi_1\, d\varphi_2 < \infty,$$

then any $x(\varphi_1, \varphi_2) \in E(\| \ \|_1)$ belongs to $L_{4/3}(D)$, where $D = [-\pi, \pi] \times [-\pi, \pi]$.

Proof. Let $x(\varphi_1, \varphi_2) \in E(\| \ \|_1)$. Then, we have $x(\varphi_1, \varphi_2) = (G(\varphi_1, \varphi_2))^{-1/2} b(\varphi_1, \varphi_2)$ with $b(\varphi_1, \varphi_2) \in L_2(D)$. Using the Hölder inequality, we get

$$\int\limits_{-\pi}^{\pi}\int\limits_{-\pi}^{\pi} |x(\varphi_1, \varphi_2)|^{4/3}\, d\varphi_1\, d\varphi_2 = \int\limits_{-\pi}^{\pi}\int\limits_{-\pi}^{\pi} \frac{1}{[G(\varphi_1, \varphi_2)]^{2/3}} |b(\varphi_1, \varphi_2)|^{4/3}\, d\varphi_1\, d\varphi_2 \leq$$

$$\leq \left[\int\limits_{-\pi}^{\pi}\int\limits_{-\pi}^{\pi} \frac{d\varphi_1\, d\varphi_2}{G^2(\varphi_1, \varphi_2)} \right]^{1/3} \left[\int\limits_{-\pi}^{\pi}\int\limits_{-\pi}^{\pi} |b(\varphi_1, \varphi_2)|^2\, d\varphi_1\, d\varphi_2 \right]^{2/3},$$

which implies the desired inclusion. Proposition 8.3.2 is proved.

Corollary 8.3.1. *Proposition* 8.3.2, *combined with the generalized Riesz–Fisher theorem* (see Zygmund (1965), Kachmans and Steinhaus (1958)) *implies that the Fourier coefficients of any* $x(\varphi_1, \varphi_2) \in E(\| \; \|_1)$ *belong to* l_4.

Consider the family of functions $\widehat{A}(\varphi_1, \varphi_2)$, $0 < C_1 < k/n < C_1$, satisfying conditions (8.2.9). For $|z_1| \le h_1$, $|z_2| \le h_2$, let us estimate the expression

$$\Delta A_{kn} = \int\limits_{-\pi}^{\pi}\int\limits_{-\pi}^{\pi} \widehat{A}_{kn}(\varphi_1, \varphi_2)\left[A_{kn}^*(\varphi_1 + z_1, \varphi_2 + z_2) - A_{kn}^*(\varphi_1, \varphi_2)\right]^2 d\varphi_1\, d\varphi_2, \qquad (8.3.8)$$

where

$$A_{kn}^* = \frac{1}{\widehat{A}_{kn}(\varphi_1, \varphi_2)}.$$

Using (8.2.9), we find that

$$\Delta A_{kn} = \int\limits_{-\pi}^{\pi}\int\limits_{-\pi}^{\pi} \widehat{A}_{kn}(\varphi_1, \varphi_2)\frac{\left[\widehat{A}_{kn}(\varphi_1, \varphi_2) - \widehat{A}_{kn}(\varphi_1 + z_1, \varphi_2 + z_2)\right]^2}{\widehat{A}_{kn}^2(\varphi_1 + z_1, \varphi_2 + z_2)\, \widehat{A}_{kn}^2(\varphi_1, \varphi_2)} d\varphi_1\, d\varphi_2 \le$$

$$\le C \ln^2 n\left(h_1^2 + h_2^2\right)\int\limits_{-\pi}^{\pi}\int\limits_{-\pi}^{\pi} \frac{d\varphi_1\, d\varphi_2}{\widehat{A}_{kn}^2(\varphi_1 + z_1, \varphi_2 + z_2)\, \widehat{A}_{kn}(\varphi_1, \varphi_2)} \le$$

$$\le C \ln^2 n\left(h_1^2 + h_2^2\right)\left[\int\limits_{-\pi}^{\pi}\int\limits_{-\pi}^{\pi} \frac{d\varphi_1\, d\varphi_2}{\widehat{A}_{kn}^4(\varphi_1 + z_1, \varphi_2 + z_2)}\right]^{1/2}\left[\int\limits_{-\pi}^{\pi}\int\limits_{-\pi}^{\pi} \frac{d\varphi_1\, d\varphi_2}{\widehat{A}_{kn}^2(\varphi_1, \varphi_2)}\right]^{1/2} \le$$

$$\le C n \ln^{5/2} n\left(h_1^2 + h_2^2\right). \qquad (8.3.9)$$

Consider Nikolskii's kernels (see *Nikolskii* (1977))

$$K_\nu(\varphi) = \left(\frac{\sin\frac{\lambda\varphi}{2}}{\sin\frac{\varphi}{2}}\right)^{2\sigma}\frac{1}{a_\nu}, \qquad (8.3.10)$$

where λ and σ are positive integers and the constant a_ν is determined by the relation

$$\int\limits_{-\pi}^{\pi} K_\nu(\varphi)\, d\varphi = 1.$$

For σ fixed, the asymptotic behavior of a_ν is known,

$$a_\nu \sim \lambda^{2\sigma-1}.$$

Moreover, $K_\nu(\varphi)$ is a trigonometric polynomial of order $\nu = \sigma(\lambda - 1)$. Therefore,

$$\frac{1}{a_\nu} \sim \frac{\sigma^{2\sigma-1}}{(\nu + \sigma)^{2\sigma-1}}.$$

Let us fix $\varepsilon > 0$ and consider the set

$$B_\varepsilon = \left\{\varphi : -\pi \le \varphi \le \frac{1}{\nu^{1-\varepsilon}} \quad \text{or} \quad \frac{1}{\nu^{1-\varepsilon}} \le \varphi \le \pi\right\}.$$

On this set, the Nikolskii kernels can be estimated as follows:

$$|K_\nu(\varphi)| \le C(\sigma) \frac{\nu^{2\sigma(1-\varepsilon)}\sigma^{2\sigma-1}}{(\nu+\sigma)^{2\sigma-1}} \le C(\sigma)\frac{\sigma^{2\sigma-1}}{\nu^{2\sigma\varepsilon-1}} \,. \tag{8.3.11}$$

One can always find σ_0 such that $2\sigma_0\varepsilon - 1 \ge 1$.

Denote the product of Nikolskii's kernels by

$$A_\nu(\varphi_1, \varphi_2) = K_\nu(\varphi_1)K_\nu(\varphi_2) \tag{8.3.12}$$

and consider the trigonometric polynomial

$$P_{kn}(\varphi_1, \varphi_2) = \int\limits_{-\pi}^{\pi}\int\limits_{-\pi}^{\pi} \frac{A_\nu(z_1, z_2)}{\widehat{A}_{kn}^2(\varphi_1 + z_1, \varphi_2 + z_2)}\, dz_1\, dz_2. \tag{8.3.13}$$

We have

$$I = \int\limits_{-\pi}^{\pi} \widehat{A}_{kn}(\varphi_1, \varphi_2)\left[A_{kn}^*(\varphi_1, \varphi_2) - P_{kn}(\varphi_1, \varphi_2)\right] d\varphi_1\, d\varphi_2 =$$

$$= \int\limits_{-\pi}^{\pi}\int\limits_{-\pi}^{\pi} \widehat{A}_{kn}(\varphi_1, \varphi_2)\left[\int\limits_{-\pi}^{\pi}\int\limits_{-\pi}^{\pi} \left[A_{kn}^*(\varphi_1, \varphi_2) - A_{kn}^*(\varphi_1 + z_1, \varphi_2 + z_2)\right] A_\nu(z_1, z_2)\, dz_1\, dz_2\right]^2 d\varphi_1\, d\varphi_2 \le$$

$$= \int\limits_{-\pi}^{\pi}\int\limits_{-\pi}^{\pi} \widehat{A}_{kn}(\varphi_1, \varphi_2)\left[\int\limits_{-\pi}^{\pi}\int\limits_{-\pi}^{\pi} \left[A_{kn}^*(\varphi_1, \varphi_2) - A_{kn}^*(\varphi_1 + z_1, \varphi_2 + z_2)\right]^2 A_\nu(z_1, z_2)\, dz_1\, dz_2\right] d\varphi_1\, d\varphi_2 \le$$

$$= \int\limits_{-\pi}^{\pi}\int\limits_{-\pi}^{\pi} \widehat{A}_{kn}(\varphi_1, \varphi_2)\left[\int\limits_{-\delta}^{\delta}\int\limits_{-\delta}^{\delta} \left[A_{kn}^*(\varphi_1, \varphi_2) - A_{kn}^*(\varphi_1 + z_1, \varphi_2 + z_2)\right]^2 A_\nu(z_1, z_2)\, dz_1\, dz_2\right] d\varphi_1\, d\varphi_2 +$$

$$+ \int\limits_{-\pi}^{\pi}\int\limits_{-\pi}^{\pi} \widehat{A}_{kn}(\varphi_1, \varphi_2)\left[\int\limits_{D\backslash\rho_\delta} \left[A_{kn}^*(\varphi_1, \varphi_2) - A_{kn}^*(\varphi_1 + z_1, \varphi_2 + z_2)\right]^2 A_\nu(z_1, z_2)\, dz_1\, dz_2\right] d\varphi_1\, d\varphi_2 =$$

$$= I_1 + I_2, \tag{8.3.14}$$

where $D = [-\pi \le \varphi_1 \le \pi] \times [-\pi \le \varphi_2 \le \pi]$, $\rho_\delta = [-\delta \le \varphi_1 \le \delta] \times [-\delta \le \varphi_2 \le \delta]$.

Let $\nu = [\theta\min(k, h)]$, $0 < \theta < 1$, where $[a]$ denotes the integer part of a. Since $0 < C_1 < k/n < C_2$, we have $\nu = An$, $A > 0$. Let $\delta = 1/\nu_{1-\varepsilon}$. Using (8.3.9), we estimate the term I_1 as follows:

$$I_1 = \int\limits_{-\delta}^{\delta}\int\limits_{-\delta}^{\delta} A_\nu(z_1, z_2)\left[\int\limits_{-\pi}^{\pi}\int\limits_{-\pi}^{\pi} \widehat{A}_{kn}(\varphi_1, \varphi_2)\left[A_{kn}^*(\varphi_1, \varphi_2) - A_{kn}^*(\varphi_1 + z_1, \varphi_2 + z_2)\right]^2 d\varphi_1\, d\varphi_2\right] dz_1\, dz_2 \le$$

$$\le C_1(\ln n)^{5/2} n\left(\frac{1}{n^{2(1-\varepsilon)}}\right) = C_1\frac{(\ln n)^{5/2}}{n^{1-2\varepsilon}} \,. \tag{8.3.15}$$

Let us estimate the term I_2. We have

$$
I_2 = \int\limits_{-\pi}^{\pi}\int\limits_{-\pi}^{\pi} \widehat{A}_{kn}(\varphi_1,\varphi_2)\Bigg\{ \int\limits_{-\delta}^{\delta} K_\nu(z_1)\Bigg[\int\limits_{\delta}^{\pi} \big[A_{kn}^*(\varphi_1,\varphi_2) - A_{kn}^*(\varphi_1+z_1,\varphi_2+z_2)\big]^2 K_\nu(z_2)\, dz_2 +
$$

$$
+ \int\limits_{-\pi}^{-\delta} \big[A_{kn}^*(\varphi_1,\varphi_2) - A_{kn}^*(\varphi_1+z_1,\varphi_2+z_2)\big]^2 K_\nu(z_2)\, dz_2 \Bigg]\, dz_1 \Bigg\}\, d\varphi_1\, d\varphi_2 +
$$

$$
+ \int\limits_{-\pi}^{\pi}\int\limits_{-\pi}^{\pi} \widehat{A}_{kn}(\varphi_1,\varphi_2)\Bigg\{ \int\limits_{-\pi}^{\delta} K_\nu(z_1)\, dz_1 \times
$$

$$
\times \Bigg[\int\limits_{-\pi}^{\pi} \big[A_{kn}^*(\varphi_1,\varphi_2) - A_{kn}^*(\varphi_1+z_1,\varphi_2+z_2)\big]^2 K_\nu(z_2)\, dz_2 \Bigg]\Bigg\}\, d\varphi_1\, d\varphi_2 +
$$

$$
+ \int\limits_{-\pi}^{\pi}\int\limits_{-\pi}^{\pi} \widehat{A}_{kn}(\varphi_1,\varphi_2)\Bigg[\int\limits_{\delta}^{\pi} K_\nu(z_1)\, dz_1 \int\limits_{-\pi}^{\pi} \big[A_{kn}^*(\varphi_1,\varphi_2) - A_{kn}^*(\varphi_1+z_1,\varphi_2+z_2)\big]^2 K_\nu(z_2)\, dz_2 \Bigg]\, d\varphi_1\, d\varphi_2 =
$$

$$
= \int\limits_{-\delta}^{\delta} K_\nu(z_1)\, dz_1 \int\limits_{\delta}^{\pi} K_\nu(z_2)\, dz_2 \int\limits_{-\pi}^{\pi}\int\limits_{-\pi}^{\pi} \big[A_{kn}^*(\varphi_1,\varphi_2) - A_{kn}^*(\varphi_1+z_1,\varphi_2+z_2)\big]^2\, d\varphi_1\, d\varphi_2 +
$$

$$
+ \int\limits_{-\pi}^{-\delta} K_\nu(z_1)\, dz_1 \int\limits_{-\pi}^{\pi} K_\nu(z_2)\, dz_2 \int\limits_{-\pi}^{\pi}\int\limits_{-\pi}^{\pi} \widehat{A}_{kn}\big[A_{kn}^*(\varphi_1,\varphi_2) - A_{kn}^*(\varphi_1+z_1,\varphi_2+z_2)\big]^2\, d\varphi_1\, d\varphi_2 +
$$

$$
+ \int\limits_{\delta}^{\pi} K_\nu(z_1)\, dz_1 \int\limits_{-\pi}^{\pi} K_\nu(z_2)\, dz_2 \int\limits_{-\pi}^{\pi}\int\limits_{-\pi}^{\pi} \widehat{A}_{kn}(\varphi_1,\varphi_2)\big[A_{kn}^*(\varphi_1,\varphi_2) - A_{kn}^*(\varphi_1+z_1,\varphi_2+z_2)\big]^2\, d\varphi_1\, d\varphi_2 +
$$

$$
+ \int\limits_{-\delta}^{\delta} K_\nu(z_1)\, dz_1 \int\limits_{-\pi}^{-\delta} K_\nu(z_2)\, dz_2 \int\limits_{-\pi}^{\pi}\int\limits_{-\pi}^{\pi} \widehat{A}_{kn}\big[A_{kn}^*(\varphi_1,\varphi_2) - A_{kn}^*(\varphi_1+z_1,\varphi_2+z_2)\big]^2\, d\varphi_1\, d\varphi_2 . \quad (8.3.16)
$$

From (8.2.9), (8.3.11), and (8.3.16), it follows that

$$
I_2 \le \frac{C\ln n}{\nu} \le \frac{C_1 \ln n}{n}, \qquad (8.3.17)
$$

and (8.3.15), (8.3.17) imply the estimate

$$
I \le \frac{C_1(\ln n)^{5/2}}{n^{1-2\varepsilon}}. \qquad (8.3.18)
$$

Consider the family of symbols $\widehat{A}_{kn}(\varphi_1,\varphi_2)$ and finite sets ε_{kn} for which the following conditions hold:

$$
\widehat{A}_{kn}(\varphi_1,\varphi_2) = \sum_{j=-pn}^{pn} \sum_{m=-pk}^{pk} a_{jm}\, e^{i(j\varphi_1 + m\varphi_2)}, \quad p \in \mathbb{Z}, \quad p > 0,
$$

$$
k = \max(m_1, n_1), \quad n_1 = \max_{(j,m)\in\varepsilon_{kn}}\{m\}, \quad m_1 = \max_{(j,m)\in\varepsilon_{kn}}\{-m\},
$$

$$
n = \max(m_2, n_2), \quad n_2 = \max_{(j,m)\in\varepsilon_{kn}}\{j\}, \quad m_2 = \max_{(j,m)\in\varepsilon_{kn}}\{-j\},
$$

$$
\varepsilon_{kn} \supset \varepsilon_{kn}^1, \quad \varepsilon_{kn}^1 = \{(j,m): \ -[\theta n] \le j \le [\theta n]; \ -[\theta k] \le m \le [\theta k]\}, \quad 0 < \theta < 1.
$$

where $[a]$ denotes the integer part of a. Assume also that the following estimates hold:

$$0 < A_1 \le \frac{n}{k} \le A_2, \quad 0 < A_1 \le \frac{n_1}{m_1} \le A_2, \quad 0 < A_1 \le \frac{n_2}{m_2} \le A_2.$$

Denote by

$$x_{kn}(\varphi_1, \varphi_2) = \sum_{(k_1,k_2) \in \varepsilon_{kn}} x_{k_1 k_2}^{kn} e^{i(k_1 \varphi_1 + k_2 \varphi_2)}$$

the solution of the restriction problem

$$\int_{-\pi}^{\pi} \int_{-\pi}^{\pi} \widehat{A}_{kn}(\varphi_1, \varphi_2) x_{kn}(\varphi_1, \varphi_2) D_{\varepsilon_{kn}}(\alpha - \varphi_1, \beta - \varphi_2) \, d\varphi_1 \, d\varphi_2 = 1. \tag{8.3.19}$$

Theorem 8.3.1. *Suppose that condition (8.2.9) holds for $\widehat{A}_{kn}(\varphi_1, \varphi_2)$ uniformly in k and the above assumptions hold for the restriction problem (8.3.19). Then the solution $x_{kn}(\varphi_1, \varphi_2)$ of the problem can be represented in the form*

$$x_{kn}(\varphi_1, \varphi_2) = z_{kn}(\varphi_1, \varphi_2) + P_{kn}(\varphi_1, \varphi_2)$$

and for any $\varepsilon > 0$, the following estimates hold:

$$\left| z_{em}^{kn} \right| \le (\ln n)^{5/4} n^{\varepsilon - 1/2} |z'_{emkn}|, \qquad \left| p_{em}^{kn} \right| \le \left| a_{em}^{kn} \right|, \qquad \sum_{(e,p) \in \varepsilon_{kn}} |z'_{emkn}|^4 \le C \ln n, \tag{8.3.20}$$

where z_{em}^{kn}, p_{em}^{kn}, a_{em}^{kn} are the Fourier coefficients of $z_{kn}(\varphi_1, \varphi_2)$, $P_{kn}(\varphi_1, \varphi_2)$, and $1/\widehat{A}_{kn}(\varphi_1, \varphi_2)$.

Proof. Let us seek the solution in the form

$$x_{kn}(\varphi_1, \varphi_2) = z_{kn}(\varphi_1, \varphi_2) + P_{kn}(\varphi_1, \varphi_2),$$

$$P_{kn}(\varphi_1, \varphi_2) = \int_{-\pi}^{\pi} \int_{-\pi}^{\pi} \frac{A_\nu(z_1, z_2) \, dz_1 \, dz_2}{\widehat{A}_{kn}(\varphi_1 + z_1, \varphi_2 + z_2)}, \qquad A_\nu(z_1, z_2) = K_\nu(z_1) K_\nu(z_2),$$

where $K_\nu(z)$ is the Nikolskii kernel of order $\nu = [\theta n]$, $0 < \theta < 1$. In order to estimate the norm

$$\|z_{kn}(\varphi_1, \varphi_2)\|_1^2 = \int_{-\pi}^{\pi} \int_{-\pi}^{\pi} \widehat{A}_{kn}(\varphi_1, \varphi_2) |z_{kn}(\varphi_1, \varphi_2)| \, d\varphi_1 \, d\varphi_2,$$

we use the fact that the operator

$$T(z_{kn}) = \int_{-\pi}^{\pi} \int_{-\pi}^{\pi} \widehat{A}_{kn}(\varphi_1, \varphi_2) z_{kn}(\varphi_1, \varphi_2) D_{\varepsilon_{kn}}(\alpha - \varphi_1, \beta - \varphi_2) \, d\varphi_1 \, d\varphi_2$$

from $E_{\varepsilon_{kn}}(\| \cdot \|_1)$ to $E_{\varepsilon_{kn}}(\| \cdot \|_2)$ is isometric. We have

$$T(z_{kn}) = \int_{-\pi}^{\pi} \int_{-\pi}^{\pi} \widehat{A}_{kn}(\varphi_1, \varphi_2) [x_{kn}(\varphi_1, \varphi_2) - P_{kn}(\varphi_1, \varphi_2)] D_{\varepsilon_{kn}}(\alpha - \varphi_1, \beta - \varphi_2) \, d\varphi_1 \, d\varphi_2 =$$

$$= 1 - \int_{-\pi}^{\pi} \int_{-\pi}^{\pi} \widehat{A}_{kn}(\varphi_1, \varphi_2) P_{kn}(\varphi_1, \varphi_2) D_{\varepsilon_{kn}}(\alpha - \varphi_1, \beta - \varphi_2) \, d\varphi_1 \, d\varphi_2 =$$

$$= \int_{-\pi}^{\pi} \int_{-\pi}^{\pi} \widehat{A}_{kn}(\varphi_1, \varphi_2) \left[\frac{1}{\widehat{A}_{kn}(\varphi_1, \varphi_2)} - P_{kn}(\varphi_1, \varphi_2) \right] D_{\varepsilon_{kn}}(\alpha - \varphi_1, \beta - \varphi_2) \, d\varphi_1 \, d\varphi_2.$$

In view of (8.3.18), we obtain

$$
\|T(z_{kn})\|_2 = \max_{\|y\|_1=1} \left| \int\limits_{-\pi}^{\pi}\int\limits_{-\pi}^{\pi} T(z_{kn})\overline{y(\alpha,\beta)}\, d\alpha\, d\beta \right| =
$$

$$
= \max_{\|y\|_1=1} \left| \int\limits_{-\pi}^{\pi}\int\limits_{-\pi}^{\pi} \overline{y(\alpha,\beta)} \left\{ \int\limits_{-\pi}^{\pi}\int\limits_{-\pi}^{\pi} \widehat{A}_{kn}(\varphi_1,\varphi_2)\left[\frac{1}{\widehat{A}_{kn}(\varphi_1,\varphi_2)} - \right.\right.\right.
$$

$$
\left.\left.\left. - P_{kn}(\varphi_1,\varphi_2)\right] D_{\varepsilon_{kn}}(\alpha-\varphi_1,\beta-\varphi_2)\, d\varphi_1\, d\varphi_2 \right\} d\alpha\, d\beta \right| =
$$

$$
= \max_{\|y\|_1=1} \left| \int\limits_{-\pi}^{\pi}\int\limits_{-\pi}^{\pi} \widehat{A}_{kn}(\varphi_1,\varphi_2)\left[\frac{1}{\widehat{A}_{kn}(\varphi_1,\varphi_2)} - P_{kn}(\varphi_1,\varphi_2)\right] \times \right.
$$

$$
\times \left[\int\limits_{-\pi}^{\pi}\int\limits_{-\pi}^{\pi} \overline{y(\alpha,\beta)} D_{\varepsilon_{kn}}(\alpha-\varphi_1,\beta-\varphi_2)\right] d\varphi_1\, d\varphi_2 =
$$

$$
= \max_{\|y\|_1=1} \left| \int\limits_{-\pi}^{\pi}\int\limits_{-\pi}^{\pi} \widehat{A}_{kn}(\varphi_1,\varphi_2)\left[\frac{1}{\widehat{A}_{kn}(\varphi_1,\varphi_2)} - P_{kn}(\varphi_1,\varphi_2)\right]\overline{y(\varphi_1,\varphi_2)}\, d\varphi_1\, d\varphi_2 \right| \le
$$

$$
\le \max_{\|y\|_1=1} \left[\int\limits_{-\pi}^{\pi}\int\limits_{-\pi}^{\pi} \widehat{A}_{kn}(\varphi_1,\varphi_2)\left[\frac{1}{\widehat{A}_{kn}(\varphi_1,\varphi_2)} - P_{kn}(\varphi_1,\varphi_2)\right]^2 \right]^{1/2} \times
$$

$$
\times \left[\int\limits_{-\pi}^{\pi}\int\limits_{-\pi}^{\pi} \widehat{A}_{kn}(\varphi_1,\varphi_2)|y(\varphi_1,\varphi_2)|^2\, d\varphi_1\, d\varphi_2 \right]^{1/2} \le A(\ln n)^{5/4} n^{\varepsilon-1/2}.
$$

Hence, we obtain the estimate

$$
\|z_{kn}(\varphi_1,\varphi_2)\|_1 \le A(\ln n)^{5/4} n^{\varepsilon-1/2}. \tag{8.3.21}
$$

Let us represent the function $z_{kn}(\varphi_1,\varphi_2)$ in the form

$$
z_{kn}(\varphi_1,\varphi_2) = z_{1kn}(\varphi_1,\varphi_2)\|z_{kn}(\varphi_1,\varphi_2)\|_1. \tag{8.3.22}
$$

According to Proposition 8.3.2, we have $z_{kn} \in L_{4/3}(D)$, $D = [-\pi,\pi] \times [-\pi,\pi]$, and the following estimate holds:

$$
\|z_{1kn}(\varphi_1,\varphi_2)\|_{L_{4/3}(D)} \le \left[\iint\limits_D \frac{d\varphi_1\, d\varphi_2}{\widehat{A}_{kn}(\varphi_1,\varphi_2)} \right]^{1/4} \le C(\ln n)^{1/4}. \tag{8.3.23}
$$

Since $z_{1kn}(\varphi_1,\varphi_2) \in L_{4/3}(D)$, it follows from the generalized Riesz–Fisher theorem that its Fourier coefficients $\{z'_{emkn}\}$ belong to l_4 and

$$
\sum_{(l,m)\in\varepsilon_{kn}} |z'_{emkn}|^4 < C(\ln n)^{1/4}. \tag{8.3.24}
$$

The Fourier coefficients c^ν_{em} of the Nikolskii kernel satisfy the inequality

$$
|c^\nu_{em}| = \left| \frac{1}{4\pi^2} \int\limits_{-\pi}^{\pi}\int\limits_{-\pi}^{\pi} K_\nu(\varphi_1)K_\nu(\varphi_2)\, e^{i(l\varphi_1+m\varphi_2)}\, d\varphi_1\, d\varphi_2 \right| \le
$$

$$
\le \frac{1}{4\pi^2} \int\limits_{-\pi}^{\pi}\int\limits_{-\pi}^{\pi} K_\nu(\varphi_1)K_\nu(\varphi_2)\, d\varphi_1\, d\varphi_2 = \frac{1}{4\pi^2}.
$$

Using the relation

$$
P_{kn}(\varphi_1, \varphi_2) = \int\limits_{-\pi}^{\pi} \int\limits_{-\pi}^{\pi} \frac{K_\nu(z_1) K_\nu(z_2)}{\widehat{A}_{kn}(\varphi_1 + z_1, \varphi_2 + z_2)} \, dz_1 \, dz_2 =
$$

$$
= \sum_{(l,m)} 4\pi^2 a_{em}^{kn} c_{-l,-m}^\nu \, e^{i(l\varphi_1 + m\varphi_2)} = \sum_{(l,m)} p_{lm}^{kn} \, e^{i(l\varphi_1 + m\varphi_2)},
$$

we obtain the following estimate for the Fourier coefficients of the polynomial $P_{kn}(\varphi_1, \varphi_2)$:

$$
\left| p_{lm}^{kn} \right| \le \left| a_{em}^{kn} \right|,
$$

where a_{em}^{kn} are the Fourier coefficients of the symbol $1/\widehat{A}_{kn}(\varphi_1, \varphi_2)$. Theorem 8.3.1 is proved.

Remark 8.3.1. The Fourier coefficients of the function $x_{kn}(\varphi_1, \varphi_2)$ coincide with values of the discrete Green function, and Theorem 8.3.1 yields asymptotic estimates for these values.

Proposition 8.3.3. *Suppose that conditions* (8.3.25) *hold for the function* $\widehat{A}_{kn}(\varphi_1, \varphi_2)$. *Then the Fourier coefficients* a_{ep}^{kn} *of the function* $1/\widehat{A}_{kn}(\varphi_1, \varphi_2)$ *satisfy the inequalities*

$$
\left| a_{ep}^{kn} \right| \le \frac{C \ln^2 n}{|l| + |p| + 1}. \tag{8.3.25}
$$

Proof. For $l = 0$, $p = 0$, the inequality (8.3.25) follows from (8.2.9). Let $l = \ne 0$. Then, using (8.2.9), we obtain

$$
|a_{ep}| = \frac{1}{4\pi^2} \left| \int\limits_{-\pi}^{\pi} \int\limits_{-\pi}^{\pi} \frac{1}{\widehat{A}_{kn}(\varphi_1, \varphi_2)} \, e^{-i(l\varphi_1 + p\varphi_2)} \, d\varphi_1 \, d\varphi_2 \right| =
$$

$$
= \left| \frac{1}{4\pi^2} \int\limits_{-\pi}^{\pi} e^{-ip\varphi_2} \left[\frac{e^{-il\varphi_1}}{-il} \frac{1}{\widehat{A}_{kn}(\varphi_1, \varphi_2)} \Big|_{-\pi}^{\pi} \right] d\varphi_2 - \right. \tag{8.3.26}
$$

$$
\left. - \frac{1}{il} \int\limits_{-\pi}^{\pi} \int\limits_{-\pi}^{\pi} \frac{1}{\widehat{A}_{kn}^2(\varphi_1, \varphi_2)} \left[\widehat{A}_{kn}^2(\varphi_1, \varphi_2) \right]'_{\varphi_1} \frac{e^{-i(l\varphi_1 + p\varphi_2)}}{4\pi^2} \, d\varphi_1 \, d\varphi_2 \right| \le \frac{C \ln^2 n}{|l| + 1}.
$$

If $|l| \ge |p|$, then

$$
|a_{ep}| \le \frac{2C \ln^2 n}{|l| + |p| + 1}. \tag{8.3.27}
$$

The inequality (8.3.25) for the other cases is proved by similar arguments.

8.4. Asymptotic Estimates of the Discrete Green Function in Rectangular Domains

Consider the linear space of 2π-periodic functions of two variables equipped with the norm

$$
\| f(\varphi_1, \varphi_2) \|_1^2 = \int\limits_{-\pi}^{\pi} \int\limits_{-\pi}^{\pi} \widehat{A}_n(\varphi_1, \varphi_2) | f(\varphi_1, \varphi_2) |^2 \, d\varphi_1 \, d\varphi_2, \tag{8.4.1}
$$

where $\widehat{A}_n(\varphi_1, \varphi_2)$ is a function satisfying condition (8.2.9). Let us estimate the integral

$$
J = \|G_n(\varphi_1, \varphi_2) - C_n(\varphi_1 + z_1, \varphi_2 + z_2)\|_1^2 =
$$

$$
= \int\limits_{-\pi}^{\pi}\int\limits_{-\pi}^{\pi} \widehat{A}_n(\varphi_1, \varphi_2)|G_n(\varphi_1, \varphi_2) - G_n(\varphi_1 + z_1, \varphi_2 + z_2)|^2 \, d\varphi_1 \, d\varphi_2, \qquad (8.4.2)
$$

where $G_n(\varphi_1, \varphi_2)$ is the Green function of the form

$$
G_n(\varphi_1, \varphi_2) = \frac{1}{A_n^{++}(\varphi_1, \varphi_2)A_n^{+-}(\varphi_1, \varphi_2)} =
$$

$$
= \left[\widehat{A}_n(\varphi_1, \varphi_2)\right]^{-1/2} \exp\left[-\frac{1}{4\pi i}\int\limits_{-\pi}^{\pi} \cot\frac{\theta - \varphi_1}{2} \ln \widehat{A}_n(\theta, \varphi_2) \, d\theta - \frac{1}{4\pi}\int\limits_{-\pi}^{\pi} \ln \widehat{A}_n(\theta, \varphi_2) \, d\theta\right] =
$$

$$
= K_1(\varphi_1, \varphi_2)K_2(\varphi_1, \varphi_2). \qquad (8.4.3)
$$

From (8.2.9), (8.2.12), and (8.2.13), we obtain the estimate

$$
\int\limits_{-\pi}^{\pi}\int\limits_{-\pi}^{\pi} \left|K_{2\varphi_i}'(\varphi_1, \varphi_2)\right|^2 \, d\varphi_1 \, d\varphi_2 \le C(\ln n)^3, \qquad i = 1, 2, \qquad (8.4.4)
$$

and from (8.2.9), (8.2.12), (8.2.13), the estimate

$$
\int\limits_{-\pi}^{\pi}\int\limits_{-\pi}^{\pi} \left|G_{n\varphi_i}'(\varphi_1, \varphi_2)\right|^2 \, d\varphi_1 \, d\varphi_2 \le Cn(\ln n)^3, \qquad i = 1, 2. \qquad (8.4.5)
$$

In a similar way, for the function

$$
F_n(\varphi_1, \varphi_2) = \qquad (8.4.6)
$$

$$
= \left[\widehat{A}_n(\varphi_1, \varphi_2)\right]^{-1/2} \exp\left[\frac{1}{4\pi i}\int\limits_{-\pi}^{\pi} \cot\frac{\theta - \varphi_1}{2} \ln \widehat{A}_n(\theta, \varphi_2) \, d\theta + \frac{1}{4\pi}\int\limits_{-\pi}^{\pi} \ln \widehat{A}_n(\theta, \varphi_2) \, d\theta\right]
$$

we obtain the inequalities

$$
\int\limits_{-\pi}^{\pi}\int\limits_{-\pi}^{\pi} \left|F_{n\varphi_i}'(\varphi_1, \varphi_2)\right| \, d\varphi_1 \, d\varphi_2 \le Cn(\ln n)^3. \qquad (8.4.7)
$$

Proposition 8.4.1. *If* $f_{\varphi_i}'(\varphi_1, \varphi_2) \in L_2(D)$, $i = 1, 2$. $D = [-\pi, \pi] \times [-\pi, \pi]$, *then*

$$
\int\limits_{-\pi}^{\pi}\int\limits_{-\pi}^{\pi} |f(\varphi_1, \varphi_2) - f(\varphi_1 + z_1, \varphi_2 + z_2)|^2 \, d\varphi_1 \, d\varphi_2 \le C\left(z_1^2 + z_2^2\right)\left(\|f_{\varphi_1}'\|_{L_2(D)}^2 + \|f_{\varphi_2}'\|_{L_2(D)}^2\right). \qquad (8.4.8)
$$

Proof. Using Parseval's identity we find that

$$
\int\limits_{-\pi}^{\pi}\int\limits_{-\pi}^{\pi} |f(\varphi_1, \varphi_2) - f(\varphi_1 + z_1, \varphi_2 + z_2)|^2 \, d\varphi_1 \, d\varphi_2 =
$$

$$
= 4\pi^2 \sum_k |a(k_1, k_2)|^2 \left|1 - e^{i(k_1 z_1 + k_2 z_2)}\right|^2 \le
$$

$$
\le 8\pi^2 \sum_k |a(k_1, k_2)|^2 |k_1 z_1 + k_2 z_2|^2 \le 8\pi^2\left(z_1^2 + z_2^2\right)\sum_k |a(k_1, k_2)|^2\left(k_1^2 + k_2^2\right) \le
$$

$$
\le C\left(z_1^2 + z_2^2\right)\left(\|f_{\varphi_1}'(\varphi_1, \varphi_2)\|_{L_2(D)}^2 + \|f_{\varphi_2}'(\varphi_1, \varphi_2)\|_{L_2(D)}^2\right),
$$

where $a(k_1, k_2)$ are the Fourier coefficients of the function $f(\varphi_1, \varphi_2)$. Proposition 8.4.1 is proved.

Next, we estimate the expression (8.4.2) for $|z_1| \leq h_1$, $|z_2| \leq h_2$. We have

$$J = \int\limits_{-\pi}^{\pi}\int\limits_{-\pi}^{\pi} \widehat{A}_n(\varphi_1, \varphi_2) |K_1(\varphi_1 + z_1, \varphi_2 + z_2) K_2(\varphi_1, \varphi_2) - K_1(\varphi_1, \varphi_2) K_2(\varphi_1, \varphi_2)|^2 \, d\varphi_1 \, d\varphi_2 \leq$$

$$\leq 2 \int\limits_{-\pi}^{\pi}\int\limits_{-\pi}^{\pi} \widehat{A}_n(\varphi_1, \varphi_2) |K_1(\varphi_1 + z_1, \varphi_2 + z_2) - K_1(\varphi_1, \varphi_2)|^2 |K_2(\varphi_1 + z_1, \varphi_2 + z_2)|^2 \, d\varphi_1 \, d\varphi_2 +$$

$$+ 2 \int\limits_{-\pi}^{\pi}\int\limits_{-\pi}^{\pi} \widehat{A}_n(\varphi_1, \varphi_2) |K_1(\varphi_1, \varphi_2)| |K_2(\varphi_1 + z_1, \varphi_2 + z_2) - K_2(\varphi_1, \varphi_2)|^2 \, d\varphi_1 \, d\varphi_2 =$$

$$= 2I_1 + 2I_2. \tag{8.4.9}$$

Using (8.2.9), we estimate I_1 as follows:

$$I_1 \leq C \int\limits_{-\pi}^{\pi}\int\limits_{-\pi}^{\pi} \widehat{A}_n(\varphi_1, \varphi_2) \left[\left(\widehat{A}_n(\varphi_1 + z_1, \varphi_2 + z_2) \right)^{-1/2} - \left(\widehat{A}_n(\varphi_1, \varphi_2) \right)^{-1/2} \right]^2 \, d\varphi_1 \, d\varphi_2 \leq$$

$$\leq C \int\limits_{-\pi}^{\pi}\int\limits_{-\pi}^{\pi} \frac{\left[\widehat{A}_n(\varphi_1, \varphi_2) - \widehat{A}_n(\varphi_1 + z_1, \varphi_2 + z_2) \right]^2 \, d\varphi_1 \, d\varphi_2}{\widehat{A}_n(\varphi_1 + z_1, \varphi_2 + z_2) \left[\left(\widehat{A}_n(\varphi_1 + z_1, \varphi_2 + z_2) \right)^{1/2} + \left(\widehat{A}_n(\varphi_1, \varphi_2) \right)^{1/2} \right]} \leq$$

$$\leq C(\ln n)^3 \left(h_1^2 + h_2^2 \right). \tag{8.4.10}$$

By virtue of (8.4.4) and (8.4.8), we obtain

$$I_2 \leq \int\limits_{-\pi}^{\pi}\int\limits_{-\pi}^{\pi} |K(\varphi_1 + z_1, \varphi_2 + z_2) - K_2(\varphi_1, \varphi_2)|^2 \, d\varphi_1 \, d\varphi_2 \leq C \left(h_1^2 + h_2^2 \right) (\ln n)^3. \tag{8.4.11}$$

From (8.4.9)–(8.4.11), it follows that

$$I \leq C_1 \left(h_1^2 + h_2^2 \right) (\ln n)^3. \tag{8.4.12}$$

Next, consider the function

$$W(y_n, z) = \int\limits_{-\pi}^{\pi}\int\limits_{-\pi}^{\pi} \widehat{A}_n(\varphi_1, \varphi_2) |y_n(\varphi_1, \varphi_2) - y_n(\varphi_1 + z_1, \varphi_2 + z_2)|^2 \, d\varphi_1 \, d\varphi_2, \tag{8.4.13}$$

where $y_n(\varphi_1, \varphi_2) = y_n(k_1^0, k_2^0, \varphi_1, \varphi_2)$, $|k_1^0| < n$, is the function defined by (8.2.19). Let us estimate $W(y_n, z)$ for $|z_1| \leq h_1$, $|z_2| \leq h_2$.

For $|k_1^0| < n$, we have

$$|b_n(\varphi_1, \varphi_2)| = \frac{1}{2\pi} \left| \int\limits_{-\pi}^{\pi} F_n(\psi_1, \varphi_2) D_{k_1^0}(\psi_1 - \varphi_1) \, d\psi_1 \right| \leq$$

$$\leq \frac{1}{2\pi} \max_{\psi_1, \varphi_2} |F_n(\psi_1, \varphi_2)| \int\limits_{-\pi}^{\pi} \left| D_{k_1^0}(\psi_1 - \varphi_1) \right| \, d\psi_1 \leq C \sqrt{n} \, \ln n, \tag{8.4.14}$$

where $F_n(\varphi_1, \varphi_2)$ is the function defined by (8.4.6); $D_{k_0^0}(\varphi_1) = \sum_{p=0}^{k_1^0} e^{ip\varphi_1}$. Let us transform (8.4.13) as follows:

$$
W(y_n, z) = \int\limits_{-\pi}^{\pi}\int\limits_{-\pi}^{\pi} \widehat{A}_n(\varphi_1, \varphi_2)|y_n(\varphi_1 + z_1, \varphi_2 + z_2) - y_n(\varphi_1, \varphi_2)|^2\, d\varphi_1\, d\varphi_2 \le
$$

$$
\le 2\int\limits_{-\pi}^{\pi}\int\limits_{-\pi}^{\pi} \widehat{A}_n(\varphi_1, \varphi_2)|G_n(\varphi_1 + z_1, \varphi_2 + z_2) - G_n(\varphi_1, \varphi_2)|^2|b_n(\varphi_1, \varphi_2)|^2\, d\varphi_1\, d\varphi_2 +
$$

$$
+ 2\int\limits_{-\pi}^{\pi}\int\limits_{-\pi}^{\pi} \widehat{A}_n(\varphi_1, \varphi_2)|G_n(\varphi_1, \varphi_2)|^2|b_n(\varphi_1, \varphi_2) - b_n(\varphi_1 + z_1, \varphi_2 + z_2)|^2\, d\varphi_1\, d\varphi_2 =
$$

$$
= 2W_1 + 2W_2. \tag{8.4.15}
$$

By (8.4.12) and (8.414), we get

$$
W_1 \le C\left(h_1^2 + h_2^2\right)n(\ln n)^5. \tag{8.4.16}
$$

From (8.4.7), (8.4.8), it follows that

$$
W_2 \le C\int\limits_{-\pi}^{\pi}\int\limits_{-\pi}^{\pi} |b_n(\varphi_1, \varphi_2) - b_n(\varphi_1 + z_1, \varphi_2 + z_2)|^2\, d\varphi_1\, d\varphi_2 \le C_1 n(\ln n)^3\left(h_1^2 + h_2^2\right). \tag{8.4.17}
$$

Applying (8.4.16) and (8.4.17), we obtain the estimate

$$
W(y_n, z) \le C\left(h_1^2 + h_2^2\right)n(\ln n)^5 \quad \text{for} \quad |z_1| \le h_1, \quad |z_2| \le h_2. \tag{8.4.18}
$$

Definition 8.4.1. Let ε be a set of indices $k = (k_1, k_2)$ with integer components. We say that ε *is a rectangular set of indices*, if it is defined by the inequalities $n_1 \le k_1 \le n_2$, $m_1 \le k_2 \le m_2$.

Consider the following problem of restriction for pseudodifference operators A_n with symbols $\widehat{A}_n$:

$$
\int\limits_{-\pi}^{\pi}\int\limits_{-\pi}^{\pi} \widehat{A}(\varphi_1, \varphi_2)\widehat{x}_{nc}(\varphi_1, \varphi_2)D_{\varepsilon_n}(\psi_1 - \varphi_1, \psi_2 - \varphi_2)\, d\varphi_1\, d\varphi_2 = 1, \tag{8.4.19}
$$

where $\widehat{A}_n(\varphi_1, \varphi_2)$ satisfies conditions (8.2.9),

$$
D_{\varepsilon_n}(\psi_1, \psi_2) = \frac{1}{4\pi^2}\sum_{k\in\varepsilon_n} e^{i(k_1\psi_1 + k_2\psi_2)}, \qquad \widehat{x}_{nc}(\varphi_1, \varphi_2) = \sum_{k\in\varepsilon_n} x_n(k_1, k_2)e^{i(k_1\varphi_1 + k_2\varphi_2)}.
$$

Here ε_n is a rectangular set of indices defined by the relations

$$
-n_1^{(n)} \le k_1 \le n_2^{(n)}, \quad -m_1^{(n)} \le k_2 \le m_2^{(n)}, \quad n_i^{(n)} \ge 0, \quad m_i^{(n)} \ge 0, \quad i = 1, 2;
$$
$$
\lim_{n\to\infty} n_2^{(n)} = \infty, \qquad \lim_{n\to\infty} m_i^{(n)} = \infty.
$$

We seek a solution of problem (8.4.19) in the form

$$
\widehat{x}_{nc}(\varphi_1, \varphi_2) = z_n(\varphi_1, \varphi_2) + P_n^\nu(\varphi_1, \varphi_2), \tag{8.4.20}
$$

where

$$P_n^\nu(\varphi_1, \varphi_2) = \int\limits_{-\pi}^{\pi} \int\limits_{-\pi}^{\pi} y_n(\varphi_1 + z_1, \varphi_2 + z_2) A_\nu(z_1, z_2)\, dz_1\, dz_2. \tag{8.4.21}$$

The function $y_n(\varphi_1, \varphi_2)$ has the form

$$y_n(\varphi_1, \varphi_2) = \frac{\widehat{y}_n(\varphi_1, \varphi_2)}{e^{n_1^{(n)}\varphi_1}},$$

where $\widehat{y}_n(\varphi_1, \varphi_2)$ is the solution of the restriction problem for pseudodifference operators A_n with symbols $\widehat{A}_n(\varphi_1, \varphi_2)$; the restriction set is

$$\varepsilon_0 = \{k = (k_1, k_2) : \ 0 \le k_1 \le \infty, \ -\infty < k_2 < \infty\},$$

and the right-hand side $f(k)$ is such that $f(n_1^{(n)}, 0) = 1$ and $f(k) = 0$ elsewhere.

The function $A_\nu(z_1, z_2)$ is the product of Nikolskii's kernels (see (8.3.10)),

$$A_\nu(z_1, z_2) = K_\nu(z_1) K_\nu(z_2), \tag{8.4.22}$$

where $\nu = \sigma(\lambda - 1)$.

For a fixed $\varepsilon > 0$, consider the set

$$B_\varepsilon = \left\{ \varphi : \ -\pi \le \varphi \le \frac{1}{\nu^{1-\varepsilon}} \quad \text{or} \quad \frac{1}{\nu^{1-\varepsilon}} \le \varphi \le \pi \right\}.$$

The Nikolskii kernels satisfy the inequality (8.3.11) on B_ε. There always exists σ_0 such that $2\sigma_0\varepsilon - 1 > 1$. We choose the order of the Nikolskii kernels, $\nu = \min\left(m_1^{(n)}, m_2^{(n)}, n_2^{(n)}\right)$. It should be observed that the trigonometric polynomial $P_n^\nu(\varphi_1, \varphi_2)$ has the form

$$P_n^\nu(\varphi_1, \varphi_2) = \sum_{k \in \varepsilon_n} p_n^\nu(k_1, k_2)\, e^{i(k_1\varphi_1 + k_2\varphi_2)},$$

since $y_n(\varphi_1, \varphi_2)$ is defined by (8.2.19).

Theorem 8.4.1. *If conditions (8.2.9) hold for the symbols $\widehat{A}_m(\varphi_1, \varphi_2)$ and the restriction set $\varepsilon = \left\{ (k_1, k_2) \in \mathbb{Z}_2 : \ -n_1^{(n)} \le k_1 \le n_2^{(n)}, \ -m_1^{(n)} \le k_2 \le m_2^{(n)} \right\}$ is that specified above, then the solution $\widehat{x}_{nc}(\varphi_1, \varphi_2)$ of problem (8.4.19) can be represented in the form*

$$\widehat{x}_{nc}(\varphi_1, \varphi_2) = z_n(\varphi_1, \varphi_2) + P_n^\nu(\varphi_1, \varphi_2).$$

Moreover, for any ε, the following estimates hold with $\nu = \min\left(m_1^{(n)}, m_2^{(n)}, n_2^{(n)}\right)$:

$$\|z_n\|_1 \le \left(W(y_n, \nu^{\varepsilon-1})\right)^{1/2} + C\nu^{\sigma\varepsilon-1/2}\left(\|y_n\|^2| + \|y_n\|_{L_2}^2\right)^{1/2}, \tag{8.4.23}$$
$$|p_n^\nu(k_1, k_2)| \le |y_n(k_1, k_2)|,$$

where $W(y_n, \nu^{\varepsilon-1})$ is given by (8.4.13); the norm $\|z_n\|_1$ is defined by (8.3.2); the function $P_n^\nu(\varphi_1, \varphi_2)$ is defined by (8.4.21); $p_n^\nu(k_1, k_2)$ and $y_n(k_1, k_2)$ are the Fourier coefficients of the functions $P_n^\nu(\varphi_1, \varphi_2)$ and $y_n(\varphi_1, \varphi_2)$, respectively.

Proof. With the help of Jensen's inequality (see *Zygmund* (1965)), we find that

$$
J = \int_{-\pi}^{\pi}\int_{-\pi}^{\pi} \widehat{A}_n(\varphi_1,\varphi_2)|y_n(\varphi_1,\varphi_2) - P_n^{\nu}(\varphi_1,\varphi_2)|^2 \, d\varphi_1 \, d\varphi_2 -
$$

$$
= \int_{-\pi}^{\pi}\int_{-\pi}^{\pi} \widehat{A}_n(\varphi_1,\varphi_2)\left|\int_{-\pi}^{\pi}\int_{-\pi}^{\pi} [y_n(\varphi_1,\varphi_2) - y_n(\varphi_1+z_1,\varphi_2+z_2)]A_\nu(z_1,z_2)\,dz_1\,dz_2\right|^2 d\varphi_1\,d\varphi_2 \le
$$

$$
\le \int_{-\pi}^{\pi}\int_{-\pi}^{\pi} \widehat{A}_n(\varphi_1,\varphi_2)\left[\int_{-\pi}^{\pi}\int_{-\pi}^{\pi} [y_n(\varphi_1,\varphi_2) - y_n(\varphi_1+z_1,\varphi_2+z_2)]^2 A_\nu(z_1,z_2)\,dz_1\,dz_2\right] d\varphi_1\,d\varphi_2 =
$$

$$
= \int_{-\pi}^{\pi}\int_{-\pi}^{\pi} \widehat{A}_n(\varphi_1,\varphi_2)\left[\iint_{D_\delta} [y_n(\varphi_1,\varphi_2) - y_n(\varphi_1+z_1,\varphi_2+z_2)]^2 A_\nu(z_1,z_2)\,dz_1\,dz_2\right] d\varphi_1\,d\varphi_2 +
$$

$$
+ \int_{-\pi}^{\pi}\int_{-\pi}^{\pi} \widehat{A}_n(\varphi_1,\varphi_2)\left[\iint_{D\backslash D_\delta} [y_n(\varphi_1,\varphi_2) - y_n(\varphi_1+z_1,\varphi_2+z_2)]^2 A_\nu(z_1,z_2)\,dz_1\,dz_2\right] d\varphi_1\,d\varphi_2 =
$$

$$
= J_1 + J_2, \tag{8.4.24}
$$

where $D = [-\pi,\pi]\times[-\pi,\pi]$, $D_\delta = [-\delta,\delta]$, $\delta = \nu^{\varepsilon-1}$. Let us estimate J_1 and J_2. With the help of Jensen's inequality, we get

$$
J_1 = \int_{-\delta}^{\delta}\int_{-\delta}^{\delta} A_\nu(z_1,z_2)\left[\int_{-\pi}^{\pi}\int_{-\pi}^{\pi} \widehat{A}_n(\varphi_1,\varphi_2)|y_n(\varphi_1,\varphi_2) - y_n(\varphi_1+z_1,\varphi_2+z_2)|^2\,d\varphi_1\,d\varphi_2\right] dz_1\,dz_2 \le
$$

$$
\le W\left(y_n\nu^{\varepsilon-1}\right), \tag{8.4.25}
$$

$$
J_2 \le C\nu^{1-2\sigma\varepsilon}\left[\int_{-\pi}^{\pi}\int_{-\pi}^{\pi} \widehat{A}_n(\varphi_1,\varphi_2)|y_n(\varphi_1,\varphi_2)|^2\,d\varphi_1\,d\varphi_2 + \int_{-\pi}^{\pi}\int_{-\pi}^{\pi} |y_n(\varphi_1,\varphi_2)|^2\,d\varphi_1\,d\varphi_2\right]. \tag{8.4.26}
$$

From (8.4.25) and (8.4.26), it follows that

$$
J \le W\left(y_n,\nu^{\varepsilon-1}\right) + C\nu^{1-2\sigma\varepsilon}\left(\|y_n\|_1^2 + \|y_n\|_{L_2}^2\right). \tag{8.4.27}
$$

Using (8.4.19) and (8.4.20), we find that

$$
T(z_n) = \int_{-\pi}^{\pi}\int_{-\pi}^{\pi} \widehat{A}_n(\varphi_1,\varphi_2)z_n(\varphi_1,\varphi_2)D_{\varepsilon_n}(\psi_1-\varphi_1,\psi_2-\varphi_2)\,d\varphi_1\,d\varphi_2 =
$$

$$
= 1 - \int_{-\pi}^{\pi}\int_{-\pi}^{\pi} \widehat{A}_n(\varphi_1,\varphi_2)P_n^{\nu}(\varphi_1,\varphi_2)D_{\varepsilon_n}(\psi_1-\varphi_1,\psi_2-\varphi_2)\,d\varphi_1\,d\varphi_2 = \tag{8.4.28}
$$

$$
= \int_{-\pi}^{\pi}\int_{-\pi}^{\pi} \widehat{A}_n(\varphi_1,\varphi_2)\left[y_n(\varphi_1,\varphi_2) - P_n^{\nu}(\varphi_1,\varphi_2)\right]D_{\varepsilon_n}(\psi_1-\varphi_1,\psi_2-\varphi_2)\,d\varphi_1\,d\varphi_2.
$$

From (8.4.28), (8.4.24), and (8.3.3) we obtain

$$\|T(z_n)\|_2 = \max_{\|b\|_1=1} \left| \int\limits_{-\pi}^{\pi} \int\limits_{-\pi}^{\pi} T(z_n) \overline{b(\psi_1, \psi_2)} \, d\psi_1 \, d\psi_2 \right| = \max_{\|b\|_1=1} \left| \int\limits_{-\pi}^{\pi} \int\limits_{-\pi}^{\pi} \overline{b(\psi_1, \psi_2)} \times \right.$$

$$\times \left[\int\limits_{-\pi}^{\pi} \int\limits_{-\pi}^{\pi} \left[y_n(\varphi_1, \varphi_2) - P_n^{\nu}(\varphi_1, \varphi_2) \right] D_{\varepsilon_n}(\psi_1 - \varphi_1, \psi_2 - \varphi_2) \, d\varphi_1 \, d\varphi_2 \right] d\psi_1 \, d\psi_2 \Bigg| =$$

$$= \max_{\|b\|_1=1} \left| \int\limits_{-\pi}^{\pi} \int\limits_{-\pi}^{\pi} \widehat{A}_n(\varphi_1, \varphi_2) \left[y_n(\varphi_1, \varphi_2) - P_n^{\nu}(\varphi_1, \varphi_2) \right] \times \right.$$

$$\times \int\limits_{-\pi}^{\pi} \int\limits_{-\pi}^{\pi} \overline{b(\psi_1 \psi_2) \overline{D_{\varepsilon_n}(\psi_1 - \varphi_1, \psi_2 - \varphi_2)}} \, d\psi_1 \, d\psi_2 \, d\varphi_1 \, d\varphi_2 \Bigg| =$$

$$= \max_{\|b\|_1=1} \left| \int\limits_{-\pi}^{\pi} \int\limits_{-\pi}^{\pi} \widehat{A}_n(\varphi_1, \varphi_2) |y_n(\varphi_1, \varphi_2) - P_n^{\nu}(\varphi_1, \varphi_2)| \overline{b(\varphi_1, \varphi_2)} \, d\varphi_1 \, d\varphi_2 \right| \leq$$

$$\leq \max_{\|b\|_1=1} \left[\int\limits_{-\pi}^{\pi} \int\limits_{-\pi}^{\pi} \widehat{A}_n(\varphi_1, \varphi_2) |y_n(\varphi_1, \varphi_2) - P_n^{\nu}(\varphi_1, \varphi_2)|^2 \right]^{1/2} \times$$

$$\times \left[\int\limits_{-\pi}^{\pi} \int\limits_{-\pi}^{\pi} \widehat{A}_n(\varphi_1, \varphi_2) |b(\varphi_1, \varphi_2)|^2 \, d\varphi_1 \, d\varphi_2 \right]^{1/2} \leq$$

$$\leq \left[W\left(y_n, \nu^{\varepsilon-1}\right) \right]^{1/2} + C_1 \nu^{1/2 - \sigma\varepsilon} \left[\|y_n\|_1^2 + \|y_n\|_{L_2}^2 \right]^{1/2}. \tag{8.4.29}$$

Proposition 8.3.1, combined with the estimate (8.4.29), implies that

$$\|z_n\| \leq \left[W\left(y_n, \nu^{\varepsilon-1}\right) \right]^{1/2} + C_1 \nu^{1/2 - \sigma\varepsilon} \left[\|y_n\|_1^2 + \|y_n\|_{L_2}^2 \right]^{1/2}. \tag{8.4.30}$$

Just as in Theorem 8.3.1, one can prove the inequality

$$|p_n^{\nu}(k_1, k_2)| \leq |y_n(k_1, k_2)|, \tag{8.4.31}$$

where $p_n^{\nu}(k_1, k_2)$ and $y_n(k_1, k_2)$ are the Fourier coefficients of the functions $P^{\nu}(\varphi_1, \varphi_2)$ and $y_n(\varphi_1, \varphi_2)$, respectively. Now, the result of Theorem 8.4.1 follows from (8.4.30) and (8.4.31).

8.5. Asymptotic Estimates of Special Matrices

Consider a hypersingular integral equation of Prandtl type,

$$\iint_G \frac{Q(x_1, x_2)}{r^3} \, dx_1 \, dx_2 = f\left(x_1^0, x_2^0\right), \tag{8.5.1}$$

where G is a plane bounded domain, $r = \left[\left(x_1^0 - x_1\right)^2 + \left(x_2^0 - x_2\right)^2 \right]^{1/2}$, $M_0 = (x_1^0, x_2^0) \in G$; the integral is understood in the sense of Hadamard (see (5.2.9)).

Consider a canonical partition of $\mathbb{R}_2$ into cells $D(k, h)$ with constant vector-valued step $h = (h_1, h_2)$ such that

$$0 < C_1 < \frac{h_1}{h_2} = \theta < C_2. \tag{8.5.2}$$

Denote by $\varepsilon_h = \varepsilon(G, h)$ the set of all indices $k = (k_1, k_2)$ such that $D(k, h) \subset G$. In each cell $D(k, h)$, consider a point $M_{0k}(h_1 k_1, h_2 k_2)$ and let $r_{0k} = |MM_{0k}|$, where M is the point with coordinates $x = (x_1, x_2)$. Consider the discrete convolution operator of Prandtl type,

$$\Pi a(k) = \sum_k a(k) \gamma(k, m), \tag{8.5.3}$$

where $a(k) \in l_2$ is a grid function and the coefficients $\gamma(k, m)$ are given by

$$\gamma(k, m) = \int_{D(k,h)} \frac{dx}{r_{0m}^3} . \tag{8.5.4}$$

The above discrete convolution operator can be represented as a pseudodifference operator,

$$\Pi a(k) = \frac{1}{4\pi^2} \int\limits_{-\pi}^{\pi} \int\limits_{-\pi}^{\pi} \widehat{A}(\varphi_1, \varphi_2, h) \widehat{a}_c(\varphi_1, \varphi_2) e^{-i(m_1\varphi_1 + m_2\varphi_2)} \, d\varphi_1 \, d\varphi_2 , \tag{8.5.5}$$

where $\widehat{a}_c(\varphi_1, \varphi_2)$ is the discrete Fourier transform of the function $a(k)$ (see Definition 5.1.4); the function $\widehat{A}(\varphi_1, \varphi_2, h)$ is the symbol of this pseudodifference operator and

$$\widehat{A}(\varphi_1, \varphi_2, h) = \sum_m \gamma(0, m) e^{i(m_1\varphi_1 + m_2\varphi_2)} . \tag{8.5.6}$$

Let $G(\varphi_1, \varphi_2) = -h_1 \widehat{A}(\varphi_1, \varphi_2, h)$. Consider the restriction problem (8.3.1) for the pseudodifference operator Π_1 with the symbol $G(\varphi_1, \varphi_2)$ and the restriction set ε_h. Let $n = \max_{k \in \varepsilon_h} |k_1|$ (n depends on h). Since the domain G is bounded, there is $p > 0$ such that the restriction problems (8.3.1) for the operator Π_1 and the operator Π_{1n} with the symbol $G_n(\varphi_1, \varphi_2)$ of the form

$$G_n(\varphi_1, \varphi_2) = \sum_{m_2=-pn}^{pn} \sum_{m_1=-pn}^{pn} (-h_1) \gamma(0, m) e^{i(m_1\varphi_1 + m_2\varphi_2)} \tag{8.5.7}$$

are equivalent and the number p does not depend on h.

For our further exposition, we examine two restriction problems,

$$\int\limits_{-\pi}^{\pi} \int\limits_{-\pi}^{\pi} G_n(\varphi_1, \varphi_2) \widehat{x}_c(\varphi_1, \varphi_2) D_{\varepsilon_h}(\alpha - \varphi_1, \beta - \varphi_2) \, d\varphi_1 \, d\varphi_2 = e^{i(m_1\alpha + m_2\beta)} \left(e^{i\alpha} - 1 \right), \tag{8.5.8}$$

$$\int\limits_{-\pi}^{\pi} \int\limits_{-\pi}^{\pi} G_n(\varphi_1, \varphi_2) \widehat{x}_c(\varphi_1, \varphi_2) D_{\varepsilon_h}(\alpha - \varphi_1, \beta - \varphi_2) \, d\varphi_1 \, d\varphi_2 = e^{i(m_1\alpha + m_2\beta)} \left(e^{i\beta} - 1 \right), \tag{8.5.9}$$

where $(m_1 + 1, m_2) \in \varepsilon_h$, $(m_1, m_2 + 1) \in \varepsilon_h$, $(m_1, m_2) \in \varepsilon_h$.

The function $A_n(\varphi_1, \varphi_2) = G_n(\varphi_1, \varphi_2)/h_1$ satisfies the inequality (5.2.23) with $N = n$. Direct calculations show that for the symbol $G_n(\varphi_1, \varphi_2)$ conditions (8.2.9) hold.

Consider the function

$$B_n(\varphi_1, \varphi_2) = \frac{e^{i\varphi_1} - 1}{G_n(\varphi_1, \varphi_2)} . \tag{8.5.10}$$

Let

$$\Delta_z^2 B_n = B_n(\varphi_1 + 2z_1, \varphi_2 + 2z_2) - 2B_n(\varphi_1 + z_1, \varphi_2 + z_2) + B_n(\varphi_1, \varphi_2).$$

In order to estimate $\Delta_z^2 B_n$, we transform it as follows:

$$\Delta_z^2 B_n = \frac{(B-D)G_n(\varphi_1, \varphi_2) + D[G_n(\varphi_1, \varphi_2) - G_n(\varphi_1 + 2z_1, \varphi_2 + 2z_2)]}{G_n(\varphi_1, \varphi_2)G_n(\varphi_1 + z_1, \varphi_2 + z_2)G_n(\varphi_1 + 2z_1, \varphi_2 + 2z_2)} = I_1 + I_2, \qquad (8.5.11)$$

where

$$D = \left(e^{i(\varphi_1 + z_1)} - 1\right)G_n(\varphi_1, \varphi_2) + \left(e^{i\varphi_1} - 1\right)G_n(\varphi_1 + z_1, \varphi_2 + z_2) =$$
$$= \left[\left(e^{i\varphi_1} - 1\right)G_n(\varphi_1, \varphi_2) - \left(e^{i(\varphi_1 + z_1)} - 1\right)G_n(\varphi_1 + z_1, \varphi_2 + z_2)\right] +$$
$$+ \left(e^{i(\varphi_1 + z_1)} - e^{i\varphi_1}\right)[G_n(\varphi_1 + z_1, \varphi_2 + z_2) + G_n(\varphi_1, \varphi_2)], \qquad (8.5.12)$$

$$B = \left[\left(e^{i(\varphi_1 + z_1)} - 1\right)G_n(\varphi_1 + z_1, \varphi_2 + z_2) - \left(e^{i(\varphi_1 + 2z_1)} - 1\right)G_n(\varphi_1 + 2z_1, \varphi_2 + 2z_2)\right] +$$
$$+ \left(e^{i(\varphi_1 + 2z_1)} - e^{i(\varphi_1 + z_1)}\right)[G_n(\varphi_1 + 2z_1, \varphi_2 + 2z_2) + G_n(\varphi_1 + z_1, \varphi_2 + z_2)]. \qquad (8.5.13)$$

Using condition (8.2.9), together with (8.5.10) and (8.5.11), we obtain

$$|I_2| = \left| \frac{e^{i(\varphi_1 + z_1)} - e^{i\varphi_1}(G_n(\varphi_1, \varphi_2) - G_n(\varphi_1 + 2z_1, \varphi_2 + 2z_2))}{G_n(\varphi_1, \varphi_2)G_n(\varphi_1 + z_1, \varphi_2 + z_2)G_n(\varphi_1 + 2z_1, \varphi_2 + 2z_2)} \times \right.$$
$$\times (G_n(\varphi_1 + z_1, \varphi_2 + z_2) + G_n(\varphi_1, \varphi_2)) + \left(e^{i\varphi_1} - 1\right) \times$$
$$\times \frac{(G_n(\varphi_1, \varphi_2) - G_n(\varphi_1 + z_1, \varphi_2 + z_2))(G_n(\varphi_1, \varphi_2) - G_n(\varphi_1 + 2z_1, \varphi_2 + 2z_2))}{G_n(\varphi_1, \varphi_2)G_n(\varphi_1 + z_1, \varphi_2 + z_2)G_n(\varphi_1 + 2z_1, \varphi_2 + 2z_2)} +$$
$$\left. + \frac{\left(e^{i\varphi_1} - e^{i(\varphi_1 + z_1)}\right)G_n(\varphi_1 + z_1, \varphi_2 + z_2)(G_n(\varphi_1, \varphi_2) - G_n(\varphi_1 + 2z_1, \varphi_2 + 2z_2))}{G_n(\varphi_1, \varphi_2)G_n(\varphi_1 + z_1, \varphi_2 + z_2)G_n(\varphi_1 + 2z_1, \varphi_2 + 2z_2)} \right| \le$$
$$\le C \ln^2 n \left(z_1^2 + z_2^2\right) \left(\frac{1}{G_n(\varphi_1 + z_1, \varphi_2 + z_2)G_n(\varphi_1 + 2z_1, \varphi_2 + 2z_2)} + \right.$$
$$\left. + \frac{1}{G_n(\varphi_1, \varphi_2)G_n(\varphi_1 + 2z_1, \varphi_2 + 2z_2)} \right). \qquad (8.5.14)$$

Next, we carry out the transformation

$$B - D = \left\{ \left[\left(e^{i(\varphi_1 + z_1)} - 1\right)G_n(\varphi_1 + z_1, \varphi_2 + z_2) - \left(e^{i(\varphi_1 + 2z_1)} - 1\right)G_n(\varphi_1 + 2z_1, \varphi_2 + 2z_2)\right] - \right.$$
$$- \left[\left(e^{i\varphi_1} - 1\right)G_n(\varphi_1, \varphi_2) - \left(e^{i(\varphi_1 + z_1)} - 1\right)G_n(\varphi_1 + z_1, \varphi_2 + z_2)\right] \right\} +$$
$$+ \left\{ \left(e^{i(\varphi_1 + 2z_1)} - e^{i(\varphi_1 + z_1)}\right)(G_n(\varphi_1 + 2z_1, \varphi_2 + 2z_2) + G_n(\varphi_1 + z_1, \varphi_2 + z_2)) - \right.$$
$$\left. - \left(e^{i(\varphi_1 + z_1)} - e^{i\varphi_1}\right)(G_n(\varphi_1 + z_1, \varphi_2 + z_2) + G_n(\varphi_1, \varphi_2)) \right\} = I_3 + I_4. \qquad (8.5.15)$$

It follows that

$$|I_4| \le \left| \left(e^{i(\varphi_1 + 2z_1)} - e^{i(\varphi_1 + z_1)}\right)(G_n(\varphi_1 + 2z_1, \varphi_2 + 2z_2) - G_n(\varphi_1, \varphi_2)) + \right.$$
$$\left. + \left(e^{i(\varphi_1 + 2z_1)} - 2e^{i(\varphi_1 + z_1)} + e^{i\varphi_1}\right)(G_n(\varphi_1 + z_1, \varphi_2 + z_2) - G_n(\varphi_1, \varphi_2)) \right| \le$$
$$\le C(\ln n)\left(z_1^2 + z_2^2\right). \qquad (8.5.16)$$

Let us transform $|I_3|$ as follows:

$$|I_3| = \left| \left(e^{i(\varphi_1+z_1)} - 1 \right)(G_n(\varphi_1 + z_1, \varphi_2 + z_2) - G_n(\varphi_1 + 2z_1, \varphi_2 + 2z_2)) + \right.$$

$$+ G_n(\varphi_1 + 2z_1, \varphi_2 + 2z_2)\left(e^{i(\varphi_1+z_1)} - e^{i(\varphi_1+2z_1)} \right) -$$

$$- \left[\left(e^{i\varphi_1} - 1 \right)(G_n(\varphi_1, \varphi_2) - G_n(\varphi_1 + z_1, \varphi_2 + z_2)) + \right.$$

$$\left. \left. + \left(e^{i\varphi_1} - e^{i(\varphi_1+z_1)} \right)G_n(\varphi_1 + z_1, \varphi_2 + z_2) \right] \right| =$$

$$= \left| \left(e^{i(\varphi_1+z_1)} - e^{i\varphi_1} \right)(G_n(\varphi_1 + z_1, \varphi_2 + z_2) - G_n(\varphi_1 + 2z_1, \varphi_2 + 2z_2)) + \right.$$

$$+ \left(e^{i\varphi_1} - 1 \right)(2G_n(\varphi_1 + z_1, \varphi_2 + z_2) - G_n(\varphi_1, \varphi_2) - G_n(\varphi_1 + 2z_1, \varphi_2 + 2z_2)) +$$

$$+ G_n(\varphi_1 + z_1, \varphi_2 + z_2)\left(2\,e^{i(\varphi_1+z_1)} - e^{i(\varphi_1+2z_1)} - e^{i\varphi_1} \right) +$$

$$\left. + \left(e^{i(\varphi_1+z_1)} - e^{i(\varphi_1+2z_1)} \right)(G_n(\varphi_1 + 2z_1, \varphi_2 + 2z_2) - G_n(\varphi_1 + z_1, \varphi_2 + z_2)) \right| \le$$

$$\le C\left(z_1^2 + z_2^2 \right) \ln n + \left| (1 - e^{i\varphi_1}) \Delta_z^2 G_n \right|. \tag{8.5.17}$$

We further have

$$\widehat{D} = \left(e^{i\varphi_1} - 1 \right)(G_n(\varphi_1, \varphi_2) - 2G_n(\varphi_1 + z_1, \varphi_2 + z_2) + G_n(\varphi_1 + 2z_1, \varphi_2 + 2z_2)) =$$

$$= \left[\left(e^{i(\varphi_1+2z_1)} - 1 \right)G_n(\varphi_1 + 2z_1, \varphi_2 + 2z_2) - \right.$$

$$\left. - 2\left(e^{i(\varphi_1+z_1)} - 1 \right)G_n(\varphi_1 + z_1, \varphi_2 + z_2) + \left(e^{i\varphi_1} - 1 \right)G_n(\varphi_1, \varphi_2) \right] +$$

$$+ \left[\left(e^{i\varphi_1} - e^{i(\varphi_1+2z_1)} \right)G_n(\varphi_1 + 2z_1, \varphi_2 + 2z_2) - \right.$$

$$\left. - 2\left(e^{i\varphi_1} - e^{i(\varphi_1+z_1)} \right)G_n(\varphi_1 + z_1, \varphi_2 + z_2) \right] = D_1 + D_2, \tag{8.5.18}$$

$$|D_2| = \left| \left(e^{i\varphi_1} - e^{i(\varphi_1+2z_1)} \right)(G_n(\varphi_1 + 2z_1, \varphi_2 + 2z_2) - G_n(\varphi_1 + z_1, \varphi_2 + z_2)) + \right.$$

$$\left. + G_n(\varphi_1 + z_1, \varphi_2 + z_2)\left(e^{i\varphi_1} + 2\,e^{i(\varphi_1+z_1)} - e^{i(\varphi_1+2z_1)} \right) \right| \le$$

$$\le C\left(z_1^2 + z_2^2 \right) \ln n. \tag{8.5.19}$$

Consider the function

$$\Phi_n(\varphi_1, \varphi_2) = \left(e^{i\varphi_1} - 1 \right)G_n(\varphi_1, \varphi_2) =$$

$$= \left(e^{i\varphi_1} - 1 \right) \sum_{m_2=-pn}^{pn} \sum_{m_1=-pn}^{pn} (-h_1 \gamma(0, 0, m_1, m_2))\, e^{i(m_1\varphi_1 + m_2\varphi_2)} =$$

$$= - \sum_{m_2=-pn}^{pn} (-h_1)\gamma(0, 0, -pn, m_2)\, e^{i(-pn\varphi_1 + m_2\varphi_2)} +$$

$$+ \sum_{m_2=-pn}^{pn} \sum_{m_1=-pn}^{pn} (-h_1)[\gamma(0, 0, m_1 + 1, m_2) - \gamma(0, 0, m_1, m_2)]\, e^{i(m_1+1)\varphi_1}\, e^{im_2\varphi_2} +$$

$$+ \sum_{m_2=-pn}^{pn} (-h_1)\gamma(0,0,pn,m_2) \, e^{i(pn+1)\varphi_2+im_2\varphi_1} =$$

$$= F_{1n}(\varphi_1,\varphi_2) + F_{2n}(\varphi_1,\varphi_2) + F_{3n}(\varphi_1,\varphi_2).$$

From (8.5.4), it follows that

$$0 < \sum_{m_2=-pn}^{pn} |h_1\gamma(0,0,pn,m_2)| \le h_1 \int_{-\pi/2}^{\pi/2} \int_{(pn-1/2)h_1/\cos\varphi}^{(pn+1/2)h_1/\cos\varphi} \frac{1}{r^2} \, dr \, d\varphi \le \frac{C_1}{p^2 n^2}. \qquad (8.5.20)$$

Similarly,

$$0 < \sum_{m_2=-pn}^{pn} |h_1\gamma(0,0,-pn,m_2)| \le \frac{C_1}{p^2 n^2}. \qquad (8.5.21)$$

From (8.5.20), (8.5.21), it follows that the functions $F_{1n}(\varphi_1,\varphi_2)$ and $F_{3n}(\varphi_1,\varphi_2)$ have bounded second derivatives. Let us estimate the Fourier coefficients of $F_{2n}(\varphi_1,\varphi_2)$.

Changing the variables to $y_1 = h_1^{-1}[x_1 - (m_1 - 1/2)h_1]$, $y_2 = h_2^{-1}[x_2 - (m_2 - 1/2)h_1]$, we obtain the relation

$$h_1^2\gamma(0,0,m_1,m_2) = h_1 \int_{(m_1-1/2)h_1}^{(m_1+1/2)h_1} \int_{(m_2-1/2)h_2}^{(m_2+1/2)h_2} (x_1^2 + x_2^2)^{-3/2} \, dx_1 \, dx_2 =$$

$$= \theta \int_0^1 \int_0^1 \left[\left(y_1 + m_1 - \frac{1}{2}\right)^2 + \theta^2 \left(y_2 + m_2 - \frac{1}{2}\right)^2 \right]^{-3/2} dy_1 \, dy_2, \qquad (8.5.22)$$

where $\theta = h_2/h_1$. Using (8.5.22), we obtain the inequality

$$b_{m_1 m_2} = h_1 |\gamma(0,0,m_1+1,m_2) - \gamma(0,0,m_1,m_2)| =$$

$$= \theta \left| \int_0^1 \int_0^1 \left[\left(y_1 + m_1 + \frac{1}{2}\right)^2 + \theta^2 \left(y_2 + m_2 - \frac{1}{2}\right)^2 \right]^{-3/2} dy_1 \, dy_2 - \right.$$

$$\left. - \int_0^1 \int_0^1 \left[\left(y_1 + m_1 - \frac{1}{2}\right)^2 + \theta^2 \left(y_2 + m_2 - \frac{1}{2}\right)^2 \right]^{-3/2} dy_1 \, dy_2 \right| \le$$

$$\le C \left(m_1^2 + m_2^2 + 1\right)^{-2}. \qquad (8.5.23)$$

From (8.5.23) it follows that

$$\left| (F_{2n})''_{\varphi_1\varphi_1} \right| \le C \sum_{m_2=-pn}^{pn} \sum_{m_1=-pn}^{pn-1} b_{m_1 m_2} m_1^2 \le C_1 \ln n,$$

$$\left| (F_{2n})''_{\varphi_1\varphi_2} \right| \le C \sum_{m_2=-pn}^{pn} \sum_{m_1=-pn}^{pn-1} b_{m_1 m_2} m_1 m_2 \le C_1 \ln n, \qquad (8.5.24)$$

$$\left| (F_{2n})''_{\varphi_2\varphi_2} \right| \le C \sum_{m_2=-pn}^{pn} \sum_{m_1=-pn}^{pn-1} b_{m_1 m_2} m_2^2 \le C_1 \ln n.$$

Hence, in view of (8.5.18), we obtain the estimate

$$|D_1| \le C(z_1^2 + z_2^2) \ln n. \tag{8.5.25}$$

From (8.5.14), (8.5.16), (8.5.17), (8.5.19), and (8.5.25), we obtain the inequality

$$|\Delta_z^2 B_n| \le C(z_1^2 + z_2^2) \ln n \left[\frac{1}{G_n(\varphi_1 + z_1, \varphi_2 + z_2)G_n(\varphi_1 + 2z_1, \varphi_2 + 2z_2)} + \tag{8.5.26}$$

$$+ \frac{1}{G_n(\varphi_1, \varphi_2)G_n(\varphi_1 + 2z_1, \varphi_2 + 2z_2)} \right].$$

Let $B_{1n}(\varphi_1, \varphi_2) = (e^{i\varphi_2} - 1)/G_n(\varphi_1, \varphi_2)$. Then, for $\Delta_z^2 B_{1n}$ the estimate (8.5.26) holds. Using (8.5.26), we obtain the inequality

$$\int_{-\pi}^{\pi}\int_{-\pi}^{\pi} G_n(\varphi_1, \varphi_2)|\Delta_z^2 B_n|^2 \, d\varphi_1 \, d\varphi_2 \le C(z_1^2 + z_2^2)^2 \ln^2 n \times$$

$$\times \left[\int_{-\pi}^{\pi}\int_{-\pi}^{\pi} \frac{[(G_n(\varphi_1, \varphi_2) - G_n(\varphi_1 + z_1, \varphi_2 + z_2)) + G_n(\varphi_1 + z_1, \varphi_2 + z_2)] \, d\varphi_1 \, d\varphi_2}{|G_n(\varphi_1 + z_1, \varphi_2 + z_2)|^2 |G_n(\varphi_1 + 2z_1, \varphi_2 + 2z_2)|^2} + \right.$$

$$+ 2\int_{-\pi}^{\pi}\int_{-\pi}^{\pi} \frac{d\varphi_1 \, d\varphi_2}{|G_n(\varphi_1 + z_1, \varphi_2 + z_2)|^2 |G_n(\varphi_1 + 2z_1, \varphi_2 + 2z_2)|^2} +$$

$$+ \left. \int_{-\pi}^{\pi}\int_{-\pi}^{\pi} \frac{d\varphi_1 \, d\varphi_2}{G_n(\varphi_1, \varphi_2)|G_n(\varphi_1 + 2z_1, \varphi_2 + 2z_2)|^2} \right] \le$$

$$\le C_1(z_1^2 + z_2^2)^2 \ln^2 \left[n^2(z_1^2 + z_2^2)^{1/2} \ln n + 4n \ln n \right]. \tag{8.5.27}$$

In a similar way, we obtain the estimate

$$\int_{-\pi}^{\pi}\int_{-\pi}^{\pi} G_n(\varphi_1, \varphi_2)|\Delta_z^2 B_{1n}|^2 \, d\varphi_1 \, d\varphi_2 \le C(z_1^2 + z_2^2)^2 \ln^2 n \left[n^2(z_1^2 + z_2^2)^{1/2} \ln n + 4n \ln n \right]. \tag{8.5.28}$$

Let us estimate the Fourier coefficients of the functions $B_n(\varphi_1, \varphi_2)$ and $B_{1n}(\varphi_1, \varphi_2)$. The second derivatives of $B_n(\varphi_1, \varphi_2)$ have the form

$$(B_n)''_{\varphi_2\varphi_2} = -(e^{i\varphi_1} - 1)\frac{[G_n(\varphi_1, \varphi_2)]''_{\varphi_2\varphi_2}G_n(\varphi_1, \varphi_2) - 2[G_n(\varphi_1, \varphi_2)]'^2_{\varphi_2}}{|G_n(\varphi_1, \varphi_2)|^3}.$$

$$(B_n)''_{\varphi_1\varphi_2} = -\frac{i\,e^{i\varphi_1}[G_n(\varphi_1, \varphi_2)]'_{\varphi_2} + (e^{i\varphi_1} - 1)[G_n(\varphi_1, \varphi_2)]''_{\varphi_1\varphi_2}}{|G_n(\varphi_1, \varphi_2)|^2} +$$

$$+ \frac{2(e^{i\varphi_1} - 1)[G_n(\varphi_1, \varphi_2)]'_{\varphi_2}[G_n(\varphi_1, \varphi_2)]'_{\varphi_1}}{|G_n(\varphi_1, \varphi_2)|^3}, \tag{8.5.29}$$

$$(B_n)''_{\varphi_1\varphi_1} = -\frac{-i\,e^{i\varphi_1}G_n(\varphi_1, \varphi_2) - i\,e^{i\varphi_1}[G_n(\varphi_1, \varphi_2)]'_{\varphi_1}}{|G_n(\varphi_1, \varphi_2)|^2} - \frac{2(1 - e^{i\varphi_1})[G_n(\varphi_1, \varphi_2)]'^2_{\varphi_1}}{|G_n(\varphi_1, \varphi_2)|^3} +$$

$$+ \frac{-i\,e^{i\varphi_1}[G_n(\varphi_1, \varphi_2)]'_{\varphi_1} + (1 - e^{i\varphi_1})[G_n(\varphi_1, \varphi_2)]''_{\varphi_1\varphi_1}}{|G_n(\varphi_1, \varphi_2)|^2}.$$

It has been shown above that the absolute values of the second derivatives of the functions $\Phi_n(\varphi_1, \varphi_2) = \left(e^{i\varphi_1} - 1\right)G_n(\varphi_1, \varphi_2)$ are bounded by $C \ln n$ (see (8.5.24), and since conditions (8.2.9) hold for $G_n(\varphi_1, \varphi_2)$, we have

$$
\begin{aligned}
\left| \left(e^{i\varphi_1} - 1\right)[G_n(\varphi_1, \varphi_2)]''_{\varphi_1\varphi_1} \right| &\leq c \ln n, \\
\left| \left(e^{i\varphi_1} - 1\right)[G_n(\varphi_1, \varphi_2)]''_{\varphi_1\varphi_2} \right| &\leq c \ln n, \\
\left| \left(e^{i\varphi_1} - 1\right)[G_n(\varphi_1, \varphi_2)]''_{\varphi_2\varphi_2} \right| &\leq c \ln n.
\end{aligned}
\tag{8.5.30}
$$

Using (8.2.9), (8.5.29), and (8.5.30), we obtain the inequalities

$$
\int_{-\pi}^{\pi}\int_{-\pi}^{\pi} \left| [B_n(\varphi_1, \varphi_2)]''_{\varphi_1\varphi_1} \right| d\varphi_1 \, d\varphi_2 \leq c \ln^3 n,
$$

$$
\int_{-\pi}^{\pi}\int_{-\pi}^{\pi} \left| [B_n(\varphi_1, \varphi_2)]''_{\varphi_1\varphi_2} \right| d\varphi_1 \, d\varphi_2 \leq c \ln^3 n,
$$

$$
\int_{-\pi}^{\pi}\int_{-\pi}^{\pi} \left| [B_n(\varphi_1, \varphi_2)]''_{\varphi_2\varphi_2} \right| d\varphi_1 \, d\varphi_2 \leq c \ln^3 n,
$$

Hence follow the estimates for the Fourier coefficients $b_n(m_1, m_2)$ of the function $B_n(\varphi_1, \varphi_2)$,

$$
\begin{aligned}
|b_n(m_1, m_2)| &\leq \frac{C \ln^3 n}{(|m_1| + 1)(|m_2| + 1)}, \\
|b_n(m_1, m_2)| &\leq \frac{C \ln^3 n}{m_1^2 + m_2^2 + 1}.
\end{aligned}
\tag{8.5.31}
$$

Remark 8.5.1. Estimates similar to (8.5.31) hold for the Fourier coefficients $b_{1n}(m_1, m_2)$ of the function $B_{1n}(\varphi_1, \varphi_2)$.

Let ε_h be the restriction set for problems (8.5.8) and (8.5.9). Denote by ε_h^m the set of vectors $\widehat{k} \in \mathbb{Z}_2$ such that $\widehat{k} = (k_1 - m_1, k_2 - m_2)$, where $k = (k_1, k_2) \in \varepsilon_h$. Then problems (8.5.8) and (8.5.9) are equivalent to the problems

$$
\int_{-\pi}^{\pi}\int_{-\pi}^{\pi} G_n(\varphi_1, \varphi_2)y_m(\varphi_1, \varphi_2)D_{\varepsilon_h^m}(\alpha - \varphi_1, \beta - \varphi_2) d\varphi_1 \, d\varphi_2 = e^{i\alpha} - 1, \tag{8.5.32}
$$

$$
\int_{-\pi}^{\pi}\int_{-\pi}^{\pi} G_n(\varphi_1, \varphi_2)y_m(\varphi_1, \varphi_2)D_{\varepsilon_h^m}(\alpha - \varphi_1, \beta - \varphi_2) d\varphi_1 \, d\varphi_2 = e^{i\beta} - 1, \tag{8.5.33}
$$

where

$$
y_m(\varphi_1, \varphi_2) = \frac{\widehat{x}_c(\varphi_1, \varphi_2)}{e^{i(m_1\varphi_1 + m_2\varphi_2)}}, \qquad D_{\varepsilon_h^m} = \frac{1}{4\pi^2} \sum_{k \in \varepsilon_h^m} e^{i(k_1\varphi_1 + k_2\varphi_2)}.
$$

Suppose that for the points $M(m_1h_1, m_2h_2)$ the following condition holds

$$
\lim_{|h| \to 0} \frac{\rho_M}{h_1} = \infty, \tag{8.5.34}
$$

where ρ_M is the distance from $M(m_1h_1, m_2h_2)$ to the boundary of the domain G.

Let us seek a solution of problem (8.5.32) in the form

$$y_m(\varphi_1, \varphi_2) = P_m^\nu(\varphi_1, \varphi_2) + z_m(\varphi_1, \varphi_2), \tag{8.5.35}$$

where

$$P_m^\nu(\varphi_1, \varphi_2) = \int_{-\pi}^{\pi}\int_{-\pi}^{\pi} \Big[B_n(\varphi_1 + z_1, \varphi_2 + z_2) - \big(B_n(\varphi_1 + 2z_1, \varphi_2 + 2z_2) -$$

$$- B_n(\varphi_1 + z_1, \varphi_2 + z_2)\big) \Big] K_\nu(z_1) K_\nu(z_2)\, dz_1\, dz_2, \tag{8.5.36}$$

$K_\nu(z)$ is the Nikolskii kernel (see (8.3.10)); $\nu = \min\{[\rho_m/(2h_1)], [\rho_m/(2h_2)]\}$, with $[a]$ being the integer part of a.

Let us show that $P_m^\nu(\varphi_1, \varphi_2)$ is a trigonometric polynomial. Indeed, consider the expression

$$I(\varphi_1, \varphi_2) = \int_{-\pi}^{\pi}\int_{-\pi}^{\pi} B_n(\varphi_1 + 2z_1, \varphi_2 + 2z_2) K_\nu(z_1) K_\nu(z_2)\, dz_1\, dz_2 =$$

$$= \int_{0}^{4\pi}\int_{0}^{4\pi} B_n(\varphi_1 + z_1, \varphi_2 + z_2) K_\nu\Big(\frac{z_1}{2}\Big) K_\nu\Big(\frac{z_2}{2}\Big) dz_1\, dz_2 =$$

$$= \int_{0}^{2\pi} dz_1 \int_{0}^{4\pi} B_n(\varphi_1 + z_1, \varphi_2 + z_2) K_\nu\Big(\frac{z_1}{2}\Big) K_\nu\Big(\frac{z_2}{2}\Big) dz_2 +$$

$$+ \int_{2\pi}^{4\pi} dz_1 \int_{0}^{4\pi} B_n(\varphi_1 + z_1, \varphi_2 + z_2) K_\nu\Big(\frac{z_1}{2}\Big) K_\nu\Big(\frac{z_2}{2}\Big) dz_2 =$$

$$= \int_{0}^{2\pi}\int_{0}^{2\pi} B_n(\varphi_1 + z_1, \varphi_2 + z_2) \Big[K_\nu\Big(\frac{z_2}{2}\Big) + K_\nu\Big(\frac{z_2 + 2\pi}{2}\Big)\Big] K_\nu\Big(\frac{z_1}{2}\Big) dz_1\, dz_2 +$$

$$+ \int_{0}^{2\pi}\int_{0}^{2\pi} B_n(\varphi_1 + z_1, \varphi_2 + z_2) \Big[K_\nu\Big(\frac{z_2}{2}\Big) + K_\nu\Big(\frac{z_2 + 2\pi}{2}\Big)\Big] K_\nu\Big(\frac{z_1 + 2\pi}{2}\Big) dz_1\, dz_2 =$$

$$= \int_{0}^{2\pi}\int_{0}^{2\pi} B_n(\varphi_1 + z_1, \varphi_2 + z_2) \Big[K_\nu\Big(\frac{z_2}{2}\Big) + K_\nu\Big(\frac{z_2 + 2\pi}{2}\Big)\Big] \times$$

$$\times \Big[K_\nu\Big(\frac{z_1}{2}\Big) + K_\nu\Big(\frac{z_1 + 2\pi}{2}\Big)\Big] dz_1\, dz_2. \tag{8.5.37}$$

Let $K_\nu(z_1) = \sum_p a_p e^{iz_1 p}$. Then

$$\Big[K_\nu\Big(\frac{z_2}{2}\Big) + K_\nu\Big(\frac{z_2 + 2\pi}{2}\Big)\Big]\Big[K_\nu\Big(\frac{z_1}{2}\Big) + K_\nu\Big(\frac{z_1 + 2\pi}{2}\Big)\Big] =$$

$$= \sum_{p,r} a_p a_r\, e^{iz_1 p/2 + iz_2 r/2}(1 + \cos\pi p)(1 + \cos\pi r). \tag{8.5.38}$$

Relations (8.5.37) and (8.5.38) show that $I(\varphi_1, \varphi_2)$ is a trigonometric polynomial, and therefore, $P_m^\nu(\varphi_1, \varphi_2)$ is also a trigonometric polynomial.

Next, we estimate the expression

$$E^\nu_{nm} = \int\limits_{-\pi}^{\pi}\int\limits_{-\pi}^{\pi} G_n(\varphi_1,\varphi_2)|B_n(\varphi_1,\varphi_2) - P^\nu(\varphi_1,\varphi_2)|^2\, d\varphi_1\, d\varphi_2 = \tag{8.5.39}$$

$$= \int\limits_{-\pi}^{\pi}\int\limits_{-\pi}^{\pi} G_n(\varphi_1,\varphi_2)\left|\int\limits_{-\pi}^{\pi}\int\limits_{-\pi}^{\pi} \Delta_z^2 B_n(\varphi_1,\varphi_2)K_\nu(z_1)K_\nu(z_2)\, dz_1\, dz_2\right|^2 d\varphi_1\, d\varphi_2 \le$$

$$\le \int\limits_{-\pi}^{\pi}\int\limits_{-\pi}^{\pi} G_n(\varphi_1,\varphi_2)\left[\int\limits_{-\pi}^{\pi}\int\limits_{-\pi}^{\pi} |\Delta_{z_1 z_2}^2 B_n(\varphi_1,\varphi_2)|^2 K_\nu(z_1)K_\nu(z_2)\, dz_1\, dz_2\right] d\varphi_1\, d\varphi_2 =$$

$$= \int\limits_{-\delta}^{\delta}\int\limits_{-\delta}^{\delta}\left[\int\limits_{-\pi}^{\pi}\int\limits_{-\pi}^{\pi} G_n(\varphi_1,\varphi_2)|\Delta_z^2 B_n(\varphi_1,\varphi_2)|^2\, d\varphi_1\, d\varphi_2\right] K_\nu(z_1)K_\nu(z_2)\, dz_1\, dz_2 +$$

$$+ \int\limits_{-\pi}^{\pi} dz_1 \int\limits_{\delta}^{\pi}\left[\int\limits_{-\pi}^{\pi}\int\limits_{-\pi}^{\pi} G_n(\varphi_1,\varphi_2)|\Delta_z^2 B_n(\varphi_1,\varphi_2)|^2\, d\varphi_1\, d\varphi_2\right] K_\nu(z_1)K_\nu(z_2)\, dz_2 +$$

$$+ \int\limits_{-\pi}^{\pi} dz_1 \int\limits_{-\pi}^{-\delta}\left[\int\limits_{-\pi}^{\pi}\int\limits_{-\pi}^{\pi} G_n(\varphi_1,\varphi_2)|\Delta_z^2 B_n(\varphi_1,\varphi_2)|^2\, d\varphi_1\, d\varphi_2\right] K_\nu(z_1)K_\nu(z_2)\, dz_2 +$$

$$+ \int\limits_{-\pi}^{-\delta} dz_1 \int\limits_{-\delta}^{\delta}\left[\int\limits_{-\pi}^{\pi}\int\limits_{-\pi}^{\pi} G_n(\varphi_1,\varphi_2)|\Delta_z^2 B_n(\varphi_1,\varphi_2)|^2\, d\varphi_1\, d\varphi_2\right] K_\nu(z_1)K_\nu(z_2)\, dz_2 +$$

$$+ \int\limits_{\delta}^{\pi} dz_1 \int\limits_{-\delta}^{\delta}\left[\int\limits_{-\pi}^{\pi}\int\limits_{-\pi}^{\pi} G_n(\varphi_1,\varphi_2)|\Delta_z^2 B_n(\varphi_1,\varphi_2)|^2\, d\varphi_1\, d\varphi_2\right] K_\nu(z_1)K_\nu(z_2)\, dz_2 = \sum_{i=1}^{5} I_i.$$

Fix $\varepsilon > 0$ and let M be the set of all φ such that either $-\pi \le \varphi \le -\nu^{\varepsilon-1}$ or $\nu^{\varepsilon-1} \le \varphi \le \pi$. The Nikolskii kernels on the set M can be estimated as follows:

$$K_\nu(\varphi) \le \frac{C(\sigma)}{\nu^{2\sigma\varepsilon-1}}. \tag{8.5.40}$$

For any $\varepsilon > 0$ there is σ_0 such that

$$2\sigma_0\varepsilon - 1 \ge 7. \tag{8.5.41}$$

Let $\delta = \nu^{\varepsilon-1}$. Then, (8.5.27) yields

$$I_1 \le \frac{C}{\nu^{4-4\varepsilon}} \ln^2 n\left[\frac{n^2}{\nu^{1-\varepsilon}} + 4n\ln n\right]. \tag{8.5.42}$$

It follows from (8.2.9), (8.5.10) (recall that (8.2.9) holds for $G_n(\varphi_1,\varphi_2)$), that $|B_n(\varphi_1,\varphi_2)| < C$. This, together with (8.5.40), (8.5.41), implies that

$$I_i \le \frac{c}{\nu^7}, \quad i = 2,3,4,5. \tag{8.5.43}$$

Relations (8.5.39), (8.5.42), and (8.5.43) imply the estimate

$$E^\nu_{nm} \le \frac{C_1}{\nu^{4-4\varepsilon}} \ln^2 n\left[\frac{n^2}{\nu^{1-\varepsilon}} + 4n\ln n\right] + \frac{C_1}{\nu^7}. \tag{8.5.44}$$

Let us estimate the Fourier coefficients of the function $P_m^\nu(\varphi_1, \varphi_2)$. Since

$$\int_{-\pi}^{\pi} K_\nu(z)\, dz = 1,$$

the Fourier coefficients of the Nikolskii kernels are bounded. This observation, together with (8.5.36)–(8.5.38), (8.5.31), implies the following estimates for the Fourier coefficients $k_{mn}^\nu(p, r)$ of the function $P_m^\nu(\varphi_1, \varphi_2)$:

$$|k_{mn}^\nu(p, r)| \le \frac{c \ln^3 n}{(|p| + 1)(|r| + 1)}, \qquad |k_{mn}^\nu(p, r)| \le \frac{c \ln^3 n}{p^2 + r^2 + 1}. \tag{8.5.45}$$

Let us estimate the norm $\|z_m(\varphi_1, \varphi_2)\|_1$ of the function $z_m(\varphi_1, \varphi_2)$ defined by (8.5.35) (see (8.3.2)). For this purpose, consider the expression

$$T(z_m) = \int_{-\pi}^{\pi}\int_{-\pi}^{\pi} G_n(\varphi_1, \varphi_2)\big[y_m(\varphi_1, \varphi_2) - P_m^\nu(\varphi_1, \varphi_2)\big] D_{\varepsilon_h^m}(\alpha - \varphi_1, \beta - \varphi_2)\, d\varphi_1\, d\varphi_2 =$$

$$= e^{i\alpha} - 1 - \int_{-\pi}^{\pi}\int_{-\pi}^{\pi} G_n(\varphi_1, \varphi_2) P_m^\nu(\varphi_1, \varphi_2) D_{\varepsilon_h^m}(\alpha - \varphi_1, \beta - \varphi_2)\, d\varphi_1\, d\varphi_2 =$$

$$= \int_{-\pi}^{\pi}\int_{-\pi}^{\pi} G_n(\varphi_1, \varphi_2)\big[B_n(\varphi_1, \varphi_2) - P_m^\nu(\varphi_1, \varphi_2)\big] D_{\varepsilon_h^m}(\alpha - \varphi_1, \beta - \varphi_2)\, d\varphi_1\, d\varphi_2. \tag{8.5.46}$$

Using (8.3.3) and (8.5.46), we obtain the inequality

$$\|T(z_m)\|_2 = \max_{\|y\|_1=1} \left| \int_{-\pi}^{\pi}\int_{-\pi}^{\pi} T(z_m)\overline{y(\alpha\beta)}\, d\alpha\, d\beta \right| = \max_{\|y\|_1=1} \left| \int_{-\pi}^{\pi}\int_{-\pi}^{\pi} \overline{y(\alpha, \beta)} \times \right.$$

$$\times \int_{-\pi}^{\pi}\int_{-\pi}^{\pi} G_n(\varphi_1, \varphi_2)\big[B_n(\varphi_1, \varphi_2) - P_m^\nu(\varphi_1, \varphi_2)\big] D_{\varepsilon_h^m}(\alpha - \varphi_1, \beta - \varphi_2)\, d\varphi_1\, d\varphi_2\, d\alpha\, d\beta \bigg| =$$

$$= \max_{\|y\|_1=1} \left| \int_{-\pi}^{\pi}\int_{-\pi}^{\pi} G_n(\varphi_1, \varphi_2)\big[B_n(\varphi_1, \varphi_2) - P_m^\nu(\varphi_1, \varphi_2)\big] \times \right.$$

$$\times \left[\int_{-\pi}^{\pi}\int_{-\pi}^{\pi} \overline{y(\alpha, \beta) D_{\varepsilon_h^m}(\varphi_1 - \alpha, \varphi_2 - \beta)} \right] d\varphi_1\, d\varphi_2 \bigg| =$$

$$= \max_{\|y\|_1} \left| \int_{-\pi}^{\pi}\int_{-\pi}^{\pi} G_n(\varphi_1, \varphi_2)\big[B_n(\varphi_1, \varphi_2) - P_m^\nu(\varphi_1, \varphi_2)\big] \overline{y(\varphi_1, \varphi_2)}\, d\varphi_1\, d\varphi_2 \right| \le$$

$$\le \max_{\|y\|_1=1} \left[\int_{-\pi}^{\pi}\int_{-\pi}^{\pi} G_n(\varphi_1, \varphi_2)|B_n(\varphi_1, \varphi_2) - P_m^\nu(\varphi_1, \varphi_2)|^2\, d\varphi_1\, d\varphi_2 \right]^{1/2} \times$$

$$\times \left[\int_{-\pi}^{\pi}\int_{-\pi}^{\pi} G_n(\varphi_1, \varphi_2)|y(\varphi_1, \varphi_2)|^2\, d\varphi_1\, d\varphi_2 \right]^{1/2}. \tag{8.5.47}$$

Proposition 8.3.1 and formulas (8.5.39), (8.5.47) imply the estimate

$$\|z_m\|_1 \leq \sqrt{E_{nm}^{\nu}} \ . \tag{8.5.48}$$

Let us represent $z_m(\varphi_1, \varphi_2)$ in the form

$$z_m(\varphi_1, \varphi_2) = z_{1m}(\varphi_1, \varphi_2)\|z_m(\varphi_1, \varphi_2)\|_1,$$

where $\|z_m(\varphi_1, \varphi_2)\|_1 = 1$. By Proposition 8.3.2, we have $z_{1m}(\varphi_1, \varphi_2) \in L_{4/3}(D)$, $D = [-\pi, \pi] \times [-\pi, \pi]$ and the following estimate holds

$$\|z_{1m}(\varphi_1, \varphi_2)\|_{L_{4/3}(D)} \leq \left[\int\limits_{-\pi}^{\pi} \int\limits_{-\pi}^{\pi} \frac{d\varphi_1 \, d\varphi_2}{|G_n(\varphi_1, \varphi_2)|^2} \right]^{1/2} \leq c(\ln n)^{1/4}.$$

Corollary 8.3.1 shows that the Fourier coefficients $z_{1m}^n(\varphi_1, \varphi_2)$ of the function $z_{1m}(\varphi_1, \varphi_2)$ belong to l_4, and

$$\sum_{k_1, k_2} |z_{1m}^n(k_1, k_2)|^4 \leq c \ln n. \tag{8.5.49}$$

The relationship between the restriction problems (8.5.32) and (8.5.8) ensures that the Fourier coefficients $x_m^n(k_1, k_2)$ of the solutions $\widehat{x}_c(\varphi_1, \varphi_2)$ of problem (8.5.8) admit the representation

$$x_m^n(k_1, k_2) = p_m^n(k_1, k_2) + x_{1m}^n(k_1, k_2). \tag{8.5.50}$$

The inequalities (8.5.45) imply that

$$|p_m^n(k_1, k_2)| \leq \frac{C(\ln n)^3}{(|k_1 - m_1| + 1)(|k_2 - m_2| + 1)} \, ,$$
$$|p_m^n(k_1, k_2)| \leq \frac{C(\ln n)^3}{(k_1 - m_1)^2 + (k_2 - m_2)^2 + 1} \ . \tag{8.5.51}$$

From (8.5.48) and (8.5.49), it follows that

$$|x_{1m}^n(k_1, k_2)| \leq C\sqrt{E_{nm}^{\nu}}|x_{2m}(k_1, k_2)|, \qquad \sum_{k_1, k_2} |x_{2m}(k_1, k_2)|^4 \leq C \ln n,$$
$$E_{nm}^{\nu} \leq C \left[\frac{1}{\nu^{4-4\varepsilon}} \ln^2 n \left[\frac{n^2}{\nu^{1-\varepsilon}} + 4n \ln n \right] + \frac{1}{\nu^7} \right]. \tag{8.5.52}$$

Remark 8.5.2. Relations (8.5.50)–(8.5.52) also hold for the Fourier coefficients of the solution $\widehat{x}_c(\varphi_1, \varphi_2)$ of the restriction problem (8.5.9).

Chapter 9

Quadrature Formulas for Singular and Hypersingular Integrals

In this chapter, some quadrature formulas for one-dimensional singular integrals are listed (without proof) for the sake of completeness of our exposition. The convergence of these integrals is established in *Belotserkovskii and Lifanov* (1985) and *Lifanov* (1996).

9.1. Integrals over a Closed Smooth Curve; Hilbert Integrals

First, we consider some quadrature formulas for singular integrals on a smooth closed curve and Hilbert integrals. These formulas pertain to the method of discrete vortices.

Consider a singular integral

$$I(t_0) = \int_L \frac{\varphi(t)\, dt}{t - t_0} \tag{9.1.1}$$

on the unit circle L with center at the origin, where $\varphi(t)$ is a function of class H on L.

Let $E = \{t_k : k = 1, \ldots, n\}$ and $E_0 = \{t_{0k} : k = 1, \ldots, n\}$ be two sets of nodes on L such that the points t_k, $k = 1, \ldots, n$, divide the circle into n equal parts and t_{0k} is the middle point of the arc $t_k t_{k+1}$; $t_{n+1} = t_1$. In what follows, the sets E and E_0 are called *the canonical partition* of the circle L.

In the special case of $\varphi(t) \equiv 1$ in (9.1.1), the following result holds.

Lemma 9.1.1. *For any $t_{0j} \in E_0$, we have*

$$\left| \int_L \frac{dt}{t - t_{0j}} - \sum_{k=1}^{n} \frac{\Delta t_k}{t_k - t_{0j}} \right| \leq O\left(\frac{1}{n}\right), \tag{9.1.2}$$

where $\Delta t_k = t_{k+1} - t_k$, $k = 1, \ldots, n$.

Here and in the sequel, by $O(\delta(n))$ we denote a quantity of the same order as $\delta(n)$, i.e., $|O(\delta(n))| \leq B|\delta(n)|$, where $B > 0$ is a constant independent of n.

Remark 9.1.1. The following inequality holds:

$$\sum_{k=1}^{n} \frac{|\Delta t_k|}{|t_{0j} - t_k|} \leq O(\ln n), \quad j = 1, \ldots, n. \tag{9.1.3}$$

Now, it is not difficult to prove the following result.

Theorem 9.1.1. *Let* $\varphi(t) \in H(\alpha)$ *on* L. *Then*

$$|I(t_{0j}) - S_n(t_{0j})| \leq \theta(t_{0j}), \quad j = 1, \ldots, n, \qquad S_n(t_{0j}) = \sum_{k=1}^{n} \frac{\varphi(t_k)\Delta t_k}{t_k - t_{0j}},$$

$$\theta(t_{0j}) = O\left(\frac{\ln n}{n^\alpha}\right) + |\varphi(t_{0j})| O\left(\frac{1}{n}\right). \tag{9.1.4}$$

Definition 9.1.1. We say that a function $\varphi(t)$ is of *class* Π (resp., Π^*) on a piecewise smooth curve L, if

$$\varphi(t) = \frac{\psi(t)}{q - t},$$

where $\psi(t) \in H$ (resp., $\psi(t) \in H^*$) on L, and q is a fixed point on L different from the nodes.

Note that we can write

$$\int_L \frac{\varphi(t)\,dt}{t - t_0} = \frac{1}{q - t_0}\left[\int_L \frac{\psi(t)\,dt}{t - t_0} - \int_L \frac{\psi(t)\,dt}{t - q}\right]. \tag{9.1.5}$$

The following statement is obtained from Theorem 9.1.1 and relation (9.1.5).

Theorem 9.1.2. *Let* $\varphi(t) \in \Pi$ *on the circle* L *and let the sets* E, E_0 *form a canonical partition of* L, $q \in E_0$ *for* $j = j_q$. *Then*

$$|I(t_{0j}) - S_n(t_{0j})| \leq \theta(t_{0j}), \quad j \neq q_j, \quad j = 1, \ldots, n, \tag{9.1.6}$$

where

$$\theta(t_{0j}) = \frac{1}{|t_{0j} - q|} O\left(\frac{\ln n}{n^\alpha}\right), \quad j \neq q_j.$$

It is not difficult to show that $\theta(t_{0j})$ satisfies the inequalities

$$\theta(t_{0j}) \leq O_l\left(\frac{1}{n^{\lambda_1}}\right), \quad \lambda_1 > 0, \tag{9.1.7}$$

for all $t_{0j} \in L \setminus l$, where l is an arbitrarily small neighborhood of the point q, and

$$\sum_{j=1,\, j \neq j_q} \theta(t_{0j})|\Delta t_{0j}| \leq O\left(\frac{1}{n^{\lambda_2}}\right), \quad \lambda_2 > 0. \tag{9.1.8}$$

Clearly, the inequality (9.1.7) can be made more precise, if we write $O_l(n^{-\alpha}\ln n)$ on its right-hand side, and also the inequality (9.1.7) becomes more precise with $O\left(n^{-\alpha}\ln^2 n\right)$ on its right-hand side.

Remark 9.1.2. The inequality (9.1.4) remains valid for the integral $\int_L \varphi(t, t_0)(t - t_0)^{-1}\,dt$ if $\varphi(t) \in H(\alpha)$ on the circle L, i.e.,

$$\left|\int_L \frac{\varphi(t, t_{0j})\,dt}{t - t_{0j}} - \sum_{k=1}^{n} \frac{\varphi(t_k, t_{0j})\Delta t_k}{t_k - t_{0j}}\right| \leq O\left(\frac{\ln n}{n^\alpha}\right), \quad j = 1, \ldots, n. \tag{9.1.9}$$

Remark 9.1.3. Let L_1 be a closed Lyapunov curve. Then there is a one-to-one correspondence $\tau = \tau(t)$ between the points $\tau \in L_1$ and the points $t \in L$ of the standard unit circle L (with center at the origin), with the derivative $\tau'(t) = d\tau/dt$ belonging to $H(\beta)$ and non-vanishing on L.

For $\varphi(\tau) \in H(\alpha)$ on L_1, using the formula for changing variables in singular integrals (1.1.7), we obtain

$$\int_{L_1} \frac{\varphi(\tau)\, d\tau}{\tau - \tau_0} = \int_L \frac{\psi(t, t_0)\, dt}{t - t_0}, \tag{9.1.10}$$

where $\psi(t, t_0) \in H$ on L.

On the circle L, consider the canonical partition formed by the sets E, E_0. The sets

$$\tau_k = \tau(t_k), \quad t_k \in E, \quad \text{and} \quad \tau_{0k} = \tau(t_{0k}), \quad t_{0k} \in E_0$$

will be called a *canonical partition of the curve* L_1. The inequality (9.1.9) and formula (9.1.10) imply that

$$\left| \int_{L_1} \frac{\varphi(\tau)\, d\tau}{\tau - \tau_{0j}} - \sum_{k=1}^{n} \frac{\varphi(\tau_k)\Delta\tau_k}{\tau_k - \tau_{0j}} \right| \le O\left(\frac{1}{n^\lambda} \right), \qquad \lambda > 0. \tag{9.1.11}$$

Now, let L be the union of p mutually disjoint Lyapunov curves $L_1, \ldots, L_p$. For each $m = 1, \ldots, p$, let

$$E_m = \{\tau_k, \quad k = n_{m-1} + 1, \ldots, n_m\}, \qquad E_{0m} = \{\tau_{0k}, \quad k = n_{m-1} + 1, \ldots, h_m\}$$

be a canonical partition of L_m into $N_m = n_m - n_{m-1}$ parts. Let $N = \min_{m=1,\ldots,p} N_m$ and assume that $N_m/N \le R < \infty$. Let us also introduce the quantities

$$S_{n_p}(\tau_{0j}) = \sum_{k=1}^{n_p} \frac{\varphi(\tau_k)\Delta\tau_k}{\tau_k - \tau_{0j}}, \quad j = 1, \ldots, n_p,$$

$$\Delta\tau_k = \tau_{k+1} - \tau_k, \quad k = 1, \ldots, n_p - 1, \quad k \ne n_1, \ldots, n_{p-1};$$

$$\Delta\tau_{n_m} = \tau_{n_{m-1}+1} - \tau_{n_m}, \quad m = 1, \ldots, p.$$

Theorem 9.1.3. *Let* $\varphi(\tau)$ *be of class* H *on* L. *Then, for any* $\tau_{0j} \in \bigcup_{m=1}^{p} E_{0m}$, *the following inequality holds:*

$$\left| I(\tau_{0j}) - S_{n_p}(\tau_{0j}) \right| \le O\left(\frac{1}{N^\lambda} \right), \quad \lambda > 0. \tag{9.1.12}$$

If the functions $\varphi(\tau)$ and $\tau = \tau(t)$ possess additional regularity properties, one can obtain a quadrature sum of a higher order than (9.1.11) for the singular integral on a closed smooth curve.

Indeed, suppose that $\varphi(t)$ is of class $H_r(\alpha)$ on the unit circle L with center at the origin. Thus, $\varphi^{(r)}(t) \in H(\alpha)$. Consider the polynomial

$$\varphi_n(t) = \frac{1}{2n+1} \sum_{k=0}^{2n} \varphi(t_k) \frac{t^{2n+1} - t_k^{2n+1}}{(t - t_k) t^n t_k^n}, \tag{9.1.13}$$

with the points t_k splitting the circle L into $2n + 1$ equal parts. Using the polynomial division formula, one can show that

$$\varphi_n(t_k) = \varphi(t_k), \quad k = 0, 1, \ldots, 2n, \tag{9.1.14}$$

since

$$\frac{1}{2n+1} \frac{t^{2n+1} - t_k^{2n+1}}{(t - t_k) t^n t_k^n} = \begin{cases} 1 & \text{for } t = t_k, \quad k = 0, 1, \ldots, 2n, \\ 0 & \text{for } t = t_m, \quad m \ne k, \ m = 0, 1, \ldots, 2n. \end{cases} \tag{9.1.15}$$

The last relation can also be obtained from the formulas

$$\frac{\sin(2n+1)\frac{\theta}{2}}{\sin\frac{\theta}{2}} = 1 + (\cos\theta + \cdots \cos n\theta) = \tag{9.1.16}$$

$$= e^{-in\theta} + e^{-i(n-1)\theta} + \cdots e^{-i\theta} + 1 + e^{i\theta} + \cdots e^{in\theta} = \frac{t^{-n} - t^{n+1}}{1-t}, \qquad t = e^{i\theta},$$

and

$$\lim_{\alpha \to 0} \frac{\sin(2n+1)\frac{\alpha}{2}}{\sin\frac{\alpha}{2}} = 2n+1. \tag{9.1.17}$$

From (9.1.16), we get

$$\frac{\sin(2n+1)\frac{(\theta-\theta_k)}{2}}{\sin\frac{(\theta-\theta_k)}{2}} = \frac{t^{2n+1} - t_k^{2n+1}}{(t-t_k)t^n t_k^n}, \qquad t_k = e^{i\theta_k}. \tag{9.1.18}$$

Now, consider the singular integral (9.1.1) for $\varphi(t) \in H_r(\alpha)$ on the circle L. Set

$$S(t_0) = \int_L \frac{\varphi_n(t)\,dt}{t-t_0} = \frac{1}{2n+1} \sum_{k=0}^{2n} \varphi(t_k) \int_L \frac{t^{2n+1} - t_k^{2n+1}}{(t-t_k)t^n t_k^n} \frac{dt}{t-t_0}. \tag{9.1.19}$$

Using the formula (see *Gakhov* (1966))

$$\frac{1}{\pi i} \int_L \frac{t^n\,dt}{t-t_0} = \begin{cases} t_0^n & \text{for } n \geq 0, \\ -t_0^n & \text{for } n < 0, \end{cases} \qquad t = e^{i\theta}, \quad t_0 = e^{i\theta_0}, \tag{9.1.20}$$

and the obvious identity

$$\frac{1}{(t-t_k)(t-t_0)} - \frac{1}{t_0 - t_k}\left(\frac{1}{t-t_0} - \frac{1}{t-t_k}\right), \tag{9.1.21}$$

we obtain

$$S(t_0) = \sum_{k=0}^{2n} \frac{\varphi(t_k)}{t_k - t_0} \frac{\pi i}{2n+1}\left[2t_k - \frac{t_0^{2n+1} + t_k^{2n+1}}{t_0^n t_k^n}\right]. \tag{9.1.22}$$

As shown in *Belotserkovskii and Lifanov* (1985) and *Lifanov* (1996),

$$|I(t_0) - S(t_0)| \leq O\left(\frac{n + \ln n}{n^{r+\alpha}}\right). \tag{9.1.23}$$

Since

$$\frac{t_0^{2n+1} + t_k^{2n+1}}{t_0^n t_k^n (t_0 - t_k)} = i\frac{\cos(2n+1)\frac{(\theta_0-\theta_k)}{2}}{\sin\frac{(\theta_0-\theta_k)}{2}}, \tag{9.1.24}$$

we see that the roots of the function in the left-hand side of (9.1.24) coincide with the points

$$t_{0m} = \exp\left\{\theta_m + \frac{\pi}{2n+1}\right\}, \qquad m = 0, 1, \ldots, 2n,$$

which means that t_{0k} is the middle point of the arc $t_k t_{k+1}$. Therefore, for the points t_{0m} we have

$$S(t_m) = \sum_{k=0}^{2n} \frac{\varphi(t_k)}{t_k - t_{0m}} \frac{2\pi i t_k}{2n+1}, \qquad m = 0, 1, \ldots, 2n, \tag{9.1.25}$$

and (see *Belotserkovskii and Lifanov* (1985) and *Lifanov* (1996))

$$|I(t_{0m}) - S(t_{0m})| \leq O\left(\frac{\ln n}{n^{r+\alpha}}\right). \qquad (9.1.26)$$

Singular integrals over a circle are closely related to integrals with Hilbert kernel,

$$I(\theta_0) = \int_0^{2\pi} \cot\frac{\theta - \theta_0}{2}\varphi_1(\theta)\,d\theta, \qquad (9.1.27)$$

and this relation is due to the well known identity (see *Muskhelishvili* (1968))

$$\int_L \frac{\varphi(t)\,dt}{t - t_0} = \frac{1}{2}\int_0^{2\pi}\cot\frac{\theta - \theta_0}{2}\varphi_1(\theta)\,d\theta + \frac{i}{2}\int_0^{2\pi}\varphi_1(\theta)\,d\theta, \qquad (9.1.28)$$

where L is the unit circle with center at the origin, $t = e^{i\theta}$, $t_0 = e^{i\theta_0}$, $\varphi_1(\theta) = \varphi(e^{i\theta})$.

Let $\varphi(t) \in H$ on L and let $E = \{t_k = e^{i\theta_k},\ k = 1,\ldots,n\}$, $E_0 = \{t_{0k} = e^{i\theta_{0k}},\ k = 1,\ldots,n\}$ be a canonical partition of L, i.e., $\Delta\theta_k = \theta_{k+1} - \theta_k = 2\pi/n$, $k = 1,\ldots,n$, $\theta_{n+1} = \theta_1 + 2\pi$. Then, the following inequality holds:

$$|I(\theta_{0j}) - S_n(\theta_{0j})| \lesssim \left(\frac{\ln n}{n^\lambda}\right), \qquad (9.1.29)$$

where

$$S_n(\theta_{0j}) = \sum_{k=1}^n \cot\frac{\theta_k - \theta_{0j}}{2}\varphi_1(\theta_k)\frac{2\pi}{n}.$$

Now, if $\varphi_1(\theta) \in H_r(\alpha)$ on $[0, 2\pi]$ is periodic and the number of the points θ_k and θ_{0k} in the canonical partition is odd, then, taking into account (9.1.26), we obtain (see *Belotserkovskii and Lifanov* (1985) and *Lifanov* (1996))

$$|I(\theta_{0j}) - S^*_{2n+1}(\theta_{0j})| \leq O\left(\frac{\ln n}{n^{r+\lambda}}\right),$$

$$S^*_{2n+1}(\theta_{0j}) = \sum_{k=0}^{2n}\cot\frac{\theta_k - \theta_{0j}}{2}\varphi_1(\theta_k)\frac{2\pi}{2n+1},\quad j = 0, 1,\ldots,2n. \qquad (9.1.30)$$

9.2. Integrals over an Open-Ended Smooth Curve

Consider the singular integral (9.1.1) with $L = [a, b]$ being an interval on the real axis and let $\varphi(t) \in H^*$ on L,

$$\varphi(t) = \frac{\psi(t)}{(t - a)^\nu (b - t)^\mu}, \qquad (9.2.1)$$

where $\psi(t) \in H$ on $[a, b]$, $0 \leq \nu, \mu < 1$.

Let $t_0 = a < t_1 < \cdots < t_n < t_{n+1} = b$ be the points dividing the segment $[a, b]$ into $n + 1$ equal parts of length $h = (b - a)/(n + 1)$, and let t_{0j} be the middle point of the segment $[t_j, t_{j+1}]$, $j = 1,\ldots,n$. We say that the points of the sets $E = \{t_k,\ k = 1,\ldots,n\}$ and $E_0 = \{t_{0j},\ j = 1,\ldots,n\}$ form a canonical partition of the segment $[a, b]$ with step h.

Lemma 9.2.1. *For any point $t_{0j} \in E_0$, the following inequality holds:*

$$\left| \int_a^b \frac{dt}{t - t_{0j}} - \sum_{k=1}^n \frac{h}{t_k - t_{0j}} \right| \leq \frac{Bh}{(t_{0j} - a)(b - t_{0j})}, \tag{9.2.2}$$

where B is a constant.

Note that in this case we also have the inequality

$$\sum_{k=1}^n \frac{h}{|t_k - t_{0j}|} \leq O(|\ln h|). \tag{9.2.3}$$

This inequality, combined with the Hölder continuity properties, allows us to obtain the following result.

Lemma 9.2.2. *Let $\varphi(t) \in H(\alpha)$ on $[a, b]$. Then, for any point $t_{0j} \in E_0$, we have*

$$I_1 = \left| \int_a^b \frac{\varphi(t) - \varphi(t_{0j})}{t - t_{0j}} - \sum_{k=1}^n \frac{\varphi(t_k) - \varphi(t_{0j})}{t_k - t_{0j}} \right| \leq O(h^\alpha |\ln h|). \tag{9.2.4}$$

Inequalities (9.2.2) and (9.2.4) imply the following statement.

Theorem 9.2.1. *Let $\varphi(t) \in H(\alpha)$ on $[a, b]$ and let the sets E, E_0 form the canonical partition of $[a, b]$. Then, for any $t_{0j} \in E_0$, we have*

$$|I(t_{0j}) - S_n(t_{0j})| \leq O\left(\frac{h^\alpha |\ln h| + |\varphi t_{0j})|h}{(t_{0j} - a)(b - t_{0j})} \right), \tag{9.2.5}$$

where

$$S(t_{0j}) = \sum_{k=1}^n \frac{\varphi(t_k)h}{t_k - t_{0j}}, \quad j = 0, 1, \ldots, n.$$

However, in the general case, the solutions of singular integral equations on a segment belong to the class H^* on that segment, i.e., have the form (9.2.1). For such functions, the following result holds (see *Belotserkovskii and Lifanov* (1985) and *Lifanov* (1996)).

Theorem 9.2.2. *Let $\varphi(t) \in H^*$ on $[a, b]$ and let E, E_0 be the canonical partition of the segment. Then,*

$$\left| \int_a^b \frac{\varphi(t)\, dt}{t - t_{0j}} - \sum_{k=1}^n \frac{\varphi(t_k)h}{t_k - t_{0j}} \right| \leq \theta(t_{0j}), \quad j = 0, 1 \ldots, n, \tag{9.2.6}$$

with $\theta(t_{0j})$ satisfying the inequalities

$$\theta(t_{0j}) \leq O_\delta\left(h^{\lambda_1}\right), \quad 0 < \lambda_1 < 1, \tag{9.2.7}$$

for $t_{0j} \in [a + \delta, b - \delta]$ and any small enough $\delta > 0$; and

$$\sum_{j=0}^n \theta(t_{0j})|\Delta t_{0j}| \leq O\left(h^{\lambda_2}\right), \quad 0 < \lambda_2 < 1, \tag{9.2.8}$$

for all $t_{0j} \in [a, b]$, where $|\Delta t_{0j}| = h$, $j = 0, 1, \ldots, n$.

According to *Belotserkovskii and Lifanov* (1985) and *Lifanov* (1996), the value $\theta(t_{0j})$ in (9.2.6) can be refined, namely,

$$\theta(t_{0j}) = \theta_0\big(t_{0j}, \alpha, \nu, \mu\big)O(|\ln h|), \qquad (9.2.9)$$

where

$$\theta_0(t_{0j}, \alpha, \nu, \mu) = \eta(t_{0j}, \nu, \mu)h^\alpha + \eta(t_{0j}, 2\nu, \mu)h^\nu + \eta(t_{0j}, \nu, 2\mu)h^\mu +$$

$$+ \eta(t_{0j}, 1, \mu)h^{1-\nu} + \eta(t_{0j}, \nu, 1)h^{1-\mu} + \eta(t_{0j}, 1 + \nu, 1 + \mu)h,$$

$$0 < \nu, \quad \mu < 1, \qquad \eta(t_{0j}, \alpha, \beta) = (t_{0j} - a)^{-\alpha}(b - t_{0j})^{-\beta}.$$

For $\nu \neq 0$, $\mu = 0$, the expression for $\theta_0(t_{0j}, \alpha, \nu, 0)$ is

$$\theta_0(t_{0j}, \alpha, \nu, 0) = \eta(t_{0j}, \nu, 0)h^\alpha + \eta(t_{0j}, 2\nu, 0)h^\nu + \eta(t_{0j}, 1, 0)h^{1-\nu} + \eta(t_{0j}, 1 + \nu, 1)h. \quad (9.2.10)$$

For $\nu = \mu = 0$, the value of $\theta_0(t_{0j})$ coincides with the right-hand side of the inequality (9.2.5).

Remark 9.2.1. If the function $\varphi(t)$ has the form

$$\varphi(t) = \frac{(b - t)^\mu}{(t - a)^\nu}\psi(t),$$

where $0 < \nu < 1$, $0 < \mu$, and $\psi(t) \in H$ on $[a, b]$, then (9.2.7) holds for all $t_{0j} \in [a + \delta, b]$.

Remark 9.2.2. In problems of aerodynamics, the partition points t_k and t_{0j} are often chosen in the following way. Let us divide the segment $[a, b]$ into n equal parts of length h and denote these by Δ_k, $k = 1, \ldots, n$. Let t_k be the point of the segment Δ_k whose distance from its left end-point is equal to $h/4$, and let t_{0k} be the point of Δ_k distanced from its left end-point by $3h/4$, $k = 1, \ldots, n$ (the so-called "one fourth — three fourths" scheme; see *Belotserkovskii* (1965)). Numerical analysis shows that this scheme yields better results in model examples than the canonical partition.

A more general statement can be made. The inequalities (9.2.2), (9.2.4)–(9.2.6) still hold if instead of forming a canonical partition, the sets $E = \{t_k, \ k = 1, \ldots, n\}$ and $E = \{t_{0j}, \ j = 1, \ldots, n\}$ satisfy the following conditions:

$$|t_{k+1} - t_k| = h, \quad k = 1, \ldots, n - 1, \qquad |t_{0j+1} - t_{0j}| = h, \quad j = 1, \ldots, n - 1,$$

$$t_1 - t_{00} = \frac{h}{2}, \quad t_{0k} = t_k + \frac{h}{2}, \quad k = 1, \ldots, n, \quad t_{00} - a = hq_h^a,$$

$$b - t_{0n} = hq_h^b, \quad 0 < P_1 \le q_h^a, \quad q_h^b \le P_2 < \infty,$$

where P_1 and P_2 are given numbers.

This situation arises (see *Belotserkovskii and Lifanov* (1985)) when it is required that some fixed point $q \in [a, b]$ be in a given position relative to the partition points, for instance, q should belong either to E or E_0 for any n. Thus, in the problem of flow past a wing-flap profile, the sets E and E_0 are chosen such that the point q is exactly in the middle between the points of E and E_0 nearest to q.

The above results for singular integrals on the segment remain valid if the points of E change their roles with those of E_0, i.e., the integrand is taken at the points of E_0 and the integrals are calculated at the points of E.

Remark 9.2.3. The above results hold for the function

$$\varphi(t, \tau) = \frac{\psi(t, \tau)}{(t - a)^\nu(b - t)^\mu}, \qquad (9.2.11)$$

provided that $\psi(t, \tau) \in H$ on $[a, b]$ with respect to both variables. The above results are also true if the sum is constructed only on the basis of the points of either E or E_0. Thus, if we take

$$S'_j = \sum_{k=1,\, k \neq j}^{n} \frac{\varphi(t_k)h}{t_k - t_j}, \tag{9.2.12}$$

$$I'_j = \int_a^b \frac{\varphi(t)\, dt}{t - t_j}, \tag{9.2.13}$$

then an inequality similar to (9.2.6) holds for $\left| I'_j - S'_j \right|$.

Now let us formulate a result for a function $\varphi(t) \in \Pi^*$ (see Definition 9.1.1).

Theorem 9.2.3. *Let $\varphi(t) \in \Pi^*$ and let the sets E and E_0 be such that the point q belongs to E_0 for any n; $q = t_{0j_q}$. Then*

$$|I(t_{0j}) - S_n(t_{0j})| \leq \theta_q(t_{0j}), \quad j \neq j_q, \tag{9.2.14}$$

where $\theta_q(t_{0j}) = |q - t_{0j}|^{-1}\theta(t_{0j})$ and $\theta(t_{0j})$ has the same properties as $\theta(t_{0j})$ in (9.2.6), namely, condition (9.2.7) holds for all $t_{0j} \in [a + \delta, q - \delta] \cup [q + \delta, b - \delta]$, and in the inequality (9.2.8) the sum should be taken over all $j \neq j_q$.

Remark 9.2.4. Let L be a open-ended Lyapunov curve, i.e., there is a one-to-one correspondence $\tau = \tau(t)$ between the points $\tau \in L$ and the points $t \in [a, b]$ such that the derivative $\tau'(t) = d\tau/dt$ is of class $H(\beta)$ on $[a, b]$ and does not vanish on this segment. Consider a canonical partition on $[a, b]$. Then, the mapping $\tau = \tau(t)$ induces a canonical partition of the curve L. Now, for $\varphi(\tau) \in H^*$ on L, take the quadrature sum $S(\tau_{0j})$ of the same structure as in (9.1.4) (with t replaced by τ). Then, for $|I(\tau_{0j}) - S_n(\tau_{0j})|$ a statement similar to Theorem 9.2.2 holds. Moreover, a similar result can be formulated for any piecewise smooth curve (see *Belotserkovskii and Lifanov* (1985)).

Just as for a singular integral on a circle, quadrature sums which are more precise than those of Theorem 9.2.2 can constructed for singular integrals on a segment, provided that the density of the singular integral has additional regularity properties.

Let us represent a function $\varphi(t) \in H^*$ on $[a, b]$ in the form $\varphi(t) = \omega(t)\psi(t)$, where $\omega(t) = (t - a)^\nu (b - t)^\mu$, $0 \leq |\nu|, |\mu| \leq 1$, and denote the integral over the segment by $I(\omega, \psi, t_0)$. Most frequently, this integral is considered for the segment $L = [-1, 1]$. An effective method for the construction of quadrature formulas for $I(\omega, \psi, t_0)$ amounts to replacing the function $\psi(t)$ by an interpolation polynomial $\psi_n(t)$ with respect to a system of polynomials $P_n(t)$, $n = 0, 1, \ldots$, mutually orthogonal on the segment $[-1, 1]$ with weight $\omega(t)$. Denote by t_k, $k = 1, \ldots, n$, the mutually distinct roots of the polynomial $P_n(t)$ on $(-1, 1)$, and let

$$\psi_n(t) = \sum_{k=1}^{n} \psi(t_k) \frac{P_n(t)}{(t - t_k)P'_n(t_k)}. \tag{9.2.15}$$

Obviously, this polynomial satisfies the conditions

$$\psi_n(t_k) = \psi(t_k), \quad k = 1, \ldots, n, \tag{9.2.16}$$

since

$$\lim_{t \to t_k} \frac{P_n(t)}{(t - t_k)P'_n(t_k)} = 1.$$

Denoting the function $I(\omega, \psi_n, t_0)$ by $S(t_0)$, we obtain

$$S(t_0) = \int_{-1}^{1} \frac{\omega(t)\psi_n(t)\,dt}{t - t_0} = \sum_{k=1}^{n} \frac{\psi(t_k)}{t_k - t_0}\left[\frac{Q_n(t_k)}{P_n'(t_k)} - \frac{Q_n(t_0)}{P_n'(t_k)}\right], \qquad (9.2.17)$$

where we have taken

$$Q_n(t_0) = \int_{-1}^{1} \frac{\omega(t)P_n(t)\,dt}{t - t_0}. \qquad (9.2.18)$$

Let t_{0j}, $j = 1, \ldots, R$, be the roots of the function $Q_n(t_0)$. At these points, formula (9.2.17) for $S(t_0)$ becomes

$$S(t_{0j}) = \sum_{k=1}^{n} \frac{\psi(t_k)a_k}{t_k - t_{0j}}, \quad j = 1, \ldots, R, \quad a_k = \frac{Q_n(t_k)}{P_n'(t_k)}, \quad k = 1, \ldots, n. \qquad (9.2.19)$$

As shown by *Korneichuk* (1964), the following inequality holds:

$$\left|I(\omega, \psi, t_{0j}) - S(t_{0j})\right| \le \frac{2}{\pi}E_{n-1}\alpha(1), \qquad (9.2.20)$$

where $\alpha(1) = \int_{-1}^{1} \omega(\tau)\,d\tau$, and E_{n-1} is the best approximation of the function $\psi'(t)$ by polynomials of degree $n - 1$. Therefore, if $\psi(t) \in H_r(\alpha)$ on $[-1, 1]$, then $E_{n-1} \le O(n^{-r-\alpha+1})$. Moreover, exact constants in the inequality for E_{n-1} are given in *Natanson* (1949). It has been shown by *Stark* (1971) that $S(t_{0m}) = I(\omega, \psi, t_{0m})$ if $\psi(t)$ is a polynomial of a degree $\le 2n$.

Recently, more precise estimates have been obtained for these quadrature sums. Thus, as shown by *Musaev* (1985), for $\psi(t) \in H_r(\alpha)$, $t_0 \in (-1, 1)$ and sufficiently large n, the following estimate holds:

$$\left|I(\omega, \psi, t_0) - S(t_0)\right| = \omega^*(t_0)O\left(\frac{\ln n}{n^{r+\alpha}}\right), \qquad (9.2.21)$$

where $\omega^*(t_0) = \omega_\mu(t_0)$ for $t_0 \in [0, 1)$; $\omega^*(t_0) = \omega_\nu(t_0)$ for $t_0 \in (-1, 0]$;

$$\omega_\mu(t_0) = \begin{cases} 1 & \text{for} \quad \mu \ge \dfrac{1}{2}, \\[2mm] (1 - t_0)^{(2\mu-1)/4} & \text{for} \quad -\dfrac{1}{2} < \mu < \dfrac{1}{2}, \\[2mm] (1 - t_0)^\mu & \text{for} \quad -1 < \mu \le -\dfrac{1}{2}; \end{cases}$$

$$\omega_\nu(t_0) = \begin{cases} 1 & \text{for} \quad \nu \ge \dfrac{1}{2}, \\[2mm] (1 + t_0)^{(2\nu-1)/4} & \text{for} \quad -\dfrac{1}{2} < \nu < \dfrac{1}{2}, \\[2mm] (1 + t_0)^\nu & \text{for} \quad -1 < \nu \le -\dfrac{1}{2}. \end{cases}$$

Previously, some other estimates were obtained for this quadrature sum by *Sheshko* (1976). Namely, if $-1 < \nu, \mu < 0$, then for all $t_0 \in (-1, 1)$, we have

$$\left|I(\omega, \psi, t_0) - S(t_0)\right| = \omega(t_0)\begin{cases} O\left(\dfrac{\ln n}{n^{r+\alpha-\gamma-1/2a}}\right), & \gamma \ge -\dfrac{1}{2}, \\[4mm] O\left(\dfrac{\ln^2 n}{n^{r+\alpha}}\right), & \gamma \le -\dfrac{1}{2}, \end{cases} \qquad (9.2.22)$$

$$r + \alpha - \gamma - \frac{1}{2} > 0,$$

where $\gamma = \max\{\nu, \mu\}$. If $0 < \nu$, $0 < \mu$, then for all $t_0 \in [-1, 1]$ relation (9.2.22) holds without $\omega(t_0)$ in the right-hand side.

If the nodes of the partition have the form

$$t_k = \cos\frac{2k-1}{2n}\pi, \quad k = 1, \ldots, n, \quad \text{or} \quad t_k = \cos\frac{k\pi}{n}, \quad k = 0, \ldots, n,$$

then we have the following estimate:

$$|I(\omega, \psi, t_0) - S(t_0)| = \begin{cases} \omega(t_0)O\left(\dfrac{\ln n}{n^{r+\alpha}}\right), & -1 < \mu, \mu < 0, \\ O\left(\dfrac{\ln^2 n}{n^{r+\alpha}}\right), & 0 < \nu, \mu. \end{cases} \quad (9.2.23)$$

Note that formula (9.2.21) is more convenient than (9.2.22), since it does not require that ν and μ be both negative or both positive. For $-1 < \nu$, $\mu \le -1/2$, formula (9.2.21) yields a more accurate estimate. However, for $0 > \nu$, $\mu \ge -1/2$, formula (9.2.22) is more convenient, since in many problems it is desirable that the function $\omega^{-1}(t_0)|I(\omega, \psi, t_0) - S(t_0)|$ could be uniformly estimated on $[-1, 1]$. Formula (9.2.22) gives such an estimate for suitable α, while (9.2.21) does not. Indeed, let $\nu = \mu = -1/4$. Then, we see that in (9.2.21), the multiplication by $\omega^{-1}(t_0)$ yields the estimate

$$(1 + t_0)^{(-1-2\nu)/4}(1 - t_0)^{(-1-2\mu)/4}O\left(\frac{\ln n}{n^{r+\alpha}}\right),$$

which is unbounded on $[-1, 1]$. For $\nu, \mu \ge 1/2$, formula (9.2.21) is preferable.

Let us examine separately the cases $\nu, \mu = \pm 1/2$, which most often occur in applications.

1. Let $\nu = \mu = -1/2$, i.e.,

$$\omega(t) = \left(1 - t^2\right)^{-1/2}. \quad (9.2.24)$$

The polynomials mutually orthogonal with this weight coincide with the Chebyshev polynomials of the first kind, $T_n(t) = \cos(n \arccos t)$, $n = 0, 1, \ldots$, and the functions $Q_n(t)$ are the Chebyshev polynomials of the second kind,

$$U_n(t) = \frac{\sin((n+1)\arccos t)}{\sin(\arccos t)}, \quad n = 1, \ldots, n,$$

which means that in this case, we have

$$P_n(t) = T_n(t), \quad Q_n(t) = \pi U_{n-1}(t), \quad (9.2.25)$$

and in (9.2.19),

$$t_k = \cos\frac{2k-1}{n}\pi, \quad k = 1, \ldots, n, \quad t_{0j} = \cos\frac{j}{n}\pi, \quad j = 1, \ldots, n-1,$$
$$a_k = \frac{\pi}{n} \quad k = 1, \ldots, n. \quad (9.2.26)$$

2. Let $\nu = \mu = 1/2$, i.e.,

$$\omega(t) = \left(1 - t^2\right)^{1/2}. \quad (9.2.27)$$

In this case,

$$P_n(t) = U_n(t), \quad Q_n(t) = \pi T_{n+1}(t), \quad t_k = \cos\frac{k\pi}{n+1}, \quad a_k = \frac{\pi}{n+1}\sin^2\frac{k}{n+1}$$
$$k = 1, \ldots, n, \quad t_{0j} = \cos\frac{2j-1}{2(n+1)}\pi, \quad j = 1, \ldots, n+1. \quad (9.2.28)$$

3. Let $\nu = -1/2$, $\mu = 1/2$, i.e.,

$$\omega(t) = \left(\frac{1-t}{1+t}\right)^{1/2}. \tag{9.2.29}$$

In this case, we have

$$P_n(t) = \frac{T_{n+1}(t) - T_n(t)}{1 - t}, \quad Q_n(t) = \pi\{U_n(t) - U_{n-1}(t)\}, \quad t_k = \cos\frac{2k\pi}{2n+1},$$

$$a_k = \frac{4\pi}{2n+1}\sin^2\frac{k\pi}{2n+1}, \quad k = 1, \ldots, n, \quad t_{0j} = \cos\frac{2j-1}{2(n+1)}\pi, \quad j = 1, \ldots, n. \tag{9.2.30}$$

Note that in case 1, for $I(\omega, \psi, t_0)$ the following quadrature sum was suggested by *Sanikidze* (1974):

$$S(t_0) = \frac{1}{n}\sum_{k=1}^{n}\frac{(-1)^k\left(1 - t_k^2\right)^{1/2}U_{n-1}(t_0) - 1}{t_0 - t_k}\varphi(t_k), \tag{9.2.31}$$

together with a useful estimate for all $t_0 \in [-1, 1]$,

$$|I(\omega, \psi, t_0) - S(t_0)| = O\left(\frac{\ln n}{\left((1 - \eta^2)^{1/2} + t_0 n^{-1}\right)n^{r+\alpha}}\right), \quad \eta = \frac{1 + |t_0|}{2}. \tag{9.2.32}$$

9.3. Integrals Arising in Boundary Value Problems for the Laplace and the Helmholtz Equations

As mentioned in Section 1.1, the problem of circulation-free flow past a segment can be reduced to a strongly singular integral (in the sequel, the term "hypersingular" is preferred)

$$I_1(t_0) = \int_a^b \frac{g(t)\,dt}{(t - t_0)^2}, \quad t_0 \in (a, b), \tag{9.3.1}$$

where $g'(t) = \varphi(t) \in H^*$ on $[a, b]$. And, as observed in Section 1.2, if $g(-1) = g(1) = 0$, integration by parts reduces the integral $I_1(t_0)$ to a singular integral on the segment $[a, b]$,

$$I(t_0) = I_1(t_0) = \int_a^b \frac{g'(t)\,dt}{t - t_0}, \quad t_0 \in (a, b). \tag{9.3.2}$$

For this reason, we are going to obtain a quadrature formula for the integral $I_1(t_0)$ which can be transformed into a quadrature formula for $I(t_0)$. To that end, we consider the following two sets on the segment $[a, b]$:

$$E = \{t_k : k = 1, \ldots, n+1\}, \quad E_0 = \{t_{0k} : k = 1, \ldots, n\};$$

$$t_k = a + (k-1)h, \quad h = \frac{b-a}{n}, \quad k = 1, \ldots, n+1; \quad t_{0k} = t_k + \frac{h}{2}, \quad k = 1, \ldots, n,$$

which form a canonical partition of the segment $[a, b]$. Now, let us replace the integral $I_1(t_{0j})$ by the quadrature sum

$$S_{n,1}(t_{0j}) = \sum_{k=1}^{n}g(t_{0k})\int_{t_k}^{t_{k+1}}\frac{dt}{(t - t_{0j})^2}, \quad j = 1, \ldots, n. \tag{9.3.3}$$

Using (1.1.14) and (9.3.3), we can write

$$S_{n,1}(t_{0j}) = \sum_{k=1}^{n}g(t_{0k})\left(\frac{1}{t_k - t_{0j}} - \frac{1}{t_{k+1} - t_{0j}}\right), \quad j = 1, \ldots, n, \tag{9.3.4}$$

or

$$S_{n,1}(t_{0j}) = \frac{g(t_{01})}{t_1 - t_{0j}} + \sum_{k=2}^{n}\frac{g(t_{0k}) - g(t_{0k-1})}{t_k - t_{0j}} + \frac{-g(t_{0n})}{t_{n+1} - t_{0j}}, \quad j = 1, \ldots, n. \tag{9.3.5}$$

Now, from Theorem 9.2.2 we obtain the following result.

Theorem 9.3.1. *If $g'(t) \in H^*$ on $[a, b]$, then*

$$|I_1(t_{0j}) - S_{n,1}(t_{0j})| \leq \theta(t_{0j}), \quad j = 1, \ldots, n, \tag{9.3.6}$$

where $\theta(t_{0j})$ has the same properties as in Theorem 9.2.2.

On the basis of the double layer potential, the Neumann problems (both the exterior and the interior problems, if the curve is closed) for the Laplace and the Helmholtz equations are reduced to hypersingular integrals. Indeed, since the normal derivative of the double layer potential is continuous across the boundary curve, we can write

$$\frac{\partial \Phi_L(M_0)}{\partial n_{M_0}} = \frac{1}{2\pi} \int_L \frac{\partial}{\partial n_{M_0}} \frac{\partial}{\partial n_M} \left(\ln \frac{1}{r_{MM_0}} \right) g(M) \, dl_M, \qquad M_0 \in L, \tag{9.3.7}$$

or

$$\frac{\partial \Phi_N(M_0)}{\partial n_{M_0}} = -\frac{i}{4} \int_L \frac{\partial}{\partial n_{M_0}} \frac{\partial}{\partial n_M} H_0^{(2)}(\varkappa r_{MM_0}) g(M) \, dl_M, \qquad M_0 \in L, \tag{9.3.8}$$

where $\Phi_M(M_0)$ and $\Phi_N(M_0)$ are the double layer potentials for the Laplace and the Helmholtz equations, respectively; $H_0^{(2)}(z)$ is the zero order Hankel function of the first kind. Quadrature formulas for integrals (9.3.7) and (9.3.8) are constructed in the same way as for the integral (9.3.1). Indeed, let $X = X(t)$, $Y = Y(t)$, $t \in [a, b]$, be a parametrization of the curve L. Just as we have done for the integral (9.3.1), consider the canonical partition of $[a, b]$ by the sets $E = \{t_k \ k = 1, \ldots, n+1\}$ and $E_0 = \{t_{0k}, \ k = 1, \ldots, n\}$. We say that the sets

$$\widetilde{E} = \{M_k, \ M(x(t_k), y(t_k)), \ k = 1, \ldots n+1\}, \quad \widetilde{E}_0 = \{M_{0k}, \ M(x(t_{0k}), y(t_{0k})), \ k = 1, \ldots n\}$$

form a canonical partition of the curve L. We replace the integral (9.3.8) by the quadrature sum

$$\widetilde{S}_n(M_{0j}) = \sum_{k=1}^n g(M_{0k}) \int_{L_{M_k M_{k+1}}} -\frac{i}{4} \frac{\partial}{\partial n_{M_0}} \frac{\partial}{\partial n_M} H_0^{(2)}(\varkappa r_{MM_{0j}}) \, dl_M, \qquad j = 1, \ldots, n, \tag{9.3.9}$$

where $L_{M_k M_{k+1}}$ is the part of the curve L between the points M_k and M_{k+1}.

For our purposes, it is convenient to transform the integral (9.3.9) in the following way. Recall that

$$\frac{\partial}{\partial n_M} f(M) = n_M \cdot \mathrm{grad}\, f(M). \tag{9.3.10}$$

Therefore, we can write

$$-\frac{i}{4} \int_{L_{M_k M_{k+1}}} \frac{\partial}{\partial n_{M_0}} \frac{\partial}{\partial n_M} H_0^{(2)}(\varkappa r_{MM_{0j}}) \, dl_M =$$

$$= -\frac{i}{4} \frac{\partial}{\partial n_{M_0}} \int_{L_{M_m M_{k+1}}} \frac{\partial}{\partial n_M} H_0^{(2)}(\varkappa r_{MM_{0j}}) \, dl_M = n_{M_0} \cdot \mathrm{grad}\, \Phi_N(M_{0j}). \tag{9.3.11}$$

By definition, we have

$$\mathrm{grad}_{M_0} \Phi_N(M_0) = \frac{\partial}{\partial x_0} \Phi_N(M_0) i + \frac{\partial}{\partial y_0} \Phi_N(M_0) j, \qquad M_0 \notin L, \tag{9.3.12}$$

where i and j are orthogonal unit vectors on the plane.

Consider the component with $\partial/\partial x_0$ in (9.3.12). We have $(g \equiv 1)$

$$\frac{\partial}{\partial x_0} \Phi_N(M_0) = -\frac{i}{4} \int_L \frac{\partial}{\partial x_0} \frac{\partial}{\partial n_M} H_0^{(2)}(\varkappa r_{MM_0})\, dl_M = \frac{i}{4} \int_L \frac{\partial}{\partial n_M} \frac{\partial}{\partial x} H_0^{(2)}(\varkappa r_{MM_0})\, dl_M =$$

$$= \frac{i}{4} \int_L \left[\frac{\partial^2}{\partial x^2} \left(H_0^{(2)}(\varkappa r_{MM_0}) \right) \cdot y' - \frac{\partial}{\partial y}\left(\frac{\partial}{\partial x} H_0^{(2)}(\varkappa r_{MM_0}) \right) \cdot x' \right] dl_M . \tag{9.3.13}$$

Here y' and x' stand for the derivatives along the arc length of the curve L. Using the fact that the Hankel function is a solution of the Helmholtz equation and setting $z = \varkappa r_{MM_0}$, we find that

$$\frac{\partial}{\partial x_0} \Phi_N(M_0) = \frac{i}{4} \int_L \left[-\frac{\partial}{\partial y}\left(\frac{\partial}{\partial y} H_0^{(2)}(z) \right) y' - \frac{\partial}{\partial x}\left(\frac{\partial}{\partial y} H_0^{(2)}(z) \right) x' \right] dl_M - \tag{9.3.14}$$

$$- \frac{i}{4} \int_L \varkappa^2 H_0^{(2)}(z) y'\, dl_M = -\frac{i}{4} \int_L d\left(\frac{\partial}{\partial y} H_0^{(2)}(z) \right) - \frac{i}{4} \varkappa^2 \int_L H_0^{(2)}(z) y'\, dL_M =$$

$$= -\frac{i}{4} \left[\frac{\partial}{\partial y} H_0^{(2)}(\varkappa r_{M_B M_0}) - \frac{\partial}{\partial y} H_0^{(2)}(\varkappa r_{M_A M_0}) \right] - \frac{1}{4} \int_L \varkappa^2 H_0^{(2)}(\varkappa r_{MM_0}) y'\, dl_M,$$

where M_A and M_B are the end-points of L.

Similar transformations of the term with $\partial/\partial y_0$ yield the relation

$$\frac{\partial}{\partial y_0} \Phi_N(M_0) = \frac{i}{4} \left[\frac{\partial}{\partial y_0} H_0^{(2)}(\varkappa r_{M_B M_0}) - \frac{\partial}{\partial x} H_0^{(2)}(\varkappa r_{M_A M_0}) \right] + \frac{i}{4} \int_L \varkappa^2 H_0^{(2)}(\varkappa r_{MM_0}) x'\, dl_M . \tag{9.3.15}$$

Replacing the terms with $\partial/\partial x_0$ and $\partial/\partial y_0$ in (9.3.12) by the expressions just obtained and taking into account that (see *Brychkov and Prudnikov* (1977))

$$\frac{\partial}{\partial y} H_0^{(2)}(\varkappa r_{MM_0}) = -\varkappa H_1^{(2)}(\varkappa r_{MM_0}) \frac{y - y_0}{r_{MM_0}} ,$$

$$\frac{\partial}{\partial x} H_0^{(2)}(\varkappa r_{MM_0}) = -\varkappa H_1^{(2)}(\varkappa r_{MM_0}) \frac{x - x_0}{r_{MM_0}} ,$$

we rewrite (9.3.12) in the form

$$\operatorname{grad}_{M_0} \Phi_N(M_0) = -\frac{i}{4} \left[H_1^{(2)}(\varkappa r_{M_B M_0}) \frac{(y_B - y_0)\boldsymbol{i} - (x_B - x_0)\boldsymbol{j}}{r_{M_B M_0}^2} \varkappa r_{M_B M_0} - \right. \tag{9.3.16}$$

$$\left. - H_1^{(2)}(\varkappa r_{M_A M_0}) \frac{(y_A - y_0)\boldsymbol{i} - (x_A - x_0)\boldsymbol{j}}{r_{M_A M_0}^2} \varkappa r_{M_A M_0} \right] + \varkappa^2 \frac{i}{4} \int_L H_0^{(2)}(\varkappa r_{MM_0}) \boldsymbol{n}_M\, dl_M .$$

Now, letting $\boldsymbol{n}_M = y_0' \boldsymbol{i} - x_0' \boldsymbol{j}$, we can rewrite (9.3.9) as

$$\widetilde{S}_n(M_{0j}) = -\frac{i}{4} \sum_{k=1}^{n} g(M_{0k}) \left\{ \left[H_1^{(2)}(\varkappa r_{M_{k+1} M_{0j}}) \frac{(y_{k+1} - y_{0j})y_{0j}' + (x_{k+1} - x_{0j})x_{0j}'}{r_{M_{k+1} M_{0j}}^2} \varkappa r_{M_{k+1} M_{0j}} - \right. \right.$$

$$- H_1^{(2)}(\varkappa r_{M_k M_{0j}}) \frac{(y_k - y_{0j})y_{0j}' + (x_k - x_{0j})x_{0j}'}{r_{M_k M_{0j}}^2} \varkappa r_{M_k M_{0j}} \right] +$$

$$\left. + \varkappa^2 \int_{L_{M_{k+1} M_k}} H_0^{(2)}(\varkappa r_{MM_0}) \boldsymbol{n}_M\, dl_M \right\}. \tag{9.3.17}$$

Passing to the limit as $\varkappa \to 0$ in (9.3.12), we obtain a quadrature formula for the normal derivative of the double layer potential for the Laplace equation (a formula for discrete vortex pairs),

$$\widetilde{S}_{n,L}(M_{0j}) = \tag{9.3.18}$$

$$= -\frac{1}{2\pi} \sum_{k=1}^{n} g(M_{0k}) \left[\frac{(y_{k+1} - y_{0j})y'_{0j} + (x_{k+1} - x_0)x'_{0j}}{r^2_{M_{k+1}M_{0j}}} - \frac{(y_k - y_{0j})y'_{0j} + (x_k - x_{0j})x'_{0j}}{r^2_{M_k M_{0j}}} \right].$$

Theorem 9.3.1 yields estimates for the approximation of hypersingular integrals (9.3.7) and (9.3.8) by quadrature sums (9.3.18) and (9.3.17), respectively.

9.4. Integrals on Smooth Surfaces with Border

Consider the following integral (understood in the sense of Hadamard; see (5.2.9)):

$$I(v) = \int_{\sigma} \frac{v(M_0)\, d\sigma_{M_0}}{|MM_0|^3}, \tag{9.4.1}$$

where $\sigma \subset \mathbb{R}_3$ is a surface (a differentiable manifold with border) of class C^4, $M_0, M \in \sigma$. It is assumed there is a Cartesian coordinate system $OX_0Y_0Z_0$ such that the plane OX_0Y_0 is non-orthogonal to the tangential planes to σ at $M \in \sigma$. Let $z_0 = f(x_0, y_0)$ be the equation of σ in these coordinates. We introduce the notation $v(M_0) = v(x_0, y_0, f(x_0, y_0)) = g(x_0, y_0)$ and denote by D the projection of the surface σ to the plane OX_0Y_0.

Condition 9.4.1. We assume that D is a domain whose boundary Γ is of class C^3. Let $\rho(x_0, y_0)$ be the distance from the point $P(x_0, y_0) \in D$ to Γ. Then, there is a neighborhood U of Γ in which $\rho(x_0, y_0)$ is twice differentiable. The neighborhood U is called *normal*. It is assumed that the function $g(x_0, y_0)$ in U has the form

$$g(x_0, y_0) = \rho^{\beta} l(x_0, y_0), \quad 0 < \beta \le 1, \tag{9.4.2}$$

and $l'_{x_0}(x_0, y_0), l'_{y_0}(x_0, y_0) \in H_{\alpha}(\overline{D})$, $g'_{x_0}(x_0, y_0), g'_{y_0}(x_0, y_0) \in H_{\alpha}(\overline{U}_1)$, where U_1 is a neighborhood such that $U \cup U_1 \supset D$ and $\overline{U}_1$ is separated from the boundary Γ by a positive distance.

Let $(x, y, f(x, y))$ and $(x_0, y_0, f(x_0, y_0))$ be the coordinates of the points M and M_0, respectively. Then

$$|M_0 M| = \left[(x - x_0)^2 + (y - y_0)^2 + (f(x, y) - f(x_0, y_0))^2 \right]^{1/2},$$

$$d\sigma_{M_0} = \left[1 + f'^2_{x_0}(x_0, y_0) + f'^2_{y_0}(x_0, y_0) \right]^{1/2} dx_0\, dy_0.$$

Therefore, the integral (9.4.1) can be represented in the form

$$I(v) = A(x, y) = \frac{1}{4\pi} \iint_{D} K(x, y, x_0, y_0) g(x_0, y_0)\, dx_0\, dy_0, \tag{9.4.3}$$

with

$$K(x, y, x_0, y_0) = \frac{1}{r^3} \left[1 + f'^2_{x_0}(x_0, y_0) + f'^2_{y_0}(x_0, y_0) \right]^{1/2} \left[1 + \left(\frac{f(x, y) - f(x_0, y_0)}{r} \right)^2 \right]^{-3/2} =$$

$$= \frac{1}{r^3} B(x, y, x_0, y_0), \qquad r = \left[(x - x_0)^2 + (y - y_0)^2 \right]^{1/2}. \tag{9.4.4}$$

From Taylor's expansion of the function $B(x, y, x_0, y_0)$ at the point $x_0 = x$, $y_0 = y$, we obtain

$$K(x, y, x_0, y_0) = \frac{K_1(x, y, \varphi)}{r^3} + \frac{K_2(x, y, \varphi)}{r^2} + \frac{K_3(x, y, r, \varphi)}{r}, \tag{9.4.5}$$

where φ is the polar angle of the polar coordinate system whose origin coincides with $P(x, y)$ and the polar axis is parallel to OX_0. The functions $K_1(x, y, \varphi)$, $K_2(x, y, \varphi)$, $K_3(x, y, r, \varphi)$ satisfy the conditions

$$K_1(x, y, \varphi + \pi) = K_1(x, y, \varphi), \quad \int_0^{2\pi} K_2(x, y, \varphi)\, d\varphi = 0, \quad |K_3(x, y, r, \varphi)| \le C. \qquad (9.4.6)$$

Now, the integral in (9.4.3) can be written in the form

$$A(x, y) = \frac{1}{4\pi} \sum_{p=1}^{2} \iint_D \frac{b(x, y, r, \varphi) K_p(x, y, \varphi)}{r^{4-p}}\, dx_0\, dy_0 +$$

$$+ \frac{1}{4\pi} \iint_D b(x, y, r, \varphi) \frac{K_3(x, y, r, \varphi)}{r}\, dx_0\, dy_0 = \sum_{p=1}^{3} A_p(x, y), \qquad (9.4.7)$$

where $b(x, y, r, \varphi) = g(x + r\cos\varphi, y + r\sin\varphi)$; $A_2(x, y)$ is a singular integral and the integral $A_1(x, y)$ is defined in Section 1.3.

Consider a canonical partition of $\mathbb{R}_2$ (see Definition 5.1.1) with cells $D(k, h)$ and step $h = (h_1, h_2)$ for which the inequality (8.5.2) holds. Let $\varepsilon_h = \varepsilon(D, h)$ be the set of all indices $k = (k_1, k_2)$ such that $D(k, h) \subset D$. In each cell $D(k, h)$, consider a point $M_{0k}(k_1 h_1, k_2 h_2)$ and let $r_{0k} = |M M_{0k}|$, where M is the point with coordinates (x_0, y_0). Then the quadrature formula for the integral (9.4.7) is constructed as follows:

$$S(m_1 h_1, m_2 h_2) = \frac{1}{4\pi} \sum_{p=1}^{2} \sum_{k \in \varepsilon_h} g(k_1 h_1, k_2 h_2) \iint_{D(k,h)} \frac{K_p(m_1 h_1, m_2 h_2, \varphi)\, dx_0\, dy_0}{r_{0m}^{4-p}} + \qquad (9.4.8)$$

$$+ \frac{1}{4\pi} \sum_{k \in \varepsilon_h} g(k_1 h_1, k_2 h_2) \iint_{D(k,h)} \frac{K_3(m_1 h_1, m_2 h_2, r_{0m}, \varphi)}{r_{0m}}\, dx_0\, dy_0 = \sum_{p=1}^{3} S_p(m_1 h_1, m_2 h_2).$$

Denote by M_h the interior of the set $\overline{\bigcup_{k \in \varepsilon_h} D(k, h)}$. Let us find the distance ρ from a point $P \in D \setminus M_h$ to the boundary of D. It can be shown that $\rho \le 2|h|$. Indeed, let $L_h \subset D$ be the set of all $P \in D$ such that $\rho(L_h, \partial D) = 2|h|$. For any $P \in L_h$, there is a cell $D(k, h)$ such that $P \in \overline{D(k, h)}$. Let us show that $D(k, h) \subset D$. Assuming the contrary, we can find a point $M \in D(k, h) \cap \partial D$, and therefore, $\rho(M, p) \ge 2|h|$, which is in contradiction with the condition that $\rho(M_1, M_2) \le |h|$ for any $M_1, M_2 \in D(k, h)$.

Let us estimate the quadrature error,

$$\Delta(m_1 h_1, m_2 h_2) = |A(m_1 h_1, m_2 h_2) - S(m_1 h_1, m_2 h_2)| \le$$

$$\le \sum_{p=1}^{3} |A_p(m_1 h_1, m_2 h_2) - S_p(m_1 h_1, m_2 h_2)| \le \sum_{p=1}^{3} \Delta_p(m_1 h_1, m_2 h_2). \qquad (9.4.9)$$

(I) For $\Delta_3(m_1 h_1, m_2 h_2)$, we have

$$\Delta_3(m_1 h_1, m_2 h_2) \le$$

$$\le \frac{1}{4\pi} \left| \iint_{D \setminus M_h} b(m_1 h_1, m_2 h_2, r_{0m}, \varphi) \frac{K_3(m_1 h_1, m_2 h_2, r_{0m}, \varphi)}{r_{0m}}\, dx_0\, dy_0 \right| +$$

$$+ \frac{1}{4\pi} \sum_{k \in \varepsilon_h} \iint_{D(k,h)} |b(m_1 h_1, m_2 h_2, r_{0m}, \varphi) - g(k_1 h_1, k_2 h_2)| \times$$

$$\times \frac{|K_3(m_1 h_1, m_2 h_2, r_{0m}, \varphi)|}{r_{0m}}\, dx_0\, dy_0 = I_3' + I_3''. \qquad (9.4.10)$$

Using the inequality for K_3 in (9.4.6), together with Condition 9.4.1 on $g(x_0, y_0)$ and the inequality $\rho(M, \partial D) \leq 2|h|$ for $M(x_0, y_0) \in D \setminus M_h$ established above, we find that $|g(x_0, y_0)| \leq C|h|^\beta$ for $M(x_0, y_0) \in D \setminus M_h$. Therefore,

$$I_3' \leq C_1|h|^\beta, \qquad 0 < \beta < 1. \tag{9.4.11}$$

Condition 9.4.1 implies that $g(x_0, y_0) \in H(\beta)$ on $\overline{D}$, and therefore,

$$I_3'' \leq C_2|h|^\beta. \tag{9.4.12}$$

From (9.4.1)–(9.4.12), we obtain the estimate

$$\delta_3(m_1h_1, m_2h_2) \leq C_3|h|^\beta, \qquad 0 < \beta \leq 1. \tag{9.4.13}$$

(II) For $\Delta_2(m_1h_1, m_2h_2)$ we obtain

$$\Delta_2(m_1h_1, m_2h_2) \leq$$

$$\leq \frac{1}{4\pi} \left| \iint_{D \setminus M_h} b(m_1h_1, m_2h_2, r_{0m}, \varphi) \frac{K_2(m_1h_1, m_2h_2, \varphi)}{r_{0m}^2} \, dx_0 \, dy_0 \right| +$$

$$+ \frac{1}{4\pi} \sum_{k \in \varepsilon_h, \, k \neq m} \left| \iint_{D(k,h)} \frac{b(m_1h_1, m_2h_2, r_{0m}, \varphi) - g(k_1h_1, k_2h_2)}{r_{0m}^2} K_2(m_1h_1, m_2h_2, \varphi) \, dx_0 \, dy_0 \right| +$$

$$+ \frac{1}{4\pi} \left| \iint_{D(m,h)} \frac{b(m_1h_1, m_2h_2, r_{0m}, \varphi) - g(m_1h_1, m_2h_2)}{r_{0m}^2} K_2(m_1h_1, m_2h_2, \varphi) \, dx_0 \, dy_0 \right| =$$

$$= \lambda_1 + \lambda_2 + \lambda_3. \tag{9.4.14}$$

From (9.4.1) and (9.4.6), it follows that

$$\lambda_3 \leq \frac{C}{\pi} \int_0^{2\pi} \left(\int_0^{|h|} \frac{r_{0m}^{\beta+1}}{r_{0m}^2} \right) d\varphi \leq \frac{C_1}{\beta} |h|^\beta, \qquad 0 < \beta \leq 1. \tag{9.4.15}$$

We further have

$$\lambda_2 \leq c|h|^\beta \int_0^{2\pi} \int_{|h|/2}^R \frac{dr}{r} \leq C_2|h|^\beta |\ln|h||, \tag{9.4.16}$$

where R is the radius of a sufficiently large circle containing the domain D and having its center at $M(m_1h_1, m_2h_2)$.

Taking into account Condition 9.4.1 for $f(x_0, y_0)$, we get

$$\lambda_1 \leq c|h|^\beta \int_0^{2\pi} \int_{|h|/2}^R \frac{1}{r} \, dr \leq C_1|h|^\beta |\ln|h||. \tag{9.4.17}$$

From (9.4.14)–(9.4.17), we obtain the inequality

$$\Delta_2(m_1h_1, m_2h_2) \leq C_3|h|^\beta |\ln|h||. \tag{9.4.18}$$

(III) Let us estimate $\Delta_1(m_1h_1, m_2h_2)$. We have

$$\Delta_1(m_1h_1, m_2h_2) \leq \frac{1}{4\pi} \left| \iint_{D \setminus M_h} b(m_1h_1, m_2h_2, r_{0m}, \varphi) \frac{K_1(m_1h_1, m_2h_2, \varphi)}{r_{0m}^3} \, dx_0 \, dy_0 \right| +$$

$$+ \frac{1}{4\pi} \left| \sum_{k \in \varepsilon_h} \iint_{D(k,h)} \frac{b(m_1h_1, m_2h_2, r_{0m}, \varphi) - g(k_1h_1, k_2h_2)}{r_{0m}^3} K_1(m_1h_1, m_2h_2, \varphi) \, dx_0 \, dy_0 \right| =$$

$$= \psi_1(m_1h_1, m_2h_2) + \psi_2(m_1h_1, m_2h_2). \tag{9.4.19}$$

Let the distance from the point $M_0(m_1h_1, m_2h_2)$ to the boundary Γ of the domain D be equal to $\delta_m \geq 5|h|$. Using the estimate $|g(x_0, y_0)| \leq C\rho^\beta(x_0, y_0)$, where $\rho(x_0, y_0) = \rho(M, \Gamma)$, we obtain

$$\psi_1(m_1h_1, m_2h_2) \leq c|h|^\beta \int\limits_0^{2\pi} d\varphi \int\limits_{\delta_{m/2}}^\infty \frac{dr}{r^2} \leq C_1 \frac{|h|^\beta}{\delta_m}. \tag{9.4.20}$$

Consider the circle K of radius $R = \delta_m/2$ with center at M_0, and let ε_h' be the set of indices $k = (k_1, k_2)$ such that $D(k, h) \subset K$. The following relation holds:

$$\psi_2(m_1h_1, m_2h_2) \leq$$

$$\leq \left| \sum_{k \in \varepsilon_h \setminus \varepsilon_h'} \frac{1}{4\pi} \iint_{D(k,h)} \frac{b(m_1h_1, m_2h_2, r_{0m}, \varphi) - g(k_1h_1, k_2h_2)}{r_{0m}^3} K_1(m_1h_1, m_2h_2, \varphi)\, dx_0\, dy_0 \right| +$$

$$+ \left| \sum_{k \in \varepsilon_h'} \frac{1}{4\pi} \iint_{D(k,h)} \frac{b(m_1h_1, m_2h_2, r_{0m}, \varphi) - g(k_1h_1, k_2h_2)}{r_{0m}^3} K_1(m_1h_1, m_2h_2, \varphi)\, dx_0\, dy_0 \right| =$$

$$= \psi_{21}(m_1h_1, m_2h_2) + \psi_{22}(m_1h_1, m_2h_2). \tag{9.4.21}$$

Using Condition 9.4.1, we get

$$\psi_{21}(m_1h_1, m_2h_2) \leq C|h|^\beta \int\limits_0^{2\pi} \int\limits_{\delta_{m/2}-2|h|}^\infty \frac{dr}{r^2} \leq C_1 \frac{|h|^\beta}{\delta_m}. \tag{9.4.22}$$

In order to estimate $\psi_{22}(m_1h_1, m_2h_2)$, we carry out the following transformation:

$$\frac{1}{4\pi} \sum_{k \in \varepsilon_h'} \iint_{D(k,h)} \frac{g_{x_0}'(m_1h_1, m_2h_2)(x_0 - k_1h_1) + g_{y_0}'(m_1h_1, m_2h_2)(y_0 - k_2h_2)}{r_{0m}^3} \times$$

$$\times K_1(m_1h_1, m_2h_2, \varphi)\, dx_0\, dy_0 +$$

$$+ \frac{1}{4\pi} \sum_{k \in \varepsilon_h'} \int_{D(k,h)} \frac{g_{x_0}'(m_1h_1, m_2h_2)(m_1h_1 - k_2h_1) + g_{y_0}'(m_1h_1, m_2h_2)(m_2h_2 - k_2h_2)}{r_{0m}^3} \times$$

$$\times K_1(m_1h_1, m_2h_2, \varphi)\, dx_0\, dy_0 = A_1 + A_2. \tag{9.4.23}$$

It is easy to see that $A_1 = A_2 = 0$, because of (9.4.6) and the symmetry of the circle K with center at the point $M(m_1h_1, m_2h_2)$ and the symmetry of the set of cells $D(k, h) \in M_h'$ with respect to that point.

From (9.4.23), it follows that

$$\psi_{22}(m_1h_1, m_2h_2) =$$

$$= \frac{1}{4\pi} \left| \sum_{k \in \varepsilon_h'} \int_{D(k,h)} \left(\frac{g(x_0, y_0) - g(k_1h_1, k_2h_2) - g_{x_0}'(m_1h_1, m_2h_2)(x_0 - k_1h_1)}{r_{0m}^3} - \right. \right.$$

$$\left. \left. - \frac{g_{y_0}'(m_1h_1, m_2h_2)(y_0 - k_2h_2)}{r_{0m}^3} \right) K_1(m_1h_1, m_2h_2, \varphi)\, dx_0\, dy_0 \right|. \tag{9.2.24}$$

Next, we perform the following transformations:

$$
\begin{aligned}
\lambda &= g(x_0, y_0) - g(k_1 h_1, k_2 h_2) - g'_{x_0}(m_1 h_1, m_2 h_2)(x_0 - k_1 h_1) - g'_{y_0}(m_1 h_1, m_2 h_2)(y_0 - k_2 h_2) = \\
&= g(x_0, y_0) - g(k_1 h_1, y_0) + g(k_1 h_1, y_0) - g(k_1 h_1, k_2 h_2) - \\
&\quad - g'_{x_0}(m_1 h_1, m_2 h_2)(x_0 - k_1 h_1) - g'_{y_0}(m_1 h_1, m_2 h_2)(y_0 - k_2 h_2) = \\
&= g'_{x_0}(x_{1k_1}, y_0)(x_0 - k_1 h_1) + g'_{y_0}(k_1 h_1, y_{1k_2})(y_0 - k_2 h_2) - \\
&\quad - g'_{x_0}(m_1 h_1, m_2 h_2)(x_0 - k_1 h_1) - g'_{y_0}(m_1 h_1, m_2 h_2)(y_0 - k_2 h_2) = \\
&= (x_0 - k_1 h_1)\left[g'_{x_0}(x_{1k_1}, y_0) - g'_{x_0}(m_1 h_1, m_2 h_2)\right] + \\
&\quad + (y_0 - k_2 h_2)\left[g'_{y_0}(k_1 h_1, y_{1k_2}) - g'_{y_0}(m_1 h_1, m_2 h_2)\right] = \\
&= (x_0 - k_1 h_1)A_{km} + (y_0 - k_2 h_2)B_{km}.
\end{aligned}
\tag{9.4.25}
$$

Let us estimate the coefficients

$$
A_{km}(x_0, y_0, k_1 h_1, k_2 h_2, m_1 h_1, m_2 h_2), \qquad B_{km}(x_0, y_0, k_1 h_1, k_2 h_2, m_1 h_1, m_2 h_2).
$$

for $(x_0, y_0) \in D(k, h)$.

Case 1. Let $M_0(m_1 h_1, m_2 h_2) \in U_1$ (see Condition 9.4.1). Then, the distance δ_m from the point $M_0(m_1 h_1, m_2 h_2)$ to the boundary Γ of the domain D is greater than $C > 0$ for all m_1, m_2 such that $M_0(m_1 h_1, m_2 h_2) \in U_1$. Since $D(k, h) \subset K$, the points $M(x_0, y_0)$ and $M_0(m_1 h_1, m_2 h_2) \in D(k, h) \subset K$ are separated from the boundary Γ by a distance larger than $C/2$. Therefore, $g'_{x_0}(x_0, y_0), g'_{y_0}(x_0, y_0) \in H_\alpha(\overline{K})$. It follows that

$$
\begin{aligned}
|A_{km}| &\le C_1 \left[(k_1 h_1 - m_1 h_1)^2 + (k_2 h_2 - m_2 h_2)^2\right]^{\alpha/2}, \\
|B_{km}| &\le C_1 \left[(k_1 h_1 - m_1 h_1)^2 + (k_2 h_2 - m_2 h_2)^2\right]^{\alpha/2}.
\end{aligned}
\tag{9.4.26}
$$

Case 2. Let $M_0(m_1 h_1, m_2 h_2) \in U$ (see Condition 9.4.1). Then, there is $\delta_0 > 0$ such that all points $M(x, y) \in D$ for which $\rho(x, y) < \delta_0$ belong to U. Then, for all $M(x, y)$ such that $\rho(x, y) \le \delta_0/4$, we have $K \subset U$, where K is the circle of radius $R = \rho(x, y)/2 \le \delta_0/8$ with center at $M(x, y)$. Indeed, take $M_0(x_0, y_0) \in K$. Then, the triangle inequality ensures that $\rho(x_0, y_0) \le \rho(x, y) + R \le \delta_0/4 + \delta_0/8 < \delta_0$, and therefore, $M_0(x_0, y_0) \in U$.

Take $\delta_1 = \delta_0/4$ and let $\delta_m > \delta_1$. Then it follows from Condition 9.4.1 that for A_{km} and B_{km} the inequalities (9.4.26) hold.

Now, suppose that for the point $M_0(m_1 h_1, m_2 h_2)$ we have $\delta_m \le \delta_1$. Then the circle K of radius $R = \delta_m/2$ with center at $M_0(m_1 h_1, m_2 h_2)$ belongs to V and thus, in view of Condition 9.4.1, we have $g(x_0, y_0) = \rho^\beta(x_0, y_0)l(x_0, y_0)$. Therefore,

$$
\begin{aligned}
|A_{km}| &= \left|g'_{x_0}(x_{1k_1}, y_0) - g'_{x_0}(m_1 h_1, m_2 h_2)\right| = \\
&= \left|\beta\rho^{\beta_1}(x_{1k_1}, y_0)\rho'_{x_0}(x_{1k_1}, y_0)l(x_{1k_1}, y_0) + \rho^\beta(x_{1k_1}, y_0)l'_{x_0}(x_{1k_1}, y_0) - \right. \\
&\quad - \beta\rho^{\beta-1}(m_1 h_1, m_2 h_2)\rho'_{x_0}(m_1 h_1, m_2 h_2)l(m_1 h_1, m_2 h_2) - \\
&\quad \left. - \rho^\beta(m_1 h_1, m_2 h_2)l'_{x_0}(m_1 h_1, m_2 h_2)\right| \le \\
&\le \left|\beta\rho^{\beta-1}(x_{1k_1}, y_0) - \beta\rho^{\beta-1}(m_1 h_1, m_2 h_2)\right|\left|\rho'_{x_0}(x_{1k_1}, y_0)l(x_{1k_1}, y_0)\right| + \\
&\quad + \beta\rho^{\beta-1}(m_1 h_1, m_2 h_2)\left|\rho'_{x_0}(x_{1k_1}, y_0)l(x_{1k_1}, y_0) - \rho'_{x_0}(m_1 h_1, m_2 h_2)l(m_1 h_1, m_2 h_2)\right| + \\
&\quad + \left|\rho^\beta(x_{1k_1}, y_0) - \rho^\beta(m_1 h_1, m_2 h_2)\right|\left|l'_{x_0}(x_{1k_1}, y_0)\right| + \\
&\quad + \rho^\beta(m_1 h_1, m_2 h_2)\left|l'_{x_0}(x_{1k_1}, y_0) - l'(m_1 h_1, m_2 h_2)\right| = K_1 + K_2 + K_3 + K_4.
\end{aligned}
\tag{9.4.27}
$$

Using the properties of the functions $\rho(x_0, y_0)$ and $l(x_0, y_0)$, we obtain the inequalities

$$
\begin{aligned}
&K_4 \le C_4 \big[(k_1 h_1 - m_1 h_1)^2 + (k_2 h_2 - m_2 h_2)^2\big]^{\alpha/2} = C_4 r_{km}^\alpha, \\
&K_3 \le C_3 r_{km}^\alpha, \qquad K_2 \le C_2 \frac{1}{\delta_m^{1-\beta}} r_{km}, \qquad K_1 \le C_1 \frac{1}{\delta_m^{2-2\beta}} r_{km}^{1-\beta}.
\end{aligned} \tag{9.4.28}
$$

From (9.4.28), we deduce that

$$
|A_{km}| \le \left[\frac{1}{\delta_m^{1-\beta}} r_{km}^\gamma + \frac{1}{\delta_m^{2-2\beta}} r_{km}^{1-\beta} \right], \tag{9.4.29}
$$

where $\gamma = \min(\alpha, \beta)$. A similar inequality holds for B_{km}.

Let us separately estimate λ (see (9.4.25)) for $k = m$.

For $M_0(m_1 h_1, m_2 h_2) \in U_1$, it follows from Condition 9.4.1 that

$$
|\lambda| \le C \big[(x_0 - m_1 h_1)^2 + (y_0 - m_2 h_2)^2\big]^{(\alpha+1)/2} \le C r_{0m}^{1+\alpha}. \tag{9.4.30}
$$

For $M_0(m_1 h_1, m_2 h_2) \in U$, using the relations

$$
x_{1m_1} = x_0 + \theta_1(x_0 - m_1 h_1), \qquad y_{1m_2} = y_0 + \theta_2(y_0 - m_2 h_2),
$$

we obtain, in the same way as (9.4.28), the following estimates:

$$
K_1 \le \frac{C_1}{\delta_m^{2-2\beta}} r_{0m}^{1-\beta}, \qquad K_2 \le C_1 \frac{1}{\delta_m^{1-m}} r_{0m}, \qquad K_3 \le C_1 r_{0m}^\beta, \qquad K_4 \le C_1 r_{0m}^\alpha.
$$

The above inequalities imply that

$$
|A_{mm}|, |B_{mm}| \le C_1 \left[\frac{r_{0m}^\gamma}{\delta_m^{1-\beta}} + \frac{r_{0m}^{1-\beta}}{\delta_m^{2-2\beta}} \right], \qquad \gamma = \min(\alpha, \beta). \tag{9.4.31}
$$

From (9.4.24) and (9.4.25), we obtain the relation

$$
\psi_{22}(m_1 h_1, m_2 h_2) \le
$$

$$
\le \frac{1}{4\pi} \left| \sum_{k \in \varepsilon_h', \, k \ne m} \iint_{D(k,h)} \frac{(x_0 - k_1 h_1) A_{km} + (y_0 - k_2 h_2) B_{km}}{r_{0m}^3} K_1(m_1 h_1, m_2 h_2, \varphi) \, dx_0 \, dy_0 \right| +
$$

$$
+ \frac{1}{4\pi} \left| \iint_{D(m,h)} \frac{(x_0 - m_1 h_1) A_{mm} + (y_0 - m_2 h_2) B_{mm}}{r_{0m}^3} K_1(m_1 h_1, m_2 h_2, \varphi) \, dx_0 \, dy_0 \right| =
$$

$$
= \psi_{122}(m_1 h_1, m_2 h_2) + \psi_{222}(m_1 h_1, m_2 h_2). \tag{9.4.32}
$$

From (9.4.31), we find that

$$
\psi_{222}(m_1 h_1, m_2 h_2) =
$$

$$
= \frac{1}{4\pi} \left| \lim_{\varepsilon \to 0} \iint_{D(k,h)} \frac{(x_0 - m_1 h_1) A_{mm} + (y_0 - m_2 h_2) B_{mm}}{r_{0m}^3} K_1(m_1 h_1, m_2 h_2, \varphi) \, dx_0 \, dy_0 \right| \le
$$

$$
\le C_1 \lim_{\varepsilon \to 0} \left| \left[\frac{1}{\delta_m^{1-\beta}} \int_0^{2\pi} d\varphi \int_\varepsilon^{f(\varphi)} r^{\gamma-1} \, dr + \frac{1}{\delta_m^{2-2\beta}} \int_0^{2\pi} d\varphi \int_\varepsilon^{f(\varphi)} r^{-\beta} \, dr \right] \right| \le
$$

$$
\le C_2 \left[\frac{|h|^\gamma}{\delta_m^{1-\beta}} + \frac{|h|^{1-\beta}}{\delta_m^{2-2\beta}} \right]. \tag{9.4.33}
$$

Here we have used the inequality $|f(\varphi)| \leq C|h|$. With the help of the estimate (9.4.29), we find that

$$\psi_{122}(m_1 h_1, m_2 h_2) \leq C|h| \sum_{k \in \varepsilon_h', \, k \neq m} \left[\frac{r_{km}^{\gamma}}{\delta_m^{1-\beta}} + \frac{r_{km}^{1-\beta}}{\delta_m^{2-2\beta}} \right] \iint_{D(k,h)} \frac{dx_0 \, dy_0}{r_{0m}^3} \leq$$

$$\leq C|h| \left[\frac{1}{\delta_m^{1-\beta}} \int_0^{2\pi} d\varphi \int_{|h|}^{R_0} r^{\gamma-2} \, dr + \frac{1}{\delta_m^{2-2\beta}} \int_0^{2\pi} d\varphi \int_{|h|}^{R_0} r^{-1-\beta} \, dr \right] \leq$$

$$\leq C_1 \left[\frac{|h|^{\gamma}}{\delta_m^{1-\beta}} + \frac{|h|^{1-\beta}}{\delta_m^{2-2\beta}} \right], \tag{9.4.34}$$

where R_0 is the radius of a circle with center at $M_0(m_1 h_1, m_2 h_2)$ containing the domain D.

Thus, for $\delta_m \geq 5|h|$, it follows from (9.4.13), (9.4.18), (9.4.20), (9.4.22), (9.4.32), (9.4.33) that

$$\Delta(m_1 h_1, m_2 h_2) \leq C \left[|h|^{\beta} |\ln|h|| + \frac{|h|^{\beta}}{\delta_m} + \frac{|h|^{\gamma}}{\delta_m^{1-\beta}} + \frac{|h|^{1-\beta}}{\delta_m^{2-2\beta}} \right]. \tag{9.4.35}$$

Consider the case $h_3/2 \leq \delta_m \leq 5|h|$, where $h_3 = \min(h_1, h_2)$. For $\Delta_2(m_1 h_1, m_2 h_2)$, $\Delta_3(m_1 h_1, m_2 h_2)$ the estimates (9.4.13), (9.4.18) are still valid. Now we should estimate $\Delta_1(m_1 h_1, m_2 h_2)$.

Keeping the notation of (9.4.19), we note that in this case the inequality (9.4.20) holds. Let us represent $\psi_2(m_1 h_1, m_2 h_2)$ in the form

$$\psi_2(m_1 h_1, m_2 h_2) \leq C|h|^{\beta} \left| \sum_{k \in \varepsilon_h, \, k \neq m} \iint_{D(k,h)} \frac{dx_0 \, dy_0}{r_{0m}^3} \right| +$$

$$+ \frac{1}{4\pi} \left| \iint_{D(k,h)} \frac{g(x_0, y_0) - g(m_1 h_1, m_2 h_2)}{r_{0m}^3} K_1(m_1 h_1, m_2 h_2, \varphi) \, dx_0 \, dy_0 \right| =$$

$$= \alpha_1(m_1 h_1, m_2 h_2) + \alpha_2(m_1 h_1, m_2 h_2). \tag{9.4.36}$$

We further have

$$\alpha_1(m_1 h_1, m_2 h_2) \leq C_1 |h|^{\beta} \int_0^{2\pi} d\varphi \int_{|h|}^{\infty} \frac{dr}{r^2} \leq C_2 \frac{|h|^{\beta}}{|h|} \leq C_3 \frac{|h|^{\beta}}{\delta_m}. \tag{9.4.37}$$

Since the function $K_1(m_1 h_1, m_2 h_2, \varphi)$ is symmetric with respect to the point $M_0(m_1 h_1, m_2 h_2)$ and

$$g^*(x_0, y_0) = g_{x_0}'(m_1 h_1, m_2 h_2)(x_0 - m_1 h_1) + g_{y_0}'(m_1 h_1, m_2 h_2)(y_0 - m_2 h_2) =$$

$$= g^{**}(r, \varphi) = g_{x_0}'(m_1 h_1, m_2 h_2) r \cos \varphi + g_{y_0}'(m_1 h_1, m_2 h_2) r \sin \varphi,$$

we have $g^{**}(r, \varphi + \pi) = -g^{**}(r, \varphi)$, and therefore,

$$\iint_{D(m,h)} \frac{g^*(x_0, y_0)}{r_{0m}^3} K_1(m_1 h_1, m_2 h_2, \varphi) \, dx_0 \, dy_0 = 0. \tag{9.4.38}$$

From (9.4.31) and (9.4.38), we obtain

$$\alpha_2(m_1 h_1, m_2 h_2) =$$

$$= \frac{1}{4\pi} \left| \iint_{D(m,h)} \frac{g(x_0, y_0) - g(m_1 h_1, m_2 h_2) - g^*(x_0, y_0)}{r_{0m}^3} K_1(m_1 h_1, m_2 h_2, \varphi) \, dx_0 \, dy_0 \right| =$$

$$= \frac{1}{4\pi} \left| \lim_{\varepsilon \to 0} \int_{D(m,h) \setminus U_\varepsilon} \frac{g(x_0, y_0) - g(m_1 h_1, m_2 h_2) - g^*(x_0, y_0)}{r_{0m}^3} K_1(m_1 h_1, m_2 h_2, \varphi) \, dx_0 \, dy_0 - 0 \right| \leq$$

$$\leq C \left| \left[\frac{1}{\delta_m^{1-\beta}} \int_0^{2\pi} d\varphi \int_{\varepsilon}^{f(\varphi)} r^{\gamma-1} \, dr + \frac{1}{\delta_m^{2-2\beta}} \int_{\varepsilon}^{f(\varphi)} r^{-\beta} \, dr \right] \right| \leq$$

$$\leq C_1 \left[\frac{|h|^{\gamma}}{\delta_m^{1-\beta}} + \frac{|h|^{1-\beta}}{\delta_m^{2-2\beta}} \right], \tag{9.4.39}$$

where we have used the inequality $|f(\varphi)| \le C_2|h|$.

From (9.4.13), (9.4.18), (9.4.36), (9.4.37), (9.4.39), it follows that for $h_3/2 \le \delta_m \le 5|h|$ the estimate (9.4.35) is preserved. Thus, the following result holds regarding the error estimate.

Theorem 9.4.1. *Suppose that Condition 9.4.1 holds for the domain D and the function $g(x_0, y_0)$. Then at any point $M(m_1h_1, m_2h_2) \in D$, $D(m, h) \subset D$, the estimate (9.4.35) holds for the difference between the quadrature sum $S(m_1h_1, m_2h_2)$ and the exact value of the integral (9.4.7).*

Consider quadrature sums for $g(x_0, y_0)$ of another class.

CONDITION 9.4.2. Suppose that for the domain D and its boundary Γ the assumptions are the same as in Condition 9.4.1 and in the normal neighborhood U (see Condition 9.4.1), the function $g(x_0, y_0)$ has the form

$$g(x_0, y_0) = \rho^\beta[\ln \rho(x_0, y_0)]^n l(x_0, y_0) = \rho_1(x_0, y_0)l(x_0, y_0), \tag{9.4.40}$$

where $\beta > 1$, $n \ge 1$, $n \in \mathbb{Z}$ is an integer, and the second derivatives of the function $l(x_0, y_0)$ are bounded in D.

In the neighborhood U_1 (see Condition 9.4.1), the second derivatives of $g(x_0, y_0)$ are bounded.

The quadrature formula for the integral (9.4.7) is constructed in accordance with (9.4.8), with $g(x_0, y_0)$ satisfying Condition 9.4.2. Let us estimate the error $\Delta(m_1h_1, m_2h_2)$ of the quadrature sum defined by (9.4.9). We proceed in the same way as we have done for $g(x_0, y_0)$ satisfying Condition 9.4.1.

The derivatives of $\rho_1(x_0, y_0) = \rho^\beta(x_0, y_0)[\ln \rho(x_0, y_0)]^n$ have the form

$$(\rho_1(x_0, y_0))'_{x_0} = \left[\beta\rho^{\beta-1}(x_0, y_0)\ln^n \rho(x_0, y_0) + n\rho^{\beta-1}(x_0, y_0)\ln^{n-1} \rho(x_0, y_0)\right]\rho'_{x_0}(x_0, y_0) =$$

$$= g_1(x_0, y_0)\rho'_{x_0}(x_0, y_0),$$

$$(\rho_1(x_0, y_0))'_{y_0} = g_1(x_0, y_0)\rho'_{y_0}(x_0, y_0),$$

$$(\rho_1(x_0, y_0)))''_{x_0x_0} = \left(\rho'_{x_0}(x_0, y_0)\right)^2\left[\beta(\beta-1)\rho^{\beta-2}(x_0, y_0)\ln^n \rho(x_0, y_0)+\right.$$

$$+ \beta n\rho^{\beta-2}\ln^{n-1} \rho(x_0, y_0) + n(\beta-1)\rho^{\beta-2}(x_0, y_0)\ln^{n-1} \rho(x_0, y_0)+$$

$$\left. + n(n-1)\rho^{\beta-2}(x_0, y_0)\ln^{n-2} \rho(x_0, y_0)\right] + g_1(x_0, y_0)\rho''_{x_0x_0}(x_0, y_0) =$$

$$= g_2(x_0, y_0)\left(\rho'_{x_0}(x_0, y_0)\right)^2 + g_1(x_0, y_0)\rho''_{x_0x_0}(x_0, y_0),$$

$$(\rho_1(x_0, y_0))''_{y_0y_0} = g_2(x_0, y_0)\left(\rho'_{y_0}(x_0, y_0)\right)^2 + g_1(x_0, y_0)\rho''_{y_0y_0}(x_0, y_0),$$

$$(\rho_1(x_0, y_0))''_{x_0y_0} = g_2(x_0, y_0)\rho'_{x_0}(x_0, y_0)\rho'_{y_0} + g_1(x_0, y_0)\rho''_{x_0y_0}(x_0, y_0). \tag{9.4.41}$$

Hence, taking into account Condition 9.4.2, we obtain the estimates

$$|g(x_0, y_0)| \le C\rho(x_0, y_0), \tag{9.4.42}$$

$$\left|g'_{x_0}(x_0, y_0)\right|, \ \left|g'_{y_0}(x_0, y_0)\right| \le C,$$
$$\left|g''_{x_0x_0}(x_0, y_0)\right|, \ \left|g''_{y_0y_0}(x_0, y_0)\right|, \ \left|g''_{x_0y_0}(x_0, y_0)\right| \le C(\rho(x_0, y_0))^{-1}. \tag{9.4.43}$$

From (9.4.10), (9.4.42), (9.4.43), it follows that

$$\Delta_3(m_1h_1, m_2h_2) \le C|h|. \tag{9.4.44}$$

From (9.4.14), (9.4.42), (9.4.43), we obtain

$$\lambda_3 \le C|h|, \qquad \lambda_2 \le C|h| \, |\ln|h||, \qquad \lambda_1 \le C|h| \, |\ln|h||. \tag{9.4.45}$$

From (9.4.14) and (9.4.45), we find that

$$\Delta_2(m_1 h_1, m_2 h_2) \le C_1 |h| \, |\ln|h||. \tag{9.4.46}$$

In order to estimate $\Delta_1(m_1 h_1, m_2 h_2)$, consider the case of $M_0(m_1 h_1, m_2 h_2)$ lying at a distance $\delta_m \ge 5|h|$ from the boundary Γ of the domain D. From (9.4.19), (9.4.20), (9.4.42), we obtain the inequality

$$\psi_1(m_1 h_1, m_2 h_2) \le C \frac{|h|}{\delta_m}. \tag{9.4.47}$$

From (9.4.21), (9.4.22), (9.4.43), we have

$$\psi_{21}(m_1 h_1, m_2 h_2) \le C \frac{|h|}{\delta_m}. \tag{9.4.48}$$

Formulas (9.4.25) and (9.4.43) yield estimates for A_{km}, B_{km} with $k \ne m$,

$$|A_{km}| \le C \frac{r_{km}}{\delta_m}, \qquad |B_{km}| \le C \frac{r_{km}}{\delta_m}, \tag{9.4.49}$$

and for $k = m$, we obtain the inequalities

$$|A_{mm}| \le C \frac{r_{0m}}{\delta_m}, \qquad |B_{mm}| \le C \frac{r_{0m}}{\delta_m}, \tag{9.4.50}$$

From (9.4.32), (9.4.33), (9.4.50), we get

$$\psi_{222}(m_1 h_1, m_2 h_2) \le C \frac{|h|}{\delta_m}. \tag{9.4.51}$$

Using (9.4.34) and (9.4.49), we obtain the estimate

$$\psi_{122}(m_1 h_1, m_2 h_2) \le C \frac{|h| \, |\ln|h||}{\delta_m}. \tag{9.4.52}$$

In the same way as we have proved (9.4.35), from (9.4.44), (9.4.46)–(9.4.48), (9.4.51), (9.4.52), we obtain the following inequality for $\delta_m \ge 5|h|$:

$$\Delta(m_1 h_1, m_2 h_2) \le C_1 \frac{|h| \, |\ln|h||}{\delta_m}. \tag{9.4.53}$$

Now, consider the case $h_3/2 \le \delta_m \le 5|h|$, $h_3 = \min\{h_1, h_2\}$.

For the terms $\Delta_i(m_1 h_1, m_2 h_2)$, $i = 2, 3$, the estimates (9.4.44) and (9.4.46) hold. Thus, it remains to estimate $\Delta_1(m_1 h_1, m_2 h_2)$. Keeping the notation of (9.4.19), we note that in this case the inequality (9.4.47) is valid. The same estimate holds for $\alpha_1(m_1 h_1, m_2 h_2)$,

$$\alpha_1(m_1 h_1, m_2 h_2) \le C \frac{|h|}{\delta_m}. \tag{9.4.54}$$

Since

$$0 < C_1 < \frac{h_1}{h_2} < C_2, \tag{9.4.55}$$

there exists θ independent of $h = (h_1, h2)$, $0 < \theta < 1$, such that the ball ρ_h of radius $R = \theta|h|$ with center at $M_0(m_1h_1, m_2h_2)$ belongs to the cell $D(m, h)$. Therefore, $\alpha_2(m_1h_1, m_2h_2)$ can be estimated as follows:

$$\alpha_2(m_1h_1, m_2h_2) \leq$$

$$\leq \frac{1}{4\pi} \left| \iint_{\rho_h} \frac{g(x_0, y_0) - g(m_1h_1, m_2h_2) - g^*(x_0, y_0)}{r_{0m}^3} K_1(m_1h_1, m_2h_2, \varphi) \, dx_0 \, dy_0 \right| +$$

$$+ \frac{1}{4\pi} \left| \iint_{D(m,h) \backslash \rho_h} \frac{g(x_0, y_0) - g(m_1h_1, m_2h_2)}{r_{0m}^3} K_1(m_1h_1, m_2h_2, \varphi) \, dx_0 \, dy_0 \right| =$$

$$= \alpha_{12}(m_1h_1, m_2h_2) + \alpha_{22}(m_1h_1, m_2h_2). \tag{9.4.56}$$

Similarly to (9.5.54), we obtain

$$\alpha_{22}(m_1h_1, m_2h_2) \leq C \frac{|h|}{\delta_m}. \tag{9.4.57}$$

With the help of (9.4.43), we can show that

$$\alpha_{12}(m_1h_1, m_2h_2) \leq C \frac{|h|}{\delta_m}. \tag{9.4.58}$$

From (9.4.54), (9.4.56)–(9.4.58), we obtain the estimate

$$\Delta_1(m_1h_1, m_2h_2) \leq C_1 \frac{|h|}{\delta_m}.$$

Hence, it follows that for $h_3/2 \leq \delta_m \leq 5|h|$, the estimate (9.4.53) for the quadrature error is preserved. Thus, the following result about quadrature error holds.

Theorem 9.4.2. *For the domain D and the function $g(x_0, y_0)$ satisfying Condition 9.4.2, the difference between the quadrature sum $S(m_1h_1, m_2h_2)$ and the exact value of the integral (9.4.7) at any point $M(m_1h_1, m_2h_2)$, $D(m, h) \subset D$, satisfies the inequality (9.4.53).*

Remark 9.4.1. Consider the integral (see (4.1.4))

$$\Pi(\nu) = \frac{1}{4\pi} \int_\sigma \frac{|\overline{M_0M}|^2 (n_{M_0}, n_M) - 3(\overline{M_0M}, n_M)(\overline{M_0M}, n_{M_0})}{|\overline{M_0M}|^5} \, d\sigma_{M_0}, \tag{9.4.59}$$

Suppose that the surface σ and the function $\nu(M_0)$ satisfy the conditions assumed in the integral (9.4.1). Let us construct quadrature formulas for the integral (9.4.59), just as we have done for the integral (9.4.1), (9.4.3), (9.4.7). Then, taking into account the inequality

$$|C(M_0M)| \leq \frac{K_0}{|\overline{M_0M}|}, \qquad K_0 = \text{const},$$

for the function

$$C(M_0M) = \frac{[(n_{M_0}, n_M) - 1]|\overline{M_0M}|^2 - 3(\overline{M_0M}, n_M)(\overline{M_0M}, n_{M_0})}{|\overline{M_0M}|^5},$$

we obtain the estimate (9.4.35) (resp., (9.4.53)) for the quadrature errors, provided that the function $\nu(x_0, y_0, f(x_0, y_0)) = g(x_0, y_0)$ satisfies Condition 9.4.1 (resp., Condition 9.4.2).

Chapter 10

Numerical Analysis of Hypersingular Integral Equations

10.1. Convergence in Quotient Spaces for Equations on a Smooth Surface with Border

In the study of circulation-free flow about a surface σ in aerodynamics, one obtains the following equation for the dipole density $\nu(M)$ on the surface σ:

$$\int_\sigma \frac{\nu(M)}{|\overline{MM_0}|^3} \left[\frac{|\overline{MM_0}|^2 (n_{M_0}, n_M)) - 3(\overline{MM_0}, n_M)(\overline{MM_0}, n_{M_0})}{|\overline{MM_0}|^2} \right] d\sigma_M = g(M_0), \quad (10.1.1)$$

where $M, M_0 \in \sigma$, n_{M_0} and n_M are the unit normals to σ at M and M_0, respectively, and $g(M_0)$ is a given function.

We assume that there exist surfaces σ_1 and σ_2 such that $\sigma \subset \sigma_1 \subset \sigma_2$, and there is a Cartesian coordinate system $OXYZ$ in which the surface σ_2 is given by the equation $z = f(x, y)$. Let Ω, Φ, and Λ be the projections of σ, σ_1, and σ_2 to the plane XOY. We assume that $f(x, y) \in C_4(\overline{\Lambda})$ and the boundaries $\Gamma\Phi$ and $\Gamma\Lambda$ of the domains Φ and Λ are separated by a positive distance from Ω and Φ, respectively. The domain Λ is such that every function $f(x, y) \in C_4(\overline{\Lambda})$ admits an extension $lf \in C_4(\mathbb{R}_2)$. The boundary $\Gamma\Lambda = L$ of σ_2 is assumed convex (see *Dvorak* (1986)) in the sense that the orientation of L is chosen such that L is traversed counterclockwise, when viewed from the end-point of the vector n_M; the vectors n_{M_0}, $\overline{MM_0}$, τ_M form a right-hand triple for any $M_0 \in \sigma_2$, $M \in L$, where τ_M is the tangential vector to L whose direction coincides with the orientation of L. Thus, we write

$$(n_{M_0}, \overline{MM_0}, \tau_M) > 0. \tag{10.1.2}$$

We choose a continuous field of normals on the surface σ. From (10.1.2), it follows that for any $M_0 \in \sigma_2$, resp., $M \in \sigma_2$, we have

$$\int_{\sigma_2} G(M, M_0)\, d\sigma_M < 0, \qquad \text{resp.,} \qquad \int_{\sigma_2} G(M, M_0)\, d\sigma_{M_0} < 0. \tag{10.1.3}$$

where $G(M, M_0) = G(M, M_0)$ is the kernel of the integral in (10.1.1).

Assume that for any $M, M_0 \in \sigma_2$, the following inequality holds:

$$0 < C_1 < G_2(M, M_0) = \frac{|\overline{MM_0}|^2 (n_{M_0}, n_M) - 3(\overline{MM_0}, n_M)(\overline{MM_0}, n_{M_0})}{|\overline{MM_0}|^2} < C_2. \tag{10.1.4}$$

As mentioned above, for any function $z = f(x, y) \in C_4(\overline{\Lambda})$, there is an extension $lf \in C_4(\mathbb{R}_2)$. Without loss of generality, we assume that lf has a compact support, and we keep the notation $f(x, y)$ for lf.

Consider a partition of the plane XOY into cells $D(k, h)$ with step $h = (h, h)$ (the step $h > 0$ is the same along each axis); see Definition 5.1.1. Let

$$a(k, m) = \iint_{D(k,h)} \frac{dx\, dy}{\left[(x - hm_1)^2 + (y - hm_2)^2 + (f(x, y) - f(hm_1, hm_2))^2\right]^{3/2}}. \qquad (10.1.5)$$

Proposition 10.1.1. *Under the above assumptions, the following inequality holds:*

$$\Delta a(k, m) = |a(k, m) - a(m, k)| \leq Ch\gamma^*(k, m), \qquad (10.1.6)$$

where C is a constant independent of k, h, m;

$$\gamma^*(k, m) = \iint_{D(k,h)} \frac{dx\, dy}{\left[(x - hm_1)^2 + (y - hm_2)^2\right]^{3/2}}, \qquad k \neq m.$$

Proof. Since $k \neq m$ by assumption, the integral in (10.1.5) has no singularities. Changing the variables in (10.1.5) by letting $x - hm_1 = hk_1 - z_0$, $y - hm_2 = hk_2 - t_0$, we obtain

$$a(k, m) = \iint_{D(m,h)} \left[(x - k_1 h)^2 + (y - k_2 h)^2 + \right.$$

$$\left. + \left(f(h(k_1 + m_1) - x,\ h(k_2 + m_2) - y) - f(hm_1, hm_2)\right)^2\right]^{-3/2} dx\, dy.$$

Using the fact that the function $g(z) = (1 + z)^{-3/2}$ has a bounded derivative for $z \geq 0$, and taking

$$z = \left(\frac{f(x, y) - f(x_0, y_0)}{r(x_0, y_0)}\right)^2, \qquad r(x_0, y_0) = \left[(x - x_0)^2 + (y - y_0)^2\right]^{1/2},$$

we obtain the following relation for some $z_1 > 0$:

$$\Delta a(k, m) = \left| \iint_{D(m,k)} \frac{1}{r^3_{k_1 k_2}} g'(z_1) \Delta z\, dx\, dy \right|, \qquad (10.1.7)$$

where $r_{k_1 k_2} = \left[(x - hk_1)^2 + (y - hk)^2\right]^{1/2}$, $(x, y) \in D(m, h)$.

Let us estimate Δz. We have

$$|\Delta z| \leq \qquad\qquad\qquad\qquad\qquad\qquad\qquad\qquad\qquad\qquad\qquad (10.1.8)$$

$$\leq C \left| \frac{f(x, y) - f(hk_1, hk_2)}{r_{k_1 k_2}} - \frac{f(hm_1, hm_2) - f(h(k_1 + m_1) - x,\ h(k_2 + m_2) - y)}{r_{k_1 k_2}} \right|.$$

Here, we have used the inequality

$$\left| \frac{f(x, y) - f(hk_1, hk_2)}{r_{k_1 k_2}} + \frac{f(hm_1, hm_2) - f(h(k_1 + m_1) - x,\ h(k_2 + m_2) - y)}{r_{k_1 k_2}} \right| < C,$$

where $(x, y) \in D(m, h)$, $(k_1, k_2) \neq (m_1, m_2)$.

Representing the functions in the right-hand side of (10.1.8) by Taylor's formula at the point (hm_1, hm_2), we obtain

$$|\Delta z| \le C \left| \frac{f'_x(hm_1, hm_2) - f'_x(hk_1, hk_2)}{r_{k_1 k_2}} (x - hm_1) + \right.$$

$$\left. + \frac{f'_y(hm_1, hm_2) - f'_y(hk_1, hk_2)}{r_{k_1 k_2}} (y - hm_2) + \frac{O(h)}{r_{k_1 k_2}} \right| \le C_1 h, \qquad (10.1.9)$$

where $O(h)$ is the remainder term in Taylor's formula, $|O(h)| \le C_2 h^2$ for $(x, y) \in D(m, h)$.
Relations (10.1.7), (10.1.9) prove Proposition 10.1.1.

Consider the quantities

$$\beta_1(k, m) = \iint_{D(k,h)} J(M'_{0m}, M') q(x, y)\, dx\, dy,$$

$$\beta_2(m) = \frac{1}{h^2} \iint_{D(m,h)} q(x, y)\, dx\, dy,$$

$$\gamma(k, m) = \beta_2(m) \beta_1(k, m), \qquad (10.1.10)$$

$$q(x, y) = \frac{1}{2} \left[1 + |f'_x(x, y)|^2 + |f'_y(x, y)|^2 \right]^{1/2},$$

where $M'_{0m} = (hm_1, hm_2)$, $M' = M'(x, y)$, $J(M'_0, M') = G(M_0, M)$, $M'_0 = M'_0(x_0, y_0)$.

Proposition 10.1.2. *For $k \ne m$, the following inequality holds:*

$$|\Delta\gamma(k, m)| = |\gamma(k, m) - \gamma(m, k)| \le Ch\gamma^*(k, m), \qquad (10.1.11)$$

where C is a constant that does not depend on k, h, m.

Proof. Set

$$G_1(M, M_0) = \frac{1}{|\overline{MM_0}|^3} = J_1(M', M'_0),$$

$$G_2(M, M_0) = \frac{G(M, M_0)}{C_1(M, M_0)} = J_2(M', M'_0). \qquad (10.1.12)$$

Since $f(x, y) \in C_0^4(\mathbb{R}_2)$, the function $J_2(M', M'_0)$ of the arguments x, y, x_0, y_0 has uniformly bounded first derivatives in these variables for $(x, y) \ne (x_0, y_0)$, which can be verified directly with the help of the Taylor formula.

For $k \ne m$, the integrals in (10.1.10) involve no singularities, all functions are continuous and positive. Hence, using the mean value theorem, we obtain

$$\gamma(k, m) = q(M'_1) J_2(M'_{0m}, M'_2) q(M'_3) a(k, m),$$

$$\gamma(m, k) = q(M'_4) J_2(M'_{0k}, M'_5) q(M'_6) a(m, k) = \qquad (10.1.13)$$

$$= q(M'_4) J_2(M'_5, M'_{0k}) q(M'_6) [a(k, m) + \Delta a(k, m)],$$

where $M'_1 \in D(m, h)$; $M'_2, M'_3 \in D(k, h)$; $M'_4 \in D(k, h)$; $M'_5, M'_6 \in D(m, h)$.

The above observation regarding the function $J_2(M', M'_0) = J_2(M'_0, M')$ and condition (10.1.4), combined with (10.1.13) and Proposition 10.1.1, yield the inequality (10.1.11), as required. Proposition 10.1.2. is proved.

For $f(x, y) \in C_0^4(\mathbb{R}_2)$, the function $Q(M, M_0)$ satisfies the estimate

$$|Q(M, M_0)| = \left| \frac{1}{|\overline{M_0 M}|^3} \frac{|\overline{M_0 M}|^2 \left[(n_{M_0}, n_M) - 1 \right] - 3 \left(\overline{M_0 M}, n_{M_0} \right) \left(\overline{M_0 M}, n_M \right)}{|\overline{M_0 M}|^2} \right| \leq$$

$$\leq \frac{C}{|\overline{M_0 M}|}, \tag{10.1.14}$$

for $|\overline{M_0 M}| < \delta$, with C independent of M_0, M.

In what follows, we use the notation $Q^*(M', M_0') = Q(M, M_0)$. Consider the expression

$$I = \iint_{\mathbb{R}_2} J(M', M_0') q(x_0, y_0)\, dx_0\, dy_0 = \int_{\mathbb{R}_2} J_1(M', M_0') q(x_0, y_0)\, dx_0\, dy_0 +$$

$$+ \int_{\mathbb{R}_2} Q^*(M', M_0') q(x_0, y_0)\, dx_0\, dy_0 = I_1(M') + I_2(M'). \tag{10.1.15}$$

From (10.1.15), (10.1.14), it follows that

$$|I_2(M')| \leq C_1 \tag{10.1.16}$$

with C_1 independent of M'.

Let us represent $I_1(M')$ in the form

$$I_1(M') = \iint_{S(R,M')} J_1(M', M_0') q(x_0, y_0)\, dx_0\, dy_0 + \iint_{\mathbb{R}_2 \setminus S(R,M')} J_1(M', M_0') q(x_0, y_0)\, dx_0\, dy_0 =$$

$$= I_{11}(M') + I_{12}(M'), \tag{10.1.17}$$

where $S(R, M')$ is the circle of radius R with center at M'.

The following estimate holds:

$$|I_{12}(M')| \leq \frac{C_2}{R}. \tag{10.1.18}$$

For the integral $I_{11}(M')$, understood in the sense of Hadamard, the following representation is proved by *Lifanov and Poltavskii* (1998):

$$I_{11}(M') = \lim_{\varepsilon \to 0} \left[\iint_{S(R,M') \setminus S(\varepsilon,M')} \frac{k_1(x, y, \varphi)}{r^2}\, dr\, d\varphi - \frac{M_1(x, y)}{\varepsilon} \right] +$$

$$+ \lim_{\varepsilon \to 0} \iint_{S(R,M') \setminus S(\varepsilon,M')} \frac{k_2(x, y, \varphi)}{r^2}\, dr\, d\varphi + \iint_{S(R,M')} k_3(x, y, r, \varphi)\, dr\, d\varphi =$$

$$= I_{21}(M') + I_{31}(M') + I_{41}(M'). \tag{10.1.19}$$

where

$$|k_i(x, y, r, \varphi)| < C, \quad i = 1, 2, 3; \qquad M_1(x, y) = q(x, y) \int_0^{2\pi} \frac{d\psi}{L^{3/2}(x, y, \psi)};$$

$$L(x, y, \psi) = 1 + \cos^2 \psi A^2(x, y); \qquad A^2(x, y) = (f_x^1)^2(x, y) + (f_y^1)^2(x, y);$$

$$\int_0^{2\pi} k_2(x, y, \varphi)\, d\varphi = 0.$$

Consequently, $I_{31}(M') = 0$, and we also have $I_{21}(M') = -R^{-1} M_1(x, y)$, as shown by *Lifanov and Poltavskii* (1998). The above observations yield the inequality

$$|I_{11}(M')| \leq C \left(R + \frac{1}{R} \right). \tag{10.1.20}$$

Since R is arbitrary, it follows from (10.1.16), (10.1.18), (10.1.20) that

$$|I(M')| \leq C_0 \tag{10.1.21}$$

with C_0 independent of M'.

Remark 10.1.1. From (10.1.21), it follows that for any bounded domain $D \subset \mathbb{R}_2$, there is a bounded domain $D_1 \supset D$ such that

$$\left| \iint_{D_1} J(M'mM_0')q(x_0, y_0)\, dx_0\, dy_0 \right| \leq K \tag{10.1.22}$$

for any $M' \in D$, where K is a constant independent of M', D, D_1.

Let us introduce a family of discrete operators T_h defined by

$$\pi \mathrm{T}_h u(x, h) = \sum_k u(k)\beta_1(k, m), \tag{10.1.23}$$

where $\beta_1(k, m)$ is defined by (10.1.12); the operator π is that of Definition 5.1.2; $u(k) = \pi u(x, h)$.

Proposition 10.1.3. *For $f(x, y) \in C_0^4(\mathbb{R}_2)$, the family of operators $\mathrm{T}_h : M(r, h) \to M(r-1, h)$, $0 < r < 1$, $|h| \leq \delta$, is bounded.*

Proof. In view of Remark 5.4.1, it suffices to prove this statement for $u(x, h) \in M_h^\infty$, $u(x, h) \in M(r, h)$, $0 < r < 1$. Consider the linear functional $I = (4\pi)^{-1}l(\nu(x, h)))$ on $M(1-r, h)$ associated with the element $\mathrm{T}_h u(x, h)$. In order to prove Proposition 10.1.3, it suffices to show that

$$|I(\nu(x, h)| \leq C\|\nu(x, h)\|_{M(1-r,h)}\|u(x, h)\|_{M(r,h)}, \qquad \forall\, \nu(x, h) \in M(1 - r, h). \tag{10.1.24}$$

In view of Remark 5.4.1, it suffices to prove (10.1.24) for $\nu(x, h) \in M_h^\infty$, $\nu(x, h) \in M(1-r, h)$. Thus, let $u(x, h), \nu(x, h) \in M_h^\infty$. Then, according to Remark 10.1.1, there is a bounded domain D_1 such that $D_1 \supset \operatorname{supp} u(x, h)$, $D_1 \supset \operatorname{supp} \nu(x, h)$, and for any $D(m, h)$ such that either $D(m, h) \subset \operatorname{supp} u(x, h)$ or $D(m, h) \subset \operatorname{supp} \nu(x, h)$, we have

$$\left| \sum_{k \in \Lambda(h)} \beta_1(k, m) \right| \leq C_0, \tag{10.1.25}$$

where $\Lambda(h)$ is the set of all k for which $D(k, h) \subset D_1$; C_0 is a constant independent of m, D_1.

Next, from (5.5.5) we get

$$I(\nu(x, h)) = h^2 \sum_{m \in \Lambda(h)} \nu(m) \sum_{k \in \Lambda(h)} \overline{u}(k)\beta_1(k, m) =$$

$$= h^2 \sum_{m \in \Lambda(h)} \beta_2(m)\nu_1(m) \sum_{k \in \Lambda(h)} \overline{u}(k)\beta_1(k, m) =$$

$$= h^2 \sum_{m \in \Lambda(h)} \beta_2(m)\nu_1(m) \left[\sum_{k \in \Lambda(h),\, k \neq m} \overline{u}(k)\beta_1(k, m) - \beta_1(m, m)\overline{u}(m) \right] =$$

$$= h^2 \sum_{m \in \Lambda(h)} \beta_2(m)\nu_1(m) \left[\sum_{k \in \Lambda(h)} \beta_1(k, m)(\overline{u}(k) - \overline{u}(m)) \right] +$$

$$+ h^2 \sum_{m \in \Lambda(h)} \beta_2(m)\beta(m)\nu_1(m)\overline{u}(m) =$$

$$= I_1(\nu(x, h)) + I_2(\nu(x, h)), \tag{10.1.26}$$

where

$$\nu_1(m) = \frac{\nu(m)}{\beta(m)}, \qquad \nu(m) = \pi\nu(x, h), \qquad \beta(m) = \sum_{k \in \Lambda(h)} \beta_1(k, m).$$

It is easy to check that the step-functions $\psi(x, h)$, $\pi\psi(x, h) = 1/\beta_2(m)$ form a Lipschitz family (see Definition 7.3.1). Proposition 7.3.1 implies that $\nu_1(x, h) \in M(1-r, h)$, $\pi\nu_1(x, h) = \nu_1(m)$. The above reasoning and (10.1.25) show that

$$|I_2(\nu(x, h))| \leq C_0 \|\nu(x, h)\|_{L_2(\mathbb{R}_2)} \|u(x, h)\|_{L_2(\mathbb{R}_2)} \leq C_1 \|\nu(x, h)\|_{M(1-r,h)} \|u(x, h)\|_{M(r,h)}. \quad (10.1.27)$$

Interchanging the indices k and m in $I_1(\nu(x, h))$, we obtain

$$I_1 = h^2 \sum_{k \in \Lambda(h)} \beta_2(k)\nu_1(k) \left[\sum_{m \in \Lambda(h)} \beta_1(m, k)[\overline{u}(m) - \overline{u}(k)] \right] =$$

$$= h^2 \sum_{k \in \Lambda(h)} \sum_{m \in \Lambda(h)} \beta_2(m)\beta_1(k, m)\nu_1(k)[\overline{u}(m) - \overline{u}(k)] +$$

$$+ h^2 \sum_{k \in \Lambda(h)} \sum_{m \in \Lambda(h)} [\beta_2(k)\beta_1(m, k) - \beta_2(m)\beta_1(k, m)]\nu_1(k)[\overline{u}(m) - \overline{u}(k)].$$

Comparing this relation with (10.1.26), we find that

$$I_1(\nu(x, h)) = \frac{h^2}{2} \sum_{k \in \Lambda(h)} \sum_{m \in \Lambda(h)} \beta_2(m)\beta_1(k, m)[\nu_1(k) - \nu_1(m)][\overline{u}(m) - \overline{u}(k)] +$$

$$+ \frac{h^2}{2} \sum_{k \in \Lambda(h)} \sum_{m \in \Lambda(h)} \Delta\gamma(k, m)\nu_1(k)[\overline{u}(m) - \overline{u}(k)] =$$

$$= A_1(\nu(x, h)) + A_2(\nu(x, h)). \quad (10.1.28)$$

Consider the quantities $a_r(k, m) = 1/(|k - m|^{2+2r}|h|^{2r})$, with $k \neq m$, $0 < r < 1$. We have $a_{1/2}(k, m) = [a_r(k, m)a_{1-r}(k, m)]^{1/2}$. It follows from Proposition 5.3.4 that

$$h^2 \sum_{\substack{m \in \Lambda(h) \\ k \neq m}} \sum_{k \in \Lambda(h)} a_r(k, m)|u(m) - u(k)|^2 \leq C \|u(x, h)\|_{M_1(r,h)}^2 \leq C_1 \|u(x, h)\|_{M(r,h)}^2. \quad (10.1.29)$$

Using the definition of $\beta_2(m)$ and $\beta_1(k, m)$ (see (10.1.10)), together with Proposition 10.1.2, we obtain

$$|\beta_2(m)\beta_1(k, m)| \leq C_1\alpha_{1/2}(k, m), \qquad |\Delta\gamma(k, m)| \leq C_1 h[\alpha_r(k, m)\alpha_{1-r}(k, m)]^{1/2}, \quad (10.1.30)$$

for $k \neq m$, where C_1 does not depend on h, k, m.

From (10.1.28)–(10.1.30), we obtain the inequalities

$$|A_1(\nu(x, h))| \leq C_2 \left[h^2 \sum_{m \in \Lambda(h),\, k \neq m} \sum_{k \in \Lambda(h)} \alpha_{1-r}(k, m)|\nu_1(k) - \nu_1(m)|^2 \right]^{1/2} \times$$

$$\times \left[h^2 \sum_{m \in \Lambda(h),\, k \neq m} \sum_{k \in \Lambda(h)} \alpha_r(k, m)|u(m) - u(k)|^2 \right]^{1/2} \leq$$

$$\leq C_3 \|\nu(x, h)\|_{M(1-r,h)} \|u(x, h)\|_{M(r,h)}; \quad (10.1.31)$$

$$
|A_2(\nu(x,h))| \le C_2 \left[h^2 \sum_{m \in \Lambda(h),\, k \ne m} \sum_{k \in \Lambda(h)} h^2 \alpha_{1-r}(k,m) |\nu_1(k)|^2 \right]^{1/2} \times
$$

$$
\times \left[h^2 \sum_{m \in \Lambda(h),\, k \ne m} \sum_{k \in \Lambda(h)} a_r(k,m) |u(m) - u(k)|^2 \right]^{1/2} \le \quad (10.1.32)
$$

$$
\le C_3 \| u(x,h) \|_{M(r,h)} \left[h^2 \sum_{k \in \Lambda(h)} h^{2r} |\nu(k)|^2 \sum_{m \in \Lambda(h),\, k \ne m} h^{2-2r} \alpha_{1-r}(k,m) \right].
$$

From (10.1.32), taking into account the inequality $\sum_{k \ne m} h^{2-2r} \alpha_{1-r}(k,m) \le C_0$ with C_0 independent of h and k, we get

$$
|A_2(\nu(x,h))| \le C_4 \| u(x,h) \|_{M(r,h)} h^r \| \nu(x,h) \|_{L_2(\mathbb{R}_2)}. \quad (10.1.33)
$$

Combining (10.1.27), (10.1.31), and (10.1.33), we obtain the estimate (7.2.4), which proves Proposition 10.1.3.

Proposition 10.1.4. *The bounded family of operators* T_h *acting from* $\mathring{M}(1/2, h, \Omega_{1h})$ *to* $M(-1/2, h, \Omega_{1h})$ *is stable.*

Proof. Let $A(h)$, resp., $A_1(h)$, be the set of all indices $k = (k_1, k_2)$ such that $D(k,h) \in \Omega$, resp., $G(k,h) \in \Phi$. Because of the above assumptions about σ, we have

$$
\beta_1(m,m) < 0, \quad \forall\, m \in A_1(h); \qquad \sum_{k \in A(h)} \beta_1(k,m) < 0,
$$

$$
\sum_{k \in A_1(h)} \beta_1(k,m) = \beta(m) < 0, \quad \forall\, m \in A(h); \qquad 0 < C_1 \le |\beta(m)| \le C_2, \quad (10.1.34)
$$

where C_1 and C_2 are independent of h and m, $m \in A(h)$.

Consider the linear functional $I = (4\pi^2)^{-1} l(\nu(x,h))$ on $\mathring{M}(1/2, h, \Omega_{1h})$ associated with the element $T_h u(x,h)$, $u(x,h) \in \mathring{M}(1/2, h, \Omega_{1h})$. Let us find its values on the family of functions $\nu(x,h) = q(x,h)u(x,h)$, where $\pi q(x,h) = \beta_2(k)$. The functions $q(x,h)$ form a Lipschitz family, and therefore, $\nu(x,h) \in \mathring{M}(1/2, h, \Omega_{1h})$ by Proposition 7.3.1.

Similarly to (10.1.26) and (10.1.28), we obtain

$$
I(\nu(x,h)) = h^2 \sum_{m \in A(h)} \beta_2(m) u(m) \sum_{k \in A(h)} \overline{u}(k) \beta_1(k,m) =
$$

$$
= h^2 \sum_{m \in A_1(h)} \beta_2(m) u(m) \sum_{k \in A_1(h)} \overline{u}(k) \beta_1(k,m) =
$$

$$
= -\frac{h^2}{2} \sum_{k,m \in A_1(h)} \beta_2(m) \beta_1(k,m) |u(m) - u(k)|^2 + \quad (10.1.35)
$$

$$
+ \frac{h^2}{2} \sum_{k,m \in A_1(h)} \Delta\gamma(k,m) u(k) (\overline{u}(m) - \overline{u}(k)) - h^2 \sum_{m \in A(h)} \beta_2(m) |\beta(m)| |u(m)|^2.
$$

From (10.1.4) and (10.1.35), we get

$$
I(\nu(x,h)) \ge C_1 \left[\frac{h^2}{2} \sum_{k,m \in A_1(h)} \gamma^*(k,m) |u(m) - u(k)|^2 + h^2 \sum_{m \in A(h)} \beta_2(m) |\beta(m)| |u(m)|^2 \right] -
$$

$$
- \left| \frac{h^2}{2} \sum_{k,m \in A_1(h)} \Delta\gamma(k,m) u(k) [\overline{u}(m) - \overline{u}(k)] \right| = I_1(\nu(x,h)) - I_2(\nu(x,h)). \quad (10.1.36)
$$

Let $T_{h,1/2}$ be a family of vortex operators defined by the relation

$$\pi T_{h,1/2} u(x, h) = \sum_{k \in A(h)} u(k) \gamma^*(k, m),$$

for $u(x, h) \in \overset{\circ}{M}(1/2, h, \Omega_{1h})$. Let $l_1 = (4\pi^2)^{-1} l(\nu(x, h))$ be the linear functional on $\overset{\circ}{M}(1/2, h, \Omega_{1h})$ associated with the element $T_{h,1/2} u(x, h)$, $\nu(x, h) \in \overset{\circ}{M}(1/2, h, \Omega_{1h})$. Consider its values on the family of functions $u(x, h)$. By analogy with (10.1.35) we obtain

$$l_1(u(x, h)) = -\frac{h^2}{2} \sum_{k, m \in A_1(h)} \gamma^*(k, m) |u(m) - u(k)|^2 - h^2 \sum_{m \in A(h)} |\beta^*(m)| |u(m)|^2, \quad (10.1.37)$$

where $\Delta \gamma^*(k, m) = 0$ for all k, m; $\beta^*(m) = \sum_{k \in A_1(h)} \gamma^*(k, m) < 0$; and the inequality $0 < C_1 < |\beta^*(m)| < C_2$ holds for all $m \in A(h)$ with constants C_1 and C_2 independent of h, m.

Proposition 7.2.1 implies the inequality

$$|l_1(u(x, h))| \geq K \|u(x, h)\|_{\overset{\circ}{M}(1/2, h, \Omega_{1h})}, \quad (10.1.38)$$

with K independent of h.

From (10.1.34), (10.1.36) and (10.1.38), it follows that

$$|I_1(\nu(x, h))| \geq K_1 \|u(x, h)\|^2_{\overset{\circ}{M}(1/2, h, \Omega_{1h})}, \quad (10.1.39)$$

with K_1 independent of h.

From (10.1.33) we get

$$|I_2(\nu(x, h))| \leq K_1 h^{1/2} \|u(x, h)\|^2_{\overset{\circ}{M}(1/2, h, \Omega_{1h})}, \quad (10.1.40)$$

with K_2 independent of h.

From (10.1.40), it follows that there is $\delta > 0$ such that the inequality

$$|I(\nu(x, h))| \geq K_3 \|u(x, h)\|^2_{\overset{\circ}{M}(1/2, h, \Omega_{1h})}$$

holds for $|h| < \delta$ with a constant K_3 independent of h. This inequality implies the stability of the operator family T_h. Proposition 10.1.4 is proved

Proposition 10.1.5. *For $f(x, y) \in C_0^4(\mathbb{R}_2)$, the operator*

$$T(u) = \iint_{\mathbb{R}_2} J(M', M_0') q(x_0, y_0) u(x_0, y_0) \, dx_0 \, dy_0$$

from $H_{1/2}(\mathbb{R}_2)$ to $H_{-1/2}(\mathbb{R}_2)$ is bounded.

Proof. Let $I = (4\pi^2)^{-1} l(\nu(x, h))$ be the linear functional on $H_{1/2}(\mathbb{R}_2)$ associated with the element $T(u)$. Then, our statement will be proved if we show that

$$|I(\nu)| \leq C \|u(x, y)\|_{H_{1/2}(\mathbb{R}_2)} \|\nu(x, y)\|_{H_{1/2}(\mathbb{R}_2)} \quad (10.1.41)$$

for all $u(x, y), \nu(x, y) \in H_{1/2}(\mathbb{R}_2)$.

Since $C_0^\infty(\mathbb{R}_2)$ is dense in $H_{1/2}(\mathbb{R}_2)$, it suffices to show that the inequality (10.1.41) holds for $u(x, y), \nu(x, y) \in C_0^\infty(\mathbb{R}_2)$. Moreover, since the operator Q of the form $Q\nu = q(x, y)\nu(x, y)$ is a one-to-one mapping from $H_{1/2}(\mathbb{R}_2)$ to $H_{1/2}(\mathbb{R}_2)$, it suffices to prove (10.1.41) for $I(Q\nu)$, where $q(x, y) = \left[1 + f_x'^2 + f_y'^2\right]^{1/2}$.

Remark 10.1.1 shows that there is a domain $\Omega_1 \supset \Omega$ such that

$$\left| \iint_{\Omega_1} J(M', M_0')q(x_0, y_0)\, dx_0\, dy_0 \right| < K, \tag{10.1.42}$$

for any $M' \in \Omega$, where K is a constant independent of M', Ω_1, Ω; and $\Omega \supset \operatorname{supp} u(x, y)$, $\Omega \supset \operatorname{supp} \nu(x, y)$. We have

$$I(Q\nu) = \iint_{\Omega} q(x,y)\nu(x,y) \left[\iint_{\Omega} J(M', M_0')\overline{u(x_0, y_0)}q(x_0, y_0)\, dx_0\, dy_0 \right] dx\, dy =$$

$$= \iint_{\Omega_1} q(x,y)\nu(x,y) \left[\iint_{\Omega_1} J(M', M_0')\overline{u(x_0, y_0)}q(x_0, y_0)\, dx_0\, dy_0 \right] dx\, dy =$$

$$= \iint_{\Omega_1} q(x,y)\nu(x,y) \left[\iint_{\Omega_1} J(M', M_0')\overline{u(x_0, y_0)}q(x_0, y_0)\, dx_0\, dy_0 - \right.$$

$$\left. - \overline{u(x, y)} \iint_{\Omega_1} J(M', M_0)q(x_0, y_0)\, dx_0\, dy_0 + \overline{u(x, y)}\beta(x, y) \right] dx\, dy =$$

$$= \iint_{\Omega_1} q(x,y)\nu(x,y) \left[\iint_{\Omega_1} J(M', M_0) \left[\overline{u(x_0, y_0)} - \overline{u(x, y)} \right] q(x_0, y_0)\, dx_0\, dy_0 \right] dx\, dy +$$

$$+ \iint_{\Omega} q(x,y)\beta(x,y)\nu(x,y)\overline{u(x,y)}\, dx\, dy, \tag{10.1.43}$$

where

$$\beta(x, y) = \iint_{\Omega_1} J(M', M_0')q(x_0, y_0)\, dx_0\, dy_0$$

satisfies the inequality $|\beta(x, y)| < K$ for $(x, y) \in \Omega$, because of (10.1.42).

Interchanging the variables x and x_0, as well as y and y_0 in (10.1.43), and taking into account that $J(M', M_0')$ is symmetric, we find that

$$I(Q\nu) = \iint_{\Omega_1} \left[\iint_{\Omega_1} J(M', M_0')q(x, y)q(x_0, y_0)\nu(x_0, y_0) \left[\overline{u(x, y)} - \overline{u(x_0, y_0)} \right] dx\, dy \right] dx_0\, dy_0 +$$

$$+ \iint_{\Omega} q(x,y)\beta(x,y)\nu(x,y)\overline{u(x,y)}\, dx\, dy. \tag{10.1.44}$$

From (10.1.43) and (10.1.44), it follows that

$$I(Q\nu) =$$

$$= -\frac{1}{2} \iint_{\Omega_1} \iint_{\Omega_1} J(M', M_0')q(x,y)q(x_0, y_0)[\nu(x, y) - \nu(x_0, y_0)] \left[\overline{u(x, y)} - \right.$$

$$\left. - \overline{u(x_0, y_0)} \right] dx_0\, dy_0\, dx\, dy + \iint_{\Omega} q(x,y)\beta(x,y)\nu(x,y)\overline{u(x,y)}\, dx\, dy =$$

$$= I_1(Q\nu) + I_2(Q\nu). \tag{10.1.45}$$

From (10.1.42), we obtain

$$|I_2(Q\nu)| \leq C\|u(x, y)\|_{L_2(\Omega)}\|\nu(x, y)\|_{L_2(\Omega)}. \tag{10.1.46}$$

The integrals in $I_1(Q\nu)$ are absolutely convergent, and therefore, using (10.1.4) and the Cauchy inequality, we get

$$|I_1(Q\nu)| \leq \left[\iint_{\Omega_1} \iint_{\Omega_1} \frac{|\nu(x,y) - \nu(x_0,y_0)|^2 \, dx \, dy \, dx_0 \, dy_0}{[(x-x_0)^2 + (y-y_0)^2]^{3/2}} \right]^{1/2} \times$$

$$\times \left[\iint_{\Omega_1} \iint_{\Omega_1} \frac{|u(x,y) - u(x_0,y_0)|^2 \, dx \, dy \, dx_0 \, dy_0}{[(x-x_0)^2 + (y-y_0)^2]^{3/2}} \right]^{1/2} \leq$$

$$\leq C_2 \|\nu(x,y)\|_{\dot{H}_{1/2}(\Omega)} \|u(x,y)\|_{\dot{H}_{1/2}(\Omega)}. \tag{10.1.47}$$

The inequality (10.1.41) follows from (10.1.46) and (10.1.47). Proposition 10.1.5 is proved.

Let us examine the properties of the function

$$\nu(x,y) = \iint_{\Omega} J(M', M_0') u(x_0, y_0) \, dx_0 \, dy_0, \qquad u(x_0, y_0) \in C_0^3(\Omega).$$

We represent $\nu(x,y)$ in the form

$$\nu(x,y) = \nu_1(x,y) + \nu_2(x,y),$$

where

$$\nu_1(x,y) = \iint_{\Omega} J_1(M', M_0') u(x_0, y_0) \, dx_0 \, dy_0,$$

$$\nu_2(x,y) = \iint_{\Omega} Q_1^*(M', M_0') u(x_0, y_0) \, dx_0 \, dy_0,$$

and $Q^*(M', M_0')$ is defined by (10.1.14).

Since $f(x,y) \in C_0^4(\mathbb{R}_2)$, the function

$$\alpha(x, y, r, \varphi) = \frac{f(x,y) - f(x + r\cos\varphi, \ y + r\sin\varphi)}{r}$$

has the derivatives $\partial\alpha/\partial r \in C^1(\mathbb{R}_2 \times \mathbb{R}_2)$ and $\partial^2\alpha/\partial r^2 \in C(\mathbb{R}_2 \times \mathbb{R}_2)$. Therefore, introducing the variables

$$r = \left[(x-x_0) + (y-y_0)^2\right]^{1/2}, \qquad x - x_0 = -r\cos\varphi, \qquad y - y_0 = -r\sin\varphi,$$

we see that the function $F(x, y, r, \varphi) = 1/(1+\alpha^2)^{3/2}$ has its derivatives $\partial F/\partial r$ and $\partial^2 F/\partial r^2$ continuous in all variables x, y, r, φ. Thus, the function $J_1(M', M_0')$ can be represented in the form

$$J_1(M', M_0') = \frac{1}{r^3} A_0(x, y, \varphi) + A_1(x, y, \varphi) \frac{1}{r^2} + A_2(x, y, r, \varphi) \frac{1}{r}, \tag{10.1.48}$$

with

$$A_0(x, y, \varphi) = F(x, y, \varphi, 0), \qquad A_1(x, y, \varphi) = \left. \frac{\partial F}{\partial r} \right|_{r=0}, \qquad A_2(x, y, r, \varphi) = \frac{1}{r} O(x, y, r, \varphi),$$

where $O(x, y, r, \varphi)$ is the remainder in the Taylor formula, and $A_2(x, y, r, \varphi)$ is continuous in x, y, r, φ.

As shown by *Lifanov and Poltavskii* (1998), we have

$$\int_0^{2\pi} A_1(x, y, \varphi)\, d\varphi = \int_0^{2\pi} A_0(x, y, \varphi)\cos\varphi\, d\varphi = \int_0^{2\pi} A_0(x, y, \varphi)\sin\varphi\, d\varphi = 0. \qquad (10.1.49)$$

The function $A_0(x, y, \varphi)$ is thrice continuously differentiable in all its arguments, and the function $A_1(x, y, \varphi)$ is twice continuously differentiable.

Let $R > 0$ be so large that any circle $S(R, M')$ of radius R with center at $M'(x, y)$ contains Ω. Then the function $\nu_1(x, y)$ can be represented in the form

$$\nu_1(x, y) =$$
$$= \lim_{\varepsilon \to 0}\left[\iint_{S(R,M')\setminus S(\varepsilon,M')} \frac{u(x + r\cos\varphi,\ y + r\sin\varphi)}{r^2} A_0(x, y, \varphi)\, dr\, d\varphi - \frac{u(x, y)B(x, y)}{\varepsilon} \right] +$$
$$+ \lim_{\varepsilon \to 0}\left[\iint_{S(R,M')\setminus S(\varepsilon,M')} \frac{u(x + r\cos\varphi,\ y + r\sin\varphi)}{r} A_1(x, y, \varphi)\, dr\, d\varphi + \right.$$
$$\left. + \iint_{S(R,M')\setminus S(\varepsilon,M')} u(x + r\cos\varphi,\ y + r\sin\varphi)A_2(x, y, r, \varphi)\, dr\, d\varphi \right] =$$
$$= \lambda_1(x, y) + \lambda_2(x, y) + \lambda_3(x, y), \qquad (10.1.50)$$

where

$$B(x, y) = \int_0^{2\pi} \frac{d\psi}{L^{3/2}(x, y)}, \qquad L(x, y, \psi) = 1 + \cos^2\psi A^2(x, y),$$
$$A^2(x, y) = f_x'^2(x, y) + f_y'^2(x, y).$$

The theorem about continuous dependence of integrals on a parameter, in view of the properties of the functions $A_2(x, y, r, \varphi)$ and $u(x_0, y_0)$, ensures that $\lambda_3(x, y)$ is continuous in x and y.

Using (10.1.50) and the fact that $u(x + R\cos\varphi,\ y + R\sin\varphi) = 0$ and integrating by parts, we get

$$\lambda_2(x, y) = \int_0^{2\pi}\int_0^R A_1(x, y, \varphi)\left[\frac{\partial u(x + r\cos\varphi,\ y + r\sin\varphi)}{\partial x}\cos\varphi + \right.$$
$$\left. + \frac{\partial u(x + r\cos\varphi,\ y + r\sin\varphi)}{\partial y}\sin\varphi\right]\ln r\, dr\, d\varphi.$$

Hence, reasoning as we have done for $\lambda_3(x, y)$, we conclude that $\lambda_2(x, y)$ is continuous in x and y.

Using (10.1.49), (10.1.5) and integrating by parts, we represent $\lambda_1(x, y)$ in the form

$$\lambda_1(x, y) = -\int_0^{2\pi}\int_0^R A_0(x, y, \varphi)\left[\frac{\partial^2 u(x + r\cos\varphi,\ y + r\sin\varphi)}{\partial x^2}\cos^2\varphi + \right.$$
$$\left. + \sin 2\varphi\frac{\partial^2 u(x + r\cos\varphi,\ y + r\sin\varphi)}{\partial x\partial y} + \sin^2\varphi\frac{\partial^2 u(x + r\cos\varphi,\ y + r\sin\varphi)}{\partial y^2}\right]\ln r\, dr\, d\varphi.$$

Hence, in the same way as for $\lambda_3(x, y)$, $\lambda_2(x, y)$, we conclude that $\lambda(x, y) \in C(\mathbb{R}_2)$. Thus, $\nu_1(x, y) \in C(\mathbb{R}_2)$.

Let us represent the function $\nu_2(x, y)$ as

$$\nu_2(x, y) = \iint_\Omega \frac{(\boldsymbol{n}_{M_0}, \boldsymbol{n}_M) - 1}{|\overline{M_0 M}|^3} u(x_0, y_0)\, dx_0\, dy_0 - \qquad (10.1.51)$$
$$- 3\iint_\Omega \frac{(\overline{M_0 M}, \boldsymbol{n}_M)\,(\overline{M_0 M}, \boldsymbol{n}_{M_0})}{|\overline{M_0 M}|^5} u(x_0, y_0)\, dx_0\, dy_0 = \beta_1(x, y) + \beta_2(x, y).$$

Introducing the variables

$$r = \left[(x - x_0)^2 + (y - y_0)^2\right], \quad x_0 - x = r\cos\varphi, \quad y - y_0 = r\sin\varphi,$$

we obtain in coordinate notation

$$\frac{1}{r^2}\left(\overline{M_0 M}, \boldsymbol{n}_{M_0}\right) = \gamma_1(x, y, r, \varphi), \qquad \frac{1}{r^2}\left(\overline{M_0 M}, \boldsymbol{n}_M\right) = \gamma_2(x, y, r, \varphi),$$

$$\frac{1}{r^2}\left(\boldsymbol{n}_{M_0}, \boldsymbol{n}_M\right) = \frac{1}{r^2} + \gamma_3(x, y, r, \varphi),$$

$$\frac{1}{\left|\overline{M_0 M}\right|^3} = \frac{1}{r^3} F_1(x, y, r, \varphi), \qquad \frac{1}{\left|\overline{M_0 M}\right|^5} = \frac{1}{r^5} F_2(x, y, r, \varphi), \qquad (10.1.52)$$

where the functions $\gamma_i(x, y, r, \varphi)$, $i = 1, 2, 3$, and $F_j(x, y, r, \varphi)$, $j = 1, 2$, are continuous in (x, y, r, φ).

From (10.1.51) and (10.1.52), it follows that $\beta_1(x, y), \beta_2(x, y) \in C_0^3(\mathbb{R}_2)$, and therefore, $\nu_2(x, y) \in C(\mathbb{R}_2)$.

The above considerations allow us to state the following result.

Proposition 10.1.6. *For* $f(x, y) \in C_0^4(\mathbb{R}_2)$, $u(x, y) \in C_0^3(\Omega)$, *the function*

$$\nu(x, y) = \iint_\Omega J(M' M_0') u(x_0, y_0) q(x_0, y_0)\, dx_0\, dy_0$$

is continuous in x *and* y.

Let us show that the family of discrete operators $\mathbf{T}_h$ defined by (10.1.23) yield a Π_h^*-approximation in the domain Ω for the bounded operator

$$\mathbf{T}(u) = \iint_\Omega J(M', M_0') u(x_0, y_0) q(x_0, y_0)\, dx_0\, dy_0$$

acting from $\mathring{H}_{1/2}(\Omega)$ to $H_{-1/2}(\Omega)$.

For $u(x_0, y_0) \in C_0^3(\Omega)$, let us calculate the approximate value $S(M_{0m}')$ of the integral $\mathbf{T}(u)$ by the formula

$$S(M_{0m}') = \sum_{k \in \Lambda(h)} u(hk_1, hk_2)\beta_1(k, m),$$

where $\Lambda(h)$ is the set of all indices $k = (k_1, k_2)$ such that

$$D(k, h) \subset \Omega_{1h}, \qquad M_{0m}' = M(hm_1, hm_2), \qquad M_{0m}' \in D(m, h) \subset \Omega_{1h}.$$

From the results of *Lifanov and Poltavskii* (1998), it follows that for

$$\nu(M_{0m}') = \iint_\Omega J(M_{0m}', M_0') u(x_0, y_0) q(x_0, y_0)\, dx_0\, dy_0,$$

we have

$$\left|S(M_{0m}') - \nu(M_{0m}')\right| = O(h), \qquad (10.1.53)$$

uniformly with respect to $M_{0m}' \in \Omega_{1h}$, where $O(h) \to 0$ as $|h| \to 0$.

Relation (10.1.53) and Proposition 10.1.6 imply that

$$\int_{\Omega_{1h}} \left|S((x, h) - \nu(x, y)\right|^2 dx\, dy = O_1(h),$$

where $O_1(h) \to 0$ as $h \to 0$, $\pi S(x,h) = S(M'_{0m})$ (π is the operator from Definition 5.1.2).

Thus, the family of operators T_h yields a Π_h^*-approximation of T in Ω (see Definition 6.2.7). Consider the equation

$$PT(u) = g(x,y), \qquad (10.1.54)$$

where $u(x,y) \in \mathring{H}_{1/2}(\Omega)$, $g(x,y) \in H_{-1/2}(\Omega)$, P is the operator of restriction to the domain Ω.

In order to find an approximate solution $u(x,y,h)$ of equation (10.1.54), consider the system of equations

$$P_h T_h u(x,y,h) = g(x,y,h), \qquad (10.1.55)$$

where P_h is the restriction of T_h to Ω_{1h}; $u(x,y,h) \in \mathring{M}(1/2,h,\Omega_{1h})$, $g(x,y,h) \in M(-1/2,h,\Omega_{1h})$, and there is an extension $\widetilde{g}(x,y,h) \in M(-1/2,h)$ of the function $g(x,y,h)$ such that $\widetilde{g}(x,y,h) \to lg(x,y)$ in the $H_{-1/2}(\mathbb{R}_2)$-norm as $h \to 0$, where $lg(x,y) \in H_{-1/2}(\mathbb{R}_2)$ is some extension of $g(x,y)$.

Theorem 10.1.1. *Let σ be the surface specified in the beginning of Section* 10.1. *Then, for any $g(x,y) \in H_{-1/2}(\Omega)$, equation* (10.1.54) *admits one and only one solution $u(x,y) \in \mathring{H}_{1/2}(\Omega)$. Moreover, for some $\delta > 0$ and all h, $0 < h < \varphi$, system* (10.1.55) *admits a solution $u(x,y,h)$, which converges to the solution $u(x,y)$ in $\mathring{H}_r(\Omega)$ for any $r \in (0, 1/2)$, so that*

$$\|\Pi_{1h}u(x,y) - u(x,y,h)\|_{\mathring{M}(1/2,h,\Omega_{1h})} \to 0 \quad \text{as} \quad h \to 0.$$

This result follows from Propositions 10.1.3–10.1.6, and Proposition 7.1.3.

10.2. Neumann Problem for the Helmholtz Equation: Convergence in Quotient Spaces for the Corresponding Hypersingular Integral Equation

Let σ be the surface specified at the beginning of Section 10.1. Consider the following problem for the Helmholtz equation:

$$\Delta u + \lambda^2 u = 0, \quad x \in \mathbb{R}_3 \setminus \sigma; \qquad \left.\frac{\partial u}{\partial n}\right|_\sigma = g(M), \qquad (10.2.1)$$

with the Sommerfeld radiation conditions

$$u(x) = O(|x|^{-1}), \qquad \frac{\partial u(x)}{\partial |x|} + i\lambda u(x) = o(|x|^{-1}) \quad \text{as} \quad |x| \to \infty, \qquad (10.2.2)$$

where the parameter λ is such that $\operatorname{Re}\lambda > 0$, $\operatorname{Im}\lambda \le 0$.

If one tries to solve problem (10.2.1), (10.2.2) in terms of the double layer potential, one obtains the following equation for the density $\nu(M_0)$ of the double layer potential:

$$\int_{\sigma_{M_0}} G(M,M_0)\nu(M_0)\,d\sigma_{M_0} + \int_{\sigma_{M_0}} K_1(M,M_0)\nu(M_0)\,d\sigma_{M_0} +$$

$$+ \int_{\sigma_{m_0}} K_2(M,M_0)\nu(M_0)\,d\sigma_{M_0} = g(M), \qquad (10.2.3)$$

where the function $G(M,M_0)$ is the kernel of the integral (10.1.1),

$$K_1(M_0,M) = \frac{G_1(M_0,M)\left(\overline{M_0M}, n_{M_0}\right)\left(\overline{M_0M}, n_M\right)}{|\overline{M_0M}|^5},$$

$$K_2(M_0,M) = -\frac{G_2(M_0,M)\left(n_{M_0}, n_M\right)}{|\overline{M_0M}|^3}, \qquad (10.2.4)$$

$$G_1(M_0,M) = e^{-i\lambda|\overline{M_0M}|}\left[-\lambda^2|\overline{M_0M}|^2 + 3i\lambda|\overline{M_0M}| + 3\right] - 3,$$

$$G_2(M_0,M) = e^{-i\lambda|\overline{M_0M}|} + i\lambda\,e^{-i\lambda|\overline{M_0M}|} - 1$$

Passing to Cartesian coordinates, just as in Section 10.1, and preserving the notation for the integral kernels, we obtain an equation for the density $\nu(x, y)$ of the double layer potential,

$$\iint_\Omega G(M', M_0')\nu(x_0, y_0)q(x_0, y_0)\, dx_0\, dy_0 + \iint_\Omega K_1(M', M_0')\nu(x_0, y_0)q(x_0, y_0)\, dx_0\, dy_0 +$$

$$+ \iint_\Omega K_2(M', M_0')\nu(x_0, y_0)q(x_0, y_0)\, dx_0\, dy_0 = F(x, y), \tag{10.2.5}$$

where $q = \left[f_x'^2(x, y) + f_y'^2(x, y) \right]^{1/2}$, $M' = M'(x, y)$, $M_0' = M_0'(x_0, y_0)$.

Similarly to Proposition 10.2.6, we prove the following result.

Proposition 10.2.1. *Let σ be the surface satisfying the assumptions of Section* 10.1, *and let $\nu(x_0, y_0) \in C_0^3(\Omega)$. Then $\psi(x, y) \in C(\mathbb{R}_2)$, where*

$$\psi(x, y) = \iint_\Omega G(M', M_0')\nu(x_0, y_0)q(x_0, y_0)\, dx_0\, dy_0 + \tag{10.2.6}$$

$$+ \iint_\Omega P(M', M_0')\nu(x_0, y_0)q(x_0, y_0)\, dx_0\, dy_0 = \psi_1(x, y) + \psi_2(x, y),$$

$$P(M'M_0') = K_1(M', M_0') + K_2(M', M_0').$$

Let us split the plane XOY into elementary cells $D(k, h)$ with vector-valued step $h = (h, h)$ (see Definition 5.1.1) and denote the family of discrete operators $\mathbf{B}_h$ by

$$\pi \mathbf{B}_h u(x, h) = \sum_{k \in \Lambda(h)} u(k)\beta_1(k, m) + \sum_{k \in \Lambda(h)} u(k)\beta_2(k, m) = \pi \mathbf{T}_h u(x, h) + \pi \mathbf{P}_h u(x, h), \tag{10.2.7}$$

where $\Lambda(h)$ is the set of indices $k = (k_1, k_2)$ such that $D(k, h) \subset \Omega_{1h}$; the operator $\mathbf{T}_h$ is the same as in Section 10.1; π is the operator from Definition 5.1.2, and $\beta_2(k, m)$ are defined by

$$\beta_2(k, m) = \iint_{D(k,h)} P(M_{0m}', M_0')a(x_0, y_0)\, dx_0\, dy_0, \tag{10.2.8}$$

$$M_{0m}' = M(hm_1, hm_2) \in D(m, h) \subset \Omega_{1h},$$

$$P(M', M_0') = K_1(M', M_0') + K_2(M', M_0'), \quad \pi u(x, h) = u(k).$$

The kernel $P(M', M_0')$ satisfies the inequality

$$|P(M', M_0')| \leq \frac{C}{|M'M_0'|}, \tag{10.2.9}$$

which implies that

$$\sum_{k \in \Lambda(h)} |\beta_2(k, m)| \leq C, \tag{10.2.10}$$

with C independent of h, m such that $D(m, h) \subset \Omega_{1h}$.

To obtain an approximation of the function $\psi(x, y)$ (see (10.2.6)) we use the quadrature formula

$$S(M_{0m}') = S_1(M_{0m}') + S_2(M_{0m}') = \sum_{k \in \Lambda(h)} \nu(hk_1, hk_2)\beta_1(k, m) + \sum_{k \in \Lambda(h)} \nu(hk_1, hk_2)\beta_2(k, m),$$
$$\tag{10.2.11}$$

where $\nu(x, y) \in C_0^3(\Omega)$. In view of (10.1.53), we have

$$|S_1(M_{0m}') - \psi_1(M_{0m}')| = O_1(h) \to 0 \quad \text{as} \quad h \to 0,$$

and this relation holds uniformly with respect to m such that $D(m, h) \subset \Omega_{1h}$. From (10.2.10), it follows that

$$|S_2(M_{0m}') - \psi_2(M_{0m}')| = O_1(h) \to 0 \quad \text{as} \quad h \to 0,$$

uniformly with respect to m such that $D(m, h) \in \Omega_{1h}$. This, together with Proposition 10.1.1 implies that

$$\|S(x, y, h) - \psi(x, y)\|_{L_2(\Omega_{1h})} \to 0 \quad \text{as} \quad h \to 0.$$

The above considerations give us the following result.

Proposition 10.2.2. *Let* B *be the operator defined by the left-hand side of* (10.2.5). *The family of operators* B_h *yields a* Π_h^*-*approximation of* B *in the domain* Ω.

Consider a function $\theta(x) \in C^\infty[0,\infty)$, $\theta(x) \geq 0$, $\theta(x) = 1$ for $0 \leq x \leq R$, where R is so large that the interior of the circle $S(R, M')$ of radius R with center at M' contains $\overline{\Omega}$; $\theta(x) = 0$ for $x > R_0 > R$. Let

$$P_1(M'M_0') = P(M', M_0')\theta|M'M_0'|.$$

We have $P_1(M', M_0') = P(M', M_0')$ for $M', M_0' \in \Omega$. It follows that the equation

$$B_1(\nu) = \iint_\Omega G(M', M_0')\nu(x_0, y_0)q(x_0, y_0)\, dx_0\, dy_0 +$$

$$+ \iint_\Omega P_1(M', M_0')\nu(x_0, y_0)q(x_0, y_0)\, dx_0\, dy_0 = F(x, y), \qquad (10.2.12)$$

where $(x, y) \in \Omega$, is equivalent to equation (10.2.5). Denote the first term in (10.2.12) by $T(\nu)$ and the second term by $P_1(\nu)$.

Remark 10.2.1. The structure of the function $P_1(M', M_0')$ shows that Proposition 10.2.1 is also valid for the operator $B_1(\nu)$.

Proposition 10.2.3. *The operator* $B_1(\nu) : \mathring{H}_{1/2}(\Omega) \to H_{-1/2}(\mathbb{R}_2)$ *is bounded.*

Proof. By proposition 10.1.5, this statement is valid for the operator $T(\nu)$. Thus, it suffices to prove Proposition 10.2.3 for the operator $P_1(\nu)$. Let $\psi_2(x, y) = P_1(\nu)$. Using the Jensen inequality (see *Zygmund* (1965)), we obtain

$$\iint_{\mathbb{R}_2} |\psi_2(x, y)|^2\, dx\, dy = \iint_{\mathbb{R}_2} \left| \iint_\Omega P_1(M', M_0')q(x_0, y_0)\nu(x_0, y_0)\, dx_0\, dy_0 \right| dx\, dy \leq$$

$$\leq \iint_{\mathbb{R}_2} \left(\left[\iint_\Omega |P_1(M', M_0')|q(x_0, y_0)|\, dx_0\, dy_0 \right] \times \right. \qquad (10.2.13)$$

$$\left. \times \iint_\Omega |P_1(M', M_0')q(x_0, y_0)||\nu(x_0, y_0)|^2\, dx_0\, dy_0 \right) dx\, dy.$$

In view of the properties of $\theta(|M_0'M'|)$, the inequalities

$$\iint_\Omega |P_1(M', M_0')q(x_0, y_0)|\, dx_0\, dy_0 \leq C_1, \qquad \iint_{\mathbb{R}_2} |P_1(M', M_0')|\, dx\, dy \leq C_2 \qquad (10.2.14)$$

hold with C_1 independent of $M'(x, y)$ and C_2 independent of $M_0'(x_0, y_0)$.

From (10.2.13) and (10.2.14), it follows that

$$\iint_{\mathbb{R}_2} |\psi_2(x, y)|^2\, dx\, dy \leq C_1 \iint_\Omega q(x_0, y_0)|\nu(x_0, y_0)|^2 \left[\iint_{\mathbb{R}_2} |P_1(M', M_0')|\, dx\, dy \right] dx_0\, dy_0 \leq$$

$$\leq C_3 \|\nu(x_0, y_0)\|_{L_2(\Omega)}^2.$$

This inequality shows that the operator $P_1(\nu)$ from $\mathring{H}_{1/2}(\Omega)$ to $H_{-1/2}(\mathbb{R}_2)$ is bounded. Proposition 10.2.1 is proved.

Next, we define a family of discrete operators B_{1h} by the relation

$$\pi B_{1h} u(x, y, h) = \pi T_h u(x, y, h) + \pi P_{1h}(x, y, h) =$$

$$= \sum_{k \in \Lambda(h)} u(k)\beta_1(k, m) + \sum_{k \in \Lambda(h)} u(k)\beta_2'(k, m), \qquad (10.2.15)$$

where $u(k) = \pi u(h)$, and $\beta'_2(k, m)$ has the form

$$\beta'_2(k, m) = \iint_{D(k,h)} P_1(M'_{0m}, M'_0) q(x_0, y_0) \, dx_0 \, dy_0, \qquad M'_{0m}(hm_1, hm_2) \in D(m, h).$$

Remark 10.2.2. For $k \in \Lambda(h)$, $m \in \Lambda(h)$, we have $\beta'_2(k, m) = \beta_2(k, m)$. Therefore, according to Proposition 10.2.2, the family of operators B_{1h} is a Π^*_h-approximation of B_1 in Ω.

Proposition 10.2.4. *The family of operators* P_{1h} : $\overset{\circ}{M}(r, h, \Omega_{1h}) \to L_2(\Omega_{1h})$ *is uniformly bounded for any h such that $0 \le r \le 1/2$.*

Proof. The following inequalities, similar to (10.2.14), hold:

$$\sum_{k \in \Lambda(h)} |\beta'_2(k, m)| \le C_1, \qquad \sum_m |\beta'_2(k, m)| \le C_2, \tag{10.2.16}$$

where C_1 does not depend on $m = (m_1, m_2)$; the sum in the second inequality is over all m; the constant C_2 is independent of $k \in \Lambda(h)$; $\Lambda(h)$ consists of all k such that $D(k, h) \subset \Omega_{1h}$.

Let $d(m) = \pi P_{1h} u(x, h)$. Using (10.2.16) and the Jensen inequality, we get

$$h^2 \sum_m |d(m)|^2 = h^2 \sum_m \left| \sum_{k \in \Lambda(h)} u(k) \beta'_2(k, m) \right|^2 \le$$

$$\le h^2 \sum_m \left(\sum_{k \in \Lambda(h)} |\beta'_2(k, m)| \right) \left(\sum_{k \in \Lambda(h)} |\beta'_2(k, m)| |u(k)|^2 \right) \le$$

$$\le C_1 h^2 \sum_{k \in \Lambda(h)} |u(k)|^2 \left(\sum_{m \in \Lambda(h)} |\beta'_2(k, m)| \right) \le C_3 \|u(x, h)\|^2_{L_2}.$$

This inequality proves Proposition 10.2.4.

Consider the equation

$$PB_1(u) = F(x, y), \tag{10.2.17}$$

where P is the restriction of the operator B_1 to the domain Ω, $u(x, y) \in \overset{\circ}{H}_{1/2}(\Omega)$., $F \in H_{-1/2}(\Omega)$. The parameter λ (see equation (10.2.1) is assumed such the kernel of the operator PB_1 : $\overset{\circ}{H}_{1/2}(\Omega) \to H_{-1/2}(\Omega)$ is trivial.

In order to find an approximate solution of equation (10.2.17), consider the system of equations

$$P_h B_{1h} u(x, y, h) = F(x, y, h), \tag{10.2.18}$$

where $u(x, y, h) \in \overset{\circ}{M}(1/2, h, \Omega_{1h})$, $F(x, y, h) \in M(-1/2, h, \Omega_{1h})$; P_h is the restriction of B_{1h} to Ω_{1h}; and there exist extensions $\widetilde{F}(x, y, h) \in M(-1/2, h)$ of $F(x, y, h)$ such that $\widetilde{F}(x, y, h) \to lF(x, y)$ in the norm of $H_{-1/2}(\mathbb{R}_2)$ as $h \to 0$, where $lF(x, y) \in H_{-1/2}(\mathbb{R}_2)$ is some extension of $F(x, y)$.

Theorem 10.2.1. *Suppose that the assumptions of Section 10.1 hold for the surface σ. Then, for any $F(x, y) \in H_{-1/2}(\Omega)$, equation (10.2.17) admits one and only one solution $u(x, y)$. There exists $\delta > 0$ such that for $0 < h < \delta$, system (10.2.18) has a solution $u(x, y, h)$, which converges to the exact solution $u(x, y)$ in $\overset{\circ}{H}_r(\Omega)$, for any $r, 0 \le r < 1/2$. Moreover,*

$$\|\Pi_{1h} u(x, y) - u(x, y)\|_{\overset{\circ}{M}(1/2, h, /\Omega_{1h})} \to 0 \quad \text{as} \quad h \to 0,$$

where Π_{1h} is the restriction of the integral operator Π_h to Ω_{1h}.

This result follows from Theorem 10.1.1, Remarks 10.2.1 and 10.2.2, Propositions 10.2.3, 10.2.4, 7.1.3, and 7.1.5.

10.3. Weak Convergence for Equations in a Plane Domain

Consider the hypersingular integral equation

$$\iint_G \frac{Q(x_1, x_2)}{r^3}\, dx_1\, dx_2 = f(x_1^0, x_2^0), \tag{10.3.1}$$

in the rectangle $G = (0 < x_1 < l) \times (0 < x_2 < b)$, where $r = \left[(x_1^0 - x_1)^2 + (x_2^0 - x_2)^2\right]^{1/2}$, $M_0 = M_0(x_1^0, x_2^0) \in G$, and the integral is understood in the sense of Hadamard (see (5.2.9)).

Let $\left\{x_{k_1}^1, x_{0m_1}^2\right\}$, $\left\{x_{k_2}^2, x_{0m_2}^2\right\}$ be the points splitting the segments

$$0 \le x_1 \le l, \qquad 0 \le x_2 \le b$$

and satisfying the relations

$$x_{k_1}^1 = k_1 h_1, \quad k_1 = 0, 1, \ldots, n, \quad nh_1 = l; \qquad x_{0m_1}^1 = (m_1 + 2^{-1}) h_1, \quad m_1 = 0, = \ldots, n-1,$$

$$x_{k_2}^2 = k_2 h_2, \quad k_2 = 0, 1, \ldots, N, \quad Nh_2 = b, \qquad x_{0m_2}^2 = (m_2 + 2^{-1}) h_2, \quad m_2 = 0, 1, \ldots, N-1.$$

We seek an approximate solution $Q_h(x_{0m_1}^1, x_{0m_2}^2)$ from the system of linear algebraic equations

$$\sum_{k_1=0}^{n-1} \sum_{k_2=0}^{N-1} Q_h(x_{0m_1}^1, x_{0m_2}^2) \gamma(k_1, k_2, m_1, m_2) = f(x_{0m_1}^1, x_{0m_2}^2), \tag{10.3.2}$$

$$m_1 = 0, \ldots, n-1; \quad m_2 = 0, 1, \ldots, N-1,$$

where

$$\gamma(k_1, k_2, m_1, m_2) = \int_{x_{k_1}^1}^{x_{k_1}^1 + 1} dx_1 \int_{x_{k_2}^2}^{x_{k_2}^2 + 1} \frac{dx_2}{r^3}. \tag{10.3.3}$$

The exact values $Q(x_{0k_1}^1, x_{0k_2}^2)$ satisfy the relation

$$\sum_{k_1=0}^{n-1} \sum_{k_2=0}^{N-1} Q(x_{0k_1}^1, x_{0k_2}^2) \gamma(k_1, k_2, m_1, m_2) = f(x_{0m_1}^1, x_{0m_2}^2) - \Delta f(x_{0m_1}^1, x_{0m_2}^2), \tag{10.3.4}$$

where $\Delta f(x_{0m_1}^1, x_{0m_2}^2)$ is the error of the quadrature formula.

Subtracting combinations of (10.3.4) from (10.3.2), we obtain

$$\sum_{k_1=0}^{n-1} \sum_{k_2=0}^{N-1} \left[Q_h(x_{0k_1}^1, x_{0k_2}^2) - Q(x_{0k_1}^1, x_{0k_2}^2)\right] \gamma(k_1, k_2, m_1, m_2) = \Delta f(x_{0m_1}^1, x_{0m_2}^2), \tag{10.3.5}$$

$$m_1 = 0, \ldots, n-1; \quad m_2 = 0, 1, \ldots, N-1.$$

For the function $G(\varphi_1, \varphi_2) = -h_1 \widehat{A}(\varphi_1, \varphi_2, h)$, where $\widehat{A}(\varphi_1, \varphi_2, h)$ is defined by (8.5.6), conditions (8.2.9) are fulfilled, and therefore, for the operator defined by the left-hand side of (10.3.2) there exists a discrete Green function $y_h(k_1, k_2, m_1, m_2)$ (see Definition 8.3.2). Now, from (10.3.5) we obtain

$$\Delta_h(k_1, k_2) = Q_h(x_{0k_1}^1, x_{0k_2}^2) - Q(x_{0k_1}^1, x_{0k_2}^2) =$$

$$= \sum_{k_1=0}^{n-1} \sum_{k_2=0}^{N-1} \Delta f(x_{0m_1}^1, x_{0m_2}^2) y_h(m_1, m_2, k_1, k_2), \tag{10.3.6}$$

$$k_1 = 0, \ldots, n-1; \quad k_2 = 0, 1, \ldots N-1.$$

Suppose that the properties of the function $f(x_1^0, x_2^0)$ ensure a solution $Q(x_1, x_2)$ of equation (10.3.1) with the following properties: $Q(x_1, x_2) = 0$ on the boundary of the rectangle G; $Q(x_1, x_2)$ has continuous second derivatives inside G, the derivative $Q''_{x_1 x_2}(x_1, x_2)$ has the form

$$Q''_{x_1 x_2}(x_1, x_2) = \frac{a(x_1, x_2)}{[x_1(l - x_1)x_2(b - x_2)]^{1/2}}, \qquad a(x_1, x_2) \in H_\alpha(\overline{G}), \alpha = 1.$$

As shown in *Belotserkovskii and Lifanov* (1985), the quadrature error for functions $Q(x_1, x_2)$ of this type has the form $(n = l/h_1)$

$$\left| \Delta f(x_{0m_1}^1, x_{0m_2}^2) \right| \leq \tag{10.3.7}$$

$$\leq \frac{Ch_1^{1/2} \ln(n+1)}{x_{0m_1}^1 (l - x_{0m_1}^1) \left[x_{0m_2}^2 (b - x_{0m_2}^2) \right]^{1/2}} + \frac{Ch_1^{1/2} \ln(n+1)}{x_{0m_2}^2 (b - x_{0m_2}^2) \left[x_{0m_1}^1 (b - x_{0m_1}^1) \right]^{1/2}}.$$

Let $E_{p_1 p_2} = [0 \leq k_1 \leq p_1] \times [0 \leq k_1 \leq p_2]$ with integer p_1, p_2 By a linear transformation one can always transform any vertex of the rectangle G to its bottom left corner. Moreover, the estimate (10.3.7) is symmetric with respect to the vertices of G. Therefore, it suffices to establish an estimate for $\Delta_h(k_1, k_2)$ on the set $E_{p_1^0 p_2^0}$ with $p_1^0 = [n/2] + 1$, $p_2^0 = [N/2] + 1$, where $[a]$ denotes the integer part of a.

Theorem 10.3.1. *Suppose that the above assumptions hold for the right-hand side of equation (10.3.1). Then, the following estimates are valid:*

(I) $|\Delta_h(k_1, k_2)| \leq C_1 (\ln n)^{5/4} n^{1/4}$ $(k_1, k_2) \in E_{l_1, l_2}$, $l_1 = l_2 = \left[n^{\varepsilon + 1/2} \right]$, $0 < \varepsilon < \dfrac{1}{4}$;

(II) $|\Delta_h(k_1, k_2)| \leq \dfrac{C_1 (\ln n)^5 n^{3/4}}{\nu^{1 - \varepsilon_1}}$, $\varepsilon_1 > 0$, $\nu = \max\{k_1, k_2\}$, $(k_1, k_2) \in E_{p_1^0 p_2^0} \setminus E_{l_1 l_2}$.

Proof. For a given $0 < \varepsilon < 1/4$, let $M(x_{0k_1}^1, x_{0k_2}^2)$ be some points for which $0 \leq k_1 \leq \left[n^{\varepsilon + 1/2} \right]$, $0 \leq k_2 \leq \left[n^{\varepsilon + 1/2} \right]$. According to Corollary 8.3.1, for the discrete Green function $y_h(m_1, m_2, k_1, k_2)$, the following relations hold:

$$y_h(m_1, m_2, k_1, k_2) = h \widehat{y}_h(m_1, m_2, k_1, k_2), \qquad \sum_m |\widehat{y}_h(m, k)| \leq C \ln n. \tag{10.3.8}$$

Let us transform the expression for the error $\Delta_h(k_1, k_2)$,

$$\Delta_h(k_1, k_2) = \sum_{m_1=0}^{p_1^0} \sum_{m_2=0}^{p_2^0} \Delta f(x_{0m_1}^1, x_{0m_2}^2) y_h(m_1, m_2, k_1, k_2) +$$

$$+ \sum_{m_1=p_1^0+1}^{n-1} \sum_{m_2=0}^{p_2^0} \Delta f(x_{0m_1}^1, x_{0m_2}^2) y_h(m_1, m_2, k_1, k_2) +$$

$$+ \sum_{m_1=0}^{p_1^0} \sum_{m_2=p_2^0+1}^{N-1} \Delta f(x_{0m_1}^1, x_{0m_2}^2) y_h(m_1, m_2, k_1, k_2) +$$

$$+ \sum_{m_1=p_1^0+1}^{n-1} \sum_{m_2=p_2^0+1}^{N-1} \Delta f(x_{0m_1}^1, x_{0m_2}^2) y_h(m_1, m_2, k_1, k_2) = \sum_{i=1}^{4} I_i. \tag{10.3.9}$$

Using (10.3.7), (10.38), and the Hölder inequality, we find that

$$|I_1| \le C \ln(n+1) \sum_{m_1=0}^{p_1^0} \sum_{m_2=0}^{p_2^0} \frac{|\widehat{y}_h(m_1, m_2, k_1, k_2)|}{(m_1+1)(m_2+1)^{1/2}} \le$$

$$\le C \ln(n+1) \left[\sum_{m_1=0}^{p_1^0} \sum_{m_2=0}^{p_2^0} |\widehat{y}_h(m_1, m_2, k_1, k_2)|^4 \right]^{1/4} \left[\sum_{m_1=0}^{p_1^0} \sum_{m_2=0}^{p_2^0} \frac{1}{(m_1+1)^{4/3}(m_2+1)^{2/3}} \right]^{3/4} \le$$

$$\le C_1 [\ln(n+1)]^{5/4} n^{1/4} \tag{10.3.10}$$

$$|I_2| \le C \ln(n+1) \sum_{m_1=p_1^0+1}^{n-1} \sum_{m_2=0}^{p_2^0} \frac{|\widehat{y}_h(m_1, m_2, k_1, k_2)|}{(n+1-m_1)(m_2+1)^{1/2}} \le$$

$$\le C \ln(n+1) \left[\sum_{m_1=p_1^0+1}^{n-1} \sum_{m_2=0}^{p_2^0} |y_h(m_1, m_2, k_1, k_2)|^4 \right]^{1/4} \times$$

$$\times \left[\sum_{m_1=p_1^0+1}^{n-1} \sum_{m_2=0}^{p_2^0} \frac{1}{(n+1-m_1)^{4/3}(m_2+1)^{2/3}} \right]^{3/4} \le C_1 [\ln(n+1)]^{5/4} n^{1/4}. \tag{10.3.11}$$

Inequalities similar to (10.3.10) and (10.3.11) can be obtained for I_3 and I_4. Thus, we conclude that

$$|\Delta_h(k_1, k_2)| \le C_1 [\ln(n+1)]^{5/4} n^{1/4}. \tag{10.3.12}$$

Now, consider the points $M(x^1_{0k_1}, x^2_{0k_2})$ for which $(k_1, k_2) \in E_{p_1^0 p_2^0} \setminus E_{l_1 l_2}$. In order to estimate the components of the discrete Green function, we utilize Theorem 8.4.1 and the inequality (8.2.22). From (8.2.19) and (8.2.20), it follows that

$$\frac{\left(\|y_h\|_1^2 + \|y_h\|_{L_2}^2 \right)}{\nu^{\sigma \varepsilon_1 - 1/2}} \le \frac{C[\ln(n+1)]^{1/2}}{\nu}; \tag{10.3.13}$$

and from (8.4.18), we have

$$\left[W\left(y_h, \nu^{\varepsilon_1 - 1} \right) \right] \le \frac{C[\ln(n+1)]^{5/2} \varepsilon^{1/2}}{\nu^{1-\varepsilon}}, \tag{10.3.14}$$

where $\nu = \max \{k_1, k_2\}$, $\varepsilon_1 > 0$.

Relations (10.3.13), (10.3.14), (8.2.22) show that $y_h(m_1, m_2, k_1, k_2)$ can be represented in the form

$$y_h(m_1, m_2, k_1, k_2) = h p_h^\nu(m_1, m_2, k_1, k_2) + h \delta_{\nu n} z_h(m_1, m_2, k_1, k_2),$$

$$|h p_h^\nu(m_1, m_2, k_1, k_2)| \le \frac{C h \ln^2 n}{|m_1 - k_1| + |m_2 - k_2| + 1}, \tag{10.3.15}$$

$$|\delta_{\nu n}| \le \frac{C[\ln(n+1)]^{5/2} n^{1/2}}{\nu^{1-\varepsilon_1}}, \qquad \sum_m |z_h(m_1, m_2, k_1, k_2)|^4 \le C \ln n.$$

Using (10.3.6), we get

$$\Delta_h(k_1, k_2) = \sum_{m_1=0}^{n-1} \sum_{m_2=0}^{N-1} \Delta f(x^1_{0m_1}, x^2_{0m_2}) \left[h p_h^\nu(m_1, m_2, k_1, k_2) + h \delta_{\nu n}(m_1, m_2, k_1, k_2) \right] =$$

$$= \Delta_h^1(k_1, k_2) + \Delta_h^2(k_1, k_2). \tag{10.3.16}$$

With the help of (10.3.15), similarly to (10.3.12), we obtain the estimate

$$\left|\Delta_h^2(k_1, k_2)\right| \le \frac{C[\ln(n+1)]^{15/4} n^{3/4}}{\nu^{1-\varepsilon_1}}.$$

(10.3.17)

Let us represent $\Delta_h^1(k_1, k_2)$ in the form

$$\Delta_h^1(k_1, k_2) = \sum_{i=1}^{4} I_i'.$$

(10.3.18)

The term I_1' is estimated as follows:

$$|I_1'| \le C[\ln(n+1)]^3 \sum_{m_1=0}^{p_1^0} \sum_{m_2=0}^{p_2^0} \left(\frac{1}{(m_1+1)(m_2+1)^{1/2}} + \frac{1}{(m_1+1)^{1/2}(m_2+1)}\right) \times$$

$$\times \frac{1}{|m_1 - k_1| + |m_2 - k_2| + 1}.$$

(10.3.19)

Let $\lambda = [\nu/2]$ be the integer part of $\nu/2$. Denote by $M(\lambda)$ the set of all indices $k = (k_1, k_2)$ such that $|m_1 - k_1| \le \lambda$, $|m_2 - k_2| \le \lambda$, $0 \le k_1 \le n-1$; $0 \le k_2 \le N-1$. Now, from (10.3.19) we get

$$|I_{11}'| = C[\ln(n+1)]^3 \sum_{m \in M(\lambda)} \left(\frac{1}{(m_1+1)(m_2+1)^{1/2}} + \frac{1}{(m_1+1)^{1/2}(m_2+1)}\right) \times$$

$$\times \frac{1}{|m_1 - k_1| + |m_2 - k_2| + 1} \le$$

$$\le C_1[\ln(n+1)]^5 \left[\frac{1}{\sqrt{\nu}} + \frac{\sqrt{n}}{\nu}\right] \le C_2 \frac{\sqrt{n}\ln^5(n+1)}{\nu}.$$

(10.3.20)

Similarly to (10.3.19), we obtain the inequality

$$|I_{21}'| = C[\ln(n+1)]^3 \sum_{m \in E_{p_1^9 p_2^0} \setminus M(\lambda)} \left(\frac{1}{(m_1+1)(m_2+1)^{1/2}} + \frac{1}{(m_1+1)^{1/2}(m_2+1)}\right) \times$$

$$\times \frac{1}{|m_1 - k_1| + |m_2 - k_2| + 1} \le C_1[\ln(n+1)]^4 \frac{\sqrt{n}}{\nu}.$$

(10.3.21)

From (10.3.10–(10.3.21), it follows that

$$|I_1'| \le |I_{11}'| + |I_{21}'| \le C_3 \frac{\sqrt{n}\ln^5(n+1)}{\nu}.$$

(10.3.22)

The estimate (10.3.22) for the other terms in (10.3.18) is established in a similar way. The statements of Theorem 10.3.1 follow from the estimates (10.3.12), (10.3.17), and (10.3.22).

Definition 10.3.1. We say that *a sequence of approximate solutions* $Q_h(x_{0k_1}^1, x_{0k_2}^2)$ *weakly converges to the exact solution of equation* (10.3.1) *as* $|h| \to 0$, if there is a sequence of domains $G_p \subset G$, $p = 1, 2, \ldots$, such that for any $M_0 \in G$ there is p_0 such that $M_0 \in G_p$ for $p \ge p_0$ and in each G_p the approximate solutions $Q_h(x_{0k_1}^1, x_{0k_2}^2)$ converge uniformly to the exact solution $Q(x_{0k_1}^1, x_{0k_2}^2)$ as $|h| \to 0$.

Corollary 10.3.1. *Theorem 10.3.1 implies that the solutions of system* (10.3.2) *weakly converge to the exact solution of equation* (10.3.1).

Now, let G be a bounded domain with boundary Γ of class C^∞.

Consider a canonical partition of $\mathbb{R}_2$ into cells $D(k, h)$ with constant vector-valued step $h = (h_1, h_2)$ (see Definition 5.1.1) satisfying condition (8.5.2). Denote by ε_h the set of all indices $k = (k_1, k_2)$ for which $D(k, h) \subset G$. In each cell $D(k, h)$, $k \in \varepsilon_h$, consider a point $M_{0k}(k_1 h_1, k_2 h_2)$ and let $x^1_{0k_1} = k_1 h_1$, $x^2_{0k_2} = k_2 h_2$. Let us seek an approximate solution $Q_h(x^1_{0k_1}, x^2_{0k_2})$ of equation (10.3.1) in the form

$$\sum_{k \in \varepsilon_h} Q_h(x^1_{0k_1}, x^2_{0k_2}) \gamma(k_1, k_2, m_1, m_2) = f(x^1_{0m_1}, x^2_{0m_2}), \qquad (m_1, m_2) \in \varepsilon_h, \qquad (10.3.23)$$

where

$$\gamma(k, m) = \iint_{D(k,h)} \frac{dx_1 \, dx_2}{r^3_{0M}}, \qquad r_{0M} = \left[(x_1 - m_1 h_1)^2 + (x_2 - m_2 h_2)^2 \right]^{1/2}. \qquad (10.3.24)$$

The following result can be proved similarly to Theorem 10.3.1, on the basis of Theorems 9.4.1 and 9.4.2, combined with the results of Section 8.3.

Theorem 10.3.2. *Let G be a bounded domain with the boundary Γ of class C^∞. Then, for $f(x^0_1, x^0_2) \in C^\infty(\overline{G})$, the following inequality holds for the solution of system (10.3.23) and the solution $Q(x_1, x_2)$ of equation (10.3.1):*

$$\left| Q_h(x^1_{0k_1}, x^2_{0k_2}) - Q(x^1_{0k_1}, x^2_{0k_2}) \right| \le C(\varepsilon) |h|^{-\varepsilon + 1/4},$$

where $\varepsilon > 0$ is arbitrary and $M(x^1_{0k_1}, x^2_{0k_2})$ is an arbitrary point separated from the boundary by the distance $\ge \delta$.

10.4. Convergence for the Multhopff Equation

Consider the equation (see *Nekrasov* (1947))

$$\int_a^b dx_2 \int_0^l \frac{\Gamma(x_1, x_2)}{(x^0_2 - x_2)^2} \left[1 + \frac{x^0_1 - x_1}{\left[(x^0_1 - x_1)^2 + (x^0_2 - x_2)^2 \right]} \right] dx_1 = f(x^0_1, x^0_2),$$

$$\int_0^l \Gamma(x_1, x_2) \, dx_1 = 0, \qquad \forall \, x_2 \in (0, b), \qquad (10.4.1)$$

where $M_0(x^0_1, x^0_2) \in G$, $G = (0 < x_1 < l) \times (0 < x_2 < b)$, and the integral in the first relation is understood in the sense of Hadamard.

Consider points $\{x^1_{k_1}, x^1_{0m_1}\}$ forming a partition of the axis OX_1: $x^1_{k_1} = k_1 h_1$; $(n + 1)h_1 = l$; $x^1_{0m_1} = h_1(2^{-1} + m_1)$. Similarly, let $\{x^2_{k_2}, x^2_{0m_2}\}$ form a partition of the axis OX_2: $x^2_{k_2} = k_2 h_2$; $(N + 1)h_2 = b$, $x^2_{m_2} = h_2(2^{-1} + m_2)$. It is assumed that

$$0 < \theta_1 < \frac{h_1}{h_2} < \theta_2. \qquad (10.4.2)$$

We seek an approximate solution $\Gamma_h(x^1_{k_1}, x^2_{0k_2})$ from the system of linear algebraic equations

$$\sum_{k_1 = l}^{n} \sum_{k_2 = 0}^{N} \Gamma_h(x^1_{k_1}, x^2_{0k_2}) \alpha(k_1, k_2, m_1, m_2) h_1 = f(x^1_{0m_1}, x^2_{0m_2}),$$

$$m_1 = 1, 2, \ldots, n - 1; \qquad m_2 = 0, 1, \ldots, N; \qquad (10.4.3)$$

$$\sum_{k_1 = 0}^{n} \Gamma_h(x^1_{k_1}, x^2_{0k_2}) = 0, \qquad k_1 = 0, 1, \ldots, N,$$

where $\alpha(k_1, k_2, m_1, m_2)$ have the form

$$\alpha(k_1, k_2, m_1, m_2) = \frac{x_{0m_1}^1 - x_{k_1}^1 + \left[\left(x_{0m_1}^1 - x_{k_1}^1\right)^2 + \left(x_{0m_2}^2 - x_{k_2+1}^2\right)^2\right]^{1/2}}{\left(x_{0m_1}^1 - x_{0k_1}^1\right)\left(x_{0m_2}^2 - x_{k_2+1}^2\right)} -$$

$$- \frac{x_{0m_1}^1 - x_{k_1}^1 + \left[\left(x_{0m_1}^1 - x_{k_1}^1\right)^2 + \left(x_{0m_2}^2 - x_{k_2}^2\right)^2\right]^{1/2}}{\left(x_{0m_1}^1 - x_{k_1}^1\right)\left(x_{0m_2}^2 - x_{k_2}^2\right)}. \qquad (10.4.4)$$

In order to write system (10.4.3) in a more compact form, set

$$F(m_1, m_2) = \begin{cases} f(x_{0m_1}^1, x_{0m_2}^2), & m_1 = 1, \ldots, n-1; \ m_2 = 0, \ldots, N; \\ 0, & m_1 = n; \ m_2 = 0, \ldots, N; \end{cases}$$

$$d(k_1, k_2, m_1, m_2) = \begin{cases} \alpha(k_1, k_2, m_1, m_2)h_1, & 1 \le k_1 \le n; \ 0 \le k_2, m_2 \le N, ; 1 \le m_1 \le n-1; \\ 1, & 1 \le k_1 \le n, \ k_2 = m_2, \ 0 \le m_2 \le N, m_1 = n; \\ 0, & 1 \le k_1 \le n, \ k_2 \ne m_2, \ m_1 = n. \end{cases}$$

$$(10.4.5)$$

Now, the algebraic system (10.4.3) reads

$$\sum_{k_1=1}^{n} \sum_{k_2=0}^{N} \Gamma_h(x_{k_1}^1, x_{0k_2}^2) d(k_1, k_2, m_1, m_2) = F(m_1, m_2), \qquad (10.4.6)$$

$$m_1 = 1, \ldots, n; \ m_2 = 0, 1, \ldots, N.$$

In order to find the elements of the discrete Green function for problem (10.4.6), consider the system of equations

$$\sum_{k_1=1}^{n} \sum_{k_2=0}^{N} x_n(k_1, k_2, p_1, p_2) \, d'(k_1, k_2, m_1, m_2) = \delta(m_1, m_2, p_1, p_2), \qquad (10.4.7)$$

$$m = 1, \ldots, n; \ m_2 = 0, 1, \ldots, N,$$

where p_1, p_2 are fixed; $d'(k_1, k_2, m_1, m_2) = d(m_1, m_2, k_1, k_2)$, $x_n(k_1, k_2, p_1, p_2)$ are unknown quantities; $\delta(m_1, m_2, p_1, p_2)$ is the Kronecker delta, equal to 1 for $m_1 = p_1$, $m_2 = p_2$, and equal to 0 otherwise.

System (10.4.7) can be written in the form (see (10.4.5))

$$x_n(n, m_2, p_1, p_2) + \sum_{k_1=1}^{n-1} \sum_{k_2=0}^{N} x_n(k_1, k_2, p_1, p_2) d(m_1, m_2, k_1, k_2) = \delta(m_1, m_2, p_1, p_2), \qquad (10.4.8)$$

$$1 \le m_1 \le n, \quad 0 \le m_2 \le N.$$

We introduce the functions

$$F_n(z_1, z_2) = \sum_{m_1=-2n}^{2n} \sum_{m_2=-2N}^{2N} d'(0, 0, m_1, m_2) z_1^{m_1} z_2^{m_2},$$

$$y_n(z_1, z_2) = \sum_{k_1=1}^{n-1} \sum_{k_2=0}^{N} x_n(k_1, k_2, p_1, p_2) z_1^{k_1} z_2^{k_2},$$

$$(10.4.9)$$

$$y_{1n}(z_1, z_2) = \sum_{m_1=1}^{n} \sum_{m_2=0}^{N} x_n(n, m_2, p_1, p_2) z_1^{m_1} z_2^{m_2} =$$

$$= \sum_{m_2=0}^{N} x_n(n, m_2, p_1, p_2) \frac{z_1 - z_1^{n+1}}{1 - z_1} z_2^{m_2}.$$

In view of (10.4.4) and (10.4.9), according to the rules of multiplication of power series, we can write system (10.4.8) in the form

$$y_{1n}(z_1, z_2) + F_n(z_1, z_2)y_n(z_1, z_2) = z_1^{p_1} z_2^{p_2} + \sum_{i=1}^{4} \varphi_{in}(z_1, z_2), \qquad (10.4.10)$$

where $\varphi_{in}(z_1, z_2)$ are unknown functions of the form

$$\varphi_{1n}(z_1, z_2) = \sum_{k_1=-\infty}^{0} \sum_{k_2=-\infty}^{\infty} a_{k_1 k_2}^{1n} z_1^{k_1} z_2^{k_2},$$

$$\varphi_{2n}(z_1, z_2) = \sum_{k_1=n+1}^{\infty} \sum_{k_2=-\infty}^{\infty} a_{k_1 k_2}^{2n} z_1^{k_1} z_2^{k_2},$$

$$\varphi_{3n}(z_1, z_2) = \sum_{k_1=1}^{n} \sum_{k_2=N+1}^{\infty} a_{k_1 k_2}^{3n} z_1^{k_1} z_2^{k_2},$$

$$\varphi_{4n}(z_1, z_2) = \sum_{k_1=1}^{n} \sum_{k_2=-\infty}^{-1} a_{k_1 k_2}^{4n} z_1^{k_1} z_2^{k_2},$$

Multiplying both sides of (10.4.10) by $z_1^{-n}\left(1 - z_1^{-1}\right)$, we obtain

$$\sum_{m_2=0}^{N} x_n(n, m_2, p_1, p_2) z_2^{m_2} \left(1 - z_1^{-n}\right) + z_1^{-n}\left(1 - z_1^{-1}\right) F_n(z_1, z_2) y_n(z_1, z_2) =$$

$$= (z_1 - 1) z_1^{p_1-n-1} z_2^{p_2} + \sum_{i=1}^{4} z_1^{-n}\left(1 - z_1^{-1}\right) \varphi_{in}(z_1, z_2). \qquad (10.4.11)$$

Let us examine the coefficients of $z_1^l z_2^k$ in each function entering (10.4.11) with $1 - n \le l \le -1$, $0 \le k \le N$.

For the functions

$$\psi_n(z_1, z_2) = \sum_{m_2=0}^{N} x(n, m_2, p_1, p_2) z_2^{m_2} \left(1 - z_1^{-n}\right),$$

$$\psi_{in}(z_1, z_2) = z_1^{-n}\left(1 - z_1^{-1}\right) \varphi_{in}(z_1, z_2), \quad i = 1, \ldots, 4,$$

the said coefficients are all equal to zero.

Let p_1, p_2 in (10.4.8) satisfy the inequalities

$$\theta_0 < \frac{p_1}{n} < \theta_1, \quad \theta_0 < \frac{p_2}{N} < \theta_1, \quad 0 < \theta_0, \quad \theta_1 < 1. \qquad (10.4.12)$$

From (10.4.2), (10.4.12), it follows that there is h_{10} such that for all $h_1 < h_{10}$, we have

$$\frac{1}{4\pi^2} \int_{-\pi}^{\pi} \int_{-\pi}^{\pi} \left(1 - e^{i\varphi_1}\right) F_n\left(e^{i\varphi_1}, e^{i\varphi_2}\right) \widehat{y}_n\left(e^{i\varphi_1}, e^{i\varphi_2}\right) D_{nN}(\alpha - \varphi_1, \beta - \varphi_2) \, d\varphi_1 \, d\varphi_2 =$$

$$= \left(e^{i\alpha} - 1\right) e^{i\alpha(p_1-n-1)} e^{i\beta p_2}, \qquad (10.4.13)$$

where

$$\widehat{y}_n(z_1, z_2) = z_1^{-n} y_n(z_1, z_2), \qquad D_{nN}(\alpha, \beta) = \sum_{k_2=0}^{N} \sum_{k_1=1-n}^{-1} e^{ik_1\alpha} e^{ik_2\beta}.$$

Next, we perform the transformations

$$K_n(z_1, z_2) = \left(1 - z_1^{-1}\right) F_n(z_1, z_2) =$$

$$= \sum_{k_1=-2n}^{2n-1} \sum_{k_2=-2N}^{2N} \left[d'(0, 0, k_1, k_2) - d'(0, 0, k_1 + 1, k_2) \right] z_1^{k_1} z_2^{k_2} +$$

$$+ z_1^{2n} \sum_{k_2=-2N}^{2N} d'(0, 0, 2n, k_2) z_2^{k_2} - z_1^{-2n-1} \sum_{k_2=-2N}^{2N} d'(0, 0, -2n, k_2) z_2^{k_2} =$$

$$= \sum_{i=1}^{3} K_{in}(z_1, z_2). \tag{10.4.14}$$

The functions $K_{2n}(z_1, z_2)$ and $K_{3n}(z_1, z_2)$ are unimportant for problem (10.4.13). Let us find the coefficients $a(k_1, k_2) = d'(0, 0, k_1, k_2) - d'(0, 0, k_1 + 1, k_2)$, making use of the following formula (see *Belotserkovskii and Lifanov* (1985)):

$$\int_{x_i}^{x_{i+1}} \int_{z_k}^{z_{k+1}} \frac{dx\, dz}{\left[(x_{0j} - x)^2 + (z_{0m} - z)^2 \right]^{3/2}} =$$

$$= -\frac{\left[(x_{0j} - x_{i+1})^2 + (z_{0m} - z_{k+1})^2 \right]^{1/2}}{(x_{0j} - x_{i+1})(z_{0m} - z_{k+1})} + \frac{\left[(x_{0j} - x_i)^2 + (z_{0m} - z_{k+1})^2 \right]^{1/2}}{(x_{0j} - x_i)(z_{0m} - z_{k+1})} +$$

$$+ \frac{\left[(x_{0j} - x_{i+1})^2 + (z_{0m} - z_k)^2 \right]^{1/2}}{(x_{0j} - x_{i+1})(z_{0m} - z_k)} - \frac{\left[(x_{0j} - x_i)^2 + (z_{0m} - z_k)^2 \right]^{1/2}}{(x_{0j} - x_i)(z_{0m} - z_k)}. \tag{10.4.15}$$

The relation $d'(0, 0, k_1, k_2) = d(k_1, k_2, 0, 0)$, together with (10.4.4), (10.4.5), (10.5.15), (8.5.4), implies that

$$a(k_1, k_2) = d'(0, 0, k_1, k_2) - d'(0, 0, k_1 + 1, k_2) = -h_1 \gamma(k, 0) = -\gamma(0, k) h_1. \tag{10.4.16}$$

It follows that problem (10.4.13) is equivalent to the restriction problem (8.5.8),

$$\int_{-\pi}^{\pi} \int_{-\pi}^{\pi} G_n(\varphi_1, \varphi_2) \widehat{x}_c(\varphi_1, \varphi_2) D_{\varepsilon h}(\alpha - \varphi_1, \beta - \varphi_2)\, d\varphi_1\, d\varphi_2 = \left(e^{i\alpha} - 1 \right) e^{i\alpha(p_1-1)} e^{i\beta p_2}, \tag{10.4.17}$$

where $G_n(\varphi_1, \varphi_2)$ is the function defined by (8.5.7) and

$$\widehat{x}_c(\varphi_1, \varphi_2) = \sum_{k_1=1}^{n-1} \sum_{k_2=0}^{N} x_n(k_1, k_2, p_1, p_2)\, e^{i(\varphi_1 k_1 + \varphi_2 k_2)}, \qquad D_{\varepsilon h} = \frac{1}{4\pi^2} \sum_{k_1=1}^{n-1} \sum_{k_2=0}^{N} e^{i(\alpha k_1 + \beta k_2)}.$$

As shown in Section 8.5, problem (8.5.8), and therefore, problem (10.4.17), admits a unique solution. The method used for the construction of the above quantities makes it clear that system (10.4.8) admits one and only one solution. Hence, we infer the existence and the uniqueness of a solution of system (10.4.3).

The coefficients $x_n(k_1, k_2, p_1, p_2)$ can be represented in the form (8.5.50) (the role of $x_m^n(k_1, k_2)$ is played by $x_n(k_1, k_2, p_1, p_2)$), and the estimates (8.5.51) and (8.5.52) hold. In view of (10.4.12), we have $\nu = \theta n$, $0 < \theta < 1$. Therefore, E_{np}^{ν} are estimated as follows:

$$E_{np}^{\nu} \leq C(\varepsilon) \frac{\ln^3 n}{n^{3-5\varepsilon}}, \tag{10.4.18}$$

where $\varepsilon > 0$ can be chosen arbitrary.

Let us estimate $\sum_{k_2}^{N} |x_n(n, m_2, p_1, p_2)|$. Take $m_1 = n$ in (10.4.8). Using (8.5.50), we get

$$|x_n(n, m_2, p_1, p_2)| \leq$$

$$\leq \sum_{k_1=1}^{n-1} \sum_{k_2=0}^{N} |d(n, m_2, p_1, p_2)| |x_{1p_1p_2}^n(k_1, k_2)| + \sum_{k_1=1}^{n-1} \sum_{k_2=0}^{N} |d(n, m_2, k_1, k_2)| |p_{p_1p_2}^n(k_1, k_2)| =$$

$$= I_1(n, m_2, p_1, p_2) + I_2(n, m_2, p_1, p_2). \tag{10.4.19}$$

From (10.4.4), by elementary transformations we obtain for $k_1 < 0$

$$|d(0, 0, k_1, k_2)| \leq \frac{C}{\left|k_1 - \frac{1}{2}\right| \left(\left|k_1 - \frac{1}{2}\right| + \left|k_2 + \frac{1}{2}\right|\right)} .$$

This inequality implies that for $k_1 < n$ we have

$$|d(n, m_2, k_1, k_2)| \leq \frac{C}{\left|k_1 - n - \frac{1}{2}\right| \left(\left|k_1 - n - \frac{1}{2}\right| + \left|k_2 - m_2 + \frac{1}{2}\right|\right)} . \tag{10.4.20}$$

With the help of (8.5.51), (8.5.52), (10.4.18), (10.4.20), we obtain the estimate

$$\sum_{m_2=0}^{N} I_1(n, m_2, p_1, p_2) \leq$$

$$\leq \frac{C(\varepsilon) \ln^{3/2} n}{n^{3/2 - 5\varepsilon/2}} \sum_{m_2=0}^{N} \sum_{k_1=1}^{n-1} \sum_{k_2=0}^{N} \frac{|x_{2p_1, p_2}^n(k_1, k_2)|}{\left|k_1 - n - \frac{1}{2}\right| \left(\left|k_1 - n - \frac{1}{2}\right| + \left|k_2 - m_2 + \frac{1}{2}\right|\right)} \leq$$

$$\leq \frac{C_1(\varepsilon) \ln^{3/2} n \ln N}{n^{3/2 - 5\varepsilon/2}} \sum_{k_1=1}^{n-1} \sum_{k_2=0}^{N} \frac{|x_{2p_1, p_2}^n(k_1, k_2)|}{\left|k_1 - n - \frac{1}{2}\right|} \leq$$

$$\leq \frac{C_1(\varepsilon) \ln^{3/2} n \ln N}{n^{3/2 - 5\varepsilon/2}} \left(\sum_{k_1=1}^{n-1} \sum_{k_2=0}^{N} \frac{1}{\left|k_1 - n - \frac{1}{2}\right|} \right)^{3/4} \left(\sum_{k_1=1}^{n-1} \sum_{k_2=0}^{N} |x_{2p_1, p_2}^n(k_1, k_2)|^4 \right)^{1/4} \leq$$

$$\leq \frac{C_2(\varepsilon)(\ln n)^{11/4}}{n^{3/4 - 5\varepsilon/2}} . \tag{10.4.21}$$

In view of (10.4.12), there exist β_1, β_2, β_1', β_2' such that

$$p_1 - \beta_1 n \geq \beta_1' n, \quad \beta_2 n - p_1 \geq \beta_2' n, \quad 0 < \beta_1, \beta_2, \beta_1', \beta_2' < 1. \tag{10.4.22}$$

Utilizing (8.5.51), (10.4.20), and (10.4.22), we obtain the inequality

$$I_2(n, m_2, p_1, p_2) = \sum_{k_1=1}^{[\beta_1 n]} \sum_{k_2=0}^{N} |d(n, m_2, k_1, k_2)| |p_{p_1, p_2}^n(k_1, k_2)| +$$

$$+ \sum_{k_1=[\beta_1 n]+1}^{[\beta_2 n]} \sum_{k_2=0}^{N} |d(n, m_2, k_1, k_2)| |p_{p_1, p_2}^n(k_1, k_2)| +$$

$$+ \sum_{k_1=[\beta_2 n]+1}^{n-1} \sum_{k_2=0}^{N} |d(n, m_2, k_1, k_2)| |p_{p_1, p_2}^n(k_1, k_2)| \leq$$

$$\leq \lambda_1^n \sum_{k_1=1}^{[\beta_1 n]} \sum_{k_2=0}^{N} |d(n, m_2, k_1, k_2)| + \tag{10.4.23}$$

$$+ \lambda_2^n \sum_{k_1=[\beta_1 n]+1}^{[\beta_2 n]} \sum_{k_2=0}^{N} |p_{p_1, p_2}^n(k_1, k_2)| + \lambda_3^n \sum_{k_1=[\beta_2 n]+1}^{n-1} \sum_{k_2=0}^{N} |d(n, m_2, k_1, k_2)|,$$

where $[a]$ is the integer part of a, and

$$\lambda_1^n = \max \left| p_{p_1,p_2}^n(k_1, k_2) \right| \quad \text{for} \quad 0 \le k_2 \le N, \quad 1 \le k_1 \le [\beta_1 n];$$

$$\lambda_3^n = \max \left| p_{p_1,p_2}^n(k_1, k_2) \right| \quad \text{for} \quad 0 \le k_2 \le N, \quad [\beta_2 n] + 1 \le k_1 \le n - 1;$$

$$\lambda_2^n = \max \left| d(n, m_2, k_1, k_2) \right| \quad \text{for} \quad 0 \le k_2 \le N, \quad [\beta_1 n] + 1 \le k_1 \le [\beta_2 n].$$

The following estimates hold:

$$\lambda_1^n < \frac{C(\ln n)^3}{n^2}, \qquad \lambda_3^n < \frac{C(\ln n)^3}{n^2}, \qquad \lambda_2^n < \frac{C}{n^2}. \tag{10.4.24}$$

From (10.4.23), (10.4.24) we obtain

$$I_2(n, m_2, p_1, p_2) \le \frac{C(\ln n)^5}{n^2},$$

which implies that

$$\sum_{m_2=0}^N I_2(n, m_2, p_2, p_2) \le \frac{C_1(\ln n)^5}{n}. \tag{10.4.25}$$

Combining (10.4.19), (10.4.21), and (10.4.25), we obtain

$$\sum_{m_2=0}^N x_n(n, m_2, p_1, p_2) \le \frac{C_3(\ln n)^{11/4}}{n^{3/4-5\varepsilon/2}}. \tag{10.4.26}$$

Suppose that the properties of $f(x_1^0, x_2^0)$ on the right-hand side of equation (10.4.1) ensure that its solution $\Gamma(x_1, x_2)$ has the form

$$\Gamma(x_1, x_2) = \left(\frac{x_2(b - x_2)}{x_1(l - x_2)} \right)^{1/2} \gamma(x_1, x_2),$$

and $\gamma_{x_1}'(x_1, x_2) \in H_\alpha[\overline{G}]$, $\gamma_{x_2}'(x_1, x_2) \in H_\alpha[\overline{G}]$, $\alpha = 1$.

Denote the value of the integral on the left-hand side of (10.4.1) by $A(x_1^0, x_2^0)$. We seek an approximate solution of (10.4.1) in terms of the quadrature formula

$$S(x_{0m_1}^1, x_{0m_2}^2) = \sum_{k_1=1}^n \sum_{k_2=0}^N \Gamma(x_{k_1}^1, x_{0k_2}^2) \alpha(k_1, k_2, m_1, m_2) h_1.$$

As shown by *Belotserkovskii and Lifanov* (1985), the error of quadrature sums of this type has the form

$$\left| \Delta f(x_{0m_1}^1, x_{0m_2}^2) \right| \le$$

$$\le \frac{Ch_1^{1/2} \ln(n+1)}{x_{0m_1}^1 (l - x_{0m_1}^1) \left[x_{0m_2}^2 (b - x_{0m_2}^2) \right]^{1/2}} + \frac{Ch_1^{1/2} \ln(n+1)}{x_{0m_2}^2 (b - x_{0m_2}^2) \left[x_{0m_1}^1 (l - x_{0m_1}^1) \right]^{1/2}} =$$

$$= \Delta_1(m_1, m_2) + \Delta_2(m_1, m_2). \tag{10.4.27}$$

Theorem 10.4.1. *Let $f(x_1^0, x_2^0)$ satisfy the assumptions made above for equation* (10.4.1). *Then the following estimate holds for the solution of system* (10.4.3) *and the solution $\Gamma(x_1, x_2)$ of equation* (10.4.1)*:*

$$|\Delta\Gamma_{k_1 k_2}| = \left| \Gamma_h(x_{k_1}^1, x_{0k_2}^2) - \Gamma(x_{k_1}^1, x_{0k_2}^2) \right| \le C(\varepsilon)|h|^{-\varepsilon+1/4}, \quad \varepsilon > 0, \quad |h| = \left[h_1^2 + h_2^2 \right]^{1/2},$$

for all points $M(x^1_{k_1}, x^2_{0k_2})$ *whose coordinates satisfy the inequalities*

$$\delta < x^1_{k_1} < l - \delta, \qquad \delta < x^2_{0k_2} < b - \delta, \qquad \delta > 0.$$

Proof. For the exact values $\Gamma(x^1_{k_1}, x^2_{0k_2})$, we have

$$\sum_{k_1=1}^{n} \sum_{k_2=1}^{N} \Gamma(x^1_{k_1}, x^2_{0k_2}) d(k_1, k_2, m_1, m_2) = F(m_1, m_2) - \Delta F(m_1, m_2), \qquad (10.4.28)$$

$$m_1 = 1, \ldots, n; \quad m_2 = 0, 1, \ldots, N,$$

where $\Delta F(m_1, m_2)$ is the quadrature error. Subtracting (10.4.28) from (10.4.6), we obtain

$$\sum_{k_1=0}^{n} \sum_{k_2=0}^{N} \left[\Gamma_h(x^1_{k_1}, x^2_{0k_2}) - \Gamma(x^1_{k_1}, x^2_{0k_2}) \right] d(k_1, k_2, m_1, m_2) = \Delta F(m_1, m_2), \qquad (10.4.29)$$

$$m_1 = 1, \ldots, n; \quad m_2 = 0, 1, \ldots, N.$$

For $m = 1, \ldots, n-1$; $m_2 = 0, 1, \ldots,$ the error $\Delta(m_1, m_2)$ satisfies the inequality (10.4.27), and for $m_1 = n$, $m_2 = 0, 1, \ldots, N$, the error $\Delta F(m_1, m_2)$ satisfies the inequality

$$|\Delta F(m_1, m_2)| \le C h_1^{-1/2}. \qquad (10.4.30)$$

From (10.4.29), it follows that

$$\Delta \Gamma_{k_1 k_2} = \sum_{m_1=1}^{n} \sum_{m_2=0}^{N} x_n(m_1, m_2, k_1, k_2) \Delta F(m_1, m_2), \qquad (10.4.31)$$

where $x_n(m_1, m_2, k_1, k_2)$ are the components of the discrete Green function; these are solutions of system (10.4.7). On the basis of the representation (8.5.50), we transform (10.4.31) as follows:

$$\Delta \Gamma_{k_1 k_2} = \sum_{m_2=0}^{N} x_n(n, m_2, k_1, k_2) \Delta F(n, m) + \sum_{m_1=1}^{n-1} \sum_{m_2=0}^{N} x^n_{1k_1 k_2}(m_1, m_2) \Delta F(m_1, m_2) +$$

$$+ \sum_{m_1=1}^{n-1} \sum_{m_2=0}^{N} p^n_{k_1 k_2}(m_1, m_2) \Delta F(m_1, m_2) = I_1(n) + I_2(n) + I_3(n). \qquad (10.4.32)$$

From (10.4.26) and (10.4.30), we obtain the estimate

$$|I_1(n)| \le C(\varepsilon_0) |h|^{1/4 - \varepsilon_0},$$

with arbitrary $\varepsilon_0 > 0$. We further have

$$I_2(n) = \sum_{m_1=1}^{n-1} \sum_{m_2=0}^{N} x^n_{1k_1 k_2}(m_1, m_2) [\Delta_1(m_1, m_2) + \Delta_2(m_1, m_2)] = I'_2(n) + I''_2(n). \qquad (10.4.33)$$

Let us write the estimate for $\Delta_1(m_1, m_2)$ as follows: for $l - x^1_{0m_1} > C_1$, we have

$$\Delta_1(m_1, m_2) \le \frac{C h_1^{-\varepsilon_1 + 1/4} \ln(n+1)}{\left(x^1_{0m_1}\right)^{-\varepsilon_1 + 3/4} (l - x^1_{0m_1}) \left[x^2_{0m_2}(b - x^2_{0m_2})\right]}; \qquad (10.4.34)$$

and for $x^1_{om_1} \geq C_2$, we have

$$\Delta_1(m_1, m_2) \leq \frac{Ch_1^{-\varepsilon+1/4}\ln(n+1)}{x^1_{0m_1}(l - x_{0m_1})^{-\varepsilon+3/4}\left[x^2_{0m_2}(b - x^2_{0m_2})\right]^{1/2}} . \tag{10.4.35}$$

Using (8.5.51), (8.5.52), (10.4.18), (10.4.34), (10.4.35), and the Hölder inequality, we obtain

$$|I_2'(n)| \leq C_3(\varepsilon)\frac{\ln^{3/2}(n+1)}{n^{3/2-5\varepsilon/2}}\left|\sum_{m_1=1}^{n-1}\sum_{m_2=0}^{N}x^n_{2k_1k_2}(m_1, m_2)\Delta_1(m_1, m_2)\right| \leq$$

$$\leq C_3(\varepsilon)\frac{\ln^{3/2}(n+1)}{n^{3/2-5\varepsilon/2}}\left[\sum_{m_1=1}^{n-1}\sum_{m_2=0}^{N}\left|x^n_{2k_1k_2}(m_1, m_2)\right|^4\right]^{1/4}\left[\sum_{m_1=1}^{n-1}\sum_{m_2=0}^{N}|\Delta_1(m_1, m_2)|^{4/3}\right]^{3/4} \leq$$

$$\leq C_3(\varepsilon)\frac{\ln^{7/4}(n+1)}{n^{3/2-5\varepsilon/2}}\left[\sum_{\substack{m_1,m_2, \\ (l-x^1_{0m_1})>C_1}}\frac{h_1^{1/3-4\varepsilon_1/3}\ln^{4/3}(n+1)}{(x^1_{0m_1})^{1-4\varepsilon_1/3}(l - x^1_{0m_1})^{4/3}\left[x^2_{0m_2}(b - x^2_{0m_2})\right]^{2/3}} + \right.$$

$$\left. + \sum_{\substack{m_1,m_2, \\ x^1_{0m_1}>C_2}}\frac{h_1^{1/3-4\varepsilon_1/3}\ln^{4/3}(n+1)}{(x^1_{0m_1})^{4/3}(l - x^1_{0m_1})^{1-4\varepsilon_1/3}\left[x^2_{0m_2}(b - x^2_{0m_2})\right]^{2/3}}\right]^{3/4} \leq$$

$$\leq \frac{C_3(\varepsilon)[\ln(n+1)]^{11/4}}{h_1^{1/4-5\varepsilon/2-\varepsilon_1}}\left(\sum_{\substack{m_1,m_2 \\ (l-x^1_{0m_1})>C_1}}\frac{h_1 h_2}{(x^1_{0m_1})^{1-4\varepsilon_1/3}(l - x^1_{0m_1})^{4/3}[x^2_{0m_2}(b - x^2_{0m_2})]^{2/3}} + \right.$$

$$\left. + \sum_{\substack{m_1,m_2 \\ x^1_{0m_1}>C_2}}\frac{h_1 h_2}{(x^1_{0m_1})^{4/3}(l - x^1_{0m_1})^{1-4\varepsilon_1/3}[x^2_{0m_2}(b - x^2_{0m_2})]^{2/3}}\right)^{3/4} \leq$$

$$\leq A(\varepsilon_3)|h|^{1/4-\varepsilon_3} \tag{10.4.36}$$

In a similar way it can be shown that the inequality (10.4.36) holds for $|I_2''(n)|$ and $I_3(n)$. Theorem 10.4.1 is proved.

10.5. Convergence of the Numerical Solution in the C-Norm

In this section, we examine convergence with respect to the norm in the space C for numerical solutions of the hypersingular integral equation

$$\iint_G \frac{Q(x_1, x_2)\, dx_1\, dx_2}{r^3_{M_0}} = f(x^0_1, x^0_2), \tag{10.5.1}$$

where G is a plane bounded domain with the boundary Γ of class C^∞;

$$M_0(x^0_1, x^0_2) \in G, \qquad r^3_{M_0} = \left[(x^0_1 - x_1)^2 + (x^0_2 - x_2)^2\right]^{1/2}.$$

Lemma 10.5.1. *Let G be a plane bounded domain with the boundary Γ of class C^2. Then,*

$$\left|A(x_1^0, x_2^0)\right| = \left|\iint_G \frac{dx_1\, dx_2}{r_{M_0}^3}\right| \geq \frac{K}{\rho(M_0)}, \tag{10.5.2}$$

where $\rho(M_0)$ is the distance from the point $M_0(x_1^0, x_2^0)$ to the boundary Γ; $K > 0$ is a constant independent of M_0.

Proof. Let $M_1 \in \Gamma$ be the point whose distance from M_0 is equal to the distance from M_0 to Γ. We introduce a Cartesian coordinate system $M_0 Y_1 Y_2$, with M_0 being the origin, the axis $M_0 Y_1$ being the line passing through M_0 and M_1 and directed along the vector $\overrightarrow{M_0 M_1}$; the axis $M_0 Y_2$ is orthogonal to $M_0 Y_1$. Let l be the tangential line to the curve Γ at the point M_1. Since Γ is of class C^2, l is parallel to the axis Y_2, and therefore, can be described by the equation $y_1 = \rho(M_0)$. Since the domain G is convex, for the points $M(y_1, y_2) \in G$ we have $y_1 < \rho(M_0)$. Let D be the domain that consists of the points $P(y_1, y_2)$ such that $y_1 \geq \rho(M_0)$. Passing to polar coordinates, we obtain the relation

$$\iint_D \frac{dx_1\, dx_2}{r_{M_0}^3} = \int_{-\pi/2}^{\pi/2} d\varphi \int_{\rho(M_0)/\cos\varphi}^{\infty} \frac{dr}{r^2} = \frac{2}{\rho(M_0)}. \tag{10.5.3}$$

The integral $\iint r_{M_0}^{-3}\, dx_1\, dx_2$ is understood in the sense of Hadamard and has the following properties:

$$\iint_{\mathbb{R}_2} \frac{dx_1\, dx_2}{r_{M_0}^3} = 0, \quad \forall\, M_0 \in \mathbb{R}_2,$$
$$\iint_{G_1} \frac{dx_1\, dx_2}{r_{M_0}^3} > 0 \quad \text{for} \quad M_0 \notin G_1; \qquad \iint_{G_1} \frac{dx_1\, dx_2}{r_{M_0}^3} < 0 \quad \text{for} \quad M_0 \in G_1, \tag{10.5.4}$$

where G_1 is an arbitrary domain that does not coincide with the plane $\mathbb{R}_2$.

From (10.5.3), (10.5.4), it follows that

$$\left|\iint_G \frac{dx_1\, dx_2}{r_{M_0}^3}\right| \geq \frac{2}{\rho(M_0)}.$$

This inequality proves Lemma 10.5.1.

Consider a canonical partition of the plane into calls $D(k, h)$ (see Definition 5.1.1) with vector-valued step $h = (h_1, h_2)$ satisfying the inequality (8.5.2). Let $\varepsilon_h = \varepsilon(G, h)$ be the set of all indices $k = (k_1, k_2)$ for which $D(k, h) \subset G$. In each cell $D(k, h)$ with $k \in \varepsilon_h$, take a point $M_{0k}(h_1 k_1, h_2 k_2)$. We say that *the points M_{0k} form a grid in the domain G*. By C_h we denote the set of grid functions $a(k)$ equipped with the norm

$$\|a(k)\|_{C_h} = \max_{k \in \varepsilon_h} |a(k)|. \tag{10.5.5}$$

By $C_{\rho h}$ we denote the space of grid functions $a(k)$ with the norm

$$\|a(k)\|_{C_{\rho h}} = \max_{k \in \varepsilon_h} |\rho(k) a(k)|, \tag{10.5.6}$$

where $\rho(k)$ is the distance from the point $M(k_1 h_1, k_2 h_2)$ to the boundary Γ.

Let $r_{0k} = |M M_{0k}|$ for M with coordinates $x = (x_1, x_2)$. Consider a discrete operator Π defined by the relation

$$\Pi a(k) = \sum_{k \in \varepsilon_h} a(k)\gamma(k, m), \tag{10.5.7}$$

where the coefficients $\gamma(k, m)$ have the form

$$\gamma(k, m) = \iint_{D(k,h)} \frac{dx_1 \, dx_2}{r_{0m}^3}. \tag{10.5.8}$$

Proposition 10.5.1. *The operator Π is a stable mapping from C_h onto $C_{\rho h}$.*

Proof. Let $a(k)$ be an arbitrary grid function and let $M(h_1 k_1^0, h_2 k_2^0)$ be a point in which $\|a(k)\|_{C_h}$ is attained. Without loss of generality, we may assume that

$$a(k_1^0, k_2^0) = \|a(k)\|_{C_h}.$$

Let $f(m_1, m_2) = \Pi a(k)$. We have

$$\|f(m_1, m_2)\|_{C_{\rho h}} = \max_{m \in \varepsilon_h} |\rho(m_1, m_2) f(m_1, m_2)|.$$

For $m_1 = k_1^0$, $m_2 = k_2^0$, it follows from Lemma 10.5.1, combined with (10.5.4), that

$$\left| \rho(k_1^0, k_2^0) f(k_1^0, k_2^0) \right| = \left| \rho(k_1^0, k_2^0) \right| \left| \sum_{k \in \varepsilon_h} a(k)\gamma(k, k_1^0, k_2^0) \right| =$$

$$= \left| \rho(k_1^0, k_2^0) \right| \left| a(k_1^0, k_2^0)\gamma(k_1^0, k_2^0, k_1^0, k_2^0) + \sum_{k \neq (k_1^0, k_2^0)} a(k)\gamma(k, k_1^0, k_2^0) \right| \geq$$

$$\geq \rho(k_1^0, k_2^0)\|a(k)\|_{C_h} \left(\left| \gamma(k_1^0, k_2^0, k_1^0, k_2^0) \right| - \sum_{k \neq (k_1^0, k_2^0)} \left| \gamma(k, k_1^0, k_2^0) \right| \right) \geq$$

$$\geq \rho(k_1^0, k_2^0)\|a(k)\|_{C_h} \left| \iint_G \frac{dx_1 \, dx_2}{r_{0 k_1^0 k_2^0}^3} \right| \geq 2\|a(k)\|_{C_h}. \tag{10.5.9}$$

Hence, we obtain the estimate

$$\|\Pi a(k)\|_{C_{\rho h}} \geq 2\|a(k)\|_{C_h}.$$

The last inequality proves Proposition 10.5.1.

The approximate solution $Q_h(x_{1m_1}, x_{2m_2})$ of equation (10.5.1) is sought as a solution of the system

$$\sum_{k \in \varepsilon_h} Q_h(x_{1k_1}, x_{2k_2})\gamma(k_1, k_2, m_1, m_2) = f(x_{1m_1}, x_{2m_2}), \qquad (m_1, m_2) \in \varepsilon_h. \tag{10.5.10}$$

Theorem 10.5.1. *Let G be a bounded convex domain with the boundary Γ of class C^∞, and let $f(x_1^0, x_2^0) \in C^\infty(\overline{G})$. Then*

$$|Q_h(x_{1k_1}, x_{2k_2}) - Q(x_{1k_1}, x_{2k_2})| \leq C|h|^{1/2}, \tag{10.5.11}$$

where $Q(x_1, x_2)$ is the solution of equation (10.5.1) and Q_h is the solution of system (10.5.10).

Proof. For the exact values $Q(x_{1k_1}, x_{2k_2})$ we have

$$\sum_{k \in \varepsilon_h} Q(x_{1k_1}, x_{2k_2})\gamma(k_1, k_2, m_1, m_2) = f(x_{1m_1}, x_{2m_2}) - \Delta f(x_{1m_1}, x_{2m_2}), \qquad (m_1, m_2) \in \varepsilon_h,$$

$$(10.5.12)$$

where $\Delta f(x_{1m_1}, x_{2m_2})$ is the quadrature error.

Subtracting (10.5.12) from (10.5.10), we obtain

$$\sum_{k \in \varepsilon_h} \left[Q_h(x_{1k_1}, x_{2k_2}) - Q(x_{1k_1}, x_{2k_2}) \right]\gamma(k_1, k_2, m_1, m_2) = \Delta f(x_{1m_1}, x_{2m_2}), \qquad (m_1, m_2) \in \varepsilon_h.$$

$$(10.5.13)$$

As shown by *Eskin* (1973), under our assumptions regarding $f(x_1^0, x_2^0)$ and Γ, the solution $Q(x_1, x_2)$ of equation (10.5.1) can be represented in the form

$$Q(x_1, x_2) = y_1(x_1, x_2) + \sum_{k=1}^{p} y_k(x_1, x_2),$$

where $y_1(x_1, x_2)$ satisfies the conditions of Theorem 9.4.1 with $\alpha = 1$, $\beta = 1/2$, and each term $y_k(x_1, x_2)$, $2 \le k \le p$, satisfies the conditions of Theorem 9.4.2.

Thus, Theorems 9.4.1 and 9.4.2 yield the following estimate for the quadrature error:

$$|\Delta f(x_{1m_1}, x_{2m_2})| \le \frac{C|h|^{1/2}}{\delta_{m_1 m_2}}, \qquad (10.5.14)$$

where $\delta_{m_1 m_2}$ is the distance from $M(m_1 h_1, m_2 h_2)$ to the boundary Γ.

From (10.5.6) and (10.5.4), it follows that

$$\|\Delta f(x_{1m_1}, x_{2m_2})\|_{C_{\rho h}} \le C|h|^{1/2}. \qquad (10.5.15)$$

This inequality, combined with Proposition 10.5.1, proves Theorem 10.5.1.

The results of the present section have been communicated to the authors by A. V. Setukha.

10.6. Convergence of Difference Ratios for the Numerical Solution

In this section, we examine the convergence of finite difference ratios constructed for the solution of the hypersingular integral equation (10.3.1).

Let G be a bounded domain with the boundary Γ of class C^∞. Consider a canonical partition of $\mathbb{R}_2$ into cells $D(k, h)$ (see Definition 5.1.1) with constant vector-valued step $h = (h_1, h_2)$ satisfying the inequality (8.5.2). Denote by $\varepsilon_h = \varepsilon(G, h)$ the set of all indices $k = (k_1, k_2)$ such that $D(k, h) \subset G$. In each cell $D(k, h)$, $k \in \varepsilon_h$, we take a point $M_{0k}(k_1 h_1, k_2 h_2)$. Let $r_{0k} = |MM_{0k}|$ for the point M with coordinates $x = (x_1, x_2)$, and let $x_{1k_1} = k_1 h_1$, $x_{2k_2} = k_2 h_2$. The approximate solution $Q_h(x_{1k_1}, x_{2k_2})$ of the integral equation is sought as a solution of the system

$$\sum_{k \in \varepsilon_h} Q_h(x_{1k_1}, x_{2k_2})\gamma(k_1, k_2, m_1, m_2) = f(x_{1m_1}, x_{2m_2}), \qquad (m_1, m_2) \in \varepsilon_h, \qquad (10.6.1)$$

where $\gamma(k, m)$ is defined by (10.3.24).

We introduce the following quantities

$$\widehat{Q}^1_{x_1 h}(x_{1k_1}, x_{2k_2}) = \frac{Q_h(x_{1k_1+1}, x_{2k_2}) - Q_h(x_{1k_1}, x_{2k_2})}{h_1},$$

$$\widehat{Q}^1_{x_2 h}(x_{1k_1}, x_{2k_2}) = \frac{Q_h(x_{1k_1}, x_{2k_2+1}) - Q_h(x_{1k_1}, x_{2k_2})}{h_2},$$

$$Q^1_{x_1 h}(x_{1k_1}, x_{2k_2}) = \frac{Q(x_{1k_1+1}, x_{2k_2}) - Q(x_{1k_1}, x_{2k_2})}{h_1}, \qquad (10.6.2)$$

$$Q^1_{x_2 h}(x_{1k_1}, x_{2k_2}) = \frac{Q(x_{1k_1}, x_{2k_2+1}) - Q(x_{1k_1}, x_{2k_2})}{h_2},$$

$$\Delta Q^1_{x_1 h}(x_{1k_1}, x_{2k_2}) = \widehat{Q}^1_{x_1 h}(x_{1k_1}, x_{2k_2}) - Q^1_{x_1 h}(x_{1k_1}, x_{2k_2}),$$

$$\Delta Q^1_{x_2 h}(x_{1k_1}, x_{2k_2}) = \widehat{Q}^1_{x_2 h}(x_{1k_1}, x_{2k_2}) - Q^1_{x_2 h}(x_{1k_1}, x_{2k_2}).$$

For the exact values $Q(x_{1k_1}, x_{2k_2})$ we have

$$\sum_{k \in \varepsilon_h} Q(x_{1k_1}, x_{2k_2}) \gamma(k_1, k_2, m_1, m_2) = f(x_{1m_1}, x_{2m_2}) - \Delta f(m_1, m_2), \qquad (m_1, m_2) \in \varepsilon_h, \quad (10.6.3)$$

where $\Delta f(m_1, m_2)$ is the quadrature error.

From (10.6.1), (10.6.3), (10.6.2), by means algebraic transformations, we obtain

$$\Delta Q^1_{x_1 h}(x_{1k_1}, x_{2k_2}) = \sum_{m \in \varepsilon_h} \frac{y_h(m_1, m_2, k_1 + 1, k_2) - y_h(m_1, m_2, k_1, k_2)}{h_1} \Delta f(m_1, m_2), \quad (10.6.4)$$

$$\Delta Q^1_{x_2 h}(x_{1k_1}, x_{2k_2}) = \sum_{m \in \varepsilon_h} \frac{y_h(m_1, m_2, k_1, k_2 + 1) - y_h(m_1, m_2, k_1, k_2)}{h_2} \Delta f(m_1, m_2), \quad (10.6.5)$$

where $y_h(m_1, m_2, k_1, k_2)$ are the values of the discrete Green function for system (10.6.1).
The quantities

$$x_h(m_1, m_2, k_1, k_2) = \frac{y_h(m_1, m_2, k_1 + 1, k_2) - y_h(m_1, m_2, k_1, k_2)}{h_1}, \qquad (10.6.6)$$

$$\widehat{x}_h(m_1, m_2, k_1, k_2) = \frac{y_h(m_1, m_2, k_1, k_2 + 1) - y_h(m_1, m_2, k_1, k_2)}{h_2}, \qquad (10.6.7)$$

are the Fourier coefficients of the solutions of problems (8.5.8) and (8.5.9) respectively. The role of the parameters m_1, m_2 in problems (8.5.8) and (8.5.9) is played by k_1 and k_2 in (10.6.6) and (10.6.7). The quantities $x_h(m_1, m_2, k_1, k_2)$ and $\widehat{x}_h(m_1, m_2, k_1, k_2)$ can be represented in the form (8.5.50) and satisfy the estimates (8.5.51), (8.5.52). Let $\rho(x_1, x_2)$ be the distance from the point $M(x_1, x_2)$ to the boundary Γ of the domain G. Consider the points $M(x_{1k_1}, x_{2k_2})$ for which $\rho(x_{1k_1}, x_{2k_2}) \geq A_0 |h|^\alpha$, with $0 \leq \alpha < 1$ and the constant A_0 independent of α, h. For such points, we can take $\nu = C[|h|^{\alpha-1}]$ in (8.5.52), where $C > 0$ is independent of $|h|$, and $[m]$ is the integer part of m. These observations allow us to obtain the estimate

$$E^\nu_{nk} = C(\varepsilon) |\ln^3 |h|| |h|^{3-5\alpha-5\varepsilon(1-\alpha)}, \qquad (10.6.8)$$

and we have $K_1 |h| \leq 1/n \leq K_2 |h|$ with K_1, K_2 independent of $|h|$.

In order to estimate $\Delta Q^1_{x_1 h}(x_{1k_1}, x_{2k_2})$, $\Delta Q^1_{x_2 h}(x_{1k_1}, x_{2k_2})$, we need the following result.

Lemma 10.6.1. *Let $D_{\delta_1\delta_2}$ be the set of all points $M(x_{1k_1}, x_{2k_2})$, $(k_1, k_2) \in \varepsilon_h$, which satisfy the inequality $\delta_1 \le \rho(x_{1k_1}, x_{2k_2}) \le \delta_2$. Then, there exist $h_0 > 0$ and $B_0 > 0$, $\delta_0 > 0$ independent of $h = (h_1, h_2)$ such that for $|h| < h_0$, $\delta_2 < \delta_0$, we have*

$$N_{\delta_1\delta_2} \le \left[\frac{B_0 l(\delta_2 - \delta_1 + |h|)}{|h|^2}\right],$$

where $N_{\delta_1\delta_2}$ is the number of points of $D_{\delta_1\delta_2}$; l is the length of Γ; $[m]$ is the integer part of m.

Proof. Since the boundary Γ is of class C^∞, there is a neighborhood of Γ, denoted by V, in which we can introduce curvilinear coordinates with coordinate lines coinciding with the curves $\Gamma_\varepsilon:\ r = r_M + \varepsilon n_M^0$, where $M \in \Gamma$, n_M^0 is the unit inward normal to Γ at the point M. The coordinates of a point $A \in V$ are S, ε, where S is the length of the segment $M_0 M \subset \Gamma$ and M_0 is a fixed point of Γ. We have

$$J = \left\|\begin{array}{cc} \dfrac{\partial x}{\partial S} & \dfrac{\partial x}{\partial \varepsilon} \\ \dfrac{\partial x}{\partial S} & \dfrac{\partial y}{\partial \varepsilon} \end{array}\right\| = |1 + \varepsilon k|,$$

where k is the curvature of Γ.

Let $d > 0$ be such that the set D formed by the points $M(x_1, x_2)$ for which $0 \le \rho(x_1, x_2) \le d$ belongs to V. Now, we take $\delta_0 = d/2$ and choose $h_0 > 0$ such that for $|h| < h_0$, the cells $D(k, h)$, $h \in \varepsilon_h$, with centers at $M(x_{1k_1}, x_{2k_2}) \in D_{0\delta_0}$ belong to D. Consider the domain $M_{\delta_1\delta_2} \subset G$ consisting of the points $M(x_1, x_2)$ such that

$$\left(\delta_1 - \frac{|h|}{2}\right) \le \rho(x_1, x_2) \le \left(\delta_2 + \frac{|h|}{2}\right).$$

Obviously, each cell $D(k, h)$ with center at $M(x_{1k_1}, x_{2k_2}) \in D_{\delta_1\delta_2}$ belongs to $M_{\delta_1\delta_2}$. Let us find the area of $M_{\delta_1\delta_2}$, denoted by $\sigma(M_{\delta_1\delta_2})$. We have

$$\sigma(M_{\delta_1\delta_2}) = \int_0^l \int_{\delta-|h|/2}^{\delta_2+|h|/2} |J|\, dS\, d\varepsilon \le B_0 l(\delta_2 - \delta_1 + |h|).$$

Hence, we obtain the desired result. Lemma 10.6.1. is proved.

Now, let us estimate $\Delta Q_{x_1 h}^1(x_{1k_1}, x_{2k_2})$ at the points $M(x_{1k_1}, x_{2k_2})$ for which $\rho(x_{1k_1}, x_{2k_2}) \ge A_0 |h|^\alpha$, $0 \le \alpha < \alpha_0 < 1$.

Suppose that $f(x_1^0, x_2^0) \in C^\infty(\overline{G})$ in (10.3.1). Then, according to (10.5.14), the quadrature error can be estimated as follows:

$$|\Delta f(m_1, m_2)| \le \frac{C|h|^{1/2}}{\rho(x_{1m_1}, x_{2m_2})}. \tag{10.6.9}$$

From (10.6.4), (10.6.6), (8.5.50), we obtain

$$\Delta Q_{x_1 h}^1(x_{1k_1}, x_{2k_2}) = \sum_{m\in\varepsilon_h} p_{k_1 k_2}^n(m_1, m_2)\Delta f(m_1, m_2) + \sum_{m\in\varepsilon_h} x_{1k_1 k_2}^n(m_1, m_2)\Delta f(m_1, m_2) =$$
$$= \Delta_{1h}(k_1, k_2) + \Delta_{2h}(k_1, k_2). \tag{10.6.10}$$

Denote by $D_1(k_1, k_2)$ the set of all indices $m = (m_1, m_2)$ such that the distance from the point $M(x_{1m_1}, x_{2m_2})$ to $M(x_{1k_1}, x_{2k_2})$ is less than $\rho(x_{1k_1}, x_{2k_2})/2$, and let $D_2 = \varepsilon_h \setminus D_1$. Then, using

(8.5.51) and (10.6.9), we obtain the estimate

$$|\Delta_{1h}(k_1, k_2)| = \sum_{m \in D_1} \frac{C|\ln|h||^3|h|^{1/2}}{|m_1 - k_1|^2 + |m_2 - k_2|^2 + 1} \frac{1}{\rho(x_{1m_1}, x_{2m_2})} +$$

$$+ \sum_{m \in D_2} \frac{C|\ln|h||^3|h|^{1/2}}{|m_1 - k_1|^2 + |m_2 - k_2|^2 + 1} \frac{1}{\rho(x_{1m_1}, x_{2m_2})} =$$

$$= \Delta'_{1h}(k_1, k_2) + \Delta''_{1h}(k_1, k_2). \tag{10.6.11}$$

In view of the definition of D_1, we have

$$|\Delta'_{1h}(k_1, k_2)| \le C_1 |\ln|h||^3|h|^{1/2-\alpha}. \tag{10.6.12}$$

For $M = (m_1, m_2) \in D_2$, the following inequality holds:

$$(m_1 - k_1)^2 + (m_2 - k_2)^2 \ge C_1|h|^{2\alpha-2}.$$

Therefore, for $|h| \le h_0$, with h_0 specified in Lemma 10.6.1, we obtain the estimate

$$|\Delta''_{1k}(k_1, k_2)| \le C_2 \sum_{m \in D_2} \frac{|h|^{5/2-2\alpha}|\ln|h||^3}{\rho(x_{1m_1}, x_{2m_2})} \le C_3|h|^{1/2-2\alpha}|\ln|h||^4. \tag{10.6.13}$$

From (10.6.11)–(10.6.13), it follows that

$$|\Delta_{1h}(k_1, k_2)| \le C_4|h|^{1/2-2\alpha}|\ln|h||^4. \tag{10.6.14}$$

Let $D_{0\delta_0}$ be the set of indices $m = (m_1, m_2)$ from Lemma 10.6.1, with $\delta_1 = 0$, $\delta_2 = \delta_0$, and $D'_{0\delta_0} = \varepsilon_h \setminus D_{0\delta_0}$. Then, using (8.5.52), (10.6.8), and (10.6.9), we find that

$$|\Delta_{2h}(k_1, k_2)| \le$$

$$\le C(\varepsilon, \alpha)|\ln|h||^{3/2}|h|^{2-5\alpha/2-5\varepsilon(1-\alpha)/2} \left| \sum_{m \in D_{0\delta_0}} \frac{x^n_{2k_1k_2}(m_1, m_2)}{\rho(x_{1m_1}, x_{2m_2})} + \sum_{m \in D'_{0\delta_0}} \frac{x^n_{2k_1k_2}(m_1, m_2)}{\rho(x_{1m_1}, x_{2m_2})} \right| \le$$

$$\le \quad C(\varepsilon, \alpha)|\ln|h||^{3/2}|h|^{2-5\alpha/2-5\varepsilon(1-\alpha)/2} \left[\sum_{m \in \varepsilon_h} \left| x^n_{2k_1k_2}(m_1, m_2) \right|^4 \right]^{1/4} \times$$

$$\times \left(\left[\sum_{m \in D_{0\delta_0}} [\rho(x_{1m_1}, x_{2m_2})]^{-4/3} \right]^{3/4} + + \left[\sum_{m \in D'_{0\delta_0}} [\rho(x_{1m_1}, x_{2m_2})]^{-4/3} \right]^{3/4} \right) \le$$

$$\le C_1(\varepsilon, \alpha)|\ln|h||^{7/4}|h|^{1/4-5\alpha/2-5\varepsilon(1-\alpha)/2}. \tag{10.6.15}$$

From (10.6.14) and (10.6.15), we obtain the estimate

$$\left| \Delta Q^1_{x_1h}(x_{1k_1}, x_{2k_2}) \right| \le C_2(\varepsilon, \alpha)|\ln|h||^{7/4}|h|^{1/4-5\alpha/2-5\varepsilon(1-\alpha)/2}. \tag{10.6.16}$$

where $|C_2(\varepsilon, \alpha)| \le C_3(\varepsilon)$ for $0 \le \alpha \le \alpha_0 < 1$.

In a similar manner, we obtain the estimate (10.6.16) for $\Delta Q^1_{x_2h}(x_{1k_1}, x_{2k_2})$.

Proposition 10.6.1. *Let G ba a bounded domain with the boundary Γ of class C^∞, and let $f(x_1^0, x_2^0) \in C^\infty(\overline{G})$. Then, for all $M(x_{1k_1}, x_{2k_2})$ whose distance form Γ is $\delta > 0$, we have*

$$\left| \widehat{Q}_{x_i h}^1(x_{1k_1}, x_{2k_2}) - Q_{x_i h}^1(x_{1k_1}, x_{2k_2}) \right| \leq C(\varepsilon) |h|^{1/4-\varepsilon}, \qquad \varepsilon > 0, \quad i = 1, 2. \tag{10.6.17}$$

This statement follows from the inequality (10.6.16) with $\alpha = 0$.

Remark 10.6.1. From (10.6.16) with $0 < \alpha < 1/11$, $0 < \varepsilon < 1/200$, it follows that

$$\lim_{|h| \to 0} \left| \Delta Q_{x_i h}^1(x_{1k_1}, x_{2k_2}) \right| = 0, \tag{10.6.18}$$

and this convergence is uniform with respect to $M(x_{1k_1}, x_{2k_2})$ for which $\rho(x_{1k_1}, x_{2k_2}) \geq A_1 |h|^\alpha$, $0 < \alpha < 1/11$.

Let $\Delta f(x, h)$ be a step-function such that $\pi f(x, h) = \Delta f(m_1, m_2)$ for $x \in D(m, h) \subset G$, where π is the standard mapping from Definition 5.1.2.

Proposition 10.6.2. *Let G be a bounded convex domain with the boundary Γ of class C^∞, and let $f(x_1^0, x_2^0) \in C^\infty(\overline{G})$. Then the following estimate holds for the quadrature error:*

$$\|\Delta f(x, h)\|_{M(-1/2, h, \Omega_{1h})} \leq C |h|^{1/2} |\ln |h||^{1/2}, \tag{10.6.19}$$

where $C > 0$ does not depend on $h = (h_1, h_2)$; $M(-1/2, h, \Omega_{1h})$ is a quotient space (see Definition 5.5.4).

Proof. The properties of linear functionals allow us to write

$$\|\Delta f(x, h)\|_{M(-1/2, 1, \Omega_{1h})} = \sup_{\substack{\|u\|=1 \\ u \in \overset{\circ}{M}(1/2, h, \Omega_{1h})}} |(\Delta f(x, h), u(x, h))|, \tag{10.6.20}$$

where $(\Delta f(x, h), u(x, h))$ is the value of the linear functional associated with the element $u(x, h) \in \overset{\circ}{M}(1/2, h, \Omega_{1h})$.

Using Lemma 10.5.1, from (7.2.5), (10.6.9), and (10.6.20), we obtain the inequality

$$\|\Delta f(x, h)\|_{M(-1/2, h\Omega_{1h})} = C \sup_{\|u\|=1} \sum_{m \in \varepsilon_h} \frac{|h|^{1/2} |u(m_1, m_2)|}{\rho(x_{1m_1}, x_{2m_2})} h_1 h_2 \leq$$

$$\leq C |h|^{1/2} \sup_{\|u\|=1} \left(\sum_{m \in \varepsilon_h} \frac{|u(m_1, m_2)|^2}{\rho(x_{1m_1}, x_{2m_2})} h_1 h_2 \right)^{1/2} \left(\sum_{m \in \varepsilon_h} \frac{h_1 h_2}{\rho(x_{1m_1}, x_{2m_2})} \right)^{1/2} \leq$$

$$\leq C_1 |h|^{1/2} |\ln |h||^{1/2} \sup_{\|u\|=1} \|u(x, h)\|_{\overset{\circ}{M}(1/2, h, \Omega_{1h})} = C_1 |h|^{1/2} |\ln |h||^{1/2}.$$

Proposition 10.6.2 is proved.

Corollary 10.6.1. *Let $\Delta Q(x, h)$ be a step-function such that*

$$\pi Q(x, h) = Q_h(x_{1k_1}, x_{2k_2}) - Q(x_{1k_1}, x_{2k_2}) \quad \text{for} \quad x \in D(k, h).$$

Then, Propositions 7.2.1 and 10.6.2 imply the estimate

$$\|\Delta Q(x, h)\|_{\overset{\circ}{M}(1/2, h, \Omega_{1h})} \leq C_1 |h|^{1/2} |\ln |h||^{1/2}. \tag{10.6.21}$$

Denote by $\Delta Q^1_{x_i h}$ the step-functions defined by the relations

$$\pi \Delta Q^1_{x_i h}(x, h) = \Delta Q^1_{x_i h}(x_{1k_1}, x_{2k_2}) \quad \text{for} \quad x \in D(k, h), \quad i = 1, 2.$$

Proposition 10.6.3. *Let G be a bounded convex domain with the boundary Γ of class C^∞, and let $f(x_1^0, x_2^0) \in C^\infty(\overline{G})$. Then*

$$\lim_{|h| \to 0} \int_{\mathbb{R}_2} \left| \Delta Q^1_{x_i h}(x, h) \, dx_1 \, dx_2 \right| = 0. \tag{10.6.22}$$

Proof. Let us show that for any $u(x, h) \in M(1, h)$ (see Definition 5.3.1), the following inequality holds:

$$\|u(x, h)\|_{M(1,h)} \le \frac{C}{|h|^{1/2}} \|u(x, h)\|_{M(1/2,h)}. \tag{10.6.23}$$

To that end, we utilize Propositions 5.3.4 and 5.3.5. We have

$$h_1 h_2 \sum_k \frac{|u(k_1 + 1, k_2) - u(k_1, k_2)|^2}{h_1^2} \le \sum_{|l| \ne 0} \frac{1}{|l|^3 h_1^2} h_1 h_2 \sum_k |u(k + l) - u(k)|^2 =$$

$$= \frac{1}{h_1} h_1 h_2 \sum_{|l| \ne 0} \frac{1}{|l|^3 h_1^2 h_2} h_1 h_2 \sum_k |u(k + l) - u(k)|^2 \le \frac{C}{|h|} \|u(x, h)\|^2_{M(1/2,h)}. \tag{10.6.24}$$

Similarly,

$$h_1 h_2 \sum_k \frac{|u(k_1, k_2 + 1) - u(k_1, k_2)|^2}{h_2^2} \le \frac{C}{|h|} \|u(x, h)\|^2_{M(1/2,h)}. \tag{10.6.25}$$

The inequality (10.6.23) follows from (10.6.24) and (10.6.25). From (10.6.2), we find that

$$\Delta Q^1_{x_1 h}(x_{1k_1}, x_{2k_2}) = \frac{\Delta Q(x_{1k_1+1}, x_{2k_2}) - \Delta Q(x_{1k_1}, x_{2k_2})}{h_1},$$

$$\Delta Q^1_{x_2 h}(x_{1k_1}, x_{2k_2}) = \frac{\Delta Q(x_{1k_1}, x_{2k_2+1}) - \Delta Q(x_{1k_1}, x_{2k_2})}{h_1}, \tag{10.6.26}$$

where $\Delta Q(x_{1k_1}, x_{2k_2}) = Q_h(x_{1k_1}, x_{2k_2}) - Q(x_{1k_1}, x_{2k_2})$.

Denote by $G_i(h)$ the support of the function $\Delta Q^1_{x_i h}(x, h)$, and let $G'(h)$ be the set consisting of the points of $D(k, h)$ for which $\rho(x_{1k_1}, x_{2k_2}) \ge A_0 |h|^\alpha$, $0 < \alpha < 1/11$, where $A_0 > 0$ does not depend on $h = (h_1, h_2)$. Let $G''_i(h) = G_i(h) \setminus G'(h)$. Note that the measure of $G''_i(h)$ satisfies the inequality

$$\text{meas } G''_i(h) \le C_1 |h|^\alpha, \tag{10.6.27}$$

with C_1 independent of $h = (h_1, h_2)$.

We further have

$$I_i(h) = \int_{G_i(h)} \left| \Delta Q^1_{x_i h}(x, h) \right| dx_1 \, dx_2 =$$

$$= \int_{G'(h)} \left| \Delta Q^1_{x_i h}(x, h) \right| dx_1 \, dx_2 + \int_{G''_i(h)} \left| \Delta Q^1_{x_i h}(x, h) \right| dx_1 \, dx_2 =$$

$$= I'_i(h) + I''_i(h). \tag{10.6.28}$$

From (10.6.18), it follows that

$$\lim_{|h| \to 0} I'_i(h) = 0. \tag{10.6.29}$$

Combining (10.6.21), (10.6.23), (10.6.26), and (10.6.27), we obtain

$$I_i''(h) \leq C_2 \left(\int_{G_i''(h)} |h|^{\alpha/2} \left| \Delta Q_{x_i h}^1(x, h) \right|^2 dx_1 \, dx_2 \right)^{1/2} \leq C_3 \| \Delta Q(x, h) \|_{M(1,h)} |h|^{\alpha/2} \leq$$

$$\leq C_4 |h|^{\alpha/2} \frac{1}{|h|^{1/2}} |h|^{1/2} |\ln|h||^{1/2} = C_4 |h|^{\alpha/2} |\ln|h||^{1/2}. \tag{10.6.30}$$

Convergence (10.6.22) follows from (10.6.29) and (10.6.30).

Chapter 11

Problems in Aerodynamics

11.1. Mathematical Modelling of Flow Past an Airfoil with Suction and Pseudodifferential Operators

Let L be the curve delimiting an airfoil in potential incompressible flow whose velocity is

$$U_0(M) = \nabla\Phi_0(M) = \frac{\partial}{\partial x}\Phi_0(M)\boldsymbol{i} + \frac{\partial}{\partial x}\Phi_0(M)\boldsymbol{j},$$

where $\boldsymbol{i}$, $\boldsymbol{j}$ are orthogonal unit vectors on the axes OX, OY, respectively; M is a point on the plane OXY; $\Phi_0(M)$ is a harmonic function defined on the entire plane. We assume that the curve L admits a parametric representation $x = x(t)$, $y = y(t)$, where $t \in [0, l]$ is the arc length. The curve L is oriented clockwise, if L is closed. The functions $x'(t)$ and $y'(t)$ are supposed to satisfy the Hölder condition on $[0, l]$ with exponent α, i.e., $x'(t), y'(t) \in H(\alpha)$ on $[0, l]$. Since t is the arc length, we have $x'^2(t) + y'^2(t) \equiv 1$ for $t \in [0, l]$. If L is a closed curve , the functions $x(t)$, $y(t)$ and their derivatives are periodic with period l, which is assumed equal to 2π for simplicity.

The airfoil is modeled by a vortex layer (see *Belotserkovskii* (1965), *Belotserkovskii and Lifanov* (1985)) of intensity $\gamma(t) = \gamma(M(t)$ at the point $M(x(t), y(t)) = M(t)$ of curve L. At any point $M_0 = M(x_0, y_0)$ of the plane, the velocity $V_\gamma(M_0)$ produced by that layer is determined by the formula

$$V_\gamma(M_0) = \frac{1}{2\pi} \int_L \frac{y_1(M, M_0)\boldsymbol{i} - x_1(M, M_0)\boldsymbol{j}}{r_{MM_0}^2}\gamma(t)\, dt, \qquad M_0 \neq L, \qquad (11.1.1)$$

$$x_1(M, M_0) = x_0 - x(t), \qquad y_1(M, M_0) = y_0 - y(t),$$

$$r_{MM_0} = |\boldsymbol{r}_{MM_0}| = |x_1(M, M_0)\boldsymbol{i} + y_1(M, M_0)\boldsymbol{j}|.$$

At each point $M \in L$, there exists the unit tangential vector $\boldsymbol{\tau}_M = x'(t)\boldsymbol{i} + y'(t)\boldsymbol{j}$. We choose the normal vector $\boldsymbol{n}_M = -y'(t)\boldsymbol{i} + x'(t)\boldsymbol{j}$ at this point and denote by L^+ the side of the curve in the direction of $\boldsymbol{n}_M$ and by L^- its opposite side. It is known that the velocity $V_\gamma^\pm$ produced by the vortex layer at the points $M_0 \in L$ approached from the corresponding side is given by the formula

$$V_\gamma^\pm(M_0) = \frac{1}{2\pi} \int_L \frac{y_1(M, M_0)\boldsymbol{i} - x_1(M, M_0)\boldsymbol{j}}{r_{MM_0}^2}\gamma(t)\, dt \pm \frac{1}{2}\boldsymbol{\tau}_{M_0}\gamma(t_0), \qquad M_0 \in L. \quad (11.1.2)$$

The total velocity of the flow, $U(M_0)$, at the points of the plane is

$$U(M_0) = V_\gamma(M_0) + U_0(M_0). \qquad (11.1.3)$$

It is assumed that the flow past the profile L is smooth, in the sense that on both sides of L, the direction of the flow velocity is tangential to L. Thus, at the points of L the following *condition of non-penetration* holds:

$$U(M_0) \cdot \boldsymbol{n}_{M_0} = 0, \qquad M_0 \in L,$$

or

$$V_\gamma(M_0) \cdot \boldsymbol{n}_{M_0} = -U_0(M_0) \cdot \boldsymbol{n}_{M_0}, \quad M_0 \in L, \tag{11.1.4}$$

where $V_\gamma(M_0)$ is the integral in (11.1.2).

Thus, the problem of finding the velocity field $U(M_0)$ reduces to the following integral equation

$$-\frac{1}{2\pi} \int_L \frac{x'(t_0)x_1(t, t_0) + y'(t_0)y_1(t, t_0)}{r^2_{MM_0}} \gamma(t)\, dt = f_{U_0}(t_0), \qquad t_0 \in [0, l], \tag{11.1.5}$$

where $f_{U_0}(t_0) = -U_0(M_0) \cdot \boldsymbol{n}_{M_0}$, $x_1(t, t_0) = x_1(M, M_0)$, $y_1(t, t_0) = y_1(M, M_0)$, $M, M_0 \in L$.

Equation (11.1.5) holds for a closed curve L, as well as for an open-ended L. For an open-ended curve L, equation (11.1.5) can be written in the form

$$-\frac{1}{2\pi} \int_0^l \frac{\gamma(t)\, dt}{t_0 - t} + \int_0^t K_1(t_0, t)\gamma(t)\, dt = f_{U_0}(t_0), \qquad t_0 \in (0, l); \tag{11.1.6}$$

and for a closed curve L, this equation can be written as

$$-\frac{1}{4\pi} \int_0^{2\pi} \cot \frac{t_0 - t}{2} \gamma(t)\, dt + \int_0^{2\pi} K_2(t_0, t)\gamma(t)\, dt = f_{U_0}(t_0), \qquad t_0 \in [0, 2\pi]. \tag{11.1.7}$$

If $x''(t), y''(t) \in H(\alpha)$ on $[0, l]$ ($[0, 2\pi]$ for a closed curve), then it can be shown (on the basis of the results of *Sobolev* (1947)) that the kernels $K_1(t_0, t)$ and $K_2(t_0, t)$ in equations (11.1.6) and (11.1.7) are also of class $H(\alpha)$ on the corresponding sets.

Suppose now that at a point $M_q(x_q, y_q) \in L$, $x_q = x(t_q)$, $y_q = y(t_q)$, $t_q \in (0, l)$, there is suction of the outer flow into the airfoil shell. Suction of the flow at M_q is modeled by an outlet, according to the experiments described by *Koenig and Falarski* (1971). The velocity field produced by the outlet at the point M_q is given by the formula

$$V_q(M_0) = \frac{Q}{2\pi} \frac{r_{M_0 M_q}}{r^2_{M_0 M_q}}. \tag{11.1.8}$$

Suction into the airfoil shell means that the total velocity field $U(M_0)$ arising in this problem has a singularity of type (11.1.8) in a neighborhood of the point M_q on the outer side (L^+) of the curve L, and has no singularities (is smooth) on the other side of the profile (L^-). In this case, the total velocity of the flow, $U(M_0)$, at the point M_0 has the form

$$U(M_0) = V_\gamma(M_0) + U_0(M_0) + V_q(M_0),$$

and in order to satisfy the above physical conditions on the field $U(M_0)$, let us consider more closely each of the terms in $U(M_0)$ in a neighborhood of the curve L. Recall, that by assumption, the velocity field $U(M_0)$ is continuous on the entire plane. If the function $\gamma(t)$ is Hölder continuous on L, it is known that the velocity field $V_\gamma(M_0)$ is continuous on each side of L. The field $V_1(M_0)$ is unbounded on each side of L, and therefore, it is clear that $\gamma(t)$ must have a jump at M_q, in order that the required physical conditions be fulfilled. Now, we only have to obtain a suitable integral equation for the function $\gamma(t)$, on the basis of the above physical conditions on the behavior of the total velocity field on both sides of L.

Traditional Approach

The traditional approach to these problems (see *Belotserkovskii* (1965), *Belotserkovskii and Lifanov* (1985), *Poltavskii* (1993)) can be described as follows. Under the non-penetration condition (11.1.4), one uses the exterior velocity field for a closed curve L or the velocity field on the side L^+ for an open-ended curve L. Thus, in equation (11.1.4) one takes $\boldsymbol{V}_\gamma^+(M_0) \cdot \boldsymbol{n}_{M_0}$ and $\boldsymbol{U}_0^+(M_0)$ for any point M_0 of curve L. However, since

$$\boldsymbol{V}_\gamma^+(M_0) \cdot \boldsymbol{n}_{M_0} = \boldsymbol{V}_\gamma^-(M_0)\boldsymbol{n}_{M_0} = \boldsymbol{V}_\gamma(M_0) \cdot \boldsymbol{n}_{M_0},$$

$$\boldsymbol{U}_0^+(M_0) \cdot \boldsymbol{n}_{M_0} = \boldsymbol{U}_0^-(M_0) \cdot \boldsymbol{n}_{M_0} = \boldsymbol{U}_0(M_0) \cdot \boldsymbol{n}_{M_0},$$

one drops the index $+$ in equation (11.1.4). However, in the problem with suction according to the physical requirements on the flow velocities on L^+, the non-penetration condition on L^+ can be written for any point $M_0 \neq M_q$. Thus, the traditional non-penetration condition on L^+ yields the equation

$$\boldsymbol{V}_\gamma^+(M_0) \cdot \boldsymbol{n}_{M_0} = -\boldsymbol{U}_0^+(M_0) \cdot \boldsymbol{n}_{M_0} - \boldsymbol{V}_q^+(M_0)\boldsymbol{n}_{M_0}, \qquad M_0 \in L, \quad M_0 \neq M_q. \qquad (11.1.9)$$

For $M_0 \in L$, $M_0 \neq M_q$ we have $\boldsymbol{V}_q^+(M_0) \cdot \boldsymbol{n}_{M_0} = \boldsymbol{V}_q^-(M_0) \cdot \boldsymbol{n}_{M_0} = \boldsymbol{V}_q(M_0) \cdot \boldsymbol{n}_{M_0}$, and therefore, equation (11.1.9) becomes

$$\boldsymbol{V}_\gamma(M_0) \cdot \boldsymbol{n}_{M_0} = -\boldsymbol{U}_0(M_0) \cdot \boldsymbol{n}_{M_0} - \boldsymbol{V}_q(M_0) \cdot \boldsymbol{n}_{M_0}, \qquad M_0 \in M, \quad M_0 \neq M_q, \qquad (11.1.10)$$

where

$$\boldsymbol{V}_q^+(M_0) \cdot \boldsymbol{n}_{M_0} = \frac{Q}{2\pi} \frac{\left(\boldsymbol{r}_{M_0 M_q}, \boldsymbol{n}_{M_0}\right)}{r_{M_0 M_q}^2} = \frac{Q}{2\pi}\omega_1(M_0, M_q), \qquad M_0 \in L, \quad M_0 \neq M_q,$$

and the function $\omega_1(M_0, M_q)$ is the kernel of the double layer potential for $M_0, M_q \in L$. As shown in traditional courses in mathematical physics, if $x''(t), y''(t) \in H(\alpha)$ on $[0, l]$, then the function $\omega_1(M_0, M_q)$, extended by continuity to the point $M_0 = M_q$, belongs to $H(\alpha)$ on $[0, l]$, uniformly in both variables. For example, if L is a unit circle with center at the origin, then

$$\omega_1(M_0, M_q) = -\frac{1}{2}, \qquad M_0 \in L, \quad M_0 \neq M_q. \qquad (11.1.11)$$

In this case, let us take $U_0(M_0) \equiv 0$ for $M_0 \in \mathbb{R}_2$. Then, by virtue of (11.1.7), equation (11.1.10) becomes

$$-\frac{1}{4\pi} \int_0^{2\pi} \cot \frac{t_0 - t}{2} \gamma(t)\, dt = \frac{Q}{4\pi}, \qquad t_0 \in [0, 2\pi], \quad t_0 \neq t_q. \qquad (11.1.12)$$

The type of the solution to be sought cannot be determined merely on the basis of equation (11.1.12) or (11.1.10), especially since it is known that for $Q \neq 0$ this equation has no solutions in L_2. However, from physical considerations, it appears that the total velocity field of the flow cannot have singularities at M_q on the side L_- opposite to the side of the profile on which there is a source. Therefore, the tangential components of the flow, too, cannot have singularities on that side. Thus, the following equation should hold:

$$\boldsymbol{V}_\gamma^- \cdot \boldsymbol{\tau}_{M_0} + \boldsymbol{V}_q^- \cdot \boldsymbol{\tau}_{M_0} = \psi(t_0), \qquad t_0 \in (0, l), \qquad (11.1.13)$$

with a function $\psi(t_0)$ having no singularities near the point t_q. Let us examine the behavior of each term in (11.1.13) near the point M_q. For $M_0 \in L$, $M_0 \neq M_q$, we have

$$\boldsymbol{V}_q^- \cdot \boldsymbol{\tau}_{M_0} = \frac{Q}{2\pi} \frac{x'(t_0)x_1(t_0, t_q) + y'(t_0)t_1(t_0, t_q)}{r_{M_0 M_q}^2}. \qquad (11.1.14)$$

We see that the function $V_q^- \cdot \tau_{M_0}$ has the same structure as the kernel of equation (11.1.5). Therefore, if L is an open-ended curve, this function has the form

$$V_q^- \cdot \tau_{M_0} = \frac{Q}{2\pi} \frac{1}{t_q - t_0} + D_1(t_0, t_q); \tag{11.1.15}$$

and if L is a smooth closed curve, then

$$V_q^- \cdot \tau_{M_0} = \frac{Q}{4\pi} \cot \frac{t_q - t_0}{2} + D_2(t_0, t_q). \tag{11.1.15'}$$

The functions $D_i(t_0, t_q)$ in (11.1.15) and (11.1.15') have no singularities at M_q, provided that $x''(t), y''(t) \in H(\alpha)$ on $[0, l]$.

Remark 11.1.1. If L is a unit circle with center at the origin, then $D_2(t_0, t_q) \equiv 0$, $M_0 \in L$.

Thus, the tangential velocity components due to the source have a singularity at M_q, and this singularity is integrable in the sense of the Cauchy principal value. The function $V_\gamma^- \cdot \tau_{M_0}$ can be expressed by (see *Belotserkovskii* (1965)),

$$V_\gamma^- \cdot \tau_{M_0} = -\frac{\gamma(t_0)}{2} + \frac{1}{2\pi} \int_L \frac{x'(t_0) y_1(t, t_0) - y'(t_0) x_1(t, t_0)}{r_{MM_0}^2} \gamma(t) \, dt. \tag{11.1.16}$$

We see that the kernel of the integral in (11.1.16) has no singularities for any t_0. Moreover, for $x''(t), y''(t) \in H(\alpha)$ on $[0, l]$, this kernel is of the same Hölder class. Therefore, if

$$\gamma(t) = \frac{Q}{\pi} \frac{1}{t_q - t} + \eta_1(t) \tag{11.1.17}$$

for an open-ended curve, or

$$\gamma(t) = \frac{Q}{2\pi} \cot \frac{t_q - t}{2} + \eta_2(t) \tag{11.1.18}$$

for a closed curve L, with $\eta_i(t) \in H(\alpha)$ on $[0, l]$, $i = 1, 2$, then the integral in (11.1.13) has no singularities in a neighborhood of M_q. It follows that relation (11.1.13) holds.

Remark 11.1.2. If L is the unit ball, then the kernel of the integral in (11.1.16) is equal to $1/2$, and therefore, for $\gamma(t) = \frac{Q}{2\pi} \cot \frac{t_q - t}{2}$, we obtain

$$w V_\gamma^- \tau_{M_0} = -\frac{Q}{4\pi} \cot \frac{t_q - t_0}{2}. \tag{11.1.19}$$

Thus, $\psi(t_0) \equiv 0$ in (11.1.13).

This remark implies that if L is the unit ball, the solution of equation (11.1.12) must coincide with the function involved therein, which means that the relation

$$-\frac{1}{4\pi} \int_0^{2\pi} \cot \frac{t_0 - t}{2} \frac{Q}{2\pi} \cot \frac{t_q - t}{2} \, dt = \frac{Q}{4\pi} \tag{11.1.20}$$

must hold in the sense of the Cauchy principal value.

Let us give an analytical proof of (11.1.20) understood in the sense of the Cauchy principal value.

Lemma 11.1.1. *For any 2π-periodic $\gamma(t) \in H(\alpha)$, the relation*

$$\int_0^{2\pi} \cot\frac{y_0 - t}{2} \cot\frac{t - t_q}{2}\gamma(t)\,dt = \tag{11.1.21}$$

$$= \cot\frac{t_0 - t_q}{2}\left[\int_0^{2\pi} \cot\frac{t - t_q}{2}\gamma(t)\,dt - \int_0^{2\pi} \cot\frac{t_q - t}{2}\gamma(t)\,dt\right] + \int_0^{2\pi}\gamma(t)\,dt,$$

$$t_0 \neq t_q, \quad t_0, t_q \in [0, 2\pi].$$

holds in the sense of the Cauchy principal value.

Proof. Elementary trigonometric transformations show that

$$\cot\frac{t_0 - t}{2} \cot\frac{t - t_q}{2} = \cot\frac{t_0 - t_q}{2}\left(\cot\frac{t_0 - t}{2} - \cot\frac{t_q - t}{2}\right) + 1, \tag{11.1.22}$$

for $t_0, t_q \in [0, 2\pi]$, $t_0 \neq t_q$. Multiplying (11.1.12) by $\gamma(t)$ and integrating in t (in the sense of the Cauchy principal value), we obtain (11.1.21). Lemma 11.1.1 is proved.

Let

$$\psi(t_0) = \int_0^{2\pi} \cot\frac{t_0 - t}{2}\gamma(t)\,dt. \tag{11.1.23}$$

Since $\gamma(t) \in H(\alpha)$ on $[0, 2\pi]$, we have $\psi(t_0 \in H(\alpha)$. Therefore, in the square brackets in (11.1.21) we have the difference $\psi(t_0) - \psi(t_q)$ which tends to zero at the same rate as $|t_0 - t_q|^\alpha$ as $t_0 \to t_0$. Therefore, the function

$$\cot\frac{t_0 - t_q}{2}(\Psi(t_0) - \psi(t_q))$$

has an integrable singularity at $t = t_q$. In particular, if we take $\gamma(t) \equiv Q/(2\pi)$ in (11.1.21), we obtain (11.1.20). Thus, we have established (11.1.20) for any $t_0, t_q \in [0, 2\pi]$, with the integral for $t_0 = t_q$ understood in the sense of Hadamard finite value.

From (11.1.20) and the Poincaré–Bertrand formula for integrals with Hilbert kernel

$$\int_0^{2\pi} \cot\frac{t_0 - t}{2}\,dt \int_0^{2\pi} \cot\frac{t - t_1}{2}f(t_1)\,dt_1 = -4\pi^2 f(t_0) + 2\pi\int_0^{2\pi} f(t)\,dt, \tag{11.1.24}$$

it follows that if we have an arbitrary function $f(t_0)$ in equation (11.1.12) and t_q is given, then all its solutions in the class of functions integrable in the sense of the Cauchy principal value are given by the formula

$$\gamma(t) = \frac{1}{\pi}\int_0^{2\pi} \cot\frac{t - t_0}{2}f(t_0)\,dt_0 + C + \frac{1}{\pi}\cot\frac{t_q - t}{2}\int_0^{2\pi} f(t)\,dt. \tag{11.1.25}$$

Remark 11.1.3. It is interesting to note that on the right-hand side of (11.1.21) we have a function with integrable singularity. Therefore, integrating both sides of (11.1.21) and using (11.1.24), we get

$$\int_0^{2\pi} dt_0 \int_0^{2\pi} \cot\frac{t_0 - t}{2} \cot\frac{t - t_0}{2}\gamma(t)\,dt = 4\pi^2\gamma(t_q). \tag{11.1.26}$$

Non-Traditional Approach

The physical statement of the problem with suction shows that on the side L_- of the curve L, the flow should be smooth at any point $M_0 \in L$. Therefore, the relation

$$V_\gamma^-(M_0) \cdot n_{M_0} = -U_0^-(M_0) \cdot n_{M_0} - V_q^-(M_0) \cdot n_{M_0} \qquad (11.1.27)$$

should hold at any point of L. But here $V_\gamma^-(M_0) \cdot n_{M_0}$ and $U_0^-(M_0) \cdot n_{M_0}$ are the same as in (11.1.9). Therefore, it makes sense to examine more closely the function $V_q^-(M_0) \cdot n_{M_0}$ at the points of L. From fluid dynamics, it is clear that this function describes the distributed density of the fluid quantity passing through the points of the curve L. Therefore, for $M_0 \in L$, $M_0 \neq M_q$, we have

$$V_q^-(M_0) \cdot n_{M_0} = V_q(M_0) \cdot n_{M_0} = \frac{Q}{2\pi} \omega_1(M_0, M_q). \qquad (11.1.28)$$

If there exists a tangential line to the curve L at M_q, the nature of the source suggests that the quantity of the fluid passing through M_q from the side L_- is equal to $Q/2$ (the direction of motion of fluid particles passing through M_q from the side L_- forms an acute angle with the vector n_{M_q}). Therefore, for the points of L we can write

$$V_q^-(M_0) \cdot n_{M_0} = \frac{Q}{2\pi} \omega_1(M_0, M_q) + \frac{Q}{2} \delta(M_0 - M_q), \qquad (11.1.29)$$

where M_0 is an arbitrary point of L and $\delta(M_0 - M_q)$, in terms of physicists, is defined by

$$\delta(M_0 - M_q) = \begin{cases} 0 & \text{for} \quad M_0 \in L, ; M_0 \neq M_q, \\ +\infty & \text{for} \quad M_0 = M_q \in L, \end{cases} \qquad (11.1.30)$$

$$\int_L \delta(M_0 - M_q) \, dt_0 = 1. \qquad (11.1.31)$$

The function $\omega_1(M_0, M_q)$ is extended to M_q by continuity. Now, it is clear that for a smooth closed curve L, we have

$$\int_L V_q^-(M_0) \cdot n_{M_0} \, dt_0 = \frac{Q}{2\pi} \int_L \omega_1(M_0, M_q) \, dt_0 + \frac{Q}{2} = -\frac{Q}{2} + \frac{Q}{2} = 0. \qquad (11.1.32)$$

Indeed, all fluid to the right of the tangential line at M_q (the vector n_{M_q} is directed to the left) passes through the points $M_0 \in L$, $M_0 \neq M_q$, and then enters the point M_q.

Thus, equation (11.1.27) becomes

$$V_\gamma(M_0) \cdot n_{M_0} = -U_0(M_0) \cdot n_{M_0} - \frac{Q}{2\pi} \omega_1(M_0, M_q) - \frac{Q}{2} \delta(M_0 - M_q), \qquad M_0 \in L. \quad (11.1.33)$$

This equation differs from (11.1.10) in that it is written for any point $M_0 \in L$, in particular, for M_q, and there is the delta-function (a generalized function) on the right-hand side. Now, the integral operator in (11.1.33) should be understood in the sense of the theory of pseudodifferential operators (see *Saranen and Vainikko* (1999), *Vainikko* (1997). For instance, if L is the unit circle and $U_0(M_0) \equiv 0$, then equation (11.1.33) becomes

$$-\frac{1}{4\pi} \int_0^{2\pi} \cot \frac{t_0 - t}{2} \gamma(t) \, dt = \frac{Q}{4\pi} - \frac{Q}{2} \delta(t_0 - t_q). \qquad (11.1.34)$$

Merely on the basis of the theory of pseudodifferential operators, without examining the behavior of tangential velocities on L_-, we can immediately say that the function

$$\gamma(t) = \frac{Q}{2\pi} \cot \frac{t_q - t}{2}$$

is a solution of equation (11.1.34), as well as of equation (11.1.12).

Remark 11.1.4. Comparing equations (11.1.12) and (11.1.34) and taking into account relation (11.1.20), one can see the difference between understanding an integral with Hilbert kernel as an integral operator in the sense of the Cauchy principal value and as a pseudodifferential operator. These operators coincide on the space L_2 but have different properties if considered on the set of functions integrable in the sense of the Cauchy principal value. Indeed, if we take m different points t_{q_i}, $i = 1, \ldots, m$, on the unit circle and place a source of intensity Q in each point, then the Hilbert operator in the sense of the Cauchy principal value maps different functions $Q(2\pi)^{-1} \cot(2^{-1}(t_{q_i} - t)2)$, $i = 1, \ldots, m$, to a single point, whereas the Hilbert operator understood as a pseudodifferential operator, maps these into different δ-functions. Moreover, in the latter case, there will always be the term $f(t_0)$ in equation (11.1.34) such that

$$\int_0^{2\pi} f(t)\, dt = 0.$$

This can be explained as follows. The pseudodifferential operator corresponding to the integral with Hilbert kernel is constructed on the basis of the following considerations (see *Saranen and Vainikko* (1999), *Vainikko* (1997)). Any periodic generalized functions is represented by a Fourier series. This series is convergent in a suitable Sobolev space related to the space of periodic functions in L_2. Substituting the Fourier series into the integral with Hilbert kernel, integrating it termwise, and taking into account how exponents are transformed by this integral, we obtain a Fourier series in the same Sobolev space. Since this integral maps all constants to zero, the constant in the resulting Fourier series will always be equal to zero, and therefore, the integral of the resulting Fourier series is also equal to zero.

This simple observation allows us to introduce weighted spaces H_ρ^λ (of Sobolev type) related to the spaces $L_{2,p}$ consisting of functions which are square summable with weight ρ. Then, we can introduce linear operators from $H_{\rho_1}^{\lambda_1}$ to $H_{\rho_2}^{\lambda_2}$ on arbitrary curves or surfaces.

Remark 11.1.5. To give a more precise mathematical explanation of what has been said in Remark 11.1.4, let us once again turn to the comparison of equations (11.1.10), (11.1.33) or, for simplicity, equations (11.1.12), (11.1.34) with Hilbert kernel.

For $Q \neq 0$, the right-hand side of equation (11.1.12) does not satisfy the condition of solvability in L_2 (the integral of the right-hand side over $[0, 2\pi]$ should be equal to zero). But, having extended the Hilbert operator to the set of functions integrable in the sense of the Cauchy principal value, we have found a solution of equation (11.1.12), namely, $\gamma(t) = Q(2\pi)^{-1} \cot((t_q - t)/2)$, i.e., we have proved (11.1.20). This function does not belong to L_2 but is integrable in the sense of the principal value.

The right-hand side of equation (11.1.34) contains the δ-function and cannot be integrated in the sense of the Cauchy principal value on $[0, \pi]$. The integral of the right-hand side should be regarded as a linear functional on the basic space of functions. Although the integral of the right-hand side in this sense is equal to zero, i.e., the solvability condition holds, the solution of equation (11.1.34) (if it exists) does not belong to L_2. Now, it is natural to extend the Hilbert operator by continuity to the corresponding Sobolev space and then find a solution of equation (11.1.34). For this purpose, we utilize a construction described by *Vainikko* (1997).

For an arbitrary $\lambda \in \mathbb{R}$, denote by H^λ the Sobolev space of 1-periodic (generalized) functions of the form

$$u(t) = \sum_{n \in \mathbb{Z}} \widehat{u}(n) e_n, \qquad e_n = e^{in2\pi t},$$

satisfying the condition

$$\|u\|_\lambda = \left(\sum_{z \in \mathbb{Z}} \underline{n}^{2\lambda} |\widehat{u}(n)|^2 \right)^{1/2} < \infty, \tag{11.1.35}$$

where $\underline{n} = \max\{1, |n|\}$, $\hat{u}(n) = \int_0^1 u(t) e^{-in2\pi t} \, dt = \langle u, e_{-n} \rangle$ are the Fourier coefficients of u with respect to the orthogonal basis $\{e^{in2\pi t}\}_{n\in\mathbb{Z}}$.

Consider the Hilbert operator

$$(H^* u)(t) = \frac{1}{i} \int_0^1 \cot \pi(s - t) u(s) \, ds, \qquad u \in \bigcap_{\lambda\in\mathbb{R}} H^\lambda.$$

It is known that

$$H e_n = \text{sign}(n) e_n,$$

which means that

$$\frac{1}{i} \int_0^1 \cot \pi(s - t) e^{in2\pi s} \, ds = \text{sign}(n) e^{in2\pi t}, \qquad (11.1.36)$$

where

$$\text{sign}(n) = \begin{cases} 1, & n > 0, \\ 0, & n = 0, \\ -1, & n < 0. \end{cases}$$

From (11.1.36), it follows that the operator H^* is bounded in each H^λ, and therefore, admits a unique extension as a bounded operator $H^{*(\lambda)} \in L(H^\lambda)$ defined on the whole of H^λ. These extensions are consistent with one another in the sense that for $\lambda_1 < \lambda_2$ and $u \in H^{\lambda_2}$, we have $H^{(\lambda_1)} u = H^{(\lambda_2)} u$. The operators $H^{(\lambda)}$ are Fredholm operators of index zero with kernel $N(H^{*(\lambda)}) = \text{span}(1)$ (the set of all constants), and their range is $R(H^{*(\lambda)}) = \{v \in H^\lambda : \langle v, 1 \rangle = 0\}$. In what follows, we write H^* instead of $H^{*(\lambda)}$, keeping in mind that $H^* \in L(H^\lambda)$ for each $\lambda \in \mathbb{R}$.

For $t = 0$, using (11.1.36), we obtain

$$\frac{1}{i} \langle \cot \pi s, e_{-n} \rangle = \text{sign}(-n) = -\text{sign}(n).$$

Thus,

$$\frac{1}{i} \cot \pi s = -\sum_{n\in\mathbb{Z}} \text{sign}(n) e^{in2\pi s}, \qquad (11.1.37)$$

and therefore,

$$\frac{1}{i} \cot \pi(s - s_0) = -\sum_{n\in\mathbb{Z}} \text{sign}(n) e^{in2\pi(s-s_0)} = -\sum_{n\in\mathbb{Z}} \left[\text{sign}(n) e^{-in2\pi s_0} \right] e^{in2\pi s}. \qquad (11.1.38)$$

These series are convergent in each space H^λ, $\lambda < -1/2$. Since $H \in L(H^\lambda)$, we can apply this operator to each term of the series, which yields

$$\frac{1}{i} \int_0^1 \cot \pi(s - t) \frac{1}{i} \cot \pi(s - s_0) \, ds = -\sum_{n\in\mathbb{Z}} \text{sign}(n) e^{-in2\pi s_0} \frac{1}{i} \int_0^1 \cot \pi(s - t) e^{in2\pi s} \, ds. \quad (11.1.39)$$

Taking into account (11.1.36), we obtain

$$\int_0^1 \cot \pi(s - t) \cot \pi(s - s_0) \, ds = \sum_{n\in\mathbb{Z},\, n\neq0} e^{-in2\pi s_0} e^{in2\pi t} = \sum_{n\in\mathbb{Z},\, n\neq0} e^{in2\pi(t-s_0)}. \qquad (11.1.40)$$

Note that the 1-periodic δ-function $\delta(t)$ can be represented by the Fourier series

$$\delta(t) = \sum_{n\in\mathbb{Z}} e^{in2\pi t}, \qquad (11.1.41)$$

which is also convergent in all H^λ, $\lambda < 1/2$. Accordingly,

$$\delta(t - s_0) = \sum_{n \in \mathbb{Z}} e^{in2\pi(t-s_0)}. \tag{11.1.42}$$

Comparing (11.1.40) and (11.1.42), we obtain

$$\int_0^1 \cot \pi(s - t) \cot \pi(s - s_0)\, ds = \delta(t - s_0) - 1. \tag{11.1.43}$$

Relation (11.1.34) can be rewritten in the form

$$\int_0^{2\pi} \cot \frac{t - s}{2} \cot \frac{s_0 - s}{2}\, ds = 4\pi^2 \delta(t - s_0) - 2\pi$$

or, in the notation of (11.1.34), as

$$-\frac{1}{4\pi} \int_0^{2\pi} \cot \frac{t_0 - t}{2} \frac{Q}{2\pi} \cot \frac{t_q - t}{2}\, dt = \frac{Q}{4\pi} - \frac{Q}{2}\delta(t_0 - t_q). \tag{11.1.44}$$

We see that in (11.1.44) (as well as in (11.1.34)), as compared with (11.1.20), there is a new term $-2^{-1}Q\delta(t_0 - t_q)$ with 2π-periodic δ-function. This means that the extensions of the Hilbert operator considered above are different. Indeed, the extension constructed on the basis of the Cauchy principal value "does not take into account" the δ-function that arises when the Hilbert operator is extended by continuity in H^λ with $\lambda < -1/2$. The extension in the sense of the principal value has another unpleasant property in H^λ, $\lambda < -1/2$, namely, the square of the operator, $H \cdot H = H^2$, maps the continuum of functions $\cot(\pi(t_q - t)/2)$, $0 \le t_q \le 2$, to zero. Nevertheless, on the space span $\{H^0, \cot((t_q - t)/2)\}$ with t_q fixed, the behavior of the extension in the sense of the Cauchy principal value is "good". Moreover, the solution obtained on the basis of this (traditional) approach coincides with the solution obtained by the continuous extension of the operator H to H^λ, $\lambda < -1/2$. In each case, we should always remain within the formalism of the adopted extension. In the first case, both singularities in the integral (11.1.20) should be interpreted in the sense of the Cauchy principal value. In the second case, the Cauchy principal value is used only when applying the Hilbert integral to smooth functions, and then one obtains formula (11.1.36), which is then used for the continuous extension of the Hilbert operator to the Sobolev spaces H^λ, $\lambda < -1/2$, and after that, one deals only with this extension.

11.2. Elements of the Potential Theory in the Plane Case

As mentioned in the previous section, the non-traditional approach to the problem with suction involves an equation with the delta-function concentrated at the source, while the right-hand side of that equation coincides with the kernel of the double layer potential. In the present section, we examine this problem in more detail.

First, consider the source function and penetration of fluid through the points of a curve.

Suppose that there is a source of intensity Q at a point $M_0(x_0, y_0)$ on the plane OXY. This source creates a vector field

$$V_0(M) = \frac{Q}{2\pi} \frac{r_{MM_0}}{r_{MM_0}^2}. \tag{11.2.1}$$

The vector field $V_0(M)$ will be regarded as the field of velocities of an ideal incompressible flow.

Let L be a curve on the plane OXY (sometimes denoted by $\mathbb{R}_2$). The curve L is defined by the same parametric relations as in Section 11.1. In the present section, L^+ will be called the right-hand side of L, and L^- is its left-hand side (see Section 11.1). If L splits the plane $\mathbb{R}_2$ into two disjoint domains (for example, L is a straight line or a simple closed curve), then the domain whose direction coincides with that of n_M we denote by D^+, and its complement is denoted by D^-. If L is a simple closed curve, D^+ is the exterior domain, i.e., the orientation of L coincides with the clockwise direction. If L is an open-ended curve, then for any of its points, M, we assume the existence of a neighborhood $O(M)$ such that $O(M)$ is split into two parts by L; the parts adjacent to L^+ and L^- are denoted by $O^+(M)$ and $O^-(M)$, respectively. The set $O(M)$ is called *an admissible neighborhood*.

Consider the following function defined for $M \in L$:

$$V_0(M) \cdot n_M = \frac{Q}{2\pi} \frac{(r_{MM_0}, n_M)}{r_{MM_0}^2} = Q\omega_1(M, M_0), \qquad (11.2.2)$$

where M_0 is an arbitrary point of the plane. Physically, the function $\omega_1(M, M_0)$ is the penetration density of the fluid passing through the points of L due to the unit source $Q = 1$ at M_0.

First, let L be a smooth closed curve. If $M_0 \in D^+$, then it is physically clear that the quantity of the fluid passing through the points of the curve is equal to zero,

$$\int_L \omega_1(M, M_0)\, dt = 0. \qquad (11.2.3)$$

Indeed, each fluid particle moves along a ray passing through M_0. Along this ray, the particle enters the curve L and then abandons it (see Fig. 11.2.1). This is a physical justification of (11.2.3).

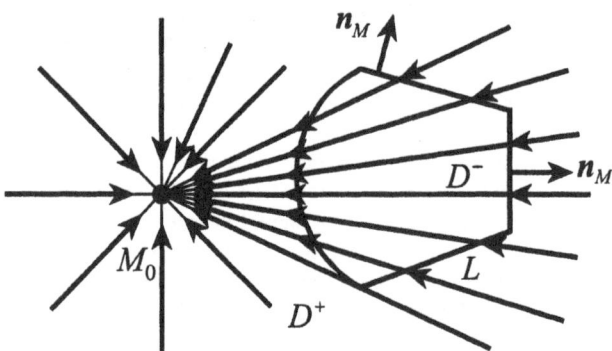

Figure 11.2.1

In order to give a mathematical justification to (11.2.3), we recall the Green formula

$$\int_L P\, dx + Q\, dy = \iint_D \left(\frac{\partial Q}{\partial x} - \frac{\partial P}{\partial y} \right) dx\, dy, \qquad (11.2.4)$$

where D is the region bounded by L, and the orientation of L is such that D remains to the right of L.

The functions P, Q, $\partial P/\partial y$, $\partial Q/\partial x$ are assumed continuous in $\overline{D} = D \cup L$.

By the definition of the normal derivative, we have

$$\int_L \omega_1(M, M_0)\, dt = \frac{1}{2\pi} \int_L \left[-\frac{\partial}{\partial x} \left(\ln \frac{1}{r_{MM_0}} \right) dy + \frac{\partial}{\partial y} \left(\ln \frac{1}{r_{MM_0}} \right) dx \right]. \qquad (11.2.5)$$

Letting

$$P = \frac{\partial}{\partial y}\left(\ln\frac{1}{r_{MM_0}}\right), \qquad Q = \frac{\partial}{\partial x}\left(\ln\frac{1}{r_{MM_0}}\right)$$

and taking into account (11.2.4), we obtain

$$\int_L \omega_1(M, M_0)\, dt = \frac{1}{2\pi}\iint_D \left[-\frac{\partial^2}{\partial x^2}\left(\ln\frac{1}{r_{MM_0}}\right) - \frac{\partial^2}{\partial y^2}\left(\ln\frac{1}{r_{MM_0}}\right)\right] dx\, dy = 0, \qquad (11.2.6)$$

since at each point of the plane the following relation holds:

$$\Delta_M\left(\ln\frac{1}{r_{MM_0}}\right) = 0, \quad M \neq M_0. \qquad (11.2.7)$$

If $M_0 \in D^-$ (i.e., M_0 is inside of the curve L), then the representation of the fluid motion into the source shows that (see Fig. 11.2.2)

$$\int_L \omega_1(M, M_0)\, dt = -1. \qquad (11.2.8)$$

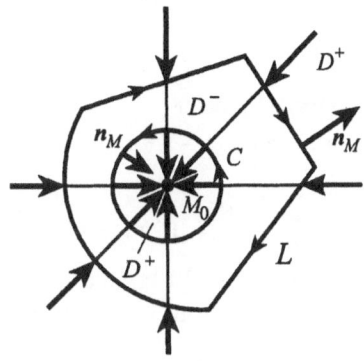

Figure 11.2.2

In order to give a mathematical proof of (11.2.8), consider a circular line $C_r \subset D^-$ of radius r with center at M_0 and choose its orientation counterclockwise. Thus oriented, C_r is denoted by $\widetilde{C}_r$. Then the region between L and $\widetilde{C}_r$ remains to the right, as we move in the positive direction along $\widetilde{C}_r$. Therefore, for the points of $\widetilde{C}_r$, we have $x - x_0 = -r\cos\varphi$, $y_0 - y = -r\sin\varphi$, $n_M = -i\cos\varphi - j\sin\varphi$, and thus

$$\int_{\widetilde{C}_r} \omega_1(M, M_0)\, dt = \frac{1}{2\pi}\int_0^{2\pi}\frac{r}{r^2}r\, d\varphi = 1. \qquad (11.2.9)$$

Now, applying the Green formula in the region between $\widetilde{C}_r$ and L (its boundary is $L \cup \widetilde{C}_r$) and taking into account (11.2.6), we obtain

$$\int_L \omega_1(M, M_0)\, dt = -\int_{\widetilde{C}_r} \omega_1(M, M_0)\, dt = -1.$$

Suppose now that L coincides with the circular line C_r with center at M_0, and let $C_{r,\theta}$ be its arc corresponding to the central angle θ. Then from (11.2.9), keeping in mind Fig. 11.2.2, we obtain

$$\int_{C_{r,\theta}} \omega_1(M, M_0)\, dt = -\frac{\theta}{2\pi}. \tag{11.2.10}$$

Consider more closely the function $\omega_1(M, M_0)$, where $M = M(x(t), y(t))$, $M_0 = M(x(t_0), y(t_0))$ are points on L. By the definition of $\omega_1(M, M_0)$, we can write

$$\omega_1(M, M_0) = \frac{1}{2\pi} \frac{\left(r_{MM_0}, n_M\right)}{r_{MM_0}^2} = \frac{1}{2\pi} \frac{-y'(t)x_3(t, t_0) + x'(t)y_3(t, t_0)}{x_2^2(t, t_0) + y_2^2(t, t_0)}, \tag{11.2.11}$$

where

$$x_3(t, t_0) = \frac{x_2(t, t_0) - x'(t)}{t_0 - t}, \qquad y_3(t, t_0) = \frac{y_2(t, t_0) - y'(t)}{t_0 - t}, \qquad x_2(t, t_0) = \frac{x_1(t, t_0)}{t_0 - t},$$

$$y_2(t, t_0) = \frac{y_1(t, t_0)}{t_0 - t}, \qquad x_1(t.t_0) = x(t_0) - x(t), \qquad y_1(t, t_0) = y(t - 0) - y(t),$$

where the parameter t is the length of curve L.

Theorem 11.2.1. *Let* $x(t)$, $y(t)$ *be defined on* $[0, l]$, *and* $x''(t), y''(t) \in H_r(\alpha)$ *on* $[0, l]$; $r \in \mathbb{N}_0 = \{0, 1, 2, \ldots\}$ *(i.e., the derivatives* $x^{(2+r)}$ *and* $y^{(2+2)}$ *are of class* $H(\alpha)$ *on* $[0, l]$). *Then,* $x_3(t, t_0), y_3(t, t_0)$ *are also of class* $H_r(\alpha)$; *and* $x_2(t, t_0), t_2(t, t_0) \in H_{r+1}(\alpha)$ *on* $[0, l]$, *uniformly in both variables.*

Proof. Since $x(t)$ is continuously differentiable, we have

$$x(t_0) - x(t) = \int_t^{t_0} x'(\sigma)\, d\sigma. \tag{11.2.12}$$

Let us change the variables in this integral by taking

$$\sigma = t + u(t_0 - t), \qquad d\sigma = (t_0 - t)\, du, \qquad u_a = 0, \qquad u_b = 1.$$

Then, we get

$$x_2(t, t_0) = \frac{x(t_0) - x(t)}{t_0 - t} = \int_0^1 x'(t + u(t_0 - t))\, du. \tag{11.2.13}$$

Hence, we obtain the statement of Theorem 11.2.1 regarding $x_2(t, t_0)$ and $y_2(t, t_0)$. Moreover, $x_2(t, t_0)\big|_{t=t_0} = x'(t_0)$. In a similar way, we obtain the statement for the function $y_2(t_0, t_0)$.

On the basis of (11.2.13), we obtain

$$x_3(t, t_0) = \int_0^1 \frac{x'(t + u(t_0 - t)) - x'(t)}{t_0 - t}\, du. \tag{11.2.14}$$

Integrating by parts in (11.2.14) and taking

$$\tilde{u} = \frac{x'(t + u(t_0 - t)) - x'(t)}{t_0 - t}, \qquad d\tilde{v} = du,$$

we get

$$x_3(t, t_0) = \frac{x'(t_0) - x'(t)}{t_0 - t} - \int_0^1 u x''(t + u(t_0 - t))\, du. \tag{11.2.15}$$

By analogy with (11.2.13), we find that

$$\frac{x'(t_0) - x'(t)}{t_0 - t} = \int_0^1 x''(t + u(t_0 - t))\, du. \tag{11.2.16}$$

From (11.2.15) and (11.2.16), it follows that

$$x_3(t, t_0) = \int_0^1 (1 - u)x''(t(1 - u) + t_0 u)\, du. \tag{11.2.17}$$

Hence, we obtain the statement of Theorem 11.2.1 regarding the functions $x_3(t, t_0)$ and $y_3(t, t_0)$. Moreover, we see that $x_3(t_0, t_0) = x''(t_0)/2$, $y_3(t_0, t_0) = y''(t_0)/2$, and the proof is complete.

Corollary 11.2.1. *Let L be a simple smooth curve, i.e., $x'^2(t) + y'^2(t) \equiv 1$, $t \in [0, l]$ (if L is closed, then $x(t)$, $y(t)$ and their derivatives are periodic functions). Suppose that $x''(t), y''(t) \in H_r(\alpha)$ on $[0, l]$. Then the function $\omega(M, M_0) = \omega_1(t, t_0)$ is of the same class on $[0, l] \times [0, l]$.*

This result is ensured by the fact that the denominator in (11.2.11) is a continuous non-vanishing function on $[0, l] \times [0, l]$.

Corollary 11.2.2. *For a piecewise smooth curve L, the function $\omega_1(M, M_0) = \omega_1(t, t_0)$ is piecewise smooth, and therefore, integrable. If $\widetilde{L}$ is a part of L of length $\lambda(\widetilde{L})$, then there is a constant $M > 0$ such that the following inequality holds:*

$$\int_{\widetilde{L}} |\omega_1(M, M_0)| \, dt \leq M \lambda(\widetilde{L}). \tag{11.2.18}$$

Let L be a piecewise smooth curve, i.e., there are finitely many points on L at which the tangential lines do not exist, but at these points there exist one sided limits of the unit normal and tangential vectors, while the second derivatives $x''(t)$ and $y''(t)$ have finite jumps at these points. On the smooth parts of L the functions $x''(t)$ and $y''(t)$ are of class $H_r(\alpha)$. From (11.2.11), (11.2.13), (11.2.17), we obtain the following result.

Let us calculate the value of the integral of $\omega_1(M, M_0)$ over a piecewise smooth curve L,

$$V_L(M_0) = \int_L \omega_1(M, M_0) \, dt, \qquad M_0 \in L. \tag{11.2.19}$$

Previously in this section, we have shown (see (11.2.3) and (11.2.8)) that for a closed piecewise smooth curve L and $M_0 \notin L$, we have $V_L(M_0) = 0$ for $M_0 \in D^+$, and $V_L(M_0) = -1$ for $M_0 \in D^-$.

Now, let $M_0 \in L$. Assume first that M_0 belongs to a smooth piece of L (see Fig. 11.2.3). The physical meaning of the function $\omega_1(M, M_0)$ (it is the penetration density of the fluid passing through the points M of the curve L, $M \neq M_0$) suggests the following approach. Let P be the tangential line at M_0. Clearly, the fluid particles moving along the rays entering M_0 to the right of P cross the curve L. Thus, the quantity of the fluid that enters M_0 to the right of the line P, passes through points $M \neq M_0$ of L. This quantity is equal to $-1/2$. Therefore, if M_0 belongs to a smooth part of L, then

$$V_L(M_0) = \int_L \omega_1(M, M_0) \, dt = -\frac{1}{2}. \tag{11.2.20}$$

This is a physical justification of (11.2.20)

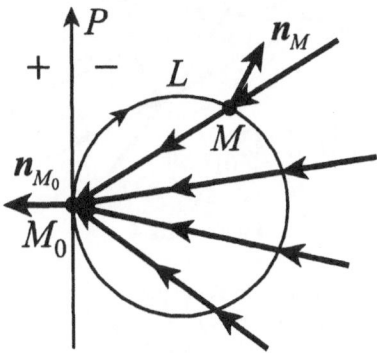

Figure 11.2.3

Let us give a rigorous proof of (11.2.20). To that end, take a circle C_ε of radius ε with center at M_0 and denote by $L_{M_0,\varepsilon}$ the part of L inside C_ε. Let C_ε^- be the part of C_ε belonging to D^-, i.e., lying to the right of L. Clearly, the length $\lambda(L_{M_0,\varepsilon})$ of the piece $L_{M_0,\varepsilon}$ tends to zero as $\varepsilon \to 0$. The angle of C_ε^- is equal to $\pi + O(\varepsilon)$, since there is a tangential line at M_0. Now, we can write

$$V_L(M_0) = \int_L \omega_1(M, M_0)\, dt = \int_{L_{M_0,\varepsilon}} \omega_1(M, M_0)\, dt + \int_{L \setminus L_{M_0,\varepsilon}} \omega_1(M, M_0)\, dt = I_1 + I_2 . \quad (11.2.21)$$

From (11.2.18), it follows that $|I_1| \to 0$ as $\varepsilon \to 0$. In order to calculate I_2, consider the integral of $\omega_1(M, M_0)$ over the curve $\widetilde{L} = (L \setminus L_{M_0,\varepsilon}) \cup C_\varepsilon^-$, choosing the orientation of C_ε^- that agrees with the orientation of $L \setminus L_{M_0,\varepsilon}$. Now, M_0 is an exterior point relative to region bounded by the closed piecewise smooth curve $\widetilde{L}$, and according to (11.2.3), we get

$$\int_{\widetilde{L}} \omega_1(M, M_0)\, dt = 0 = \int_{C_\varepsilon^-} \omega_1(M, M_0)\, dt + \int_{L \setminus L_{M_0,\varepsilon}} \omega_1(M, M_0)\, dt. \quad (11.2.22)$$

Now, using (11.2.10) and taking into account the orientation of C_ε^- and the chosen direction of the normal, we obtain

$$\int_{C_\varepsilon^-} \omega_1(M, M_0)\, dt = \frac{\pi + O(\varepsilon)}{2\pi} . \quad (11.2.23)$$

It follows from (11.2.22) and (11.2.23) that

$$\int_{L \setminus L_{M_0,\varepsilon}} \omega_1(M, M_0)\, dt = -\frac{1}{2} + O(\varepsilon). \quad (11.2.24)$$

This, together with (11.2.21) and (11.2.24), implies that

$$\int_L \omega_1(M, M_0)\, dt = \lim_{\varepsilon \to 0} \int_{L \setminus L_{M_0,\varepsilon}} \omega_1(M, M_0)\, dt = -\frac{1}{2} , \quad (11.2.25)$$

and the proof of (11.2.20) is complete.

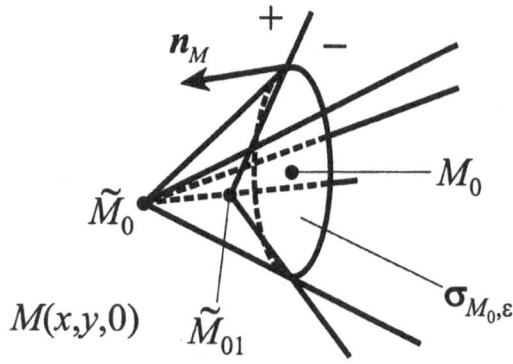

Figure 11.2.4

Now, suppose the curve L is non-smooth at the point $M_0 \in L$. Then, the one-sided tangential unit vectors $\tau_{M_0}^+$ and $\tau_{M_0}^-$ are different (see Fig. 11.2.4). Let θ, $0 \le \theta \le 2\pi$, be the angle between the vectors $\tau_{M_0}^+$ and $-\tau_{M_0}^-$ (if we turn $-\tau_{M_0}^-$ within the domain D^- by the angle θ, we obtain $\tau_{M_0}^+$).

For $\theta = \pi$, the curve L is smooth at M_0. In this case $(\theta = \tau_{M_0}^+ \widehat{(-\tau_{M_0}^-)})$, the fluid particles moving along the rays between the vectors $\tau_{M_0}^+$ and $-\tau_{M_0}^-$ (the rays are within the angle θ) enter the domain D^- through the curve L and remain in D^-. Consider a particle which moves along a ray entering M_0 outside that angle and its first crossing of this curve occurs at a point $M_1 \neq M_0$ and its last crossing of this curve is at $M_2 \neq M_0$. Clearly, this particle can be neglected. Thus the quantity of the fluid which passes through the curve L at points M different from M_0 is equal to the quantity of the fluid entering M_0 within the angle between the vectors $\tau_{M_0}^+$ and $-\tau_{M_0}^-$. This means that

$$V_L(M_0) = \int_L \omega_1(M, M_0)\, dt = -\frac{\Theta}{2\pi}. \tag{11.2.26}$$

Formula (11.2.20) is a special case of (11.2.26). The mathematical proof of (11.2.26) is similar to that of (11.2.20).

Next, we examine the functions $\omega_1^\pm(M, M_0)$ for various types of curves L.

Consider the function $\omega_1(M, \widetilde{M_0})$, where $M \in L$ and $\widetilde{M_0}$ is an arbitrary point of $\mathbb{R}_2$. The value of $\omega_1(M, M_0)$ for $M_0 \in L$ is called *the direct value* of this function on L. As shown above (see Corollary 11.2.2), if L is a piecewise smooth curve and $x''(t)$, $y''(t)$ are of class $H(\alpha)$ on smooth pieces of L and have finite jumps at the points of juncture of smooth pieces of L, then $\omega_1(M, M_0)$ is a piecewise smooth function on $[0, l] \times [0, l]$ and is integrable on L with respect to each variable. Moreover, relation (11.2.18) holds for $\omega_1(M, M_0)$.

Denote by $\omega_1^\pm(M, M_0)$ the limit values of $\omega_1(M, \widetilde{M_0})$ as the point $\widetilde{M_0} \notin L$ approaches $M_0 \in L_0$ from the left or from the right, respectively. Thus,

$$\omega_1^\pm(M, M_0) = \lim_{\widetilde{M_0} \to M_0^\pm} \omega_1(M, \widetilde{M_0}). \tag{12.2.27}$$

Let us see how the values of $\omega_1^\pm(M, M_0)$ are related to the values of $\omega_1(M, M_0)$.

Assume first that M_0 belongs to a smooth piece of L. Then, the function

$$\omega_1(M, M_0) = \frac{1}{2\pi} \frac{(r_{MM_0}, n_M)}{r_{MM_0}^2}$$

is smooth with respect to M on that piece and the value $\omega_1(M_0, M_0)$ is well-defined. The function $\omega_1(M, \widetilde{M_0})$ for $\widetilde{M_0} \notin L$ is also smooth on the said piece and is piecewise smooth on the entire curve L.

Consider the function

$$w(M, M_0, \widetilde{M_0}) = \omega_1(M, \widetilde{M_0}) - \omega_1(M, M_0) = \frac{1}{2\pi} \left[\frac{(r_{M\widetilde{M_0}}, n_M)}{r_{M\widetilde{M_0}}^2} - \frac{(r_{MM_0}, n_M)}{r_{MM_0}} \right] \tag{11.2.28}$$

for $M \in L$. Let us show that

$$w(M, M_0, M_0^\pm) = \lim_{\widetilde{M_0} \to M_0^\pm} w(M, M_0, \widetilde{M_0}) = \omega_1^\pm(M, M_0) - \omega_1(M, M_0) = \lambda \delta(M - M_0), \tag{11.2.29}$$

where $\delta(M - M_0)$ is the delta-function and λ is a constant.

Consider the space of basic functions on curve L. This space consists of infinitely smooth functions $\varphi(M)$, since the curve L is assumed bounded. Let us show that for any such $\varphi(M)$, we have

$$\lim_{\widetilde{M_0} \to M_0^\pm} \int_L w(M, M_0, \widetilde{M_0}) \varphi(M)\, dt = \pm \frac{1}{2}\varphi(M_0), \qquad M_0 \in L. \tag{11.2.30}$$

Denote by $L_{M_0,\varepsilon}$ the piece cut out of L by the ε-neighborhood of M_0 in $\mathbb{R}_2$. Then, we can write

$$\lim_{\widetilde{M}_0 \to M_0^{\pm}} \int_L w(M, M_0, \widetilde{M}_0)\varphi(M)\, dt =$$

$$= \lim_{\widetilde{M}_0 \to M_0^{\pm}} \int_{L \setminus L_{m_0,\varepsilon}} w(M, M_0, \widetilde{M}_0)\varphi(M)\, dt + \lim_{\widetilde{M}_0 \to M_0^{\pm}} \int_{L_{M_0,\varepsilon}} w(M, M_0, \widetilde{M}_0)\varphi(M)\, dt =$$

$$= \lim_{\widetilde{M}_0 \to M_0^{\pm}} I_1(M_0, \widetilde{M}_0) + \lim_{\widetilde{M}_0 \to M_0^{\pm}} I_2(M_0, \widetilde{M}_0). \tag{11.2.31}$$

From (11.2.28), it is easy to see that the integrand in I_1 tends to zero as $\widetilde{M}_0 \to M_0^{\pm}$, uniformly with respect to $\widetilde{M}_0$ (recall that ε is fixed). Therefore, we can pass to the limit under the integral, which yields

$$\lim_{\widetilde{M}_0 \to M_0^{\pm}} I_1(M_0, \widetilde{M}_0) = \int_{L \setminus L_{M_0,\varepsilon}} 0 \cdot \varphi(M)\, dt = 0. \tag{11.2.32}$$

In order to examine the second limit, assume first that $L_{M_0,\varepsilon}$ is a segment on a straight line P. Therefore, $w_1(M, M_0) = 0$ on $L_{M_0,\varepsilon}$, and it remains to calculate the limit

$$\lim_{\widetilde{M}_0 \to M_0^{\pm}} \int_{L_{M_0,\varepsilon}} \omega_1(M, \widetilde{M}_0)\, dt =$$

$$= \varphi(M_0) \lim_{\widetilde{M}_0 \to M_0^{\pm}} \int_{L_{M_0,\varepsilon}} \omega_1(M, \widetilde{M}_0)\, dt + \lim_{\widetilde{M}_0 \to M_0^{\pm}} \int_{L_{M_0,\varepsilon}} \omega_1(M, \widetilde{M}_0)[\varphi(M) - \varphi(M_0)]\, dt =$$

$$= \varphi_0 \lim_{\widetilde{M}_0 \to M_0^{\pm}} I_{2,\varepsilon}^*(\widetilde{M}_0) + \lim_{\widetilde{M}_0 \to M_0^{\pm}} I_{2,\varepsilon}^{**}(M_0, \widetilde{M}_0). \tag{11.2.33}$$

In order to find these two limits, consider a coordinate system OXY on the plane $\mathbb{R}_2$ with the origin at M_0, the axis OX directed along the line P with a given direction, and OY along n_{M_0} (see Fig. 11.2.5). Then, for the integral $I_{2,\varepsilon}^*(\widetilde{M}_0)$ in (11.2.33) with $|x_0| < \varepsilon$ we obtain

$$\int_{L_{M_0,\varepsilon}} \omega_1(M, \widetilde{M}_0)\, dt = \frac{1}{2\pi} \int_{-\varepsilon}^{\varepsilon} \frac{y_0\, dx}{(x - x_0)^2 + y_0^2} = \frac{1}{2\pi}\left(\arctan\frac{\varepsilon - x_0}{y_0} + \arctan\frac{\varepsilon + x_0}{y_0}\right). \tag{11.2.34}$$

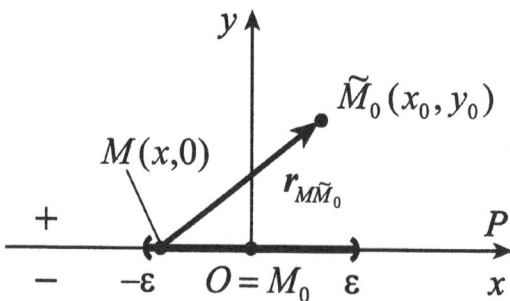

Figure 11.2.5

Since $|x_0| < \varepsilon$, both parameters $\varepsilon - x_0$ and $\varepsilon + x_0$ are positive, and therefore, can be both assumed larger than $\varepsilon/2$ for $\widetilde{M}_0 \to M_0^{\pm}$. For $\widetilde{M}_0 \to M_0^{\pm}$, we have $y_0 \to 0^{\pm}$, and the sum in parentheses in

(11.2.34) tends to $\pm\pi$. Thus, we obtain

$$\lim_{\widetilde{M}_0 \to M_0^{\pm}} I_{2,\varepsilon}^*(\widetilde{M}_0) = \pm\frac{1}{2}. \tag{11.2.35}$$

Let us estimate the term $I_{2,\varepsilon}^{**}(M_0, \widetilde{M}_0)$. We have

$$\left| I_{2,\varepsilon}^{**}(M_0, \widetilde{M}_0) \right| \le A \int_{L_{M_0,\varepsilon}} \left| \omega_1(M, \widetilde{M}_0) \right| r_{MM_0} \, dt,$$

where $|\varphi(M) - \varphi(M_0)| \le Ar_{MM_0}$ for $M, M_0 \in L$. Using the above coordinate system OXY, just as for $I_{2,\varepsilon}^*(\widetilde{M}_0)$, we obtain

$$\left| I_{2,\varepsilon}^{**}(M_0, \widetilde{M}_0) \right| \le \frac{A}{2\pi} \int_{-\varepsilon}^{\varepsilon} \frac{|y_0||x - x_0| \, dx}{(x - x_0)^2 + y_0^2} \le$$

$$\le \frac{2A\varepsilon}{2\pi} \int_{-\varepsilon}^{\varepsilon} \frac{|y_0| \, dx}{(x - x_0)^2 + y_0^2} = A\varepsilon\pi \arctan \left. \frac{x - x_0}{|y_0|} \right|_{-\varepsilon}^{\varepsilon}. \tag{11.2.36}$$

Therefore,

$$\lim_{\widetilde{M}_0 \to M_0^{\pm}} \left| I_{2,\varepsilon}^{**}(M_0, \widetilde{M}_0) \right| = O(\varepsilon). \tag{11.2.37}$$

Thus, formulas (11.2.31)–(11.2.37) imply (11.2.30). Consequently, using the definition of the delta-function $\delta(M - M_0)$, we can write

$$\omega_1^{\pm}(M, M_0) = \omega_1(M, M_0) \pm \frac{1}{2}\delta(M - M_0), \tag{11.2.38}$$

$$\int_L \omega_1^{\pm}(M, M_0)g(M) \, dt = \int_L \omega_1(M, M_0)g(M) \, dt \pm \frac{1}{2}g(M_0), \tag{11.2.39}$$

$$\lim_{\widetilde{M}_0 \to M_0^{\pm}} \int_L \omega_1(M, \widetilde{M}_0) \, dt = \int_L \omega_1(M, M_0) \, dt \pm \frac{1}{2}, \tag{11.2.40}$$

for any point $M_0 \in L$ on a smooth piece of L.

Let us give a physical justification to formula (11.2.35). Since the function $\omega_1(M, \widetilde{M}_0)$ describes the distributed penetration density on the segment $[-\varepsilon, \varepsilon]$ of the axis OX (see Fig. 11.2.5), or on a segment I, due to the source of intensity 1 at $\widetilde{M}_0$, it is clear that the fluid quantity that passes through this segment is equal to $\pm\theta/2\pi$, where θ is the value of the angle at which I is seen from the point $\widetilde{M}_0$ and the signs $\pm$ correspond to whether the source is to the right or to the left of the segment I. Thus,

$$\int_I \omega_1(M, \widetilde{M}_0) \, dt = \pm\frac{\theta}{2\pi}. \tag{11.2.41}$$

Figure 11.2.5 shows that as $\widetilde{M}_0$ approaches I, we have $\theta \to \pi$, and therefore,

$$\lim_{\widetilde{M}_0 \to M_0^{\pm}} \int_I \omega_1(M, \widetilde{M}_0) \, dt = \pm\frac{1}{2}. \tag{11.2.42}$$

These physical arguments show that relations (11.2.41) and (11.2.42) are preserved to within ε, if the integral over the segment is replaced by the integral over a smooth piece $L_{M_0,\varepsilon}$ containing M_0, which means that relation (11.2.35) remains valid to within ε. Therefore, (11.2.38) holds.

Let us give a mathematical proof to these statements. Suppose that the piece $L_{M_0,\varepsilon}$ of curve L is smooth but is not a linear segment. We introduce a coordinate system on the plane, OXY, with the origin at M_0, the axis OX directed along the tangent to $L_{M_0,\varepsilon}$ at M_0, and the axis OY directed to the left of this curve (see Fig. 11.2.6). Then, the equation of the curve can be written as $y = f(x)$, $y(0) = f(0) = 0$, $y'(0) = f'(0) = 0$. In this case,

$$n_M = \frac{-f'(M)i + j}{[1 + f'^2(M)]^{1/2}}, \qquad r_{M\widetilde{M}_0} = (0 - x)i + (y_0 - y)j = xi + (y_0 - f(x))j.$$

Therefore,

$$\omega_1(M, \widetilde{M}_0) = \frac{1}{2\pi} \frac{\left(r_{M\widetilde{M}_0}, n_M\right)}{r_{M\widetilde{M}_0}^2} = \frac{1}{2\pi} \frac{-xf'(x) + y_0 - f(x)}{[x^2 + (y_0 - f(x))^2][1 + f'^2(x)]^{1/2}}. \qquad (11.2.43)$$

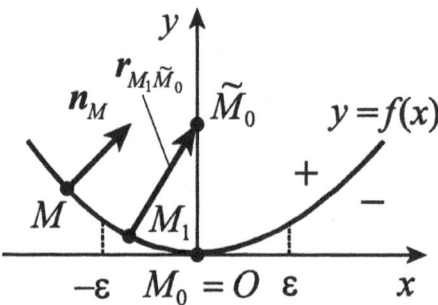

Figure 11.2.6

The above conditions on the function $y = f(x)$ allow us to write

$$y(x) = \frac{1}{2} f''(\theta_1) x^2, \qquad y'(x) = f''(\theta_2) x, \qquad (11.2.44)$$

where θ_1, θ_2 are some values of x on the interval $(-\varepsilon, \varepsilon)$. Taking into account (11.2.44), we can rewrite (11.2.43) in the form

$$\omega_1(M, \widetilde{M}_0) = \frac{-f''(\theta_2)x^2 + y_0 - 2^{-1}f''(\theta_1)x^2}{\left[x^2 + \left(y_0 - 2^{-1}f''(\theta_1)x^2\right)^2\right][1 + f'^2(x)]^{1/2}}. \qquad (11.2.45)$$

Assume that on the curve $L_{M_0,\varepsilon}$, we have $|f''(x)| \le C$ for any $x \in (-\varepsilon, \varepsilon)$. Therefore, for any $x \in (-\varepsilon, \varepsilon)$, the following inequality holds:

$$\frac{\left|-x^2\left[2^{-1}f''(\theta_1) + f''(\theta_2)\right]\right|}{\left[x^2 + \left(y_0 - 2^{-1}f''(\theta_1)x^2\right)^2\right][1 + f'^2(x)]^{1/2}} \le \frac{3}{2}C. \qquad (11.2.46)$$

Therefore,

$$\left|\int_{-\varepsilon}^{\varepsilon} \frac{-x^2\left[2^{-1}f''(\theta_1) + f''(\theta_2)\right]}{\left[x^2 + \left(y_0 - 2^{-1}f''(\theta_1)x^2\right)^2\right][1 + f'^2(x)]^{1/2}}\right| \le 3C\varepsilon. \qquad (11.2.47)$$

We further have

$$\frac{y_0}{\left[x^2 + \left(y_0 - 2^{-1}f''(\theta_1)x^2\right)^2\right]\left[1 + f'^2(x)\right]^{1/2}} = \frac{y_0}{x^2 + y_0^2} + \lambda(x, y_0, \theta_1),\qquad(11.2.48)$$

and therefore, taking into account that $f''(x)$ is bounded on $(-\varepsilon, \varepsilon)$, we obtain

$$\left|\int_{-\varepsilon}^{\varepsilon} \lambda(x, y, \theta_1)\, dx\right| \le O(\varepsilon).\qquad(11.2.49)$$

Now, using (11.2.35), (11.2.47), and (11.2.49), we find that

$$\lim_{\widetilde{M}_0 \to M_0^{\pm}} \int_{L_{M_0,\varepsilon}} \omega_1(M, \widetilde{M}_0)\, dt = \pm\frac{1}{2} + O(\varepsilon).\qquad(11.2.50)$$

Consequently, from (11.2.31), (11.2.32), and (11.2.50), we have

$$\lim_{\widetilde{M}_0 \to M_0^{\pm}} \int_L W(M, M_0, \widetilde{M}_0)\, dt = \pm\frac{1}{2}.\qquad(11.2.51)$$

Thus, we have established the following result.

Theorem 11.2.2. *For M_0 on a smooth piece of L, the following relations hold:*

$$\omega_{1,M_0}^{\pm}(M, M_0) = \omega_1(M, M_0) \pm \frac{1}{2}\delta(M - M_0),\qquad(11.2.52)$$

$$\int_L \omega_{1,M_0}^{\pm}(M, M_0)g(M)\, dt = \int_L \omega_1(M, M_0)g(M)\, dt \pm \frac{1}{2}g(M_0).\qquad(11.2.53)$$

Now, let M_0 be an angular point of curve L. Thus, at the point M_0, there exist one-sided tangential vectors $\tau_{M_0}^+$ and $\tau_{M_0}^-$ of unit length. Let θ be the angle between these vectors. Suppose that there is a source of intensity 1 at $\widetilde{M}_0$, and the point $\widetilde{M}_0$ approaches M_0 along a path between the rays λ_1 and λ_2, and this path is non-tangential to the rays. In this case, we write $\widetilde{M}_0 \to M_0^+$. The rays λ_1 and λ_2 issue from M_0 in the direction of the vectors $\tau_{M_0}^+$ and $-\tau_{M_0}^-$. Then, the following relations hold:

$$\int_L \omega_{1,M_0}^{\pm}(M, M_0)\, dt = \int_L \omega_1(M, M_0)\, dt + \begin{cases} \dfrac{\theta}{2\pi} & \text{for } \omega_{1,M_0}^+, \\[2mm] -\dfrac{2\pi - \theta}{2\pi} & \text{for } \omega_{1,M_0}^-, \end{cases}\qquad(11.2.54)$$

$$\omega_{1,M_0}^{\pm}(M, M_0) = \omega_1(M, M_0) + \begin{cases} \dfrac{\theta}{2\pi}\delta(M - M_0) & \text{for } \omega_{1,M_0}^+, \\[2mm] -\dfrac{2\pi - \theta}{2\pi}\delta(M - M_0) & \text{for } \omega_{1,M_0}^-, \end{cases}\qquad(11.2.55)$$

Let us give a mathematical proof of (11.2.54) and (11.2.55). Assume first that L is an angle that consists of two segments, L_1 and L_2. The motion along L is chosen in the direction from L_1 to L_2. Let OXY be the coordinate system on the plane such that OX contains L_1 and has the same direction as L_1, the origin coincides with the vertex of the angle, and the axis OY is orthogonal to L_1 and is directed to the left of the angle (see Fig. 11.2.7).

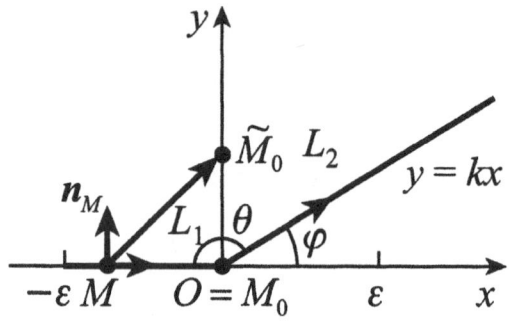

Figure 11.2.7

Let $L_1 = [-\varepsilon, 0]$, $\varepsilon > 0$, be a segment on the axis OX, and let L_2 be a piece of the straight line $y = kx$, $k = \tan \varphi$. Then,

$$\boldsymbol{n}_M = \boldsymbol{j}, \quad \boldsymbol{r}_{M\widetilde{M}_0} = -x\boldsymbol{i} + y_0\boldsymbol{j} \quad \text{for} \quad x \in [-\varepsilon, 0];$$

$$\boldsymbol{n}_M = \left[1 + k^2\right]^{1/2}(-k\boldsymbol{i} + \boldsymbol{j}), \quad \boldsymbol{r}_{M\widetilde{M}_0} = (-x)\boldsymbol{i} + (y_0 - kx)\boldsymbol{j} \quad \text{for} \quad x \in [0, \varepsilon].$$

Since $dt = \left[1 + k^2\right]^{1/2} dx$ on L_2, we have

$$\int_L \omega_1(M, M_0)\, dt = \frac{1}{2\pi} \int_{-\varepsilon}^0 \frac{y_0\, dx}{x^2 + y_0^2} + \frac{1}{2\pi} \int_0^\varepsilon \frac{kx + y_0 - kx}{\left[1 + k^2\right]^{1/2}\left(x^2 + (y_0 - kx)\right)^2} \left[1 + k^2\right]^{1/2} dx.$$

$$(11.2.56)$$

Simple calculations show that

$$\int_L \omega_1(M, \widetilde{M}_0)\, dt = \frac{1}{2\pi}\left[\arctan \frac{\varepsilon}{y_0} + \arctan \frac{\varepsilon(1 + k^2) - ky_0}{y_0} - \arctan k\right]. \qquad (11.2.57)$$

Now, it is clear that for $y_0 \to 0^\pm$ (i.e., $y_0 \to 0$, $y_0 > 0$; or $y_0 \to 0$, $y_0 < 0$), we have

$$\lim_{\widetilde{M}_0 \to M_0^\pm} \int_L \omega_1(M, \widetilde{M}_0)\, dt = \begin{cases} \dfrac{\pi - \varphi}{2\pi} & \text{for } M_0^+, \\[2mm] \dfrac{-\pi - \varphi}{2\pi} & \text{for } M_0^-, \end{cases} \qquad (11.2.58)$$

Setting $\theta = \pi - \varphi$, we obtain

$$\lim_{\widetilde{M}_0 \to M_0^\pm} \int_L \omega_1(M, \widetilde{M}_0)\, dt = \begin{cases} \dfrac{\theta}{2\pi} & \text{for } M_0^+, \\[2mm] -\dfrac{2\pi - \theta}{2\pi} & \text{for } M_0^-, \end{cases} \qquad (11.2.59)$$

Just as above, when we have replaced the segment I by a smooth curve (see Fig. 11.2.6), in the present case we can replace each segment I_1 and I_2 by smooth curves L_1 and L_2 forming an angle θ at the point M_0 in the above sense. In this way, we obtain (11.2.54).

Relations (11.2.55) can be established on the basis of the representation (11.2.31) and the fact that M_0 is an angular point.

Next, we examine the double layer potential and its properties.

By definition, *the double layer potential with density* $g(M) = g(t)$ *on a piecewise smooth curve* L is the function

$$\Phi(M_0) = \frac{1}{2\pi} \int_L \frac{\partial}{\partial n_M} \left(\ln \frac{1}{r_{MM_0}} \right) g(M)\, dt = \tag{11.2.60}$$

$$= \frac{1}{2\pi} \int_L \frac{(r_{MM_0}, n_M)}{r_{MM_0}^2} g(M)\, dt = \int_L \omega_1(M, M_0) g(M)\, dt, \qquad M_0 \in L.$$

As shown in Remark 11.2.1, for a piecewise smooth curve L and $M_0 \in L$, the function $\omega_1(M, M_0)$, for $M, M_0 \in L$, is piecewise smooth on $L \times L$. For this reason, we call the function

$$\Phi(M_0) = \int_L \omega_1(M, M_0) g(M)\, dt, \qquad M_0 \in L, \tag{11.2.61}$$

the direct value of the double layer potential on L.

It can be easily verified that

$$\Delta\Phi(M_0) = 0, \quad M_0 \notin L, \tag{11.2.62}$$

i.e., $\Phi(M_0)$ is a harmonic function outside L.

Let us examine the limit values of the function $\Phi(M_0)$. Let

$$\Phi^\pm(M_0) = \lim_{\widetilde{M}_0 \to M_0^\pm} \int_L \omega_1(M, \widetilde{M}_0) g(M)\, dt. \tag{11.2.63}$$

It follows from (11.2.55) that

$$\omega_{1,M_0}^\pm(M, M_0) g(M) = \omega_1(M, M_0) g(M) + \begin{cases} \dfrac{\theta}{2\pi} \delta(M - M_0) g(M) & \text{for} \quad \omega_{1,M_0}^+, \\[2mm] -\dfrac{2\pi - \theta}{2\pi} \delta(M - M_0) g(M) & \text{for} \quad \omega_{1,M_0}^-, \end{cases} \tag{11.2.64}$$

which implies that

$$\Phi^\pm(M_0) = \int_L \omega_1(M, M_0) g(M)\, dt + \begin{cases} \dfrac{\theta}{2\pi} g(M) & \text{for} \quad \Phi^+, \\[2mm] -\dfrac{2\pi - \theta}{2\pi} g(M) & \text{for} \quad \Phi^-. \end{cases} \tag{11.2.65}$$

Proposition 11.2.1. *For points* M_0 *at which* L *is smooth, we have*

$$\Phi^\pm(M_0) = \int_L \omega_1(M, M_0) g(M)\, dt \pm \frac{1}{2} g(M_0), \tag{11.2.66}$$

and for angular points, formula (11.2.65) is valid.

Proposition 11.2.2. *For points* M_0 *at which* L *is smooth, formulas of Sokhotski–Plemelj type hold* (see Muskhelishvili (1968)),

$$\frac{1}{2} \left[\Phi^+(M_0) + \Phi^-(M_0) \right] = \int_L \omega_1(M, M_0) g(M)\, dt, \tag{11.2.67}$$

$$\Phi^+(M_0) - \Phi^-(M_0) = g(M_0). \tag{11.2.68}$$

For angular points we have

$$\frac{1}{2}\left[\Phi^+(M_0) + \Phi^-(M_0)\right] = \int_L \omega_1(M, M_0)g(M)\,dt - \frac{\pi - \theta}{2\pi}g(M_0), \qquad (11.2.69)$$

$$\Phi^+(M_0) - \Phi^-(M_0) = g(M_0). \qquad (11.2.70)$$

Next, consider the derivatives of the double layer potential. We start with its normal derivative. For the points of a smooth piece of L, using (11.2.66), we obtain the relation

$$\frac{\partial}{\partial n_{M_0}}\Phi^\pm(M_0) = \frac{\partial}{\partial n_{M_0}}\int_L \omega_1(M, M_0)g(M)\,dt, \qquad M_0 \in L, \qquad (11.2.71)$$

since the function $g(M)$ is defined only on the curve L and for shifts along the normal, the value $g(M_0)$ remains the same.

For the tangential derivatives at regular points of L, we get

$$\frac{\partial}{\partial \tau_{M_0}}\Phi^\pm(M_0) = \frac{\partial}{\partial \tau_{M_0}}\int_L \omega_1(M, M_0)g(M)\,dt \pm \frac{1}{2}\frac{\partial}{\partial \tau_{M_0}}g(M_0). \qquad (11.2.72)$$

In particular, if L is a segment of a straight line, we have $\omega_1(M, M_0) \equiv 0$ for $M, M_0 \in L$, and therefore,

$$\frac{\partial}{\partial \tau_{M_0}}\Phi^\pm(M_0) = \pm\frac{\partial}{\partial \tau_{M_0}}g(M_0). \qquad (11.2.73)$$

The gradient of this function is invariant with respect to the coordinate systems, and therefore, for regular points we can write

$$\text{grad}\,\Phi^\pm(M_0) = \text{grad}\,\Phi_0 \pm \frac{1}{2}\frac{\partial}{\partial \tau_{M_0}}g(M_0)\tau_{M_0}. \qquad (11.2.74)$$

In order to write (11.2.74) avoiding the coordinate notation (in the form of $\text{grad}(\cdots)$), we introduce the tangential component of the gradient on the curve L,

$$\frac{\partial}{\partial \tau_{M_0}}g(M_0)\tau_{M_0} = \text{Grad}\,g(M_0). \qquad (11.2.75)$$

Then, (11.2.74) becomes

$$\text{grad}\,\Phi^\pm(M_0) = \text{grad}\,\Phi(M_0) \pm \frac{1}{2}\text{Grad}\,g(M_0), \qquad (11.2.76)$$

in particular, if L is a segment on a straight line, we have

$$\text{grad}\,\Phi^\pm(M_0) = \pm\frac{1}{2}\text{Grad}\,g(M_0), \qquad M_0 \in L. \qquad (11.2.77)$$

Thus, we have obtained a well-known formula for the double layer potential.

Finally, consider the simple layer potential and its derivatives.

By definition, *the simple layer potential on the curve L* is the function

$$\varphi(M_0) = \frac{1}{2\pi}\int_L \ln\frac{1}{r_{MM_0}}\mu(M)\,dt, \qquad (11.2.78)$$

where $\mu(M) = \mu(t)$ is *the simple layer density*. For $M_0 \notin L$, $\varphi(M_0)$ is a harmonic function, i.e., equation (11.2.62) holds for $\varphi(M_0)$.

Since the kernel of the integral in (11.2.78) has only a logarithmic singularity for $M_0 \in L$, the function $\varphi(M_0)$ can be continuously extended to L.

Let us examine the derivatives of this function. For $M_0 \notin L$, the function $\varphi(M_0)$ is harmonic in sufficiently small neighborhood of M_0, and therefore, is infinitely differentiable for $\mu(t) \in L_2$ on L.

First, consider the normal derivative of this function at a point $\widetilde{M} \notin L$ close to L. We have

$$\frac{\partial}{\partial n_{M_0}} \varphi(\widetilde{M}_0) = \frac{1}{2\pi} \int_L \frac{\partial}{\partial n_{M_0}} \left(\ln \frac{1}{r_{M\widetilde{M}_0}} \right) \mu(M) \, dt, \tag{11.2.79}$$

where M_0 is a point on L which is approached by $\widetilde{M}_0$ in what follows. Using (11.2.5), we obtain

$$\frac{1}{2\pi} \frac{\partial}{\partial n_{M_0}} \left(\ln \frac{1}{r_{M\widetilde{M}_0}} \right) = -\frac{1}{2\pi} \frac{\left(r_{M\widetilde{M}_0}, n_{M_0} \right)}{r^2_{M\widetilde{M}_0}} = \omega_{1,1}(M, \widetilde{M}_0). \tag{11.2.80}$$

The integral

$$\frac{\partial}{\partial n_{M_0}} \varphi(M_0) = \int_L \omega_{1,1}(M, \widetilde{M}_0)\mu(M) \, dt \tag{11.2.81}$$

is called *the direct value of the normal derivative of the simple layer*. Let us find the relation between this direct value and the limit values of the normal derivative,

$$\frac{\partial}{\partial n_{M_0}} \varphi^{\pm}(M_0) = \lim_{\widetilde{M}_0 \to M_0^{\pm}} \int_L \omega_{1,1}(M, \widetilde{M}_0)\mu(M) \, dt. \tag{11.2.82}$$

To that end, we note that

$$\omega_{1,1}(M, \widetilde{M}_0) = -\frac{1}{2\pi} \left[\frac{\left(r_{M\widetilde{M}_0}, n_{M_0} \right)}{r^2_{M\widetilde{M}_0}} + \frac{\left(r_{M\widetilde{M}_0}, n_{M_0} - n_M \right)}{r^2_{M\widetilde{M}_0}} \right] =$$
$$= -\omega_{1,1}(M, \widetilde{M}_0) + \lambda(M, M_0, \widetilde{M}_0), \tag{11.2.83}$$

where $\omega_{1,1}(M, \widetilde{M}_0)$ is the function defined in (11.2.2).

Consider the function $\lambda(M, M_0, \widetilde{M}_0)$. Assume that L is a piecewise Lyapunov curve, i.e., the inequality

$$|n_{M_0} - n_M| \leq A r_{MM_0} \tag{11.2.84}$$

holds on its smooth pieces. To ensure (11.2.84), it suffices to assume that the Lipschitz condition holds for $x''(t)$ and $y''(t)$ on smooth pieces of L. Let us transform the function $\lambda(M, M_0, \widetilde{M}_0)$ as follows:

$$\lambda(M, M_0, \widetilde{M}_0) = \frac{1}{2\pi} \frac{\left(r_{M\widetilde{M}_0} - r_{MM_0}, n_{M_0} - n_M \right)}{r^2_{M\widetilde{M}_0}} + \frac{1}{2\pi} \frac{\left(r_{M\widetilde{M}_0}, n_{M_0} - n_M \right)}{r^2_{M\widetilde{M}_0}} =$$
$$= \frac{1}{2\pi} \frac{\left(r_{M_0\widetilde{M}_0}, n_{M_0} - n_M \right)}{r^2_{M\widetilde{M}_0}} + \frac{1}{2\pi} \frac{\left(r_{MM_0}, n_{M_0} - n_M \right)}{r^2_{MM_0}} +$$
$$+ \frac{1}{2\pi} \frac{\left(r_{MM_0}, n_{M_0} - n_M \right)\left(r^2_{MM_0} - r^2_{M\widetilde{M}_0} \right)}{r^2_{M\widetilde{M}_0} r^2_{MM_0}} =$$
$$= \lambda_1(M, M_0, \widetilde{M}_0) + \lambda_2(M, M_0) + \lambda_3(M, M_0, \widetilde{M}_0). \tag{11.2.85}$$

For $\lambda_1(M, M_0, \widetilde{M_0})$ we have

$$\lim_{\widetilde{M_0} \to M_0} \lambda_1(M, M_0, \widetilde{M_0}) = \begin{cases} 0, & M \neq M_0, \\ 0, & M = M_0, \end{cases} \qquad (11.2.86)$$

since

$$\lambda_1(M_0, M_0, \widetilde{M_0}) = \frac{\left(r_{M_0\widetilde{M_0}}, 0\right)}{r^2_{M_0\widetilde{M_0}}} = 0, \qquad M_0 \neq \widetilde{M_0}.$$

For $\lambda(M, M_0)$, in view of (11.2.84) and Remark 11.2.1, we have

$$|\lambda_2(M, M_0)| \leq A, \qquad \lambda_2(M_0, M_0) = 0, \qquad (11.2.87)$$

for any $M, M_0 \in L$.

For $\lambda_3(M, M_0, \widetilde{M_0})$, we have

$$\lambda_3(M, M_0, \widetilde{M_0}) = \lambda_2(M, M_0) \frac{\left(r_{MM_0}, r_{\widetilde{M_0}M_0}\right) + \left(r_{\widetilde{M_0}M_0}, r_{M\widetilde{M_0}}\right)}{r^2_{M\widetilde{M_0}}} =$$

$$= \lambda_2(M, M_0)\lambda_{3,1}(M, M_0, \widetilde{M_0}). \qquad (11.2.88)$$

We see that

$$\lim_{\widetilde{M_0} \to M_0} \lambda_{3,1}(M, M_0, \widetilde{M_0}) = \begin{cases} 0, & M \neq M_0, \\ -1, & M = M_0. \end{cases} \qquad (11.2.89)$$

Thus, the function $\lambda(M, M_0, \widetilde{M_0})$ is continuous in $\widetilde{M_0}$ on the entire plane. In view of the properties of $\omega_1(M, \widetilde{M_0})$ (see (11.2.55)), we obtain

$$\omega_{1,1}^{\pm}(M, \widetilde{M_0})\mu(M) = \omega_{1,1}(M, M_0)\mu(M) + \begin{cases} -\dfrac{\theta}{2\pi}\delta(M - M_0)\mu(M), \\ \dfrac{2\pi - \theta}{2\pi}\varphi(M - M_0)\mu(M). \end{cases} \qquad (11.2.90)$$

Proposition 11.2.3. *For the limit values of the normal derivative of the simple layer potential, the following formula holds:*

$$\frac{\partial}{\partial n_{M_0}}\varphi^{\pm}(M_0) = \frac{1}{2\pi}\int_L \frac{\partial}{\partial n_{M_0}}\left(\ln\frac{1}{r_{MM_0}}\right)\mu(M)\,dt + \begin{cases} -\dfrac{\theta}{2\pi}\mu(M), \\ \dfrac{2\pi - \theta}{2\pi}\mu(M). \end{cases} \qquad (11.2.91)$$

Therefore, at regular points,

$$\frac{\partial}{\partial n_{M_0}}\varphi^{\pm}(M_0) = \frac{\partial}{\partial n_{M_0}}\varphi(M_0) \mp \frac{1}{2}\mu(M_0). \qquad (11.2.92)$$

Now, consider the tangential derivative of the simple layer potential,

$$\frac{\partial\varphi(\widetilde{M_0})}{\partial\tau_{M_0}} = -\frac{1}{2\pi}\int_L \frac{\left(r_{MM_0}, \tau_{M_0}\right)}{r^2_{M\widetilde{M_0}}}\mu(M)\,dt, \qquad \widetilde{M_0} \notin L. \qquad (11.2.93)$$

By definition, *the direct value of the tangential derivative* is the value of the integral in (11.2.93) for $\widetilde{M}_0 = M_0$, $M_0 \in L$. First, consider the function

$$\omega_\tau(M, M_0) = \omega_\tau(t, t_0) = -\frac{1}{2\pi} \frac{(r_{MM_0}, \tau_{M_0})}{r^2_{MM_0}} =$$

$$= -\frac{1}{2\pi} \frac{x'(t_0)x_2(t, t_0) + y'(t_0)y_2(t, t_0)}{(t_0 - t)\left[x_2^2(t, t_0) + y_2^2(t, t_0)\right]} = -\frac{1}{2\pi} \frac{1}{t_0 - t} K(t, t_0), \qquad (11.2.94)$$

where $x_2(t, t_0)$ and $y_2(t, t_0)$ are defined in (11.2.11). From the proof of Theorem 11.2.1, it follows that if $x'^2(t) + y'^2(t) \neq 0$ and $x'(t), y'(t) \in H(\alpha)$ on the smooth part of L, then the function $K(t, t_0)$ is of class $H(\alpha)$ in each variable, uniformly with respect to the other, and $K(t_0, t_0) = 1$. Now we can write

$$\omega_\tau(M, M_0) = -\frac{1}{2\pi} \frac{1}{t_0 - t} - \frac{1}{2\pi} \frac{x_2(t, t_0)\widetilde{x}_3(t, t_0) + y_2(t, t_0)\widetilde{y}_3(t, t_0)}{x_2^2(t, t_0) + y_2^2(t, t_0)} =$$

$$= -\frac{1}{2\pi} \frac{1}{t_0 - t} + K_1(t, t_0), \qquad (11.2.95)$$

$$\widetilde{x}_3(t, t_0) = \frac{x'(t_0) - x_2(t, t_0)}{t_0 - t}, \qquad \widetilde{y}_3(t, t_0) = \frac{y'(t_0) - y_2(t, t_0)}{t_0 - t}.$$

Again, reasoning as in the proof of Theorem 11.2.1, one can show that if $x''(t), y''(t) \in H_r(\alpha)$, then $\widetilde{x}_3(t, t_0), \widetilde{y}_3(t, t_0) \in H_r(\alpha)$. Thus, on a smooth part of L in a neighborhood of $M_0 = M(t_0)$, we obtain the representation (11.2.95), provided that $x''(t), y''(t) \in H(\alpha)$ in that neighborhood. In angular points of L, the function $K_1(t, t_0)$ has one-sided limits, $\lim\limits_{t_0 \to \widehat{t}_0} \lim\limits_{t \to \widehat{t}_0} K_1(t, t_0)$, where $\widehat{t}_0$ is the arc length corresponding to an angular point or an end-point.

Consider the limit values of the function $\omega_\tau(M, \widetilde{M}_0)$ as $\widetilde{M}_0 \to M_0^\pm$,

$$\omega^\pm_{\tau, M_0}(M, \widetilde{M}_0) = \lim\limits_{\widetilde{M}_0 \to M_0^\pm} \omega_\tau(M, \widetilde{M}_0) = \lim\limits_{\widetilde{M}_0 \to M_0^\pm} -\frac{1}{2\pi} \frac{(r_{M\widetilde{M}_0}, \tau_{M_0})}{r^2_{M\widetilde{M}_0}}. \qquad (11.2.96)$$

For $M \neq M_0$ and $M_0 \in L$, we have

$$\omega^\pm_{\tau, M_0} = \omega_\tau(M, M_0). \qquad (11.2.97)$$

For $M = M_0$ the above limits do not exist, since $\omega^\pm_{\tau, M_0}(M, M_0) = 0$ if $r_{M\widetilde{M}_0}$ is parallel to n_{M_0}. If the point $\widetilde{M}_0$ approaches M_0 along a ray, i.e., $\left(r^0_{M\widetilde{M}_0}, n_{M_0}\right) = \cos\theta$, then $\left(r^0_{M\widetilde{M}_0}, \tau_{M_0}\right) = \sin\theta$ for $\widetilde{M}_0 \to M_0^+$. Therefore, $\omega^\pm_{\tau, M_0}(M, \widetilde{M}_0) = \pm\infty$ in this case. The above arguments show that at the point M_0 the functions $\omega^\pm_{\tau, M_0}(M, M_0)$ cannot have a singularity like that of the δ-function, but their singularities are like that of $1/x$ as $x \to 0$.

Thus, formula (11.2.97) shows that the following relation is valid:

$$\frac{\partial \varphi^\pm(M_0)}{\partial \tau_{M_0}} = -\frac{1}{2\pi} \int_L \frac{(r_{MM_0}, \tau_{M_0})}{r^2_{MM_0}} \mu(M)\, dt, \qquad (11.2.98)$$

where $M_0 \in L$ is a point in which L is smooth.

From (11.2.92) and (11.2.98) we make the following conclusion.

Proposition 11.2.4. *For the points $M_0 \in L$ at which L is smooth, we have*

$$\operatorname{grad} \varphi^\pm(M_0) = \operatorname{grad} \varphi(M_0) \mp \frac{1}{2} n_{M_0}\mu(M_0). \qquad (11.2.99)$$

11.3. Mathematical Modelling of Flow Past an Airfoil with Suction and Jet Discharge

Here we study the nonlinear problem of finding aerodynamical characteristics of an airfoil in potential flow of an ideal incompressible weightless fluid. We are dealing with a smooth flow past an impenetrable profile (an airfoil of infinite span), which is suddenly set in motion with constant velocity U_0 from the state of rest. Fluid perturbations caused by the airfoil and the discharged jet are supposed to decay at large distances from the surface of the airfoil and the jet. Suction of the external flow is modelled by a sink of intensity Q_1 at a point M_q on the surface of the airfoil. Let Q_2 be a given net flow of the fluid discharged through the surface of the exhaust port and forming a jet bounded by curves σ_1 and σ_2 (recall that we consider a plane nonstationary problem).

Let us give a mathematical statement of the above problem of aerodynamics.

Since the incident flow is assumed potential on the entire plane, the velocity field excited by the contour L and the jet is also assumed potential in the domain D^+ which is exterior to L and the jet (see *Loitsyanskii* (1978)). The problem of describing the resulting the velocity consists in finding a function $\Phi(M, t)$ satisfying the following conditions:

(i) In the domain D^+, the Laplace equation holds for $\Phi(M, t)$,

$$\Delta\Phi(M, t) = 0, \qquad M \in D^+. \tag{11.3.1}$$

(ii) The perturbed potential and the velocity field become infinitely small at large distances from the curves L, σ_1, σ_2,

$$\Phi(M, t) \to 0, \quad \nabla\Phi(M, t) \to 0 \quad \text{as} \quad \rho(M, L \cup \sigma_1 \cup \sigma_2) \to \infty. \tag{11.3.2}$$

(iii) The sink of intensity Q_1 at the point M_q produces the velocity field (see (11.1.8))

$$\boldsymbol{V}_q(M, t) = \frac{Q_1(t)}{2\pi} \frac{\boldsymbol{r}_{MM_q}}{r^2_{MM_q}}. \tag{11.3.3}$$

(iv) The condition of non-penetration on L,

$$\frac{\partial}{\partial \boldsymbol{n}_{M_0}} \Phi(M_0, t)\bigg|_L = -\boldsymbol{U}_0(M_0) \cdot \boldsymbol{n}_{M_0} - \boldsymbol{V}_q(M_0, t) \cdot \boldsymbol{n}_{M_0}, \quad M_0 \in L, \qquad M_0 \neq M_q. \tag{11.3.4}$$

(v) On the boundary of the jet, $\sigma_1 \cup \sigma_2$, there should be no pressure drop and no jump of the normal derivatives,

$$P^+(M_0) = P^-(M_0), \qquad \frac{\partial \Phi^+(M_0, t)}{\partial \boldsymbol{n}_{M_0}} = \frac{\partial \Phi^-(M_0, t)}{\partial \boldsymbol{n}_{M_0}}, \qquad M_0 \in \sigma_1 \cup \sigma_2, \tag{11.3.5}$$

where the symbols '+' and '−' have the same meaning as in Section 11.1.

(vi) The Chaplygin–Zhukovskii–Kutta condition: the velocities should be finite near the sharp edges of the nozzle from which the vortex sheets σ_1 and σ_2 start, i.e., the gradient $\nabla\Phi(M, t)$ is bounded near the edge-points M_1^* and M_2^* for any $t \geq 0$.

(vii) The net flow through the surface (curve) σ of the discharge jet is given, namely,

$$\int_\sigma \left[\frac{\partial \Phi(M, t)}{\partial \boldsymbol{n}_M} + \left(\boldsymbol{U}_q(M) + \boldsymbol{V}_q(M) \right) \cdot \boldsymbol{n}_M \right] ds_M = Q_2(t) \tag{11.3.6}$$

for $t \geq 0$, where ds_M is the element of arc length on the curve σ at the point M.

We seek the function $\Phi(M, t)$ in the form

$$\Phi(M_0, t) = \frac{1}{2\pi} \int_L \frac{\partial}{\partial n_M} \ln \frac{1}{r_{MM_0}} g(M, t)\, ds_M + \qquad (11.3.7)$$

$$+ \frac{1}{2\pi} \sum_{i=1}^{2} \int_{\sigma_i} \frac{\partial}{\partial n_{M_i}} \ln \frac{1}{r_{M_0 M_i(t,\tau)}} g_i(M_i(t, \tau), t)\, ds_{i, M_i(t,\tau)}, \quad 0 \leq \tau \leq t,$$

where $g(M, t)$ is the potential density at the point $M \in L$ at the instant t; the curve L does not change in time; $g_i(M_i(t, \tau), t)$ is the potential density at the point $M_i(t, \tau)$ of the curve σ_i at time t. The point $M_i(t, \tau)$ is the position at time t of a fluid particle which occupied the position M_i^* at time $t = \tau$, i.e.,

$$M_i(\tau, \tau) = M_i^*, \qquad i = 1, 2. \qquad (11.3.8)$$

For studying two-dimensional problems of aerodynamics, it is convenient to introduce the notion of *distributed vortex intensity*, which is the derivative of the potential density with respect to arc length, and is denoted by γ for L and δ_i for σ_i, $i = 1, 2$,

$$\gamma(M, t) = g_s'(M, t), \qquad \delta_i(M_i(t, \tau), t) = g_{s_i}'(M_i(t, \tau), t). \qquad (11.3.9)$$

It is also convenient to consider the discontinuity surface for tangential velocities as the limit case of a vortex layer. Thus, the behavior of the airfoil contour L is modelled by a vortex layer of intensity $\gamma(M, t)$, $M \in L$, $M \neq M_q$. According to the above assumptions and physical considerations, the pressure and the normal velocities on the jet boundaries σ_1 and σ_2 are continuous (see conditions (11.3.5)) and the fluid density is the same inside and outside the jet. Therefore, it is natural to replace the curves σ_1 and σ_2 by free vortex layers in which free vortices that separate from the points M_1^* and M_2^* are moving along the trajectories of fluid particles without changing intensity (see *Belotserkovskii and Lifanov* (1985)) and coinciding with these particles. Thus, for the vortex intensity on the vortex sheets δ_i, $i = 1, 2$, we have

$$\delta_i(M_i(t, \tau), t) = \delta_i(\tau). \qquad (11.3.10)$$

For the velocity of the free vortex $\delta_i(\tau)$ located at the point $M_i(t, \tau)$ with radius-vector $r_i(t, \tau) = x_i(t, \tau)\boldsymbol{i} + y_i(t, \tau)\boldsymbol{j}$ and moving with the same velocity as the fluid particle at this point, we obtain

$$\frac{dr_i(t, \tau)}{dt} = \int_L \boldsymbol{V}_\gamma(M, M_i(t, \tau))\gamma(M, t)\, ds_M + \qquad (11.3.11)$$

$$+ \sum_{j=1}^{2} \int_0^t \boldsymbol{V}_{\delta_\gamma}\left(M_j(t, \tau'), M_i(t, \tau)\right)\delta_j(\tau')\left[x_{j,\tau'}'^2(t, \tau') + y_{j,\tau'}'^2(t, \tau')\right]^{1/2} d\tau' +$$

$$+ \boldsymbol{U}_0(M_i(t, \tau)) + \boldsymbol{V}_q(M_i(t, \tau), t)), \quad i = 1, 2, \qquad 0 \leq \tau \leq t,$$

with the initial condition (see (11.3.8))

$$r_i(\tau, \tau) = r(M_i^*), \quad i = 1, 2, \qquad (11.3.12)$$

where, $r(M_i^*)$ is the radius-vector of the point M_i^*; the vector $\boldsymbol{V}_\gamma$ is determined by the Biot–Savart law (see *Prandtl* (1939), *Loitsyanskii* (1978))

$$\boldsymbol{V}_\gamma(M, M_0) = \frac{1}{2\pi} \frac{y_1(M, M_0)\boldsymbol{i} - x_1(M, M_0)\boldsymbol{j}}{r_{MM_0}^2},$$

and the notation is the same as in (11.1.1).

The vectors $V_{\delta_i}(M, M_0)$ are determined in a similar way. The condition of non-penetration (11.3.4) becomes

$$n_{M_0} \cdot \int_L V_\gamma(M, M_0)\gamma(M, t)\, ds_M = -U_0(M_0) \cdot n_{M_0} - V_q(M_0, t) \cdot n_{M_0} -$$

$$- n_{M_0} \cdot \sum_{i=1}^{2} \int_0^t V_{\delta_i}(M_i(t, \tau), M_0)\, \delta_i(\tau) \left[x_{i,\tau}'^2(t, \tau) + y_{i,\tau}'^2 \right]^{1/2} d\tau, \quad (11.3.13)$$

$$M_0 \in L, \quad M_0 \neq M_q \in L.$$

where the notation is the same as in (11.3.11).

Since the airfoil starts its motion from the state of rest, Thomson's theorem about the preservation of velocity circulation on any curve embracing vortex layers yields the following relation

$$\int_L \gamma(M, t)\, ds_M + \sum_{i=1}^{2} \int_0^t \delta_i(\tau) \left[x_{i,\tau}'^2(t, \tau) + y_{i,\tau}'^2(t, \tau) \right]^{1/2} d\tau = 0, \quad t \geq 0. \quad (11.3.14)$$

Now, the condition of given net flow (11.3.6) becomes

$$\int_\sigma \left[n_{M_0} \cdot \int_L V_\gamma(M, M_0)\gamma(M, t)\, ds_M + U_0(M_0) \cdot n_{M_0} + V_q(M_0, t) \cdot n_{M_0} + \quad (11.3.15) \right.$$

$$\left. + n_{M_0} \cdot \sum_{i=1}^{2} \int_0^t V_{\delta_i}(M_i(t, \tau), M_0)\delta_i(\tau) \left[xx_{i,\tau}'^2(t, \tau) + y_{i,\tau}'^2(t, \tau) \right]^{1/2} d\tau \right] ds_M = Q_2(t).$$

As shown by *Belotserkovskii and Lifanov* (1985), at the instant τ when a free vortex separates from M_i^*, $i = 1, 2$, its intensity $\delta_i(\tau)$ is defined by

$$\delta_i(\tau) = \gamma(M_i^*, \tau). \quad (11.3.16)$$

Thus, we have a system of integro-differential equations (11.3.11)–(11.3.16) which should be solved with respect to the vortex layer intensity $\gamma(M, t)$ on L and the unknown curves δ_i, $i = 1, 2$, forming the boundary of the jet. It should be mentioned that, as shown by *Lifanov, Mikhailov, and Titskii* ((1990), the function $\gamma(M, t)$ should be sought in the class of functions having a singularity of the type $1/(s - q)$ at M_q, at any time t, if $Q_1 \neq 0$.

System (11.3.11)–(11.3.16) cannot be solved analytically and has to be approached by numerical methods. In order to solve such systems arising aerodynamics, the numerical method of discrete vortices was developed in *Belotserkovskii and Nisht* (1978), *Belotserkovskii and Lifanov* (1985) (see also *Belotserkovskii* (1988)).

Consider the nonstationary process at discrete time instants $t_\nu = \nu \Delta t$ with step Δt, where ν is called *the calculated time instant*. At each calculated time instant ν, the vortex layer that models the profile L is replaced by a system of N discrete vortices $\Gamma_{i,\nu} = \gamma(M_i, t_\nu)$. The lateral boundaries of the jet, σ_1 and σ_2, are modelled by two systems of free discrete vortices $\Delta_{1,m}$, $\Delta_{2,m}$, $m = 1, \ldots, \nu$, which separate from the discharge edges at the calculated instants $m = 1, \ldots, \nu$. Since the intensity of the free discrete vortices does not change in time, it follows that the only unknown quantities at the instant ν are the circulations of the vortices $\Gamma_{i,\nu}$, $i = 1, \ldots, N$, and $\Delta_{1,\nu}$, $\Delta_{2,\nu}$, because the circulations of the vortices $\Delta_{1,m}$, $\Delta_{2,m}$, $m = 1, \ldots, \nu - 1$, have been found on the previous steps. The discrete vortices $\Gamma_{i,\nu}$, $i = 1, \ldots, N$, are distributed on the boundary of the profile according to the law that ensures greater density of vortices on the parts of the boundary with greater curvature (see *Belotserkovskii* (1988)). The calculation points M_{0j}, $j = 1, \ldots, N + 1$, in which the non-penetration condition (11.3.4) holds, are located between neighboring discrete vortices on the profile. The points M_i of discrete vortices and the calculation points M_{0j} are chosen in such a way that M_q coincides with some M_{0j_q} and the corresponding quadrature sums converge to the precise value of the integrals. In order to fulfill condition (11.3.6), the cross-section σ of the discharge nozzle is divided into p equal parts of length σ_j, $j = 1, \ldots, p$. Consider calculation points A_{0j}, $j = 1, \ldots, p$, located in geometrical centers of the pieces $\Delta\sigma_j$. For each point A_{0j}, we know the velocity component orthogonal to $\Delta\sigma_j$ and arising due to the incident flow and the system of singularities. In order to determine the unknown circulations at each calculated time instant, we have to solve the following system of linear algebraic equations corresponding to integral equations (11.3.11)–(11.3.16):

$$\sum_{i=1}^{N} \omega(M_i, M_{0j})\Gamma_{i,\nu} + \sum_{k=1}^{2} \omega(M_{k,\nu}^{\nu}, M_{0j})\Delta_{k,\nu} = \tag{11.3.17}$$

$$= -2\pi U_0 \cdot n_{M_{0j}} - Q_{1,\nu}\omega(M_{0j_q}, M_{0j}) - \sum_{i=1}^{2} \left[\sum_{m=1}^{\nu-1} \omega(M_{k,m}^{\nu}, M_{0j})\Delta_{k,m} \right],$$

$$j = 1, \ldots, N+1; \quad j \neq j_q, \quad \nu = 1, 2, \ldots;$$

$$\sum_{j=1}^{p} \left[\sum_{i=1}^{N} \omega(M_i, A_{0j})\Gamma_{1,\nu} + \sum_{k=1}^{2} \omega(M_{k,\nu}^{\nu}, A_{0j})\Delta_{k,\nu} \right] \Delta\sigma_j = \tag{11.3.18}$$

$$= -2\pi Q_{2,\nu} - \sum_{j=1}^{p} \left[\sum_{k=1}^{2} \left(\sum_{m=1}^{\nu} \omega(M_{k,\nu}^{\nu}, A_{0j})\Delta_{k,m} \right) + 2\pi U_0 \cdot n_{A_{0j}} + Q_{1,\nu}\omega(M_{0j_q}, A_{0j}) \right],$$

$$A_{0j} \in \sigma, \quad \nu = 1, 2, \ldots;$$

$$\sum_{i=1}^{N} \Gamma_{i,\nu} + \sum_{k=1}^{2} \left(\sum_{m=1}^{\nu} \Delta_{k,m} \right) = 0, \quad \nu = 1, 2, \ldots, \tag{11.3.19}$$

where $Q_{1,\nu}$ and $Q_{2,\nu}$ are respectively the dimensionless coefficients of suction intensity and jet flow rate at the instants $\nu = 1, 2, \ldots$; $\omega(M_i, M_{0j})$ and $\omega(M_i, A_{0j})$ are respectively the normal components of dimensionless velocities produced by the discrete vortices $\Gamma_{i,\nu}$ at the point M_{0j} of the profile and A_{0j} of the discharge nozzle; $\omega(M_{k,m}^{\nu}, M_{0j})$ and $\omega(M_{k,m}^{\nu}, A_{0j})$ are respectively the normal components of dimensionless velocities produced at M_{0j} and A_{0j} by the discrete vortices at time ν; $\omega(M_{0j_q}, M_{0j})$ and $\omega(M_{0j_q}, A_{0j})$ are the normal components of dimensionless velocities produced at M_{0j} and A_{0j} by the source of intensity $Q_{1,\nu}$; U_0 is the velocity of the incident flow.

Having determined all circulations $\Gamma_{i,\nu}$, $i = 1, \ldots, N$, and $\Delta_{k,m}$, $k = 1, 2$; $m = 1, 2, \ldots, \nu$, at time $t = t_\nu$, we can express the velocity at each spatial point M_0 as a superposition of the known velocities, U_0, $V_q(M_0, t_\nu)$, and the velocities induced by the discrete vortices. Thus, we have

$$V(M_0, t) = U_0 + V_q(M_0, t) + \sum_{i=1}^{N} V_\gamma(M_i, M_0)\Gamma_{i,t} + \sum_{k=1}^{2} \left(\sum_{m=1}^{\nu} V_{\delta_k}(M_{k,m}^{\nu}, M_0)\Delta_{k,m} \right). \tag{11.3.20}$$

If we know the absolute flow velocity at an arbitrary point of the profile surface, then, using the Cauchy–Lagrange integral, we can find the pressure coefficient at that point,

$$P(M_0, t) = \left(P^*(M_0, t) - P_0) \right) \left(\rho \frac{|U_0|^2}{2} \right)^{-1} = 1 - |V(M_0, t)|^2 - 2\frac{\partial \Phi(M_0, t)}{\partial t}. \tag{11.3.21}$$

It should be observed that in addition to the above hydrodynamical effects related to energy control devices, the flow past a profile in our problem is accompanied by the action of the jet reaction force, with the following contribution to the moments:

$$R_j(t) = -2 \int_\sigma \mathbf{V}(M, t) \cdot \mathbf{V}_{n\sigma}(M, t) \, ds_M, \qquad M \in \sigma, \tag{11.3.22}$$

and the reaction of the suction force,

$$R_{qj}(t) = -Q_1(t) \mathbf{V}(M_q, t), \tag{11.3.23}$$

where M_q is the location of the outlet.

The method considered in this work is a logical continuation of the methods described in *Lifanov* (1989), *Lifanov, Mikhailov, and Titskii* (1990), where mathematical models, systems of integro-differential equations, and methods of their numerical analysis were proposed for problems of flow past a profile with suction. In order to refine the methods of solving such problems and verify their reliability, we have constructed an exact solution for stationary flow of an ideal incompressible fluid past a cylinder of unit radius with suction of the exterior flow on its surface. On the basis of the results obtained by *Lifanov* (1989), *Lifanov, Mikhailov, and Titskii* (1990), it can be shown that this solution has the form

$$\gamma(\varphi) = 2 \sin\varphi + \frac{Q_1}{2\pi} \cot \frac{\varphi_q - \varphi}{2} + C, \tag{11.3.24}$$

where φ_q is the position of the outlet on the unit circle. If the flow past the cylinder is circulation-free, we have $C = 0$ in (11.3.24). Using (11.3.24), one can obtain an expression for the circulation of discrete vortices modelling the surface of the cylinder,

$$\Gamma_i = 2 + 2\cos(\alpha - \psi_i) - \frac{Q}{\pi} \ln \left| \sin \frac{\psi_i - \psi_{0j_q}}{2} \right|, \tag{11.3.25}$$

where ψ_i is the angle in the clockwise direction between the positive direction of the axis OX and the radius-vector of the i-th discrete vortex; ψ_{0j_q} is the angular coordinate of the suction point of intensity Q_1; and α is the angle of attack of the incident flow.

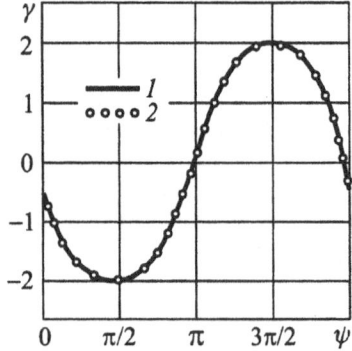

Figure 11.3.1

Figure 11.3.1 shows the distributed intensity of the vortex layer, $\gamma(\psi)$, on an infinite cylinder. Curve 1 gives the intensity values calculated by (11.3.24) and curve 2 gives the intensity values obtained by *Lifanov, Mikhailov, and Titskii* ((1990) for the flow past a two-dimensional body with suction. Here, the angle of attack is $\alpha = 6°$, the contour of the cylinder is divided into $N = 71$ equal parts; $Q_1 = 0$ corresponds to the absence of suction (Fig. 11.3.2(a)); suction intensity $Q_1 = 0.05$ at the outlet located at the calculation point $\psi_{0j_q} = 65$ (Fig. 11.3.2(b)). We see that the solution obtained by the method of discrete vortices is in fair agreement with the exact solution.

Next, we are going to describe a method of numerical analysis of nonstationary flows past an airfoil when the air is drawn off through the nozzle into the jet and acts as a jet flap. As a demonstration of this method, we have calculated aerodynamic characteristics of a symmetric NASA-0012 profile in the following three cases: without energy control devices; only with suction on the upper surface; with a jet–suction combination. Our investigation pertains to subcritical values of the attack angle and its aim is to estimate the efficiency of the device and determine the lift coefficient C_y in velocity coordinates, for a given suction point x_{j_q} of intensity Q_1, and the exhaust angle θ near the rear edge, and the given exhaust rate coefficient Q_2. Figure 11.3.2(a) shows the dependence $C_y = f(x)$ for a profile without energy control devices (curve 1), with suction of the outer flow at $x_{0j_q} = 0.94$ with intensity $Q_1 = 0.025$ (curve 1), and suction combined with jet exhaust on the rear edge at the angle $\theta = 30°$, for $Q_1 = 0.05$ and $Q = 0.05$ (curve 3). We see that the suction–jet device is the most efficient. The largest lift coefficient in this case is achieved by the joint favorable action of suction and the jet on the flow past the upper surface.

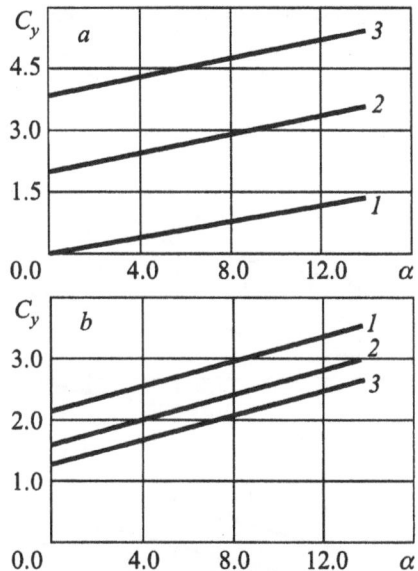

Figure 11.3.2

The flow slows down on the surface behind the suction point and its positive pressure gradient sharply increases. The outflow of the jet from the rear edge changes the main flow and modifies the local attack angle due to the jet ejection accompanied by partial inhibition of the flow behind the suction point on the upper surface, and increases the pressure on the lower surface of the airfoil.

The action of suction and the jet on the flow past the NASA-0012 profile can be seen from the structure of pressure distribution on its surface (Fig. 11.3.3). Curve 1 gives the pressure distribution in the presence of both suction and jet discharge, curve 2 corresponds to the presence of only suction, and curve 3 is the pressure distribution without energy control devices. It is clear that a favorable position of the suction point for subcritical values of the attack angle would be near the rear edge, because favorable action of suction extends to the largest part of the upper airfoil surface

(Fig. 11.3.2(b); curves 1, 2, and 3 correspond to $x_{0j_q} = 0.94$, $x_{0j_q} = 0.43$, and $x_{0j_q} = 0.09$). An increase of the suction rate coefficient and the jet discharge improve the performance of the system, since the flow becomes faster on the upper surface.

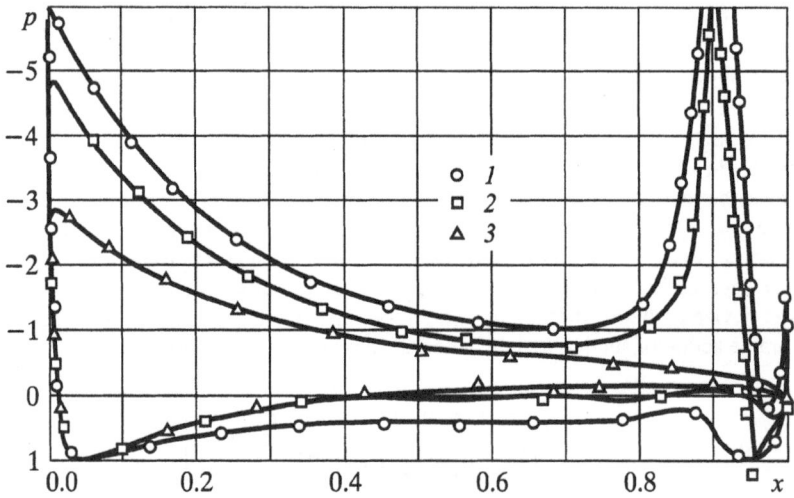

Figure 11.3.3

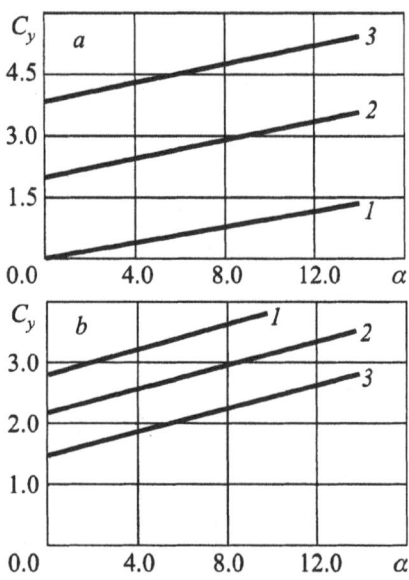

Figure 11.3.4

Moreover, an increase of Q_2 improves the energy properties of the jet and changes the shape of the jet, causing thereby a substantial change of the main flow, especially in the rear part of the airfoil. Figure 11.3.4 represents the dependence $C_y = f(x)$ for the flux coefficients $Q_1 = 0.0$ and $Q_2 = 0.0$; $Q_1 = 0.05$ and $Q_2 = 0.05$; $Q_1 = 0.1$ and $Q_2 = 0.1$ (curves 1, 2, 3, respectively).

As mentioned above, jet discharge acts near the rear edge like a jet flap with its typical effect on the structure of flow past a body. The results regarding the dependence of bearing properties of the airfoil on the discharge jet angle are given in Fig. 11.3.4(b) (curves 1, 2, and 3 correspond to the angles $\theta = 45°$, $\theta = 30°$, and $\theta = 15°$).

Thus, the mathematical model of two-dimensional flow of an ideal fluid past an airfoil with energy control devices and the corresponding numerical method allow us to estimate, with sufficient accuracy, the efficiency of the devices for increasing the lift force and to determine geometrical and kinetic parameters of an airfoil in a wide range of flight regimes.

11.4. Numerical Analysis of 3D Flows Past Bodies of Arbitrary Shape

Consider a bluff body moving in a fluid that fills up the entire space. Suppose that the surface of the body is non-smooth along the ribs L_p, $p = 1, \ldots, N$. As a rule, flow past such bodies is accompanied by the phenomenon of separation. We assume that the fluid is ideal and incompressible, and there is separation on some of the ribs L_p.

Let us introduce orthogonal coordinate systems usually considered in aerodynamics (see *Belotserkovskii and Nisht* (1978)), namely, a moving frame $OXYZ$ fixed to the body, and a frame $O_0X_gY_gZ_g$ which is fixed in relation to the unperturbed medium (see Fig. 11.4.1). The position of the points of the surface of the body in the fixed frame is described by the equation

$$F(M, t) = 0,$$

where t is time, $M = M(x, y, z)$ is a point on the surface of the body.

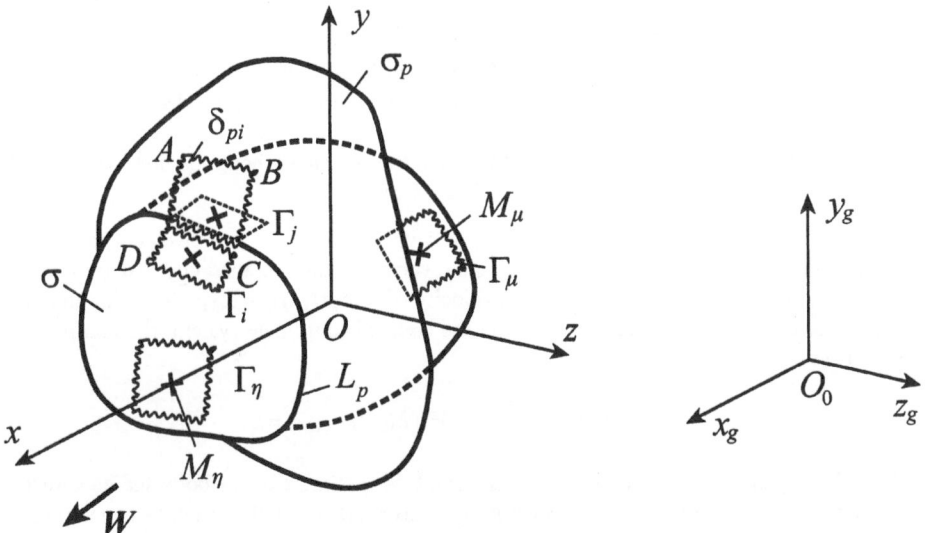

Figure 11.4.1

The lines L_p, $p = 1, \ldots m$, $m \le N$, on the surface σ from which vortex sheets σ_p separate are assumed known. The velocity field excited by the surfaces σ and σ_p is assumed potential, i.e.,

$$V_\Sigma(M, t) = \nabla\Phi_\Sigma(M, t),$$

where $\Phi_\Sigma(M, t)$ is the potential of the excited velocities.

From the continuity equation, it follows that the velocity potential is a harmonic function outside the surface of the body σ and the vortex surfaces σ_p, i.e., equation (11.3.1) holds for the potential, where D^+ is the domain outside these surfaces in the three dimensional space $\mathbb{R}_3$ with the coordinate system $OXYZ$. Then, pressure at an arbitrary point is again expressed by the Cauchy–Lagrange integral (see (11.3.21), which in this case can be written as

$$p(M_0, t) = p_\infty - \rho \left[\frac{\partial \Phi_\Sigma(M_0, t)}{\partial t} + \frac{1}{2} V_\Sigma(M_0, t) \right],$$

where p_∞ is the pressure of the unperturbed flow at infinity. It is assumed that the perturbations decay at infinity, i.e., conditions (11.3.2) hold for $\rho\left(M, \sigma \cup \cup_{p=1}^m \sigma_p\right) \to 0$.

Our aim is to describe the action of the fluid on the body in terms of the integral of pressure over the surface σ,

$$R = -\int_\sigma p_+(M, t) n_M(t) \, d\sigma,$$

where $p_+(M, t)$ is the value of the pressure at the point $M \in \sigma$ approached from the outside; n_M is the unit vector of the exterior normal to σ at the point M at time t.

Thus, the problem of flow past a body again reduces to finding the potential of excited velocities, $\Phi_\Sigma(M, t)$, satisfying the Laplace equation (11.3.1) in the domain D^+, and also the following boundary conditions:

(i) condition of non-penetration on the surface σ,

$$\frac{\partial \Phi_\Sigma(M_0, t)}{\partial n_M(t)} = W(M_0, t) \cdot n_M(t), \qquad M_0 \in \sigma, \qquad (11.4.1)$$

where $W(M_0, t)$ is the velocity of the points on σ;

(ii) decay of perturbations (11.3.2) at infinity, as $\rho\left(M, \sigma \cup \cup_{p=1}^m \sigma_p\right) \to \infty$;

(iii) the kinematic compatibility condition and no pressure drop on the free vortex sheets σ_p, $p = 1, \ldots, m$;

(iv) Chaplygin–Zhukovskii–Kutta condition that the velocities should be finite near the ribs of the surface σ from which the vortex sheets separate.

In order to solve the above problem of flow with separation, it is convenient to introduce a coupled coordinate system corresponding to the reverse motion, i.e., the body is regarded as fixed and the flow is moving about the body with velocity $W(M, t)$, $M \in \mathbb{R}_3$, $t \geq 0$. Assume that the shape of the body does not change. Then, in the coupled coordinate system, the non-penetration condition (11.4.1) reads

$$V_\Sigma(M_0, t) \cdot n_{M_0} + W(M_0, t) \cdot n_{M_0} = 0, \qquad (11.4.2)$$

where $V_\Sigma(M_0, t)$ is the velocity excited at the point M_0 at time t by the body and its vortex trace.

The surface of the body σ and the separating vortex sheets σ_p will be modelled by vortex layers. The velocity vectors at a point M induced by these vortex formations are denoted by $V(M, t)$ and $V_p(M, t)$, $p = 1, \ldots, m$. Then, equation (11.4.2) at $M_0 \in \sigma$ can be written as

$$V(M_0, t) \cdot n_{M_0} + \sum_{p=1}^m V_p(M_0, t) n_{M_0} = -W(M_0, t) \cdot n_{M_0}. \qquad (11.4.3)$$

Let r_M be the radius-vector of the point M. Clearly, in order to ensure the kinematic condition that there should be no jump of normal velocities at the points of the vortex sheets σ_p it suffices to assume that these points are moving with local velocities. Thus, for the point $M_p(t, \tau)$ that separates from the curve L_p at time t, the following condition should hold:

$$\frac{d\boldsymbol{r}_{M_p}(t, \tau)}{dt} = \boldsymbol{V}(M_p(t, \tau), t) + \sum_{p=1}^{m} \boldsymbol{V}_p(M_p(t, \tau), t) + \boldsymbol{W}(M_o(t, \tau), t), \qquad (11.4.4)$$

and also the boundary condition

$$\boldsymbol{r}_{M_p}(\tau, \tau) = \boldsymbol{r}_{M_p}, \qquad (11.4.5)$$

where M_p is the point of $L_p \subset \sigma$ from which a vortex particle separates at time τ and occupies the position $M_p(t, \tau)$ on the vortex surface σ_p at time t.

We seek the velocities in (11.4.3) and (11.4.4) as the gradients of the double layer potential concentrated on the surfaces σ and σ_p, $p = 1, \ldots, m$, namely,

$$\boldsymbol{V}(M_0, t) = \nabla_{M_0} \left[\frac{1}{4\pi} \int_{\sigma} \frac{\partial}{\partial n_M} \left(\frac{1}{r_{MM_0}} \right) g(M, t) \, d\sigma_M \right], \qquad (11.4.6a)$$

$$\boldsymbol{V}_p(M_0, t) = \nabla_{M_0} \left[\frac{1}{4\pi} \int_{\sigma_p} \frac{\partial}{\partial n_{M_p(t, \tau')}} \left(\frac{1}{r_{M_p(t, \tau')M_0}} \right) g_p(M_p, \tau') \, d\sigma_{M_p(t, \tau')} \right], \qquad (11.4.6b)$$

where $g(M, t)$ is the density of the double layer potential at $M \in \sigma$ at time t; $g_p(M_p, \tau')$ is its value at the point $M_p(t, \tau') \in \sigma_p$ at time t. Thus, the value $g_p(M_p, \tau')$ depends only on the time of separation, τ', and does not depend on the current time t (see *Lifanov* (1996)).

The expressions for $\boldsymbol{V}(M_p(t, \tau), t)$ and $\boldsymbol{V}_p(M_p(t, \tau), t)$ in equation (11.4.4) are obtained from the respective formulas (11.4.6) by replacing M_0 with $M_p(t, \tau)$.

Thus, the problem is reduced to solving equations (11.4.3), (11.4.4) with the unknown functions $g(M, t)$ and $g_p(M_p, \tau)$, $0 \le \tau \le t$, and finding free vortex surfaces σ_p. Within this approach to finding the velocities $\boldsymbol{V}(M_0, t)$ and $\boldsymbol{V}_p(M_0, t)$, equation (11.4.3) is a two-dimensional hypersingular integral equation of the first kind with the unknown functions $g(M, t)$ and $g_p(M_p, \tau)$, $0 \le \tau \le t$, and singularity of the type $1/r_{MM_0}^3$. As mentioned in Chapter 1, the integral in this equation should be understood either in the sense of Hadamard's finite value (see *Lifanov and Poltavskii* (1998)) or as an integral in the sense of distributions. Since σ, being the surface of a body, is a closed surface and $\boldsymbol{V}(M_0, t) \cdot \boldsymbol{n}_{M_0}$ is the normal derivative of a harmonic function on a piecewise smooth closed surface, it follows that (see *Vladimirov* (1976))

$$\int_{\sigma_p} \boldsymbol{V}(M_0, t) \cdot \boldsymbol{n}_{M_0} \, d\sigma = 0. \qquad (11.4.7)$$

In practice, a numerical solution of system (11.4.3), (11.4.4) is found by the method of discrete closed vortex frameworks (see *Aparinov et al.* (1988), *Lifanov* (1996)). To that end, this system of integro-differential equations is reduced to a linear system of algebraic equations. From (11.4.7), it follows that the linear algebraic system is degenerate. In the problem under consideration, it suffices to know the values of $g(M, t)$ to within a constant, and therefore, the intensity of one of the N vortex frames modelling the surface of the body is taken equal to zero (see *Lifanov* (1996)), and in order to solve the resulting system of algebraic equations, (which happens to be overdetermined and incompatible, in general), one employs the method of regularization variables (see *Aparinov et al.* (1988), *Lifanov* (1996)).

In the numerical analysis of the above problem, we assume that at time t the fluid, previously at rest, starts to move with constant velocity $\boldsymbol{W}$ and the formation of vortex sheets behind the body. We introduce the dimensionless time variable

$$t^* = \frac{|\boldsymbol{W}|t}{b},$$

where b is the characteristic linear size of the body, t is the actual time since the body has started its motion.

At the initial instant, $t^* = 0$, there is no vortex trace, and one has to solve the problem of circulation-free flow past the body. In this case, the system of linear algebraic equations expressing the condition of non-penetration on σ and written for N calculation points M_ν has the form

$$\lambda_N + \sum_{\mu=1,\,\mu\neq q}^{N} \Gamma_m \mu a_{\nu\mu} = -\omega_{n\nu}, \qquad \nu = 1, \ldots, N. \tag{11.4.8}$$

Here, λ is a regularization variable; $\Gamma_\mu = \Gamma_{\mu+}/|W|b$ is the dimensionless circulation of the μ-th vortex frame ($\Gamma_{\mu+}$ is the dimensional circulation); q is the index of the frame with zero circulation; $a_{\nu\mu}$ is a dimensionless function determined by the mutual position of the points M_ν and M_μ on σ (see *Belotserkovskii, Lifanov, and Mikhailov* (1985)); $\omega_{n\nu} = W_{n\nu}/W$ is the dimensionless normal velocity component for the unperturbed flow at the point M_ν.

For $t^* > 0$, on step k of our calculations, a vortex frame separates from the i-th piece of the line $L_p \subset \sigma$ (see Fig. 11.4.1). The dimensionless circulation of this frame, δ_{pi}^k, is taken equal to the difference of the circulations of the vortex frames adjacent to the i-th piece of line L_p (these circulations have been calculated on the previous step),

$$\delta_{pi}^k = \Gamma_m^{k-1} - \Gamma_j^{k-1}.$$

This expression means that the vortex layer on the surface σ near the separation line L_p consists of all free vortices that have entered the flow during the time of calculation Δt^*.[1] Here, the direction in which the frames enter the flow is not assumed beforehand, but is calculated on the basis of the vector of local velocity. This means that the position of angular points A and B of a vortex frame δ_{pi}^k entering the flow is determined as follows:

$$r_A = r_D + V^*(D)\Delta t^*, \qquad r_B = r_C + V^*(C)\Delta t^*,$$

where $V^*(D) = V^*(D)/W$, $V^*(C) = V^*(C)/W$ are dimensionless velocities calculated at the points D and C.

On the k-th calculation step, as a result of a free vortex sheet entering the flow, system (11.4.8) becomes

$$\lambda_N + \sum_{\mu-1,\,\mu\neq q}^{N} \Gamma_\mu^k a_{\nu\mu} = -\omega_{n\nu} - \sum_{p=1}^{m}\sum_{i=1}^{N_p}\sum_{s=1}^{k} \delta_{0i}^s a_{pi}^{k-s+1}, \qquad \nu = 1, \ldots, N, \tag{11.4.9}$$

where N_p is the number of longitudinal strips in the p-th vortex sheet.

On each step, the unknown circulations of the vortex frames modelling the surface of the body are determined from system (11.4.9). Note that in this case, there is no need to include into the calculation scheme the condition that the circulation on closed surfaces embracing the body and its vortex trace should be constant. This condition is satisfied automatically, due to the structure of the scheme. As a result, the dimension of the algebraic system to be solved is reduced (see *Aparinov et al.* (1988), *Lifanov* (1996)).

Having found the circulations Γ_μ^k, $\mu = 1, \ldots, N$, one can calculate the nonstationary aerodynamic load acting on the body, and also find instantaneous values of aerodynamic coefficients. This is done as follows.

---

[1] This is our hypothesis in the present case. Its mathematical proof is a serious open problem worthy of investigation.

The pressure coefficient at any calculation point $M_q \in \sigma$ is defined by the Cauchy–Lagrange integral, which can be written in the form (see *Aparinov at al.* (1988), *Belotserkovskii, Lifanov, and Mikhailov* (1987))

$$p_\mu = 2\frac{p_{\mu+} - p_{\infty-}}{\rho W^2} = 1 - v_\mu^2 - 2\frac{\partial \varphi_{\mu+}}{\partial \tau}. \tag{11.4.10}$$

Here, v_μ is the dimensionless relative velocity of the flow at the point under consideration; $\varphi_+ = \Phi_+/|W|b$ is the dimensionless potential of the flow (the indices '+' and '−' refer to different sides of the surface σ). Let us write (11.4.10) as follows:

$$p_\mu = 1 - v_\mu^2 - 2\frac{\partial(\varphi_{\mu+} - \varphi_{\eta+})}{\partial \tau} - 2\frac{\partial \varphi_{\eta+}}{\partial \tau}, \tag{11.4.11}$$

where $\varphi_{\eta+}$ is the flow potential at the η-th calculation point taken as the basic point.

Denote by $\Delta\varphi$ the jump of the potential across the surface σ at some calculation point. Then we can write

$$(\varphi_{\mu+} - \varphi_{\eta+}) = (\varphi_{\mu-} - \varphi_{\eta-}) + (\Gamma_\mu - \Gamma_\eta), \tag{11.4.12}$$

since $\Delta\varphi(M) = \Gamma$.

Differentiating this expression, we obtain

$$\frac{\partial}{\partial t}(\varphi_{\mu+} - \varphi_{\eta+}) = \frac{\partial}{\partial t}(\varphi_{\mu-} - \varphi_{\eta-}) + \frac{\partial}{\partial t}(\Gamma_\mu - \Gamma_\eta).$$

According to our assumptions, the difference of the potentials $\varphi_{\mu-} - \varphi_{\eta-}$ does not depend on time, and therefore,

$$\frac{\partial}{\partial t}(\varphi_{\mu+} - \varphi_{\eta+}) = \frac{\partial}{\partial t}(\Gamma_\mu - \Gamma_\eta).$$

For the k-th calculation step, this expression has the form

$$\frac{\partial}{\partial t}(\varphi_{\mu+} - \varphi_{\eta+}) = \frac{\Gamma_\mu^k - \Gamma_\mu^{k-1} - \Gamma_\eta^k + \Gamma_\eta^{k-1}}{\Delta \tau}.$$

The value $\varphi_{\eta+}$ can be defined as

$$\varphi_{\eta+} = \varphi_\infty + \int_{+\infty}^{x_\eta} v_x \, dx, \tag{11.4.13}$$

where φ_∞ is the velocity potential for the unperturbed flow at ∞, and x_η is the coordinate of the basic point M_η. Differentiating (11.4.13), we obtain

$$\frac{\partial \varphi_{\eta+}}{\partial t} = \int_{+\infty}^{x_\eta} \frac{\partial v_x}{\partial t} \, dx.$$

This integral is found numerically.

Thus, all quantities in (11.4.11) have been determined, and at each calculated time instant we can find the distribution of the pressure coefficient p on the surface of the body σ. Then, in the coupled coordinate system, we can find the components of the aerodynamic forces and moments (see *Belotserkovskii and Nisht* (1978))

$$C_X = -\sum_{\mu=1}^{N} p_\mu \cos(m, x)_\mu \Delta\sigma_\mu \,,$$

$$C_Y = \sum_{\mu=1} p_\mu \cos(n, y)_\mu \Delta\sigma_\mu \,,$$

$$C_Z = \sum_{\mu=1}^{N} p_\mu \cos(n, z)_\mu \Delta\sigma_\mu \,,$$

$$m_X = \sum_{\mu=1}^{N} p_\mu \left[Z_\mu \cos(n, y)_\mu - Y_\mu \cos(n, z)_\mu \right] \Delta\sigma_\mu \,,$$

$$m_Y = \sum_{\mu=1}^{N} p_\mu \left[X_\mu \cos(n, z)_\mu - Z_\mu \cos(n, x)_\mu \right] \Delta\sigma_\mu \,,$$

$$m_Z = \sum_{\mu=1}^{N} p_\mu \left[Y_\mu \cos(n, x)_\mu - X_\mu \cos(n, y)_\mu \right] \Delta\sigma_\mu \,,$$

where $\Delta\sigma_\mu$ is the area embraced by the vortex frame μ; $\cos(nx)_\mu$, etc., are the direction cosines of the exterior normal to σ at the calculation point $M_\mu(x_\mu, y_\mu, z_\mu)$ with dimensionless coordinates.

Let us apply the above scheme to the problem of flow with separation past a cube, assuming that the velocity of the unperturbed flow W is orthogonal to one of its faces and there is separation from all its ribs. Let the origin of the coordinate system $OXYZ$ be at the center of the cube, and let $\Delta\tau = 0.25$ be the dimensionless time-step.

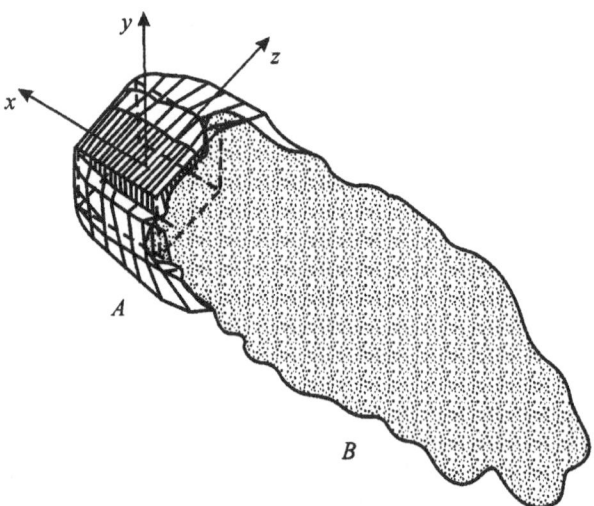

Figure 11.4.2

Figure 11.4.2 shows the vortex sheet behind the cube for $\tau = 6$. We see that stable vortex surfaces are preserved only in the immediate vicinity of the cube (zone A). Here, the vortex structure is determined mainly by stable vortex sheets that separate from the ribs on the frontal face of the cube (in Fig. 11.4.2, this vortex surface has a cut, for the sake of illustration). Vortex sheets separating from different faces are almost immediately destroyed and have no ordered structure.

Almost immediately behind the cube, all vortex sheets become completely destroyed and a chaotic alternating flow is observed (zone B in Fig. 11.4.2).

Figure 11.4.3 shows the velocity field calculated in the longitudinal plane OXY for $\tau = 6$. Behind the cube, we observe a nonstationary vortex flow with an ordered structure above and below the cube.

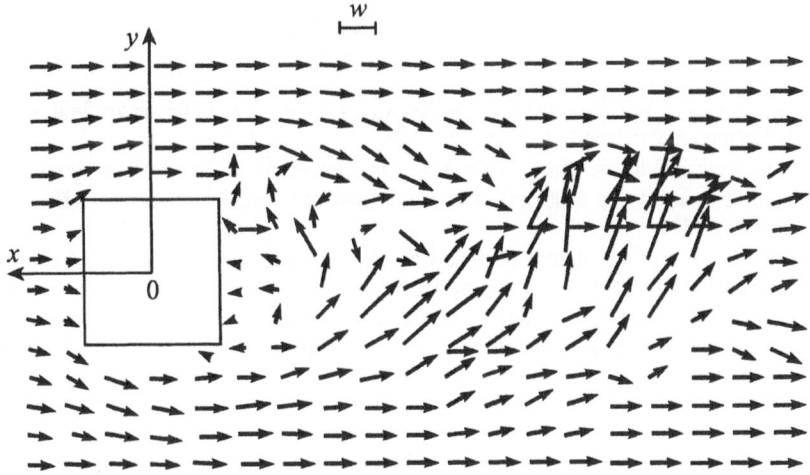

Figure 11.4.3

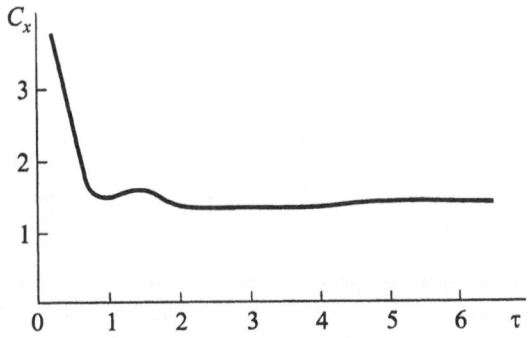

Figure 11.4.4

The curve in Fig. 11.4.4 shows the resistance coefficient C_X versus τ. And there is a point corresponding to the value of C_X obtained for a cube in a weighted experiment. We see that for $\tau > 2$, the calculated value of C_X varies but slightly and there is a satisfactory agreement with the experiment (see *Khudyakov* (1973)).

11.5. Elements of the Potential Theory in the Three-Dimensional case

In the previous section, we have used the notion of the double layer potential and its properties in the three-dimensional case. It seems that there is no adequate exposition of these topics in classical text books, and therefore, we offer our version of the theory of double layer and single layer potentials.

First, we again consider a source function. Suppose that there is a source of intensity Q at a point $M_0(x_0, y_0, z_0)$ in the $OXYZ$ space. This source can be associated with the vector field

$$V_0(M) = \frac{Q}{4\pi} \frac{r_{MM_0}}{r^3_{MM_0}},$$
(11.5.1)

where $r_{MM_0} = (x - x_0)i + (y_0 - y)j + (z_0 - z)k$, $r^2_{MM_0} = |r_{MM_0}|^2$. The vector field $V_0(M)$ will be regarded as the velocity field of an ideal fluid flow.

Let σ be a surface in the space $OXYZ$ (again denoted by $\mathbb{R}_3$) and its parametric equations have the form $x = x(u, v)$, $y = y(u, v)$, $z = z(u, v)$, where (u, v) is a point of a domain D on the Cartesian plane O^*UV.

We assume that σ is a simple piecewise smooth surface, i.e., at every point $M \in \sigma$, possibly, outside some curves on σ, there exists a tangential plane, and therefore, there is a unit normal vector n_M. The vectors n_M are all directed to one side of the surface. If σ is a closed surface, we denote by D^+ and D^- the exterior and the interior domains bounded by σ. In this case, we choose n_M directed outside D^- (Fig. 11.5.1).

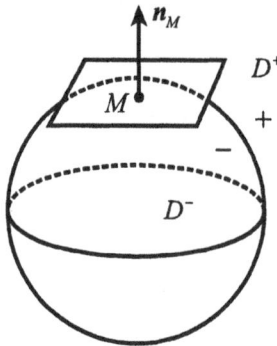

Figure 11.5.1

The side of σ corresponding to the direction of n_M is denoted by σ^+ and σ^- is the opposite side of σ. We assume that if σ is smooth at M, then there is a neighborhood $O(M)$ and its parametric representation such that the tangential vectors $\tau_{1,M} = \overline{x}'_u i + \overline{y}'_u j + \overline{z}'_u k$ and $\tau_{2,M} = \overline{x}'_v i + \overline{y}'_v j + \overline{z}'_v k$ are non-parallel at any point of that neighborhood. Then, for n_M the following formula holds

$$n_M = \frac{\tau_{1,M} \times \tau_{2,m}}{|\tau_{1,M} \times \tau_{2,m}|} = \frac{[\tau_{1,M} \times \tau_{2,m}]}{|[\tau_{1,M} \times \tau_{2,m}]|}.$$
(11.5.2)

Consider the function

$$V_0(M) \cdot n_M = \frac{Q}{4\pi} \frac{(r_{MM_0}, n_M)}{r^3_{MM_0}} = Q\omega_1^{(2)}(M, M_0)$$
(11.5.3)

defined at the points of σ. As above, the function $\omega_1^{(2)}(M, M_0)$ can be interpreted in physical terms as penetration density at the points of σ due to the unit source at M_0.

Assume first that σ is a closed piecewise smooth surface. For $M_0 \in D^+$, it is physically clear that the quantity of the fluid that passes through σ is equal to zero,

$$\int_\sigma \omega_1^{(2)}(M, M_0) \, d\sigma_M = 0, \qquad M_0 \in D^+, \tag{11.5.4}$$

where $d\sigma_M$ is the area element on σ. This formula can be justified in physical terms similarly to the plane case. In order to prove (11.5.4) in mathematical terms, we recall the Ostrogradskii divergence theorem (see *Shilov* (1972)), namely, that the flux of a vector field $a(M) = (P, Q, R)$ through a closed surface σ is equal to the volume integral of the divergence of the vector $a(M)$ over the region bounded by σ,

$$\int_\sigma a(M) \cdot n_M \, d\sigma = \int_V \operatorname{div} a(M) \, dx \, dy \, dz,$$

$$\int_\sigma P \, dy \, dz + Q \, dz \, dx + R \, dx \, dy = \int_V \left(\frac{\partial P}{\partial x} + \frac{\partial Q}{\partial y} + \frac{\partial R}{\partial z} \right) dx \, dy \, dz,$$

or

$$\int_\sigma (P \cos \lambda + Q \cos \mu + R \cos \nu) \, d\sigma = \int_V \left(\frac{\partial P}{\partial x} + \frac{\partial Q}{\partial y} + \frac{\partial R}{\partial z} \right) dx \, dy \, dz, \tag{11.5.5}$$

where $n_M = (\cos \lambda) i + (\cos \mu) j + (\cos \nu) k$. Therefore, we can write

$$\int_\sigma \omega_1^{(2)}(M, M_0) \, d\sigma_M = \int_\sigma \frac{1}{4\pi} \frac{(r_{MM_0}, n_M)}{r_{MM_0}^3} \, d\sigma_M =$$

$$= \frac{1}{4\pi} \int_\sigma \frac{\partial}{\partial n_M} \left(\frac{1}{r_{MM_0}} \right) d\sigma_M = \frac{1}{4\pi} \int_\sigma \nabla \left(\frac{1}{r_{MM_0}} \right) \cdot n_M \, d\sigma_M =$$

$$= \frac{1}{4\pi} \int_\sigma \left[\frac{\partial}{\partial x} \left(\frac{1}{r_{MM_0}} \right) \cos \lambda + \frac{\partial}{\partial y} \left(\frac{1}{r_{MM_0}} \right) \cos \mu + \frac{\partial}{\partial z} \left(\frac{1}{r_{MM_0}} \right) \cos \nu \right] d\sigma_M =$$

$$= \frac{1}{4\pi} \int_V \left[\frac{\partial^2}{\partial x^2} \left(\frac{1}{r_{MM_0}} \right) + \frac{\partial^2}{\partial y^2} \left(\frac{1}{r_{MM_0}} \right) + \frac{\partial^2}{\partial z^2} \left(\frac{1}{r_{MM_0}} \right) \right] dV =$$

$$= \frac{1}{4\pi} \int_V 0 \, dV = 0, \tag{11.5.6}$$

since the function $1/r_{MM_0}$ is harmonic in any domain that does not contain M_0.

Now, consider a point $M_0 \in D^-$ and suppose that σ is the sphere of radius R with center at M_0 (cf. Fig. 11.2.2). Physically, it is clear that

$$\int_{S_R} \omega_1^{(2)}(M, M_0) \, d\sigma_M = -1. \tag{11.5.7}$$

To give a rigorous proof of (11.5.7), we pass to the spherical coordinates

$$x = R \sin \theta \cos \varphi, \quad y = R \sin \theta \sin \varphi, \quad z = R \cos \theta, \quad 0 \le \theta \le \pi, \quad 0 \le \varphi \le 2\pi.$$

in the integral on the left-hand side. Taking into account that $d\sigma_M = R^2 \sin \theta \, d\theta \, dy$, we obtain

$$\int_{S_R} \omega_1^{(2)}(M, M_0) \, d\sigma_M = \frac{1}{4\pi} \int_0^{2\pi} d\varphi \int_0^\pi \frac{(r_{MM_0}, n_M)}{r_{MM_0}^3} R^2 \sin \theta \, d\theta =$$

$$= \frac{1}{4\pi} \int_0^{2\pi} d\varphi \int_0^\pi \frac{-RR^2}{R^3} \sin \theta \, d\theta = -\frac{1}{4\pi} 2\pi \left(-\cos \theta \Big|_0^\pi \right) = -1. \tag{11.5.8}$$

Remark 11.5.1. If n_M is directed inside the sphere, we have

$$\int_{S_R} \omega_1^{(2)}(M, M_0)\, d\sigma_M = 1. \tag{11.5.9}$$

Now, let σ be a closed piecewise smooth surface and let M_0 be a point inside σ. Take a sphere S_ε of small radius ε with center at M_0. The unit normal on S_ε is chosen in the direction of M_0, so that we obtain the exterior unit normal to the region bounded by $S_\varepsilon \cup \sigma$.

By the Ostrogradskii formula (which holds for an arbitrary piecewise smooth closed surface), we get

$$\int_{\sigma \cup S_\varepsilon} \omega_1^{(2)}(M, M_0)\, d(\sigma \cup S_\varepsilon)_M = \int_\sigma \omega_1^{(2)}(M, M_0)\, d\sigma_M + \int_{S_\varepsilon} \omega_1^{(2)}(M, M_0)\, dS_{\varepsilon, M} = 0. \tag{11.5.10}$$

From (11.5.9) and (11.5.10), we obtain

$$\int_\sigma \omega_1^{(2)}(M, M_0)\, d\sigma_M = -1. \tag{11.5.11}$$

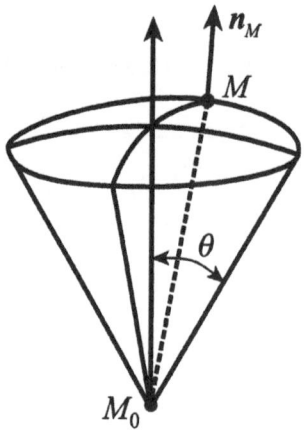

Figure 11.5.2

Remark 11.5.2. Let $S_{R,\theta}$ be the part of the sphere cut out by the cone of angle θ with vertex at M_0 (Fig. 11.5.2). And let $K_{R,\theta}$ be the union of $S_{R,\theta}$ and the part of the conical surface inside the sphere. Then

$$\int_{S_{R,\theta}} w_1^{(2)}(M, M_0)\, d\sigma_M = \int_{K_{R,\theta}} w_1^{(2)}(M, M_0)\, d\sigma_M = -\frac{\theta}{\pi},$$
$$\int_{S \setminus S_{R,\theta}} w_1^{(2)}(M, M_0)\, d\sigma_M = \frac{\pi - \theta}{\pi}. \tag{11.5.12}$$

Let $S_{R,\varphi}$ be the part of the sphere cut out by the dihedral angle of magnitude φ, whose rib passes through M_0 (Fig. 11.5.3), then

$$\int_{S_{R,\varphi}} w_1^{(2)}(M, M_0)\, d\sigma_M = -\frac{\varphi}{2\pi}. \tag{11.5.13}$$

There is another interesting case. Let $S_{R,\theta,\varphi}$ be the part of the sphere cut out by two dihedral angles of magnitudes θ and φ with orthogonal ribs (Fig. 11.5.4). Then

$$\int_{S_{R,\theta,\varphi}} w_1^{(2)}(M, M_0) \, d\sigma_M = -\frac{\varphi\theta}{2\pi^2}. \tag{11.5.14}$$

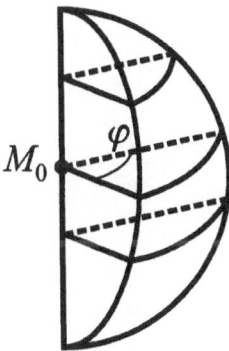

Figure 11.5.3

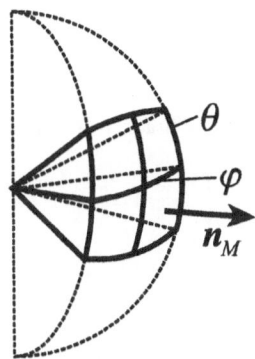

Figure 11.5.4

Consider more closely the function $\omega_1^{(2)}(M, M_0)$, assuming that σ is a piecewise smooth surface and $M_0 \in \sigma$. By definition, we have

$$\omega_1^{(2)}(M, M_0) = \frac{1}{4\pi} \frac{(\mathbf{r}_{MM_0}, \mathbf{n}_M)}{r_{MM_0}^3} = \tag{11.5.15}$$

$$= \frac{1}{4\pi} \frac{(x - x_0)[\boldsymbol{\tau}_{1,M}, \boldsymbol{\tau}_{2,M}]_x + (y_0 - y)[\boldsymbol{\tau}_{1,M}, \boldsymbol{\tau}_{2,M}]_y + (z_0 - z)[\boldsymbol{\tau}_{1,M}, \boldsymbol{\tau}_{2,M}]_z}{\left|[\boldsymbol{\tau}_{1,M}, \boldsymbol{\tau}_{2,M}]\right| \left[(x_0 - x)^2 + (y_0 - y)^2 + (z_0 - z)^2\right]^{3/2}},$$

where $x_0 = x(u_0, v_0)$, $y_0 = y(u_0, v_0)$, $z_0 = z(u_0, v_0)$, $x = x(u, v)$, $y = y(u, v)$, $z = z(u, v)$.

Denote by $m = m(u, v)$ the point on the plane O^*UV with the coordinates u, v, and let $\rho_{mm_0}^2 = |mm_0|^2 = (u - u_0)^2 + (v_0 - v)^2$. Consider the ratio

$$\frac{x(u_0, v_0) - x(u, v)}{\rho_{mm_0}} = \frac{x(u_0, v_0) - x(u, v)}{\rho_{mm_0}} + \frac{x(u, v_0) - x(u_0, v_0)}{\rho_{mm_0}}. \tag{11.5.16}$$

Now, formulas (11.2.12)–(11.2.17) can be transformed as follows. We have

$$x(u_0, v_0) - x(u, v_0) = \int_u^{u_0} x_u'(\xi, v_0) \, d\xi. \tag{11.5.17}$$

Changing the integration variable, as we have done in (11.2.12), we get

$$x(u_0, v_0) - x(u, v_0) = (u_0 - u) \int_0^1 x_u'(u + \alpha(u_0 - u), v_0) \, d\alpha,$$

or

$$x_{2,u}(u; u_0, v_0) = \frac{x(u_0, v_0) - x(u, v_0)}{\rho_{mm_0}} = \frac{u_0 - u}{\rho_{mm_0}} \int_0^1 x_u'(u + \alpha(u_0 - u), v_0) \, d\alpha. \tag{11.5.18}$$

Further, we have

$$x_{3,u}(u; u_0, v_0) = \frac{1}{\rho_{mm_0}} \left[\frac{x(u_0, v_0) - x(u, v_0)}{\rho_{mm_0}} - x_u'(u, v_0) \frac{u_0 - u}{\rho_{mm_0}} \right] =$$
$$= \frac{u_0 - u}{\rho_{mm_0}} \int_0^1 \frac{x_u'(u + \alpha(u_0 - u), v_0) - x_u'(u, v_0)}{\rho_{mm_0}} \, d\alpha. \tag{11.5.19}$$

Integrating by parts in the last integral, we find that

$$x_{3,u}(u; u_0, v_0) = \frac{u_0 - u}{\rho_{mm_0}} \left[\frac{x_u'(u_0, v_0) - x_u'(u, v_0)}{\rho_{mm_0}} - \frac{u_0 - u}{\rho_{mm_0}} \int_0^1 \alpha x_{uu}''(u + \alpha(u_0 - u), v_0) \, d\alpha \right]. \tag{11.5.20}$$

Applying (11.5.18) to the first term in square brackets in (11.5.20), we obtain

$$x_{3,u}(u; u_0, v_0) = \left(\frac{u_0 - u}{\rho_{mm_0}} \right)^2 \int_0^1 (1 - \alpha) x_{uu}''(u + za(u_0 - u), v_0) \, d\alpha. \tag{11.5.21}$$

Similar arguments show that

$$x_{2,v}(u; , v, v_0) = \frac{x(u, v_0) - x(u, v)}{\rho_{mm_0}} = \frac{v_0 - v}{\rho_{mm_0}} \int_0^1 x_v'(u, v + \beta(v_0 - v)) \, d\beta, \tag{11.5.22}$$

$$x_{3,v}(u, v; v_0) = \frac{1}{\rho_{mm_0}} \left[\frac{x(u, v_0) - x(u, v)}{\rho_{mm_0}} - x_v'(u, v) \frac{v_0 - v}{\rho_{mm_0}} \right] =$$
$$= \left(\frac{v_0 - v}{\rho_{mm_0}} \right)^2 \int_0^1 (1 - \beta) x_{vv}''(u, v + \beta(v_0 - v)) \, d\beta. \tag{11.5.23}$$

Consider more closely the product (r_{MM_0}, n_M). By the properties of the mixed product of three vectors, we have

$$(r_{MM_0}, n_M) = \frac{(r_{MM_0}, \tau_{1,M} \times \tau_{2,M})}{|\tau_{1,M} \times \tau_{2,M}|} = \frac{(r_{MM_0}, \tau_{1,M}, \tau_{2,M})}{\tau_{1,M} \times \tau_{2,M}} =$$

$$= \frac{1}{|\tau_{1,M} \times \tau_{2,M}|} \begin{vmatrix} x_0 - x & y_0 - y & z_0 - z \\ x_u' & y_u' & z_u' \\ x_v' & y_v' & z_v' \end{vmatrix} = \tag{11.5.24}$$

$$= \frac{1}{|\tau_{1,M} \times \tau_{2,M}|} \Big[x_v' \big[(y_0 - t) z_u' + (z_0 - z) y_u' \big] + y_v' \big[(z_0 - z) x_u' - (x_0 - x) z_u' \big] +$$
$$+ z_v' \big[(x_0 - x) y_u' - (y_0 - y) x_u' \big] \Big].$$

Let us transform (11.5.15) as follows:

$$\omega_1^{(2)}(M, M_0) = \qquad (11.5.25)$$

$$= \frac{(\mathbf{r}_{MM_0}, \mathbf{n}_M)}{4\pi \rho_{mm_0}^2} \left[\left(\frac{x(u_0, v_0) - x(u, v)}{\rho_{mm_0}} \right)^2 + \left(\frac{y(u_0, v_0) - y(u, v)}{\rho_{mm_0}} \right)^2 + \left(\frac{z(u_0, v_0) - z(u, v)}{\rho_{mm_0}} \right)^2 \right]^{-3/2}$$

For squared terms in square brackets, using (11.5.18), (11.5.22), we obtain

$$\frac{x(u_0, v_0) - x(u, v)}{\rho_{mm_0}} = \frac{x(u_0, v_0) - x(u, v_0)}{\rho_{mm_0}} + \frac{x(u, v_0) - x(u, v)}{\rho_{mm_0}} = \qquad (11.5.26)$$

$$= \frac{u_0 - u}{\rho_{mm_0}} \int_0^1 x_u'(u + \alpha(u_0 - u), v_0) \, d\alpha + \frac{v_0 - v}{\rho_{mm_0}} \int_0^1 x_v'(u, v + \alpha(v_0 - v)) \, d\alpha.$$

And similar expressions are obtained for

$$\frac{y(u_0, v_0) - y(y, v)}{\rho_{mm_0}}, \qquad \frac{z(u_0, v_0) - z(y, v)}{\rho_{mm_0}}.$$

Formula (11.5.26) shows that in polar coordinates on the plane O^*UV with the origin (u_0, v_0), the expression in square brackets in (11.5.25) is a continuous function; moreover, it is Hölder continuous if the derivatives of x, y, z in u, v are Hölder continuous. Assume that the said expression does not vanish in the entire domain of the functions x, y, z (i.e., on the entire surface σ, or on its pieces having common points only on their boundaries). The class of surfaces for which this condition holds is relatively wide. For example, let σ be the graph of a function $z = f(x, y)$ such that f_x' and f_y' are Hölder continuous on their domain. This surface admits the parametric representation

$$x = x, \quad y = y, \quad z = f(x, y). \qquad (11.5.27)$$

Letting $x = u$, $y = v$, we get

$$\frac{x_0 - x}{\rho_{mm_0}} = -\cos\varphi, \qquad \frac{y_0 - y}{\rho_{mm_0}} = -\sin\varphi,$$

$$\frac{z_0 - z}{\rho_{mm_0}} = -\cos\varphi \int_0^1 f_x'(x + \alpha(x_0 - x), y_0) \, d\alpha - \sin\varphi \int_0^1 f_y'(x, y + \alpha(y_0 - y)) \, d\alpha. \qquad (11.5.28)$$

From (11.5.28) we find that

$$\left(\frac{x_0 - x}{\rho_{mm_0}} \right)^2 + \left(\frac{y_0 - y}{\rho_{mm_0}} \right)^2 + \left(\frac{z_0 - z}{\rho_{mm_0}} \right)^2 = \qquad (11.5.29)$$

$$= 1 + \left[\cos\varphi \int_0^1 f_x'(x + \alpha(x_0 - x), y_0) \, d\alpha + \sin\varphi \int_0^1 f_y'(x, y + \alpha(y_0 - y)) \, d\alpha \right]^2.$$

In what follows, we assume that the second partial derivatives of the functions x, y, z are Hölder continuous in their domain, and $|\tau_{1,M} \times \tau_{2,M}| \neq 0$ in that domain.

Consider the ratio

$$\frac{(\mathbf{r}_{MM_0}, \tau_{1,M}, \tau_{2,M})}{\rho_{mm_0}^2} = \qquad (11.5.30)$$

$$= \frac{1}{\rho_{mm_0}} \left[x_v' \left(\frac{y(u_0, v_0) - y(u, v)}{\rho_{mm_0}} z_u'(u, v) - \frac{z(u_0, v_0) - z(u, v)}{\rho_{mm_0}} y_u'(u, v) \right) + \right.$$

$$+ y_v' \left(\frac{z(u_0, v_0) - z(u, v)}{\rho_{mm_0}} x_u'(u, v) - \frac{x(u_0, v_0) - x(u, v)}{\rho_{mm_0}} z_u'(u, v) \right) +$$

$$\left. + z_v' \left(\frac{x(u_0, v_0) - x(u, v)}{\rho_{mm_0}} y_u'(u, v) - \frac{y(u_0, v_0) - y(u, v)}{\rho_{mm_0}} x_u'(u, v) \right) \right].$$

Let us transform the last term in square brackets (the other terms can be transformed similarly). We have

$$z_v'(u,v)\left(\frac{x(u_0,v_0)-x(u,v)}{\rho_{mm_0}}y_u'(u,v)-\frac{y(u_0,v_0)-y(u,v)}{\rho_{mm_0}}x_u'(u,v)\right)=$$

$$= z_v'(u,v)\left[\left((1)_{x,u}+(1)_{x,v}+x_u'(u,v_0)\frac{u_0-u}{\rho_{mm_0}}+x_v'(u,v_0)\frac{v_0-v}{\rho_{mm_0}}\right)y_u'(u,v)-\right.$$

$$\left.-\left((I)_{y,u}+(1)_{y,v}+y_u'(u,v_0)\frac{u_0-u}{\rho_{mm_0}}+y_v'(u,v_0)\frac{v_0-v}{\rho_{mm_0}}\right)x_u'(u,v)\right]=$$

$$= z_v'(u,v)\left\{(1)_{x,u}y_u'(u,v)+(1)_{x,v}y_u'(u,v)-(1)_{y,u}x_u'(u,v)-(1)_{y,v}x_u'(u,v)+\right.$$

$$+\frac{u_0-u}{\rho_{mm_0}}\left[x_u'(u,v_0)y_u'(u,v)-y_u'(u,v_0)x_u'(u,v)\right]+$$

$$\left.+\frac{v_0-v}{\rho_{mm_0}}\left[x_v'(u,v)y_u'(u,v)-y_v(u,v_0)x_u'(u,v)\right]\right\},\qquad (11.5.31)$$

where

$$(1)_{x,u}=\frac{x(u_0,v_0)-x(u,v_0)}{\rho_{mm_0}}-x_u'(u,v_0)\frac{u_0-u}{\rho_{mm_0}},$$

$$(2)_{x,v}=\frac{x(u,v_0)-x(u,v)}{\rho_{mm_0}}-x_v'(u,v_0)\frac{v_0-v}{\rho_{mm_0}},$$

and similar formulas hold for $(1)_{y,u}$, $(1)_{y,v}$. Let us make the following transformation:

$$\frac{u_0-u}{\rho_{mm_0}}\left[x_u'(u,v_0)y_u'(u,v)-y_u'(u,v_0)x_u'(u,v)\right]=\qquad (11.5.32)$$

$$=\frac{u_0-u}{\rho_{mm_0}}\left[\left(x_u'(u,v_0)-x_u'(u,v)\right)y_u'(u,v)-x_u'(u,v)\left(y_u'(u,v_0)-y_u'(u,v)\right)\right].$$

Note that

$$\frac{x_u'(u_0,v_0)-x_u'(u,v)}{\rho_{mm_0}}=\frac{v_0-v}{\rho_{mm_0}}\int_0^1 x_{uv}''(u,v+\alpha(v_0-v))\,d\alpha.\qquad (11.5.33)$$

From (11.5.19), (11.5.23), and (11.5.33), we see that all terms in (11.5.31) are small quantities of the order of ρ_{mm_0}, except the term

$$\frac{v_0-v}{\rho_{mm_0}}\left[z_v'x_v'y_u'-z_v'x_v'x_u'\right].\qquad (11.5.34)$$

Therefore, in the expression for $\rho_{mm_0}^{-2}\left(r_{MM_0},\tau_{1,M},\tau_{2,m}\right)$ all terms are continuous in polar coordinates, except the term

$$\frac{v_0-v}{\rho_{mm_0}^2}\left[z_v'x_v'y_u'-z_v'y_v'x_u'+y_v'z_v'x_u'-y_v'x_v'z_u'+x_v'y_v'z_u'-x_v'z_v'y_u'\right].\qquad (11.5.35)$$

But this expression is identically equal to zero.

Proposition 11.5.1. *Suppose that the second partial derivatives of the functions x, y, z are Hölder continuous in their domain. Then, in this domain, the function $\omega_1^{(2)}(M,M_0)$ can be represented in polar coordinates with center at (u_0,v_0) as the ratio*

$$\omega_1^{(2)}(M,M_0)=\frac{\lambda_1^{(2)}(M,M_0)}{\rho_{mm_0}},\qquad (11.5.36)$$

where $\lambda_1^{(2)}(M, M_0)$ is a Hölder continuous function in the said polar coordinates.

If the surface is piecewise smooth and the second derivatives of the functions x, y, z are bounded and Hölder continuous on its smooth pieces, then the function $\lambda_1^{(2)}(M, M_0)$ for a fixed M_0 is of the same class.

Let us calculate the integral of the function $\omega_1^{(2)}$ over σ,

$$V_L^{(2)}(M_0) = \int_\sigma \omega_1^{(2)}(M, M_0)\, d\sigma_M, \qquad (11.5.37)$$

for $M_0 \in \sigma$.

As shown above, if $M_0 \notin \sigma$ and σ a closed piecewise smooth surface, then $V_L^{(2)}(M_0) = 0$ for $M_0 \in D^+$; and $V_L^{(2)}(M_0) = -1$ for $M_0 \in D^-$.

Now, let $M_0 \in \sigma$. Assume first that M_0 belongs to a smooth piece of σ (see Fig. 11.2.3). If we consider $\omega_1^{(2)}(M, M_0)$ as the penetration density of fluid passing through the points $M \in \sigma$, $M \neq M_0$, then physical considerations allow us to make the following conclusions. Let P be the tangential plane to σ at the point M_0. Clearly, the fluid particles moving along the rays entering M_0 to the right of the plane P cross σ. Thus, all fluid that enters M_0 from right side of P must pass through the points $M \in \sigma$, $M \neq M_0$. The quantity of this fluid is $-1/2$. Therefore, for M_0 on a smooth part of σ, we have

$$V_L^{(2)}(M_0) = \int_\sigma \omega_1^{(2)}(M, M_0)\, d\sigma_M = -\frac{1}{2}. \qquad (11.5.38)$$

Thus, we have obtained a physical justification of formula (11.5.38). Its mathematical proof is almost the same as the proof of a similar formula in the plane case.

Now, suppose that the surface σ at the point M_0 is non-smooth and that its structure in a neighborhood of M_0 is of one of the types described in Remark 11.5.2. Then, the value of the integral in (11.5.37) will coincide with the respective value in (11.5.12), (11.5.13), or (11.5.14).

Next, we are going to examine the function $\omega_1^{(2)\pm}(M, M_0)$ for various surfaces.

Consider the function $\omega_1^{(2)}(M, \widetilde{M_0})$, where $M \in \sigma$ and $\widetilde{M_0}$ is an arbitrary point of $\mathbb{R}_3$. The value $\omega_1^{(2)}(M, M_0)$, $M_0 \in \sigma$, will be called *the direct value* of this function on σ. The function $\omega_1^{(2)}(M, M_0)$ has the form (11.5.36) in polar coordinates in the domain of the functions x, y, z whose second derivatives are assumed to be of class $H(\alpha)$ on smooth pieces of σ. Therefore, for any piece $\widetilde{\sigma} \subset \sigma$, we have

$$\left| \int_{\widetilde{\sigma}} \omega_1^{(2)}(M, M_0)\, d\sigma_M \right| \leq M \cdot s(\widetilde{\sigma}), \qquad (11.5.39)$$

where $s(\widetilde{\sigma})$ is the measure of the set corresponding to $\widetilde{\sigma}$ in the domain of the functions x, y, z.

Let $\omega_1^{(2)\pm}(M, M_0)$ be the limit values of the function $\omega_1^{(2)}(M, \widetilde{M_0})$ as $\widetilde{M_0} \notin \sigma$ approaches the point $M_0 \in \sigma$ from the left and from the right, respectively, i.e.,

$$\omega_1^{(2)\pm}(M, M_0) = \lim_{\widetilde{M_0} \to M_0^\pm} \omega_1^{(2)}(M, \widetilde{M_0}). \qquad (11.5.40)$$

Let us find a relation between the values of the functions $\omega_1^{(2)\pm}(M, M_0)$ and those of the function $\omega_1^{(2)}(M, M_0)$.

Assume first that M_0 belongs to a smooth piece of σ. Then, the function $\lambda_1^{(2)}(M, M_0)$ in (11.5.36) is smooth in polar coordinates on that piece, and the value $\lambda_1^{(2)}(M_0, M_0)$ is defined in these coordinates. The function $\omega_1^{(2)}(M, \widetilde{M_0})$, for $\widetilde{M_0} \notin \sigma$, is also smooth on that piece and is piecewise smooth on σ.

Consider the function

$$W^{(2)}(M, M_0, \widetilde{M}_0) = \omega_1^{(2)}(M, \widetilde{M}_0) - \omega_1^{(2)}(M, M_0) =$$

$$= \frac{1}{4\pi} \left[\frac{(r_{M\widetilde{M}_0}, n_M)}{r_{M\widetilde{M}_0}^3} - \frac{(r_{MM_0}, n_M)}{r_{MM_0}^3} \right]. \tag{11.5.41}$$

Let us show that

$$W^{(2)}(M, M_0, M_0^\pm) = \lim_{\widetilde{M}_0 \to M_0^\pm} W^{(2)}(M, M_0, \widetilde{M}_0) =$$

$$= \omega_1^{(2)\pm}(M, M_0) - \omega_1^{(2)}(M, M_0) = \lambda\delta(M - M_0), \tag{11.5.42}$$

where $\delta(M - M_0)$ is the delta-function and λ is a constant. Let us show that for any test function $\varphi(M)$ infinitely differentiable on σ, we have

$$\lim_{\widetilde{M}_0 \to M_0^\pm} \int_\sigma W^{(2)}(M, M_0, \widetilde{M}_0)\varphi(M)\, d\sigma_M = \pm\frac{1}{2}\varphi(M_0), \qquad M_0 \in \sigma. \tag{11.5.43}$$

Denote by $\sigma_{M_0,\varepsilon}$ the piece of σ inside the ε-neighborhood of M_0 in $\mathbb{R}_3$. Then

$$\lim_{\widetilde{M}_0 \to M_0^\pm} \int_\sigma W^{(2)}(M, M_0, \widetilde{M}_0)\varphi(M)\, d\sigma_M =$$

$$= \lim_{\widetilde{M}_0 \to M_0^\pm} \int_{\sigma \setminus \sigma_{M_0,\varepsilon}} W^{(2)}(M, M_0, \widetilde{M}_0)\varphi(M)\, d\sigma_M +$$

$$+ \lim_{\widetilde{M}_0 \to M_0^\pm} \int_{\sigma_{M_0,\varepsilon}} W^{(2)}(M, M_0, \widetilde{M}_0)\varphi(M)\, d\sigma_M =$$

$$= \lim_{\widetilde{M}_0 \to M_0^\pm} I_1(M_0, \widetilde{M}_0) + \lim_{\widetilde{M}_0 \to M_0^\pm} I_2(M_0, \widetilde{M}_0). \tag{11.5.44}$$

From (11.5.41) we see that for $\widetilde{M}_0 \to M_0^\pm$, the integrand in I_1 tends to zero uniformly with respect $\widetilde{M}_0$. Therefore, we can pass to the limit under the integral, which yields

$$\lim_{\widetilde{M}_0 \to M_0^\pm} I_1(M_0, \widetilde{M}_0) = \int_{\sigma \setminus \sigma_{M_0,\varepsilon}} 0\varphi(M)\, d\sigma_M = 0. \tag{11.5.45}$$

In order to examine the second limit, assume first that $\sigma_{M_0,\varepsilon}$ is a circle of radius ε with center at M_0 and let P be the plane of that circle. Then, $\omega_1^{(2)}(M, M_0) = 0$ on $\sigma_{M_0,\varepsilon}$ and it remains to find the limit

$$\lim_{\widetilde{M}_0 \to M_0^\pm} \int_{\sigma_{M_0,\varepsilon}} \omega_1^{(2)}(M, \widetilde{M}_0)\varphi(M)\, d\sigma_M =$$

$$= \varphi(M_0) \lim_{\widetilde{M}_0 \to M_0^\pm} \int_{\sigma_{M_0,\varepsilon}} \omega_1^{(2)}(M, \widetilde{M}_0)\, d\sigma_M +$$

$$+ \lim_{\widetilde{M}_0 \to M_0^\pm} \int_{\sigma_{M_0,\varepsilon}} \omega_1^{(2)}(M, \widetilde{M}_0)[\varphi(M) - \varphi(M_0)]\, d\sigma_M =$$

$$= \varphi(M_0) \lim_{\widetilde{M}_0 \to M_0^\pm} I_{2,\varepsilon}^*(\widetilde{M}_0) + \lim_{\widetilde{M}_0 \to M_0^\pm} I_{2,\varepsilon}^{**}(M_0, \widetilde{M}_0). \tag{11.5.46}$$

In order to examine the last two integrals, consider a coordinate frame $OXYZ$ in $\mathbb{R}_3$ such that OXY coincides with the plane P, the axis OZ is directed along n_{M_0}, and M_0 is the origin (Fig. 11.5.5). For $\widetilde{M}_0$ on the axis OZ we have

$$n_M = n_{M_0} = k, \qquad r_{M\widetilde{M}_0} = -xi - yj + z_0 k,$$

and therefore,

$$\int_{\sigma_{M_0,\varepsilon}} \omega_1^{(2)}(M, \widetilde{M}_0) \, d\sigma_M = \frac{1}{4\pi} \int_{\sigma_{M_0,\varepsilon}} \frac{\left(r_{M\widetilde{M}_0}, n_M\right)}{r_{M\widetilde{M}_0}^3} \, d\sigma_M = \frac{1}{4\pi} \int_{\sigma_{M_0,\varepsilon}} \frac{z_0 \, dx \, dy}{\left[x^2 + y^2 + z_0^2\right]^{3/2}}.$$
$$(11.5.47)$$

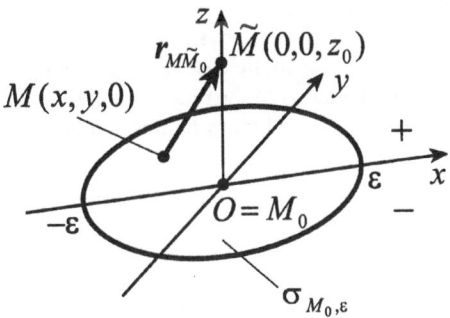

Figure 11.5.5

Introducing polar coordinates on the plane OXY, we obtain

$$\int_{\sigma_{M_0,\varepsilon}} \omega_1^{(2)}(M, \widetilde{M}_0) \, d\sigma_M = \frac{1}{4\pi} \int_0^{2\pi} d\varphi \int_0^\varepsilon \frac{z_0 \rho \, d\rho}{\left[\rho^2 + x_0^2\right]^{3/2}} = \frac{1}{2} \left(-z_0 \frac{1}{\left[\rho^2 + z_0^2\right]^{1/2}} \bigg|_0^\varepsilon \right) =$$

$$= \frac{1}{2} \left(-z_0 \frac{1}{\left[\varepsilon^2 + z_0^2\right]^{1/2}} + z_0 \frac{1}{|z_0|} \right). \tag{11.5.48}$$

From (11.5.46) and (11.5.48) we see that

$$\lim_{\widetilde{M}_0 \to M_0^{\pm}} I_{2,\varepsilon}^*(\widetilde{M}_0) = \pm \frac{1}{2}. \tag{11.5.49}$$

For $I_{2,\varepsilon}^{**}(M_0, \widetilde{M}_0)$ we obtain the estimate

$$\left| I_{2,\varepsilon}^{**}(M_0, \widetilde{M}_0) \right| \le A \int_{\sigma_{M_0,\varepsilon}} \left| \omega_1^{(2)}(M, \widetilde{M}_0) \right| r_{MM_0} \, d\sigma_M.$$

Taking the same coordinate system $OXYZ$ as in the case of $I_{2,\varepsilon}^*(\widetilde{M}_0)$ and passing to polar coordinates, we get

$$\left| I_{2,\varepsilon}^{**}(M_0, \widetilde{M}_0) \right| \le \frac{A}{4\pi} \int_0^{2\pi} d\varphi \int_0^\varepsilon \frac{|z_0| \rho \rho \, d\rho}{\left[z_0^2 + \rho^2\right]^{3/2}} \le \frac{A}{4\pi} \varepsilon 2\pi |z_0| \left(-\frac{1}{\left[\rho^2 + z_0^2\right]^{1/2}} \bigg|_0^\varepsilon \right). \tag{11.5.50}$$

It follows that

$$\lim_{\widetilde{M}_0 \to M_0^\pm} I_{2,\varepsilon}^{**}(M_0, \widetilde{M}_0) = O(\varepsilon). \tag{11.5.51}$$

Formulas (11.5.44)–(11.5.51) imply (11.5.43). Therefore, using the definition of the delta-function $\delta(M - M_0)$ and $W^{(2)}(M, M_0, M_0^\pm)$, $W^{(2)\pm}(M, M_0^\pm)$, we can write

$$\omega_1^{(2)\pm}(M, M_0) = \omega_1^{(2)}(M, M_0) \pm \frac{1}{2}\delta(M - M_0), \tag{11.5.52}$$

$$\lim_{\widetilde{M}_0 \to M_0^\pm} \int_\sigma \omega_1^{(2)}(M, \widetilde{M}_0)\, d\sigma_M = \int_\sigma \omega_1^{(2)}(M, M_0)\, d\sigma_M \pm \frac{1}{2}, \tag{11.5.53}$$

$$\int_\sigma \omega_1^{(2)\pm}(M, M_0)g(M)\, d\sigma_M = \int_\sigma \omega_1^{(2)}(M, M_0)g(M)\, d\sigma_M \pm \frac{1}{2}g(M_0), \tag{11.5.54}$$

where M_0 is a regular point of the surface σ.

Just as in the plane case, one can give a physical justification of (11.5.49) and strictly prove it not only if $\sigma_{M_0,\varepsilon}$ is a circle, but also in the case of $\sigma_{M_0,\varepsilon}$ being a smooth piece of σ. In the latter case, (11.5.49) holds to within terms of the order ε, and therefore, (11.5.53) is also valid. In the three-dimensional case, we limit ourselves to figures similar to Fig. 11.5.5. See Figs. 11.5.6 and 11.5.7.

Proposition 11.5.2. *Formulas* (11.5.53) *and* (11.5.54) *hold for any regular point* $M_0 \in \sigma$.

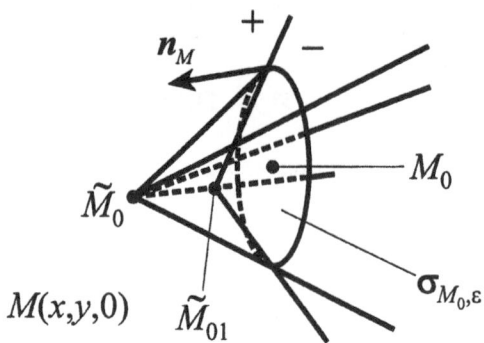

Figure 11.5.6

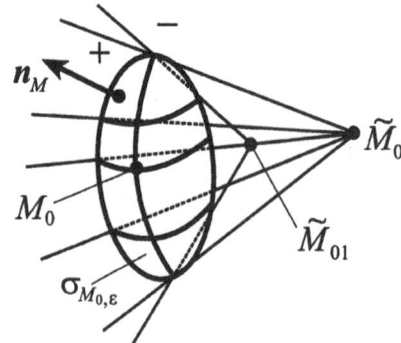

Figure 11.5.7

Now, suppose that M_0 is a non-regular point of σ and the structure of the surface σ near M is of one of the types specified in Remark 11.5.2.

Let M_0 be the vertex of the cone K_θ of angle θ (see Fig. 11.5.2). From Figs. 11.2.4 and 11.5.2 (as in the plane case) we see that for $\widetilde{M_0} \to M_0^+$ with $\widetilde{M_0}$ within the extension of the cone, the fluid quantity that passes through the cone to the source, in the limit, is equal to θ/π, and for $\widetilde{M_0} \to M_0^-$, is equal to $-(\pi - \theta)/\pi$. Thus, we have

$$
\int_{K_\theta} \omega_1^{(2)\pm}(M, M_0)\, d\sigma_M = \int_{K_\theta} \omega_1^{(2)}(M, M_0)\, d\sigma_M +
\begin{cases}
\dfrac{\theta}{\pi} & \text{for } w_1^{(2)+}, \\[2mm]
-\dfrac{\pi - \theta}{\pi} & \text{for } w_1^{(2)-},
\end{cases}
\tag{11.5.55}
$$

and

$$
\omega_1^{(2)\pm}(M, M_0) = \omega_1^{(2)}(M, M_0) +
\begin{cases}
\dfrac{\theta}{\pi}\delta(M - M_0) & \text{for } w_1^{(2)+}, \\[2mm]
-\dfrac{\pi - \theta}{\pi}\delta(M - M_0) & \text{for } w_1^{(2)-}.
\end{cases}
\tag{11.5.56}
$$

Now, let M_0 belong to the rib of the dihedral angle D_φ of the surface σ (see Fig. 11.5.3). Then, viewing Fig. 11.2.4 as the cross-section of this dihedral angle by a plane orthogonal to its rib and passing through M_0, we obtain

$$
\int_{D_\varphi} \omega_1^{(2)\pm}(M, M_0)\, d\sigma_M = \int_{D_\varphi} \omega_1^{(2)}(M, M_0)\, d\sigma_M +
\begin{cases}
\dfrac{\varphi}{2\pi} & \text{for } w_1^{(2)+}, \\[2mm]
-\dfrac{2\pi - \varphi}{2\pi} & \text{for } w_1^{(2)-},
\end{cases}
\tag{11.5.57}
$$

and

$$
\omega_1^{(2)\pm}(M, M_0) = \omega_1^{(2)}(M, M_0) +
\begin{cases}
\dfrac{\varphi}{2\pi}\delta(M - M_0) & \text{for } w_1^{(2)+}, \\[2mm]
-\dfrac{2\pi - \varphi}{2\pi}\delta(M - M_0) & \text{for } w_1^{(2)-}.
\end{cases}
\tag{11.5.58}
$$

Finally, if M_0 is the vertex of a rectangular spherical cone with angles φ, θ (see Fig. 11.5.4), then

$$
\int_{K_{\varphi,\theta}} \omega_1^{(2)\pm}(M, M_0)\, d\sigma_M = \int_{K_{\varphi,\theta}} \omega_1^{(2)}(M, M_0)\, d\sigma_M +
\begin{cases}
\dfrac{\varphi\theta}{2\pi^2} & \text{for } w_1^{(2)+}, \\[2mm]
-\dfrac{2\pi^2 - \varphi\theta}{2\pi} & \text{for } w_1^{(2)-},
\end{cases}
\tag{11.5.59}
$$

and

$$
\omega_1^{(2)\pm}(M, M_0) = \omega_1^{(2)}(M, M_0) +
\begin{cases}
\dfrac{\varphi\theta}{2\pi^2}\delta(M - M_0) & \text{for } w_1^{(2)+}, \\[2mm]
-\dfrac{2\pi^2 - \varphi\theta}{2\pi^2}\delta(M - M_0) & \text{for } w_1^{(2)-}.
\end{cases}
\tag{11.5.60}
$$

Next, we turn to the double layer potential and its properties.

By definition, *the double layer potential with density* $g(M)$ *on a piecewise smooth surface is the function*

$$
\Phi(M_0) = \frac{1}{4\pi}\int_\sigma \frac{\partial}{\partial n_M}\left(\frac{1}{r_{MM_0}}\right)g(M)\, d\sigma_M = \frac{1}{4\pi}\int_\sigma \frac{(r_{MM_0}, n_M)}{r_{MM_0}^3}g(M)\, d\sigma_M =
$$

$$
= \int_\sigma \omega_1^{(2)}(M, M_0)g(M)\, d\sigma_M, \qquad M_0 \notin \sigma.
\tag{11.5.61}
$$

As we have previously shown in this section, for a piecewise smooth σ and $M_0 \in \sigma$, the function $\omega_1^{(2)}(M, M_0)$, $M, M_0 \in \sigma$, admits the representation (11.5.36) and is therefore absolutely integrable on σ. The function

$$\Phi(M_0) = \int_\sigma \omega_1^{(2)}(M, M_0)g(M)\, d\sigma_M, \qquad M_0 \in \sigma, \tag{11.5.62}$$

is called *direct value of the double layer potential on the surface* σ.

It is easy to check that

$$\Delta\Phi(M_0) = 0, \qquad \forall M_0 \notin \sigma, \tag{11.5.63}$$

which means that the function $\Phi(M_0)$ is harmonic outside σ.

Consider the limit values of the double layer potential. Let

$$\Phi^\pm(M_0) = \lim_{\widetilde{M}_0 \to M_0^\pm} \Phi(\widetilde{M}_0) = \lim_{\widetilde{M}_0 \to M_0^\pm} \int_\sigma \omega_1^{(2)}(M, \widetilde{M}_0)g(M)\, d\sigma_M. \tag{11.5.64}$$

Formulas (11.5.54) and (11.5.62) imply that for any regular point $M_0 \in \sigma$, we have

$$\Phi^\pm(M_0) = \Phi(M_0) \pm \frac{1}{2}g(M_0), \qquad M_0 \in \sigma. \tag{11.5.65}$$

For non-regular $M_0 \in \sigma$ of any of the above three types (see Figs. 11.5.2–11.5.4), formulas similar to (11.5.55), (11.5.57), or (11.5.59) are valid.

Proposition 11.5.3. *For any regular point $M_0 \in \sigma$, the following formulas of Sokhotski–Plemelj type hold:*

$$\frac{1}{2}\left[\Phi^+(M_0) + \Phi^-(M_0)\right] = \int_\sigma \omega_1^{(2)}(M, M_0)g(M)\, d\sigma_M, \tag{11.5.66}$$

$$\Phi^+(M_0) - \Phi^-(M_0) = g(M_0). \tag{11.5.67}$$

If M_0 is a non-regular point of one of the above three types, then formula (11.5.67) is valid, while relation (11.5.66) changes according to the singularity at M_0, similarly to (11.2.74).

Consider the derivatives of the double layer potential. As in the plane case, we start with the normal derivative at regular points. Once again, we have

$$\frac{\partial}{\partial \boldsymbol{n}_{M_0}}\Phi^\pm(M_0) = \frac{\partial}{\partial \boldsymbol{n}_{M_0}}\Phi(M_0). \tag{11.5.68}$$

For the tangential component of the gradient of $\Phi^\pm(M_0)$, denoted by $\operatorname{Grad}\Phi^\pm(M_0)$, we obtain

$$\operatorname{Grad}\Phi^\pm(M_0) = \operatorname{Grad}\Phi(M_0) \pm \frac{1}{2}\operatorname{Grad} g(M_0). \tag{11.5.69}$$

Proposition 11.5.4. *For any regular point $M_0 \in \sigma$, the following relation holds for the double layer potential in $\mathbb{R}_3$ (cf. (11.2.81)):*

$$\operatorname{grad}\Phi^\pm(M_0) = \operatorname{grad}\Phi(M_0) \pm \frac{1}{2}\operatorname{Grad} g(M_0). \tag{11.5.70}$$

If σ is a piece of a plane, then

$$\operatorname{grad}\Phi^\pm(M_0) = \pm\frac{1}{2}\operatorname{Grad} g(M_0), \qquad M_0 \in \sigma. \tag{11.5.71}$$

Finally, we turn to the single layer potential and its derivatives. By definition, *the single layer potential on σ with density $\mu(M)$ is the function*

$$\varphi(M_0) = \frac{1}{4\pi} \int_\sigma \frac{1}{r_{MM_0}} \mu(M)\, d\sigma_M \,. \tag{11.5.72}$$

For $M_0 \notin \sigma$, the function $\delta(M_0)$ is harmonic in the sense of (11.5.63).

Since the kernel of the integral in (11.5.72) has a weak singularity $1/r_{MM_0}$ for $M_0 \in \sigma$, the function $\varphi(M_0)$ can be continuously extended to σ and becomes continuous across σ.

Now, let $\widetilde{M}_0 \notin \sigma$ be a point close to $M_0 \in \sigma$. We have

$$\frac{\partial}{\partial n_{M_0}} \varphi(\widetilde{M}_0) = -\frac{1}{4\pi} \int_\sigma \frac{\left(r_{M\widetilde{M}_0}, n_{M_0}\right)}{r_{M\widetilde{M}_0}^3} \mu(M)\, d\sigma_M = \int_\sigma \omega_{1,1}^{(2)}(M, \widetilde{M}_0)\mu(M)\, d\sigma_M, \tag{11.5.73}$$

for $M_0 \in \sigma$.

Similarly to (11.2.86), the function $\omega_{1,1}^{(2)}(M, \widetilde{M}_0)$ can be represented as the sum of $-\omega_1^{(2)}(M, \widetilde{M}_0)$ and $\lambda^{(2)}(M, M_0, \widetilde{M}_0)$. This function is continuous in $\mathbb{R}_3$ with respect to $\widetilde{M}_0$. Therefore, taking into account the properties of the normal derivative of the double layer potential, we obtain the following relation for the single layer potential:

$$\frac{\partial}{\partial n_{M_0}} \varphi^\pm(M_0) = \frac{\partial}{\partial n_{M_0}} \varphi(M_0) \mp \frac{1}{2}\mu(M_0),$$

where M_0 is any regular point of σ.

Consider the tangential derivative of the single layer potential,

$$\frac{\partial \varphi(\widetilde{M}_0)}{\partial \tau_{M_0}} = -\frac{1}{4\pi} \int_\sigma \frac{\left(r_{M\widetilde{M}_0}, \tau_{M_0}\right)}{r_{M\widetilde{M}_0}^3} \mu(M)\, d\sigma_M, \qquad \widetilde{M}_0 \notin \sigma. \tag{11.5.74}$$

Arguments, more or less similar to those used in the one-dimensional case, show that for regular points $M_0 \in \sigma$, we have

$$\frac{\partial \varphi^\pm(M_0)}{\partial \tau_{M_0}} = \frac{\partial \varphi(M_0)}{\partial \tau_{M_0}}\,. \tag{11.5.75}$$

Thus, we obtain a known formula (see *Colton, Kress* (1987)) for the gradient of the single layer potential at regular points of σ,

$$\operatorname{grad} \varphi^\pm(M_0) = \operatorname{grad} \varphi(M_0) \mp \frac{1}{2} n_{M_0}\mu(M_0). \tag{11.5.76}$$

11.6. Mathematical Modelling and Numerical Analysis of Nonstationary Flow past a Ship Deck

As an application of the results established in Section 11.4, we consider the problem of nonstationary flow past a ship deck (see *Lifanov, Setukha, Tsvetinskii, and Zhelannikov* (1997)). This problem is gaining more and more importance with the appearance of various types of shipborne aircraft, such as airplanes based on icebreakers and used for monitoring the ice conditions, or those on large fishing vessels following fish shoals. The aircraft landing or take off occur while the ship is moving and strong vortex sheets are produced above and behind its deck. These vortex sheets move very near to the deck and collide with deck equipment. In this connection, it is necessary to examine some special features of the numerical analysis of the velocity field near the surface.

First, we formulate our problem in physical terms. Since the speed of the ship is much less than the speed of sound, moving air will be modelled by an ideal incompressible fluid. Water surface will be modelled by a plane Π, with its wave structure neglected. The plane Π divides the space $\mathbb{R}_3$ into two domains D^+ and D^-. The part of the ship's surface (together with its deck) streamlined by the incident air flow is denoted by σ^0. We assume that σ^0 belongs to the region D^+. Let L^0 be the line on σ^0 from which a vortex sheet σ_1^0 separates. Clearly, the vortex surface σ_1^0 is above the water level and also lies in D^+. The perturbed velocity field is induced by the water surface Π, the surface of the ship σ^0, and the vortex sheet σ_1^0. The condition of non-penetration holds on the surface of the ship and on the water surface. Since the water surface is infinite and only finite surfaces can be modelled by discrete closed vortex frames, we adopt the following approach. We model the surface of the ship σ^0 and the vortex sheet σ_1^0 by a system of closed discrete vortex frames. Then, in the domain D^- we consider an airflow which is symmetrical to the incident flow with respect to the plane Π, and we consider surfaces $\widetilde{\sigma}^0$ and $\widetilde{\sigma}_1^0$ symmetric to σ^0 and σ_1^0 with respect to the same plane Π. On the surfaces $\widetilde{\sigma}^0$ and $\widetilde{\sigma}_1^0$ we take closed discrete frames symmetric to the corresponding frames on σ^0 and σ_1^0. The intensities of the frames on $\widetilde{\sigma}^0$ and $\widetilde{\sigma}_1^0$ have the same absolute values as the intensities of the respective frames on σ^0 and σ_1^0, but have opposite sign. Then, the condition of non-penetration on the plane Π will be automatically fulfilled for the velocity field in $\mathbb{R}_3$ which is formed by the constructed velocity field symmetric with respect to Π and the velocity field induced by the constructed discrete vortex frames on the surfaces $\sigma = \sigma^0 \cup \widetilde{\sigma}^0$ and $\sigma_1 = \sigma_1^0 \cup \widetilde{\sigma}_1^0$.

Thus, we have come to a nonstationary problem of infinite symmetric flow of an ideal incompressible fluid past a closed surface σ, with the above conditions of symmetry on σ and on the vortex sheet σ_1, the other conditions being the same as in the problem considered in Section 11.4.

In the initial numerical experiment pertaining to this problem we encountered the following phenomenon. The vortex sheet that separates from the front edges of the deck moves in close proximity to the deck because of the smallness of the angle of incidence. For this reason, calculations for this vortex sheet start to fail very soon. Having examined this situation for a model example described below, we formulated a condition on the region near the streamlined surface, which ensured successful calculations. This condition requires that moving free discrete vortices are prohibited to approach the streamlined surface closer than one half of the vortex frame increment.

Our model example can be described as follows. On the axis OX of the plane OXY, consider the segment $[-1, 1]$ with a vortex layer of intensity $\gamma(x) \equiv 1$, $x \in [-1, 1]$. The velocity produced by this vortex segment at any point $M_0(x_0, y_0)$ outside the segment is calculated by the formula

$$V(M_0) = \frac{1}{2\pi} \int_{-1}^{1} \frac{y_0\, \boldsymbol{i} - (x_0 - x)\boldsymbol{j}}{(x_0 - x)^2 + y + 0^2}\, dx. \tag{11.6.1}$$

On this segment, consider a system of discrete vortices located at the points $x_k = -1 + (k - 1/2)h$, $k = 1, \ldots, 2m$, $h = 1/m$. For a point M_0 whose distance from the segment $[-1, 1]$ is sufficiently large, the velocity is calculated by

$$V_m(M_0) = \frac{1}{2\pi} \sum_{k=1}^{2m} h\, \frac{y_0\, \boldsymbol{i} - (x_0 - x_k)\boldsymbol{j}}{(x_0 - x_k)^2 + y_0^2}. \tag{11.6.2}$$

For M_0 on the axis OY, the normal and the tangential components of the velocity $V_m(M_0)$ have the form

$$V_{n,m}(M_0) = \frac{1}{2\pi} \sum_{k=1}^{2m} \frac{x_k h}{x_k^2 + y_0^2},$$

$$V_{\tau,m}(M_0) = \frac{1}{2\pi} \sum_{k=1}^{2m} \frac{y_0 h}{x_k^2 + y_0^2}. \tag{11.6.3}$$

Since the points x_k, $k = 1, \ldots, 2m$, are symmetric with respect to the origin, we have

$$V_n(M_0) = V_{n,m}(M_0) = 0.$$

We see that in this case the normal velocity component can be calculated exactly.

Let us express the tangential velocity component as follows:

$$V_{\tau,m} = \frac{1}{\pi} \sum_{p=1}^{m} \frac{y_0 h}{x_p^2 + y_0^2}, \qquad x_p = \left(p - \frac{1}{2}\right) h, \qquad p = 1, \ldots, m. \qquad (11.6.4)$$

Thus, $V_{\tau,m}(M_0) \to 0$ for $y_0 \to 0$ and h fixed. We also have

$$0 \leq V_{\tau,m}(M_0) \leq \frac{y_0}{\pi} \sum_{P=1}^{m} \frac{h}{x_P^2} \leq \frac{y_0}{\pi} \left(\frac{1}{h} - \frac{1}{1 - 2^{-1}h}\right). \qquad (11.6.5)$$

In order to estimate the error of the obtained tangential velocity, we utilize the inequality

$$\frac{1}{\pi}\left(\arctan \frac{hy_0}{2(y_0^2 + 1) - h} - \arctan \frac{h}{2y_0}\right) \leq V_\tau(M_0) - V_{\tau,m}(M_0) \leq$$

$$\leq \frac{1}{\pi}\left(\arctan \frac{hy_0}{2(y_0^2 + 1) - h} + \arctan \frac{h}{2y_0}\right), \qquad (11.6.6)$$

which yields

$$|V_\tau(M_0) - V_{\tau,m}(M_0)| \leq \frac{2}{\pi} \arctan \frac{h}{2y_0}. \qquad (11.6.7)$$

The above considerations show that if the distance from the vortex layer to a free discrete vortex (more precisely, to the point at which one calculates the velocity) is much less than the distance h between the discrete vortices, then the method of discrete vortices, when applied to the calculation of tangential velocities, results in an error.

It should also be mentioned that formula (11.6.7) shows that the accuracy of the approximate tangential velocity values does not depend on $|y_0|$ but depends on the relation between h and y_0. Applications of this observation to the numerical analysis of specific problems are given in *Lifanov, Setukha, Tsvetinskii, and Zhelannikov* (1997). Here we only describe some results this work.

In our calculations it is assumed that vortex sheets separate from all edges of the deck, as well as from the ribs of structures on the deck. The angle of attack is $\theta°$ and the angle of sliding is $5°$.

Figure 11.6.1 shows a qualitative picture of excited velocity fields on the same cross-section of the deck: Fig. 11.6.1 (a) corresponds to experimental data and Fig. 11.6.1 (b) represents the results obtained numerically.

Figure 11.6.2 shows the distribution of the vertical and the horizontal components of the excited velocity field for the $A - A$ cross-section.

Experimental data and calculations based on the above method show that for a flow past a deck with a large sliding angle, the vortex sheets are twisted into powerful cord-like structures above the deck. As a rule, the sliding angle of these cords is different from that of the ship (Fig. 11.6.3).

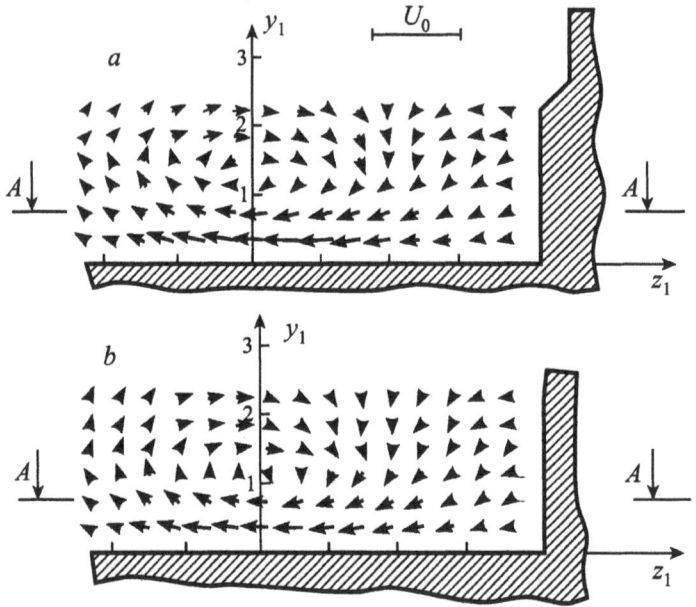

Figure 11.6.1

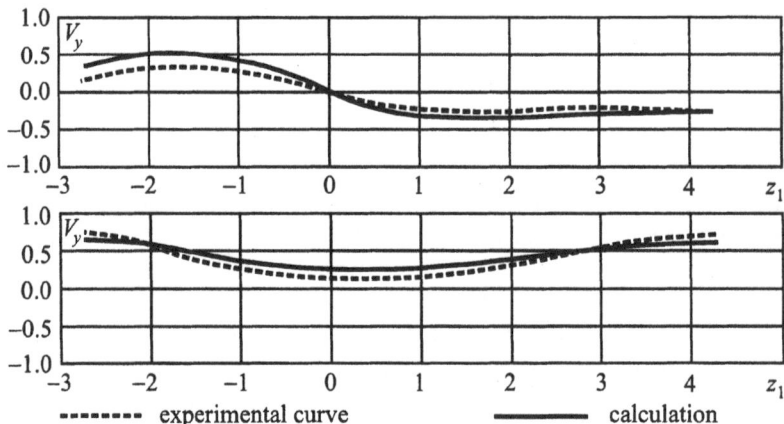

-------- experimental curve   ——————— calculation

Figure 11.6.2

If the number of steps in the calculations is large enough and the external flow conditions do not change, the vortex structures above the deck become stationary (Fig. 11.6.3).

Figure 11.6.4 demonstrates calculations when the vortex trace has the form of a curved surface. On the basis of this model, we can calculate the main aerodynamical characteristics in the first approximation. Figure 11.6.5 gives an example of calculations for an airplane about to land along a standard trajectory, and also the increments of the lift force and the lateral force, as well as the torque components along the axes, as compared to the values obtained for the unperturbed flow.

Figure 11.6.3

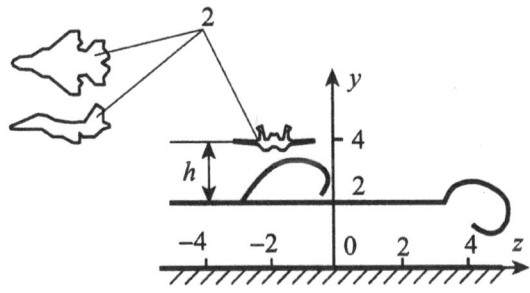

Figure 11.6.4

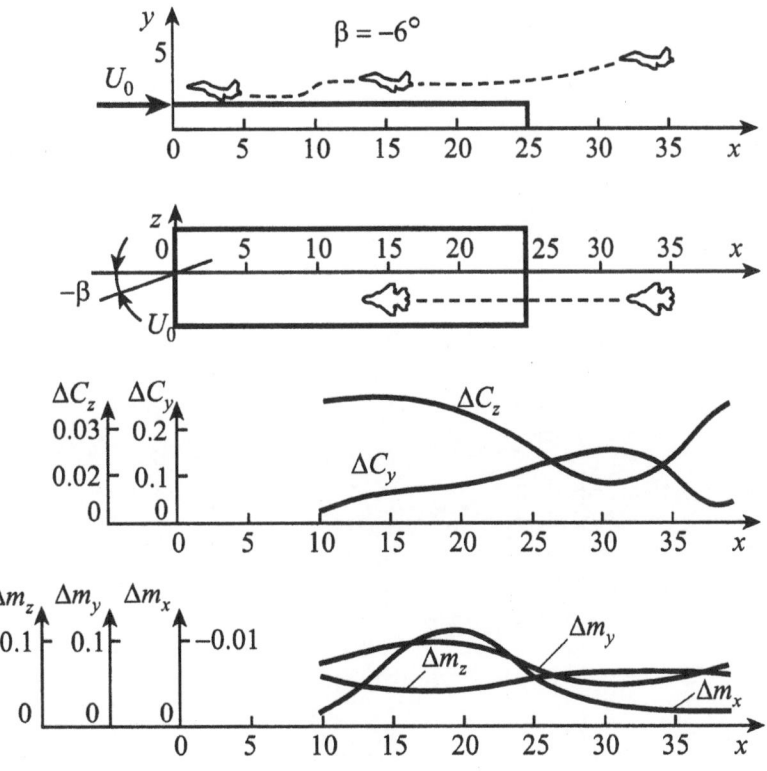

Figure 11.6.5

Chapter 12

Some Problems of Physics

In this chapter we apply the method of discrete vortices to some problems in physics, in particular, the diffraction of waves and the theory of elasticity. These applications are based on the ideas developed for the problems of aerodynamics.

12.1. Mathematical Modelling of Wide-Band Antennas

In this section, we consider electromagnetic wave radiation by an antenna excited by an H-polarized source. The problem is reduced to a hypersingular equation on a curve corresponding to the antenna geometry. The problem of electromagnetic wave radiation by the same antenna excited by an E-polarized source is reduced to a singular integral equation. The contour of the antenna may consist of finitely many open-ended or closed piecewise simple arcs and may contain angles.

First, we mention the corresponding physical model. The design of modern radar installations requires that they should be capable of wide-band operation. One of the approaches to the construction of wide-band structures formed by wide-band radiators is to increase the uncoupling between the radiators. A wide-band antenna suggested by *Lai* (1992) sheds light on the utilization of wide-band structures and the ways to connect closely located radiators of this type. The interaction between the component radiators can be calculated if we know the induced current distribution on the antenna walls. An analytical solution for the problem of current distribution in systems of the type shown in Fig. 12.1.1 is hardly possible. Numerical methods are more suitable for the investigation of electrodynamic processes in such antennas.

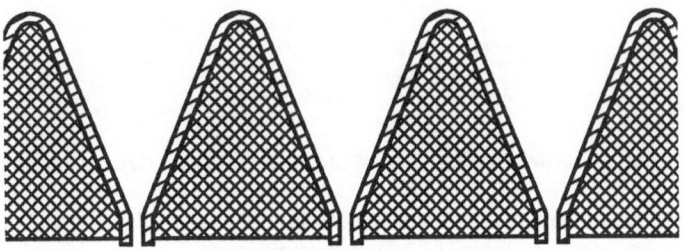

Figure 12.1.1

In order to construct a simplified mathematical model, we consider a single wide-band radiator shown in Fig. 12.1.2. The type of the source in the emitter determines the polarization of the emitted wave. For applications, it is important to know current distribution on the antenna surface. The method proposed below is effective for the description of both E and H polarizations in terms of the theory of diffraction of electromagnetic waves on cylindrical surfaces.

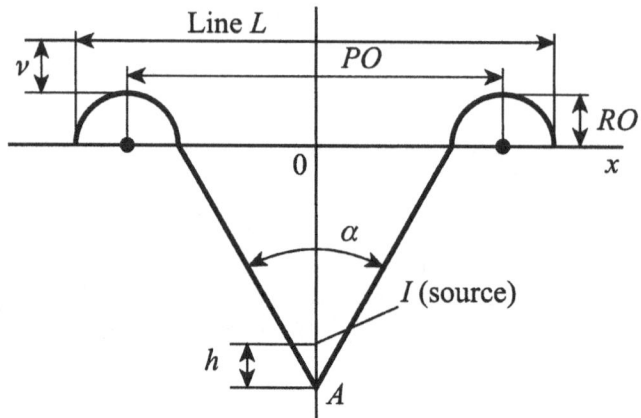

Figure 12.1.2

First, consider the E-polarization. Suppose that an antenna of the type shown in Fig. 12.1.2 is excited by an E-polarized source. The notation adopted in Fig. 12.1.2 will be explained later. The E-polarization problem can be reduced to the Dirichlet problem for the two-dimensional Helmholtz equation on the plane OXY (see *Zakharov and Pimenov* (1982), *Colton and Kress* (1987)),

$$\Delta U(M) + \varkappa^2 U(M) = 0, \qquad M \notin L, \tag{12.1.1}$$

$$U(M_0)\Big|_L = f(M_0), \qquad M_0 \in L, \tag{12.1.2}$$

together with the radiation condition at infinity. Here, $U = E_Z(x, y)$ is the Z-component of the reflected field, and L is the contour of the antenna. It is assumed that $L = \bigcup_{k=1}^{n} L_k$, where L_k are mutually disjoint simple piecewise smooth curves, which may be open-ended or closed.

We seek a solution of problem (12.1.1), (12.1.2) as a simple layer potential,

$$U(M_0) = \int_L K(M, M_0)\mu(M)\, ds_M, \qquad M_0 \in L, \tag{12.1.3}$$

$$K(M, M_0) = -\frac{i}{4} H_0^{(2)}(\varkappa r_{MM_0}),$$

where $H_0^{(2)}(z)$ is the zero order Hankel function of the second kind; ds_M is the length element on the curve L.

The simple layer potential is a continuous function on the entire plane, including the curve L, and therefore, the boundary condition (12.1.2) can be written as

$$\int_L K(M, M_0)\mu(M)\, ds_M = f(M_0), \qquad M_0 \in L. \tag{12.1.4}$$

If $\mu(M)$ is a solution of the integral equation (12.1.4), then $U(M_0)$ is a solution of the boundary value problem (12.1.1), (12.1.2) with the radiation condition at infinity. Note that the kernel $K(M, M_0)$ in (12.1.4) has a logarithmic singularity on the diagonal $M = M_0$. Therefore, it makes sense to pass to the parameters s and s_0 in (12.1.4) and differentiate the resulting equation in s_0. Thus we obtain the equation

$$\int_L K'_{s_0}(x_0, s)\mu(s)\, ds = f'(s_0), \qquad s_0 \in L, \qquad (12.1.5)$$

where L is now the range of the parameters s and s_0. This is a singular integral equation with the Cauchy kernel.

If L is a simple piecewise smooth open-ended curve, then the principal part of the kernel has a singularity of type $1/(s_0 - s)$. If L is a simple piecewise smooth closed curve, then the principal part of the kernel has a cotangent type singularity, and therefore, we have an equation with Hilbert kernel. In both cases, the solution is defined to within a constant. Thus, if L is a union of mutually disjoint curves $L_1, \ldots, L_p$ of the above type, then the solution of equation (12.1.5) will depend on p arbitrary constants and in order to have a unique solution, additional conditions should be imposed. Since equation (12.1.4) has only one solution (this follows from the uniqueness of a solution of the boundary value problem (12.1.1), (12.1.2) for any $f(M_0)$ continuous on L), it can be shown that the relations

$$\int_L k\big(s^*_{0k}, s\big)\mu(s)\, ds = f\big(s^*_{0s}\big), \quad s^*_{0k} \in L_k, \quad k = 1, \ldots, p, \qquad (12.1.6)$$

determine a unique solution of equation (12.1.5). Here s^*_{0k} is an arbitrary fixed point on L; $k = 1, \ldots, p$.

Consider a numerical approach to the problem of E-polarization. Let us construct a numerical solution of equation (12.1.5). This equation, together with (12.1.6), will be solved by the method of discrete singularities (see *Belotserkovskii and Lifanov (1985)* and *Lifanov (1996)*), which coincides with the method of discrete vortices in the present case. Consider two families of points

$$E = \bigcup_{k=1}^p E_k, \qquad E_0 = \bigcup_{k=1}^p E_{0k},$$

where

$$E_k = \big\{s_{k,m}, \ m = 1, \ldots, n_k\big\}, \qquad E_{0k} = \{s_{k.0m}, \ m = 1, \ldots, n_k + \eta(L_k)\},$$

$\eta(L_k) = 0$ if L_k is a closed curve, and $\eta(L_k) = 1$ if L_k is an open-ended curve. The points of E_k divide the curve L_k (if it is smooth) or its smooth pieces (if it is piecewise smooth) into equal parts, and E_{0k} consists of the midpoints of these pieces. The points of E_k and E_{0k} can be evenly distributed on L_k with respect to a coordinate on the curve other than the arc length. Then, we take a system of linear algebraic equations for each curve L_k according to the algorithm described by *Belotserkovskii and Lifanov* (1985) and depending on whether the curve L_k is open-ended or closed.

This method had been used for calculations, some of which were later compared with experimental data.

As an example, consider the model of antenna represented in Fig. 12.1.2. The geometrical parameters of the antenna are: $PO = 50$mm and $RO = 6$mm. The flare angle of the antenna is $\alpha = 60°$; h is the distance from the point A to the point $M_1 = M_0(0, y_1)$ in which the source $I = 4^{-1}iH_0^{(2)}(\varkappa r_{M_0 M_1})$, $M_0 = M_0(x_0, y_0)$, is located; $\varkappa = 2\pi/\lambda$ is the wave number, and $\lambda = 10$ cm is the wavelength. This mathematical model corresponds to the physical model described above. When the antenna becomes excited by this source, the initial field has the desired component $E_Z(x, y)$ corresponding to an E-wave in a waveguide, and the problem is reduced to the Dirichlet problem for the two-dimensional Helmholtz equation. It should be observed that in the case of E-polarization, the value of h depends on the wavelength of the source. In order to obtain experimental data with respect to the wavelength scale, the antenna was magnified, so that $P = K_\nu \times PO$ and $R = K_\nu \times RO$, where K_ν is the coefficient of expansion (for a constant wavelength).

Let us compare calculation results with experimental data. First, we compare the current distributions on the walls of the angle. From the mathematical standpoint, current distribution is characterized by the function $\mu(x)$ in equation (12.1.4). The solid line in Fig. 12.1.3 represents the graph of the function $\lg(\mu(x))$ for $n = 200$, where n is the number of partition points on the curve L, and $K_\nu = 3$. The deviation of the curve $\mu(x)$ for small x is explained by the fact that the source I is at a positive distance from the point A. The dots in Fig. 12.1.3 mark the experimental values of current distribution on the walls for a real physical model. We see that the numerical and the experimental data are in good agreement. Some deviation is due to the fact that in the physical experiment, we have used an open pyramid angle.

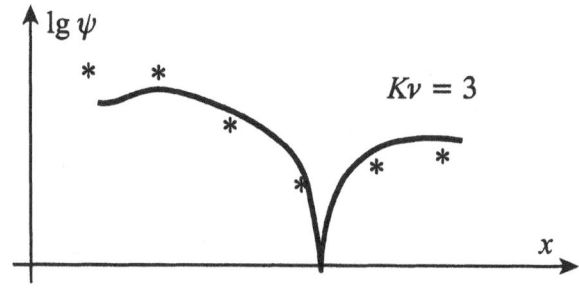

Figure 12.1.3

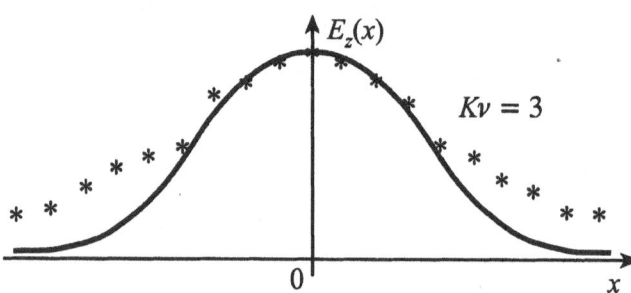

Figure 12.1.4

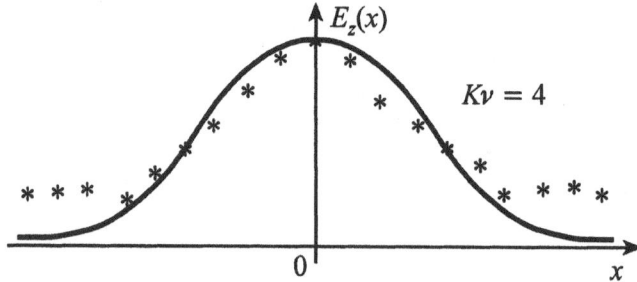

Figure 12.1.5

For applications, it is also important to know the field inside the aperture of the antenna. On the basis of the current distribution on the walls of the angle, we calculated the distribution of the component $E_Z(x)$ on the line L, which is shown in Fig. 12.1.2.

Solid lines in Figs. 12.1.4 and 12.1.5 give the numerical results for $K_\nu = 3$ and $K_\nu = 4$, respectively; and the dots correspond to the experimental data. These graphs demonstrate good agreement between numerical and experimental results and allow us to claim that the mathematical model adequately describes the physical singularity at the source.

Calculations can be performed for a wide frequency range. The solid line in Fig. 12.1.6 corresponds to the field in the aperture calculated for the original angle with $K_\nu = 8$. Of course, to ensure the desired accuracy, we have to increase the number of partition points of the curve for greater values of K_ν. The numerical method considered in this work allows us to utilize the symmetry properties present in this problem. The dashed line in Fig. 12.1.6 corresponds to the said curve for the field component $E_Z(x)$ calculated with the symmetries taken into account. The symmetries allow us to increase the accuracy of calculations for large values of K_ν, as well as to decrease the number of partition points for small values of K_ν.

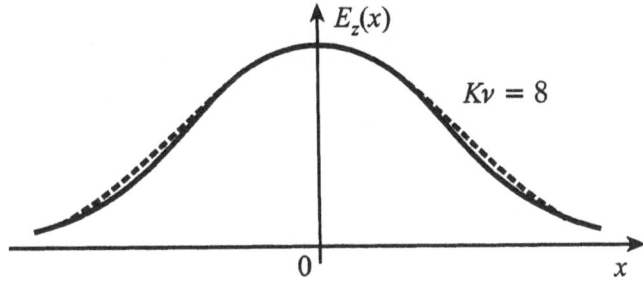

Figure 12.1.6

Finally, consider the case of H-polarization for the same antenna. Now the antenna is excited by a source of "magnetic" current required in the problem of H-polarization. This problem can be reduced to the Neumann problem for the two-dimensional Helmholtz equation on the plane OXY (see *Colton and Kress* (1987)). Thus, the Dirichlet condition (11.1.2) is replaced by the condition

$$\left. \frac{\partial U}{\partial n_{M_0}} \right|_L = f(M_0), \qquad M_0 \in \sigma, \tag{12.1.7}$$

where $U = H_Z(x, y)$ is the Z-component of the reflected field.

A solution of problem (12.1.1), (12.1.7) is sought as the double layer potential

$$U(M_0) = \int_L \frac{\partial}{\partial n_M} K(M, M_0) q(M) \, ds_M, \qquad M_0 \notin L, \tag{12.1.8}$$

where $K(M, M_0)$ is given by (12.1.3).

Since the normal derivative of the double layer potential is continuous across the curve (see Section 11.2), condition (12.1.7) can be written in the form

$$\int_L \frac{\partial}{\partial n_{M_0}} \frac{\partial}{\partial n_M} K(M, M_0) q(M) \, ds_M = f(M_0), \qquad M_0 \in L. \tag{12.1.9}$$

If $g(M)$ is a solution of the integral equation (12.1.9), then $U(M_0)$ from (12.1.8) is a solution of the corresponding boundary value problem (12.1.1), (12.1.7) and satisfies the radiation condition at infinity. Note that the kernel in (12.1.8) has a singularity of the type $1/(s_0 - s)^2$ on the diagonal $M = M_0$. Equation (12.1.9) is a hypersingular integral equation.

If L is a simple piecewise smooth open-ended curve, then the principal part of the kernel has a singularity of the type $1/(s_0 - s)^2$, and if L is closed, then the singularity of the principal part is of the type $\sin^{-2}((s_0 - s)/2)$.

Equation (12.1.9) can be solved numerically by a method similar to the method of discrete vortex pairs (see *Belotserkovskii and Lifanov (1985)* and *Lifanov (1996)*), on the basis of the same families of points E and E_0.

This approach was applied to the antenna shown in Fig. 12.1.2. The original parameters of its geometry, PO and RO, as well as the parameters α, h, λ, M_1, were the same as in the case of E-polarization. Then the antenna geometry was changed by means of the coefficient K_ν.

For the antenna excited by a source of "magnetic current", the initial field has the desired components $E_x(x, y)$ and $E_y(x, y)$ in the aperture, which corresponds to an H-wave in a waveguide. Then, the problem is reduced to the Neumann problem for the two-dimensional Helmholtz equation. It should be noted that smaller h, i.e., a smaller distance from the point A to the source, give a better approximation to the physical model. In our calculations, $h = 1.5\,\mathrm{mm}$.

Let us compare the calculated current distribution on the walls of the angle with experimental data. From the mathematical standpoint, $g(x)$ in equation (12.1.9) specifies the current distribution. The solid line in Fig. 12.1.7 is the graph of $g(x)$ for $n = 200$ and $K_\nu = 1$. As in the case of E-polarization, the distance h between the vertex A and the source should be commensurable with the lengths Δs_K of the partition segments on L. Thus, smaller values of h entail smaller values of Δs_k, which leads to a considerable increase of n, and therefore, increases the dimension of the linear algebraic system. For this reason, we have taken $h = 1.5\,\mathrm{mm}$, for which the calculation error is not very great for the chosen number of partition points.

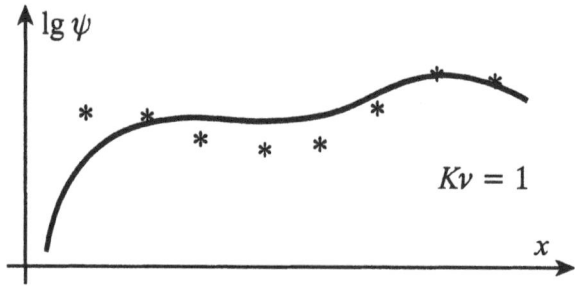

Figure 12.1.7

The asterisks in Fig. 12.1.7 mark the experimental values of the current distribution on the walls of the angle. These graphs show that the results obtained by our numerical method are in good agreement with experimental data.

12.2. Antenna-Diffraction Problems and Current Sources on the Antenna Surface

Consider the following antenna-diffraction problem. Let S be an absolutely conducting cylindrical surface with the generatrix parallel to the axis OZ and the directrix L on the plane OXY. Suppose that this surface is exposed to an incident H-polarized electromagnetic wave. The thread of the electromagnetic field is located on the surface S parallel to its generatrix. This thread crosses the curve L at a point g and generates the electromagnetic field $\boldsymbol{E}^0 = \{E_x, E_y, 0\}$, $\boldsymbol{H}^0 = \{0, 0, V\}$, where $V(M_0) = H_0^2(\varkappa r_{MM_0})$ and Q_1 is the source amplitude. This source will be used to model the current. The curve L may be either closed or open-ended. Our aim is to determine the scattering pattern (the field in the region far away from the curve) and its variation caused by the presence of a source.

We reduce this problem to a singular integral equation and indicate the class of functions in which its solution should be sought. It should be mentioned that the field will have a singularity of the same type as the source in a neighborhood of the point $q \in L$ on the side of the curve where the source is located and will have no singularities on the opposite side of L, since absolute conductor cannot be penetrated by a field.

The above physical problem amounts to finding the component $U = H_Z(x, y)$ of the excited magnetic field on the plane OXY. This field is created by the joint action of an incident H-polarized plane wave and the source on the surface S. The sought function should satisfy the Helmholtz equation (12.1.1) outside L, the Neumann boundary condition

$$\frac{\partial U(M_0)}{\partial \boldsymbol{n}_{M_0}} + \frac{\partial u_0(M_0)}{\partial \boldsymbol{n}_{M_0}} + \frac{\partial v_0(M_0)}{\partial \boldsymbol{n}_{M_0}} = 0, \qquad M_0 \in L, \qquad M_0 \neq q, \qquad (12.2.1)$$

and the radiation condition at infinity, where $u_0(M_0)$ is the function describing the incident field.

We seek the function $U(M_0)$ as the double layer potential (12.1.8). Applying the formula for the gradient of the double layer potential (see *Colton and Kress* (1987)) and the normal derivative of the function $v(M)$, and letting $f(M_0) = \partial u_0(M_0)/\partial \boldsymbol{n}_{M_0}$, we rewrite (12.2.1) as

$$-\frac{\varkappa^2 i}{4} \int_{-1}^1 H_0^{(2)}(\varkappa r_{MM_0}) \frac{y' y_0' + x' x_0'}{\left[x_0'^2 + y_0'^2\right]^{1/2}} g(t)\, dt +$$

$$+ \frac{i\varkappa}{4} \int_{-1}^1 H_1^{(2)}(\varkappa r_{MM_0}) \frac{x_0'(x_0 - x) + y_0'(y_0 - y)}{r_{MM_0} \left[x_0'^2 + y_0'^2\right]^{1/2}} g'(t)\, dt =$$

$$= -f(t_0) - Q_1 H_1^{(2)}(\varkappa r_{MM_q}) \frac{(r_{MM_q}, \boldsymbol{n}_{M_0})}{r_{M_0 M_q}}, \qquad t_0 \in (-1, 1). \qquad (12.2.2)$$

Here x', y', x_0, y_0 stand for $x'(t)$, $y'(t)$, $x(t_0)$, $y(t_0)$, respectively; $g(t) = g(x(t), y(t))$ and $x = x(t)$, $y = y(t)$ is a parametric representation of the curve on the segment $[-1, 1]$. It is assumed that $x''(t), y''(t) \in C^{r,\alpha}[-1, 1]$, where $C^{r,\alpha}[-1, 1]$ is the set of functions whose rth-order derivative satisfies the Hölder condition with exponent α. Then, the last term in the right-hand side of equation (12.2.2) is a function of the same class and has no singularity for $M_0 \to M_q$, since $\left(r_{M_0 M_q}, \boldsymbol{n}_{M_0}\right) = O\left(r_{M_0 M_q}^2\right)$ (see *Colton and Kress* (1987), Section 11.2).

Let us transform (12.2.2) into an integral equation with respect to $g'(t)$. To that end, it suffices to note that for $Z(t_0, t) = \varkappa r_{MM_0}$, we have

$$\int_{-1}^1 H_0^{(2)}(Z(t_0, t)) \frac{y' y_0' + x' x_0'}{\left[x_0'^2 + y_0'^2\right]^{1/2}} g(t)\, dt = \int_{-1}^1 \left(\int_{-1}^t H_0^{(2)}(Z(t_0, t)) \frac{y' y_0' + x' x_0'}{\left[x_0'^2 + y_0'^2\right]^{1/2}}\right) g'(t)\, dt, \quad (12.2.3)$$

since $g(-1) = g(1) = 0$.

Since the function $H_0^{(2)}(Z)$ has only a logarithmic singularity, the function in parenthesis in the right-hand side of (12.2.3) is Hölder continuous. Therefore, the integral on the right-hand side of (12.2.3) is regular. With the help of the results of *Lifanov I.I. and Lifanov I.K.* (1996), it can be shown that the second integral in the right-hand side of (12.2.2) is singular. Now, using the physical condition according to which the total field has a different representation on different sides of L

near q, we can show the following. The solution of equation (12.2.2) for the above problem should be sought in the form

$$g'(t) = \frac{A}{t_q - t} + b(t),$$

if L is an open-ended curve; and

$$g'(t) = A_1 \cot \frac{t_q - t}{2} + b_1(t),$$

if the curve L is closed. Here, A and A_1 are constants, and the functions $b(t)$, $b_1(t)$ are smooth in a neighborhood of q. Since our reasoning is of local character, we can take finitely many sources and near each source, the function $g'(t)$ will be of one of the above types.

Consider the special case of L being the segment $[-1, 1]$ on the axis OX. Then, the equation becomes

$$\int_{-1}^{1} \frac{g'(x)\, dx}{x - x_0} + \int_{-1}^{1} K(x, x_0) g'(x)\, dx = \int_{-1}^{1} \omega(x, x_0) g'(x)\, dx = f_1(x_0), \qquad x \in (-1, 1), \quad (12.2.4)$$

and $g'(x)$ near the point x_q, at which the source is located, can be represented in the form

$$g'(x) = \frac{R(x)}{x - x_q}.$$

It turns out that $R(x_q) = 4\varkappa i Q_1 / \pi$ and this representation holds near each source.

A numerical solution of equation (12.2.4) can be obtained by a method similar to that of discrete vortices. Let us indicate the system of linear algebraic equations that replaces equation (12.2.4) in each specific case.

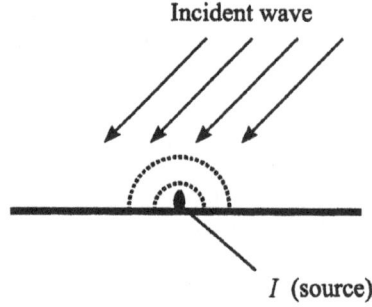

Incident wave

I (source)

Figure 12.2.1

First, consider the case of a single source on the segment $[-1, 1]$. The location of the source is at the point $x_q = 0$ (see Fig. 12.2.1) and its intensity is Q_1. Let $x_1, \ldots, x_n$ be a partition of the segment into n equal parts, and let $x_{0i} = x_1 + h/2$, $i = 1, \ldots, n-1$. We assume that $x_q = 0 = x_{0j_q}$. Then equation (12.2.4) is replaced by the following system:

$$\sum_{k=1}^{n} \omega(x_k x_{0j}) g'(x_k) h = f_1(x_{0j}), \qquad j = 1, \ldots, n-1, \quad j \neq j_q,$$

$$\sum_{k=1}^{n} g'(x_k) h = 0, \qquad g'(x_{j_q+1}) h - g'(x_{j_q}) h = \frac{16 \varkappa i}{\pi} Q_1. \qquad (12.2.5)$$

Figure 12.2.1 illustrates the model with a single source (directional diagram control).

We have taken the following parameters: the incident wavelength $\lambda = 1$ and $Q_1 = 8$; the angle of incidence $\beta = \pi/2$ corresponding to the wave directed along the axis OY; the wave number $\varkappa$ is equal to the wave number of the incident wave.

The dotted lines in Figs. 12.2.2 and 12.2.3 give the distribution of current density (see also Figs. 12.2.1 and 12.2.5). The dotted line in Fig. 12.2.4 gives the field distribution at the distance $d = \lambda$ from L. Calculations were performed with 80 points.

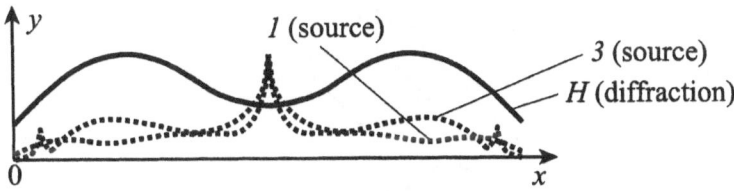

Figure 12.2.2

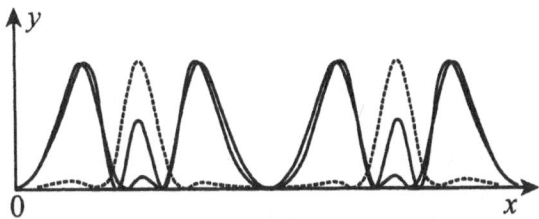

Figure 12.2.3

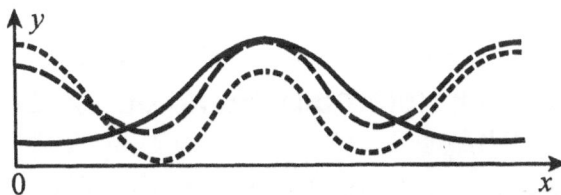

Figure 12.2.4

The above graphs show that the character of all physical quantities (current density, directional diagram, the field) differs from the case of the H-polarization without a source (solid line in Figs. 12.2.2–12.2.4). One can control the directional diagram by varying the intensity Q_1 of the source with its position on L being fixed, or by varying its position on L for Q_1 fixed. The above numerical results are in total agreement with theoretical predictions.

Finally, consider the problem of H-polarization with three sources on $L = [-1, 1]$. We take the wavelength $\lambda = 1$ and the angle of incidence $\beta = \pi/2$. The source $v_3(M)$ of intensity $Q_3 = 8$ is located at the point $q_3 = 0$, and the sources $v_1(M)$ and $v_2(M)$ of unknown intensities are located at the distance $1/8$ from the end-points of the segment $[-1, 1]$ (see Fig. 12.2.4). The values of the amplitudes Q_1 and Q_2 are found in the process of the numerical solution of equation (12.2.4) from the condition that the solution having the form (12.2.4) at the points q_1, q_2, q_3, should vanish at the points -1 and 1. Then, equation (12.2.4) is replaced by the system

$$\sum_{k=2}^{n-1} \omega(x_k, x_{0j}) g'(x) k) h = f_1(x_{0j}), \qquad j = 1, \ldots, n-1, \quad j \neq j_{q_1}, j_{q_2}, j_{q_3},$$

$$\sum_{k=2}^{n-1} g'(x_k) h = 0, \qquad g'(x_{j_{q_\lambda}+1}) h - q'(x_{j_{q_\lambda}}) h = \frac{16 \varkappa i}{\pi} Q_\lambda, \qquad \lambda = 1, 2, 3. \tag{12.2.6}$$

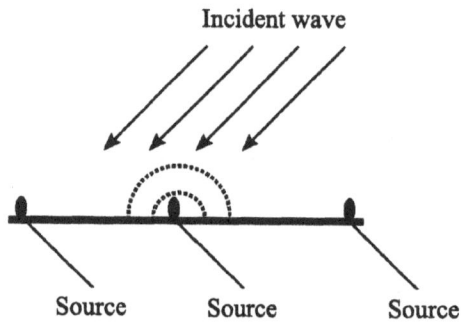

Figure 12.2.5

System (12.2.6) contains n equations and n unknown quantities $g'(x_2), \ldots, g'(x_{n-1})$, Q_1, and Q_2.

In this case we obtained the values $Q_1 = 2.6806$ and $Q_2 = 2.6805$. The dotted line in Fig. 12.2.2 is the graph of the current density distribution. The dotted line in Fig. 12.2.3 is the graph of the directional diagram, and the dotted line in Fig. 12.2.4 gives the distribution of the field at the distance $d = \lambda$ from L. Calculations were performed for $n = 80$ points.

12.3. Numerical Solution of the 3D Neumann problem for the Scalar Helmholtz Equation for Bodies of Complex Shape

As shown the previous chapter (see Sections 11.4 and 11.6) some problems of flow of an ideal incompressible fluid past bodies of a complex shape can be solved by the above methods. In fact, we have learnt how to solve three-dimensional problems for the Laplace equation in domains with boundaries of a complex structure. Since the nature of the Helmholtz equation is close to that of the Laplace equation, some ideas developed for the Laplace equation can be adapted to the study of the Helmholtz equation, as shown below.

Let σ be a piecewise smooth bounded surface in the space $OXYZ$, also denoted by $\mathbb{R}_3$. This surface may be with or without a border. Our aim is to find a function $U(M)$ satisfying the relations

$$\Delta U(M) + \varkappa^2 U(M) = 0, \qquad M \in \mathbb{R}_3 \setminus \sigma, \tag{12.3.1}$$

$$\frac{\partial U(M_0)}{\partial \boldsymbol{n}_{M_0}} = f(M_0), \qquad M_0 \in \sigma, \tag{12.3.2}$$

together with the radiation condition at infinity. If σ is a surface without a border, we consider only the exterior problem. If σ is a surface with or without a border and has ribs, we impose some additional conditions on the ribs (see *Colton and Kress* (1987)). The wave number $\varkappa$ is assumed positive. If σ is a surface without a border, we assume that $\varkappa$ does not coincide with an eigenvalue of the interior Neumann problem.

We seek a solution of problem (12.3.1), (12.3.2) as the double layer potential (see *Colton and Kress* (1987))

$$U(M_0) = \frac{1}{4\pi} \int_\sigma \frac{\partial}{\partial n_M} \left(\frac{e^{i\varkappa r_{MM_0}}}{r_{MM_0}} \right) g(M)\, d\sigma_M. \tag{12.3.3}$$

The function $U(M_0)$ in (12.3.3) satisfies equation (12.3.1) everywhere outside σ, as well as the radiation conditions at infinity (see *Colton and Kress* (1987)). Substituting (12.3.3) into the boundary condition (12.3.2) and using the continuity of the normal derivative of the double payer potential (see *Colton and Kress* (1987)), we can differentiate under the sign of the integral and obtain the following hypersingular integral equation:

$$\frac{1}{4\pi} \int_\sigma \frac{\partial}{\partial n_{M_0}} \frac{\partial}{\partial n_M} \left(\frac{e^{i\varkappa r_{MM_0}}}{r_{MM_0}} \right) g(M)\, d\sigma_M = f(M_0), \qquad M_0 \in \sigma, \tag{12.3.4}$$

where the integral is understood in the sense of Hadamard's finite part (see *Hadamard* (1932)). If $g(M)$ is a solution of equation (12.3.4), then $U(M_0)$ is a solution of the boundary value problem (12.3.1), (12.3.2).

In order to develop a numerical approach to this problem, let us find an exact solution of the hypersingular equation (12.3.4) on a sphere. First we recall the notion of *spherical functions*,

$$Y_l(\theta, \varphi) = \sum_{m=-l}^{l} a_l^{(m)} Y_l^m(\theta, \varphi), \qquad l = 1, 2, \ldots,$$

where (R, θ, φ) are the spherical coordinates of a point M in $\mathbb{R}_3$, and

$$Y_l^m(\theta, \varphi) = \begin{cases} P_l^m(\cos\theta)\cos m\varphi, & m = 0, 1, \ldots, l, \\ P_l^{|m|}(\cos\theta)\sin|m|\varphi, & m = -1, \ldots, -l. \end{cases}$$

In the above formulas

$$P_l^m(\mu) = \left(1 - \mu^2\right)^{m/2} P_l^{(m)}(\mu), \quad l = 0, 1, \ldots, \quad m = 0, 1, \ldots, l,$$

are the associated Legendre functions and

$$P_l(\mu) = \frac{1}{2^l l!} \frac{d^l}{d\mu^l} \left(\mu^2 - 1\right)^l, \qquad l = 0, 1, 2, \ldots,$$

are the Legendre polynomials.

Denote by $(Ag)(M_0)$ the left-hand side of (12.3.4), where $M_0(R_0, \theta_0, \varphi_0)$ is a point on the sphere σ of radius R_0 with center at the origin, and let

$$Z_l^{(1)}(\varkappa r) = -\frac{1}{2} r^{-3/2} H_{l+1/2}^{(1)}(\varkappa r) + r^{-1/2} \left(H_{l+1/2}^{(1)}(\varkappa r) \right)',$$

$$Z_l^{(2)}(\varkappa r) = -\frac{1}{2} r^{-3/2} G_{l+1/2}(\varkappa r) + r^{-1/2} \left(G_{l+1/2}(\varkappa r) \right)',$$

where $H_{l+1/2}^{(1)}(x)$ is the Hankel function of the first kind and $G_{l+1/2}(x)$ is the Bessel function. The following result is proved by *Lifanov*(1996).

Theorem 12.3.1. *For the operator* $(Ag)(M_0)$ *on the sphere* σ *of radius* R, *the following relations hold:*

$$(AY_l)(\theta_0, \varphi_0) = B_l^{-1}(\varkappa R) Y_l(\theta_0, \varphi_0), \qquad \theta_0 \in [0, \pi], \quad \varphi_0 \in [0, 2\pi], \quad l = 0, 1, 2, \ldots, \quad (12.3.5)$$

where

$$B_l(\varkappa R) = \frac{H_{l+1/2}^{(1)}(\varkappa R)}{R^{1/2} Z_l^{(1)}(\varkappa R)} - \frac{Y_{l+1/2}(\varkappa R)}{R^{1/2} Z_1^{(2)}(\varkappa R)}.$$

This theorem immediately implies the following result.

Theorem 12.3.2. *The function* $4^{-1} \cos \theta$ *is a solution of equation* (12.3.4) *with the right-hand side equal to* $\cos \theta_0$.

Now, we can describe a numerical approach to the solution of equation (12.3.4) based on a version of the method of discrete closed vortex frames.

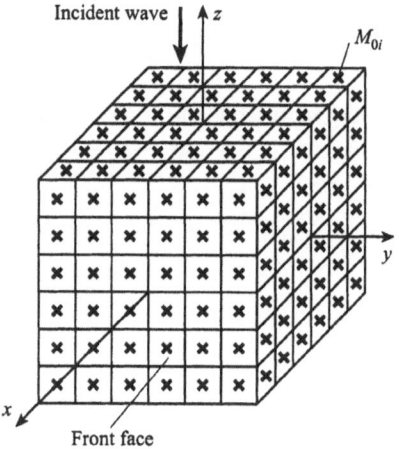

Figure 12.3.1

For a cube, we proceed as follows. Let us divide the surface of the cube into n squares σ_k, $k = 1, \ldots, n$, as shown in Fig. 12.3.1, and denote by M_{0k} the center of σ_k. Let us replace equation (12.3.4) by the following system of linear algebraic equations:

$$\frac{1}{4\pi} \sum_{k=1}^{n} g_n(M_{0k}) \int_{\sigma_k} \frac{\partial}{\partial n_{M_0}} \frac{\partial}{\partial n_M} \left(\frac{e^{i\varkappa r_{MM_{0j}}}}{r_{MM_{0j}}} \right) d\sigma_M = f(M_{0j}), \qquad j = 1, \ldots, n. \quad (12.3.6)$$

As shown in *Anfinogenov and Lifanov* (1992), we have

$$\int_{\sigma_k} \frac{\partial}{\partial n_{M_0}} \frac{\partial}{\partial n_M} \left(\frac{e^{i\varkappa r_{MM_{0j}}}}{r_{MM_{0j}}} \right) d\sigma_M = \qquad\qquad\qquad\qquad (12.3.7)$$

$$= \left[\int_{L_k} dl_M \times \nabla_M \frac{e^{i\varkappa r_{MM_{0j}}}}{r_{MM_{0j}}} \right] \cdot n_{M_{0j}} - \varkappa^2 \int_{\sigma_k} \frac{e^{i\varkappa r_{MM_{0j}}}}{r_{MM_{0j}}} (n_M, n_{M_0}) d\sigma_M,$$

where L_k is the boundary of σ_k; dl_M is the length element on L_k at the point M.

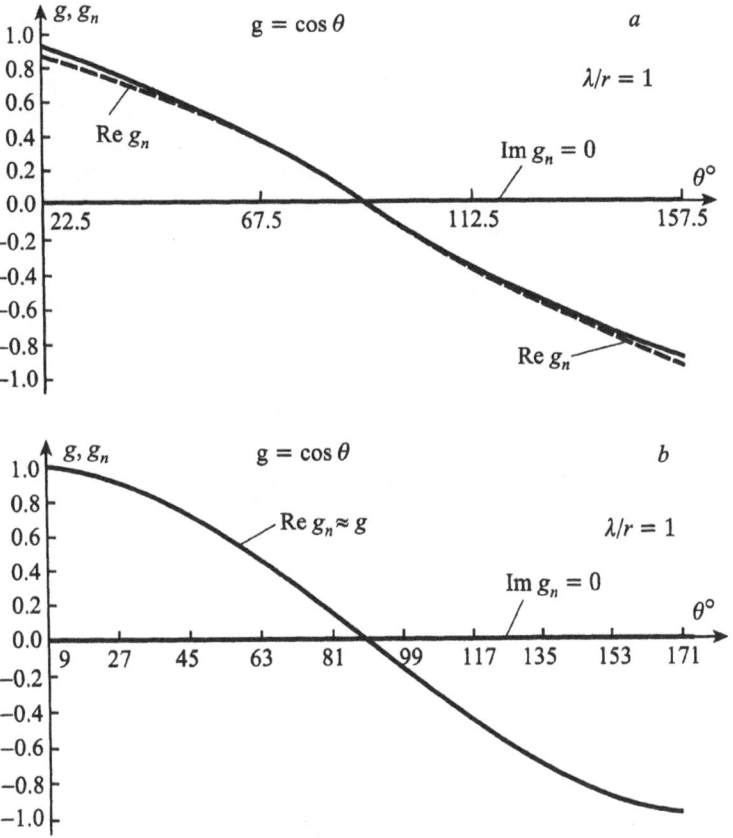

Figure 12.3.2

In a similar way, one obtains the numerical method for solving equation (12.3.4) on a sphere. In this case, σ_k are the images of the corresponding squares on the rectangle $[0, \pi] \times [0, 2\pi]$ on the plane $O^*\theta\varphi$. Note that the function $\cos\theta$ is a solution of equation (12.3.4) with the right-hand side equal to $4\pi\cos\theta_0$, in which case system (12.3.6) for the sphere turns into a system for the one-dimensional integral equation

$$\frac{1}{4\pi} \int_0^\pi K(\theta, \theta_0) g(\theta) \, d\theta = f(\theta_0), \qquad \theta_0 \in [0, \pi], \tag{12.3.8}$$

$$K(\theta, \theta_0) = \left(\int_0^{2\pi} \frac{\partial}{\partial n_{M_0}} \frac{\partial}{\partial n_M} \left(\frac{e^{i\varkappa r_{MM_{0j}}}}{r_{MM_{0j}}} \right) R^2 \, d\varphi \right) \sin\theta, \qquad r_{MM_0} = r(\varphi, \theta, \varphi_0, \theta_0).$$

Therefore, in the case of a sphere, we should take a partition only with respect to θ.

Figure 12.3.2 shows the results of comparison between the approximate solution and the exact solution on the sphere for various numbers of partition points, n, with respect to θ (the graph of the approximate solution g_n for the sphere: the exact solution is $g = \cos\theta$; the number of steps in θ is $n = 4$ (a); and $n = 10$ (b)). Relative error values for the sphere are given in Fig. 12.3.3. These graphs show that there is convergence of order $O(h)$, where h is the partition step in θ. In order to demonstrate the applicability of our method of solving equation (12.3.4) for bodies of a nontrivial structure, for instance, bodies with corners, we give graphs of the solutions of system (12.3.6) for σ being a cube. The approximate solution is constructed for the face of the cube shown in Fig. 12.3.1. The graphs of the approximate solution g_n on this cross-section are given in Fig. 12.3.4.

As a demonstration of the applicability of the method for non-closed surfaces (i.e., surfaces with border), Fig. 12.3.5 gives calculation results for equation (12.3.4) on $\sigma = [-1, 1] \times [-1, 1]$ on the plane OXY, with the incident waves moving along the axis OZ. The graphs represent the approximate solution for $y = 0$. We should also mention a known theoretical fact that this solution should vanish on the border of a non-closed surface.

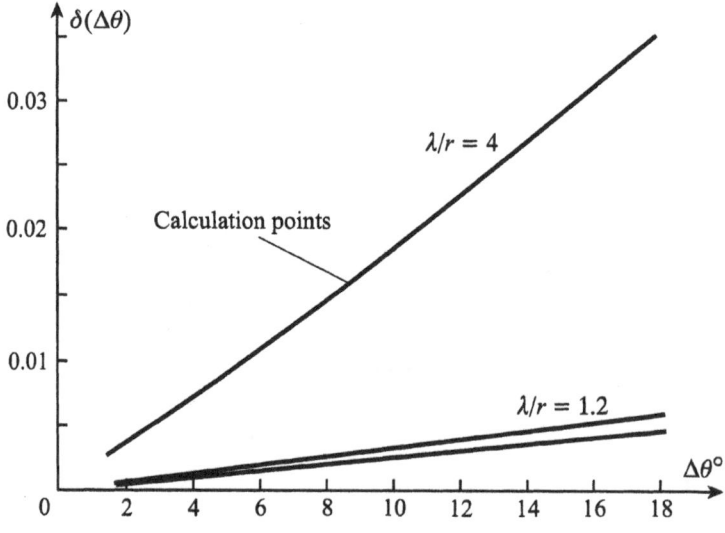

Figure 12.3.3

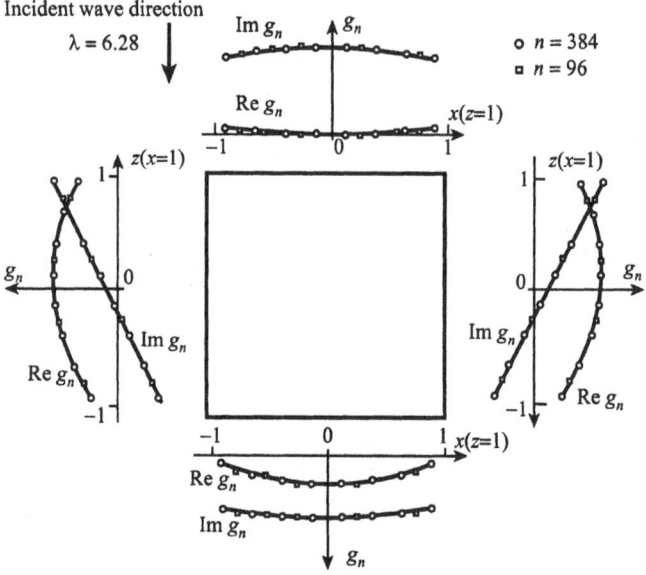

Figure 12.3.4

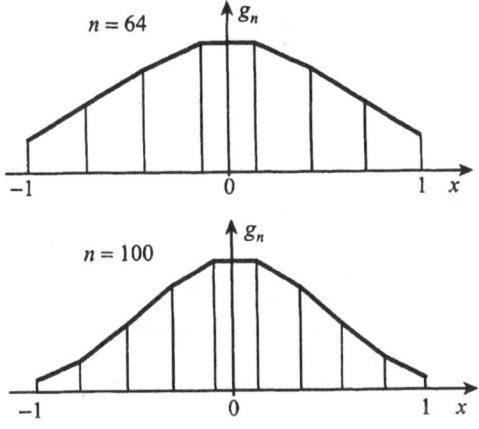

Figure 12.3.5

12.4. Contact Problem: Impression of a Uniformly Moving Punch into an Elastic Halfplane with Heat Generation

Following Saakian (see *Belotserkovskii and Lifanov* (1985)), let us show how the problem mentioned in the title can be reduced to a singular integral equation. Consider a rigid punch with an arbitrary smooth base moving along the surface of an elastic halfplane with constant velocity V_0 that does not exceed the velocity of propagation of the Rayleigh waves in a halfplane. Assume that the punch is loaded by a force P which impresses it into the halfplane (see Figure 12.4.1), and that between the punch and the halfplane there is dry friction, i.e., the tangential stresses in the contact region are proportional the normal pressure. As a result, the heat produced in the contact region will be proportional to the velocity of the punch, the friction coefficient, and the normal contact pressure (see *Belotserkovskii and Lifanov* (1985)). The parts of the halfplane and the punch outside the contact region are assumed heat-insulated.

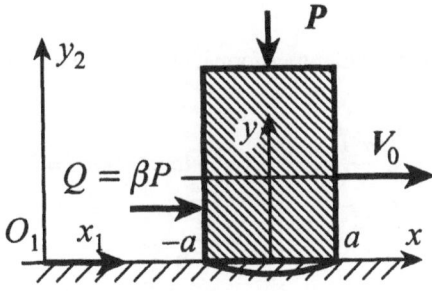

Figure 12.4.1

Consider a fixed coordinate frame $O_1 X_1 Y_1$ and a moving coordinate frame OXY rigidly fixed to the punch. Obviously, these coordinate systems are related by $X = X_1 - V_0 t$, $y = y_1$, and in the moving coordinates the sought quantities will be independent of t, i.e., we consider the quasi-static case. Heat production in the contact region causes heat flows of intensities $Q_1(x)$ and $Q_2(x)$ in the halfplane and the punch, respectively. These are related to the contact pressure $p(x)$ by

$$Q_1(x) + Q_2(x) = \beta V_0 p(x), \tag{12.4.1}$$

where β is the friction coefficient. According to the Fourier law of heat transfer, we have

$$Q_k(x) = (-1)^{k+1} \lambda_k \frac{\partial T_k(x, y)}{\partial y}, \qquad k = 1, 2,$$

where λ_1, λ_2 are the heat conduction coefficients, and $T_1(x, y)$ and $T_2(x, y)$ are the temperatures at the points of the halfplane and the punch, respectively.

It is assumed that the deformation of the elastic halfplane does not affect the thermal field. Then, the problem under consideration can be split into two problems: that of finding the thermal field and that of thermal elasticity. Consider the first problem.

We start with the construction of the influence function for this problem, i.e., we find a solution of the heat transfer problem for the elastic halfplane with heat-insulated boundary, along which a concentrated heat source is moving with constant velocity with the heat flow of unit intensity. Thus, we have the heat equation

$$\nabla^2 T(x_1, y_1, t) - \frac{\rho c_\varepsilon}{\lambda_1} \frac{\partial T(x_1, y_1, t)}{\partial t} = 0, \tag{12.4.2}$$

with the boundary conditions

$$\frac{\partial T(x_1, y_1, t)}{\partial y_1} = -\frac{1}{\lambda_1} \delta(x_1 - V_0 t - \xi) \quad \text{for} \quad y_1 = 0, \tag{12.4.3}$$

and $T(x_1, y_1, t) < \infty$ for $y_1 \to \infty$. Here $T(x_1, t_1, t)$ is the temperature at the points of the halfplane; ρ is the density of the material of the halfplane, c_ε is its specific heat, and λ_1 is its heat conduction coefficient; $\delta(x)$ is the Dirac delta-function; ξ is the coordinate of the heat source at time $t = 0$.

In the moving coordinate frame, relations (12.4.2) and (12.4.3) become

$$\nabla^2 T(x, y) + \frac{\rho c_\varepsilon}{\lambda_1} \frac{\partial T(x, y)}{\partial x} = 0, \tag{12.4.4}$$

$$\left. \frac{\partial T(x, y)}{\partial y} \right|_{y=0} = -\frac{1}{\lambda_1} \delta(x - \xi), \tag{12.4.5}$$

and $T(x, y) < \infty$ for $y \to -\infty$.

Let us apply the complex Fourier transformation in x to the boundary value problem (12.4.4), (12.4.5). We obtain

$$\frac{d^2 \widehat{T}(y, \alpha)}{dy^2} - \left(\alpha^2 + i\alpha\varkappa \right) \widehat{T}(y, \alpha) = 0, \tag{12.4.6}$$

$$\left. \frac{d\widehat{T}(y, \alpha)}{dy} \right|_{y=0} = -\frac{e^{i\alpha\xi}}{\lambda_1}. \tag{12.4.7}$$

$$\widehat{T}(y, \alpha) < \infty \quad \text{as} \quad y \to -\infty,$$

where α is the complex parameter of the transformation; $\varkappa = \rho c_\varepsilon V_0 / \lambda_1$; $\widehat{T}(y, \alpha)$ is the Fourier transform of the temperature $T(x, y)$,

$$\widehat{T}(y, \alpha) = \int_{-\infty}^{\infty} T(x, y) e^{ix\alpha} dx.$$

Solving equation (12.4.6) with the boundary conditions (12.4.7), we find that

$$\widehat{T}(y, \alpha) = -\frac{e^{i\alpha\xi}}{\lambda_1 \eta(\alpha)} e^{y\eta\alpha}, \tag{12.4.8}$$

where $\eta(\alpha) = \left[\alpha^2 + i\alpha\varkappa\right]^{1/2}$. The function $\eta(\alpha)$ has branch points $\alpha = 0$ and $\alpha = -ik$ on the complex plane $\alpha = \sigma + i\tau$. In order to select a single-valued branch of this function, it is necessary to cut the plane α along the lines joining the branch points with the point at infinity and lying in the upper and the lower halfplanes (see Fig. 12.4.2). This cut allows us to choose a single-valued branch of the root for which $\eta(\alpha) \to |\sigma|$ as $\alpha \to \pm\infty$ along the real axis.

Now, let us construct the influence function for the plane problem of elasticity with the thermal field produced by a moving concentrated source of intensity S, and also with a concentrated normal force P and a tangential force Q moving on the boundary, together with the heat source.

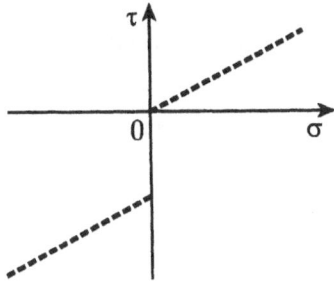

Figure 12.4.2

In this situation, the Lamé equations with thermal terms in the coordinates $O_1 X_1 Y_1$ read as follows:

$$\mu\nabla^2 u + (\lambda + \mu)\frac{\partial\theta}{\partial x_1} = \rho\frac{\partial^2 u}{\partial t^2} + \gamma\frac{\partial T}{\partial x_1},$$

$$\mu\nabla^2 v + (\lambda + \mu)\frac{\partial\theta}{\partial y_1} = \rho\frac{\partial^2 v}{\partial t^2} + \gamma\frac{\partial T}{\partial y_1}, \tag{12.4.9}$$

where $u(x_1, y_1, t)$ and $v(x_1, y_1, t)$ are the displacement components; $\theta(x_1, y_1, t) = \partial u/\partial x_1 + \partial v/\partial y_1$ is the volume strain; ρ is the density; λ and μ are the Lamé constants; $\gamma = (3\lambda + 2\mu)\alpha_t$; α_t is the linear extension coefficient; $T(x_1, y_1, t)$ is the temperature specified above.

Using the Duhamel–Neumann relations

$$\sigma_y = \lambda\theta + 2\mu\frac{\partial v}{\partial y_1} - \gamma T, \qquad \tau_{xy} = \mu\left(\frac{\partial u}{\partial y_1} + \frac{\partial v}{\partial x_1}\right),$$

we can write the boundary conditions in the form

$$\lambda\theta + 2\mu\frac{\partial v}{\partial y_1} - \gamma T - P\delta(x_1 - V_0 t - \xi),$$

$$\mu\left(\frac{\partial u}{\partial y_1} + \frac{\partial v}{\partial x_1}\right)\bigg|_{y=0} = Q\delta(x_1 - V_0 t - \xi), \tag{12.4.10}$$

$$(\sigma_y, \sigma_x, \tau_{xy}) \to 0 \quad \text{as} \quad y \to -\infty.$$

Going back to the moving coordinate frame, we can write equation (12.4.9) and the boundary conditions (12.4.10) in the form

$$\mu\nabla^2 + (\lambda + \mu)\frac{\partial\theta}{\partial x} = \rho V_0^2 \frac{\partial^2 v}{\partial x^2} + \gamma\frac{\partial T}{\partial x},$$

$$\mu\nabla^2 v + (\lambda + \mu)\frac{\partial\theta}{\partial y} = \rho V_0^2 \frac{\partial^2 v}{\partial x^2} + \gamma\frac{\partial T}{\partial y}, \tag{12.4.11}$$

$$\left(\lambda\theta + 2\mu\frac{\partial v}{\partial y} - \gamma T\right)\Bigg|_{y=0} = -P\delta(x - \xi),$$

$$\mu\left(\frac{\partial u}{\partial y} + \frac{\partial v}{\partial x}\right)\Bigg|_{y=0} = Q\delta(x - \xi). \tag{12.4.12}$$

Differentiating the first equation in (12.4.11) in x and the second in y, and taking their sum, we obtain the following equation for $\theta(x, y)$:

$$(\lambda + 2\mu)\nabla^2\theta = \rho V_0^2 \frac{\partial^2\theta}{\partial x^2} + \gamma\nabla^2 T. \tag{12.4.13}$$

Let us apply the generalized Fourier transformation

$$\widehat{\varphi}(y, \sigma) = \int_{-\infty}^{\infty} \varphi(x, y)e^{i\sigma x}\, dx$$

to (12.4.11)–(12.4.13). We obtain a system of ordinary differential equations in y for the functions $\widehat{u}(y, \sigma)$, $\widehat{v}(y, \sigma)$, and $\widehat{\theta}(y, \sigma)$ with the boundary conditions obtained from (12.4.12). Next, we find a particular solution of this system with the resulting boundary conditions, taking into account (12.4.10) and keeping in mind that the generalized Fourier transform of the temperature coincides with the complex Fourier transform (12.4.8), if we assume that on the real axis, the function $\eta(\sigma)$ coincides with the chosen single-valued branch of $\eta(\alpha)$. However, in order to solve the contact problem, it suffices to know only the displacements of the points on the boundary of the halfplane, $\widehat{u}(0, \sigma)$ and $\widehat{v}(0, \sigma)$. Applying the inverse Fourier transformation to $\widehat{u}(0, \sigma)$ and $\widehat{v}(0, \sigma)$, we obtain the influence functions $u(x, 0)$ and $v(x, 0)$ for the problem of plane elasticity with uniformly moving concentrated forces and a heat source. Now, we can use the following formulas, known from the theory of generalized functions (see *Brychkov and Prudnikov* (1977)):

$$\frac{1}{2\pi}\int_{-\infty}^{\infty}\frac{e^{i\sigma\xi}}{\sigma}e^{-i\sigma x}\, d\sigma = \frac{1}{2}\operatorname{sgn}(x - \xi), \qquad \frac{1}{2\pi}\int_{-\infty}^{\infty}\frac{i\,e^{i\sigma\xi}}{|\sigma|}e^{-i\sigma x}\, d\sigma = \frac{1}{\pi}\ln\frac{1}{|x - \xi|} + C.$$

In the theory of generalized functions, the constant C is taken to be the Euler constant, while for the problem of plane elasticity, $C = \infty$, since the system of the forces applied to the halfplane is unbalanced. Finally, we obtain the following influence functions:

$$u(x, 0) = -\pi\nu_0\nu_1\frac{P}{2\mu}\operatorname{sign}(x - \xi) + \mu_1\frac{Q}{\mu}\ln\frac{1}{|x - \xi|} + c_1 + \frac{k_2\gamma S}{\mu\lambda_1\nu_2}R_1(\xi - x),$$

$$v(x, 0) = -\pi\nu_0\nu_1\frac{Q}{2\mu}\operatorname{sign}(x - \xi) - \nu_1\frac{P}{\mu}\ln\frac{1}{|x - \xi|} + c_2 + \nu_1\nu_2\frac{\gamma S}{\lambda_1\mu}R_2(\xi - x), \tag{12.4.14}$$

where

$$R_1(\xi - x) = \int_{-\infty}^{\infty}\left[\frac{k_1}{\eta(\sigma)} - \frac{1}{|\sigma|}\right]\frac{i\,e^{i\sigma(\xi - x)}}{\sigma + i\varkappa c_1^2(V_0)^{-2}}\, d\sigma,$$

$$R_2(\xi - x) = \int_{-\infty}^{\infty} \left[\frac{k_1}{\eta(\sigma)} - \frac{1}{|\sigma|} \right] \frac{i\, e^{i\sigma(\xi - x)}}{\sigma + i\varkappa c_1^2 (V_0)^{-2}}\, d\sigma,$$

$$\nu_0 = \frac{(1 + k_2^2) - 2k_1 k_2}{k_1 (1 - k_2^2)}, \qquad \nu_1 = \frac{k_1(1 - k_2^2)}{\pi \left[4k_1 k_2 - (1 + k_2^2)^2 \right]}, \qquad \nu_2 = \frac{1 + k_2^2}{2k_1(1 - k_2^2)},$$

$$\nu_3 = \pi \left[4k_1 k_2 + (1 - k_2^2)^2 \right], \qquad k_i = \left[1 - \frac{V_0^2}{c_i^2} \right]^{1/2}, \quad i = 1, 2,$$

$$c_1 = \left[\frac{\lambda + 2\mu}{p} \right]^{1/2}, \qquad c_2 = \left[\frac{\mu}{\rho} \right]^{1/2}.$$

Now, let us find a solution of the contact problem. Obviously, the action of the punch on the halfplane is equivalent to applying to its boundary on the segment $[-a, a]$ unknown normal and tangential contact pressures p and q, together with a heat source $Q_1(\xi)$. According to the superposition principle, in order to determine the displacements of the boundary points of the halfplane, it suffices to integrate (12.4.14) in ξ, replacing P by $p(\xi)$, Q by $q(\xi)$, and S by $Q_1(\xi)$. We obtain

$$u(x, 0) = -\frac{\pi \nu_0 \nu_1}{2\mu} \int_{-a}^{a} p(\xi)\, \mathrm{sign}(x - \xi)\, d\xi + \frac{\nu_1}{\mu} \int_{-a}^{a} \ln 1 |x - \xi| q(\xi)\, d\xi +$$

$$+ \frac{k_2 \gamma}{\mu \lambda_1 \nu_3} \int_{-a}^{a} R_1(\xi - x) Q_1(\xi)\, d\xi + c_3,$$

$$v(x, 0) = -\frac{\pi \nu_0 \nu_1}{2\mu} \int_{-a}^{a} q(\xi)\, \mathrm{sign}(x - \xi)\, d\xi - \frac{\nu_1}{\mu} \int_{-a}^{a} \ln \frac{1}{|x - \xi|} p(\xi)\, d\xi +$$

$$+ \nu_1 \nu_2 \frac{\gamma}{\lambda_1 \mu} \int_{-a}^{a} R_2(\xi - x) Q_1(\xi)\, d\xi + c_4,$$

(12.4.15)

where c_3 and c_4 are some infinite constants. The conditions of contact are expressed by the relations

$$v(x, 0) = f(x) - d, \qquad T_1(x, 0) = T_2(x, 0), \qquad |x| < a, \qquad (12.4.16)$$

where $f(x)$ is the function describing the punch base; d is the immersion depth; T_1 and T_2 are the temperatures of the halfplane and the punch, respectively.

Suppose that the punch dimensions are much larger than those of the contact region. Then, in order to find the thermal field, it can be replaced by a halfplane. Then the temperature of the points on the punch boundary is given by the formula

$$T_2(x, 0) = \frac{1}{\pi \lambda_2} \int_{-a}^{a} \ln |x - \xi| Q_2(\xi)\, d\xi + c_5, \qquad (12.4.17)$$

where c_5 is also an infinite constant.

For the temperature on the boundary of the elastic halfplane, using (12.4.8), we get

$$T_1(x, 0) = -\frac{1}{\pi \lambda_1} \int_{-a}^{a} \ln |x - x| Q_1(x)\, ds - \frac{1}{\lambda_1} \int_{-a}^{a} R(s - x) Q_1(s)\, ds + c_6,$$

$$R(x - s) = \frac{1}{2\pi} \int_{-\infty}^{\infty} \left[\frac{1}{[\alpha^2 + i\alpha \varkappa]^{1/2}} - \frac{1}{|\alpha|} \right] e^{i\alpha(s - x)}\, d\alpha.$$

(12.4.18)

If instead of the stationary thermal regime in the punch and the halfplane we consider a regime harmonic in t with frequency ω and single out the wave that goes to infinity, then by equating the temperature of the punch (12.4.17) and that of the halfplane (12.4.18) in the contact region and passing to the limit as $\omega \to \infty$, we obtain the condition of thermal contact

$$\frac{1}{\pi \lambda_2} \int_{-a}^{a} \ln|x - s| Q_2(x)\, ds = \frac{1}{\pi \lambda_1} \int_{-a}^{a} \ln|x - s| Q_1(s)\, ds - \frac{1}{\pi \lambda_1} \int_{-a}^{a} R(s - x) Q_1(s)\, ds, \quad (12.4.19)$$

and also the condition under which the infinite constants c_5 and c_6 are mutually annihilated,

$$\frac{1}{\lambda_1} \int_{-a}^{a} Q_1(s)\, ds = \frac{1}{\lambda_2} \int_{-a}^{a} Q_2(s)\, ds. \tag{12.4.20}$$

Eliminating $Q_2(s)$ from (12.4.19), using (12.4.1), and differentiating this equation in x, we obtain

$$\frac{\lambda_1 + \lambda_2}{\pi \lambda_1 \lambda_2} \int_{-a}^{a} \frac{Q_1(s)}{s - x} - \frac{\beta V_0}{\pi \lambda_2} \int_{-a}^{a} \frac{p(s)\, ds}{s - x} + \frac{1}{\lambda_1} \int_{-a}^{a} \frac{\partial R(s - x)}{\partial x} Q_1(s)\, ds = 0. \tag{12.4.21}$$

Substituting the expression for $v(x, 0)$ with $q(s) = \beta p(x)$ into the first condition of contact (12.4.16) and differentiating in x, we get

$$\frac{\nu_1}{\mu} \int_{-a}^{a} \frac{p(s)}{s - x}\, ds + \frac{\beta}{\mu} \pi \nu_0 \nu_1 p(x) - \nu_1 \nu_2 \frac{\gamma}{\lambda_1 \mu} \int_{-a}^{a} \frac{\partial R_2(s - x)}{\partial x} Q_1(s)\, ds = f'(x). \tag{12.4.22}$$

This equation should be supplemented with the equilibrium condition for the punch,

$$\int_{-a}^{a} p(x)\, dx = P. \tag{12.4.23}$$

Thus, we have obtained a system of singular integral equations of the first and the second kind, (12.4.21) and (12.4.22), which has a unique solution under the conditions (12.4.20), (12.4.23).

Introducing the dimensionless quantities

$$\xi = \frac{x}{a}, \qquad \psi(\xi) = \frac{a}{P} p(x), \qquad \widehat{Q}_i(\xi) = \frac{\gamma a^2}{P \lambda_1} Q_i(x), \qquad i = 1, 2,$$

$$\varphi(\xi) = \frac{\mu a}{P} f'(x), \qquad \lambda = \frac{\lambda_1}{\lambda_2}, \qquad \zeta = \frac{\gamma \beta V_0 a}{\lambda_2}, \qquad \widehat{\varkappa} = \varkappa a = \frac{p c_\epsilon V_0 a}{\lambda_1},$$

and the new function $\chi(\xi) = (1 + \lambda)\widehat{Q}_i(\xi) - \zeta \widehat{P}(\xi)$, we finally obtain the following system of singular integral equations:

$$\int_{-1}^{1} \frac{\chi(s)\, ds}{s - \chi} + \frac{\pi}{1 + \lambda} \int_{-1}^{1} \frac{\partial R(s - \xi)}{\partial \xi}[\chi(s) + \zeta \psi(s)]\, ds = 0,$$

$$\int_{-1}^{1} \frac{\psi(s)\, ds}{s - \xi} + \pi \beta \nu_0 \psi(s) - \frac{\nu_2}{1 + \lambda} \int_{-1}^{1} \frac{\partial R_2(s - \xi)}{\partial \xi}[\chi(s) + \zeta \psi(s)]\, ds = -\frac{1}{\nu_1} \varphi(\xi),$$

$$\tag{12.4.24}$$

with the conditions

$$\int_{-1}^{1} \chi(s)\, ds = 0, \qquad \int_{-1}^{1} \psi(s)\, ds = 1. \tag{12.4.25}$$

The first relation in (12.4.24) is a singular integral equation of the first kind for the function $\chi(\xi)$, and the second is a singular integral equation of the second kind for the function $\psi(x)$. We seek a solution of index 1 of both equations in (12.4.24), and therefore, using the results of *Belotserkovskii and Lifanov* (1985), we replace (12.4.24), (12.4.25) by the following system of linear algebraic equations:

$$\frac{\pi}{n} \sum_{k=1}^{n} \left[\frac{1}{\tau_k - \xi_i} + \frac{\pi}{1+\lambda} \frac{\partial R(\tau_k - \xi_i)}{\partial \xi_i} \right] \chi_k^* + \frac{\pi\zeta}{1+\lambda} \sum_{p=1}^{m} a_p \frac{\partial R(t_p - \xi_i)}{\partial \xi_i} \psi_\pi^* = 0,$$

$$\frac{\pi\nu_2}{(1+\lambda)n} \sum_{k=1}^{n} \frac{\partial R_2(\tau_k - s_j)}{\partial s_i} \chi_k^* - \sum_{p=1}^{m} a_p \left[\frac{1}{t_p - s_j} - \frac{\zeta\nu_2}{1+\lambda} \frac{\partial R_2(t_p - s_j)}{\partial s_j} \right] \psi_p^* = \frac{\varphi(s_j)}{\nu_1}, \quad (12.4.26)$$

$$\sum_{k=1}^{n} \chi_k^* \frac{\pi}{n} = 0, \qquad \sum_{p=1}^{m} \psi_p^* a_p = 1,$$

$$j = 1, \dots, m-1, \qquad i = 1, \dots, n-1,$$

where τ_k and ξ_k are the same as in (9.2.26), and t_p, s_j, a_p are the same as in (9.2.19) with $P_n(t)$ being the Jacobi polynomial $P_n^{(\alpha, 1-\alpha)}$ for $\alpha = -\pi^{-1} \arctan(\beta\nu_0)$.

Thus, the contact problem with a moving punch has been reduced to the system of linear algebraic equations. Numerical analysis has been performed for system (12.4.26) with various values of the punch velocity and the following values of the parameters:

$$\lambda = 35, \qquad \zeta = \frac{642 V_0}{c_2}, \qquad \frac{c_2^2}{c_1^2} = 0.275, \qquad \beta = 0.27, \qquad \varkappa = 1.807 \times 10^{-5} \frac{V_0}{c_2},$$

and the punch base described by the function $f(x) = 0.1x^2$. Figure 12.4.3 shows the distribution of thermal flows in the direction of the halfplane (solid line) and inside the punch (dashed line).

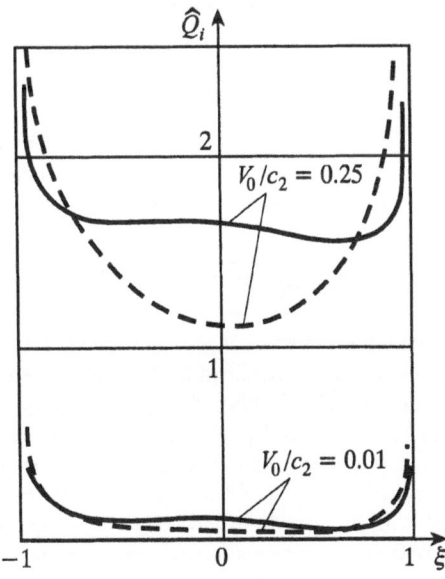

Figure 12.4.3

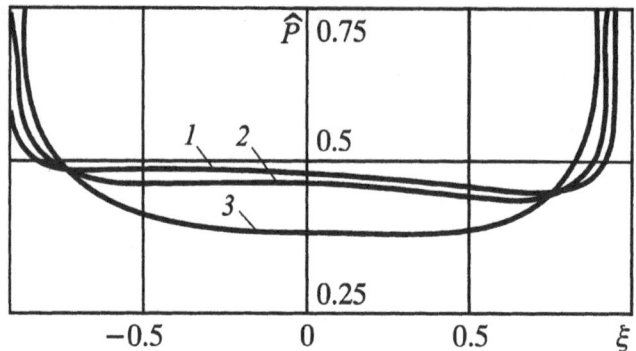

Figure 12.4.4

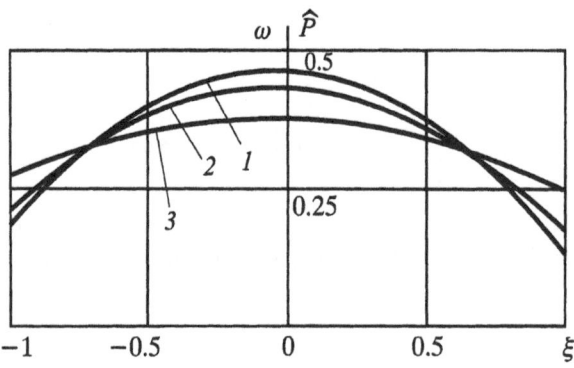

Figure 12.4.5

The values obtained for the contact pressure at the points coinciding with the roots of the polynomial $P_{10}^{(\alpha, -1-\alpha)}(x)$ were compared with the values of the function

$$\frac{a}{p} P(x) = -\omega(X) \frac{\sin \pi \alpha}{5 \pi \theta_1} \left[5\theta_1 - 2\alpha(1 + \alpha) - (1 + 2\alpha)x - x^2 \right]$$

at the same points. This function is the exact analytical expression of contact pressure for a similar problem without heat production and is obtained by the method of orthogonal polynomials. The comparison shows that the produced heat, although large in quantity (see Fig. 12.4.3), has no effect on the distribution of contact stresses (for $V_0/c_3 = 0.2$, the difference does not exceed 0.03%).

Figures 12.4.4 and 12.4.5 show the contact pressure distribution and its regular part for various values of the punch velocity (lines 1, 2, and 3 correspond to $V_0/c_2 = 0.44 \times 10^{-5}$, 0.5, and 0.8, respectively). We see that with the growth of the velocity, the pressure in the middle part of the contact region drops, whereas the coefficients of stress concentration near the edges of the contact region increase.

Conclusion

The authors are pleased to acknowledge the fact that some problems formulated by *Belotserkovskii and Lifanov* (1985) have been solved in the present monograph. In conclusion, we formulate some open problems which, in our opinion, seem interesting from the mathematical standpoint and are important for applications.

1. On a piecewise smooth curve L (with angular points), consider one-dimensional singular integral equations (see, for instance, (11.1.5) and (12.1.5)) of the type to which the Dirichlet or the Neumann problems for the Laplace and the Helmholtz equations are reduced. The problem is to prove the convergence of the numerical solutions obtained by the method of discrete vortices to the exact solutions.

2. The same problem of convergence is open for one-dimensional hypersingular integral equations of the type (12.1.9) to which the Neumann problem for the Laplace and the Helmholtz equations is reduced and the numerical solutions are obtained by the method of discrete vortex pairs.

3. On a surface σ with or without border, consider a two-dimensional hypersingular integral equation of the type to which the Neumann problem for the Laplace or the Helmholtz equations is reduced,

$$\int_\sigma \frac{\partial}{\partial n_{M_0}} \frac{\partial}{\partial n_M} K(M, M_0) g(M) \, d\sigma_M = f(M_0), \qquad M \in \sigma, \tag{1}$$

where $K(M, M_0) = (4\pi r_{MM_0})^{-1}$ or $K(M, M_0) = (4\pi)^{-1} e^{-i\varkappa r_{MM_0}}$, and $\varkappa$ does not coincide with an eigenvalue of the interior problem for the Helmholtz equation. The numerical solution of this equation is sought by the method of discrete closed vortex frames, i.e., (1) is replaced by the following linear algebraic system (see Sections 11.4 and 12.3):

$$\sum_{k=1}^{n} g_n(M_{0k}) \int_{\sigma_k} \frac{\partial}{\partial n_{M_0}} \frac{\partial}{\partial n_M} K(M, M_{0j}) \, d\sigma_M = f(M_{0j}), \qquad j = 1, \ldots, n. \tag{2}$$

In this connection it is interesting to consider the following problems:

(i) study the convergence of the quadrature sums in (2) to the exact values of the integral in the case of σ with border and ribs;

(ii) for σ being a sphere, examine the properties of this convergence near the poles of the sphere and find whether the convergence is uniform with respect to all points M_{0j}, $j = 1, \ldots, n$;

(iii) the same problem if σ is an arbitrary smooth surface without border;

(iv) prove the convergence of the solutions of system (2) to the exact solution of equation (1) for σ with ribs.

References

AGRANOVICH, M.S. (1965) Singular Elliptic Integro-Differential Operators, *Uspekhi Mat. Nauk*, Vol. 20, No. 5.

ALEXANDROV, V.M., SMETANIN, B.I. AND SOBOL, B.I. (1993) *Thin Stress Concentrators in Elastic Bodies* [In Russian], Nauka, Moscow.

ANFINOGENOV, A.YU. AND LIFANOV, I.I. (1992) On numerical solution of integral equations of planar and spatial diffraction, *Russ. J. Numer. Anal. Math. Modelling*, Vol. 7, No. 5, pp. 387–404.

APARINOV, B.A., BELOTSERKOVSKII, S.M., LIFANOV I.K. AND MIKHAILOV, A.A. (1988) Calculation of nonstationary aerodynamical characteristics of bodies in a flow with separation, *Zh. Vychisl. Matem. i Matem. Fiz.*, Vol. 28, No. 10, pp. 1558–1566.

BABKIN, V.I., BELOTSERKOVSKII, S.N., GULYAEV, V.V. AND DVORAK, A.V. (1989) *Computer Modelling of Jets and Bearing Surfaces* [In Russian], Nauka, Moscow.

BALYBIN, G.N., LIFANOV, I.I., LIFANOV, I.K. AND MOLOCHKOV, YU.B (1995) Mathematical modelling of wide-band horn antennas, *Russ. J. Numer. Anal. Math. Modelling*, Vol. 10, No. 1, pp. 1–8.

BELOTSERKOVSKII, S.M. (1965) *Thin Bearing Surfaces in Subsonic Gas Flow* [In Russian], Nauka, Moscow.

BELOTSERKOVSKII, S.M. [EDITOR] (1988) *Mathematical Modelling of Plane-Parallel Flows with Separation* [In Russian], Nauka, Moscow.

BELOTSERKOVSKII, S.M. AND LIFANOV, I.K. (1985) *Numerical Methods for Singular Integral Equations* [In Russian], Nauka, Moscow.

BELOTSERKOVSKII, S.M. AND LIFANOV, I.K. (1993) *Method of Discrete Vortices*, CRC Press.

BELOTSERKOVSKII, S.M., LIFANOV, I.K. AND MIKHAILOV, A.A. (1985) Computer simulation of flow with separation past profiles with angular points, *Dokl. Akad. Nauk SSSR*, Vol. 285, No. 6, pp. 1348–1352.

BELOTSERKOVSKII, S.M., LIFANOV, I.K. AND MIKHAILOV, A.A. (1987) Computation of irrotational flows past arbitrary bodies, *Uchenye Zapiski TSAGI*, Vol. XVIII, No. 5, pp. 1–10.

BELOTSERKOVSKII, S.M. AND NISHT, M.I. (1978) *Ideal Fluid Flow with or without Separation past an Airfoil* [In Russian], Nauka, Moscow.

BISPLINGHOFF, R.L., ASHLEY, H. AND HALFMAN, R.L. (1958) *Aeroelasticity* [Russian translation], Izd-vo. Inostr. Liter., Moscow.

BITSADZE, A.V. (1986) Singular integral equations of the first kind with Neumann kernels, *Diff. Uravnenia*, Vol. 22, No.5, pp. 823–828.

BOULIGAND, G., GIRAUD, G. AND DELENS, P. (1935) *Le Problème de la derivée oblique en thèorie de potential*, Hermann, Paris.

BRYCHKOV, YU.A. AND PRUDNIKOV, A.P. (1977) *Integral Transformations of Generalized Functions* [In Russian], Nauka, Moscow.

CHANG, P. (1970) *Separation of Flow*, Pergamon Press, London.

COLTON, D. AND KRESS, R. (1987) *Methods of Integral Equations in the Scattering Theory* [Russian translation], Mir, Moscow.

DANILENKO, N.V., ZHELANNIKOV, A.I., NISHT, M.I. AND PAVLENKO, V.F. (1983) Modelling of the immediate trace and its effect on the aerodynamical characteristics of aircraft, *Trudy VVIA imeni Zhukovskogo*, Vol. 1311.

DMITRIEV, V.I. AND ZAKHAROV, E.V. (1987) *Integral Equations in Boundary Value Problems of Electrodynamics* [In Russian], Moscow Univ. Press, Moscow.

DUNFORD, N. AND SCHWARTZ, J. (1958) *Linear Operators. Part I. General Theory,* Wiley, New York.

DVORAK, A.V. (1986) Non-degeneracy of the matrix in the discrete vortex method for problems of spatial flow, *Trudy VVIA imeni Zhukovskogo,* Vol. 1313, pp. 441–453.

DYNIN, A.S. (1961) Multi-dimensional elliptic boundary value problems with a single unknown function, *Dokl. Akad. Nauk SSSR,* Vol. 141.

ELLIOT, D., LIFANOV, I.K. AND LITVINCHUK, G.S. (1997) The solution in a class of singular functions of Cauchy type bi-singular integral equations, *J. Integral Eqs. Appl.,* Vol. 9, No. 3, pp. 237–251.

ESKIN, G.I. (1973)$_1$ On the method of variational difference solutions for elliptic pseudodifferential equations, *Uspekhi Mat. Nauk,* No. 5.

ESKIN, G.I. (1973)$_2$ *Boundary Value Problems for Pseudodifferential Equations* [In Russian], Nauka, Moscow.

FERNHOLTZ, H.H. AND KRAUSE, E. [EDITORS] (1982) *Three-Dimensional Turbulent Boundary Layers,* Springer-Verlag, Berlin.

GAKHOV, F.D. (1966) *Boundary Value Problems,* Oxford University Press, Oxford.

GELFAND, I.M. AND SHILOV, G.E. (1959) *Operations with Generalized Functions* [In Russian], Gostekhizdat, Moscow.

GINEVSKII, A.S. (1969) *Turbulence Theory for Jets and Traces* [In Russian], Mashinostroenie, Moscow.

GULYAEV, V.V., LIFANOV, I.K. AND MISKO, V.A. (1996) Mathematical Model of the flow around airfoil with energy high-lift devices, *Russ. J. Numer. Anal. Math. Modelling,* Vol. 11, No. 2, pp. 155–165.

GUREVICH, M.I. (1979) *Theory of Jets in Ideal Fluids* [In Russian], Nauka, Moscow.

HADAMARD, J. (1932) *Le problème de Cauchy et les équations aux dérivées partielles linéires hyperboliques,* Paris.

HÖRMANDER, L. (1965) *Linear Partial Differential Operators* [Russian translation], Mir, Moscow.

KACHMANS, S. AND STEINHAUS, G. (1958) *Theory of Orthogonal Series* [Russian translation], Mir, Moscow.

KAKICHEV, V.A. (1959) Boundary properties of a Cauchy type integral in several variables, *Uchen. Zapiski Shacht. Ped. Inst.,* Vol. 2, No. 6, pp. 25–90.

KAKICHEV, V.A. (1967) On regularization of singular integral equations with Cauchy kernels in bi-cylindrical domains, *Izv. Vuzov, Ser. Mat.,* Vol. 62, No. 7, pp. 54–64.

KELLOG, O.D. (1970) *Foundations of Potential Theory,* Frederick Ungar Publ. Co., New York.

KHUDYAKOV, G.E. (1973) An investigation of aerodynamical characteristics of cylinders with square cross-section, *Nauchn. Trudy Inst. Mekh. Mosc. Univ.,* No. 24, pp. 61–67.

KOCHIN, N.E., KIBEL, I.A. AND ROSE, N.V. (1963) *Theoretical Hydromechanics,* Parts 1 and 2, Fizmatgiz, Moscow.

KOENIG, D.G. AND FALARSKI, V.D. (1971) *Aerodynamic Characteristics of a Large-Scale Model with a Swept and Augmented Jet Flap,* NASA TM X62029.

KOLMOGOROV, A.N. AND FOMIN, S.B. (1972) *Theory of Functions and Elements of Functional Analysis* [In Russian], Nauka, Moscow.

KORNEICHUK, A.A. (1964) Quadrature formulas for singular integrals, *Numerical Methods for Differential and Integral Equations; Cubature Formulas* [In Russian], Nauka, Moscow, pp. 64–74.

LADYZHENSKAYA, O.A. (1973) *Boundary Value Problems of Mathematical Physics* [In Russian], Nauka, Moscow.

LAI, A.K.Y. (1992) A novel antenna for ultra-wide-band applications, *IEEE Trans., Antennas and Propagation*, AP-40, No. 7, pp. 1249–1255.

LOITSYANSKII, L.G. (1978) *Mechanics of Fluids and Gasses* [In Russian], Nauka, Moscow.

LAVRENTIEV, M.A. AND SHABAT, B.V. (1973) *Methods of the Theory of Complex Variable* [In Russian], Nauka, Moscow.

LIFANOV, I.K. (1988) Singular integral equation of the first kind for the Neumann problem, *Diff. Uravnenia*, Vol. 24, No. 1, pp. 110–115.

LIFANOV, I.K. (1989) Singular solutions of singular integral equations and flow ejecting for an arbitrary contour, *Sov. J. Numer. Anal. Math. Modelling*, Vol. 4, No. 3, pp. 239–252.

LIFANOV, I.K. (1996) *Singular Integral Equations and Discrete Vortices*, VSP, the Netherlands.

LIFANOV, I.I. AND LIFANOV, I.K. (1996) Antenna-diffraction problems and singular solutions of singular integral equations, *J. Electromagnetic Waves and Appl*, Vol. 10, pp. 925–937.

LIFANOV, I.I., LIFANOV, I.K. AND NOVIKOV, S.N. (1996) Numerical solution of the Neumann three-dimensional problem for the Helmholtz scalar equation for bodies of complex form, *Russ. J. Numer. Anal. Math. Modelling*, Vol. 11, No. 4, pp. 359–366.

LIFANOV, I.K. AND MIKHAILOV, A.A. (1986) Mathematical problems of the numerical analysis of flows without separation, *Trudy VVIA imeni Zhukovskogo*, No. 1313, pp. 454–464.

LIFANOV, I.K., MIKHAILOV, A.A. AND TITSKII, S.V. (1990) Mathematical modelling of airfoil flowing control by ejection, *Russ. J. Numer. Anal. Math. Modelling*, Vol. 5, No. 3, pp. 209–220.

LIFANOV, I.K. AND POLONSKII, YA.E. (1975) Justification for the numerical method of discrete vortices for singular integral equations, *Prikl. Mat. Mekh.*, Vol. 39, No. 4, pp. 742–746.

LIFANOV, I.K. AND POLTAVSKII, L.N. (1992) Generalized Fourier operator and its application for the justification of the method of discrete vortices, *Mat. Sbornik*, Vol. 183, No. 5, pp. 70–114.

LIFANOV, I.K. AND POLTAVSKII, L.N. (1998) Quadrature formulae for the Hadamard integral over a curvilinear surface, *Russ. J. Numer. Anal. Math. Modelling*, Vol. 13, No. 1, pp. 27–44.

LIFANOV, I.K. AND POLTAVSKII, L.N. (1999)[1] Spaces of fractional quotients. Discrete operators and their applications, I *Mat. Sbornik* Vol. 190, No. 9, pp. 41–98.

LIFANOV, I.K. AND POLTAVSKII, L.N. (1999)[2] Spaces of fractional quotients. Discrete operators and their applications, II *Mat. Sbornik* Vol. 190, No. 11, pp. 67–134.

LIFANOV, I.K., SETUKHA, A.V., TSVETINSKII, YU.G. AND ZHELANNIKOV, A.I. (1997), Mathematical modelling and the numerical analysis of a nonstationary flow around the deck of a ship, *Russ. J. Numer. Anal. Math. Modelling*, Vol. 12, No. 3, pp. 255–269.

MIKHLIN, S.G. (1962) *Multi-Dimensional Singular Integrals and Integral Equations* [In Russian], Fizmatgiz, Moscow.

MUSAEV, B.I. (1985) Approximate solution of a complete singular integral equation on a segment [In Russian]. Inst. Cybernetics AzSSR, Baku, Registered at VINITI, 23.10.85, No. 7377–85.

MUSKHELISHVILI, N.I. (1968) *Singular Integral Equations*[In Russian], Nauka, Moscow.

NATANSON, I.P. (1949) *Constructive Theory of Functions* [In Russian], GIPTL, Moscow, Leningrad.

NAZARCHUK, Z.T. (1989) *Diffraction of Waves on Cylindrical Structures: Numerical Analysis* [In Russian], Naukova Dumka, Kiev.

NEKRASOV, A.I. (1947) *Theory of Airfoil in Nonstationary Flow* [In Russian], Izd-vo AN SSSR, Moscow.

NIKOLSKII, S.M. (1977) *Approximation of Functions of Several Variables and Imbedding Theorems* [In Russian], Nauka, Moscow.

POLTAVSKII, L.N. (1993) *Mathematical Justification of some Numerical Schemes in Aerodynamics*, Doctorate Thesis, Moscow.

PRANDTL, L. (1939) Mechanics of viscous fluid [Russian translation], in *Aerodynamics*, Vol. 3, Oborongiz, Moscow, Leningrad.

RIMAN, I.S. AND KREPS, R.L.(1947) Associated masses for bodies of arbitrary horizontal projection, *Trudy TSAGI*, No. 635.

RYZHIK, I.M. AND GRADSTEYN, I.S. (1951) *Tables* [In Russian], Nauka, Moscow

SANIKIDZE, D.G. (1974) On a uniform estimate for the approximation of singular integrals with Chebyshev weight by interpolation sums, *Soobsh. AN Gruz. SSR*, Vol. 75, No. 1, pp.53–55.

SARANEN, J. AND VAINIKKO, G. (1999) Fast collocation solvers for integral equations on open arcs, *J. Integr. Eqs. Appl.*, Vol. 11, No. 1, pp. 57–102.

SCHLICHTING, G. (1974) *Theory of Boundary Layer* [Russian translation], Nauka, Moscow.

SCHWARTZ, L. (1967) *Analyse Mathématique*, Vols. 1 and 2, Hermann, Paris.

SEDOV, L.I. (1973) *Continuum Mechanics* [In Russian], Part 2, Nauka, Moscow.

SHESHKO, M.A. (1976) On the convergence of quadrature processes for singular integrals, *Izv. Vuzov, Ser. Mat.*, No. 12, pp. 108–118.

SHILOV, G.E. (1961) *Mathematical Analysis; A Special Course* [In Russian], Nauka, Moscow.

SHILOV, G.E. (1965) *Mathematical Analysis; The Second Special Course* [In Russian], Nauka, Moscow.

SHILOV, G.E. (1972) *Mathematical Analysis of Functions with Several Variable* [In Russian], Nauka, Moscow.

SHUBIN, M.A. (1978) *Pseudodifferential Operators and Spectral Theory* [In Russian], Nauka, Moscow.

SOBOLEV, S.L. (1947) *Equations of Mathematical Physics* [In Russian], GITTL, Moscow, Leningrad.

STARK, I. (1971) Generalized quadrature formula for Cauchy integrals, *Raketn. Tekhn. i Kosmon.*, No. 9, pp. 244–245.

STEIN, E. (1970) *Singular Integrals and Differential Properties of Functions*, Princeton Univ. Press, New Jersey.

TIKHONOV, A.N. AND SAMARSKII, A.A. (1966) *Equations of Mathematical Physics* [In Russian], Nauka, Moscow.

TRENOGIN, V.A. (1980) *Functional Analysis* [In Russian], Nauka, Moscow.

TSAGI REVIEWS (1987) *Three-Dimensional Flows in Boundary Layars* [In Russian], Moscow.

VAINIKKO, G. (1997) *Periodic Integral and Pseudodifferential Equations*, Research Report C13, Heksinki Univ, Technology, Institute of Mathematics.

VLADIMIROV, V.S. (1976) *Equations of Mathematical Physics* [In Russian], Nauka, Moscow.

ZAKHAROV, E.V. AND PIMENOV, YU.V. (1982) *Diffraction of Radio Waves: Numerical Analysis* [In Russian], Radio i Svyaz', Moscow.

ZYGMUND, A. (1965) *Trigonometric Series* [Russian translation], Mir, Moscow, Vols. 1 and 2.

Index